The Pituitary Gland

CLINICAL SURVEYS IN ENDOCRINOLOGY

C. R. Kannan, M.D.

Volume 1   THE PITUITARY GLAND

A Continuation Order Plan is available for this series. A continuation order will bring delivery of each new volume immediately upon publication. Volumes are billed only upon actual shipment. For further information please contact the publisher.

The Pituitary Gland

C. R. Kannan, M.D.

Chairman
Division of Endocrinology and Metabolism
Department of Medicine
Cook County Hospital
and Clinical Associate Professor of Medicine
University of Illinois at Chicago
Chicago, Illinois

PLENUM MEDICAL BOOK COMPANY
NEW YORK AND LONDON

Library of Congress Cataloging in Publication Data

Kannan, C. R. (Charkravarthy R.), 1943–
  The pituitary gland.

  (Clinical surveys in endocrinology; v. 1)
  Includes bibliographies and index.
  1. Pituitary gland—Diseases. I. Title. II. Series. [DNLM: 1. Pituitary Diseases. 2. Pituitary Gland. WK 550 K16p]
RC658.K36   1987   616.4'7   87-7812
ISBN-13: 978-1-4612-9032-2   e-ISBN-13: 978-1-4613-1849-1
DOI: 10.1007/978-1-4613-1849-1

© 1987 Plenum Publishing Corporation
Softcover reprint of the hardcover 1st editon 1987
233 Spring Street, New York, N.Y. 10013

Plenum Medical Book Company is an imprint of Plenum Publishing Corporation

To the perpetual student in all of us

Preface

The past two decades have witnessed an unprecedented growth in the field of neuroendocrinology. The conjoint research contributions by clinicians and basic scientists have promulgated revolutionary concepts at a breakneck speed. This first volume in *Clinical Surveys in Endocrinology, The Pituitary Gland,* has been written with but one purpose in mind: to integrate the current knowledge in this dynamic field with the existing body of information already available to the clinician. The chapters in this book attempt to portray current research information seen through the eyes of a clinician. The contributions of pioneers in each field have been placed in a perspective relevant to the practicing endocrinologist. The selection of the almost 1500 references from a bewildering body of literature has been influenced by the degree to which these articles—original as well as review papers—contributed to the growth of pituitary endocrinology. Despite the most scrutinizing attempts, it is inevitable and regrettable that works of importance must be excluded due to the practical limitations of any comprehensive work. Nevertheless, to the researcher these references are complete enough to serve as a significant resource. To the reader who wishes to gain an indepth clinical perspective of pituitary disorders, this work is written precisely from that vantage point.

The single authorship of this work notwithstanding, several friends have been instrumental in the completion of this work. I deeply appreciate the incessant zeal and excellent assistance of Ms. Gayla Blake, who transformed my handwritten thoughts into a legible manuscript. I am most grateful to my medical editor at Plenum, Ms. Janice Stern, whose ideas have been an inspiration to me. Without her persistent support this project would not have been realized. The entire production team of Plenum Publishing Corporation has my gratitude for once again producing my work with professional care and an eye for aesthetics. I also wish to express my thanks to Dr. Gerald Burke for his continued encouragement, and to my staff in the Division of Endocrinology for their support. Finally, I appreciate the patience of my family,

Molly, Ashley, Alexander, and Margaret for bearing with me—for the fourth time around.

                                                    C. R. Kannan, M.D.

Cook County Hospital
Chicago, Illinois

Contents

1

Introduction to the Pituitary and Hypothalamus

Historical Perspectives

The history of past events, when subjected to logical schema, very often permits its students to foretell the future to some extent. However, no student of the history of endocrinology could have predicted the heights to which this science would soar in this century, especially during the past 20 years. This is particularly so if one considers the paucity of developments in the field of endocrinology until the turn of the century and the consequent intellectual drought that pervaded that period. Everything has a history and a past, regardless of whether that past is richly textured or barren. A brief sojourn in the back woods of history, as it relates to the pituitary gland, is relevant—even essential—when one considers the prophetic nature of the observations by a few. For an eminently readable and a singularly enchanting account, the reader is referred to *A History of Endocrinology*[1] by Medvei, the only complete and definitive work of its kind in the English language. The landmarks—the places and the people—that punctuated that history are briefly outlined in this chapter.

The function of the pituitary gland was largely unknown and mostly misrepresented for the first 1500 years of history. For the student wandering through the vistas of the historic past, this misrepresentation would be readily evident. For instance, Galen (130–200 AD), the most celebrated physician of the Greco-Roman era, believed that the pituitary gland served as a "phlegmatic glandule" that removed waste products from the base of brain and that excreted these products by ducts through the sphenoid and ethmoid bones to the nasopharynx. (The word pituitary is derived from *pituita*, meaning nasal mucus, which represented the removed waste product!) The first glimmer into the function of the pituitary gland was provided by Conrad Schreiber[2] (1614–1680) and Richard Lower[3] (1631–1691). Of particular interest is Lower's prophetic opinion that blood and substances contained in blood pass from the brain into the pituitary, which performed the function of "distilling substances back into the circulation." Joseph Lieutaud[4] (1703–1780), a reputed anatomist and professor of medicine at the University of Aix-en-Provence, stumbled upon the pituitary–portal circulation. He perceived the "small vessels, which pass along the axis of the stalk and communicated with those of the gland." Even the anatomical facts regarding the pituitary gland were largely unknown, until Giovanni Domenico Santorini (1681–1737) dissected the gland and demonstrated an anterior part (pars anterior) that was separate from the posterior lobe. The notion that the pituitary gland had no substantial function was perpetuated when, in 1886, Sir Victor Horsley excised the pituitary gland from two dogs and showed that they survived reasonably well for 6 months.[5] However, all that changed in the same year, when the French neurologist Pierre Marie published his papers on "Marie's malady," acromegaly.[6] History credits the Bucharest physiologist Nicholas C. Paulesco (1869–1931) with providing proof that the pituitary gland is essential for life.[7] He performed total hypophysectomies in 22 dogs and 12 cats, all of whom died within 3 days following surgery. This observation in 1907, firmly entrenched the notion that the pituitary gland secreted substance(s) essential for the preservation of life. It also was becoming apparent that disorders of the pituitary gland are responsible for diverse clinical manifestations.

The technique of hypophysectomy, first perfected by Paulesco and later by Harvey Cushing and Bernard Aschner, paved the way for experimental observations on the consequences of partial hypophysectomy and stalk section in dogs. Slowly but surely, it became apparent that loss of pituitary function had far-reaching effects on the thyroid, sex glands, and adrenals. Harvey Cushing summarized the state of knowledge regarding the pituitary in an address to the American Medical Association in 1909 and coined the terms *hyper-* and *hypopituitarism*.[8] Endocrinology of the pituitary had become a science and was expanding at breakneck speed. By 1935, no less than six, and possibly eight, hormones had been extracted from a gland less than 500 mg in weight, once considered to represent nothing more than a vestigeal organ. Sir Walter Langdon Brown, in 1931, made the often-repeated remark that "the pituitary is the leader of the endocrine orchestra."[9] An orchestration of hormony that ensured the smooth functioning of the adrenal cortex, the thyroid, the gonads, growth, parturition, lactation, fuel regulation, and water conservation. The relationship of the pituitary with the brain and the other endocrine glands was beginning to be perceived as a complex one, the complexities of which were exceeded only by the central nervous system (CNS). In a riveting address, "The Fourth Langdon Brown Lecture on the Endocrine Orchestra," Sir Douglas Hubble,[10] further paraphrased the appropriateness of this analogy to an orchestra:

> An orchestra is a group of individuals harmoniously loyal to their higher control and playing with a common purpose towards a common end—the best possible interpretation of the piece of music in front of them. In the endocrine system is a group of individual organs, with much more complicated relationships and with controls which although complex, are yet unified, working with a common purpose toward a common end—the maintenance of homeostasis in the body.

The "hypothalamic connection" did not gain acceptance as a concept until the 1940s, although anatomists had been suggesting the same for a decade. G. B. Wislocki and L. S. King, two anatomists at Harvard, injected Indian ink into monkeys and showed that the direction of blood in the portal circulation was down the pituitary stalk and not up as it was previously thought. The concept that substances were secreted by the brain in general—and the hypothalamus in particular—appealed to several pioneers in the early days of neuroendocrinology; these included intellectual giants such as Harry B. Friedgood from Harvard, Georgy Pope from Romania, and Geoffrey Harris and John Green, both from Cambridge. It was Harris, however, who strongly believed that the pituitary gland was controlled by hormones secreted by the hypothalamus and transported by the portal vessels to their hypophyseal destination. This theory was to form the thesis that Harris submitted for his M.D. examination in 1944. It was also to become the preoccupation that would shape his life's work. Proving the existence of releasing hormones from the hypothalamus would become an obsession that was to be a race against time. The quest for isolating and characterizing hypothalamic releasing hormones covered nearly three decades, and involved another generation of endocrinologists working across the Atlantic ocean—Andrew Schally and Roger Guillemin. The 21-year race between these two scientists to isolate TRH (and to win the world's most coveted research prize) is impressively chronicled by Nicholas Wade in his book, *The Nobel Duel.* Geoffrey Harris died in 1971, 2 years after Shally and Guillemin had independently discovered and characterized thyrotropin-releasing hormone (TRH), publishing their results in the same year, (1969) in the same month (November), in papers that appeared 6 days apart. Both would go on to share conjointly the Nobel Price for medicine in 1977.

The field of endocrinology begin to experience a scientific harvest as the releasing factors of the hypothalamus were characterized one after another. The tripeptide TRH in 1969,[12,13] the decapeptide GnRH in 1971,[14] the 14-amino acid peptide somatostatin in 1973,[15] the 41-amino acid residue corticotropin releasing factor in 1981,[16] and the 40–44-amino acid peptide, growth hormone-releasing hormone (GHRH) the next year.[17,18] Neuroendocrinology research is now at its zenith. The dynamic nature of this field has been expressed most succinctly by Victor Cornelius Medvei in *"A History of Endocrinology"*—"The present state of the endocrine field cannot be readily presented like, for example, the Battle of Waterloo is painted, however huge a canvas one may choose. It is still in flux, is still progressing and is impossible to foresee or foretell the final outcome. All we can do is to take a still photograph, as it were, of the situation perhaps from various angles, and pick out a few of the contents, pinpoint and briefly discuss them on their present merits."[18]

Anatomical Considerations

The anatomical considerations can be viewed in terms of the pituitary gland, the hypothalamus, the median eminence, and the sella turcica.

Pituitary Gland

The term pituitary (*pituita* for "phlegm") is derived from Aristotle's belief that this was the organ through which phlegm, "one of the four vital humors," passed from the brain towards the various parts of the body. The pituitary gland is couched in the sella turcica, a part of the sphenoid body. The gland occupies 80% or more of the fossa; the dimensions of the pituitary gland are impressively small: 1.2–1.5 cm in its greatest diameter, from side to side, with an anteroposterior diameter approximating 1 cm and a thickness no greater than 0.5 cm. The pituitary gland weighs 500–600 mg, being slightly larger in women, especially during pregnancy. The pituitary gland in humans is encased in a dural covering except for a small opening in the diaphragm, through which the infundibulum passes.

Divisions

The pituitary gland consists of the anterior lobe (anterior pituitary) and the neural lobe (posterior pituitary and the infundibular process). A third lobe, the intermediate lobe, is rudimentary and ill defined in humans; this lobe makes up less than 0.8% of the pituitary gland. The pituitary gland is connected to the hypothalamus and the brain by the pituitary stalk. This highly strategic corridor contains several important structures; at the hypothalamic end the stalk originates at the neurohypophyseal portion of the hypothalamus. This is continued as the funnel-shaped infundibulum, a wide structure, and is enveloped from below by the pars distalis, an extension of the anterior lobe. The infundibulum is penetrated by capillary loops of the hypophyseal portal circulation. As it proceeds downward through the aperture in the diaphragma sella, the infundibulum narrows ("stem"), blending into the neural lobe.

Blood Supply

The pituitary enjoys a dual blood supply: arterial supply from the inferior and superior hypophyseal arteries and from a highly specialized hypophyseal portal vascular network. The inferior hypophyseal artery is a single vessel that originates in the cavernous portion of the internal carotid artery on each side and supplies the infundibular process and neural lobe. The superior hypophyseal arteries arise from the supraclinoidal portion of the internal carotid arteries, as well as the anterior (and posterior) cerebral arteries. These vessels form an arterial ring around the infundibulum and serve as the main providers of arterial supply to the median eminence and the anterior pituitary. The anterior pituitary also receives blood from the portal system, which is physiologically vital for transport of hypothalamic factors to the anterior pituitary. The portal system originates from specialized vascular structures that are located in the region of the median eminence and are termed the "gomitoli". These structures are comprised of short terminal arterioles surrounded by a dense capillary network. These capillaries drain into long portal veins that run along the surface of the pituitary stalk and terminate into sinusoidal capillaries of the anterior pituitary. The capillary network that surrounds the median eminence, infundibular stalk and the neurohypophysis consists of a distinct external plexus and an internal plexus. The external plexus, called the mantle plexus of Romeis,[20] is supplied by the superior hypophyseal arteries and is drained by the long portal vessels. This network forms the outer shell of the median eminence. The internal plexus, often referred to as the deep plexus of Duvernoy,[21] is formed by coils and loops derived from the vessels of the external plexus. The coils and loops of the internal plexus are continuous posteriorly with the capillary bed of the infundibular stalk. Angioarchitectural studies of the vascular casts of the median eminence resemble a bowl or a funnel, with the inner surface of the cast facing the third ventricle. The direction of flow in the portal system is predominantly directed from the median eminence into the anterior pituitary, serving as a vascular avenue of transport for the hypothalamic peptides. However, retrograde passage of pituitary hormones from the pituitary to the median eminence has been proposed.[22,23] Page and Bergland[24] studied the angioarchitecture of the pituitary-median eminence complex extensively by examining corrosion casts of injected specimens with the aid of the scanning electron microscope. Their studies suggest that the capillary bed of the neurohypophysis may be actively and selectively involved in bidirectional transport of hypothalamic and pituitary secretions. Additional work by Bergland and Page[25] is suggestive of the fact that the portal system of blood vessels can be best viewed as a loop. The general direction of flow is down the stalk to the anterior pituitary to and then across to the posterior lobe. From the posterior pituitary, some blood escapes into the general circulation and flows back into the anterior lobe, and some flows upward to the hypothalamus.

The venous drainage of the anterior pituitary is mediated by a number of inferior hypophyseal veins draining sequentially into the cavernous sinuses, the superior and inferior petrosal sinuses, and eventually the jugular bulb and jugular veins.

The clinical implications of the intricate manner by which the pituitary gland is vascularized are as follows:

1. Interruption of the stalk results in loss of hypothalamic drive to the anterior pituitary.
2. Although the pituitary gland is relatively resistant to vascular damage, there are two situations in which the gland becomes vulnerable to vascular damage: the gestational pituitary, which is highly susceptible to ischemic necrosis precipitated by hypotension, and the tumor-bearing pituitaries, which are vulnerable to the hemorrhagic necrosis caused by apoplexy.
3. The venous drainage of the pituitary into ipsilateral inferior petrosal sinuses (with little or no cross-communications between the two sides) has permitted cannulation of these sinuses. Thus, sampling of the venous effluents for pituitary hormones can accurately predict both the site and side of hormone-secreting microtumors of the pituitary gland.

Relationships

Anatomically, the pituitary gland is related to several important structures (see Fig. 1). These relationships assume importance when the gland enlarges secondary to tumorous or other disease processes. The gland is related to the optic chiasm and the hypothalamus superiorly and to the sphenoid sinuses inferiorly. Laterally on either side, the gland is related to the cavernous sinuses and the oculomotor nerves contained within. The suprasellar area is an extremely strategic location lying in the vicinity of the hypothalamus, optic chiasm, and the ventricular system. If one views the base of the brain from below, the crucial nature of this strategic area will become evident. The hypothalamus is limited anteriorly by the optic chiasm and posteriorly by the mammillary bodies and interpenduncular fossa. The posterior communicating arteries form the lateral borders of the hypothalamus. Inferiorly, the floor of the hypothalamus is formed by the suprasellar cistern. Superiorly, the roof of the hypothalamus is formed by the third ventricle, and the thalamus. The hypothalamus assumes a funnel-like shape inferiorly as it forms the pituitary stalk. The stalk is closely surrounded by the lateral walls of the third ventricle and its infundibular recess. Obviously, an expanding tumor above the sella can play extensive havoc by damaging several vital structures.

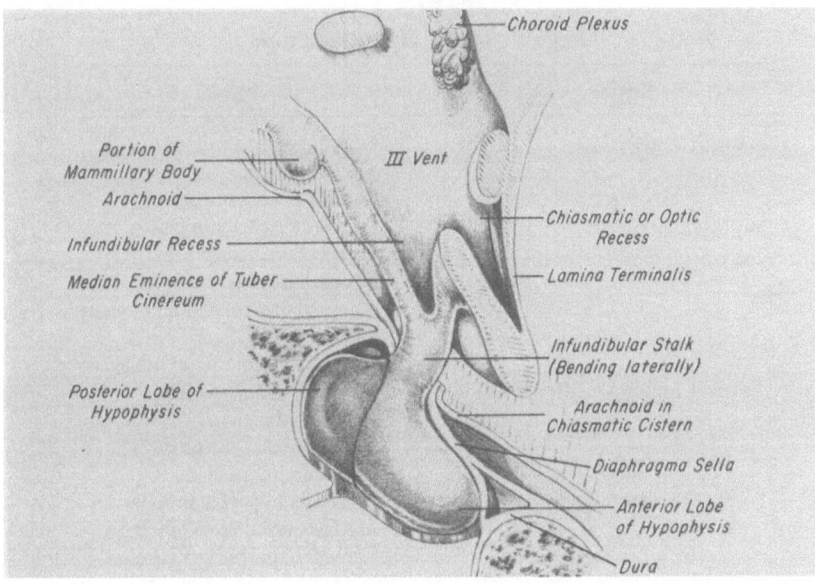

FIGURE 1. Sagittal view of the human hypothalamic pituitary unit. (From Reichlin.[43])

Hypothalamus

The hypothalamus represents only 4 g of brain tissue, yet plays a phenomenol role in endocrine regulation. The anatomical boundaries of the hypothalamus extend anteriorly up to the optic chiasm and posteriorly as far as the mammillary bodies. The hypothalamus lies ventral to the thalamus, forming the floor and lower lateral walls of the third ventricle. The base of the hypothalamus is represented by a prominence, the tuber cinerum, the center of which is constituted by the median eminence. This physiologically important site represents the originating point for the pituitary stalk. The hypothalamus is divided into several parts. In a transverse direction, the hypothalamus can be divided into preoptic, supraoptic, tuberal (infundibular), and posterior regions. Each of these areas can be further subdivided longitudinally into medial or lateral regions. Each region contains several important nuclei (Table 1) that are involved in the secretion of specific hypothalamic peptides. The nature of these releasing and inhibitory peptides are discussed separately.

Median Eminence and Stalk

The median eminence represents a crucial point in the hypothalamic hypophyseal conduit (see Fig. 2). Three types of neurons are present in the median eminence; the supraopticohypophyseal and paraventriculohypophy-

TABLE 1
Regions of the Hypothalamus

Region	Nuclei
Preoptic region	Periventricular nucleus
	Medial preoptic nucleus
	Lateral preoptic nucleus
Supraoptic region	Paraventricular nucleus
	Periventricular nucleus
	Suprachiasmatic nucleus
	Anterior hypothalamic nucleus
	Supraoptic nucleus
	Lateral hypothalamic nucleus
Tuberal region (Infundibular)	Periventricular nucleus
	Arcuate nucleus
	Ventromedial nucleus
	Dorsomedial nucleus
Posterior (mammillary) region	Posterior hypothalamic nucleus
	Medial mammillary nucleus
	Lateral mammillary nucleus

seal neurons that transport vasopressin and oxytocin to the posterior pituitary; the tuberoinfundibular neurons that transport the hypothalamic releasing factors and terminate here; the tuberohypophyseal neurons that secrete and transport peptidergic hormones (e.g., TRH, GnRH) and bioamines, particularly dopamine to the intermediate lobe. In addition to this rich neural element, the median eminence contains vascular and epithelial components

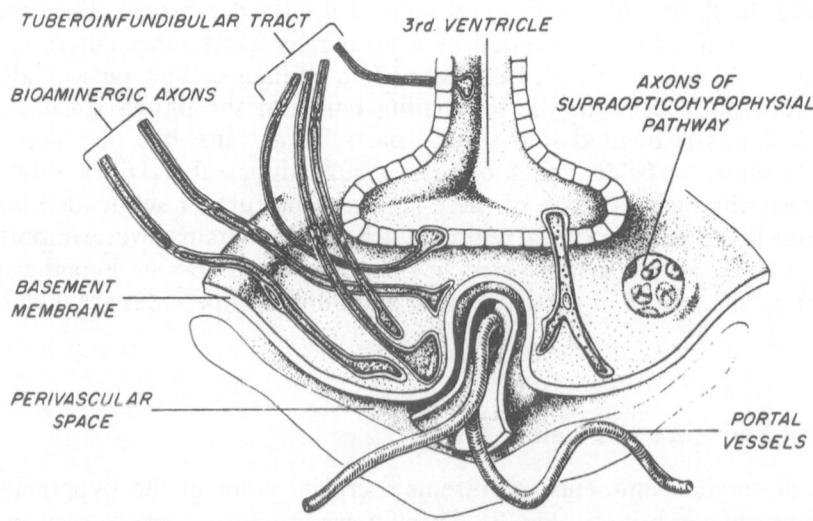

FIGURE 2. The median eminence with its relationship to nerve endings and bioaminergic tracts. (From Reichlin.[43])

as well. The vascular element of the median eminence consists of the primary capillary plexus and the portal veins. The median eminence is a specialized structure containing neurovascular anastamosis characterized by nerve terminals that end on the small capillaries of the portal vessels. There appears to be free communication between the extracellular space of the median eminence and substances in the blood. The median eminence is permeable to molecular substances such as thyroxine (T_4) or growth hormone in contrast to the brain tissue. It is believed that the nerve terminals are bathed in interstitial fluid rich in neurotransmitters. The epithelial component of the median eminence consists of the pars tuberalis of the anterior pituitary.

Sella Turcica

The lateral view of the sella turcica is represented diagramatically in Fig. 3. The following landmarks should be systematically identified on the lateral film: the planum sphenoidale, the chiasmatic sulcus, the tuberculum sellae, the anterior clinoid, the posterior clinoid, the dorsum sellae, the floor, and the sphenoid sinus. The length L and the depth D of the sella should be measured using the following definitions. The length of the sella is the greatest anteroposterior diameter. The depth of the sella is measured by drawing a line from the tuberculum sellae to the top of the dorsum sellae and then a perpendicular line to the deepest point in the floor. The width is measured on a PA film as the distance between the highest points of the lateral edges of the plateau of the floor of the sella. Figure 4 demonstrates an enlarged sella turcica with a thinned-out dorsum sella. The mean sella length is 11.9 mm (range: 9–16.2), the mean depth is 8 mm (range: 6.5–10.5), and the mean width is 13.1 mm (range: 9–19). The volume of the sella can be calculated using the formula

$$(V) = L \times D \times W/2 \text{ mm}$$

Using the mean linear values and formula stated above the mean volume is calculated to be 624 mm. The sellar size varies with age and body height. When evaluating the sella turcica attention should be paid to changes such as the shape of the sella, pathologic double contour of the floor, thinning of the cortical bone, erosions (especially asymmetrical ones), and calcifications. Regarding the size of the sella, volumetric measurements are superior to linear measurements, since a sella could appear normal on the lateral view yet be increased in width. It should be remembered that a normal sella volume does not necessarily mean a normal-sized pituitary gland, since the pituitary is bound laterally and superiorly by soft tissues and the gland can enlarge without a concomitant alteration in sellar volume. It should be noted that quantitative volume changes are not as significant as qualitative changes in the sellar contour when attempting to determine the presence or absence of pituitary lesions.

Double contouring of the floor is a sign usually associated with intrasellar lesions unevenly expanding the floor of the sella. It has been emphasized,

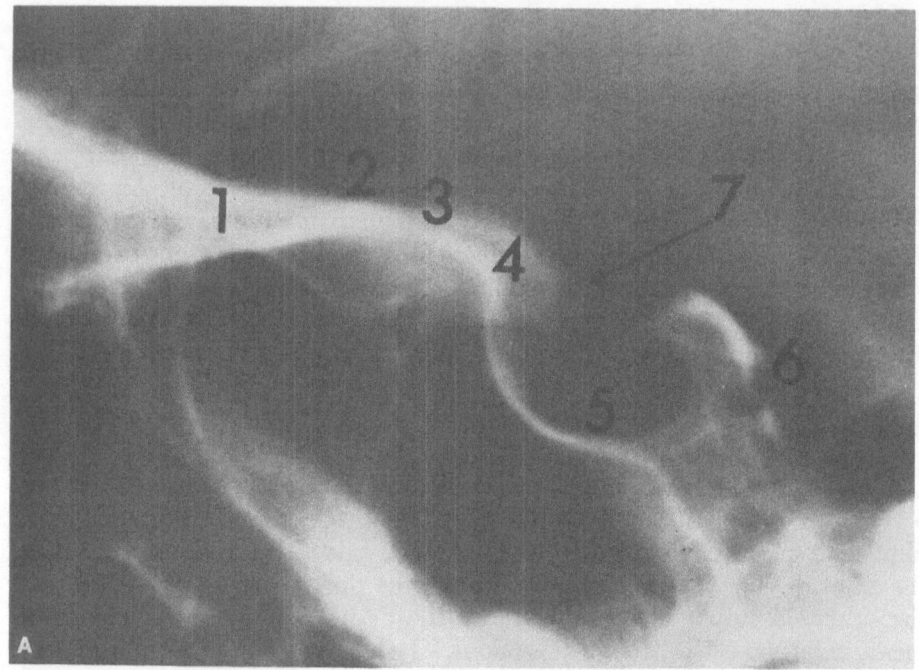

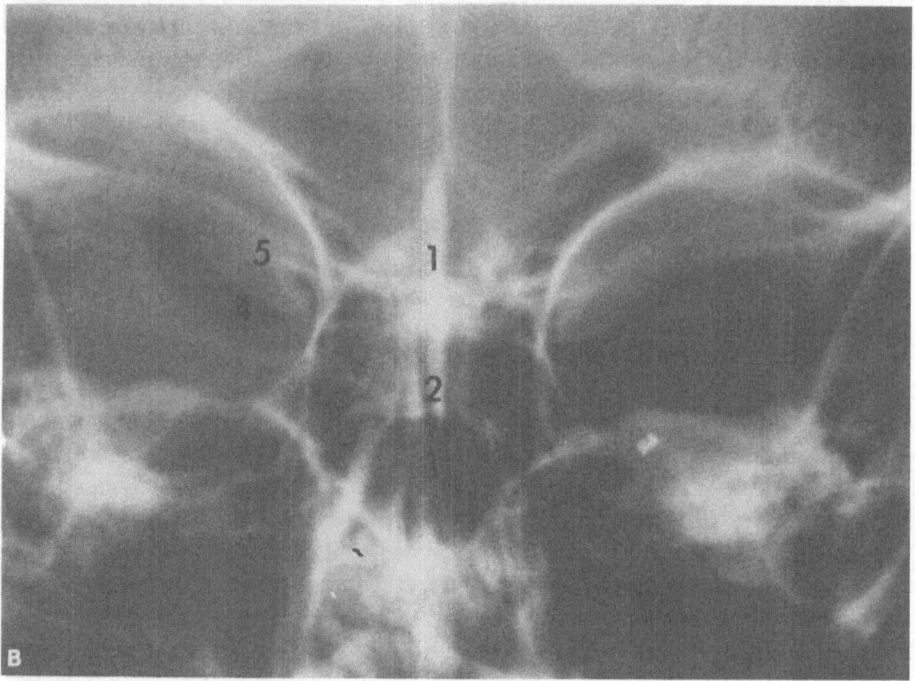

FIGURE 3. (A) Skull radiograph, lateral projection. Note the planum sphenoidale (1), limbus sphenoidale (2), sulcus chiasmaticus (3), tuberculum sellae (4), sellar floor with distinct lamina dura (5), dorsum sellae (6), anterior clinoid processes (7), and sphenoid sinus (8). (B) Caldwell projection. Note the planum sphenoidale (1), floor of sella (2), sphenoid sinus (3), sphenoidal fissure (4), and inner one-third of the sphenoidal ridge (5). (From Wolpert.[44])

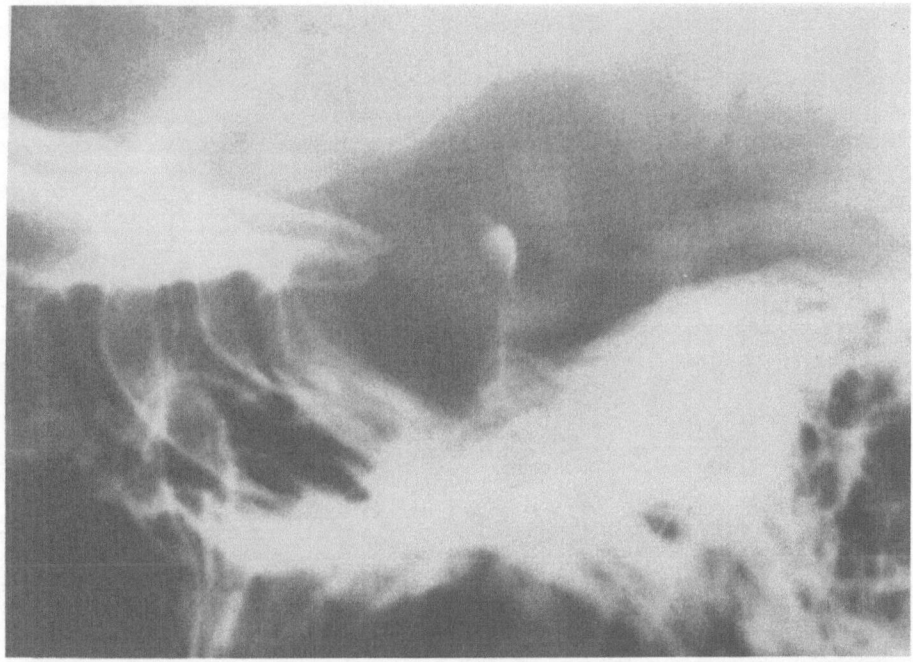

FIGURE 4. Lateral skull radiograph demonstrating an enlarged sella turcica with a thinned-out dorsum sellae. (From Wolpert.[44])

however, that minor duplications of the floor or anterior wall can frequently occur in normal people. Thinning of the cortical bone can occur with osteoporosis. Erosions, especially asymmetrical ones, should be regarded as indicative of microadenomas. Calcifications in the sella can be due to vascular lesions, granulomas, and degenerative changes. Suprasellar calcifications, especially if flocculent, are highly suggestive of craniopharyngiomas (Fig. 5).

Embryology

The anterior and the posterior pituitary share different embryological origins. The anterior pituitary develops from neurectoderm, as a downward projection of the neurectoderm, ending inferiorly as a blind pouch at the roof of the mouth. This pouch, the Rathke's pouch with its infundibulum, develops, elongates, and migrates caudally toward the base of brain. The Rathke's pouch has several segments, each one giving rise to different parts of the adenohypophysis. Thus, the pars intermedia develops from the portion of the Rathke's pouch that is in close proximity with the neurectoderm; the pars distalis is derived from the anterior fold of the residual buccal ectoderm, and the pars tuberalis from the lateral folds of the developing anterior pituitary. Of these

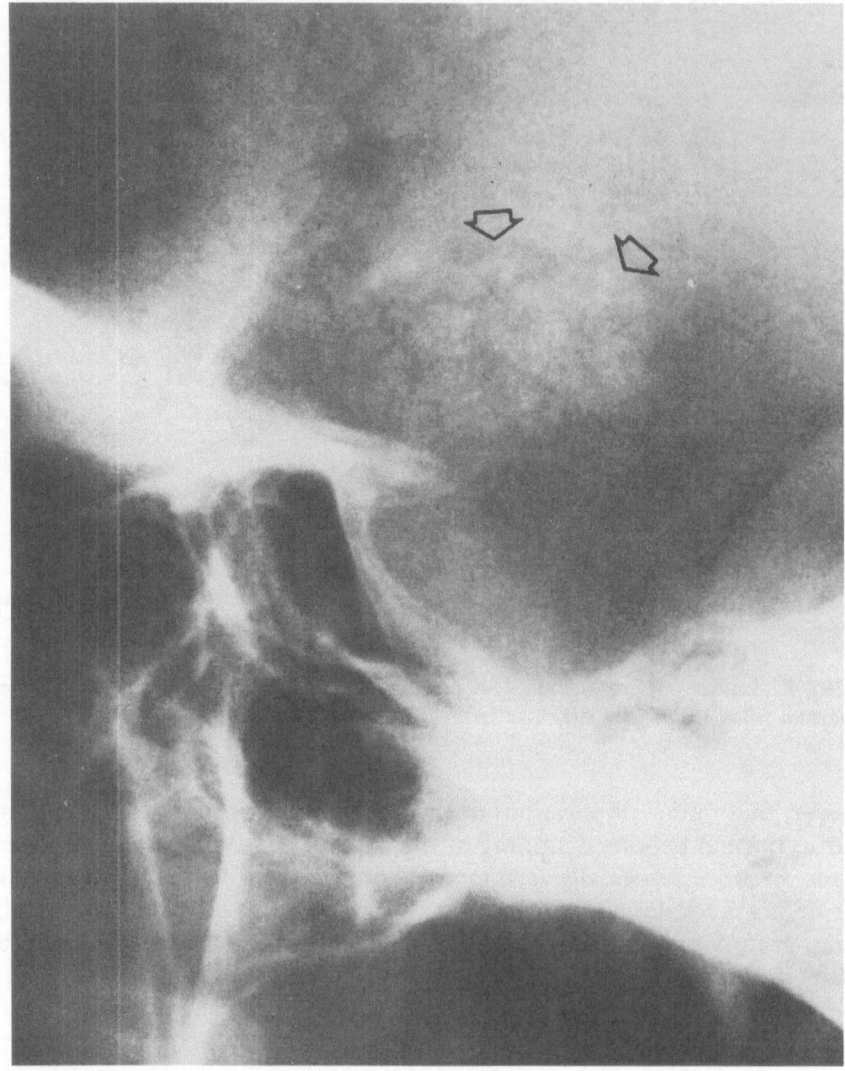

FIGURE 5. Lateral skull radiograph demonstrating enlargement of the sella, absence of dorsum sella, and impressive suprasellar calcification. (From Post.[45])

three parts, in man, the pars distalis is the part conferred with secretory potential.

The neural lobe develops as a downgrowth of the ventral diencephalon, and hence is part of the central nervous system. The posterior pituitary contains unmyelinated nerve tracts originating in the hypothalamus.

In the embryo, the neural lobe, growing downward, fuses with its anterior counterpart, growing upward from the roof of the stomadeum, to form the complete intrasellar hypophysis cerebri. Throughout adult life, the posterior pituitary retains its neural origin from the brain, while the anterior pituitary retains and revels in its neurectodermal secretory potency.

Aberrations of migration can result in the occurrence of adenohypophyseal tissue in ectopic locations, such as the nasopharynx, or in the sphenoid sinus. Remnants of the Rathke's pouch during its suprasellar course may become the seat for the development of craniopharyngiomas.

Histology

Histologically the pituitary gland consists of five varieties of cells, based on the hormone secreted. Thus, the pituitary is populated by somatotropes, lactotropes, corticototropes, thyrotropes, and gonadotropes. The somatotropes constitute approximately 50% of the adenohypophyseal cell population. These cells primarily populate the lateral wings of the pars distalis. Scattered somatotropes can also be found in the median wedge. The three characteristics of the normal somatotropes are the presence of evenly electron dense granules (250–500 mm), prominent rough endoplasmic reticulum (RER), and positive immunoperoxidase staining for GH.

The lactotropes constitute 10–25% of the pituitary cell population. The pars distalis is the favored location of these cells—clusters of cells usually visible at the posterior aspect of the pars distalis. One notable feature of normal lactotopes is the differential staining of these cells with prolactin antisera; some cells are densely granulated and show intense immunopositivity, reflected as diffuse brown deposits throughout the cytoplasm, while other lactotropes are only mildly immunopositive, demonstrating a ringlike immunopositive reaction with prolactin antisera, especially discernible over the Golgi complex.

In the normal pituitary gland, corticotropes constitute 10–20% of pituicytes. These cells characteristically appear oval or angular, and are mostly found in the median wedge of the pars distalis. Corticotropes can be identified by their immunopositivity for ACTH, β-LPH, and endorphins.

The thyrotropes are medium to large sized, angular or polyhedral cells with characteristic long cytoplasmic processes. These cells contain fine granules that are immunopositive for TSH. Thyrotropes constitute approximately 10% of the pituicyte population.

The gonadotropes are randomly distributed within the central wedge and the lateral wings of the gland. They account for approximately 10% of the total cell population. Immunoperoxidase stain localizes both FSH and LH to the same cell.

The lactotropes and somatotropes take eosinophilic stains, the corticotropes appear basophilic while the thyrotropes and gonadotropes do not stain well with either. The shortcomings of the conventional classification of pituitary tumor cells into eosinophilic, basophilic, and chromophobe adenomas are discussed in Chapter 16.

TABLE 2
Hypothalamic Hormones

Hormone	Chemical characterization
Thyrotropin-releasing hormone (TRH)	Pyroglutamyl histidylprolinamide
Gonadotropin-releasing hormone (GnRH)	Decapeptide
Corticotropin-releasing hormone	41 amino acids
Growth hormone-releasing hormone (GHRH)	40–44 amino acids, linear sequence
Somatostatin	Tetra decapeptide (somatostatin 14) and a larger 28-amino acid peptide
Prolactin-inhibiting factor	Dopamine
Prolactin-releasing factor	? Serotonin ? Vasoactive intestinal polypeptide (VIP)
Arginine vasopressin	Octapeptide
Oxytocin	Octapeptide

Physiological Considerations

Hypothalamic Factors/Hormones

The hypothalamus exerts its control over the pituitary gland by secreting several peptidergic hormones. In the case of TSH, LH, FSH, and ACTH, the hypothalamic control is represented by exclusively trophic influences from the hypothalamus. Thus TRH, GnRH, and corticotropin-releasing factor (CRF) are intimately involved in the secretion of TSH, the gonadotropins, and ACTH, respectively. In the case of growth hormone (GH) and prolactin (PRL), the hypothalamus exerts dual influences, both trophic as well as inhibitory. Thus, GH regulation is governed by growth hormone-releasing hormone (GHRH), or somatocrinin, as well as by growth hormone-inhibiting hormone, (GHIH) or somatostatin. Prolactin secretion is controlled by the tonic restraining effect of prolactin-inhibiting factor (PIF), which is in fact dopamine, as well as by the provocative influences of prolactin-releasing factors (PRF) secreted by the hypothalamus. In addition to these factors and hormones, the hypothalamus secretes arginine vasopressin (AVP), which acts on the renal tubules to conserve water, as well as oxytocin, which is essential for contraction of the pregnant uterus. Table 2 outlines the numerous peptides secreted by the hypothalamus. Each of these is briefly commented on. Figure 6 outlines the structure of TRH, LRH, and somatostatin.

Thyrotropin-Releasing Hormone

Historically, TRH was the first hypothalamic factor to be isolated and characterized from ovine and porcine hypothalami. TRH is a tripeptide and is often referred to by its chemical structure, pyroglutamyl histidylprolinam-

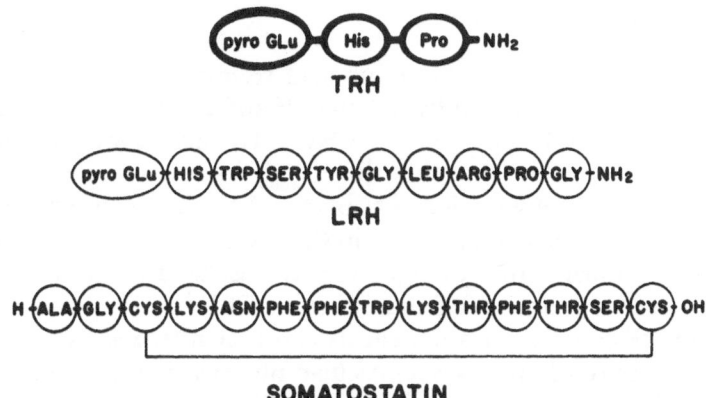

FIGURE 6. Structures of some hypothalamic-releasing hormones.

ide.[26,27] The synthesis of TRH is carried out in several hypothalamic nuclei, from which it is secreted into the portal venous system to reach its target cells, the thyrotropes of the anterior pituitary. Distribution studies, using immunohistochemical techniques, have provided evidence in delineating the preoptic suprachiasmatic nucleus, and the periparaventricular area as the thyrotropic area in the rat hypothalamus.[28] Jackson and Reichlin[29] demonstrated that ablation of this area in the rat is accompanied by depletion of nearly all hypothalamic TRH, with the consequent development of hypothalamic hypothyroidism. The TRH, thus secreted, is transported via the stalk to the thyrotropes of the anterior pituitary. Although the pituitary thyrotrope can be regarded as the major target cell for the action of TRH, distribution studies demonstrating the presence of TRH have yielded intriguing results regarding the ubiquitous nature of this peptide both within the CNS and elsewhere (see Chapter 5).

Gonadotropin-Releasing Hormone

Hypothalamic control over the secretion of LH and FSH by the pituitary gonadotropes is mediated by GnRH. This peptide is synthesized by hypothalamic neurons located primarily in two regions of the hypothalamus—the preoptic region and the arcuate region. The GnRH neurons at the preoptic region possess long axons that project into the median eminence; the GnRH neurons in the arcuate region are located more caudally and possess shorter axons that also project into the median eminence. Experiments with female rats have implied that GnRH neurons in the preoptic region are stimulated by estrogens to induce preovulatory surges of GnRH (positive feedback), while those neurons at the arcuate region are suppressed by estrogens (negative feedback). In addition to these two major sites, GnRH is also found in the lamina terminalis, a specialized structure located at the rostral tip of the third ventricle. Biochemically, GnRH is a decapeptide.[30,31]

Corticotropin-Releasing Hormone

The characterization of a 41-amino acid residue peptide with ACTH-releasing properties from sheep hypothalami[32] and the sequential analysis of its cDNA[33] led to the isolation of human CRF.[34] The homology between ovine and human CRF is so close that they differ only by seven amino acids. This fact, coupled with the observation that ovine CRF is a potent stimulator of ACTH release in humans, has resulted in the use of ovine CRF for extensive investigation in humans. Interestingly, the CRF isolated from the rat hypothalamus is closer to human CRF than to ovine CRF.

Human CRF contains 41 amino acids and is synthesized by neurons located in the vicinity of the paraventricular nucleus of the hypothalamus. Extensive studies in the rat brain have disclosed CRF immunoreactive neurons in the hindbrain and the limbic systems.[35,36] The significance of finding CRF receptors in these locations is unclear, but the proximity to the central sympathetic system suggests a relationship of CRF release in response to stress.

GHRH and Somatostatin

Dual influences from the hypothalamus regulate the amount of GH secreted. Peptidergic neurons located at specific anatomic areas of the hypothalamus secrete two important peptides that regulate GH secretion: somatotropin release-inhibiting factor, or somatostatin, and GHRH. The anatomical areas of the hypothalamus involved in the production of these peptidergic hormones are located in the medial hypothalamus. A whole body of animal experiments and clinical situations have pointed to the medial hypothalamus as the area governing the control of GH secretion. Electrical stimulation of certain parts of the hypothalamus results in the augmentation or abolition of GH release, depending on the site stimulated. Destruction of the medial basal hypothalamus (by tumors or disease) or interruption of the stalk, results in lowering of basal as well as stimulated GH secretory responses. To the contrary, lesions in certain parts of the hypothalamus may be associated with excess growth hormone secretion, presumably due to loss of the inhibitory influences of somatostatin.

The synthesis of somatostatin and GHRH is carried out by several important nuclei located in the medial hypothalamus. Two regions are particularly important in the functional regulation of GH release: the medial preoptic area (MPOA), which is primarily involved in the synthesis of somatostatin, and the ventromedial and arcuate area, which contains the ventromedial nucleus (VMN) and the arcuate nucleus (ARC). These two nuclei are primarily involved in the synthesis of GHRH. The lateral hypothalamus, which is in close proximity with the medial hypothalamus serves as a relay center connecting the medial hypothalamus to the limbic system and the brain stem of the CNS. There are demonstrable interconnections between the neurons of the medial and lateral portions of the hypothalamus. The MPOA consists of three important nuclei: medial preoptic, suprachiasmatic, and the periventricular nuclei. Of the three, the periventricular nucleus is the one with the heaviest concentration of cells that stain positive for somatostatin. By contrast, the ventromedial and arcuate nuclei contain GHRH but no somatostatin. The fibers from the arcuate nuclei project directly into the median eminence.

Although the site of production of somatostatin is believed to be in the medial preoptic area of the hypothalamus (and perhaps the D cells of the pancreatic islets), the distribution of somatostatin is impressively wide. The presence of somatostatin in such diverse tissues as the CNS, peripheral nervous system, and the GI tract suggests an important role for this hormone as a neurotransmitter or neuromodulator. Somatostatinergic neurons are found in high density within the medial preoptic area of the hypothalamus. Neural fibers containing somatostatin are found in the entire nervous system in the limbic area, cerebral cortex, brainstem, and spinal cord. The somatostatinergic nerve fibers that originate from the medial hypothalamus, particularly from the periventricular region, terminate in the median eminence. Next to the hypothalamus and the CNS, the highest content of somatostatin-containing cells are found in the pancreas, in islet cells termed D cells. These cells are in extreme proximity with other types of islet cells, particularly the α- and β-cells, an anatomical feature that perhaps confers a paracrine role to somatostatin.

The chemical configuration of the hypothalamic GHRH is identical to the growth hormone-releasing factor isolated from pancreatic tumors.[37–39] Two peptide sequences, hpGHRH 1–40 and 1–44, have been characterized. Both peptides are equal in biological potency. Hypothalamic somatostatin exists in two forms, 14 and 28 amino acid residues, both of which are powerful inhibitors of GH release.

Prolactin-Inhibiting Factor

The hypothalamus exerts a predominantly inhibitory control over prolactin secretion via prolactin inhibitory factor (PIF). Such a tenet is supported by several lines of evidence: (1) when the pituitary stalk is transected, prolactin levels rise, while there is a decline in the other adenohypophyseal hormones (the pituitary isolation syndrome); (2) crude hypothalamic extracts suppress release of prolactin from normal pituitary cells maintained in culture; and (3) when the pituitary is transplanted under the renal capsule in the experimental animal, the lactotropes escape from the tonic inhibitory hypothalamic control with a rise in prolactin levels. Finally, and most importantly, PIF has been identified in portal vessel blood, a criterion that establishes it as a hypophysiotropic factor.

The following lines of evidence indicate that PIF is in fact dopamine:

1. Dopamine and its agonists prevent the release of prolactin in vitro from cultured pituitary cells.
2. The inhibitory effect of dopamine on prolactin release can be blocked or neutralized by dopamine antagonists.[40] In this regard, clinical ex-

periments with a dopamine antagonist drug, called domperidone, are of interest.[47] This drug does not cross the blood-brain barrier and therefore does not reach the brain or the hypothalamus, thereby limiting its actions to the pituitary gland. Administration of this drug leads to a dramatic increase in the prolactin levels in humans. This indicates that dopamine is, in fact, PIF and its effects are exerted directly on the pituitary gland. Similar effects are seen with metoclopramide, another dopamine antagonist.

3. Dopamine infusion in humans leads to an immediate and sustained suppression of basal and stimulated prolactin release from the pituitary gland. This drug again does not cross the blood-brain barrier, limiting its actions to the pituitary gland.

4. Drugs such as phenothiazines and butyrophenones cause galactorrhea by depleting hypothalamic dopamine, decreasing PIF, and thereby increasing prolactin. The hypothalamic mediation of phenothiazine hyperprolactinemia is suggested by the fact that in stalk-sectioned rhesus monkeys the drug fails to cause hyperprolactinemia. In addition, these drugs may also have a direct effect on the pituitary by blocking dopaminergic receptors in the lactotropes.

5. L-Dopa and bromocriptine (a dopamine agonist) cause a lowering of the prolactin level by virtue of their dopaminergic effect.

In addition to dopamine, which is the principal PIF, minor peptides exhibiting PIF activity have been isolated from the hypothalamus; two in particular that have been noted to exhibit some prolactin inhibitory activity are α-amino butyric acid (GABA) and His-Pro-diketopiperazine, a breakdown product from the metabolism of TRH.

As far as the PRFs of the hypothalamus are concerned, several substances qualify; the most notable hypothalamic peptide that possesses prolactin-releasing properties is TRH, followed by vasopressin, vasoactive intestinal polypeptide (VIP), substance P, and β-endorphin.

Arginine Vasopressin and Oxytocin

AVP and oxytocin are synthesized by highly specialized neurosecretory cells located within the supraoptic and paraventricular nuclei located in the supraoptic region of the hypothalamus. These neurons, termed magnocellular neurons, owing to their strikingly large size, are consolidated into distinct groups situated in paired nuclei above the optic tract and on each side of the ventricle. The supraopticohypophyseal and paraventricular–hypophyseal nerve tracts originate from these neurons and descend through the infundibulum to terminate in the neural lobe. The use of specific antisera directed against vasopressin, oxytocin, and their respective neurophysins has confirmed that these neurosecretory cells contain one or the other hormone, i.e., vasopressin or oxytocin.

Pituitary Hormones

The pituitary hormones can be categorized into adenohypophyseal and neurohypophyseal hormones. The adenohypophyseal hormones are further subclassified as glycoprotein hormones (TSH, LH, FSH), and peptide hormones (GH, prolactin, and the POMC-derived peptides, ACTH, β-lipotropin, and β-endorphin). The neurohypophyseal hormones are represented by AVP and oxytocin, with their respective neurophysins (Table 3).

The synthesis of pituitary hormones is carried out by their respective neuroendocrine secretory cells. Subsequent to synthesis, the hormones are concentrated into condensed granules in the Golgi cysterni. Release of hormone occurs through a process of fusion with the plasma membrane, followed by extrusion into the extracellular space. Like many other hormones, the pituitary hormones are stored as two pools—a readily releasable pool of preformed hormone, and a less readily available pool that requires de novo synthesis. The labile pool can be easily depleted, but it is from the synthetic pool that new hormone is actively synthesized, packaged, and released in response to continuous stimulation. The synthetic machinery involved in the process is exceedingly active and highly responsive to stimulation by their respective hypothalamic releasing factors. The releasing factor–pituicyte interaction is mediated by sensitive receptors located at the membrane of the secretory cell. The mechanisms underlying the secretion and release of pituitary trophic hormones are cAMP dependent; however, the theory of calcium-mediated stimulus–secretion coupling is increasingly being proposed as a mechanism by which hypothalamic releasing hormones stimulate the pituitary cells. Such a mechanism has gained acceptance as the basis for the TRH-induced TSH release.

In the basal state the levels of the pituitary hormones in the circulation

TABLE 3
Pituitary Hormones

Adenohypophyseal hormones
  Glycoprotein hormones
    Thyroid-stimulating hormone (TSH)
    Luteinizing hormone (LH)
    Follicle-stimulating hormone (FSH)
  Peptide hormones
    Growth hormone (GH)
    Prolactin
    Pro-opio-melanocortin-derived peptides
      Adrenocorticotropic hormone (ACTH), or corticotropin
      β-Lipotropin
      β-Endorphins
Neurohypophyseal hormones
  Vasopressin
  Oxytocin

are impressively low. Basal secretion is defined as the secretory activity of the pituicytes in the absence of provocative or suppressive influences. Very little is known regarding the basal secretion of hormones by the pituicytes. However, this placid basal secretory actively is punctuated by bursts of pulsatile activity, particularly influenced by biological rhythms, sleep, and exercise. These rhythmic interruptions of the basal secretory rate are probably mediated by neuroregulatory influences from the cortex or the hypothalamus.

In contrast to the basal secretory rate of pituitary hormones, the stimulated output of these hormones has been studied extensively. The stimulatory influences that govern release of pituitary hormones are impressively diverse. The three major influences modulating the synthesis and release of pituitary hormones are the hypothalamic factors, biogenic amine neurotransmitters, and sympathetic mediation. To these three can be added a fourth factor, the direct effect of target gland hormones on the pituitary gland. Very often these influences overlap, working together in a highly complex but unified fashion. For instance, GH release by the somatotropes is under the trophic influence of GHRH from the hypothalamus. The neurotransmitter catecholamines, dopamine and serotonin, stimulate the release of GHRH. Central α-adrenergic stimulation also facilitates the release of GH, probably mediated by GHRH. Thus, when a hypoglycemic challenge is used to provoke the release of GH, the provocative response is brought about by the mediation of neurogenic (adrenergic) bioaminergic (catecholamines) and hypothalamic (GHRH) influences.

Table 4 outlines the various provocative influences that facilitate the release of growth hormone. It may well be that the final common pathway for GH release is hypothalamic GHRH regardless of the stimulus used or the mechanism by which the stimulus works. This is based on the observation that neurotransmitters work at the level of the brain and hypothalamus but have no effect on releasing growth hormone when applied directly to the pituitary.[41] The same principle probably extends to all pituitary hormones, in which the dominant provocative influence is their respective hypothalamic

TABLE 4
Stimuli for Growth Hormone[a]

Stimulus	GHRH	Neurotransmitter	α-Adrenergic
Hypoglycemia	+	NE, Serotonin	+
L-Dopa	+	DA	−
Vasopressin	−	NE	?
Glucagon	−	NE	−
Arginine	−	NE, Serotonin	−
Clonidine	+	−	+
Opioids	?	NE	−
Propranolol	−	−	β-Blocking effect
Sleep	+	Serotonin	−
Exercise	+	NE	?+

[a] DA, dopamine; NE, norepinephrine.

releasing factor, which in turn can be modulated by neurotransmitters and adrenergic receptors.

Feedback Regulation and Control

All the hormones secreted by the pituitary are controlled by feedback regulation. The pituitary hormones are unique in that they are regulated by influences from above (the hypothalamus) and from below (the target glands). Consequently, the functions of the thyroid, adrenal cortices, testes, and ovaries are modulated and fine-tuned by an intimate servo feedback mechanism involving the hypothalamic–pituitary unit. The term negative feedback implies a reciprocal relationship between target gland hormone and its trophic hormone; i.e., an increase in the concentration of target gland hormone will result in suppression of its respective trophic hormone, whereas decreasing concentrations of target gland hormone cause an increase in its respective trophic hormone level. Such is the relationship between the thyroid hormones and thyrotropin (TSH), and glucocorticoids and corticotropin (ACTH), as well as between testosterone and luteinizing hormone (LH). The effects of estrogens on the hypothalamic–pituitary unit in normal females is dichotomous, consisting of both negative and positive feedback regulatory mechanisms. The term *positive feedback* implies a linear relationship between target gland hormone and its trophic hormone; i.e., an increase in the circulating levels of target gland hormone results in stimulation of its respective trophic hormone. The most illustrative physiologic example of positive feedback between hormones is the LH surge that occurs during the mid-menstrual cycle, caused by a progressive increase in circulating concentrations of 17β-estradiol. Notably, this response is restricted only to females.

Feedback loops have also been categorized as long, short, and ultrashort feedback loops (Table 5):

1. Long feedback loops are exemplified by the feedback regulation between the hypothalamus and glucocorticoids or gonadal steroids. In the case of cortisol and testosterone, this feedback loop is exclusively a negative feedback loop, whereas in the case of 17β-estradiol this long feedback loop is both negative as well as positive under different circumstances.

2. Short feedback loops denote the servo feedback relationship between the pituitary gland and the hormones secreted by several of its target glands. The most impressive example of such a short feedback loop is the relationship between the pituitary TSH and the thyroid hormones. A similar situation is seen between ACTH and cortisol as well as between LH and gonadal steroids. Cortisol and the gonadal steroids feedback on both the hypothalamus (long negative feedback) as well as the pituitary (short negative feedback), whereas the thyroid hormones exert their negative feedback effect predominantly on the pituitary gland (short negative feedback loop).

TABLE 5
The Feedback Regulatory Loops[a]

Type of loop	Target gland hormone	Trophic hormone
I. Long feedback		
A. Negative feedback	Cortisol	CRH[b]
	Testosterone or 17β-estradiol	GnRH[c]
B. Positive feedback	17β-Estradiol	GnRH
II. Short feedback		
A. Negative feedback	Thyroid hormones	TSH
	Testosterone or 17β-estradiol	LH
	Cortisol	ACTH
B. Positive feedback	None	None
III. Ultrashort feedback		
A. Negative feedback	ACTH	CRH
	?TSH	TRH
B. Positive feedback	Growth hormone	Somatostatin
	Prolactin	Hypothalamic PIF[d] (dopamine)

[a] From Kannan.[42]
[b] CRH, corticotropin-releasing hormone.
[c] GnRH, gonadotropin-releasing hormone.
[d] PIF, prolactin inhibitory factor.

3. The term ultrashort feedback refers to the intimate relationship between the pituitary gland and its hypothalamic master. These loops can be negative or positive. Thus, a negative ultrashort feedback loop exists between ACTH and corticotropin-releasing hormone (CRH) of the hypothalamus. A similar loop perhaps exists between TSH and TRH as well. By contrast, a positive ultrashort feedback loop operates between growth hormone and somatostatin and perhaps between prolactin and hypothalamic dopamine.

Disorders of Structure and Function

Disorders of the pituitary represent a substantial percentage in the endocrinologist's practice. Whereas acromegaly and pituitary-dependent Cushing's disease are still rare, the problems of hyperprolactinemia, pituitary tumors, reproductive and menstrual disorders, impotence, and abnormal radiography of the sella constitute the most commonly encountered disorders of the pituitary gland. The availability of highly specific and sensitive radioimmunoassays as well as the emergence of high-resolution CT, have facilitated definitive and early diagnoses of syndromes that involve the pituitary and hypothalamus. The following chapters discuss several of these syndromes. The chapters are arranged to showcase the individual hormones of the pituitary. Thus, the opening chapter of each hormone deals with the physiology of the hormone, followed by chapters that deal with hyper- and hypofunctional syndromes involving these hormones. The chapters that deal with hypopituitarism, pituitary apoplexy, pituitary tumors, and the empty sella syndrome in a sense, cover generalized dysfunction of the pituitary gland. The perspective provided is primarily clinical, with a focus on understanding clinical research through the eyes of a clinician.

References

1. Medvei VC: *A History of Endocrinology.* MTP Press Limited, Boston, 1982.
2. Schneider CV: *Dissertatio de osse cribriforme, et sensu ac organo odoratus.* Wittebergae, Mevii, Mevii, 1655.
3. Lower R: Dissertatio de origine catarrhi in qua ostenditur illum non provenire a cerebro. In *Tractatus de corde.* Londini, typ. J. Redmayne, 1670, p. 221.
4. Zuckerman Sir Sy: Secretions of brain; relation of hypothalamus to pituitary gland (Addison Memorial Lecture). *Lancet* 1: 739, 1954.
5. Horsley Sir Vr: Functional nervous disorders due to loss of thyroid gland and pituitary body. *Lancet* 2: 5, 1886.
6. Marie P: Sur deux cas d'acromégalie. Hypertrophie singulière noncongénitale des extrémité supérieures, inférieures et céphaliques. *Rev Med* 6:297, 1886.
7. Paulesco NC: L'hypophyse du cerveau. *J Physiol Pathol Gen* 9:441, 1907.
8. Cushing HW: The hypophysis cerebri. *JAMA* 53:250, 1909.
9. Langon-Brown Sir W: *Practitioner* 127:614, 1931.
10. Hubble Sir D: The endocrine orchestra. *Br Med J* 1:523, 1961.
11. Wade N: *The Nobel Duel.* Anchor/Doubleday, Garden City, NY, 1981.
12. Guillemin R, Sakiz E, Ward DN: Further purification of TSH releasing factor from sheep hypothalamic tissues, with observations on the amino acid composition. *Proc Soc Exp Biol Med* 118:1132, 1965.
13. Schally AV, Bowers CY, Redding TW, et al: Isolation of thyrotropin releasing factor (TRF) from porcine hypothalamus. *Biochem Biophys Res Commun* 25:165, 1966.
14. Schally AV, Arimura A, Baba Y, et al: Isolation and properties of the FSH and LH releasing hormone. *Biochem Biophys Res Commun* 43:393, 1971.
15. Brazeau P, Vale W, Burgus R, et al: Hypothalamic polypeptide that inhibits the secretion of immunoreactive pituitary growth hormone. *Science* 179:77, 1973.
16. Vale W, Spiess J, Rivier C, et al: Characterization of a 41 residue ovine hypothalamic peptide that stimulates secretion of corticotropin and β-endorphin. *Science* 213:1394, 1981.
17. Guillemin R, Brazeau P, Bohlen P, et al: Growth hormone-releasing factor from a human pancreatic tumor that caused acromegaly. *Science* 218:585, 1982.
18. Rivier J, Spiess J, Thorner M, et al: Characterization of a growth hormone-releasing factor from a human pancreatic islet tumour. *Nature (Lond)* 300:276, 1982.
19. Medvei VC: *A History of Endocrinology.* MPT Press Limited, Boston, 1982, p. 562.
20. Green JD: The comparative anatomy of the hypophysis with special references to its local blood supply and innervation. *Am J Anat* 88:225, 1951.
21. Duvernoy H: Brain–endocrine interaction. In Knigge KM, Scott DE, Weindl BA (eds): *Median Eminence: Structure and Function* Karger, Basel, 1972, p. 79.
22. Knowles F: Ependymal function of the third ventricle in relation to pituitary function. In Kapners A, Shade JP (eds): *Progress in Brain Research. Topics in Neuroendocrinology,* Vol 38 Elsevier, Amsterdam, 1972, p. 255.
23. Scott DE, Knigge KM: Ultrastructural changes in the median eminence of the rat following deafferentiation of the basal hypothalamus. *Z Zellforsch* 105:1, 1970.
24. Page RB, Bergland RM: Pituitary vasculature. In Allen MB, Mahesh VB (eds): *The Pituitary. A Current Review* Academic Press, New York, 1977, p. 9.

25. Berland RM, Page RB: Can the pituitary secrete directly to the brain? *Endocrinology* **102:**1325, 1978.
26. Nair RMG, Barrett JF, Bowers CY, et al: Structure of porcine thyrotropin releasing hormone. *Biochemistry* **9:**1103, 1970.
27. Burgus R, Dunn TF, Desiderio D, et al: Characterization of ovine hypothalamic hypophysiotropic TSH-releasing factor. *Nature (Lond)* **226:**321, 1970.
28. Johansson O, Hokfelt T: Thyrotropin release hormone, somatostatin, and enkephalin: Distribution studies using immunohistochemical techniques. *J Histochem Cytochem* **28:**364, 1980.
29. Jackson IMD, Reichlin S: Brain thyrotropin-releasing hormone is independent of the hypothalamus. *Nature (Lond)* **267:**853, 1977.
30. Matsuo H, Babo Y, Nair RMG, et al: Structure of the porcine LH- and FSH-releasing hormone. I. The proposed amino acid sequence. *Biochem Biophys Res Commun* **43:**1334, 1971.
31. Burgus R, Butcher M, Amoss M, et al: Primary structure of the ovine hypothalamic luteinizing hormone-releasing factor (LRF). *Proc Natl Acad Sci USA* **69:**278, 1972.
32. Vale W, Spiess J, Rivier C, et al: Characterization of a 41-residue ovine hypothalamic peptide that stimulates secretion of corticotropin and beta-endorphin. *Science* **213:**1394, 1981.
33. Furutani Y, Morimoto Y, Shibahara S, et al: Cloning and sequence analysis of cDNA for ovine corticotropin-releasing factor precursor. *Nature (Lond)* **301:**537, 1983.
34. Shibahara S, Morimoto Y, Furutani Y, et al: Isolation and sequence analysis of the human corticotrophin-releasing factor precursor gene. *EMBO J* **2:**775, 1983.
35. Olschowka JA, O'Donohue TL, Mueller GP, et al: The distribution of corticotropin releasing factor-like immunoreactive neurons in rat brain. *Peptides* **3:**995, 1982.
36. De Souza EB, Perrin MH, Insel TR, et al: Corticotropin-releasing factor receptors in rat forebrain: Autoradiographic identification. *Science* **224:**1449, 1984.
37. Spiess J, Rivier J, Thorner M, et al: Sequence analysis of a growth hormone releasing factor from a human pancreatic islet tumor. *Biochemistry* **21:**6037, 1982.
38. Brazeau P, Ling N, Bohlen P, et al: Growth hormone releasing factor somatocrinin, release pituitary growth hormone in vitro. *Proc Natl Acad Sci USA* **79:**7909, 1982.
39. Rosenthal SM, Schriock EA, Kaplan SL, et al: Synthetic human pancreatic growth hormone releasing factor stimulates growth hormone secretion in normal man. *J Clin Endocrinol Metab* **57:**677, 1983.
39. Caron MG, Beaulieu M: Dopaminergic receptors in the anterior pituitary gland. *J Biol Chem* **253:**2244, 1978.
40. Scanlon MF, Pourmond M, et al: Some current aspects of clinical and experimental neuroendocrinology with particular reference to growth hormone, thyrotropin and prolactin. *J Endocrinol Invest* **2:**307, 1979.
41. Martin JB, Brazeau P, Tannenbam GS: Neuroendocrine organization of growth hormone regulation. In Reichlin S, Baldessarini RJ, Martin JB (eds): *The Hypothalamus.* Vol. 56. Raven, New York, 1978, pp. 329–357.
42. Kannan C: *Essential Endocrinology: A Primer for Nonspecialists,* Plenum Press, New York, 1986, p. 6.
43. Reichlin S: Anatomical and Physiological Basis of Hypothalamic Pituitary Regulation. In Post KD, Jackson IM, Reichlin S (eds): *The Pituitary Adenoma,* Plenum Press, New York, 1980, pp. 3–28.
44. Wolpert SM: The Radiology of Pituitary Adenomas: An Update. In Post KD, Jackson IM, Reichlin S (eds): *The Pituitary Adenoma,* Plenum Press, New York, 1980, pp. 287–320.
45. Post KD, Kasdan DL: Sellar and Parasellar Lesions Mimicking Adenoma. In Post KD, Jackson IM, Reichlin S (eds): *The Pituitary Adenoma,* Plenum Press, New York, 1980, pp. 159–218.

Growth Hormone

Structure and Synthesis

Human growth hormone (GH) is secreted by the somatotropes of the anterior pituitary gland. The somatotropes are acidophilic cells resembling, in this regard, the lactotropes. On electron microscopic study, somatotropes can be distinguished from lactotropes by their uniformity in appearance and by the size of the electron-dense granules contained within. These secretory granules are spherical, measuring approximately 300–400 nm, and appear evenly electrodense. Immunoelectron microscopy localizes GH in these secretory granules. The somatotropes demonstrate a well-developed rough endoplasmic reticulum (RER) with conspicuous Golgi bodies.

Growth hormone is 'a single-chain polypeptide containing 191 amino acids, with a molecular weight of 21,500.[1] GH has a single homologous tryptophan residue at locus 85 and two homologous disulfide bonds. GH, prolactin

(PRL), and human chorionic somatomammotropin (CSM) are closely related peptides, with similar size, structure, immunochemistry, and some aspects of function. It is believed, as originally postulated by Niall et al.,[2] that these three closely related peptides have developed by duplication of a single ancestral hormone gene. In a review resonating with clarity, Miller and Eberhardt[3] traced the evolution and structure of the GH gene family. The GH gene is located on the long arm of chromosome 17. It has been shown, using molecular cloning techniques, that the sequence of the cDNA (the DNA complementary to RNA) of bovine, rat, and human GH demonstrates 75% homology of nucleotide.[4]

Extensive studies involving the analysis of bacteriophage λ recombinants containing GH genes have revealed that these are arranged in two clusters on the long arm of chromosome 17. Of these, the first linkage group (GH gene-1) appears to be closely related to GH production. The importance of GH gene-1 has assumed significance by the revelation that deletion of GH gene-1 is associated with primary GH deficiency.[5] Although several variants of GH have been extracted from the human pituitary gland, only one (the 20-K variant) seems to be encoded by a separate messenger RNA (mRNA), which is also derived from the authentic GH gene-1. This variant of GH is believed to constitute 15% of pituitary GH and 5% of the secreted hormone.[6] Interestingly, the 20-K variant does possess growth-promoting and lactogenic activities, identical to those displayed by authentic GH, but may lack effects on carbohydrate metabolism.[7] Studies on rat GH genes have indicated that glucocorticoids and triiodothyronine (T3) increase GH gene transcription.[8] In the absence of thyroid hormones or glucocorticoids, GH synthesis is extremely low; the two hormones increase GH mRNA synergistically.[9]

Following synthesis, GH is released by a process of sequential movement of the secretory granules from their site of formation, through the Golgi apparatus to the plasma membrane. Having reached this site, the secretory granule fuses with the plasma membrane, and the hormone is released by a process of extrusion. While mediation by cyclic adenosine monophosphate cAMP, and probably cyclic guanosine monophosphate (cGMP), have been demonstrated for stimulated secretion, there is no consensus on the role of these factors in the basal secretion of GH.

Secretory Patterns of Growth Hormone

Growth hormone secretion, like many other pituitary hormones, is subject to rhythmic flux. For instance, the normal basal levels during the resting phase are quite low, often indistinguishable from patients with GH deficiency. The two major physiological factors, which during the course of the day spike the flat basal secretory state, are sleep and activity. It is well established that GH secretion markedly rises during deep slumber. This *nyctohemerel* elevation of GH has been documented extensively and has been shown to be related to stages III and IV of electroencephalographically (EEG) determined deep sleep.[10–14] The sleep-associated rise in GH is absent in blind subjects,[12] in infants under 3 months of age,[13] and in some patients with acromegaly or pituitary-dependent Cushing's disease. It is remarkable that the sleep-related GH rise cannot be abolished by hyperglycemia, an observation that points to nonadrenergic mechanisms mediating such a release. GH release is often pulsatile, demonstrating numerous fluctuations in the plasma levels during the day. These spontaneous surges are more evident during adolescence and in pathological hypersecretory states. The spontaneous fluctuation in GH levels is believed to reflect the stimulatory and inhibitory influences exerted on the anterior pituitary by the hypothalamus.

Growth hormone circulates in the plasma unbound to proteins, with a half-life ($t_{\frac{1}{2}}$) of 15–45 min. A single, random determination of GH does not reflect the integrity of somatotrope function. Like other hormones of the anterior pituitary, integrity of function would require evaluation of GH reserve capacity using provocative maneuvers.

Control of Growth Hormone Secretion

The regulatory mechanisms that control GH secretion are strikingly diverse. For purposes of discussion, the control of GH secretion is viewed from four perspectives: (1) the hypothalamic regulation of GH secretion, focusing on growth hormone-releasing hormone (GHRH) and somatotropin release-inhibiting factor (SRIF), or somatostatin; (2) the role of neurotransmitter modulation of GH secretion, particularly by the bioamines, dopamine, norepinephrine, and serotonin; (3) the modulation of GH secretion by amino acids and peptide hormones; and (4) the feedback regulation of GH secretion.

Hypothalamic Regulation

The minute-to-minute regulation of GH secretion by the somatotropes is under the elegant control exerted on these cells by the hypothalamus. Dual influences from the hypothalamus regulate the amount of GH secreted. Peptidergic neurons located at specific anatomical areas of the hypothalamus secrete two important peptides that regulate GH secretion: SRIF, or somatostatin, and GHRH. The latter has recently been christened somatocrinin, from the Greek words *somatin*, for somatotropin, and *crinin*, meaning "to secrete." The anatomical areas of the hypothalamus involved in the production of these peptidergic hormones are located in the medial hypothalamus. A whole body of animal experiments and clinical situations have pointed to the medial hypothalamus as the area governing the control of GH secretion. Electrical stimulation of certain parts of the hypothalamus results in the augmentation or abolition of GH release, depending on the site stimulated. Destruction of the medial basal hypothalamus (by tumors or disease) or interruption of the stalk results in lowering of basal as well as GH-stimulated secretory responses. By contrast, lesions in certain parts of the hypothalamus may be associated with excess GH secretion, presumably due to loss of the inhibitory influences exerted by somatostatin.

The synthesis of somatostatin and GHRH are carried out by several important nuclei located in the medial hypothalamus. Two regions are particularly important in the functional regulation of GH release: the medial preoptic area (MPOA), which is primarily involved in synthesis of somatostatin, and the Ventromedial and arcuate area, containing the ventromedial nucleus (VMN) and the arcuate nucleus (ARC). These two nuclei are involved primarily in the synthesis of GHRH. The lateral hypothalamus, which is in close proximity to the medial hypothalamus, serves as a relay center connecting the medial hypothalamus to the limbic system and the brain stem of the CNS. There are demonstrable interconnections between the neurons of the medial and lateral portions of the hypothalamus. The MPOA consists of three important nuclei: medial preoptic, suprachiasmatic, and the periventricular nuclei. Of the three, the periventricular nucleus is the one having the heaviest concentration of cells that stain positive for somatostatin. By contrast, the ventromedial and arcuate nuclei contain GHRH but no somatostatin. It has recently been shown that high concentrations of immunoreactive GHRH are also present in the infundibular nucleus.[15] The arcuate nucleus is also rich in dopamine and is viewed as the primary site for the presence of these transmitters. The fibers from the arcuate nucleus project directly into the median eminence. The median eminence is a specialized structure containing

TABLE 6
Hypothalamic Regulation of GH

Anatomical site	Comment
Lateral hypothalamus	Connects the medial hypothalamus with the CNS (limbic system, brain stem); serves as a conduit for the ascending and descending fibers of medial forebrain bundle
Medial hypothalamus	
Medial preoptic area (MPOA) [inhibitory]	
Medial preoptic nucleus	Connects with other nuclei and the median eminence; contains cells that stain for somatostatin
Suprachiasmatic nucleus	Important in preserving circadian pattern of rhythms; regulates the light–dark entrainment of GH
Periventricular nucleus	Primary area for somatostatin secretion
VMN-ARC area [stimulatory]	
Ventromedial nucleus	Contain GHRH staining cells
Arcuate nucleus	Major site for GHRH synthesis; rich in dopamine; projects into median eminence
Median eminence	Major site for neurovascular anastomoses; rich in neurotransmitters, several of which regulate GH secretion

neurovascular anastomoses with nerve terminals that end on the small capillaries of the portal system of vessels supplying the pituitary. There seems to be a free communication between the extracellular space of the median eminence and substances in the blood. The median eminence is extremely rich in its content of neurotransmitters. The importance of the medial hypothalamus, the lateral hypothalamus, and the median eminence in the regulatory control of GH secretion is outlined in Table 6.

The isolation and characterization of GHRH and somatostatin has had considerable impact on our understanding of regulation of GH secretion by the hypothalamus. A discussion of the structure, dynamics, and actions of these two peptides follows.

Growth Hormone-Releasing Hormone

Although the existence of GHRH was strongly suggested by a large body of experimental and clinical evidence, the isolation and characterization of the hormone remained unsuccessful. Technology was not a problem, since several other hypothalamic peptides had already been identified and chemically synthesized—thyrotropin-releasing hormone (TRH), gonadotropin-releasing hormone (GnRH), somatostatin, and so on, while GHRH continued to remain elusive. The major hurdle, it seemed, was the exceedingly small amounts of biologically active hormone present in hypothalamic extracts.[16,17] A major breakthrough in the characterization of GHRH came from an unexpected source; in 1982, Thorner et al.[18] reported an acromegalic patient with an enlarged sella, pituitary somatotrope hyperplasia (found at surgery), persistent GH hypersecretion after partial hypophysectomy, and a pancreatic tumor. Apparently, the acromegaly had resulted from somatotrope hyperplasia secondary to chronic stimulation by the ectopic secretion of a GHRH-like substance by the pancreatic tumor. In contrast to the normal hypothalamus, the content of this GHRH-like peptide in the pancreatic tissue was extremely high, permitting the characterization and purification of this peptide. Although partial purification of such a peptide in the past had been reported by Frohman et al.[19] in patients with acromegaly caused by nonpituitary tumors, complete characterization of the peptide was yet to come. In 1982, two groups of investigators characterized a family of peptides with GH-releasing properties obtained from pancreatic tumor extracts; Rivier et al.[20] sequenced and characterized the GHRH-like peptide as a linear peptide with 40 amino acids demonstrating remarkable potency in stimulating GH release. Guillemin et al.,[21] the same year, reported the isolation of a GHRH-like peptide from pancreatic tumors; their peptide consisted of 44 amino acids, with a terminal amide group. Regardless of the minor difference in structure, the biological activity of both peptides was identical, suggesting that the secretory machinery of these tumors in different patients is identical as well. The term human pancreatic GHRF (hpGHRF) is often used to denote the origin of this peptide in contrast to native hypothalamic GHRH. The distinction between GHRH and hpGHRF is probably limited only to their origin. Antibodies raised

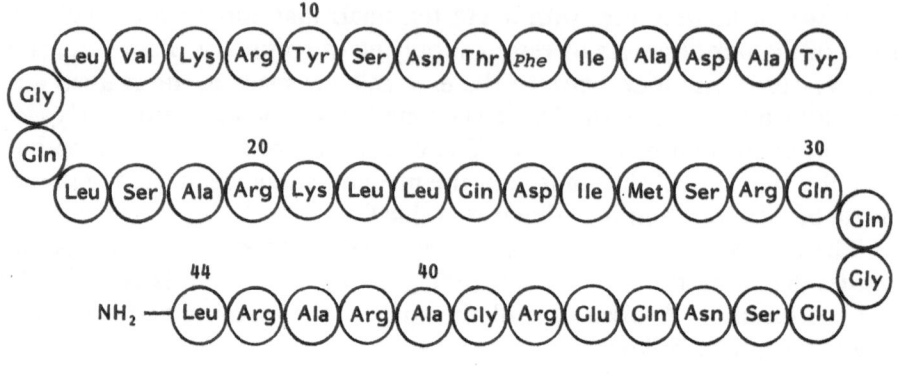

hp GRF

FIGURE 7. Structure of hpGHRH.

against hpGHRF cross-react with cells located in the ventromedial and arcuate nuclei, which are established sites for production of GHRH. More recently, Spiess et al.[22] characterized the GHRF isolated from the rat hypothalamus and showed significant homology to the human pancreatic GHRF (Fig. 7).

Actions. The availability of synthetic GHRF has established the potency of this peptide in releasing GH in normal subjects.[23–25] The selective nature of the peptide in stimulating only GH release is also well established.[23] Both peptides, the GHRH 1–40, as well as the GHRH 1–44 amide are synthetically available and have been used to study the GH response in physiological and pathological states.

The precise mechanism by which GHRH stimulates the secretion of GH by the somatotropes is not well understood. Barinaga et al.[26] showed that GHRH regulates the transcription of GH gene expression; these workers suggested that normal pituitary cells demonstrate both release as well as transcriptional responses to GHRH. Studies by Lewin et al.[27] support the notion that GHRH combines with a protein kinase to activate exocytosis and enhance GH secretion. Following an intravenous bolus of exogenous GHRH, a prompt release of GH is noticeable, beginning within 20 min, and reaching a peak at 30–40 min after the bolus. The response of GH secretion to a continuous infusion is similar, except that the GH response is not sustained during the infusion.[28,29] Also, the response to a bolus, at a dose that was previously effective, is blunted when given following the infusion, suggesting depletion of the rapidly releasable pool of hormone, or receptor desensitization. The GHRH-mediated GH release can be blunted by hyperglycemia,[30] by hypothyroidism,[31] or with advancing age.[32]

Clinical Applications. Within the short period since its discovery, synthetic GHRH has found diagnostic, and possibly therapeutic, applications. Four applications in particular deserve mention:

1. *Differential diagnosis between hypothalamic and pituitary GH deficiency:* This application can be attempted with the use of GHRH 1–40 or GHRH 1–44. Laron et al.[33] studied the effects of synthetic GHRH 1–44 administered as an intravenous bolus to children with constitutional short stature and those diagnosed as having GH deficiency. They were able to identify a subset of patients with GH deficiency who responded to GHRH administration while not responding to either insulin hypoglycemia or clonidine, or both. This is indicative of the fact that, at least in some instances, GH deficiency can occur as a result of hypothalamic deficiency of GHRH. Similar results have been reported by others.[34]

2. *Identification of ectopic GHRH production in acromegaly:* The plasma GH response to GHRH may help in identifying ectopic GHRH production in these patients. Those with untreated acromegaly caused by pituitary tumor often respond to GHRH with a further rise in the basal GH levels.[35] Schulte et al.[36] described a patient with acromegaly caused by ectopic GHRH secretion from a pancreatic tumor, who failed to demonstrate a rise in GH after GHRH. Similarly, Ch'ng et al.[37] described a female patient with acromegaly, hypercalcemia, and Zollinger-Ellison syndrome who demonstrated high concentrations of basal GH levels without a rise after administration of synthetic GHRH.

3. *GHRH as a means of ascertaining cure of acromegaly following surgery:* This application is under investigation. Losa et al.[38] demonstrated normalization of the GH response to GHRH after transphenoidal surgery. Their study, encompassing 20 patients with active acromegaly, represents the largest series of acromegalics (as of 1985) studied with GHRH. Their data indicated that while the test was not extremely useful in establishing the diagnosis of hypersomatotropism, it can be used as one additional parameter to assess normalization of GH secretion following surgery.

4. *The therapeutic use of synthetic GHRH in idiopathic GH deficiency:* This application is also under study. Borges et al.[39] studied the response to intermittent pulsatile administration of synthetic hpGHRF-40 in six adult subjects with a diagnosis of idiopathic GH deficiency. Their preliminary data revealed that 5 days of intermittent administration of the analogue augmented GH secretion in some of these patients, suggesting the presence of viable somatotropes capable of being primed by the peptide. More encouraging was the demonstration of a rise in serum levels of somatomedin C after multiple injections of hpGHRH-40. If the therapeutic use of this peptide is confirmed in long-term studies, the impact of such a finding would be significant as an alternative for GH therapy. This is especially so in view of the expense and the adverse effects of GH treatment.[40] Thorner et al.[41] treated two GH-deficient children with GHRH for 6 months and demonstrated a significant acceleration of growth in both. Similarly, Gelato et al.[42] assessed the efficacy of pulsatile administration of GHRH on

short-term linear growth in seven children with GH deficiency. These investigators showed that repeated administration of GHRH stimulated GH, somatomedin level, and linear growth in some, but not all, patients treated.

Somatostatin[43]

The discovery of growth hormone inhibitory factor (GHIF) was quite accidental, occurring during a search for the releasing factor of GH in the hypothalami of rats.[44] A resurgence of interest in this factor did not occur until 1972, when Brazeau et al.[45] from the Salk Institute isolated a peptide from sheep hypothalami that inhibited the secretion of GH. Although the term *somatostatin* is in common usage to describe this peptide, such a term is quite restrictive, in light of the bewildering array of activities possessed by this hormone. The original description of somatostatin as a tetradecapeptide (14 amino acids) has expanded to include a 28-amino acid peptide (somatostatin 28) and a larger prohormone of somatostatin-14.

Although the site of production of somatostatin is believed to be in the medial preoptic area of the hypothalamus (and perhaps the D cell of the pancreatic islets), the distribution of somatostatin is impressively wide.[46–57] The presence of somatostatin in such diverse tissues as the CNS, peripheral nervous system, and the gastrointestinal (GI) tract suggests an important role for this hormone as a neurotransmitter or neuromodulator. Table 7 outlines the anatomical distribution of somatostatin. Somatostatinergic neurons are found in high density within the medial preoptic area of the hypothalamus. Neural fibers containing somatostatin are found in the entire nervous system

TABLE 7
Anatomical Distribution of Somatostatin

Hypothalamic pituitary unit
    Medial hypothalamus
    Median eminence
    Neurohypophysis
    Anterior pituitary
CNS (extrahypothalamic)
    The limbic system[44]
    Cerebral cortex[45]
    Brain stem[46]
    Spinal cord[46]
Peripheral nervous system
    Sensory neurons[49]
    Sympathetic neurons[50]
Gastrointestinal tract
    Pancreas[51–53]
    Stomach and intestines[55]
Other sites
    Parafollicular cells of thyroid[56]
    Renal tubules[57]

in the limbic area, cerebral cortex, brain stem, and spinal cord. The somatostatinergic nerve fibers that originate in the medial hypothalamus, particularly the periventricular region, terminate in the median eminence. Next to
the hypothalamus and the CNS, the highest content of somatostatin-containing cells is found in the pancreas, in islet cells termed D cells. These cells are
in extreme proximity with other types of islet cells, particularly the α and β
cells. It is this close anatomical contact that has led to the concept of paracrine
control exerted locally by somatostatin on the secretion of insulin and glucagon.

Synthesis. The synthesis of somatostatin has been well studied in tissue
culture, especially from human pancreatic tumor tissue.[58] It is noteworthy
that tissue obtained from pancreatic tumors has provided the basis for studying the sequence of both GHRH as well as somatostatin. Somatostatin is synthesized as a pre-prohormone. After translocation into the Golgi apparatus,
the precursor peptide is cleaved into prohormone, which is further processed
into somatostatin-14 and somatostatin-28 within the secretory granules. While
both forms of somatostatin are quite potent in suppressing GH, the 28-amino
acid residue is more potent in suppressing insulin release.[59]

Actions. The actions of somatostatin are as diverse as its distribution;
they can be viewed in terms of the effects of this hormone on the pituitary
and the extrapituitary sites.

Effects on the Pituitary. The effects of somatostatin on the pituitary are
most pronounced on the somatotropes. Somatostatin inhibits the release of
GH in normals. It is also a powerful inhibitor of the GH response to a variety
of provocative stimuli. Thus, prior infusion of somatostatin inhibits the normal
rise in GH following L-dopa,[60] arginine,[60] stress,[61] hypoglycemia,[62] sleep,[63]
starvation,[64] and even GHRH.[65] Thus, somatostatin is extremely potent in its
ability to completely abolish the release of GH from the somatotropes. The
ability of somatostatin in suppressing somatotrope function extends to acromegalics, in whom marked inhibition of GH levels is seen following administration of synthetic somatostatin.[66–69] The suppressive effect of somatostatin
on normal as well as hypersecretory somatotropes is short lived, however, and
a rebound increase in GH levels is seen in normals as well as in acromegalics
following discontinuation of the infusion of somatostatin.[66,70] Curiously, this
postinhibitory rebound is not observed when somatostatin is administered
during the GH ebb in normal rats or when given to animals with lesions of
the ventromedial nucleus. The consensus of opinions is that GH secretion is
a net effect of the interaction between GHRH and somatostatin at the somatotrope level. The dominant force, however, appears to be that of somatostatin. Direct evidence for a relationship between somatostatin regulation
and GH release has been provided by experiments conducted in rats by Kasting et al.[71] By the use of a stereotactically implanted push–pull type of cannula
in the median eminence, these workers showed GH secretion to be closely

correlated to the release of somatostatin into the perifusate. The close relationship between GH and somatostatin is also illustrated by the positive feedback relationship between the two; when the hypothalamus is exposed to GH, somatostatin release is enhanced, resulting in decreased GH secretion.[72-75]

In addition to its powerful effects on inhibition of GH secretion, somatostatin also inhibits the TRH-induced release of thyroid-stimulating hormone (TSH). Concomitant administration of somatostatin and TRH significantly inhibits the release of TSH induced by TRH without affecting the TRH-induced prolactin release.[76] Somatostatin also lowers the basal TSH levels in normals as well as in patients with primary hypothyroidism.

Somatostatin does not exert any effect on the secretion of adrenocorticotropic hormone (ACTH) or prolactin under normal circumstances. However, somatostatin can lower basal levels of prolactin in some acromegalics,[77] as it can lower basal ACTH levels in some patients with Addison's disease. Although it is believed that somatostatin has no effect on gonadotropin secretion, Millar et al.[78] showed that somatostatin-28 inhibits the LHRH-stimulated gonadotropin secretion in humans.

Extrapituitary Effects.   The extrapituitary effects of somatostatin are intriguing. These effects pertain to its role in affecting the pancreatic islet cell function, as well as its effects on the GI tract at large. Somatostatin infusion causes inhibition of secretion of insulin and glucagon.[79,80] De Vane et al.,[81] among others, demonstrated that synthetic somatostatin, when infused into normal subjects, causes an acute decline in insulin levels as well as glucose levels. The dual effects of somatostatin on both α- and β-cell function have been studied extensively. Gerich et al.[82] evaluated the effects of somatostatin on plasma glucose and glucagon levels in patients with diabetes mellitus. These workers noted a significant reduction in fasting glucagon in the plasma, as well as complete abolition of postmeal hyperglycemic excursions in some patients. The effects were noticeable even in hypophysectomized diabetics, indicating that these effects were exerted independent of the inhibitory effects of somatostatin on GH. These observations have been supported by in vitro data. Gerich et al.[83] also demonstrated that somatostatin inhibits the release of glucagon and insulin from the perfused pancreas of the rat, in response to a variety stimuli such as arginine, isoproterenol, and theophylline. In vitro studies seem to indicate that somatostatin has a preferential inhibitory effect on glucagon release. The histological observation that the D cells (presumably the site of synthesis of pancreatic somatostatin) are in close anatomical proximity to the α- and β-islet cells may have important connotations. Viewed in light of the hormonal effects of somatostatin on insulin and glucagon release, this anatomical proximity suggests that somatostatin must have an important local regulatory role in the secretion of insulin and glucagon by the pancreatic islet cells. The observation that somatostatin receptors can be demonstrated on the α- and β-islet cells has added support for such a role.[84]

In addition to its effects on the secretion of insulin and glucagon, somatostatin exhibits an array of activities on the GI tract.[85,86] The major effects of somatostatin on the gastrointestinal tract, reside in its ability to inhibit secretion of a variety of gut hormones,[87-91] as well as to inhibit exocrine secretion, motor activity, absorption of nutrients, and blood flow to the gut. The gut hormones that decline following the administration of somatostatin include pancreozymin, secretin, motilin, vasoactive intestinal polypeptide (VIP), and gastric inhibiting polypeptide (GIP). The most notable effects of somatostatin on exocrine secretion are its effects on inhibition of gastric acid[92] (and pepsin) as well as its inhibitory effects on pancreatic enzyme secretion and bile. In addition to these effects on gut hormone secretion and exocrine secretion, somatostatin inhibits gastric emptying and gallbladder contraction. Finally, somatostatin impairs absorption of nutrients across the mucosa of the gut, the most notable of which are glucose, amino acids, calcium, and water.

In addition to its effects on the pituitary and the GI tract, somatostatin functions in the nervous system as a neurotransmitter or neuromodulator.[93] When somatostatin is experimentally applied to brain preparation, there is a noticable decrease in the spontaneous as well as the evoked excitability of neurons. Guillemin[94] showed that somatostatin inhibits the release of acetylcholine (ACh)-induced electrically. A similar inhibitory effect on norepinephrine release has also been reported.[95] The wide variety of roles assumed by somatostatin is outlined in Table 8.

Clinical Implications and Applications. The only clinical condition that has been clearly linked to somatostatin is the somatostatinoma of the pancreatic islets.[96-102] This tumor secretes excessive amounts of somatostatin. Somatostatinomas are rare and manifest a highly nonspecific presentation consisting of weight loss, steatorrhea, dyspepsia, cholelithiasis, diabetes, and hypochlor-

TABLE 8
Actions of Somatostatin

As a neurohormone
  On anterior pituitary function
    Inhibits GH release
    Inhibits TSH release
As a true hormone
  On islet cell function
    Inhibits insulin release
    Inhibits glucagon release
  On GI tract
    Inhibits gastric acid secretion
    Inhibits pancreatic exocrine secretion
    Inhibits release of guthormones (pancreozymin, CZK, VIP, GIP, motilin)
    Decreases gastric emptying
    Decreases blood flow
    Decreases absorption of nutrients
As a paracrine hormone (local regulation)
  Regulates α- and β-islet cell function
As a neurotransmitter/neuromodulator

TABLE 9
Therapeutic Applications of Somatostatin Analogues

Acromegaly, resistant to conventional therapy
Hypersecretory islet cell adenomas
  VIPOMA
   Carcinoid
   Insulinoma
Bleeding peptic ulcer disease
Acute pancreatitis
Brittle diabetes

hydria. The hormonal hallmark is the demonstration of markedly elevated plasma levels of somatostatin. In most instances, there is concomitant hypersecretion of other hormones by the tumor (most notably gut hormone peptides) and calcitonin. These patients also demonstrate classic abnormalities in pituitary function: impairment of GH and TSH release in response to their respective provocative maneuvers. These tumors are nearly always malignant and can be diagnosed either by measuring somatostatin levels in plasma or by extracting the peptide from tumor tissue.

As far as a role for somatostatin in the pathogenesis of any form of diabetes, this is largely undefined. Decreased D-cell populations have been described in some animal models of diabetes,[103,104] but the significance of these histological findings is far from clear.

The diagnostic indication for obtaining plasma levels of somatostatin is limited to the diagnosis of somatostatin producing tumors of the pancreatic islet cells.

Therapeutic applications for the use of somatostatin were limited, owing mostly to the transitory action of the originally synthesized compound. The availability of longer-acting analogues of somatostatin have changed that situation. There are four areas that hold promise in the therapeutic application of somatostatin analogues; management of acute, massive GI bleeding secondary to ulcer disease, management of resistant acromegaly, treatment of hyperfunctioning islet cell adenomas, and perhaps management of brittle diabetes (Table 9).

The use of the long-acting somatostatin analogue in the management of acromegaly has been investigated by several groups. These studies were prompted by the observations that the plasma half-life of the somatostatin analogue was considerably longer (10–20 times) than that of the original peptide, that the octapeptide analogue is more potent (50–100 times) than that of the natural peptide, that the administration of the analogue to normals resulted in protracted suppression of GH to provocative stimuli, and that the rebound phenomenon in GH secretion seen following discontinuation of the natural peptide was absent when the analogue was used.[105] Preliminary results from studies by Plewe et al.,[106] Ch'ng et al.,[107] and Lamberts et al.[108] support the notion that significant reductions in GH levels can be attained, lasting as long as 9 hr following the administration of a single dose of the somatostatin analogue. In a recent study by Lamberts et al.,[109] four patients with acromegaly treated for as long as 24 weeks showed rapid amelioration of symptoms as well as near normalization of the laboratory parameters, including somatomedin C levels. The only side effect noted was hyperglycemia resulting from suppression of insulin secretion, a phenomenon that improved with continuation of treatment. More importantly, there was evidence of tumor shrinkage, albeit slight, in three of four subjects. At the time of this writing, the use of somatostatin analogues for the treatment of acromegaly would seem to be reserved for patients who have not benefited from conventional modalities of therapy, such as surgery, radiation, or dopaminergic agents.

The use of somatostatin analogues in the management of patients with severely bleeding peptic ulcer disease is promising.[110] The two effects of the somatostatin analogue that are exploited here, are the potent inhibition of acid secretion and the reduction of blood flow to the stomach. In addition to these well-known effects of somatostatin, a possible local cytoprotective effect of the peptide has been proposed.[111] Several randomized multicenter trials are underway to compare the efficacy of somatostatin analogues to the more conventional forms of treatment in the management of bleeding peptic ulcer disease. The beneficial use of somatostatin analogues in the management of acute pancreatitis, particularly the hemorrhagic variety, has been reported.[112]

The therapeutic use of somatostatin analogues in the management of hypersecretory tumors of the islet cells is based on the effect of somatostatin in lowering gastroenteropancreatic hormones. Santangelo et al.[113] reported that the use of somatostatin analogues in the pancreatic cholera syndrome caused by secreting VIP was attended with a decrease—even total "cure"—of the diarrhea. A significant improvement in absorption of water and sodium, as well as reductions in plasma VIP level, were also seen. Davis et al.[114] reported improvement in the secretory diarrhea caused by malignant carcinoid syndrome. Similarly, Osei et al.[115] reported on the use of the somatostatin analogue in a patient with malignant insulinoma and showed significant reductions in the premeal insulin and C-peptide concentrations in the serum.

The therapeutic role of somatostatin in the management of brittle diabetics has not been conclusively established. Natural somatostatin has been used in conjunction with insulin to reduce the magnitude of the hyperglycemic excursions in the brittle diabetic.[116–119] The limited number of studies, the heterogeneity of patient responses, the short half-life of natural somatostatin, the adverse effects of long-term treatment with somatostatin on GH and TSH release, and the availability of other excellent methods for intensive insulin therapy (i.e., the insulin pumps) have all diminished the therapeutic impact of somatostatin in the treatment of diabetes. Therapeutic trials with the octapeptide analogue of somatostatin in treatment of type I diabetics have reinforced the observation that its use does indeed result in diminished postprandial glucose excursions.[120] However, at this time such therapy is deemed experimental at best.

Neurotransmitter Regulation

A variety of neurotransmitter substances are involved in integrating the effects of hypothalamic factors on GH release. These neurotransmitters act at the synaptic junction between the afferent neuronal connections and the peptidergic cells that synthesize the hypothalamic hormones somatostatin and GHRH. Growth hormone is rivaled only by prolactin in its proclivity to be regulated by neurotransmitter substances. The three major substances focused on in this section are dopamine (DA), norepinephrine (NE), and serotonin or 5-hydroxytryptamine (5-HT).

Dopaminergic Regulation

Dopamine is found in extremely high concentrations within the hypothalamus and the median eminence. The arcuate nucleus, the periventricular nucleus, and the periventricular preoptic nucleus contain cells that are very rich in DA content.[121] These cells send afferent projections to the median eminence, which is also very rich in dopaminergic cells. The following lines of evidence are supportive of a major role for DA in the regulation of GH.

1. Administration of L-dopa results in stimulating the secretion and release of GH in normal subjects.[122–125] This effect is brought about by virtue of its conversion to dopamine, after crossing the blood–brain barrier into the hypothalamus. It is believed that dopamine per se or its metabolite norepinephrine stimulates the release of GHRH from the medial hypothalamus or the median eminence. The prior administration of glucose blunts or abolishes the L-dopa-induced GH release.[126] This would imply that dopaminergic stimulation of GH release is superseded by the suppressive effect of hyperglycemia.
2. Dopamine agonists such as bromocriptine elicit a provocative GH response in normal subjects.
3. Apomorphine, a dopamine agonist, is a strong stimulus for GH release.[127] The provocative effect of apomorphine on GH release is more potent and prompt than L-dopa and is not antagonized by cholinergic mechanisms.
4. L-dopa is believed to have both central (hypothalamic) and peripheral (pituitary) effects, but the central action is dominant. When carbidopa (an inhibitor of dopa-decarboxylase) is given along with L-dopa, the conversion of L-dopa to dopamine is blocked peripherally, but not in the brain. This results in higher concentrations of dopamine in the hypothalamus and the brain. Such a setting has been known to result in augmentation of GH response to even suboptimal doses of L-dopa, presumably by central dopamine augmentation.[127–130]
5. The effect of dopamine (which does not cross the blood–brain barrier per se) on the pituitary release of GH is somewhat controversial. The intravenous administration of dopamine results in a blunted GH re-

sponse to hypoglycemia,[131] arginine,[132] and strangely enough to orally administered L-dopa.[133] These studies imply that dopamine, while stimulating basal GH secretion, paradoxically blunts the augmented release of GH to provocative stimuli in normals. Bansal et al.[134] demonstrated that under basal conditions dopamine and bromocriptine stimulated GH to a similar degree but also diminished the provoked GH response to hypoglycemia in an equally comparable fashion. However, when subjects were pretreated with metoclopramide, a specific dopamine antagonist, the augmented GH secretion was restored to normal. These data suggest that at the pituitary level dopamine may have a direct inhibitory effect. In vitro studies of GH modulation by dopamine have also suggested that dopamine is inhibitory at the level of the pituitary somatotropes.[135] This may underlie the so-called paradoxical inhibition of GH secretion seen in some acromegalics following the administration of dopamine receptor agonists such as L-dopa, apomorphine, or bromocriptine. It is highly possible that the hypersecretory adenoma cells directly respond to the dopaminergic agonism exerted by such drugs.

Regardless of the mechanism of dopaminergic participation in the regulation of GH secretion, modulation by this neurotransmitter provides valuable insights in diagnosis of GH reserve as well in therapy of GH hypersecretion.

Adrenergic Regulation

Norepinephrine (NE), a true neurotransmitter, is intimately involved in the regulation of GH secretion. Several hypothalmic nuclei demonstrate intense adrenergic innervation. These include the periventricular, medial preoptic, paraventricular, and supraoptic nuclei. Several lines of clinical and experimental data have strongly imparted a significant role for NE in the control of GH secretion:

1. α-Adrenergic stimulation is a potent provocative stimulus for GH release. In its most dramatic form this is observed during insulin-induced hypoglycemia. The hypothalamic glucoreceptors are exquisitely sensitive to α-adrenergic stimulation. The provocative response of several well-known stimuli for GH release is mediated by α-adrenergic stimulation.
2. The well-known provocative effect of L-dopa on GH secretion in normal subjects probably occurs as a result of conversion of L-dopa into NE within the CNS. Kansal et al.[124] showed that the provocative effect of L-dopa on GH secretion can be abolished by prior administration of phentolamine, an α-adrenergic antagonist.
3. Phentolamine not only blocks the GH rise following L-dopa but virtually blocks the GH rise following several other stimuli, including stress, hypoglycemia, and vasopressin as well.

4. Propranolol, a β-adrenergic receptor blocker, potentiates the response of GH to various stimuli. One combination in particular works exceedingly well as a provocative test to assess GH reserve; the combination of propranolol and glucagon serves as a potent and useful stimulus to study growth hormone release.[136]

5. Clonidine, a centrally mediated adrenergic receptor agonist, causes release of GH in normal subjects.[137]

6. Experimentally, microinjection of NE into the ventromedial nucleus (VMN) causes a prompt release of GH.[138] The VMN has been most strongly identified as a major site for synthesis of GHRH.

Based on these lines of evidence, it is irrefutable that NE has a strong facilitatory role in the regulation of GH release and that its action is exerted at the hypothalamic level. The most important clinical connotation is that nearly all provocative stimuli to study GH release—insulin, L-dopa, clonidine, glucagon, vasopressin, stress, and exercise—are mediated by adrenergic modulation of the hypothalamic peptide, GHRH.

Serotonergic-Mediated Regulation

In contrast to the dopaminergic and adrenergic regulation of GH secretion, serotonin has at best only a secondary and peripheral role in the regulation of GH release. Plasma GH levels rise following the administration of tryptophan, a precursor for the synthesis of serotonin (5-HT).[139] Physiologically, it is believed that the GH rise associated with stage III and IV sleep is mediated by serotonergic mechanisms. This is supported by the observation that the sleep-related GH rise is not abolished by hyperglycemia,[140] a powerful suppressant of catecholamine-mediated GH release. The observation that GH is elevated in patients with carcinoid syndrome[141] further suggests a role for serotonin in regulation of GH. The bulk of evidence for serotinergic mediation comes from studies employing serotonin-antagonists. Bivens et al.[142] showed that cyproheptadine, when given to normal subjects, decreased the GH response to hypoglycemia by 60% compared with the pretreatment response. Similar results have been reported by others.[143] It is unclear whether the inhibitory effects of serotonin antagonists are mediated by stimulating somatostatin release. The sporadic (and often anecdotal) reports of cyproheptadine-induced response in acromegaly may be due to stimulation of somatostatin release by serotonin-antagonists.

In summary, the hypothalamic hormones—GHRH and, to a lesser extent, somatostatin—are heavily modulated by neurotransmitters. Catecholamines, either NE or its precursor dopamine (DA), exert a profound influence on the release of GHRH. Understanding the neuropharmacology of such release has added to the diagnostic armamentarium in evaluation of GH dynamics in health and disease. In addition to DA, NE, and 5-HT, minor neurotransmitters such a γ-aminobutyric acid (GABA), histamine receptor antagonists, nicotinic acid, and cholinergic agents have been implicated, from time to time, in the regulation of GH secretion. These substances at best have only a marginal effect on GH release.

Peptide and Amino Acid Regulation

Several peptides and amino acids are capable of releasing GH from the pituitary. The four major ones to consider are glucagon, vasopressin, opioids, and arginine. Glucagon is a fairly consistent stimulus for the release of GH,[144] especially in conjunction with β-blocker augmentation. The stimulatory effects of glucagon are probably brought about by changes in circulating glucose level, although a direct effect on the pituitary cannot be excluded. The provocative effect of vasopressin on GH release is believed to be exerted directly on the pituitary gland, as is the effect of arginine monochloride. Opioid peptides release GH, as evidenced by the potent effect of morphine on stimulating GH release.[145] Although the major effect of opioids is exerted on the hypothalamus, a direct effect on the pituitary has not been ruled out. Table 10 outlines the diverse factors that regulate GH secretion.

TABLE 10
Factors That Regulate GH Release

Factor	Effect[a]	Comment
Hypothalamic hormones		
GHRH	+	Most provocative stimulus for GH release
Somatostatin	−	The most potent physiological suppressor of GH release
TRH	−	Does not stimulate GH release, except in acromegaly
Neurotransmitters		
Dopaminergic	+	Dopamine agonists stimulate release of GH in normals but lower GH in acromegaly
Adrenergic	+	Most tests employed to assess GH reserve work through adrenergic stimulation
Serotonergic	+	Facilitatory role in mediating GH release
Other neurotransmitters	(?)+	GABA, cholinergics, H_2-receptor antagonists play a minor role, if any
Other peptides and amino acids	+	Glucagon, vasopressin, opioid peptides, arginine, leucine, and isoleucine stimulate release
Neuroregulation		
α-Adrenergic stimulation	+	Augments GH release
β-Adrenergic stimulation	−	Suppresses GH release
α-Adrenergic blockade	−	Suppresses GH release
β-Adrenergic blockade	+	Potentiates GH release

[a] +, Stimulates; −, suppresses.

Feedback Regulation

Growth hormone secretion by the somatotropes is elegantly controlled by a feedback mechanism that involves four factors: GHRH, somatostatin, GH itself, and the somatomedins. Viewed in a simplistic fashion, GH regulates its own secretion in the following three ways (Fig. 8).

1. GH exerts a positive feedback effect on hypothalamic somatostatin.
2. The somatomedin-C generated by GH, also exerts a positive feedback effect on the release of somatostatin. Both the above phenomena result in release of somatostatin, which in turn suppresses GH.
3. Somatomedin-C, in addition to its positive feedback effect on the hypothalamic somatostatin, exerts a direct negative feedback effect on pituitary GH, resulting in suppression of GH release.

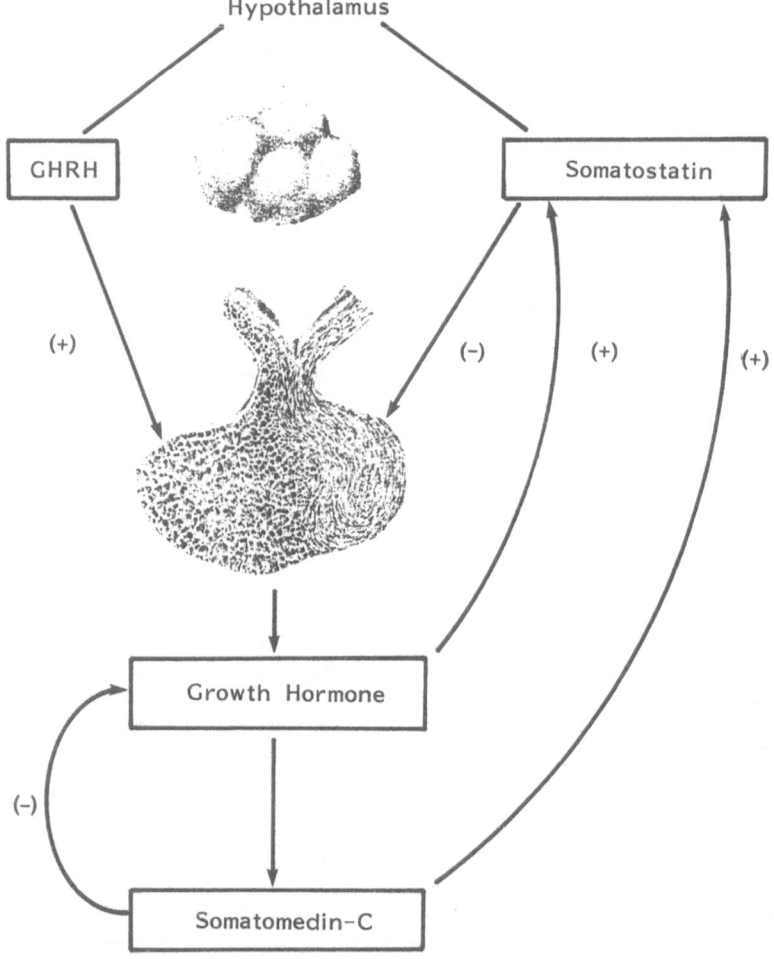

FIGURE 8. Feedback regulation of growth hormone secretion.

4. While it is established that GHRH has a trophic effect on GH release by the pituitary, the feedback loop between GH and GHRH, or the feedback loop between somatomedin C and GHRH, has not been clearly defined.

Actions of Growth Hormone

The predominant action of growth hormone is to promote growth. In addition to its growth-promoting activity, which for obvious reasons is no longer needed once linear growth is completed, growth hormone has important metabolic actions. These relate to its anabolic effects, and its effects on glucose metabolism. The anabolic effects of growth hormone are most strikingly evident when GH is administered to hypopituitary children. In addition to promoting skeletal and soft tissue growth, there is a marked positive balance in the metabolism of nitrogen, phosphorus, potassium, calcium, and even sodium. The actions of GH on glucose metabolism are clearly geared in the direction of being diabetogenic—antiinsulin like, since it represents an important counterinsulin hormone.

Although it is true that GH does not act directly on cartilage and requires the mediation of the intermediary somatomedins, it is not correct to assume that GH is incapable of any direct action. Several observations suggest that GH does indeed have a direct action on some target tissues. GH has been shown to stimulate DNA synthesis directly in vitro in thymocytes[146] as well as affect the metabolism of isolated rat diaphragm and adipose tissue in vitro.[147] Furthermore, specific receptors that bind GH have been found in several tissues, such as liver cells, adipocytes, fibroblasts, and lymphocytes. These observations assume significance in light of the paradox that somatomedins possess insulinlike properties, while their generator, GH, possesses antiinsulinlike properties. Thus, it is entirely possible that such discordance may reflect the presence of different receptors, at least in some tissues, for GH and somatomedins, resulting in differing postreceptor events.

The actions of GH will be considered in terms of growth promotion, as well as, its effects on carbohydrate, fat, and protein metabolism.

Growth-Promoting Action

The biological actions of GH are mediated by a family of peptides, the somatomedin family,[148–150] generated in the liver. As early as 1957, Salmon and Doughaday[151] reported three significant observations; that the serum from hypophysectomized rats demonstrated decreased activity in terms of stimulating the uptake of radioactive sulfate by the cartilage; that GH per se did not stimulate incorporation of sulfate; and that GH treatment of hypophysectomized rats resulted in the appearance of a serum factor that promoted sulfation of the cartilage (the sulfation factor). This factor mediated the biological effects of GH. It is now recognized that this factor in reality, represents a whole group of peptides, and is referred to as the somatomedins.[150] Members of this family of polypeptides share several common characteristics: they are low molecular weight (6,000–10,000), are mostly, but not exclusively synthesized in the liver, and circulate in the plasma bound to carrier proteins of high molecular weight. They also possess varying degrees of growth-promoting and insulinlike activity. Hence the term insulinlike growth factors is often used to describe some somatomedins. Further characterization of these peptides has resulted in categorizing these into somatomedin A, B, and C and insulin growth factors I and II.[152-154] The differences between each somatomedin reside in structure, physicochemical properties, and their potency, as it relates to the spectrum of biological effects, exerted by these peptides. For instance, somatomedin C and A are more closely correlated with promotion of growth, while somatomedin B is not. The insulinlike properties of the somatomedins are related to their effects on the liver and their structural similarity to proinsulin. Somatomedins are also referred to as having nonsuppressible insulinlike activity (NSILA)—although possessing insulinlike bioactivity, these peptides do not suppress to glucose administration. The term insulin growth factors aptly describes the anabolic effects of one class of somatomedins on muscle, adipose tissue and the liver. In addition to somatomedin A, B, and C and IGF I and II, another peptide with predominantly multiplication stimulating activity has been described.[155]

The measurement of somatomedins in plasma can be carried out by traditional bioassays, or by newer methods involving radioreceptor assay, competitive protein-binding assays, and RIAs. Each method has limitations and advantages.[156-158] For instance, bioassays will measure activity on non-GH-dependent growth factors as well as somatomedin activity. Interpretation of the radioreceptor assays can be affected by the presence of carrier proteins. The RIAs for somatomedins must take into account the significant immunological cross-reactivity among IGF-I, somatomedin A, and somatomedin C. In fact some of the confusion in terminology is a consequence of different methodologies employed by various laboratories in detecting the presence of these peptides in serum. It is important to recognize the existence of several growth factors of which the somatomedin family is just one.[159] Somatomedins, however, are the only type of growth factors that are dependent on GH. As would be expected, the levels of somatomedin in the circulation mirror the status of GH activity. Thus, somatomedin activity is markedly elevated in acromegaly and markedly diminished in GH deficiency.

The regulation of somatomedin generation is controlled by several factors, both hormonal, and nonhormonal (Table 11). While GH is the major stimulus for synthesis of somatomedins, three other hormones may promote somatomedin generation: insulin, prolactin, and chorionic somatomammotropin. Of these, the effects of insulin are the most striking. Insulin (and good nutrition) can sustain somatomedin generation even in the complete absence of GH.[160,161] Conversely, in the presence of insulin lack, as seen in poorly controlled diabetic children, growth is impaired, despite high circulating GH levels. Prolactin, a hormone with homology structurally similar to that of GH,

TABLE 11
Factors That Regulate Somatomedin Activity

Stimulation of somatomedin activity	Impairment of somatomedin activity
Growth hormone	Glucocorticoids
Insulin	Estrogens
Prolactin (in high concentrations)	Malnutrition
Chorionic somatomammotropin	Systemic illness
(human placental lactogen)	Circulating inhibitors
Nutrional factors	

can stimulate somatomedin generation, when present in high concentrations.[162] Support for the role of chorionic somatomammotropin in stimulating somatomedin comes predominantly from animal experiments, and its significance in human pregnancy is largely unclear. Amongst inhibitors of somatomedin production the two important hormones are glucocorticoids[163] and estrogens.[164] The most important nonhormonal factors that regulate somatomedin generation are nutrition and good general health. The importance of good nutrition in stimulating somatomedin synthesis has been underscored by several reports.[165–167]

The actions of somatomedins are impressively diverse, not being restricted to promotion of growth alone. In vitro, somatomedins have demonstrable effects on four different tissues, the cartilage, muscle, adipose tissue, and other cells. On cartilage, the somatomedins stimulate incorporation of sulfate into chondroitin sulfate, promote amino acid transport, and increase the synthesis of DNA and RNA. On the muscle, the actions of somatomedins are identical to insulin and appear to be equally potent. Somatomedins promote transport of amino acids and glucose into the muscle cell and stimulate the formation of glycogen as well as muscle protein. This anabolic effect is identical to that of insulin. In the adipocyte, the actions of somatomedins are weaker than insulin and consist of inhibiting lipolysis, promoting glucose transport, and glucose oxidation. The actions of somatomedins on other cells grown in culture, is to stimulate multiplication. This action has been referred to as mitogenic activity and has been shown to be dependent on GH.[168]

In addition to its replicative potential, the somatomedins have also demonstrated cytodifferentiative properties in vitro. Thus, somatomedins have been shown to promote the differentiation of widely diverse cells such as chondrocytes, osteoblasts, and even myeloblasts. It has been proposed recently that somatomedin C/IGF-I may even modulate granulosa cell ontogeny, with possibly an intraovarian role in the regulation of granulosa cell growth and function.[169]

In summary much has been learned in the past decade regarding the somatomedins. Clearly, the growth-promoting properties of GH are heavily dependent on these mediators. The most striking example of this dependency is demonstrated in Laron dwarfism, in which growth retardation occurs despite a high level of GH, which is unable to stimulate release of somatomedins. The full implication of the in vivo role of somatomedins in noncartilaginous tissues awaits elucidation.

Effects on Carbohydrate, Fat, and Protein Metabolism

The effects of GH on carbohydrate metabolism are threefold:

1. Growth hormone is a counterinsulin hormone, antagonistic to the peripheral action of insulin. Thus, GH decreases the peripheral utilization of glucose by tissues. In this sense, GH is diabetogenic.
2. Growth hormone is β-cytotropic, i.e., it stimulates the β cell to secrete more insulin. This insulinotropic effect is evidenced by the fact that prior administration of GH results in a vastly enhanced insulin response of β cell to glucose.
3. An effect of GH on glucose homeostasis is seen only when supraphysiological doses of growth hormone are administered. This glucose-lowering effect is paradoxical and resembles the action of insulin. It is believed that this early action is seen only when very high tissue levels are attained, and it is transient. It is unclear which somatomedin, if any, mediates this effect.

The effect of GH on adipose tissue is also antiinsulinlike. Growth hormone is lipolytic, in contrast to the lipogenic effects of insulin. Following the injection of GH, there is a prompt increase of free fatty acid (FFA). This lipolytic effect can be attenuated by concomitant administration of glucose or by administering insulin, which is lipolytic.

The effect of GH on protein metabolism is anabolic. Growth hormone enhances incorporation of leucine into muscle protein. This is reflected best when the muscle tissue of acromegalics is examined. Increase in muscle fiber size and increased mitochondria are strikingly obvious. The incorporation of amino acid into muscle may be mediated by somatomedins and is probably brought about by enhancing the tissue permeability.

Growth Hormone Dynamics in Health and Disease

Evaluation of GH dynamics is an integral part of the diagnostic workup when hypofunction or hyperfunction of the somatotropes are suspected. The former involves testing the adequacy of growth hormone reserve with provocative maneuvers and the latter involves assessing the suppressibility of growth hormone to physiological stimuli. Measurement of GH in a single basal sample does not establish or exclude disordered somatotrope function.

Testing Growth Hormone Reserve

There are several indications for evaluating the GH reserve: evaluation of short stature or growth retardation during the growing years; evaluation

TABLE 12.
Provocative Stimuli for Growth Hormone Release[a]

Stimulus	Comment
Insulin 0.1 U reg insulin per kg body weight	Considered the gold standard; may be hazardous, especially in presence of hypopituitarism and coronary disease
L-Dopa 500 mg by mouth	Side effect is vomiting; 85–90% of normals respond to L-dopa. The maximum GH levels are less than with insulin hypoglycemia
Glucagon 1 mg IM	Less consistent than insulin or L-dopa (10% failure rate)
Glucagon with propranolol (40 mg)	Better than glucagon alone but often requires an additional test to establish GH deficiency
Apomorphine 0.75 mg SC	GH response is as good or better than L-dopa
Arginine 300 g (600 ml of 5% solution)	Less potent then insulin or L-dopa
Vasopressin 10 U IM	Compares with arginine or glucagon in potency, can be hazardous in patients with coronary disease
Sleep; stages III and IV Exercise (15 min of vigorous stair climbing)	Cumbersome Difficult to quantify the response
GHRH	Consistently releases GH

[a]See references 170–181.

of fasting hypoglycemia, particularly in children; and evaluation of patients with hypothalamic pituitary lesions, particularly tumors.

Numerous pharmacological and physiological stimuli have been used to assess the adequacy of GH reserve (Table 12). Insulin-induced hypoglycemia has stood the test of time as an excellent stimulus for GH release. Several studies have compared the use of many other provocative agents,[170-181] none of which seem to match the potency of insulin hypoglycemia as the most effective and perhaps the most consistent stimulus for GH release. The hypoglycemia test is performed on the overnight-fasted patient by intravenously administering 0.1 U of regular insulin per kilogram body weight. The GH levels are measured basally and 30, 60, and 90 min following insulin, with simultaneous glucose determinations. The degree of reduction in the blood glucose needed to stimulate GH is variable, but in general a reduction by 50% of the baseline level is considered optimal. The reserve of growth hormone is considered normal when the level exceeds an absolute value of 10 ng/dl. In addition to hypopituitarism, causes for a blunted growth hormone response include obesity, depression, and hypothyroidism. Occasionally, a normal person who fails to respond to hypoglycemia on one occasion may demonstrate a perfectly normal response on a subsequent day. The reason for such a phenomenon is unclear.

The insulin test can be hazardous, especially in patients with hypopituitarism who may develop severe and protracted hypoglycemia, and in patients with coronary heart disease. Therefore, strict supervision is needed during the performance of the test. In view of these difficulties, alternative provocative maneuvers have been evaluated. L-dopa is a reasonably effective alternative, eliciting a GH response in 80–85% of subjects. Its advantages are that it can be given orally, and is free from major side effects. All the other tests in use—glucagon, glucagon/propranolol combination, vasopressin, arginine, and apomorphine—are not nearly as effective or as consistent in their ability to provoke GH secretion.

Evaluation of GH response to sleep is cumbersome, requiring EEG monitoring and the services of the sleep laboratory. Ostensibly there is one situation in which determining the GH response to sleep can be most valuable; children with growth failure secondary to emotional deprivation demonstrate absence of GH response to a variety of pharmacologic stimuli, while preserving the sleep-induced GH response.

In clinical practice, GH reserve is usually assessed by the use of insulin, L-dopa, or both. Studies that have compared the efficacy of several stimuli for GH release have not displaced insulin or L-dopa as the two major ones.[171,172] Admittedly, there will be a small percentage of normal patients failing to show a response to insulin hypoglycemia, or L-dopa. The demonstration of an absent response to both stimuli is strong evidence for compromised GH reserve.

The availability of synthetic hpGHRH, has added a new dimension to the evaluation of GH reserve. Both insulin and L-dopa stimulate GH release working predominantly at the hypothalamic level. Thus, failure to respond to these stimuli would not differentiate between an intrinsic pituitary lesion and a hypothalamic lesion. By contrast, synthetic hPGHRH directly stimulates GH release from the pituitary. Thus, patients with GH deficiency arising from a hypothalamic lesion will demonstrate preservation of response to GHRH, while showing no response to insulin hypoglycemia or L-dopa. The application of hpGHRH has already resulted in diagnosing several patients with hypothalamic etiologies for GH deficiency.

Testing Growth Hormone Suppressibility

The primary indication to test suppressibility of GH is when acromegaly or gigantism is suspected. The response of normal somatotropes to hyperglycemia is a decline of GH level to below 5 ng/ml after 100 g of orally administered glucose. By contrast, patients with acromegaly demonstrate failure to decrease GH completely below 5 ng/ml or may even shown a paradoxical rise in GH after ingestion of glucose. This test merely represents a screening test for acromegaly, since several other conditions may be associated with impairment in GH suppression to glucose (Chapter 3).

Disorders of Growth Hormone Secretion

The clinical disorders involving perturbations in GH secretion are hyper- and hyposomatotropism.

Hypersomatotropism results in acromegaly or gigantism disabling disorders with spectacular manifestations. While an intrinsic pituitary (or rarely hypothalamic) etiology usually underlies acromegaly, rarely it can be caused by ectopic secretion of GH or GHRH.

Hyposomatotropism or growth hormone deficiency can occur singly or in combination with deficiencies of anterior pituitary hormones. In its purest form, isolated GH deficiency can result from intrinsic loss of somatotropes, or as a consequence of deficiency of the hypothalamic GHRH. Either variety can result in pituitary dwarfism. Rarely, pituitary dwarfism can occur as a result of circulating abnormal polymers.[182]

The disorders that result from hyper- or hyposecretion are discussed in Chapters 3 and 4.

References

1. Martin JB, Brazeau P, Tannenbaum GS, et al: Neuroendocrine organization of growth hormone regulation. In Reichlin S, Baldessarini R, Martin JB (eds): *The Hypothalamus*. Raven Press, New York, 1978, p. 329.
2. Niall HD, Hogan ML, Sayer R, et al: Sequences of pituitary and placental lactogenic and growth hormones: Evolution from a primordial peptide by gene duplication. *Proc Natl Acad Sci USA* 68:866, 1971.
3. Miller WL, Eberhardt NL: Structure and evolution of the growth hormone gene family. *Endocr Rev* 4:97, 1983.
4. Martial JA, Hallewell RA, Baxter JD, et al: Human growth hormone: Complementary DNA cloning and expression in bacteria. *Science* 205:602, 1979.
5. Phillips JA III, Parks JS, Hjelle BL, et al: Genetic analysis of familial isolated growth hormone deficiency Type I. *J Clin Invest* 70:489, 1982.
6. Baumann R, MacCart JG: Growth hormone production by human pituitary glands in organ culture: Evidence for predominant secretion of the single-chain 22,000 molecular weight form (isohormone B). *J Clin Endocrinol Metab* 55:611, 1982.
7. Sigel MG, Thorpe NA, Korbin MS, et al: Binding characteristics of a biologically active variant of human growth hormone (20K) to growth hormone and lactogen receptors. *Endocrinology* 108:1600, 1981.
8. Miller WL, Eberhardt NL, Baxter JD: Growth hormone genes. Secretory tumors of the pituitary gland. In Black PM, Zervas NT, Ridgway EC, et al. (eds): *Progress in Endocrine and Therapy*, Vol. 1. Raven Press, New York, 1984, p. 135.
9. Martial JA, Baxter JD, Goodman HM, et al: Regulation of growth hormone messenger RNA by thyroid and glucocorticoid hormones. *Proc Natl Acad Sci USA* 74:1816, 1977.
10. Takahashi Y, Kipnis DM, Daughaday WH: Growth hormone secretion during sleep. *J Clin Invest* 47:2079, 1968.
11. Honda Y, Takahashi K, Takahashi S, et al: Growth hormone secretion during nocturnal sleep in normal subjects. *J Clin Endocrinol Metab* 29:20, 1969.
12. Krieger DT, Glick S: Absent sleep peak of growth hormone release in blind subjects: Correlation with sleep EEG stages. *J Clin Endocrinol Metab* 33:847, 1971.
13. Vigneri R, D'Agata R: Growth hormone release during the first year of life in relation to sleep–wake periods. *J Clin Endocrinol Metab* 33:561, 1971.
14. Parker DC, Sassin JF, Mace JW, et al: Human growth hormone release during sleep: Electroencephalographic correlation. *J Clin Endocrinol Metab* 29:871, 1969.

15. Leidy JW Jr, Robbins RJ: Regional distribution of human growth hormone-releasing hormone in the human hypothalamus by radioimmunoassay. *J Clin Endocrinol Metab* **62**:372, 1986.

16. Frohman LA, Maran JW, Dhariwal APS: Plasma growth hormone responses to intrapituitary injections of GH RF in the rat. *Endocrinology* **88**:1483, 1971.

17. Martin JB: Neural regulation of growth hormone secretion. Medical progress report. *N Engl J Med* **288**:1384, 1973.

18. Thorner MO, Perryman RL, Cronin MJ, et al: Acromegaly with somatotroph hyperplasia: Successful treatment by resection of a pancreatic tumor secreting a growth hormone (GH) releasing factor (GRF). *J Clin Invest* **70**:965, 1982.

19. Frohman LA, Szabo M, Berelowitz M, et al: Partial purification and characterization of a peptide with growth hormone-releasing activity from extrapituitary tumors in patients with acromegaly. *J Clin Invest* **65**:43, 1980.

20. Rivier J, Spiess M, Thorner M, et al: Characterization of a growth hormone-releasing factor from a human pancreatic islet tumour. *Nature (Lond)* **300**:276, 1982.

21. Guillemin R, Brazeau P, Bohlen P, et al: Growth hormone-releasing factor from a human pancreatic tumor that caused acromegaly. *Science* **218**:585, 1982.

22. Spiess J, Rivier J, Vale W: Characterization of rat hypothalamic growth hormone-releasing factor. *Nature (Lond)* **303**:532, 1983.

23. Thorner MO, Rivier J, Spiess J, et al: Human pancreatic growth-hormone releasing factor selectively stimulates growth-hormone secretion in man. *Lancet* **1**:24, 1983.

24. Gelato MC, Pescovits O, Cassorla F, et al: Effects of a growth hormone releasing factor in man. *J Clin Endocrinol Metab* **57**:674, 1983.

25. Rosenthal SM, Schriock EA, Kaplan SL, et al: Synthetic human pancreatic growth hormone-releasing factor stimulates growth hormone secretion in normal man. *J Clin Endocrinol Metab* **57**:677, 1983.

26. Barinaga M, Yamonoto G, Rivier C, et al: Transcriptional regulation of growth hormone gene expresion by growth hormone releasing factor. *Nature (Lond)* **306**:84, 1983.

27. Lewin MJ, Reyl-Desmarz F, Ling N: Somatocrinin receptor coupled with cAMP dependent protein kinase in anterior pituitary granules. *Proc Natl Acad Sci USA* **80**:6538, 1983.

28. Gelato MC, Rittmaster RS, Pescovitz OH: Growth hormone responses to continuous infusions of growth hormone-releasing hormone. *J Clin Endocrinol Metab* **61**:223, 1985.

29. Webb CB, Vance ML, Thorner MO, et al: Plasma growth hormone responses to constant infusions of human pancreatic growth hormone releasing factor: Intermittent secretion or response attenuation. *J Clin Invest* **74**:96, 1984.

30. Masuda A, Shibasaki T, Nakahara M, et al: The effect of glucose on growth hormone (GH)-releasing hormone-mediated GH secretion in man. *J Clin Endocrinol Metab* **60**:523, 1985.

31. Williams T, Maxon H, Thorner MO, et al: Blunted growth hormone (GH) response to GH-releasing hormone in hypothyroidism resolves in the euthyroid state. *J Clin Endocrinol Metab* **61**:454, 1985.

32. Shibasaki T, Shizume K, Nakara M, et al: Age-related changes in plasma growth hormone release to growth hormone-releasing factor in man. *J Clin Endocrinol Metab* **58**:212, 1984.

33. Laron Z, Keret R, Bauman B, et al: Differential diagnosis between hypothalamic and pituitary hGH deficiency with the aid of synthetic GH-RH 1–44. *Clin Endocrinol (Oxf)* **21**:9, 1984.

34. Takano K, Hizuka N, Shizume K, et al: Plasma growth hormone response to growth hormone releasing factor in normal children with short stature and patients with pituitary dwarfism. *J Clin Endocrinol Metab* **58**:236, 1984.

35. Wood SM, Ch'ng JLC, Adams EF, et al: Abnormalities of growth hormone release in response to human pancreatic growth hormone releasing factor (GRF 1–44) in acromegaly and hypopituitarism. *Br Med J* **286**:1687, 1983.

36. Schulte HM, Benker G, Windeck R, et al: Failure to respond to growth hormone releasing hormone (GHRH) in acromegaly due to a GHRH secreting pancreatic Tumor: Dynamics of multiple endocrine testing. *J Clin Endocrinol Metab* **61**:585, 1985.

37. Ch'ng JLC, Christofides ND, Kraenzlin ME, et al: Growth hormone secretion dynamics in a patient with ectopic growth hormone-releasing factor production. *Am J Med* **79**:135, 1985.

38. Losa M, Schopohl J, Stalla GK, et al: Growth hormone releasing factor-test in acromegaly: Comparison with other dynamic tests. *Clin Endocrinol (Oxf)* **23**:99, 1985.

39. Borges JLC, Blizzard RM, Evans WS, et al: Stimulation of growth hormone (GH) and somatomedin C in idiopathic GH-deficient subjects by intermittent pulsatile administration of synthetic human pancreatic tumor GH-releasing factor. *J Clin Endocrinol Metab* **59**:1, 1984.

40. Brown P, Gajdusek C, Gibbs CJ Jr, et al: Potential epidemic of Creutzfeldt-Jakob disease from human growth hormone therapy. *N Engl J Med* **313**:728, 1985.

41. Thorner MO, Reschke J, Chitwood J, et al: Acceleration of growth in two children treated with human growth hormone-releasing factor. *N Engl J Med* **312**:4, 1985.

42. Gelato MC, Ross JL, Malozowski S, et al: Effects of pulsatile administration of growth hormone (GH)-releasing hormone on short term linear growth in children with GH deficiency. *J Clin Endocrinol Metab* **61**:444, 1985.

43. Reichlin S: Somatostatin. *N Engl J Med* **309**:1495, 1983.

44. Krulich L, Dhariwal APS, McCann SM: Stimulatory and inhibitory effects of purified hypothalamic extracts on growth hormone release from rat pituitary in vitro. *Endocrinology* **83**:783, 1968.

45. Brazeau P, Vale W, Burgus R, et al: Hypothalamic peptide that inhibits the secretion of immunoreactive pituitary growth hormone. *Science* **179**:77, 1973.

46. Krisch B: Hypothalamic and extrahypothalamic distribution of somatostatin-immunoreactive elements in the rat brain. *Cell Tissue Res* **195**:499, 1978.

47. Krisch B: Differing immunoreactivities of somatostatin in the cortex and the hypothalamus of the rat: A light and electron microscope study. *Cell Tissue Res* **212**:457, 1980.

48. Krisch B: Somatostatin-immunoreactive fiber projections into the brain stem and the spinal cord of the rat. *Cell Tissue Res* **217**:531, 1981.

49. Hökfelt T, Johansson O, Ljungdahl A, et al: Peptidergic neurones. *Nature (Lond)* **284**:515, 1980.

50. Hökfelt T, Elfvin LG, Elde R, et al: Occurrence of somatostatin-like immunoreactivity in some peripheral sympathetic noradrenergic neurons. *Proc Natl Acad Sci USA* **74**:3587, 1977.

51. Gerich JE: Somatostatin and diabetes. *Am J Med* **70**:619, 1981.

52. Dubois M: Immunoreactive somatostatin is present in discrete cells of the endocrine pancreas. *Proc Natl Acad Sci USA* **72**:1340, 1975.

53. Polak J, Pearse AGE, Grimelius L, et al: Growth-hormone release-inhibiting hormone in gastrointestinal and pancreatic D cells. *Lancet* **1**:1220, 1975.

54. Arnold R, Lankisch PG: Somatostatin and the gastrointestinal tract. *Clin Gastroenterol* **9**:733, 1980.

55. Larsson L-I: Gastrointestinal cells producing endocrine, neurocrine and paracrine messengers. *Clin Gastroenterol* **9**:485, 1980.

56. Van Noorden S, Polak JM, Pearse AGE: Single cellular origin or somatostatin and calcitonin in the rat thryoid gland. *Histochemistry* **53**:243, 1977.

57. Bolaffi JL, Reichlin S, Goodman DBP, et al: Somatostatin: occurrence in urinary bladder epithelium and renal tubules of the toad, *Bufomarinus. Science* **210**:644, 1980.

58. Shen LP, Pictet RL, Rutter WJ: Human somatostatin I: Sequence of the cDNA. *Proc Natl Acad Sci USA* **79**:4575, 1982.

59. Mandarino L, Stenner D, Blanchard W, et al: Selective effects of somatostatin-14, -25 and -28 on in vitro insulin and glucagon secretion. *Nature (Lond)* **291**:76, 1981.

60. Siler TM, Vandenberg G, Yen SSC: Inhibition of growth hormone release in humans by somatostatin. *J Clin Endocrinol Metab* **37**:632, 1973.

61. Arimura A, Smith W, Schally AV: Blockade of the stress-induced decrease in blood GH by anti-somatostatin serum in rats. *Endocrinology* **98**:540, 1976.

62. Copinschi G, Virasoro E, Vanhaelst L, et al: Specific inhibition by somatostatin of growth hormone release after hypoglycaemia in normal men. *Clin Endocrinol (Oxf)* **3**:441, 1974.

63. Parker DC, Rossman LG, Siler TM, et al: Inhibition of the sleep-related peak in physiologic human growth hormone release by somatostatin. *J Clin Endocrinol Metab* **38**:496, 1974.

64. Tannenbaum GS, Epelbaum J, Colle E, et al: Antiserum to somatostatin reverse starvation-induced inhibition of growth hormone but not insulin secretion. *Endocrinology* **102**:1909, 1978.

65. Wehrenberg WB, Ling N, Böhlen P, et al: Physiological roles of somatocrinin and somatostatin in the regulation of growth hormone secretion. *Biochem Biophys Res Commun* **109**:562, 1982.

66. Besser GM, Mortimer CH, Carr D, et al: Growth hormone release inhibiting hormone in acromegaly. *Br Med J* **1**:352, 1974.

67. Dunn PJ, Donald RA, Espiner EA: A comparison of the effect of levodopa and somatostatin on the plasma levels of growth hormone, insulin, glucagon and prolactin in acromegaly. *Clin Endocrinol (Oxf)* **5**:167, 1976.

68. Oppizzi G, Botalla L, Verde G, et al: Homogeneity in growth hormone lowering effect of dopamine and somatostatin in acromegaly. *J Clin Endocrinol Metab* **51**:616, 1980.

69. Besser GM, Mortimer CH, McNeilly AS, et al: Long-term infusion of growth hormone release inhibiting hormone in acromegaly: Effects on pituitary and pancreatic hormones. *Br Med J* **1**:622, 1974.

70. Pieters GFFM, Romeijn JE, Smals AGH, et al: Somatostatin sensitivity and growth hormone response to releasing hormones and bromocriptine in acromegaly. *J Clin Endocrinol Metab* **54**:942, 1982.

71. Kasting NW, Martin JB, Arnold MA: Pulsatile somatostatin release from the median eminence of the unanesthestized rat and its relationship to plasma growth hormone levels. *Endocrinology* **109**:1739, 1981.

72. Tannenbaum GS: Evidence for autoregulation of growth hormone secretion via the central nervous system. *Endocrinology* **107**:2117, 1980.

73. Abe H, Molitch ME, Van Wyk JJ, et al: Human growth hormone and somatomedin C suppress the spontaneous release of growth hormone in unanesthetized rats. *Endocrinology* **113**:1319, 1983.

74. Berelowitz M, Firestone SL, Frohman LA: Effects of growth hormone excess and deficiency on hypothalamic somatostatin content and release and on tissue somatostatin distribution. *Endocrinology* **109**:714, 1981.

75. Sheppard MC, Kronheim S, Primstone BL: Stimulation by growth hormone of somatostatin release from the rat hypothalamus in vitro. *Clin Endocrinol (Oxf)* **9**:583, 1978.

76. Siler TM, Yen SSC, Vale W, et al: Inhibition by somatostatin on the release of TSH induced in man by thyrotropin-releasing factor. *J Endocrinol Metab* **38**:742, 1974.

77. Yen SSC, Siler TM, DeVane GW: Effects of somatostatin in patients with acromegaly: Suppression of growth hormone, prolactin insulin and glucose levels. *N Engl J Med* **290**:935, 1974.

78. Millar RP, Klaff LJ, Barron J, et al: Somatostatin-28 inhibits LHRH-stimulated gonadotrophin secretion in man. *Clin Endocrinol (Oxf)* **17**:103, 1982.

79. Koerker DJ, Ruch W, Chideckel E, et al: Somatostatin: hypothalamic inhibitor of the endocrine pancreas. *Science* **184**:482, 1974.

80. Sakurai H, Dobbs R, Unger RH: Somatostatin-induced changes in insulin and glucagon secretion in normal and diabetic dogs. *J Clin Invest* **54**:1395, 1974.

81. DeVane GW, Siler TM, Yen SSC: Acute suppression of insulin and glucose levels by synthetic somatostatin in normal human subjects. *J Clin Endocrinol Metab* **38**:913, 1974.

82. Gerich JE, Lorenzi M, Schneider V, et al: Effects of somatostatin on plasma glucose and glucagon levels in human diabetes mellitus: Pathophysiologic and therapeutic implications. *N Engl J Med* **291**:544, 1974.

83. Gerich JE, Lovinger R, Grodsky GM: Inhibition by somatostatin of glucagon and insulin release from the perfused rat pancreas in response to arginine, isoproterenol, and theophylline: Evidence for a preferential effect on glucagon secretion. *Endocrinology* **96**:749, 1975.

84. Patel YC, Amherdt M, Orci L: Quantitative electron microscopic autoradiography of insulin, glucagon, and somatostatin binding sites on islets. *Science* **217**:1155, 1982.

85. Raptis S, Schlegel W, Lehmann E, et al: Effects of somatostatin on the exocrine pancreas and the release of duodenal hormones. *Metabolism* **27**(suppl 1):1321, 1978.

86. Creutzfeldt W, Arnold R: Somatostatin and the stomach: Exocrine and endocrine aspects. *Metabolism* **27**(suppl 1):1309, 1978.
87. Vatn MH, Schrumpf E, Hanssen KF, et al: The effect of somatostatin on pentagastrin-stimulated gastric secretion and on plasma gastrin in man. *Scand J Gastroenterol* **12**:833, 1977.
88. Schlegel W, Raptis S, Harvey RF, et al: Inhibition of cholecystokinin-pancreozymin release by somatostatin. *Lancet* **2**:166, 1977.
89. Sakurai H, Dobbs RE, Unger RH: The effect of somatostatin on the response of GLI to the intraduodenal administration of glucose, protein, and fat. *Diabetologia* **11**:427, 1975.
90. Hanssen LE, Hanssen KF, Myren J: Inhibition of secretin release and pancreatic bicarbonate secretion by somatostatin infusion in man. *Scand J Gastroenterol* **12**:391, 1977.
91. Pederson RA, Dryburgh JR, Brown JC: The effect of somatostatin on release and insulinotropic action of gastric inhibitory polypeptide. *Can J Physiol Pharmacol* **53**:1200, 1975.
92. Bloom SR, Mortimer CH, Thorner MO, et al: Inhibition of gastrin and gastric-acid secretion by growth-hormone release-inhibiting hormone. *Lancet* **2**:1106, 1974.
93. Renaud L, Martin J, Brazeau P: Depressant action of TRH, LH-RH, and somatostatin on activity of central neurons. *Nature (Lond)* **255**:233, 1975.
94. Guillemin R: Somatostatin inhibits the release of acetylcholine induced electrically in the myenteric plexus. *Endocrinology* **99**:1653, 1976.
95. Cohen M, Rosing E, Wiley K, et al: Somatostatin inhibits adrenergic and cholinergic neurotransmission in smooth muscle. *Life Sci* **23**:1659, 1978.
96. Pipeleers D, Somers G, Gepts W, et al: Plasma pancreatic hormone levels in a case of somatostatinoma: Diagnostic and therapeutic implications. *J Clin Endocrinol Metab* **49**:572, 1979.
97. Ganda O, Weir G, Soeldner J, et al: Somatostatinoma: A somatostatin containing tumor of the endocrine pancreas. *N Engl J Med* **296**:963, 1977.
98. Larsson L, Holst J, Kühl C, et al: Pancreatic somatostatinoma: Clinical features and physiologic implications. *Lancet* **1**:666, 1977.
99. Alumets J, Ekelund G, Hakanson R, et al: Jejunal endocrine tumor composed of somatostatin and gastrin cells and associated with duodenal ulcer disease. *Virchows Arch [A]* **378**:17, 1978.
100. Kovacs K, Horvath E, Ezrin C, et al: Immunoreactive somatostatin in pancreatic islet-cell carcinoma accompanied by ectopic ACTH syndrome. *Lancet* **1**:1365, 1977.
101. Galmiche J, Colin R, Dubois P, et al: Calcitonin secretion by a pancreatic somatostatinoma. *N Engl J Med* **299**:1252, 1978.
102. Krejs G, Orci L, Conlon J, et al: Somatostatinoma syndrome: Biochemical, morphologic and clinical features. *N Engl J Med* **301**:286, 1979.
103. Petersson B, Elde R, Efendic S, et al: Somatostatin in the pancreas, stomach and hypothalamus of the diabetic Chinese hamster. *Diabetologia* **13**:463, 1977.
104. Stefan Y, Malaisse-Lagae F, Yoon J, et al: Virus-induced diabetes in mice: a quantitative evaluation of islet cell population by immunofluorescence technique. *Diabetologia* **15**:394, 1978.
105. del Pozo E, Schlüter K, Neufeld M, et al: Endocrine profile and pharmacokinetics of the new somatostatin analog SMS 201-995. *Acta Endocrinol (Copenh)* (in press).
106. Plewe G, Beyer J, Krause U, et al: Long-acting and selective suppression of growth hormone secretion by somatostatin analogue SMS 201-995 in acromegaly. *Lancet* **2**:782, 1984.
107. Ch'ng LJC, Sandler LM, Kraenzlin ME: Long term treatment of acromegaly with a long acting analogue of somatostatin. *Br Med J* **290**:284, 1985.
108. Lamberts SWJ, Oosterom R, Neufeld M, et al: The somatostatin analog SMS 201-995 induces long-acting inhibition of growth hormone secretion without rebound hypersecretion in acromegalic patients. *J Clin Endocrinol Metab* **60**:1161, 1985.
109. Lamberts SWJ, Uitterlinden P, Verschoor L, et al: Long-term treatment of acromegaly with the somatostatin analogue SMS 201-995. *N Engl J Med* **313**:1576, 1985.
110. Kayasseh L, Gyr K, Keller U, et al: Somatostatin and cimetidine in peptic ulcer haemorrhage: a randomised controlled trial. *Lancet* **1**:844, 1980.

111. Szabo S, Usadel KH: Cytoprotection-organoprotection by somatostatin: Gastric and hepatic lesions. *Experientia* **38**:254, 1982.
112. Limberg B, Kommerell B: Treatment of acute pancreatitis with somatostatin. *N Engl J Med* **303**:284, 1980.
113. Santangelo WC, O'Dorisio TM, Kim JG, et al: Pancreatic cholera syndrome: Effect of a synthetic somatostatin analog on intestinal water and ion treatment. *Ann Intern Med* **103**:363, 1985.
114. Davis GR, Camp RC, Raskin P, et al: Effect of somatostatin infusion on jejunal water and electrolyte transport in a patient with secretory diarrhea due to malignant carcinoid syndrome. *Gastroenterology* **78**:346, 1980.
115. Osei K, O'Dorisio TM: Malignant insulinoma: Effects of a somatostatin analog (compound 201-995) on serum glucose, growth, and gastro-enteropancreatic hormones. *Ann Intern Med* **103**:223, 1985.
116. Gerich J, Schultz T, Tsalikian E, et al: Clinical evaluation of somatostatin as a potential adjunct to insulin in the management of diabetes mellitus. *Diabetologia* **13**:537, 1977.
117. Raskin P, Unger R: Hyperglucagonemia and its suppression: Importance in the metabolic control of diabetes. *N Engl J Med* **299**:433, 1978.
118. Meissner C, Thum C, Beischer W, et al: Antidiabetic action of somatostatin assessed by the artificial pancreas. *Diabetes* **24**:988, 1975.
119. Christensen S, Hansen A, Weeke J, et al: 24 hour studies of the effects of somatostatin on the levels of plasma growth hormone, glucagon, and glucose in normal subjects and juvenile diabetics. *Diabetes* **27**:300, 1978.
120. Spinas GA, Bock A, Keller U: Reduced postprandial hyperglycemia after subcutaneous injection of a somatostatin-analogue (SMS 201-995) in insulin-dependent diabetes mellitus. *Diabetes Care* **8**:429, 1985.
121. Dahlstrom A, Fuxe K: Evidence for the existence of monoamine neurons in the central nervous system. I. Demonstration of monoamines in the cell bodies of brainstem neurons. *Acta Physiol Scand* **64**(suppl 232):1, 1964.
122. Boyd AE III, Lebovitz HE, Pfeiffer JB: Stimulation of human growth hormone by L-dopa. *N Engl J Med* **283**:1425, 1970.
123. Mims RB, Stein RB, Bethune JE: The effect of a single dose of L-dopa on pituitary hormones in acromegaly, obesity and in normal subjects. *J Clin Endocrinol Metab* **37**:34, 1973.
124. Kansal PC, Buse J, Talbert OR, et al: The effect of L-Dopa on plasma growth hormone, insulin, and thyroxine. *J Clin Endocrinol* **34**:99, 1972.
125. Luckc C, Höffken B, Morgner KD: L-Dopa induced growth hormone secretion. Comparison with insulin tolerance test, arginine infusion and sleep induced GH-secretion. *Acta Endicrinol (Copenh)* **77**:241, 1974.
126. Ettigi P, Lal S, Martin JB, et al: Effects of sex, oral contraceptives and glucose loading on apomorphine induced growth hormone secretion. *J Clin Endocrinol Metab* **40**:1094, 1975.
127. Lal S, Martin JB, De La Vega C, et al: Comparison of the effect of apomorphine and L-Dopa on serum growth hormone levels in normal men. *Clin Endocrinol (Oxf)* **4**:277, 1975.
128. Pinder RM, Brogden RN, Sawyer PR, et al: Levodopa and decarboxylase inhibitors: a review of their clinical pharmacology and use in the treatment of Parkinsonism. *Drugs* **11**:329, 1976.
129. Mars H, Genuth SM: Potentiation of levodopa stimulation of human growth hormone by systemic decarboxylase inhibition. *Clin Pharmacol Ther* **14**:390, 1973.
130. Bansal S, Lee LA, Woolf PD: Dopaminergic regulation of growth hormone (GH) secretion in normal man: Correlation of L-Dopa and dopamine levels with the GH response. *J Clin Endocrinol Metab* **53**:301, 1981.
131. Leebaw WF, Lee LA, Woolf PD: Dopamine affects basal and augmented pituitary hormone secretion. *J Clin Endocrinol Metab* **47**:480, 1978.
132. Bansal S, Lee LA, Woolf PD: Dopaminergic modulation of arginine mediated growth hormone and prolactin release in man. *Metabolism* **30**:649, 1981.
133. Woolf PD, Lantigua R, Lee LA: Dopamine inhibition of stimulated growth hormone secretion: Evidence for dopaminergic modulation of insulin and L-dopa induced growth hormone secretion in man. *J Clin Endocrinol Metab* **49**:326, 1979.

134. Bansal SA, Lee LA, Woolf PD: Dopaminergic stimulation and inhibition of growth hormone secretion in normal man: Studies of the pharmacologic specificity. *J Clin Endocrinol Metab* **53**:1273, 1981.

135. Tallo D, Malarkey WB: Adrenergic and dopaminergic modulation of growth hormone and prolactin secretion in normal and tumor bearing human pituitaries in monolayer culture. *J Clin Endocrinol Metab* **53**:1279, 1981.

136. Parks JS, Amrhein JA, Vaidya V, et al: Growth hormone responses to propranolol-glucagon stimulation: A comparison with other tests of growth hormone reserve. *J Clin Endocrinol Metab* **37**:85, 1973.

137. Lal S, Tolis G, Martin JB, et al: Effects of clonidine on growth hormone, prolactin, luteinizing hormone, follicle-stimulating hormone and thyroid-stimulating hormone in the serum of normal men. *J Clin Endocrinol Metab* **41**:1703, 1975.

138. Toivola PTK, Gale CC, Goodner CJ, et al: Central alpha-adrenergic regulation of growth hormone and insulin. *Hormones* **3**:193, 1972.

139. Woolf P, Lee L: Effect of the serotonin precursor tryptophan on pituitary hormone secretion. *J Clin Endocrinol Metab* **45**:123, 1977.

140. Schnure JJ, Raskin P, Lipman RL, et al: Growth hormone secretion during sleep: impairment in glucose tolerance and nonsuppressibility by hyperglycemia. *J Clin Endocrinol Metab* **33**:234, 1971.

141. Feldman JM, Lebovtiz HE: Control of insulin and growth hormone secretion by serotonin and dopamine. In *Fourth International Congress of Endocrinology, Washington, D.C., June 18–24, 1972.* International Congress Series No 256. Excerpta Medica, Amsterdam, 1972, p. 35.

142. Bivens CH, Lebovitz HE, Feldman JM: Inhibition of hypoglycemia-induced growth hormone secretion by the serotonin antagonists cyproheptadine and methysergide. *N Engl J Med* **289**:236, 1973.

143. Smythe GA, Lazarus L: Suppression of human growth hormone secretion by melatonin and cyproheptadine. *J Clin Invest* **54**:116, 1974.

144. Mitchell ML, Byrne MJ, Sanchez Y, et al: Detection of growth-hormone deficiency. The glucagon stimulation test. *N Engl J Med* **282**:539, 1970.

145. Martin JB, Audet J, Saunders A: Effect of somatostatin and hypothalamic ventromedial lesions on GH release induced by morphine. *Endocrinology* **96**:839, 1975.

146. Talwar GP, Pandian MR, Kumar N, et al: Mechanism of action of pituitary growth hormone. *Recent Prog Horm Res* **31**:141, 1975.

147. Ahren K, Hjalmarson A: Early and late effects of growth hormone on transport of amino acids and monosaccharides in the isolated rat diaphragm. In Pecile A, Muller EE (eds): *Growth Hormone, Proceedings of the First International Symposium, Milan, Italy, Sept. 11–13, 1967.* International Congress Series No. 158. Excerpta Medica, Amsterdam, 1968, p. 143.

148. Chochinov RH, Daughaday WH: Current concepts of somatomedin and other biologically related growth factors. *Diabetes* **25**:994, 1976.

149. Van Wyk JJ, Underwood LE: Relation between growth hormone and somatomedin. *Annu Rev Med* **26**:427, 1975.

150. Shields R: Growth hormones and serum factors. *Nature (Lond)* **267**:308, 1977.

151. Salmon WD Jr, Daughaday WH: A hormonally controlled serum factor which stimulates sulfate incorporation by cartilage in vitro. *J Lab Clin Med* **49**:825, 1957.

152. Phillips LS, Vassilopoulou-Sellin R: Somatomedins. *N Engl J Med* **302**:371, 1980.

153. Fryklund L, Uthne K, Sievertsson H: Identification of two somatomedin A active polypeptides and in vivo effects of a somatomedin A concentrate. *Biochem Biophys Res Commun* **61**:957, 1974.

154. Van Wyk JJ, Underwood LE, Baseman JB, et al: Explorations of the insulin-like and growth-promoting properties of somatomedin by membrane receptor assays. *Adv Metab Disord* **8**:127, 1975.

155. Dulak NC, Temin HM: A partially purified polypeptide fraction from rat liver cell conditioned medium with multiplication-stimulating activity for embryo fibroblasts. *J Cell Physiol* **81**:153, 1973.

156. Yalow RS, Hall K, Luft R: Radioimmunoassay of somatomedin B: Application to clinical and physiologic studies. *J Clin Invest* **55**:127, 1975.

157. Schalch DS, Heinrich UE, Koch JG, et al: Non-suppressible insulin-like activity (NSILA). I. Development of a new sensitive competitive protein-binding assay for determination of serum levels. *J Clin Endocrinol Metab* **46**:664, 1978.
158. Furlanetto RW, Underwood LE, Van Wyk JJ, et al: Estimation of somatomedin-C levels in normals and patients with pituitary disease by radioimmunoassay. *J Clin Invest* **60**:648, 1977.
159. Golde DW, Herschman HR, Lusis AJ, et al: Growth factors. *Ann Intern Med* **92**:650, 1980.
160. Kenny FM, Guyda HJ, Wright JC, et al: Prolactin and somatomedin in hypopituitary patients with "catch-up" growth following operations for craniopharyngioma. *J Clin Endocrinol Metab* **36**:378, 1973.
161. Costin G, Kogut MD, Phillips LS, et al: Craniopharyngioma: the role of insulin in promoting post-operative growth. *J Clin Endocrinol Metab* **42**:370, 1976.
162. Bala RM, Bohnet HG, Carter JN, et al: Effects of prolactin on serum somatomedin activity in hypophysectomized rats. *Clin Res* **24**:655A, 1976 (Abst.)
163. Unterman TG, Phillips LS: Glucocorticoid effects on somatomedins and somatomedin inhibitors. *J Clin Endocrinol Metab* **61**:618, 1985.
164. Wiedemann E, Schwartz E, Frantz AG: Acute and chronic estrogen effects upon serum somatomedin activity, growth hormone, and prolactin in man. *J Clin Endocrinol Metab* **42**:942, 1976.
165. Phillips LS, Orawski AT, Belosky DC: Somatomedin and nutrition. IV. Regulation of somatomedin activity and growth cartilage activity by quantity and composition of diet in rats. *Endocrinology* **103**:121, 1978.
166. Phillips LS, Orawski AT: Nutrition and somatomedin. III. Diabetic control, somatomedin, and growth in rats. *Diabetes* **26**:864, 1977.
167. Phillips LS, Fusco AC, Unterman TG: Nutrition and somatomedin. XIV. Altered levels of somatomedins and somatomedin inhibitors in rats with streptozotocin-induced diabetes. *Metabolism* **34**:765, 1985.
168. Moses AC, Cohen KL, Johnsonbaugh R, et al: Contribution of human somatomedin activity to the serum growth requirement of human skin fibroblasts and chick embryo fibroblasts in culture. *J Clin Endocrinol Metab* **46**:937, 1978.
169. Adashi EY, Resnick CE, D'Ercole AJ, et al: Insulin-like growth factors as intraovarian regulators of granulosa cell growth and function. *Endocr Rev* **6**:400, 1985.
170. Eddy RI, Gilliland PF, Ibarra JD Jr, et al: Human growth hormone release: Comparison of provocative test procedures. *Am J Med* **56**:179, 1974.
171. Lin T, Tucci JR: Provocative tests of growth-hormone release: A comparison of results with seven stimuli. *Ann Intern Med* **80**:464, 1974.
172. Roth J, Glick SM, Yalow RS, et al: Hypoglycemia: A potent stimulus to secretion of growth hormone. *Science* **140**:987, 1963.
173. Eddy RL, Jones AL, Chakmakjian ZH, et al: Effect of levodopa (L-Dopa) on human hypophyseal trophic hormone release. *J Clin Endocrinol Metab* **33**:709, 1971.
174. Hayek A, Crawford JD: L-dopa and pituitary hormone secretion. *J Clin Endocrinol Metab* **34**:764, 1972.
175. Weldon VV, Gupta SK, Haymond MW, et al: The use of L-dopa in the diagnosis of hyposomatotropism in children. *J Clin Endocrinol Metab* **36**:42, 1973.
176. Mitchell ML, Sawin CT: Growth hormone response to glucagon in diabetic and nondiabetic persons. *Isr J Med Sci* **8**:867, 1972.
177. Spathis GS, Bloom SR, Jeffcoate WJ, et al: Subcutaneous glucagon as a test of the ability of the pituitary to secrete GH and ACTH. *Clin Endocrinol* **3**:175, 1974.
178. Knopf RF, Conn JW, Fajans SS, et al: Plasma growth hormone response to intravenous administration of amino acids. *J Clin Endocrinol Metab* **25**:1140, 1965.
179. Merimee TJ, Lillicrap DA, Rabinowitz D: Effect of arginine on serum levels of human growth hormone. *Lancet* **2**:668, 1965.
180. Eddy RL: Aqueous vasopressin provocative test of anterior pituitary function. *J Clin Endocrinol Metab* **28**:1836, 1968.
181. Gagliardino JJ, Bailey JD, Martin JM: Effect of vasopressin on serum levels of human growth hormone. *Lancet* **1**:1357, 1967.
182. Valenta LJ, Sigel MB, Lesniak MA, et al: Pituitary dwarfism in a patient with circulating abnormal growth hormone polymers. *N Engl J Med* **312**:214, 1985.

3

Acromegaly and Gigantism

Acromegaly is a fascinating, awesome disorder caused by sustained hypersecretion of growth hormone. The term gigantism is used to denote the disorder when its onset precedes epiphyseal closure. Acromegaly was probably known to ancient Egyptians, as early as the thirteenth century B. C.[1] Gigantism had fascinated Greek mythologists, who promulgated the notion that those afflicted with this disorder were favored by the Gods. The biblical tale of David and Goliath has strong undercurrents of the possibility that Goliath, the giant, suffered from bitemporal hemianopsia (due to a pituitary tumor), which made him unable to see the stone from David's sling that came from the side and slew him.[2,3]

Acromegaly and gigantism hold the same fascination to clinicians today as through the centuries. In Western literature, Marie[4] provided the first descriptive and concise report of two cases with this malady and coined the term *acromegalie.* The role of growth hormone (GH) in the causation of this disorder was suspected but was not proved until Evans and Long[5] duplicated the syndrome in rats treated with extracts of GH. The first series of acromegaly in the American literature was reported by Davidoff.[6] Since then, considerable strides have been made in our understanding of this disease. The development of a radioimmunoassay for GH in the 1960s, the emergence of high-resolution computed tomography (CT) as well as the characterization of somatomedin and somatostatin in the 1970s, and the discovery of the human pancreatic growth hormone-releasing factor (GHRH) in the 1980s have all revolutionized our insights into this spectacular disease. Still, acromegaly continues to tantalize investigators, with its yet unraveled mysteries. Formidable and forboding, acromegaly continues to claim its toll on the life of its victims, despite all the modern techniques in diagnosis and treatment. Acromegaly is a rare disease. It is estimated that 300 new cases are diagnosed each year in the United States. Epidemiological studies in the Newcastle region have indicated a possibly higher prevalence rate.[7] Since acromegaly shortens the life-span of its victims, the early recognition of this disorder is mandatory.

Etiology and Pathogenesis

Acromegaly results from sustained hypersecretion of GH, most often secondary to a somatotrope adenoma of the pituitary gland. Less commonly, acromegaly can occur as a result of extrapituitary causes (Table 13). This section focuses on four aspects pertinent to the etiology of hypersomatotropism. The first issue that will be examined is whether the classic form of acromegaly is a primary pituitary disease or represents a phenomenon secondary to hypothalamic deregulation. Second, the possible basis of the origin and development of GH-cell tumors is outlined. Third, the phenomenon of acromegaly occurring as a consequence of ectopic secretion of GHRH, as well as ectopic secretion of GH, is discussed. Finally, the phenomenon of "eutopic" GH secretion by an ectopically located pituitary gland is briefly outlined.

Acromegaly as a Primary Hypothalamic Disorder

During the past decade, there has been a growing body of evidence to challenge the classic concept that acromegaly is primarily a disorder of the pituitary somatotropes. Since the hypothalamus exerts dual influences on the somatotropes—an inhibitory effect by somatostatin, and a provocative affect by GHRH—it is conceivable that hyperfunction of the somatotropes could result from either defective secretion of somatostatin or excessive secretion of GHRH. In fact, several workers have strongly favored the notion that the essential defect in acromegaly is almost always located at the hypothalamus.[8–10]

TABLE 13
Etiology of Acromegaly

Primary pituitary pathology
  Somatotrope adenoma
  Hyperplasia
  Carcinoma
  Mixed cell adenomas
Secondary to GHRH excess from hypothalamus
  Hypothalamic tumors (hamartoma, choristoma)
Secondary to ectopic secretion of GHRH
  Pancreatic tumors
  Bronchial tumors (foregut-derived tumors, carcinoids, carcinoma)
Secondary to ectopic secretion of GH
  Pancreatic tumors
  Bronchogenic carcinoma
  Breast and ovarian cancer
  Carcinoid tumors
Secondary to tumors arising from ectopic pituitary tissue
  Pharyngeal pituitary
  Sphenoid sinus

It has been suggested that perturbations in the central regulation of GH secretion may result in hyperplasia of the somatotropes with eventual adenomatous transformation, as a "tertiary" event. The data that support a central etiology for acromegaly can be approached from the following perspectives; histopathological, anatomical, and dynamic studies both in vivo and in vitro.

Histopathological and Anatomical Support for Hypothalamic Etiology

The histopathological data that strongly support a central hypothalamic etiology for the development of acromegaly stem from the repeated observation that occasionally the pituitary histology is entirely normal in acromegalics.[11,12] While this is reminiscent of the analogous situation encountered in the pituitaries of patients with pituitary-dependent Cushing's disease and hyperprolactinemia, such an observation is significantly rarer in patients with acromegaly. Although such reports provide compelling evidence for a possible central etiology, the presence of normal histology in acromegalics should be reviewed in light of the current knowledge regarding ectopic GHRH secretion by tumors.

Anatomically, it is now well recognized that acromegaly can be associated with hyperplasia of the somatotropes, in the absence of a localized pituitary tumor. Such an occurrence is usually attributable to excessive stimulation of the somatotropes by GHRH. While excess secretion of GHRH can be a result of a hypothalamic lesion, more commonly it occurs from ectopic secretion of GHRH by tumors elsewhere. Acromegaly caused by hypothalamic lesions is exemplified by hamartomas[13] and choristomas[14] within, or in the vicinity of, the hypothalamus. These tumors may directly secrete GHRH; alternatively, they may stimulate production of GHRH by the hypothalamic nuclei or impair the secretion of somatostatin by the hypothalamus. The precise mechanism that leads to the development of somatotrope adenomas in this situation is not clear, since these hypothalamic tumor cells have not been studied by the use of GHRH antiserum. More recently, Asa et al.[15] reported a clinicopathological study of six patients with acromegaly secondary to hypothalamic gangliocytomas producing GHRH. Immunocytochemical analysis employing specific antisera revealed the presence of hpGHRH in most neurons in all the gangliocytomas, while the cells of the pituitary adenomas in these patients did not possess hpGHRF. The demonstration that hpGHRF can be found in neurons of hypothalamic tumors supports a role for excess GHRH secretion by the hypothalamus in the development of acromegaly. Until GHRH or somatostatin levels in serum can be accurately measured, hypothalamic perturbations in the causation of acromegaly will continue to remain a theoretical consideration. Regarding ectopic secretion of GHRH by tumors, this repeatedly documented entity has become the model for acromegaly that develops as a consequence of stimulation by GHRH. This syndrome shares several similarities in GH dynamics with classic acromegaly.

Dynamic Data Supporting a Hypothalamic Etiology

Several facets of GH dynamics encountered in acromegaly suggest a central etiology. Even patients with documented somatotrope adenoma demonstrate several features that resemble physiological phenomena mediated by the hypothalamus. Table 14 outlines the similarities and dissimilarities between acromegalics and normals in terms of hormone dynamics. It is well known that patients with acromegaly, like normals, demonstrate preservation of pulsatile GH release,[16] as well as suppression to α-blockade and β-stimulation.[17] Furthermore, some acromegalics continue to demonstrate abnormal GH dynamics, even after removal of the somatotrope adenoma, and even after normalization of basal GH levels.[16,18] This implies that despite correction of hypersomatotropism, the intrinsic hypothalamic disturbances continue to prevail. Two recent studies[19,20] have shed considerable light in evaluating the role of hypothalamic control of GH secretion in acromegalics. Hanew et al.[19] studied GH responses to somatostatin, thyroid-releasing hormone (TRH), and luteinizing hormone-releasing hormone (LRH), and bromocriptine in acromegaly and distinguished two groups of acromegalics: one group with higher somatostatin responsiveness and another with relatively lower somatostatin responsiveness. The former group also demonstrated GH responses to hypothalamic releasing peptides and normalization of GH levels with bromocriptine. These observations were confirmed by Pieters et al.,[20] whose data also argue for the occurrence of two groups of acromegalics with significantly different patterns of sensitivity to somatostatin. The group with the higher sensitivity to somatostatin also had lower basal GH levels in comparison, showed

TABLE 14
Similarities and Differences in Growth Hormone Dynamics between Normals and Acromegalics

	Normal	Acromegaly
Similarities in GH dynamics		
Pulsatile secretion of GH	Pulsatile	Pulsatile
Response to insulin	Provocative response	Provocative response often preserved
Response to α-blockade (phentolamine)	Suppression of GH	Suppression often maintained
Response to β-stimulation	Suppression of GH	Suppression often maintained
Response to synthetic GHRF	Stimulation of GH release	Stimulation of GH release
Response to synthetic somatostatin	Suppression of GH release	Suppression often maintained
Differences in GH dynamics		
Response to glucose	Suppression of GH	No suppression
Response to L-DOPA	Stimulation of GH	Lowering of GH
Response to synthetic TRH	None	Paradoxical rise in GH
Response to synthetic LRH	None	Paradoxical rise in GH

significant "rebound" after discontinuation of somatostatin, and responded
to LRH administration. Their data suggest that the subgroup of acromegalics
with higher sensitivity to exogenous somatostatin might represent a deficiency
of endogenous somatostatin. The notable difference in the basal GH levels
between the two groups is also an intriguing finding in this study. One explanation might be that a deficiency of endogenous somatostatin would be
expected to result in smaller increments in GH than with an autonomously
functioning adenoma or in somatotropes stimulated by excess GHRH. Alternatively, it is possible that patients with milder elevations in GH respond better
to the administration of somatostatin.

In Vitro Data Supporting a Hypothalamic Etiology

In vitro studies have evaluated the behavior of adenoma tissue removed
from patients with acromegaly.[21,22] These studies show that GH secretion by
adenoma tissue is similar to normal pituitary tissue grown in culture. Thus,
dopaminergic agents inhibited GH release from adenoma tissue, a phenomenon also seen with adult and fetal pituitary tissue maintained in culture. The
stimulatory effects of theophylline and the inhibitory effects of somatostatin
observed in adenoma tissue also suggest that, at least in some respects, tumor
tissue behaves like normal pituitary grown in culture and appears not to be
due to a functionally less differentiated state.

Acromegaly as a Primary Pituitary Disorder

While the above body of clinical and experimental evidence supports a
hypothalamic origin of acromegaly, several established facts favor a primary
pituitary role in the genesis of acromegaly. Five lines of evidence constitute
the most vital links: (1) the impressive frequency with which tumors are found
within the pituitary; (2) the remarkable normalcy of the somatotropes around
the tumor; (3) the frequency with which "cures" are attained following selective removal of tumors; (4) the relative low rates of recurrence following
removal of the GH secreting adenoma; and (5) the impressive histological
spectrum of these adenomas, including their propensity for pluripotential
secretion of hormones. All these data point to the de novo development of a
GH-secreting adenoma due to a focal intrinsic problem within the pituitary,
each deserving a brief comment.

The incidence of demonstrating pituitary tumors in patients with acromegaly is impressively high, approaching 85–90%. Although hyperplasia, and
even a normal histology, can be encountered, these are extremely rare, indeed
so rare that the frequency of such a phenomenon has not been established
in the literature. The frequency with which demonstrable tumors are found
within the pituitary is much greater than in patients with Cushing's disease
or hyperprolactinemia, two closely related disorders, in which hypothalamic
influences are believed to be involved in pathogenesis. In acromegalics with
normal histology or hyperplasia the possibility of ectopic secretion of GH or

GHRH is a strong possibility. Thus, for all practical purposes, the assumption that a pituitary tumor will be found in acromegalics is statistically sound.

The remarkably normal appearance of the somatotropes in the uninvolved portion of the pituitary gland is also evidence against a hypothalamic influence, which if present would have exerted a generalized effect on all the somatotropes, not on just a focal few. The functional ability of these uninvolved somatotropes has not been adequately studied in man. Nakayama and Nickerson[23] studied the effects of transplanting GH secreting tumors into normal pituitaries of rats and showed suppression of normal somatotropes. Although the appearance of the paranodular somatotrope population is normal in acromegaly, one would expect—on the basis of the feedback effect of high GH levels—some degree of suppression of the nontumorous cells. This hypothesis awaits confirmation.

The frequency with which surgical cures are attained in acromegaly is another factor favoring a primary pituitary etiology. In many patients, following selective adenomectomy there is restoration of GH levels to normal, along with normalization of qualitative abnormal GH dynamics.[11,24-27] Such complete restoration to normalcy speaks against a primary hypothalamic defect. The effectiveness of surgery, which can approach as high as 80% cure rate, is discussed under treatment.

The relatively low recurrence rate following surgery also favors a primary pituitary etiology. This statement must be tempered by two facts—that detecting early recurrence by hormonal studies can be extremely difficult and that late recurrences are always a possibility in a disorder as insidious as acromegaly. Laws et al.,[28] reporting on 82 of their patients treated neurosurgically, point out that as many of 90% demonstrate normal GH levels 5 years after total excision. However, a case for late recurrence of acromegaly has been impressively made by Schuster et al.,[29] who have advocated more rigid criteria for defining persistence of biochemical cure.

Finally, the spectrum of histological abnormalities that may be encountered in somatotrope adenoma is so variable, with monomorphous, bimorphous, and even polymorphous forms, that it strengthens the case for de novo development. The most important piece of histological evidence to suggest an intrinsic pituitary origin of GH-secreting adenoma is the histological demonstration of plurihormonal tumors. It is now recognized that mixed GH and prolactin cell tumors can be associated with acromegaly.[30-33] This implies that the hypersecretory cells—secreting both GH and prolactin—must originate from a common stem cell. An alternative explanation is that two individually transformed cells undergo fusion. Either way, hypothalamic factors are unlikely to influence such events. The characterization of acidophil stem cell tumors (from the primitive noncommitted precursor stem cell), as well as mixed GH and prolactin cell adenomas, have clearly indicated that the process of tumorigenesis has its origins within the pituitary.[34-36]

In summary, there is evidence in the literature for both hypothalamic and pituitary mechanisms underlying the development of acromegaly (Tables 15 and 16). The existence of two subtypes of acromegalics has been postu-

TABLE 15
Acromegaly: Is It a Primary Hypothalamic Disorder?

Histopathological evidence
  Rare demonstration of normal pituitary histology[11,12]
Anatomical evidence
  Hyperplasia of somatotropes may be rarely encountered.
  Eutopic secretion of GHRH by hypothalamic tumors can cause acromegaly.[15]
  Ectopic secretion of GHRH by nonpituitary tumors can cause acromegaly.[47–49]
Hormonal evidence
  Pulsatile release of GH is often preserved in acromegaly.[16]
  Suppression to β-blockade is preserved.[17]
  Stimulation by insulin is often preserved.[16]
  Abnormal GH dynamics may persist even after normalization of GH.[16]
In vitro evidence
  In vitro behavior of tumor tissue is similar to that of normal tissue.[12,21]

lated.[8,9,20] It is unclear whether a hypothalamic abnormality sets the stage for hyperplasia of the somatotropes, followed by the development of an autonomous, tertiary type of adenomatous transformation. The issue remains unresolved. Studies using synthetic GHRH, as well as in vitro studies of the uninvolved somatotropes grown in culture, may provide meaningful insight into the nature and etiology of acromegaly.

Tumorigenesis of GH-Secreting Adenomas

Melmed et al.[37] proposed a multistage theory of GH cell tumorigenesis. These workers propose that the somatotropes must go through at least two stages of tumor induction in order to be transformed to neoplastic cells: a stage of initiation, wherein the DNA of somatotropes is modified by a short, irreversible stimulus, and a stage of promotion wherein putative factors influence transformation of the previously initiated cells. The factors that initiate this process are not well identified; genetic mutation, viral infection, and irradiation have been suggested, but remain unproved. Although the genetic tendency for inheriting acromegaly as part of the multiple endocrine adenomatosis (type I) syndrome is well recognized,[38] the significance of the genetic influences in the sporadic variety is unclear. The putative promoting factors that have been hypothesized are several, and include GHRH excess, somatostatin deficiency, and numerous peptides, such as vasoactive intestinal polypeptide (VIP), TRH, LRH endorphins, and dopamine. These speculations are based on in vitro observations that adenoma cells grown in culture

TABLE 16
Acromegaly: Is It a Primary Pituitary Disorder?

The impressive frequency of underlying pituitary tumors
Normal appearance of somatotropes around the tumor
Cure rates following selective microadenomectomy high
Relatively low recurrence rates following microadenomectomy
Propensity for plurihormonal secretion

respond to several of these peptides.[39,40] Changes in surface receptors, brought about by the neoplastic process, may account for some of the anomalous responses. The possibility that these tumor cells represent specialized clones of cells capable of responding to stimuli in a manner differing from the normal somatotropes is also unproved. The bihormonal secretory potential of some of these tumor cells has supported the theory that bihormonal cells probably represent in situ fusion of cells resulting in hybrid clones.

Acromegaly Caused by Ectopic Secretion of GH and GHRH

The elaboration of ectopic hormones by tumors is a common, perhaps an even ubiquitous, phenomenon. Ectopic GH secretion resulting in the development of clinical acromegaly is an uncommon event. Although immunoreactive GH has been demonstrated in several tumor tissues, including breast cancer[41] bronchogenic carcinoma,[42,43] bronchial carcinoid,[44] and ovarian cancer,[41] this alone does not imply ectopic GH secretion by these tumors. Five criteria need to be fulfilled unequivocally in order to establish the presence of ectopic GH secretion by a nonpituitary tumor:

1. The demonstration of an arteriovenous hormone gradient across the tumor bed
2. The rapid decline in circulating levels of GH (and somatomedin-C) following extirpation of the tumor
3. The demonstration of tumor cells that stain positive with immunoperoxidase staining for GH
4. Evidence for synthesis of GH by the tumor tissue grown in culture
5. Evidence for expression of the gene for human GH by the tumor

Currently several excellent techniques are available for studying the ability of the tumor tissue to synthesize and release GH, when maintained in monolayer culture. For instance, incubation of tumor tissue in the presence of radiolabeled leucine can provide evidence regarding the ability of the tumor cells to synthesize and release GH into the medium. Furthermore, the turnover rate of the newly synthesized hormone can be determined by comparison of levels of [^{3}H]-GH in the tissue, with the levels of hormone released into the medium. The ability of the tumor tissue to express the human GH gene can be studied by the use of cDNA human GH probes. This evaluates the ability of the cytoplasmic mRNA to hybridize with the specific [^{32}P]-cDNA for human GH. The demonstration of the presence of human GH mRNA in the cytoplasm of tumor cells is direct evidence for human GH gene expression by ectopic tissue.

Recently, Melmed et al.[45] documented ectopic GH secretion by a pancreatic tumor by fulfilling every criteria that has been set to establish ectopic hormonogenesis. Their patient, a 60-year-old man, presented with clinical evidence of hypersomatotropism, elevated and nonsuppressible GH levels, elevated somatomedin-C levels, and a normal CT scan of the pituitary, but an abnormal abdominal CT, which revealed a large mass. At laparotomy, an intramesenteric pancreatic tumor was removed. Melmed and co-workers demonstrated a 10 : 1 AV gradient of GH across the tumor, as well as marked decline in GH and somatomedin-C levels within 1 week of surgery. Studies on the tumor tissue revealed positive immunoperoxidase staining for GH, as well as its ability to synthesize GH and express the human GH gene.

Although cases of acromegaly caused by ectopic GH secretion have been reported in the past,[44,46] these cases were not as well studied as that of Melmed et al.[45] and were reported at a time when the phenomenon of ectopic secretion of GHRH was not well defined. In retrospect, some cases of acromegaly attributed to ectopic GH secretion may have been caused by ectopic GHRH secretion. The biochemical differences between the two entities are discussed in Section 3.4, on the diagnosis of acromegaly.

The notion that acromegaly can result from ectopic hypersecretion of GHRH was suggested by two case reports in the early 1980s.[47,48] Frohman et al.[47] partially purified extracts from nonpituitary tumors and showed that they contained specific GH-releasing activity. Leveston et al.[48] reported an 18-year-old with Cushing's syndrome and hypersomatotropic gigantism caused by a metastatic foregut carcinoid. The tumor extracts demonstrated GH-releasing activity. The turning point came when Thorner et al.[49] described a patient with acromegaly and a pituitary tumor that showed hyperplasia of somatotropes during surgery. Partial hypophysectomy was ineffective in rendering the patient eusomatotropic. Ultimately, a pancreatic tumor was discovered, extirpation of which cured the hypersomatotropism. It was from this patient's pancreatic tumor that Rivier et al.[50] isolated, sequenced, and characterized the human pancreatic GHRH, the hpGHRF, as we know it today. The same year also saw Guillemin et al.[51] isolate a GH-releasing peptide from an acromegalic patient with no tumor in the pituitary gland, but with two tumors in the pancreas. The peptide characterized by Guillemin et al.[51] was similar to that identified by Rivier et al.,[50] with the exception that it possessed a terminal amide group. Studies employing genomic sequencing methods have strongly suggested that these two peptides are identical to the human hypothalamic GHRH.[52,53] Both peptides have been synthesized (hpGHRH 1–40, and hpGHRH 1–44) and used for clinical evaluation in patients with disorders of the hypothalamic pituitary axis.

The phenomenon of acromegaly caused by ectopic secretion of GHRH has now become a well-recognized entity.[48,49,51,54–56] The actual incidence of ectopic GHRH secretion as a cause of acromegaly is unknown, but it must be quite rare. To assess the frequency with which acromegaly results from ectopic GHRH secretion, Thorner et al.,[57] in a multicentric international study, collected plasma samples from 177 unselected acromegalics. The immunoreactive GHRH was assayed by the use of a sensitive assay. None of the 177 samples showed elevated values for this peptide, underscoring its relative rarity.

The constellation of features of acromegaly secondary to ectopic secretion of GHRH is as follows:

1. There is clinical evidence of hypersomatotropism.
2. The underlying tumor that secretes GHRH is usually of neural crest origin—foregut tumors, carcinoids, and pancreatic tumors.

3. The radiograms of the sella may be normal or abnormal.
4. The GH levels may not show a rise when synthetic hpGHRH 1–40 or hpGHRH 1–44 is exogenously administered; this is in contrast to the positive response seen in most acromegalics.
5. The serum levels of IR GHRH (immunoreactive GHRH) are elevated.
6. The pituitary histology characteristically shows hyperplasia.
7. Amelioration—even "cure"—of the hypersomatotropism is often attained upon removal of the pancreatic tumor.

Although the syndrome of ectopic GHRH secretion is rare, it should be always kept in mind in order to avoid unnecessary pituitary surgery.

Acromegaly Caused by Tumors from Ectopic Pituitary Tissue

Embryological remnants of pituitary tissue may be present in the pharynx, parapharyngeal region, and even the sphenoid sinus.[58,59] The pituitary remnant in such "ectopic" locations is potentially functional.[60] Pituitary tumors, particularly hypersecretory, can develop in such locations.[61–64] More importantly, intrasellar and ectopic pituitary tumors can coexist in the same patient,[61,62] underscoring the clinical dictum that persistence of high hormone levels following operative removal of the intrasellar tumor should prompt a search for coexistent tumor in an ectopic pituitary.

The hypersomatotropism caused by adenomas originating from ectopic pituitaries demonstrates GH dynamics identical to those associated with the conventional disorder. It is interesting that the secretory dynamics of a tumor anatomically separated from the hypothalamus is still subject to the same neuroendocrine regulatory mechanisms. Warner et al.[63] described a mixed GH- and prolactin-secreting adenoma arising from a pharyngeal pituitary; they showed that even though spatially separate from the anterior pituitary, the tumor responded to dopamine agonists.

Documentation of tumors originating from ectopically located pituitary tissue can be obtained by high-resolution CT. Rarely, it may be necessary to resort to selective venous sampling of the inferior petrosal sinus to establish the origin of the excess hormone.

Histopathology

The application of sophisticated histopathological techniques has considerably changed the traditional view that acromegaly is caused by a single entity, the eosinophilic adenoma. Electron microscopy and immunocytochemical staining have clearly pointed out the existence of a spectrum of histological appearances in association with acromegaly. The monographs by Kovacs and Horvath and colleagues[65–68] continue to remain standard, classic treatises on the histopathology of pituitary adenomas.

Anatomically, the most favored site for the GH-secreting adenoma is the lateral wing of the anterior pituitary. On gross appearance, the tumor is well demarcated, with a rim of normal tissue surrounding it. Histologically, the tumor can be classified and subclassified into several types, but five types are important to recognize: densely granulated adenoma, sparsely granulated adenoma, mixed cell adenoma (GH and prolactin cell), acidophil stem cell adenoma, and mammosomatotrope adenoma. Rarely, the pathological lesion underlying acromegaly may be carcinoma or diffuse hyperplasia and, very rarely, no pathological change may be seen. In light of our current clinical knowledge, it is probably correct to state that when hyperplasia is seen, ectopic secretion of GHRH should be suspected; when the histology is entirely normal, ectopic secretion of GH should be suspected.

Densely and Sparsely Granulated Somatotrope Adenomas

Classification of the tumor histology into various types has important morphological connotations and may have clinical implications as well.[69–72] For instance, the densely granulated adenoma cells are characterized by nu-

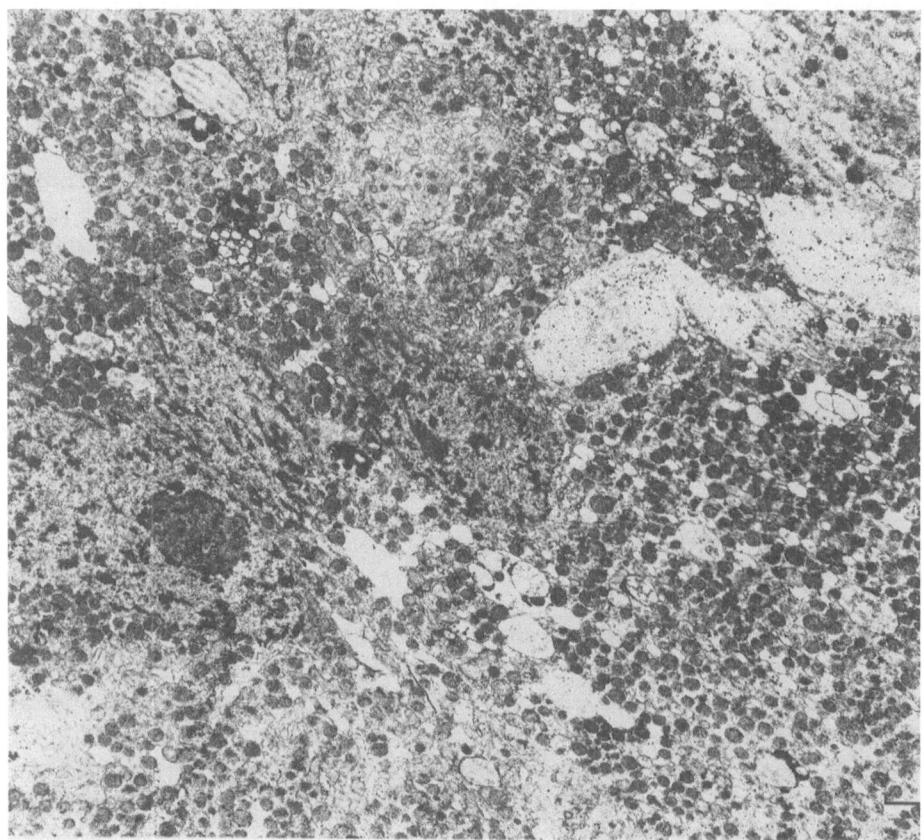

FIGURE 9. Electron-micrographic appearance of the densely granulated somatotropic adenoma demonstrating numerous GH granules. Scale bar: 1 μm. (From Adelman.[246])

merous secretory granules that are evenly dense, while the sparsely granulated adenoma cells contain highly irregular nuclei, with a scattered rough endoplasmic reticulum and smaller electrodense granules. The secretory granules of both varieties stain for GH with immunoperoxidase stains. Although these two varieties of adenoma represent variants of the same tumor, the biological behavior of the sparsely granulated variant is notably different; this variant of adenoma grows faster and more often invades the parasellar and suprasellar region, displaying an aggressive behavior. Figure 9 illustrates the electron micrographic appearance of the densely granulated somatotrope adenoma.

Mixed Cell Adenoma

The mixed cell adenoma[73,74] demonstrates both GH- and prolactin-containing cells within the tumor. Importantly, these cells are unihormonal, displaying immunoperoxidase staining for one or the other hormone. Although these tumors can be shown to synthesize both hormones in vitro, prolactin levels in the serum are not always elevated. The biological behavior of the mixed cell adenoma is not particularly aggressive.

Acidophil Stem Cell Adenoma

The acidophil stem cell adenoma,[75,76] is a monomorphous tumor composed of rather immature progenitor (stem) cells. Acromegalic patients with acidophil stem cell adenoma tend to have tumors that rapidly grow and invade neighboring tissues. The individual cells of this tumor stain for both GH and prolactin. It is in this type of variant that circulating levels of prolactin are elevated, often with clinical consequences of hyperprolactinemia. The acidophil stem cell adenoma can be recognized by virtue of its small, sparse secretory granules and abundance of mitochondria, including giant forms.

Mammosomatotropic Adenoma

Finally, the mammosomatotrope adenoma can be viewed as a mature form of a acidophil stem cell adenoma. Both GH and prolactin can be demonstrated in the same adenoma cell. The appearance of this variant is similar to the densely granulated adenoma, but in addition, characteristic large granular extrusions as well as extracellular deposits of secretory material are seen in mammosomatotropic adenoma.

Clinical Features

Chronic hypersecretion of growth hormone results in a multisystem disease with protean manifestations. The clinical features of acromegaly can be viewed from three perspectives: the clinical effects of hypersomatotropism on various organ systems, the local effects of an enlarging sellar mass caused by the GH secreting adenoma, and associated endocrine disturbances that may coexist with acromegaly.

Multisystem Involvement

While practically any organ system can be involved as a consequence of chronic hypersecretion of GH, the effects are particularly evident on somatic growth, soft tissue overgrowth, skin, and the cardiovascular system. Involvement of the neurological, musculoskeletal, respiratory systems, and other viscera occurs with varying degrees of frequency.

Growth

The effect of GH excess on somatic growth is spectacularly illustrated in pituitary gigantism, which is the expression of GH excess before epiphyseal closure. The hallmark of pituitary gigantism is an excessive rate of relatively symmetrical growth.[77] The two notable examples in the literature are the Buffalo giant, who reached an ultimate height of 96 inches,[78] and the Alton giant, who grew 99 inches tall.[79] Although the rapid growth rate usually begins during the immediate prepubertal age period (10–12 years) or adolescence, it can have its onset in infancy as in the case of the Alton giant, or in childhood, as in the case of the Minneapolis giant.[80] Obviously, when hypersomatotropism begins after epiphyseal closure, increased growth rate will not be seen. The classic features of acromegaly (acral enlargement and prognathism) may be seen only in 40–50% of patients with pituitary gigantism. The systemic effects of chronic hypersomatotropism, however, are the same regardless of whether the disease had its onset before or after epiphyseal closure.

Soft Tissue Changes

Acral enlargement is the hallmark of acromegaly and is seen in nearly all patients with the disease (Fig. 10A,B). The characteristic and insidious increase in hands and feet is responsible for the frequently positive response elicited by the questions regarding changes in the size of gloves, shoes, and rings. The massive, moist, fleshy handshake offered by the patient is unmistakable, leaving an indelible impression on the examiner. The soft tissue overgrowth is not merely limited to the hands and feet. The characteristic acromegalic facies, constituted by prognathism, enlargement of the nasal cartilage, earlobes and lips, coarsened thick leathery skin, and the overhanging hypertrophied supraorbital ridges, is in large part due to hypertrophy of the soft tissues of face (Fig. 11A,B). It is noteworthy that the changes of hypersomatotropism ensue so gradually that they literally creep under the skin of their victims. Thus, while patients may not admit to changes in their appearance, owing to the insidious nature of evolution, old photographs viewed serially would be immensely revealing.

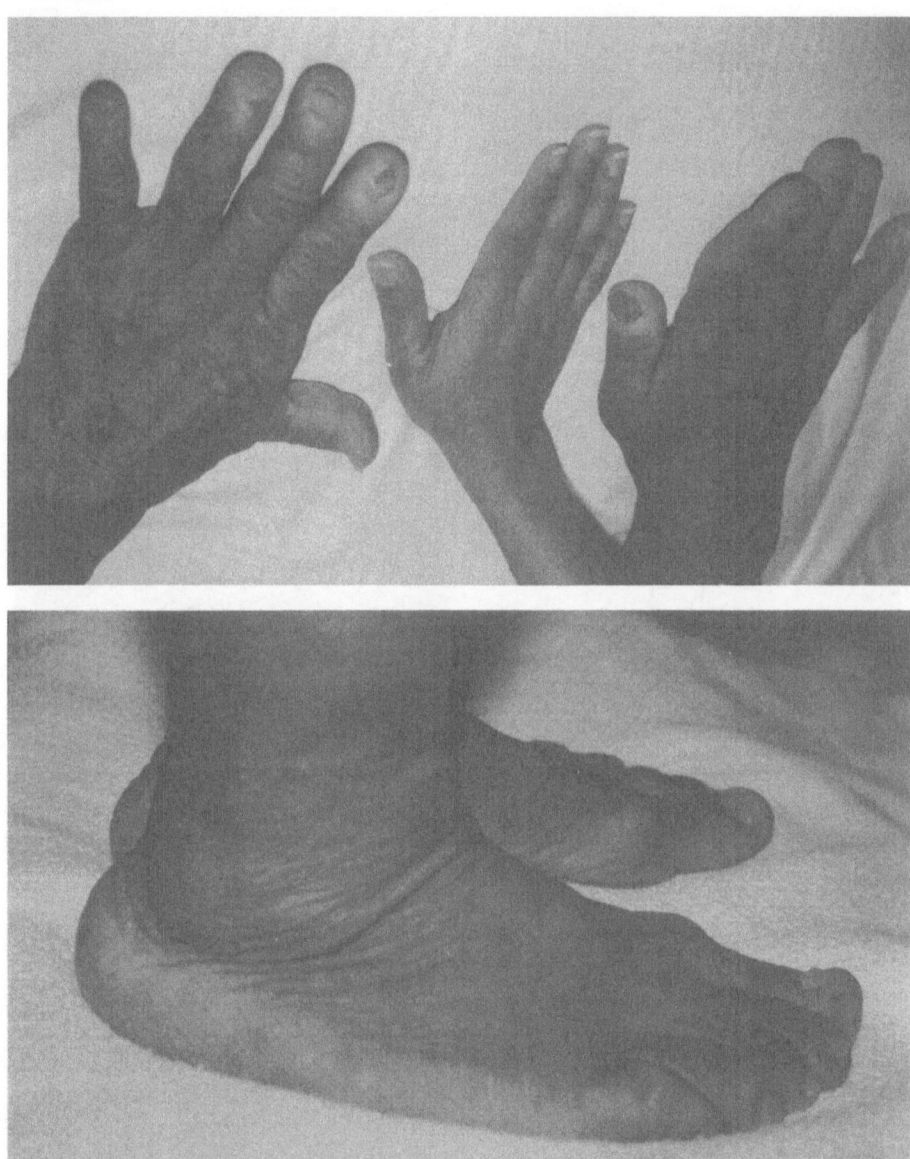

FIGURE 10. (A,B) Fifty-year-old acromegalic woman demonstrating striking soft tissue overgrowth of hands and feet.

Cutaneous Changes

Changes in the skin and integument are present in more than 90% of acromegalics. Skinfold thickening, as measured by skinfold calipers, is consistently increased in acromegaly. The thickening is most impressive in the dorsum of the hand. Hyperhidrosis is another feature noted in the skin of

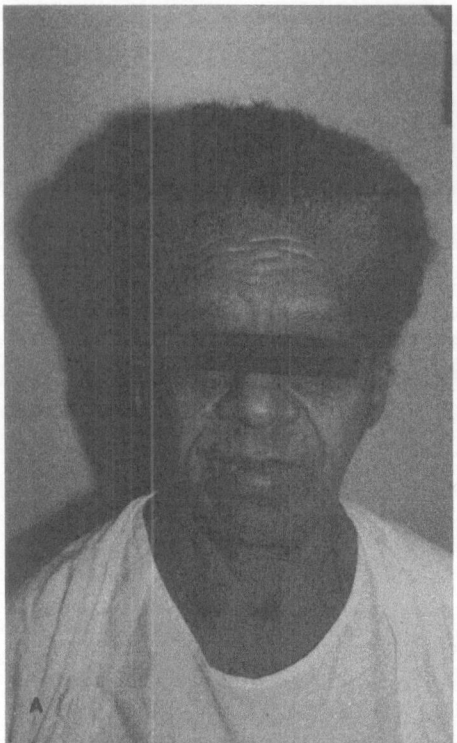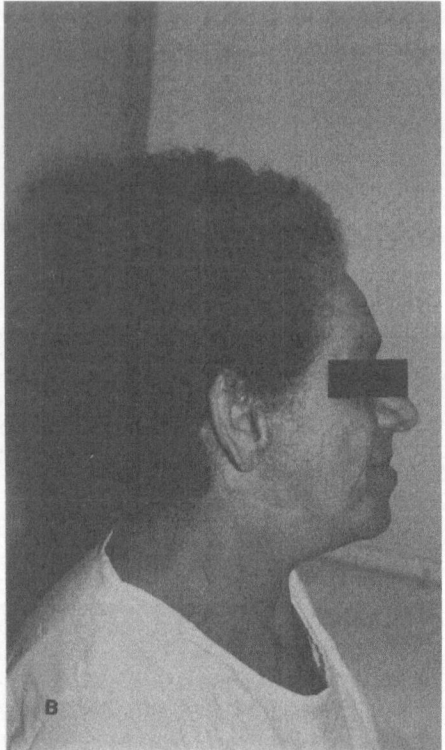

FIGURE 11. (A,B) Fifty-year-old acromegalic woman demonstrating acromegalic facies, characterized by prognathism, coarsening of features, prominent supraorbital ridges, and thickening of nasal and auricular cartilages.

acromegalics and is due to increased activity of the sweat glands. In addition, a certain degree of oiliness is imparted to the skin due to increased sebaceous secretions. Acanthosis nigricans has been associated with acromegaly,[81] but its prevalence is unknown. It has been suggested that these lesions be termed *pseudoacanthosis*, since these changes could be related to local mechanical effects on the skinfold and to increased perspiration.[82,83] The acromegalic skin often shows fibrous skin tags (molluscum fibrosum). The coarseness of the skin is one of the main reasons for the acromegalic facies.

Acral changes and the cutaneous changes represent the most consistent alterations caused by hypersomatotropism. However, these changes may not be impressive during the early phases of disease.

Cardiovascular System

The brunt of the ill effects caused by acromegaly is borne by the cardiovascular system (Table 17). In fact, the cardiac disease constitutes a major factor in contributing to the morbidity in acromegaly. At autopsy, cardiac enlargement is nearly always seen, often out of proportion in comparison with enlargement of other organs. Although the nature of acromegalic heart

TABLE 17
Cardiovascular Complications of Acromegaly

Systemic hypertension	Coronary atherosclerosis
Cardiomegaly	Arrhythmia
Congestive heart failure	Cardiomyopathy

disease is not entirely understood, the manifestations of chronic hypersomatotropism are well recognized; thus, systemic hypertension, cardiomegaly, congestive cardiac failure (often responding poorly to conventional treatment), premature coronary atherosclerosis, cardiac arrhythmias, and even a specific myocardial disease ("acromegalic cardiomyopathy") are distinct facets in the expression of the heart disease associated with chronic hypersomatotropism.

Although several reviews have stressed the manifold and deleterious effects of acromegaly on the heart,[84–86] relatively few studies have prospectively evaluated the incidence and nature of the cardiac manifestations associated with that disease. In a prospective study of the cardiovascular findings in 57 acromegalics, the findings of McGuffin et al.[87] indicate that hypertension, seen in approximately 28% of their acromegalics, was mild, often responsive to drugs, and was the only parameter that correlated with GH levels. Congestive heart failure, arrhythmias, and cardiomegaly were seen in fewer than 20% of their patients; when cardiomegaly was detected clinically, it was invariably associated with hypertension or cardiac failure. These investigators were unable to define a specific form of cardiomyopathy characteristic of acromegaly.

Hypertension is present in approximately 13–50% of patients with acromegaly. The mechanism for the development of hypertension is unclear. It is well known that an increase in the total body water and sodium is seen in many patients with acromegaly.[88] Therefore, studies of the renin-angiotensin-aldosterone (RAA) system in acromegaly are of interest. In one such study by Strauch et al.[89] the RAA system responded normally to salt deprivation and posture, while in another study of acromegalics, Cain et al.[90] showed blunting of plasma-renin activity to posture and salt deprivation, along with subnormal aldosterone-secretion rates. Occasionally, a primary aldosterone-secreting adenoma may be associated with acromegaly.[89] The association between acromegaly and pheochromocytoma is discussed separately. The hypertension associated with acromegaly is seen more often in older patients, is often mild, and responds well to drug therapy and correlates, to some extent, with circulating GH levels.

The cardiac failure encountered in acromegalics is often related to underlying problems such as hypertensive or coronary heart disease. However, in rare cases biventricular cardiac failure can occur in the absence of such factors. This rare phenomenon is characterized by rapidly progressive cardiac failure, showing little or no response to conventional remedies. The course is often punctuated by cardiac arrhythmias or pulmonary embolization and often culminates in death. In such patients, aggressive therapeutic intervention is necessary to lower GH levels rapidly.

Cardiac arrhythmias tend to occur frequently in acromegaly. Both ventricular and supraventricular arrhythmias may be encountered. It has been emphasized that the presence of supraventricular arrhythmias should raise suspicion of coexistent hyperthyroidism,[87] a not-infrequent association of acromegaly. Intraventricular conduction defects due to myocardial involvement may also be seen. Premature coronary atherosclerosis may complicate the heart disease of acromegaly.[91]

During the past decade, several observations have fostered the notion that subclinical cardiac muscle dysfunction may be present in asymptomatic acromegalics.[92] Abnormal electrocardiograms (ECGs) have been reported in 30–60% of acromegalics. The application of noninvasive technology for evaluation of cardiac muscle function in acromegaly has yielded important, and somewhat conflicting, data. Savage et al.[93] evaluated cardiac anatomy and function in 25 patients with active acromegaly. Their data indicated that abnormal echograms were seen in 80%, nearly one-half of whom were totally asymptomatic. The dominant finding noted was increased left ventricular mass, with concentric thickening of the left ventricular wall. The important observations that emerged from the study were threefold; (1) the strikingly increased frequency of abnormal findings in asymptomatic acromegalics; (2) the absence of other etiologies to explain the increased mass seen in some patients, implying that acromegaly per se can affect the myocardium; and (3) the absence of a correlation between GH levels and abnormalities in the echocardiograms.

Smallridge et al.[94] also studied the echocardiographic findings in 27 acromegalics with active disease and identified three groups; the first group consisted of six patients with asymmetrical septal hypertrophy, a finding also reported by other workers.[95,96] The second group consisted of eight patients who demonstrated concentric left ventricular hypertrophy, most showing no clinical evidence of organic heart disease. The third group, the largest one, comprising 13 patients, had normal studies or demonstrated a clearly increased thickness of the septum alone. Evaluation of left ventricular function demonstrated that the mean percentage of internal diameter shortening was greater in the first group, a finding analogous to that encountered in idiopathic hypertrophic subaortic stenosis (IHSS). Also shown in this study was a correlation between the degree of GH elevation and the presence of concentric LVH by echocardiography.

The plethora of abnormal echocardiographic findings seen in acromegalics—even when asymptomatic—has supported the notion that hypersomatotropism can directly affect cardiac structure and function in the absence of other disorders, such as hypertension or coronary atherosclerosis (Table 18). It has not been established, however, that continued disease activity progressively worsens the abnormalities in the structure and function of cardiac muscle. The frequency with which increased septal and posterior wall thickness is encountered in acromegaly highly favors a "cardiomyopathic effect" of hypersomatotropism on the heart muscle.

TABLE 18
Echocardiographic Abnormalities in Active Acromegaly

Asymmetrical septal hypertrophy
Concentric left ventricular hypertrophy
Increased thickness of septum and posterior wall
Systolic anterior mitral valve motion
Left atrial enlargement
Aortic root enlargement
Increased left ventricular transverse dimension at end of systole
Decreased left ventricular ejection fraction

Neuromuscular Involvement

The development of entrapment neuropathies and a nonspecific my-opathy represents the two major facets of neuromuscular manifestations of acromegaly.

Entrapment neuropathies, particularly involving the median nerve, are frequently encountered in acromegaly. O'Duffy et al.[97] found evidence of median neuropathy in 35 of 100 acromegalics with active disease. In their series, disease activity correlated well with median neuropathy. The incidence of neuropathy, in some series, has been reported to be as high as 70%.[98] Another rare neurological manifestation seen in severe cases is the development of root pains due to compression of spinal nerves from spinal stenosis. Epstein et al.[99] described a 54-year-old male acromegalic with a 30-year history of low back pain, progressive paraparesis, and bilateral lumbosacral radiculopathy secondary to stenosis of the spinal canal. Neurologically, the patient revealed bilateral foot drop, absent deep tendon reflexes, and bilateral L5–S1 hypesthesia. Myelography demonstrated the stenotic lumbar canal with a high-grade block. The spinal cord compression was relieved by laminectomy.

The association of weakness with acromegaly dates back to the very first cases described by Marie,[4] showing weakness, flabbiness, and wasting of proximal limb muscles. Although the myopathy is usually characterized by proximal muscle weakness, especially in the upper extremities, generalized decrease in muscle strength has been described in pituitary gigantism.[78,80] Myopathy occurs in acromegaly with an estimated incidence of 50%.[100,101] More recently, Pickett et al.[102] evaluated 17 consecutive acromegalics and found evidence of myopathy alone in five and myopathy with the carpal tunnel syndrome in four. The myopathy was chararacterized by mild, exclusively proximal, weakness with flabbiness. Electromyographic (EMG) studies showed nonspecific myopathic potentials, but the muscle enzymes and biopsies were normal. In this study there was a good correlation between duration of disease and presence of myopathy.

Articular Involvement

The joint disease of acromegaly resembles degenerative joint disease and can result in considerable disability. Like osteoarthritis, the major weight-bearing joints are involved but, unlike osteoarthritis, the joint space in acromegalic joints is enlarged. In older patients with acromegaly, joint involvement results in considerable disability. Chondrocalcinosis is said to occur more frequently in acromegalics.

Respiratory Involvement

The respiratory system can also be affected in acromegaly. Acromegalics are known to show hyperplasia and restructuring of the soft tissues of the upper airway.[103,104] As a consequence, there is a heightened tendency for narrowing of the upper airway with the risk of collapse and obstruction of airflow during sleep. The sleep apnea syndrome associated with acromegaly is being recognized with increasing frequency. Perks et al.[105] studied 11 acromegalics and demonstrated sleep apnea in three. In a larger series of 21 acromegalics, Hart et al.[106] found sleep apnea syndrome in 10 active acromegalics and in none of the cured acromegalics. They also demonstrated that while the sleep apnea syndrome was correlated with high GH levels, a high GH level does not affect the hypercapnic ventilatory drive. The endoscopic findings in acromegalics with sleep apnea syndrome were described by Cadieux et al.[107]; they demonstrated complete inspiratory collapse of the soft tissues in the posterior and lateral hypopharynx, with invagination into the laryngeal vestibule. The seriousness of the condition was underscored by the fact that both patients studied required emergency tracheostomies.

Visceral Involvement

Visceromegaly is a striking autopsy finding in acromegalics. Besides the heart, which is almost always enlarged, the liver, spleen, kidneys, large bowel, and thyroid gland show varying degrees of enlargement in postmortem studies. Clinically, thyromegaly is encountered in 20–25% of acromegalics. The enlargement can be diffuse, asymmetrical, or nodular. While hepatomegaly may be seen in acromegaly, it has been emphasized in the literature that a second cause is often found to account for the hepatomegaly.[108] Finally, sialomegaly is an infrequently recognized but fairly consistent feature of active acromegaly, most often involving the submandibular salivary glands.[109]

Local Effects of Enlarging Sellar Mass

The enlarging mass within the sella can result in significant anatomic and functional consequences. The pituitary tumor can invade the surrounding structures by suprasellar and parasellar extensions. There are five consequences of such invasiveness:

1. *Headaches:* Although it has been the traditional belief that headache is a result of upward extension of tumor, Jadresic et al.[110] found no correlation between size of the tumor and the presence or absence of

suprasellar extension. The headaches in acromegaly may be due to increased intrasellar pressure or to stretching of the capsule.

2. *Visual-field defects:* Upward extension and chiasmatic compression is the reason for developing field cuts. When these are significant, the visual-field cuts can be detected by the confrontation test at the bed side. More often, however, perimetric evaluation is required to document or confirm the extent of field cuts. While bitemporal hemianopsia is the classic field defect associated with enlarging pituitary neoplasms, superior (especially temporal) quadrantic cuts in the field are earlier lesions that can be detected by the astute observer. The sudden development of visual-field cuts that rapidly progress to even total loss of vision is the hallmark of pituitary apoplexy developing in the tumorous gland.

3. *Lateral extension into the sphenoid sinus:* This sign of extreme invasiveness is seen with the larger, more aggressive tumors, especially the chronic ones.

4. *Loss of pituitary reserve:* This may occur when large tumors encroach upon and destroy the normal pituitary tissue. As a consequence, deficiencies of gonadotropins, TSH, and even adrenocorticotropic hormone (ACTH) may eventually develop.

5. *Suprasellar extension:* This may result in compression of the pituitary stalk with resultant secondary hypopituitarism.

Several series, old and new, have compared the frequency of the various clinical features seen in acromegaly.[6,91,98,110-113] These international series encompass clinical experiences of more than 50 years, starting with Davidoff's original review of 100 cases of acromegaly in 1926,[6] and ending with the review by Jadresic et al.[110] of 145 acromegalics seen at Hammersmith Hospital, London. The most common clinical features, listed in descending order of frequency, are acral enlargement, coarsening of facial features, hyperhidrosis, carpal tunnel syndrome, headaches, altered reproductive or sexual function, and visual-field defects. Table 19 outlines the frequency with which these features occur in acromegaly, based on comparison of seven series. It is interesting to note that visual-field defects, seen with a frequency of 62% in Davidoff's series,[6] was encountered only in 6% of patients reviewed in Jadresic's series.[110] Similarly, headaches were a significant problem in 87% of Davidoff's patients,[6] while only 55% in the Hammersmith series were troubled with this symptom.[110] It is also interesting to note that the frequency of menstrual disorders, decreased libido, glucose intolerance, hyperhidrosis, acral enlargement, and soft tissue overgrowth is virtually identical in both the original review by Davidoff in 1926[6] and the current review by Jadresic in 1982.[110]

Associated Endocrine Disturbances

Acromegaly can be associated with several perturbations in endocrine function involving other hormones. Some of these perturbations involve the other pituitary hormones (e.g., prolactin, TSH, and gonadotropins); others

TABLE 19
Frequency of Clinical Features
in Acromegaly

Feature	Percent
Acral enlargement and overgrowth of soft tissue	96–100
Increased sweating	60–88
Increased skin thickness	80–90
Headache	50–87
Parasthesias	30–70
Cardiovascular problems	12–34
Hypertension	18–24
Menstrual irregularities	32–87
Decreased libido	27–46
Visual-field defects	6–24
Thyromegaly	18–30
Asthenia	30–42
Glucose intolerance and overt diabetes	25–50

are the result of the far-reaching effects of GH excess of distant endocrine glands (e.g., thyroid, islet cells, parathyroids); and finally some endocrine associations occur more frequently in association with acromegaly (multiple endocrine adenomatosis type I and pheochromocytoma). Table 20 outlines these hormonal associations. Each is briefly outlined in turn.

Prolactin Dynamics

Elevated levels of circulating prolactin levels are often encountered in acromegaly. The precise incidence of this phenomenon has been variably

TABLE 20
Endocrine Disturbances Associated with Acromegaly

Effects on other anterior pituitary hormones
    Prolactin (hyperprolactinemia)
    TSH (impaired TSH response to TRH)
    LH, FSH, ACTH (compromised pituitary reserve)
Effects on distant endocrine organs/systems
    Thyroid
        Abnormalities in thyroid function tests
        Coexistent hyperthyroidism
    Islet cells (β cells)
        Glucose intolerance
    Mineral metabolism
Associated endocrinopathies
    MEA-I
    Pheochromocytoma

reported in the literature, mostly owing to the small number of patients studied. In one of the largest series reported, De Pablo et al.[114] measured prolactin levels in 73 untreated acromegalics and found hyperprolactinemia in 40% of female and 27% of male patients. This prevalence rate is highly representative of the incidence of hyperprolactinemia in acromegaly. There are four main mechanisms for the development of hyperprolactinemia in patients with acromegaly:

1. The tumor may be composed of a mixed cell adenoma, containing individual cells that secrete either GH or prolactin.

2. The tumor may be composed of cells that are biopotential, both hormones being secreted by the same cell.[73,115] This phenomenon is particularly evident in two variants of GH-secreting adenomas: the acidophil stem cell adenoma and the mammosomatotropic cell adenoma. The demonstration that single cells of a benign tumor can secrete more than one hormone challenges the traditional "one cell, one hormone concept."

3. Two distinct and separate tumors, somatotrope adenoma and lactotrope adenoma, may coexist in the same gland. This situation has been well documented by Tolis et al.,[116] who described a patient with amenorrhea, galactorrhea, and acromegaly caused by two adenomas secreting GH and prolactin independently.

4. Hyperprolactinemia may occur as a consequence of stalk inhibition or hypothalamic involvement, resulting in impaired dopamine transport to the pituitary. As a result, the intrinsically normal lactotropes "escape" from the restraining effect of dopamine and hypersecrete prolactin; in this situation, the other trophic hormones (LH, FSH, TSH, and ACTH) suffer the consequences of diminished hypothalamic drive.

Immunocytochemically somatotrope adenomas containing GH and prolactin can be divided into three types[117]: type I, in which immunoreactive prolactin is present in single cells surrounded by immunoreactive GH cells; type II, in which immunoreactive prolactin-containing cells form clusters; and type III, in which immunoreactive prolactin and GH form a mosaic pattern. Of these three varieties, active secretion of prolactin with resultant hyperprolactinemia is most likely to be encountered in adenomas displaying the type III pattern.

Endocrine studies of prolactin dynamics in acromegalic patients with hyperprolactinemia may reveal a variety of abnormalities, none of which is considered diagnostically pathognomic. Thus, normal, subnormal, and exaggerated prolactin responses to TRH administration have been reported.[117–119] The suppressive effect of dopaminergic agents on prolactin secretion appears to be preserved in acromegalics.[119]

The clinical importance of abnormal prolactin secretion in acromegalics is fourfold. First, it should be recognized that galactorrhea may be an early manifestation of somatotrope adenomas. In general, when hyperprolactinemia occurs in acromegaly, the clinical manifestations of hypersomatotropism are readily evident. Occasionally, however, as illustrated by Tourniaire et al.,[120] patients with GH-secreting adenomas can present with galactorrhea, absence of acromegalic features, and abnormally elevated nonsuppressible GH levels. It is therefore recommended that evaluation of GH status form an integral part in the workup of hyperprolactinemia. Second, when hyperprolactinemia is associated with acromegaly, the changes in prolactin level seen after treatment often parallel the changes in GH level.[121] Third, rarely the previously normal prolactin level may slightly increase following pituitary radiation. The reason for this phenomenon is unclear. Finally, there is also evidence to suggest that following selective adenomectomy in acromegalics with hyperprolactinemia, a transient decrease in prolactin secretion from the normal gland may be observed.[122]

TSH Dynamics

Two important phenomena affect TSH dynamics in acromegaly: reversible impairment in TRH-induced TSH response, and coexistence of TSH hypersecretion with acromegaly.

The TSH response to TRH may be impaired in acromegaly even in the absence of anatomical destruction of thyrotropes by the tumor. The notion that the GH secretory status influences the response of thyrotropes to intravenous administration of TRH is a well-accepted one.[123] Thus, elevated GH levels impair the TRH-mediated TSH response, best exemplified in acromegaly, in which a blunted TSH response to exogenous TRH is often seen.[124,125] The reversible nature of this abnormality is evidenced by restoration of normal TRH-mediated TSH responses upon rendition of eusomatotropism. While the mechanism for this phenomenon is not entirely clear, the possibility that it is mediated by somatostatin is a likely one.

The concomitant hypersecretion of GH and TSH is a well-recognized event.[126–130] The hypersecretion of TSH in acromegaly is often caused by plurihormonal adenomas, which may be monomorphous or plurimorphous (i.e., composed of multiple distinct cell types, each secreting a different hormone). The cellular origin of such tumors is obscure, since the thyrotrope and somatotrope do not share a common ancestral stem cell line. The clinical, hormonal, and dynamic features of TSH-secreting tumors is discussed in Chapter 16.

Gonadotropin Dynamics

Abnormalities in pituitary reserve are present in many patients with acromegaly. Gonadotropin deficiency leads the list of trophic hormone failures associated with acromegaly, which correlates with clinical observations that decreased libido or menstrual irregularities occur with a very high frequency in that disorder. Goldfine and Lawrence[131] pointed out that there is no correlation between the hypopituitarism of acromegaly and tumor size, severity, or duration of acromegaly.

The Thyroid Gland

The most frequent effect of hypersomatotropism on the thyroid gland is promotion of growth of the gland, resulting in thyromegaly. The enlargement is usually mild to modest and often asymmetrical and can be diffuse or nodular. Some form of thyromegaly is evident in nearly 30% of acromegalics.

Thyroid function, however, is unaltered in most acromegalics. It should be pointed out that several symptoms are common to both hypersomatotropism and hyperthyroidism, a fact that can be conducive to missing concomitant hyperthyroidism in acromegaly. Thus, increased sweating, menstrual disorders, decreased libido, asthenia, hypertension, and galactorrhea may be encountered in both diseases. Two clues may point to the concomitant presence of hyperthyroidism in acromegaly; the occurrence of supraventricular arrhythmias, and the absence of the characteristic thickening of skin may point to suspicion of hyperthyroidism in association with acromegaly. In general, two mechanisms underlie the thyroidal hyperfunction associated with acromegaly. The most frequent mechanism is TSH hypersecretion by the pituitary adenoma. Less commonly, Graves' disease can be associated with acromegaly. Besides hyperfunction, secondary hypothyroidism can result when the pituitary tumor has destroyed thyrotrope reserve. Less frequently, hypothyrotropic hypothyroidism can result from interruption of TRH transport caused by interruption of the pituitary stalk.

Even though most acromegalics are euthyroid, the commonly performed thyroid function tests can be affected by hypersomatotropism. The thyroxine-binding proteins are altered in acromegaly,[132,133] resulting in decreased levels of thyroxine-binding globulin (TBG) and increased thyroxine-binding prealbumin (TBPA). Since the dominant effect is on TBG, the net result is lowering of total thyroxine and an increase in the triiodothyronine (T3)-resin uptake. The free thyroxine (T4) hormone levels are usually normal. It should also be pointed out that the blunted TRH-induced TSH release seen in acromegaly resembles that seen in secondary hypothyroidism.

Glucose Intolerance

Abnormal glucose tolerance is present in at least one-third of acromegalics.[134,135] The traditionally held concepts that overt diabetes develops in those acromegalics who have a family history of diabetes, those with longer duration of disease, and with the higher GH levels have been challenged by the studies of Linfoot et al.[136] and Wass et al.[137] The hallmark of the glucose intolerance associated with acromegaly is hyperinsulinemia and insulin resistance. Some insight into the mechanism of insulin resistance associated with acromegaly can be gained by insulin-receptor studies in acromegaly. In studying insulin binding to monocyte receptors in acromegalics with normoglycemia, Muggeo et al.[138] demonstrated two abnormalities—the total receptor concentration per cell was decreased in proportion to the hyperinsulinemia (which bore an inverse relationship) with compensatory increase in the affinity of the empty receptors. These investigators noted that the decrease in receptor concentration, as well as the increased affinity of the "empty receptors," correlated well with the magnitude of GH elevation. By contrast, studies in hyperglycemic acromegalics,[139] while also demonstrating a decrease in receptor concentration, failed to show an increase in receptor affinity. The failure in this compensatory mechanism may be the reason for the development of glucose intolerance in some acromegalics. Recent studies by Hitman et al.[140] supported the notion that there may be genetic determinants for the glucose intolerance of acromegaly.

Mineral Metabolism

Several alterations in the metabolism of calcium, phosphorus, and vitamin D are associated with the acromegalic state. There are four major perturbations in mineral metabolism caused by acromegaly:

1. Hypercalciuria is present in approximately 50–60% of patients with active acromegaly.[141]
2. The hypercalciuria is attributed to intestinal hyperabsorption of calcium.[142]
3. The increase in calcium absorption by the gut in acromegaly may be mediated by 1,25-dihydroxy D_3. The concept that GH may play a role in stimulating renal 1-α-hydroxylation of 25-hydroxy D_3 is gaining acceptance. Two independent groups[143,144] demonstrated that indeed GH possessed such activity in the rat. Clinical support for such a concept was provided when Brown et al.[145] demonstrated elevated levels of 1,25-dihydroxycholecalciferol in the sera of patients with active acromegaly. Studies evaluating the effects of GH administration on 1,25-dihydroxy D_3 to hypopituitary children have yielded conflicting data.[146–148]
4. The serum phosphorus is often elevated in acromegaly due to the stimulatory effect of GH on renal tubular reabsorption of phosphorus.[149]

In a recent study, Takamoto et al.[150] evaluated all these facets of mineral metabolism in 12 acromegalics both before and after correction of GH excess by pituitary adenomectomy. They observed significant reductions in urinary calcium excretion, serum calcium, and serum phosphorus levels. More importantly, they confirmed that serum 1,25-dihydroxy-D_3 levels, which were significantly elevated prior to treatment, declined after surgery, the decline correlating with the decrements in calcium excretion. These data suggest that the hypercalciuria of acromegaly might result from intestinal hyperabsorption of calcium, which in turn might be secondary to a GH-mediated increase in 1,25-dihydroxy D_3.

The effect of GH excess on skeletal metabolism in adults is controversial. The major effects on the bone are increased calcium turnover, increased periosteal opposition, and increased resorption of trabecular bone. Aloia et al.[151] postulated that untreated acromegaly is associated with differential bone remodeling with increased cortical bone and a paradoxically reduced trabecular bone mass. The increase in cortical bone may have a protective influence and probably prevents compression fractures.

Finally, two clinical observations are pertinent. Although acromegaly may be associated with slight increases in the serum calcium level compared with the baseline, hypercalcemia is unusual. When present, this should raise suspicion of concomitant hyperparathyroidism, often secondary to MEA-I. Second, the chronic effect of GH excess on the bone results in increased turnover and eventual osteopenia.

Endocrine Neoplasia

The association of acromegaly with either hyperparathyroidism or pancreatic islet cell adenoma, or both, is referred to as multiple endocrine neoplasia (or adenomatosis) type I.[152] While it is difficult to predict the frequency with which acromegaly is associated with pancreatic or parathyroid adenomata, it is good practice to search for these when a family history of acromegaly or islet cell tumor is present. It should also be recognized that the association of acromegaly and a pancreatic islet cell tumor need not always be on the basis of MEA-I: the acromegaly can result from ectopic secretion of GHRF by the islet cell tumor, particularly gastrinomas.[54,56]

Less well recognized is the association of acromegaly and pheochromocytoma, although several case reports[153–157] had sporadically mentioned such an association. Anderson et al.[158] reported two cases of pheochromocytomas with acromegaly and cited eight, possibly 12, other instances in the literature, suggesting that the combination may represent a nonfamilial variety of multiple endocrine neoplasia (MEN) syndrome. These patients tend to have sustained hypertension rather than the episodic hypertension characteristic of pheochromocytoma. It is noteworthy that in most of the patients reviewed by Anderson et al.[158] symptoms of pheochromocytoma had persisted for an average of 7 years before the diagnosis had been established. Even more important was the fact that four of the 10 patients died from causes directly related to pheochromocytoma. This emphasizes the importance of a careful search for pheochromocytoma in acromegalics with hypertension. Both patients reported by Anderson et al.[158] had malignant pheochromocytomas that had metastasized. Hyperparathyroidism has been variably associated with the syndrome. As with MEN, aberrant neural crest development may underlie the development of both, the pituitary tumor and the pheochromocytoma.

Diagnostic Approach

The approach to the laboratory diagnosis of acromegaly must be pursued in a methodical and stepwise fashion. The most important step is to entertain a high index of suspicion. Establishing the diagnosis of acromegaly at a stage when the classic somatic features have not yet manifested would be greatly rewarding to both patient and physician. Screening for acromegaly can be performed by measurement of basal GH levels, often with concomitant evaluation of the response of GH to an orally administered glucose load. Confirmation of the diagnosis is best achieved by measurement of the GH-dependent somatomedins (usually somatomedin C) in the circulation. The various dynamic studies (i.e., tests that evaluate the response of GH to TRH, LRH, L-dopa, and somatostatin) are at best supportive and perhaps aid in planning adjunctive medical therapy. Once the diagnosis of acromegaly has been established on the basis of demonstrating the triad of raised GH levels, non-suppressibility to glucose, and elevated circulating somatomedin levels, the diagnostic exercise is not over; it may have just begun. After having established the diagnosis of acromegaly, the workup should be aimed at answering the following five questions:

1. *Is there evidence of a tumor in the pituitary and, if so, is it invasive?* The information obtained (usually by CT studies) has enormous therapeutic and prognostic implications.
2. *Is there evidence of target organ damage and, if so, to what extent?* Documentation of the adverse effects of acromegaly on the cardiovascular, respiratory, articular, and metabolic systems would serve as guidelines to assess clinical responses to therapy.
3. *Has ectopic secretion of GHRH or GH been excluded?* Although rare, consideration must be given to this entity to avoid unnecessary treatments directed against the wrong organ.
4. *Is there evidence of hypopituitarism?* This is aimed at evaluating basal pituitary reserve before treatment in order to avoid unfair future incriminations against the modality of therapy used.
5. *Is there evidence of coexistent endocrinopathies?* The four conditions that need to be excluded are hyperprolactinemia, hyperthyroidism, MEA-I, and pheochromocytoma.

It can thus be deduced that a systematic approach is necessary to gain insights into the prognostic and therapeutic implications of this disease. Table 21 outlines the steps in the diagnostic workup of acromegaly. The discussion of the various laboratory tests follows the schema and the order outlined in Table 21.

Step 1: Screening for Acromegaly

Whereas screening for hypersomatotropism is a reflex when acral enlargement and other acromegalic features are evident, several instances may indicate screening for the disease:

1. Obvious acromegalic features are found to be present (e.g., acral enlargement, coarsened features, hyperhidrosis, prognathism).
2. All patients with pituitary neoplasms must be screened for acromegaly.
3. In patients with hyperprolactinemia with or without galactorrhea, acromegaly needs to be excluded.
4. The presence of unexplained cardiac disease—either ventricular hy-

TABLE 21
Laboratory Evaluation of Acromegaly

Step	Category of studies	Tests
I	Tests that screen for acromegaly	Basal GH level Glucose suppression test
II	Tests that confirm the diagnosis	Somatomedin-C level Supportive dynamic data
III	Localization studies to demonstrate intrasellar, suprasellar, or parasellar disease	CT scan of sella suprasellar region Visual fields by perimetry
IV	Tests that determine target organ damage by acromegaly	Glucose tolerance test Chest roentgenogram, ECG noninvasive cardiac testing (echocardiogram) Radiographs of knee, spine Nerve conduction velocity studies Pulmonary function tests
V	Tests that exclude ectopic GHRH secretion as a cause of acromegaly	Plasma level of IR GHRH GH response to exogenous administration of GHRH; if above are suggestive, CT of chest and pancreas
VI	Tests that determine pituitary reserve	Pituitary reserve testing
VII	Tests that evaluate presence of coexistent endocrinopathies	Thyroid function Prolactin studies Metanephrine, normetanephrine Calcium, PTH, gastrin

pertrophy or cardiac failure—may be a subtle clue for underlying hypersomatotropism.

5. All patients with median nerve entrapment neuropathy merit a screening for acromegaly.
6. In adolescents with extremely rapid growth rates, gigantism should be considered and excluded (this, as we shall see, may not always be easy).
7. Propositi of MEA-I syndrome need to be screened for the acromegaly.
8. Patients with repeated orthodontic problems caused by malocclusion should be screened for acromegaly.

The two simple and inexpensive studies employed for screening purposes are the measurement of basal GH levels and evaluation of GH response to an oral load of glucose.

Basal GH Levels

Tolis et al.[159] reviewed data on approximately 900 acromegalic patients (a staggering task involving analysis of 24 different studies); they showed that

TABLE 22
Causes of Elevated Growth Hormone

Pathological
    Acromegaly, gigantism
    Unstable diabetes
    Carcinoid syndrome
    Cirrhosis with portocaval shunt
    Anorexia nervosa
    Protein-calorie malnutrition
    Hypoglycemia
    Drugs: β-blockers
Physiological
    Pubertal growth spurt
    Exercise
    Stress

a basal serum GH level >10 ng/ml was present in 95.8% of acromegalics. Although the diagnostic yield of basal GH determinations is rather impressive, the major problem with the test is not so much with its sensitivity as with its specificity. Several conditions are associated with elevated basal GH concentrations (Table 22). Three important facts need to be kept in mind while interpreting the basal GH level in acromegaly. First, there may be a tremendous variation in the basal hormone level in the same patient when blood is sampled during different times of the same day or on different days. The magnitude of fluctuation may diminish the interpretive value of a single sample drawn at an isolated point in time. Second, the basal GH level, drawn under fasting conditions, does not reflect the GH secretory activity. Daggett and Nabarra[160] demonstrated impressively the lack of correlation between basal GH levels and GH secretory activity determined by 24-hr integrated plasma levels. Also, GH levels may not correlate with the severity of disease.[161] Third, and more importantly, basal GH levels can be normal or barely elevated in some patients with acromegaly. Such patients may show abnormal 24-hr integrated plasma value as well as abnormal responses of GH to various stimuli. The prognostic value of the basal GH measurement in assessment of cure is discussed in the section on treatment of acromegaly.

GH Response to an Oral Glucose Load

Administration of oral glucose to normal patients is attended by a reduction in GH levels to <5 ng/ml. This suppression of somatotropes is mediated by the glucostats located in the hypothalamus, which are highly sensitive to sudden increases in plasma glucose concentrations. Suppression of the adrenergically mediated GHRH is believed to be the mechanism underlying the lowering of GH following a glucose load. The decrease in GH levels to <5 ng/ml is usually seen at 60–120 min following the glucose load. In normal children, a rebound increase in GH levels may be seen 3–5 hr after glucose loading.[162] Lawrence et al.,[16] in a resounding article resonating with clarity, characterized the GH responses to a glucose load in acromegaly. Patients with acromegaly can be divided into three categories based on the GH response to an oral glucose load: (1) partial responders, who show some decline in GH levels after glucose, but remaining above 5 ng at all times; (2) nonresponders, who show no significant change at any time during the 2 hr after glucose; and (3) paradoxical responders, who actually demonstrated a rise in GH levels temporally related to glucose administration. It is important to remember that the interpretation of these responses can be clouded by the spontaneous fluctuation of GH levels, the magnitude of which can be quite impressive in some patients. The observations made by Lawrence et al.[16] were among the earliest that challenged the long-held concept that acromegaly represented a totally autonomous disorder.

As a screening test for acromegaly, an abnormal response to oral glucose (i.e., failure of GH levels to decline below 5 ng/ml) is encountered in approximately 93.5% of acromegalics.[159] The ease with which the test can be performed, as well as the low cost of the study, have made the oral glucose loading test a highly desirable screening procedure. However, several limitations of the test in terms of its specificity and sensitivity should be appreciated. First, several conditions besides acromegaly may be associated with varying degrees of failure of GH to suppress following an oral glucose load. Many of these situations are also characterized by an elevated basal level as well. Thus, brittle diabetes, cirrhosis, chronic renal failure, carcinoid syndrome, anorexia nervosa, and malnutrition are situations that may be associated with elevated basal GH levels as well as impairment in glucose-mediated suppression. Furthermore, some patients with primary hypothyroidism,[163] porphyria,[159] and depression[159] may demonstrate failure to suppress their GH levels normally in response to glucose. Second, in terms of its sensitivity, a small but significant number of acromegalics may demonstrate suppression of GH to below 5 ng/ml following an oral glucose load. The prevalence of such a phenomenon, which is indeed conducive for missing the diagnosis, is approximately 7–10%. For instance, in the series of 57 acromegalics reported by Clemmons et al.,[164] five with active acromegaly had "normal" GH levels after glucose suppression. In view of this incidence, it has been suggested that more stringent criteria be employed to define normal suppression (i.e., a fall in GH level below 3 ng/ml[29,165]) following an oral glucose load. This is especially important when considering the criteria for defining a "curative response" to therapy.

In summary, measurement of both basal, and postglucose GH levels are good screening tests, within the boundaries limited by specificity and sensitivity. Considering the fact that the sensitivity of these tests is rather high (93–95%), the relatively low specificity should not minimize their use as simple, first-line screening studies.

Step II: Confirmation of the Diagnosis

The diagnosis of acromegaly is usually confirmed by the measurement of fasting somatomedin-C level. The additional data provided by studying

GH dynamics to various stimuli (e.g., TRH, L-dopa) provide strong diagnostic support and yield important information on behavior of the GH secreting tumor.

Somatomedin-C Level

The knowledge that the growth-promoting actions of GH are mediated by a family of peptides is hardly new.[166] The main factor that had hampered the application of that knowledge was the lack of availability of reliable techniques that measured the level of these peptides, collectively referred to somatomedins, in the serum. The ability to measure these peptides by several different methods—bioassays, radioreceptor assays, competitive protein-binding assays, and more recently radioimmunoassays—have also generated considerable confusion in terminology as well as reporting. Four major realizations in the past decade have been instrumental in the emergence of somatomedin measurements as a valuable diagnostic tool: (1) the ability to measure specifically the GH-dependent somatomedin-C by a sensitive radioimmunoassay[164,167]; (2) the demonstration that these peptides, bound to carrier proteins, have a significantly slower rate of disappearance,[168] and hence are more stable; (3) the demonstration that somatomedin C levels reflect the GH secretory rate[169]; and the availability of data from several centers involved in the diagnosis and treatment of acromegaly.

The diagnostic superiority of measurement of circulating levels of somatomedin C in the fasted state in unequivocal. Clemmons et al.[164] measured serum concentrations of somatomedin C in 57 acromegalics and reported that the levels were increased in all patients with active acromegaly. Similar results were reported by Wass et al.[170] Several observations are pertinent as they relate to measurement of somatomedin-C as a diagnostic tool in acromegaly.

1. Somatomedin C levels are elevated in virtually all acromegalics with active disease.
2. The fasting somatomedin C level is elevated even in acromegalics with normal basal as well as glucose-suppressed GH levels. This may be due to two reasons: (1) the serum concentration of somatomedin C is stable and reflects the cumulative effects of sustained hypersomatotropism; and (2) minor increases in GH levels on a chronic basis can induce major and magnified increases in circulating somatomedin C levels.
3. In active acromegaly, somatomedin C levels, in contrast to basal GH levels, correlate better with the severity of the disease.
4. The measurement of somatomedin C level in predicting or defining cure is still somewhat controversial. This is discussed under treatment.

Thus, at least for diagnostic purposes, measurement of fasting levels of somatomedin C serves excellently well. It must be recognized that elevated levels of somatomedin C can be encountered in adolescents (especially during growth spurt) and in poorly controlled type I diabetics. While estrogens decrease the level of somatomedin C, late pregnancy is associated with elevated levels, probably as a consequence of stimulation by chorionic mammosomatotropin.

Supportive Dynamic Studies

In clinical practice, two other tests are usually performed, after the diagnosis of acromegaly has been established by measurement of somatomedin C levels: studies evaluating the response of GH to administration of TRH and L-dopa.

GH Response to Thyrotropin-Releasing Hormone in Acromegalics (Fig. 12). In normal persons, administration of TRH intravenously is not associated with a temporal rise in the release of GH. Several workers in the early 1970s demonstrated that active acromegaly was associated with a paradoxical rise in GH levels following administration of TRH.[171–174] Subsequently, a similar rise in GH was demonstrated when luteinizing hormone releasing hormone (LHRH) was administered to acromegalics.[175] This paradoxical responsiveness of adenomatous somatotropes to hypothalamic-releasing factors is believed to have represented loss of receptor specificity in the cell membrane of neoplastic cells.

The paradoxical rise in GH levels following TRH in acromegalics should be viewed from several perspectives; the definition, prevalence, and diagnostic and prognostic significance of this paradoxical response in acromegalics merit consideration. As far as the criteria to define the paradoxical response pattern, there is no unanimity of opinion. It has been argued that since normally GH does not rise following TRH administration, any rise, no matter how small, constitutes a paradoxical response. Indeed, in some studies an absolute increment of as little as 3–5 ng/ml GH above the basal levels constitutes a response.[176] Most workers, however, would accept a 50–100% increase over

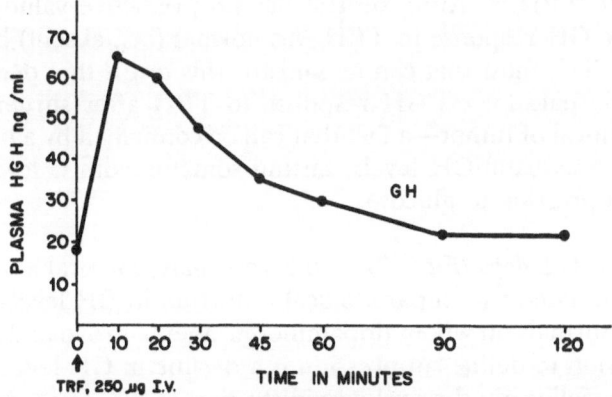

FIGURE 12. Paradoxical response of growth hormone to intravenous thyroid-releasing hormone administration in an acromegalic patient. (From Reichlin.[247])

the basal level as a response.[171,177] It should be pointed out that the spontaneous fluctuation in GH levels can make the interpretation of GH response pattern rather difficult.

The paradoxical GH response to TRH is seen in approximately 70–80% of active acromegalics. Liuzzi et al.,[177] among others,[178,179] have pointed out the homogeneity between the TRH-induced GH rise and the lowering of GH by dopaminergic agents in acromegaly. More recent, Hanew et al.[19] evaluated several dynamic parameters in acromegaly and demonstrated the existence of two groups, based on GH responsiveness to somatostatin; they were able to show that GH release following TRH (and LRH) was more prevalent in the group of acromegalics who retained a higher responsiveness to somatostatin. This group also showed normalization of GH levels after bromocriptine. It appears that all three phenomena—responsiveness of GH to hypothalamic releasing factors, heightened suppression to somatostatin, and suppression with dopamine agonists—are all linked together. It is possible that the primary pathophysiology in this group of acromegalics may indeed reside at the hypothalamic level.

As far as the diagnostic significance of the TRH-induced GH rise in acromegaly, this test is not the definitive confirmatory test for diagnosing acromegaly. It is one of the dynamic studies performed after establishing the diagnosis by other means, usually somatomedin-C. Depending on the criteria employed, 20–30% of acromegalics may not demonstrate the paradoxical rise in GH following TRH. Besides, an abnormal response may be encountered in conditions other than acromegaly. The most important of these are tall adolescent children,[180] patients with mental depression,[181] anorexia nervosa,[182] schizophrenia,[183] and primary hypothyroidism.[184,185]

Prognostically, it has been suggested by Faglia et al.[174] that normalization of a previously paradoxical GH response to TRH constitutes one parameter to define "cure" of acromegaly. Post-therapeutic assessment should take into consideration the fact that partial removal of tumor may result in normalization of several previously abnormal responses, including the paradoxical GH response to TRH.[159] Also, the test has no predictive value post-therapeutically if the GH response to TRH was normal (i.e., absent) before instituting therapy. The most that can be said for this test is that demonstration of a persistently paradoxical GH response to TRH after surgery indicates inadequate removal of tumor—a fact that can be confirmed by a host of other parameters such as basal GH levels, fasting somatomedin C levels, and abnormal GH suppression to glucose.

GH Response to L-dopa (Fig. 13).   In acromegaly, the oral administration of L-dopa is associated with a paradoxical reduction in GH levels.[186,187] This contrasts with normals, in whom dopaminergic agents provoke the release of GH. The criterion to define suppression is a decline in GH levels by 50% of the basal levels, following the oral administration of 500 mg L-dopa. Such a response is seen in 60–65% of acromegalics. Generally, patients demonstrating a paradoxical GH response to L-dopa also demonstrate a paradoxical GH

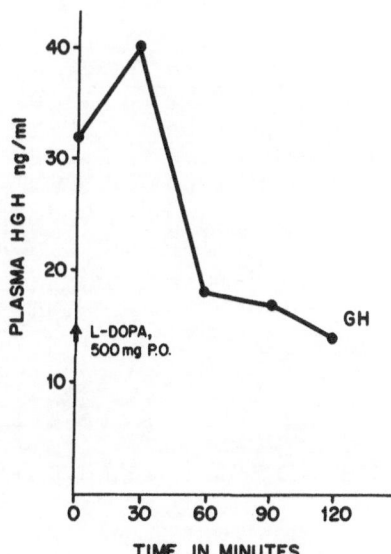

FIGURE 13. Paradoxical lowering of growth hormone level in response to L-dopa in an acromegalic patient. (From Reichlin.[247])

response to TRH. Liuzzi et al.[177] showed that suppression of GH to an oral dose of L-dopa correlates with the long-term reduction in GH levels following dopamine agonist therapy.

The role of these two tests—TRH and L-dopa—is primarily to identify the subgroup of acromegalics who might benefit from chronic dopamine agonist therapy with bromocriptine. For instance, such therapy would not be expected to work in the acromegalic patient who fails to show a paradoxical rise in GH after TRH or a paradoxical fall in GH after L-dopa. It should be noted that these tests, alone or in combination, are not likely to be the sole predictors of a therapeutic cure of the disease.

Step III: Localization

Once the diagnosis of acromegaly has been established, the next immediate step is to document the presence and extent of tumor within the pituitary gland. Abnormalities in the pituitary gland are demonstrable in more than 90% of patients with acromegaly (Fig. 14). The absence of demonstrable abnormalities in the pituitary by CT scan in a patient with acromegaly is suggestive of any one of the following possibilities, all of which are rare: acromegaly secondary to somatotrope hyperplasia; acromegaly caused by ectopic secretion of GH; acromegaly resulting from a somatotrope adenoma arising from an ectopically located pituitary, in the pharyngeal region or sphenoidal sinus; and acromegaly secondary to a microadenoma in association with empty sella syndrome. The CT study is invaluable in delineating the presence of supra- or parasellar extension of somatotrope adenomas. The more invasive the tumor, the less the likelihood of attainment of a "cure" for acromegaly. The levels of GH do not correlate with the size of the tumor or with the presence of suprasellar extension (Fig. 15).

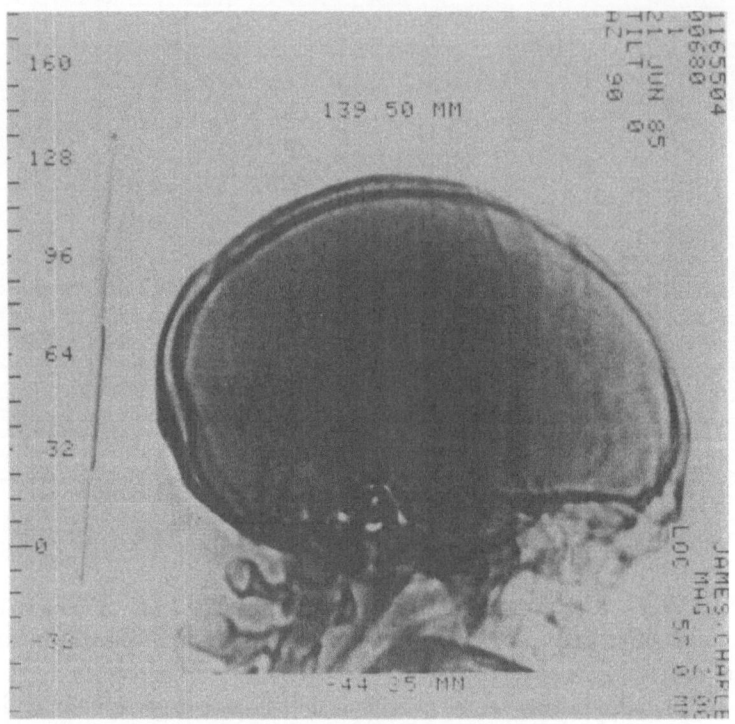

FIGURE 14. Plain skull film demonstrating enlarged sella turcica in a patient with acromegaly.

Finally, all patients with abnormalities in the pituitary fossa must have their visual fields tested by formal perimetry.

Step IV: Assessment of Target Organ Damage

The next step in the evaluation of the patient documented to have acromegaly is to determine the extent of damage caused by chronic GH excess. From the cardiac standpoint, the routine tests of chest roentgenogram, ECG, and so forth, should be coupled with echocardiography to obtain anatomical as well as functional information regarding cardiac performance. The pulmonary function tests, especially in reference to the sleep apnea syndrome, provide basic information as to the presence and degree of upper airway obstruction. The oral glucose tolerance test and the measurement of heel-pad thickness by radiography have remained classic and traditional indices of disease activity, often supplemented by nerve conduction velocity studies to detect subtle or overt median nerve entrapment. Radiographic studies of the joints and spine may be ordered when articular pain is present. These studies constitute a strong infrastructure of information when assessment of clinical response to treatment becomes an issue.

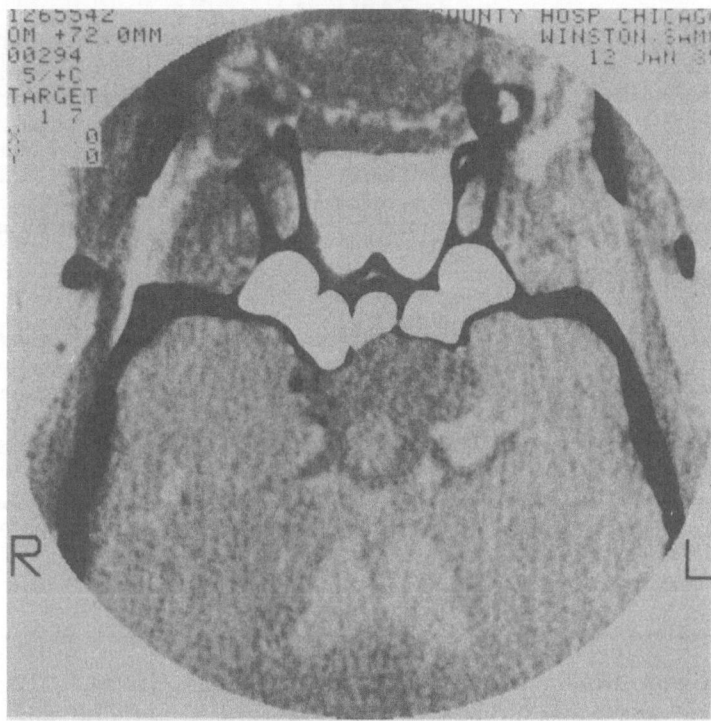

FIGURE 15. CT scan of a patient with acromegaly demonstrating a large pituitary tumor with suprasellar extension. Despite the massive tumor size, this patient's basal growth hormone levels ranged between 30 and 40 ng/ml.

Step V: Exclusion of Ectopic Source of Acromegaly

IR-GHRH Assay

A most important aspect to consider in the evaluation of every acromegalic patient is the exclusion of disease caused by ectopic secretion of GH or GHRH. Rare as these conditions are, their exclusion is crucial in order to avoid therapeutic failure. Ectopic GH secretion is rare and is most often associated with tumors of the lung or pancreas. The four notable characteristics of acromegaly caused by ectopic GH secretion are (1) a normal-appearing pituitary by CT, (2) failure of GH levels to rise following the administration of synthetic hpGHRH 1–40 (or hpGHRH 1–44), (3) absence of raised plasma levels of IR GHRH, and (3) often a lack of GH response to exogenous administration of TRH. The constellation of findings outlined reflects the fact that endogenous production of pituitary GH is suppressed by the ectopically secreted GH. It should be realized that ectopic GH possesses the ability to stimulate somatomedin-C generation.

Ectopic secretion of GHRH is also a rare cause of acromegaly and is

TABLE 23
Growth Hormone Dynamics in Normals and in Patients with Acromegaly
from Various Causes

Feature	Normal	Acromegaly from pituitary tumor	Acromegaly from ectopic GHRH	Acromegaly from ectopic GH
Growth hormone and somatomedin C	Normal	Elevated	Elevated	Elevated
GH response to glucose	Suppression	Nonsuppression	Nonsuppression	Nonsuppression
GH response to TRH	Absent	Present in 70–80%	Present	Absent
Tumor in the pituitary	—	Present	Often present	Absent
GH response to synthetic GHRH	Present	Preserved	Absent	? Absent
IR GHRH level in plasma	—	Low	Elevated	Low

associated with tumors of the foregut and pancreas. Carcinoids, carcinoma, and islet cell adenomas (apudomas) represent the histological consortium of neoplasms associated with the entity. The four notable characteristics of acromegaly caused by ectopic GHRH secretion are (1) an abnormally enlarged pituitary gland by CT, (2) an absent GH response to exogenous administration of synthetic hpGHRH 1–40 (or hpGHRH 1–44), (3) an elevated IR GHRH level in the plasma, and (4) often a paradoxical rise in GH to TRH administration. These findings reflect the fact that the pituitary gland is driven by a substance that biologically resembles native hypothalamic GHRH. The differences between normal subjects and acromegaly caused by intrinsic pituitary tumors, ectopic secretion of GH, and ectopic secretion of GHRH are outlined in Table 23.

The two diagnostic studies that permit exclusion of ectopic GHRH secretion are measurement of plasma levels of IR-GHRH and evaluation of the response of GH to exogenous administration of synthetic GHRF. Availability of both studies is limited to specialized centers. Thorner et al.[57] reported the ability of a sensitive assay in clearly separating acromegaly caused by ectopic GHRH secretion from the garden-variety acromegaly.

GH Response to hpGHRF 1–40

The evaluation of GH response to synthetic hp GHRH has added another dimension to the diagnosis of ectopic GHRH secretion. Several workers[54,55] showed that patients with acromegaly secondary to ectopic GHRH secretion fail to raise their GH level following the administration of synthetic GHRH. This contrasts with the acromegaly caused by pituitary tumors, wherein the response is usually preserved. Losa et al.[188] studied 20 patients with active acromegaly caused by pituitary tumors and showed that 16 responded with at least a 100% increase over the baseline. It has been proposed that a lack of GH response to GHRH in a patient with acromegaly is suggestive of ectopic production of GHRH.[54]

Step VI: Assessment of Pituitary Reserve

All acromegalic patients need to have their pituitary reserve tested to determine the extent of trophic hormone deficiency associated with acromegaly. While interpreting the data it should be realized that acromegaly is often associated with a blunting or even absence of a TSH response to TRH. This phenomenon is reversible upon restoration of eusomatotropism, unless anatomical destruction of thyrotropes underlies the phenomenon.

Step VII: Screening for Associated Endocrinopathies

Finally, acromegalics should be screened for the presence of concomitant endocrinopathies—hyperthyroidism, hyperprolactinemia, MEA-I, and pheochromocytoma—by the utilization of their respective screening procedures.

Differential Diagnosis

The features of acromegaly are so strikingly visual that it lends itself to "spot diagnosis." Few diseases transfigure their victims in such a predictable and obvious manner. Unmistakable as the somatic features of acromegaly are, occasionally the features of myxedema overlap those of acromegaly. The swollen face, the thickened skin, and the coarsening of features associated with myxedema can resemble the features of acromegaly. The term *myxomegaly* is often used to denote this entity.[132] The similarities to acromegaly are further extended by the development of carpal tunnel syndrome, myopathy, fatigue, headaches, thyromegaly, and acropachy—features shared by both acromegaly and primary hypothyroidism. The similarities are drawn even closer when one considers the fact that an enlarged sella and abnormal GH suppression to glucose may be seen in association with primary hypothyroidism.[163]

The most frustrating clinical encounter is the presence of acromegalic features in the absence of hormonal data to support the diagnosis. Three possibilities may account for such a combination: the presence of a growth factor other than GH,[189] the possibility of a heredofamilial disorder called Touraine–Solente–Gole syndrome, and residual (burnt-out) stigmata of past hypersomatotropism. Each of these deserves brief comment.

The term *acromegaloidism* is in current usage to denote the development of a syndrome caused by a unique (non-GH) growth factor. The hallmark of the syndrome is the presence of characteristic clinical features of acromegaly without biochemical evidence of excessive production of GH or somatomedin. Ashcraft et al.[190] described five patients with acromegaloidism, collectively manifesting nearly all the characteristic features of acromegaly: headaches, acral enlargement, hypertension, coarsening of facial features, increased sweating, decreased libido, parasthesiae, and glucose intolerance. Prognathism was not noticed in any. Some of these patients had even undergone craniotomy with no tumor detected in the pituitary gland. The GH levels and GH dynamics were normal in all patients. These workers identified the presence of a heretofore unidentified growth factor in the sera of these patients. This growth factor stimulated growth of the burst-forming unit, a primitive erythroid precursor. This growth factor had an estimated molecular weight of 70,000, heat stable at 58°C for 1 hr, but heat labile at 80°C. The role of this growth factor in the causation of this growth disorder needs further evaluation.

The Touraine–Solente–Gole syndrome (or pachydermoperiostosis) is a disorder characterized by acropachy and thickening of skin and bone, particularly in the region of the face and scalp. The GH levels, as well as GH dynamics, are normal in this entity, which is often familial; the pathogenesis underlying this disorder is far from clear.

Burnt-out acromegaly is usually a result of bleeding into the tumor (apoplectic auto cure) and not due to spontaneous resolution of the disorder. As a result, the hypersomatotropism is often replaced by hypopituitarism (including low GH levels), while the stigmata of acromegaly remain residual. The diagnosis can be suspected on the basis of a history suggestive of an episode of pituitary apoplexy—often an unforgettably painful experience— and can be confirmed by demonstrating impaired trophic hormone reserve of most or all pituitary hormones, including GH.

Finally, the diagnostic difficulties encountered in tall, rapidly growing children during adolescence can be considerable. These children may demonstrate elevated GH levels, elevated somatomedin-C levels, abnormal GH suppression to glucose, and abnormal GH response to TRH administration.[180] These children, referred to as preacromegaloid by some, show no evidence of pituitary tumor. It is not clear whether this entity in tall children represents constitutional tall stature or whether it indicates subtle perturbations in hypothalamic regulation of GH secretion. Careful follow-up studies in this group is well warranted.

Complications

Untreated acromegaly is associated with considerable morbidity and is characterized by the development of several complications. Five in particular are well recognized: pituitary apoplexy, hypopituitarism, visual-field defects, decreased life span, and development of deformities. A sixth complication— a heightened tendency for carcinogenesis—has been suggested but not proved. Each of these contributes considerably to the morbidity associated with acromegaly.

Pituitary Apoplexy

Pituitary apoplexy[191] represents a catastrophic event, characterized by bleeding into the tumorous gland, resulting in a rapidly expanding lesion within a closed space. The onset is sudden, often dramatic, manifested as excruciating headaches and rapid deterioration of vision. The presentation can be rather heterogeneous; the course is often punctuated by the development of oculomotor nerve palsies, meningeal irritation, fever, visual loss (including total blindness), lethargy, and even coma. Hormonal events that may further complicate the picture are the rapid development of secondary adrenal failure and secretion of inappropriate ADH (SIADH). Pituitary apoplexy tends to occur in acromegalics harboring relatively large tumors, especially against a background of diabetes, estrogen therapy, and following radiation to the pituitary tumor. It can, however, occur in the absence of any recognizable risk factor mentioned above. Pituitary apoplexy represents a neurosurgical emergency and should be suspected in any acromegalic presenting with sudden onset of headache, failing vision, and oculomotor palsies. Patients who have survived an apoplectic episode are often "cured" of their acromegaly, since panhypopituitarism is the usual sequel of hemorrhagic necrosis of the pituitary.[192,193] There are, however, reports of recurrent acromegaly following an apoplectic episode.[194] Pituitary apoplexy is discussed in Chapter 17.

Hypopituitarism

The development of spontaneous hypopituitarism in acromegaly was once thought to be rare. However, Goldfine and Lawrence[131] studied the pituitary reserve in 28 acromegalics both before and shortly after pituitary radiation, and observed varying degrees of compromise in trophic hormone reserve in as many as 54% of the patients studied. The most frequent deficiency noted was that of LH and FSH, followed by deficiencies of TSH and ACTH. The development of trophic hormone deficiency did not correlate with tumor size, disease duration, or the therapy given.

Visual Compromise

Field cuts and loss of vision represent the most serious local complication of GH-secreting tumors. Although the incidence is considerably lower these days, the threat to vision poses the most pressing concern when formulating therapy for acromegaly. The presence and degree of visual compromise correlate with the size of the tumor and, more importantly, the degree of suprasellar extension.

Decreased Life-Span

Acromegaly clearly decreases the life-span of its victims. Wright et al.[195] reviewed the mortality in acromegaly and showed an increase in mortality due to cardiovascular disease, myocardial infarction, pulmonary disease, and diabetes. Mortality was particularly prevalent in the fourth and fifth decades of life. This is clearly one of the major reasons for early intervention of the disease, as well as for aggressive therapies for persistent or recurrent disease.

Deformities

Deformities caused by acromegaly take a heavy toll on its victims. One of the most distressing, but not so well appreciated, problems in acromegaly is the deformity that results from overgrowth of the frontal, nasal, and malar bones. This overgrowth, when coupled with the asymmetrical prognathism of the mandible, can result in tremendous malocclusion.[196] Corrective facial surgery may be required for more than cosmetic purposes. While rhinoplasty and mandibular osteotomies may suffice for cosmetic disfigurement, more specialized reconstructive surgery may be required for severe malocculusion. Another deformity, particularly seen in pituitary gigantism, is the development of marked dorsal kyphosis.[197] The cardiovascular and respiratory restraints posed by severe kyphosis can further aggravate the morbidity and mortality associated with acromegaly.

Increased Risk of Carcinogenesis

Finally, one possible complication of acromegaly has recently received considerable attention in the literature (i.e., the increased incidence of developing cancer in acromegaly). As early as 1960, Hamwi et al.[91] noted that six of the 30 acromegalics in their series had new growths: in two, these growths were malignant. Alexander et al.,[7] while reporting on the epidemiology of acromegaly in the Newcastle region, noted that malignancies of all forms accounted for approximately 15% of deaths in acromegalics. When compared with risk for age-adjusted males, the cancer risk was fourfold greater in acromegalic males. Similarly, Klein et al.[198] detected a 25% frequency of carcinomas in a prospective study of acromegalics. Colon cancer represents the most frequent malignancy seen in association with acromegaly. In a study of 12 patients with acromegaly, Ituarte et al.[199] have confirmed the association between acromegaly and colon cancer. In addition to colon cancer, the incidence of colonic polyps is increased in patients with acromegaly. It appears that acromegalics over the age of 50—especially males with longer duration of disease and multiple skin tags—are particularly at high risk. It is to be noted that the risk extends to those with inactive disease as well. The bulk of evidence favors the recommendation that patients who have, or who have had, acromegaly be screened for colonic polyps or colon cancer starting at age 50. With the availability of better treatment modalities for the disease and a longer life expectancy of acromegalics, the issue of cancer surveillance will undoubtedly become a major one in the future.

Treatment

Despite the availability of several newer modalities of therapy (Table 24), acromegaly can be quite a frustrating disease to treat. The goals of therapy are fourfold: (1) normalization of the hypersecretory state with the attendant

TABLE 24
Treatment of Acromegaly

Surgical
   Transsphenoidal
   Transfrontal
Radiation
   Conventional
   Proton beam
Medical
   Bromocriptine
   Pergolide
   Somatostatin analogues

correction of clinical and metabolic effects of GH excess, (2) removal of the tumor mass, (3) avoidance of hypopituitarism, and (4) maintenance of eusomatotropism for long periods of time. Between the trinity of surgery, drug therapy, and radiation, no single form of therapy guarantees fulfillment of all four goals with a reasonable degree of consistency.

General Considerations

Several general considerations should be borne in mind before discussing the various therapeutic modalities available. One of the major difficulties in sorting out the numerous studies in the literature is the variability with which a "cure" is defined. There is no consensus of opinion as to the parameters that clearly constitute a cure. There seem to be as many definitions of "hormonal cure," as there are studies reporting responses to one form of therapy or another. To make matters even more confusing, the hormonal parameters that define a cure (e.g., GH, somatomedin C) may not always correlate with clinical improvement or lack of it. Aside from the hormonal definition of cure, even a clinical definition of cure can be rather difficult. In a disease such as acromegaly, in which the bony changes caused by the disorder are permanent stigmata, and in which the soft tissue changes seldom show complete reversal, defining improvement is heavily reliant on subjective data. Table 25 outlines the numerous follow-up parameters than can be used to assess response to therapy. The diverse nature of these parameters as well as the numerous variations within the themes may be the reason for conflicting reports in the literature—often to an amazing degree of contradiction. For instance, merely two examples will suffice to make the point: (1) bromocriptine, a dopamine agonist, has suffered extreme negative criticisms by some, while enjoying a thriving reputation as a remarkably effective drug by others; and (2) transsphenoidal surgery, which according to its protagonists satisfies more of the therapeutic goals than any other therapy, has not escaped bad reviews. Thus, while one group of neurosurgeons report an 80% cure rate, another group[29] reports only a 22% rate of permanent cure with the same type of procedure. Variability in treatment protocols, differences in the definition of cure pa-

TABLE 25
Follow-up Parameters to Treatment for Acromegaly

Clinical
  Subjective
    Reduction or abolition of excessive sweating
    Reduction in foot size
    Improvement in facial features
    General feeling of well-being
  Objective
    Reduction in ring circumference
    Improvement in hypertension
    Improvement in NCV
Biochemical
  Glucose tolerance test
    Reduction in area under GTT curve
    Normalization of GTT
  Phosphorus level
    Reduction of hyperphosphatemia
Hormonal
  GH level (basal)
    Basal <10 ng/ml
        vs.
    Basal <5 ng/ml
        vs.
    Undetectable basal; <1 ng in 75% of samples during the day
  Integrated GH level <5
  Dynamic tests
    Postglucose below 5 ng/ml
    Postglucose below 2.5 or 3 ng/ml
    Normalization of TRH-mediated GH rise
  Somatomedin-C level

rameters, the heterogeneity of patient responses, and finally the varying durations of follow-up in the reported series have all contributed to the conflicting data in the literature.

The choice of therapy for acromegaly depends on several factors:

1. The urgency underlying the situation plays a major role in determining the choice of therapy. The presence of visual-field cuts, severe cardiac failure, and significant suprasellar extension requires urgent intervention. Since radiation therapy takes a long time to effect a cure, these patients require surgical intervention.

2. The surgical expertise available, as well as the track record of the surgical team, play a significant role in the outcome.

3. Size of the tumor also influences the outcome. Transsphenoidal surgery offers a 80–85% cure rate for acromegaly when the underlying pathology is a microadenoma. The success rate of transsphenoidal surgery for tumors >10 mm drops to 50–52%.

4. With large and invasive tumors, avoiding hypopituitarism is an unrealistic goal, and emphasis should be more on debulking.

5. The type of radiotherapy available in a particular institution will be a decisive factor in deciding whether such therapy is to be applied at all, or applied as primary or as adjunctive therapy.

Regardless of the therapy used, the definition of a "cure" based on hormonal criteria varies among different workers. The five different hormonal parameters employed are (1) measurement of basal GH levels, (2) postglucose GH levels, (3) integrated GH levels, (4) post-TRH growth hormone, and (5) somatomedin-C levels. Measurement of basal GH levels is the conventional parameter that has long been in vogue. Tolis et al.[159] reviewed the therapeutic response to surgery in 299 acromegalics reported from 10 series and noted that the serum GH declined to <10 ng/ml in 72% and <5 ng/ml in 55% of cases. Eventual relapse occurred in some patients in whom the GH level was between 5 and 10 ng/ml.[11,200–202] This strengthens the notion that a basal level of GH > 5 ng after surgery is indicative of persistence of disease activity. Similarly, a postglucose GH level of ≤5 ng used to be considered a "cure" in the past. Schuster et al.[29] reaffirmed that stricter criteria should be employed (i.e., postglucose GH suppression to below 2.5 ng/ml) to assess therapeutic responsiveness. Besser and Wass[203] pointed out that a true cure can be present only when the GH levels are completely normalized, and this would mean demonstration of normal levels (<1 ng/ml) in the majority of samples obtained through the working day. These workers have proposed the most stringent criteria for cure and regard a patient cured of acromegaly only when the blood samples obtained four times throughout a day reveal unmeasurable GH levels in 75% of samples.

Persistence of the paradoxical GH response after TRH is a clear indication of the persistence of active disease. While normalization of a previously paradoxical GH response to TRH suggests a cure, this should be viewed in conjunction with basal GH and somatomedin-C levels. It should be noted that the TRH-induced GH response cannot be used as a predictor of cure in the minority of acromegalics who have a normal response to begin with.

It has been suggested that somatomedin C may be a superior diagnostic tool to assess response to treatment than measurement of basal GH levels.[164] There have been conflicting results in the literature regarding the correlation among somatomedin-C level, GH level, and the attainment of cure. Stonesifer et al.[204] evaluated 15 patients treated for acromegaly and presented data that showed a poor correlation between clinical improvement and measured levels of somatomedin-C as well as basal GH. Similar discrepancies have been reported by Moses et al.,[205] who noted that five of seven patients treated with bromocriptine showed a good clinical response as well as reduction in GH levels, but this was not associated with a pari passu decrease in circulating somatomedin-C level. Nortier et al.[206] studied 31 acromegalics on bromocriptine therapy and found that the clinical and metabolic improvement correlated better with changes in mean plasma GH levels than with the changes in somatomedin-C levels; these workers cautioned against the practice of using somatomedin-C level as the sole parameter for judging disease activity. The lack of correlation between GH and somatomedin-C level and clinical im-

provement may be attributable to several factors, such as differences in grad-
ing clinical response, differences in techniques of measuring somatomedin-
C levels, and errors in interpretation caused by decreased clearance of carrier
proteins for the somatomedins. With the above-mentioned generalizations,
the role of surgery, radiation, and medical therapy in the management of
acromegaly is discussed in the following sections.

Neurosurgical Management

Although the technique of transsphenoidal surgery was introduced in
the 1920s, there has been a resurgence of the technique since the 1970s. To
wit, at least nine surgical series, since 1980, have reported on the effects of
microsurgery in the treatment of acromegaly.[11,29,200-202,207-210] With the advent
of optical magnification, selective microsurgical adenomectomy with preservation of normal gland is possible through the extracranial transsphenoidal
approach. The procedure has enjoyed enormous success in the hands of Dr.
Hardy, a pioneer in the neurosurgical management of pituitary tumors.[211] A
recent series by Hardy and Somma[208] consisted of 120 patients with active
acromegaly operated on by the transsphenoidal route. Their results indicate
that when the tumor tissue was totally excised, a cure was attained in 78%,
improvement in 17%, and no change in 5% of patients. A significant observation in this series was the fact that pituitary function was preserved in 91%
of patients. Also remarkable was the fact that significant clinical improvement
was noted even in those who underwent subtotal tumor removal (i.e., debulking of tumor). The only patients who were resistant to surgery were those
with invasive lesions. The hormonal criteria of "cure" were a GH level below
5 ng/ml repeatedly and a postglucose GH level below 5 ng/ml, when the basal
was slightly high. The complication rate in this series was impressively negligible. The Mayo Clinic series of 183 acromegalic patients treated neurosurgically, reported by Laws et al.,[207] also demonstrates a 70% cure rate,
especially when microadenomas underlie the acromegaly. The success rate in
this series correlated with the size of the tumor and the degree of GH elevation.
The report by Tindall and Tindall[209] outlines the experience at Emory University and comprises the results of 50 consecutive acromegalics treated by
transsphenoidal surgery. The follow-up period ranged from 2.3 to 9.3 years.
Using the criterion of a basal GH level below 5 ng/ml as cure, 70% of patients
achieved a cure. Like the Mayo group, these workers also showed a correlation
between the size of the tumor as well as GH levels and the surgical outcome.
The poorest outcome was seen in the most invasive tumors and in the patients
with highest GH levels. In those cured, no recurrence was noted during the
follow-up period.

While the enthusiasm for transsphenoidal surgery is strengthened by
most published series, some reports in the literature have been less than
enthusiastic. Schaison et al.[210] evaluated the outcome of transsphenoidal surgery in 18 acromegalics: five of these patients were obviously not cured, and
among the remainder, despite a clinical cure that was "unequivocal," the

hormonal data were not always concordant with a cure. Notable was the fact that frequently anterior pituitary insufficiency was associated with unequivocal hormonal cures. Similarly, Schuster et al.[29] treated 11 active acromegalics by transsphenoidal selective microadenectomy and reported that four failed to show evidence of clinical improvement and an additional four, while initially improving, were found to have a recurrence upon prolonged follow-up. They noted that the three patients who had a prolonged cure demonstrated basal GH levels below 3 ng/ml. Both studies pointed out the need for more standardized and stricter criteria for defining cure.

In summary, the following statements apply to the treatment of acromegaly by the transsphenoidal route.

1. The cure rate for acromegaly by this approach is approximately 80% when microadenomas are present. The cure rate declines to 52% when the adenoma is larger than 10 mm. (The incidence of finding microadenomas in acromegaly is approximately 35–40%.)
2. The results of surgery are immediate.
3. The outcome of surgery correlates with the tumor size and preoperative GH level.
4. When larger tumors are present, the incidence of postoperative hypopituitarism proportionately increases and can be as high as 25%.[212]
5. The enthusiasm for transsphenoidal surgery, justifiable as it is, should be tempered by the possibility of late recurrences.[29]
6. Patients resistant to surgery should be treated by conventional radiation, bromocriptine, or a combination of both.

Radiation Therapy

Up until the middle 1960s and even during the early 1970s, conventional radiation therapy was considered an accepted and appropriate therapy for some patients with acromegaly. Roth et al.[213] treated 20 acromegalics with 4000 rad of external radiation to the pituitary and demonstrated a mean fall in plasma GH of 51% for the entire group by the end of 1 or 2 years. Lawrence et al.[214] also reported on 28 acromegalics who had received external radiotherapy and showed normalization (or near normalization) of GH levels in 78% of patients thus treated. Lowering of GH levels, when it occurred, correlated with disappearance of the clinical features of acromegaly. The time required for attainment of eusomatotropism in this series ranged from 1 to 2.5 years. In 1979, Eastman et al.[215] reconfirmed that conventional external megavoltage irradiation is an effective modality for the treatment of this disease (73% remission rate, based on attainment of GH level <10 ng/ml), but the arrival of cure, in some instances, may be delayed by as long as 10 years. A fourth series, by Bloom and Kramer,[216] reviewed the results of conventional radiation in 40 acromegalics treated between 1957 and 1982. Their material included patients with suprasellar extension, as well as some with disease duration for more than 10 years. Their data revealed that three-quarters of treated patients were alive with no progression of disability for at least 10 years after radiation. They also noted that complete normalization of GH levels was mostly seen when the pretreatment GH levels were only modestly elevated.

The major advantage of conventional pituitary radiation is its low complication rate. Hypopituitarism is seen only in a small number of patients.[213,214] Reports on careful endocrine testing of pituitary reserve following radiation by three different groups[217-219] have reaffirmed the low rate of this complication after radiation. Also, optic nerve damage is unusual when the total dose is kept under 4600 rad and not exceeding an individual fraction size of 180 rad.[220,221] Third, the rate of recurrence following successful radiation is less than 10%, an acceptable rate for acromegaly. The quadruple advantages of external radiation—satisfactory rate of cure in properly selected patients, low incidence of hypopituitarism, low incidence of optic nerve damage, and low recurrence rates—are outweighed by two factors: the long time taken for attainment of cure, and the unpredictability of such an occurrence.

Conventional radiotherapy is usually delivered by the use of ^{60}Co or MV X rays. The average dose is 4500 rad over a period of 4.5–5 weeks. Rotational techniques are used to minimize exposure of brain tissues, and the dose is delivered as a coronal arc with a wedge-shaped rotational technique. The only side effect is some temporal hair loss.

Another form of radiotherapy that can be employed in treatment of acromegaly is proton-beam therapy.[222] The Bragg peak method of proton-beam therapy, developed by Kjellberg at the Harvard cyclotron, is a popular source of proton beams. The differences between conventional megovoltage treatment and proton-beam therapy lie in the extremely localized effects of proton beam on the pituitary. With conventional radiation, the collimation of X-ray beams limits the field of exposure and cannot prevent the angular widening of beams. This scattering results in some loss of ionizing radiation in terms of focus on the pituitary. The proton beams, in contrast to X-ray beams, bear a single positive charge and are generated from hydrogen atoms. When the proton beams are accelerated, the result is generation of a particle beam with a phenomenal flux of 2 billion protons/sec directed from the source (emission portal). At the end of the beam, the protons produce ionizing radiation, forming a localized peak within its target, the sella turcica. With the use of 12 portals of entry—six on each side of the head—a central dose of up to 12,000 rad, can be delivered to the target. The usual dose for treating acromegaly is 9000 rad.

There are several advantages of proton beam therapy for acromegaly:

1. The entire therapeutic session is completed in one sitting, lasting approximately 2 hr.
2. The cure rates are excellent with this form of therapy. Kliman et al.[223] compiled their 20-year experience of treating more than 500 acromegalic patients with proton-beam therapy. Their data show that improvement in GH levels appears early and have remained low through-

out extended follow-up. The eventual rates of cure in their series is 97.5% (if GH levels below 10 ng/ml are considered adequate) and 87.5% if the criterion used is GH level below 5 ng/ml.
3. The treatment has been employed successfully even for patients with suprasellar extension.
4. The recurrence rate is extremely low, below 0.5%.
5. The incidence of development of hypopituitarism following proton-beam therapy is variable. In Kliman's series,[223] the overall incidence of pituitary hormone deficiency directly attributable to proton-beam therapy was 9.7%; 9.7% for TSH deficiency, 9.0% for ACTH deficiency, and 6.7% for gonadotropin efficiency. Loss of pituitary function is gradual and usually becomes evident within the first 5 years of therapy. Occasionally, it can be dramatic, when apoplexy, a rare complication, develops following proton-beam therapy.
6. The recognized complications of proton-beam therapy (e.g., oculomotor dysfunction, temporal lobe seizures) have been reduced to an acceptable minimum, with refinements in techniques.
7. Proton-beam therapy has been used successfully to treat patients who have failed to respond to surgery and conventional radiation.
8. The cost of proton-beam therapy is less than that with transsphenoidal surgery.[224]

α-Particle therapy for acromegaly without the use of the Bragg peak has been employed by Linfoot and co-workers[225,226] with considerable success. These investigators showed clinical improvement and reduction of basal GH levels below 5 ng/ml in 68% of acromegalics who received α-particle radiation. However, the incidence of hypopituitarism was 28–34%, much higher than in Kliman's series.[223]

While it is justifiable to discuss the option of proton-beam therapy for acromegaly with the patient, the limited availability of this method restrains its use considerably. Proton-beam therapy by the Bragg peak method is limited to Boston, while qualified neurosurgeons with expertise in transsphenoidal microsurgery are more readily available throughout the nation.

Drug Therapy

The purpose of medical therapy for acromegaly is to augment the effects of ablative therapy. Medical therapy is best viewed as adjunctive therapy while awaiting the results of radiotherapy to take effect or after the tumor has been surgically debulked. There are, however, investigators who have used medical therapy as the sole modality in acromegalic patients who had received neither radiation nor surgical therapy. The results of such studies are discussed separately. Medical therapy for acromegaly today is synonymous with dopamine agonist drugs. Bromocriptine is protypical of this group of drugs. A second generation of dopamine agonists, led by pergolide, was recently introduced. The newcomers to the scene are somatostatin analogues that are more potent and longer acting than native somatostatin. Thus, bromocriptine, pergolide, and somatostatin analogues dominate the drug therapy for acromegaly. It should be never forgotten that the beneficial effects of these drugs in acromegaly last only as long as drug therapy is continued. In this sense, they can hardly be regarded as curative. All the other drug therapies used during the 1960s for acromegaly—estrogens, medroxyprogesterone, and chlorpromazine—have faded into oblivion and represent no more than footnotes in the history of medical therapy for acromegaly. The discussion here focuses on the role of bromocriptine, pergolide, and somatostatin analogues in the management of acromegaly.

Bromocriptine

Bromocriptine, the first clinically available dopamine agonist, was shown to lower GH levels in acromegalics on an acute[177] as well as a chronic basis by Liuzzi et al.[227] Considering the fact that bromocriptine has received widely divergent reviews in the literature, it is critical to sort out the facts regarding the usage of bromocriptine. The five major concerns that need to be addressed are (1) the efficacy of the drug in lowering GH levels, (2) the effect of the drug in causing clinical improvement, (3) the consequences of long-term therapy on tumor size, (4) the side effects of long-term therapy, and (5) the duration as well as cost of therapy.

The effect of bromocriptine therapy on circulating GH levels has been studied extensively. A decade of experience with this drug has shown that it reduces the serum GH level in 75% of acromegalics, although normalization of GH is seen in only about 25%.[228] Maximum suppression of GH by bromocriptine varies between patients and can be seen with doses ranging from 10 to 60 mg/day. The discrepancy between the GH-lowering effects of bromocriptine and the marked clinical (and metabolic) improvements caused by the drug has intrigued many investigators. It is now accepted that bromocriptine preferentially lowers the monomeric form of GH, which is biologically more active. Since the radioimmunoassay measures both the monomeric and polymeric forms, the significant reductions in the biologically active monomeric form may go unnoticed. This may account for the fact that significant clinical as well as metabolic improvement can occur in the absence of a proportional reduction in GH levels. This preferential lowering of the monomeric form of GH, demonstrable by gel-filtration studies, is now a well-accepted notion[229–231] and may have accounted for the "negative" results reported by some investigators in earlier studies.[232,233] The alterations in levels of somatomedin C following bromocriptine have also been shrouded in controversy. Nortier et al.[206] studied the clinical, hormonal, and metabolic responses in 31 active acromegalics treated with bromocriptine, after unsuccessful surgery, radiation, or both. They were able to show that clinical and metabolic improvements correlated better with changes in the mean plasma GH levels, rather than with changes in circulating somatomedin C. Similar findings have been reported by Moses et al.,[205] who were able to demonstrate clinical responses to bromocriptine in six of seven acromegalics with significant reductions in GH levels, but with only small decreases of somatomedin C in five of these responders. These data are in conflict with those reported by Wass et al.,[234] who were able to identify a subgroup of acromegalics on bromocriptine therapy who showed marked clinical improvement with significant reduction in somatomedin-C level, without a concordant change in the mean plasma level of GH.

The effect of bromocriptine in causing clinical improvement has been well documented.[205,206,229,235,236] The series with the largest number of bromocriptine-treated acromegalics is that reported by Besser and Wass,[203] who studied the results of bromocriptine therapy on 146 active acromegalics, 49 of whom had received neither surgery nor radiation prior to initiation of therapy with bromocriptine. Clinical improvement, often dramatic, was sometimes noted as early as 72 hr after starting treatment; 95% of patients noticed improvement in headaches, hyperhidrosis, and fatigue along with reductions in facial swelling, foot size, and finger circumference. Clinical improvement was paralleled with improvement in glucose tolerance and an overall sense of well-being. Most investigators concur that the drug does indeed cause significant subjective and objective clinical improvement.

From the foregoing, it is apparent that treatment with bromocriptine does result in varying degrees of clinical and hormonal improvement. The negative side of the picture was reported by Lindholm et al.,[232] who denounced the drug as being ineffective in acromegaly. Their study has been criticized on grounds of methodology.[237] The majority of investigators on both sides of the Atlantic agree on the efficacy of bromocriptine, although the enthusiasm is more fervent on the part of the English investigators.

The effect of bromocriptine on reducing tumor size in acromegaly has also received considerable attention. Impressive reduction in tumor size following use of bromocriptine for acromegaly has been reported.[238,239] This may result in improvement of visual-field defects and even cranial nerve palsies. The reduction in tumor size, when it occurs, is seen after 12 months of therapy. The shrinkage in tumor size may even result in an "empty sella." While the most impressive reductions in tumor size are seen when concomitant hyperprolactinemia is present, pure GH-secreting tumors also demonstrate tumor shrinkage with chronic bromocriptine therapy. Even more intriguing are reports indicating that the tumor shrinking potential of the drug extends to suprasellar lesions as well.[239] Besser and Wass[203] further strengthened the case for bromocriptine by showing normalization of previously suboptimal pituitary function in acromegalics treated with bromocriptine. The mechanism cited by the authors is that the shrinkage of the tumor brings about restoration of physiological contact between the pituitary and the hypothalamus.

The side effects of bromocriptine therapy are nausea, postural hypotension, constipation, leg cramps, digital vasospasm, and hyperkinesis. These side effects are usually dose dependent. Nausea, the major adverse effect, can be minimized if the drug is taken with meals, as well as by starting with a small dose and gradually increasing it. The dose ranges from 7.5 mg/day to an average of 20 mg/day. The drug has to be given in three or four divided doses daily, owing to its short-lived effects on GH secretion. The treatment is continued for years—until other forms of definitive therapy have taken effect. When bromocriptine therapy is the sole modality, drug therapy has to be continued indefinitely. In some instances, doses greater than 20 mg/day need to be given to keep GH levels completely suppressed. The cost of an expensive drug given in large doses for protracted periods of time can be considerable.

Two main indications to use bromocriptine are (1) as adjunctive interim therapy while awaiting the results of ablative therapy to take effect; and (2) in patients who have failed to respond to surgery and radiation therapy, bromocriptine administration on a long-term basis provides an alternative method to lower GH levels and provide clinical improvement.[205] Bromocriptine as the sole agent for the treatment of acromegaly has not found wide application in the United States.

The effect of pergolide mesylate, a potent long-acting dopamine agonist, has been shown to be beneficial in acromegaly. Kendall-Taylor et al.[240] evaluated the long-term use of this agent in acromegaly. All patients treated with a single daily dose of pergolide showed statistically significant reductions in GH levels. The decreases in GH levels were dose dependent. The potential benefits of pergolide were its long-acting action and the advantage of single dose regimen. There were no adverse side effects to the drug. Experience with another ergot derivative, 8-α-amino ergoline, has also shown promising results.[241] A single dose of 8-α-amino ergoline causes powerful suppression of GH (average of 60% reduction) for more than 9 hr. It is unclear whether the second generation of dopamine agonists would replace bromocriptine. The general lack of experience with these compounds in the United States, as well as the lack of approval by the FDA for routine use, have limited the application of these agents in large clinical trials.

Somatostatin Analogues

The intravenous administration of somatostatin to acromegalics is associated with a marked inhibition of GH release.[242,243] However, three factors have limited the use of somatostatin in the treatment of acromegaly: the transitory nature of inhibition, the rebound hypersecretion of GH after discontinuation of the somatostatin infusion, and the practical inconvenience of multiple intravenous infusions of the peptide to obtain sustained inhibition of GH. The introduction of somatostatin analogues with greater potency and longer half-life has rekindled the interest in the use of these analogues in the treatment of acromegaly. Lamberts et al.[244] showed that a single subcutaneous injection of 50 μg of the octapepdide [D-Phe-Cys-Phe-D-Trp-Lys-Thr-Cys-Thr-(01)] resulted in strong inhibition of GH, without subsequent rebound hypersecretion. No adverse effects were noted on insulin secretion or glucose levels. Long-term treatment of acromegaly with somatostatin analogues immediately followed. Lamberts et al.[245] administered the same synthetic analogue to four acromegalics for a period of 8–24 weeks, in a dosage of 100–300 μg/day SC. They noted a rapid clinical improvement with normalization of mean plasma GH levels as well as circulating somatomedin-C levels in all four cases. More interestingly, the authors showed a slight decrease in tumor size in all patients during the study period. One patient with oculomotor and abducent nerve paralysis due to compression of the cavernous sinus by the tumor showed dramatic improvement in the nerve function within 4 weeks of therapy with the somatostatin analogue. While such preliminary studies have held promise for the use of somatostatin analogues in the treatment of acromegaly, the potential side effects of the peptide on insulin secretion, and the gastrointestinal hormones need to be evaluated. Additional clinical studies would be needed to determine the role of somatostatin analogues in the treatment of acromegaly.

References

1. Singer S, Hilgard HR: In *The Biology of People*. WH Freeman, San Francisco, 1978, p. 283.
2. Bollet AJ: Medical history in the bible. Part IV. Some metabolic and endocrine diseases. (Did Goliath have a pituitary tumor?) *Resident Staff Physician* **30**:52, 1984.
3. Rabin D, Rabin PL: David, Goliath, and Smiley's people. *N Engl J Med* **309**:992, 1983.
4. Marie P: On two cases of acromegaly: Marked hypertrophy of the upper and lower limbs and the head. *Rev Med* **6**:297, 1886.
5. Evans HM, Long JA: The effect of the anterior lobe of the pituitary administered intraperitoneally upon growth, maturity and oestrus cycle of the rat. *Anat Rec* **21**:62, 1921.
6. Davidoff LM: Studies in acromegaly. The anamnesis and symptomatology in one hundred cases. *Endocrinology* **10**:461, 1926.
7. Alexander L, Appleton D, Hall R, et al: Epidemiology of acromegaly in the Newcastle region. *Clin Endocrinol (Oxf)* **12**:71, 1980.
8. Besser GM, Mortimer CH, Carr D, et al: Growth hormone release inhibiting hormone in acromegaly. *Br Med J* **1**:352, 1974.
9. Daughaday WH, Cryer PE: Growth hormone hypersecretion and acromegaly. *Hosp Pract* **13**:75, 1978.
10. Cryer PE, Daughaday WH: Regulation of growth hormone secretion in acromegaly. *J Clin Endocrinol Metab* **29**:386, 1969.
11. Arafah BUM, Bradkey JS, Kaufman B, et al: Transsphenoidal microsurgery in the treatment of acromegaly and gigantism. *J Clin Endocrinol Metab* **50**:578, 1980.
12. Feingold KR, Goldfine ID, Weinstein PR: Acromegaly with normal growth hormone levels and pituitary histology. Case report. *J Neurosurg* **50**:503, 1979.
13. Asa SL, Bilbao JM, Kovacs K, et al: Hypothalamic neuronal hamartoma associated with pituitary growth hormone cell adenoma and acromegaly. *Acta Neuropathol (Berl)* **52**:131, 1980.
14. Rhodes RH, Dusseau JJ, Boyd AS Jr, et al: Intrasellar neural-adenohypophyseal choristoma. A morphological and immunocytochemical study. *J Neuropathol Exp Neurol* **41**:267, 1982.
15. Asa SL, Scheithauer BW, Bilbao JM, et al: A case for hypothalamic acromegaly: Clinicopathologic study of six patients with hypothalamic gangliocytomas producing growth hormone releasing factor. *J Clin Endocrinol Metab* **58**:796, 1984.
16. Lawrence AM, Goldfine ID, Kirstein L: Growth hormone dynamics in acromegaly. *J Clin Endocrinol Metab* **31**:239, 1970.

17. Cryer PE, Daughaday WH: Adrenergic modulation of growth hormone secretion in acromegaly: Suppression during phentolamine and phentoxamine-isoproterenol administration. *J Clin Endocrinol Metab* **39**:658, 1974.

18. Frohman LA: Diseases of the anterior pituitary. In Felig P, Baxter J, Broadus AE, Frohman LA (eds): *Endocrinology and Metabolism.* McGraw-Hill, New York, 1981, p. 204.

19. Hanew K, Kokubun M, Sasaki A, et al: The spectrum of pituitary growth hormone responses to pharmacological stimuli in acromegaly. *J Clin Endocrinol Metab* **51**:292, 1980.

20. Pieters GFFM, Romeijn JE, Smals AGH, et al: Somatostain sensitivity and growth hormone responses to releasing hormones and bromocryptine in acromegaly. *J Clin Endocrinol Metab* **54**:942, 1982.

21. Marcovitz S, Goodyer G, Guyda H, et al: Comparative study of human fetal, normal adult and somatotrophic adenoma pituitary function in tissue culture. *J Clin Endocrinol Metab* **54**:1982, 1982.

22. Tallo D, Malarkey WB: Adrenergic and dopaminergic modulation of growth hormone and prolactin secretion in normal and tumor-bearing human pituitaries in monolayer culture. *J Clin Endocrinol Metab* **53**:1278, 1981.

23. Nakayama I, Nickerson PA: Suppression of anterior pituitary in rats bearing a transplantable growth hormone and prolactin-secreting tumor (MtT.W10). *Endocrinology* **92**:516, 1973.

24. Allen JP, Cook DM, Greer MA, et al: Evidence that acromegaly is not a hypothalamic disease. *Trans Assoc Am Physicians* **80**:272, 1973.

25. Hoyte KM, Martin JB: Recovery from paradoxical growth hormone responses in acromegaly after transsphenoidal selective adenomectomy. *J Clin Endocrinol Metab* **41**:656, 1975.

26. Baskin DS, Boggan JE, Wilson CB: Transsphenoidal microsurgical removal of growth hormone-secreting pituitary adenomas. *J Neurosurg* **56**:634, 1982.

27. Faglia G, Parachi A, Ferrari C, et al: Evaluation of the results of transphenoidal surgery in acromegaly by assessment of the growth hormone response to thyrotropin-releasing hormone. *Clin Endocrinol (Oxf)* **8**:873, 1978.

28. Laws ER Jr, Piepgras DG, Randall RV, et al: Neurosurgical management of acromegaly. Results in 82 patients treated between 1972 and 1977. *J Neurosurg* **50**:454, 1979.

29. Schuster LD, Bantle JP, Oppenheimer JH, et al: Acromegaly: Reassessment of the long-term therapeutic effectiveness of transsphenoidal pituitary surgery. *Ann Intern Med* **95**:172, 1981.

30. Guyda H, Robert F, Colle E, et al: Histologic ultrastructural and hormonal characterization of a pituitary tumor secreting both hGH and prolactin. *J Clin Endocrinol Metab* **36**:531, 1973.

31. Corenblum B, Sirek AMT, Horvath E, et al: Human mixed somatotrophic and lactotrophic pituitary adenoma. *J Clin Endocrinol Metab* **42**:857, 1976.

32. Kovacs K, Horvath E, Ezrin C: Pituitary adenomas. *Pathol Annu* **12**(part 2):341, 1977.

33. Horvath E, Kovacs K: Pathology of the pituitary gland. In Ezrin C, Horvath E, Kaufman B, Kovacs K, Weiss MH (eds): *Pituitary Diseases.* CRC Press, Boca Raton, FL, 1980, pp. 1–83.

34. Horvath E, Kovacs K, Singer W, et al: Acidophil stem cell adenoma of the human pituitary. *Arch Pathol Lab Med* **101**:594, 1977.

35. Horvath E, Kovacs K, Singer W, et al: Acidophil stem cell adenoma of the human pituitary. Clinico-pathological analysis of 15 cases. *Cancer* **47**:761, 1981.

36. Kovacs K, Horvath E: Pathology of pituitary adenomas. In Givens JR (ed): *Hormone-Secreting Pituitary Tumors.* Year Book, Chicago, 1982, p. 97.

37. Melmed S, Braunstein GD, Horvath E, et al: Pathophysiology of acromegaly *Endocrinol Rev* **4**:271, 1983.

38. Ballard HS, Frame B, Hartsock RJ: Familial multiple endocrine adenomapeptic ulcer complex. *Medicine (Baltimore)* **43**:481, 1964.

39. Rivier C, Vale W, Ling M: Stimulation in vivo of the secretion of prolactin and growth hormone by β-endorphin. *Endocrinology* **100**:238, 1977.

40. Matsushita N, Kato Y, Katakami H, et al: Stimulation of growth hormone release by vasoactive intestinal polypeptide from human pituitary adenomas in vitro. *J Clin Endocrinol Metab* **53**:1297, 1981.

41. Kaganowicz A, Farkouh NH, Frantz AG, et al: Ectopic human growth hormone in ovaries and breast cancer. *J Clin Endocrinol Metab* **48:**5, 1979.
42. Sparagana M, Phillips G, Hoffman C, et al: Ectopic growth hormone syndrome associated with lung cancer. *Metabolism* **20:**730, 1971.
43. Greenberg PB, Beck C, Martin TJ, et al: Synthesis and release of human growth hormone from lung carcinoma in cell culture. *Lancet* **1:**350, 1972.
44. Dabek JT: Bronchial carcinoid tumour with acromegaly in two patients. *J Clin Endocrinol Metab* **38:**329, 1974.
45. Melmed S, Ezrin C, Kovacs K, et al: Acromegaly due to secretion of growth hormone by an ectopic pancreatic islet-cell tumor. *N Engl J Med* **312:**9, 1985.
46. Caplan RH, Koob L, Avellera RM, et al: Cure of acromegaly by operative removal of an islet cell tumor of the pancreas. *Am J Med* **64:**874, 1978.
47. Frohman LA, Szabo M, Berelowitz M, et al: Partial purification and characterization of a peptide with growth hormone-releasing activity from extrapituitary tumors in patients with acromegaly. *J Clin Invest* **65:**43, 1980.
48. Leveston SA, McKeel DW Jr, Buckley PJ, et al: Acromegaly and Cushing's syndrome associated with a foregut carcinoid tumor. *J Clin Endocrinol Metab* **53:**682, 1981.
49. Thorner MO, Perryman RL, Cronin MJ, et al: Acromegaly with somatotroph hyperplasia: Successful treatment by resection of a pancreatic tumor secreting a growth hormone (GH) releasing factor (GRF). *J Clin Invest* **70:** 965, 1982.
50. Rivier J, Spiess M, Thorner M, et al: Characterization of a growth hormone releasing factor from a human pancreatic islet tumour. *Nature (Lond)* **300:**276, 1982.
51. Guillemin R, Brazeau P, Bohlen P, et al: Growth hormone-releasing factor from a human pancreatic tumor that caused acromegaly. *Science* **218:**585, 1982.
52. Mayo KE, Vale W, Rivier J, et al: Expression cloning and sequencing of a cDNA encoding human growth hormone releasing factor. *Nature (Lond)* **306:**86, 1983.
53. Gubler U, Monahan JJ, Lomedico PT, et al: Cloning and sequence analysis of cDNA for the precursor of human growth hormone-releasing factor, somatocrinin. *Proc Natl Acad Sci USA* **80:**4311, 1983.
54. Ch'ng JLC, Christofides ND, Kraenzlin ME, et al: Growth hormone secretion dynamics in a patient with ectopic growth hormone-releasing factor production. *Am J Med* **79:**135, 1985.
55. Schulte HM, Benker G, Windeck R, et al: Failure to respond to growth hormone releasing hormone (GHRH) in acromegaly due to a GHRH secreting pancreatic tumor: Dynamics of multiple endocrine testing. *J Clin Endocrinol Metab* **61:**585, 1985.
56. Wilson DM, Ceda GP, Bostwick DG, et al: Acromegaly and Zollinger-Ellison syndrome secondary to an islet cell tumor: Characterization and quantification of plasma and tumor human growth hormone-releasing factor. *J Clin Endocrinol Metab* **59:**1002, 1984.
57. Thorner MO, Frohman LA, Leong DA, et al: Extrahypothalamic growth-hormone-releasing factor (GRF) secretion is a rare cause of acromegaly: Plasma GRF levels in 177 acromegalic patients. *J Clin Endocrinol Metab* **59:**846, 1984.
58. McGrath P: Volume and histology of the human pharyngeal hypophysis. *Aust NZ J Surg* **37:**16, 1967.
59. Melchionna RH, Moore RA: The pharyngeal pituitary gland. *Am J Pathol* **14:**763, 1938.
60. Boyd JS: Observations in the human pharyngeal hypophysis. *J Endocrinol* **14:**66, 1956.
61. Rasmussen P, Lindholm J: Ectopic pituitary adenomas. *Clin Endocrinol (Oxf)* **11:**69, 1979.
62. Corenblum B, Leblanc FE, Watanabe M: Acromegaly with an adenomatous pharyngeal pituitary. *JAMA* **243:**1456, 1980.
63. Warner BA, Santen RJ, Page RB: Growth hormone and prolactin secretion by a tumor of the pharyngeal pituitary. *Ann Intern Med* **96:**65, 1982.
64. Kammer H, George R: Cushing's disease in a patient with an ectopic pituitary adenoma. *JAMA* **246:**2722, 1981.
65. Kovacs K, Horvath E: Pituitary adenomas. Pathologic aspects. In Tolis G, Labrie F, Martin JB, Naftolin F (eds): *Clinical Neuroendocrinology. A Pathophysiological Approach.* Raven Press, New York, 1979, p. 367.

66. Kovacs K, Horvath E, Pritzker KPH, et al: Pituitary growth hormone cell adenoma with cytoplasmic tubular aggregates in the capillary endothelium. *Acta Neuropathol (Berl)* **37**:77, 1977.

67. Horvath E, Kovacs K: Ultrastructural classification of pituitary adenomas. *Can J Neurol Sci* **3**:9, 1976.

68. Kovacs K, Horvath E, Ryan N: Immunocytology of the human pituitary. In Delellis RA (ed): *Diagnostic Immunocytochemistry*. Masson USA, New York, 1981, p. 17.

69. Halmi NS, Duello T: "Acidophilic" pituitary tumor. A reappraisal with differential staining and immunocytochemical techniques. *Arch Pathol Lab Med* **100**:346, 1976.

70. Halmi NS: Immunostaining of growth hormone cells and prolactin cells in paraffin-embedded and stored or previously stained material. *J Histochem Cytochem* **26**:486, 1978.

71. Kovacs K, Horvath E, Killinger DW, et al: Growth hormone producing pituitary adenoma with giant secretory granules. *Acta Neuropathol* **46**:239, 1979.

72. Trouillas J, Girod C, Lheritier M, et al: Morphological and biochemical relationships in 31 human pituitary adenomas with acromegaly. *Virchows Arch Pathol Anat* **389**:127, 1980.

73. Zimmerman EA, Defendini R, Frantz AG: Prolactin and growth hormone in patients with pituitary adenomas. A correlative study of hormone in tumor and plasma by immunoperoxidase technique and radioimmunoassay. *J Clin Endocrinol Metab* **38**:579, 1974.

74. Corenblum B, Sirek AMT, Horvath E, et al: Human mixed somatotrophic and lactotrophic pituitary adenoma. *J Clin Endocrinol Metab* **42**:857, 1976.

75. Horvath E, Kovacs K, Singer W, et al: Acidophil stem cell adenoma of the human pituitary. *Arch Pathol Lab Med* **101**:594, 1977.

76. Horvath E, Kovacs K, Singer W, Smyth HS, et al: Acidophil stem cell adenoma of the human pituitary. Clinico-pathological analysis of 15 cases. *Cancer* **47**:761, 1981.

77. Haigler ED, Hershman JM, Meador CK: Pituitary gigantism: A case report and review. *Arch Intern Med* **132**:588, 1973.

78. Prezio JA, Griffin JE, O'Brien JJ: Acromegalic gigantism: The Buffalo giant. *Am J Med* **31**:966, 1961.

79. Humberd CD: Gigantism: Report of a case. *JAMA* **108**:544, 1937.

80. Gray H: The Minneapolis giant. *Ann Intern Med* **10**:1669, 1937.

81. Levin SR, Hofeldt FD, Becker N, et al: Hypersomatotropism and acanthosis nigricans in two brothers. *Arch Intern Med* **134**:365, 1974.

82. Curth HO: Pigmentary changes of the skin associated with internal diseases. *Postgrad Med* **41**:439, 1967.

83. Fitzpatrick TB, et al: *Dermatology in General Medicine*. McGraw-Hill, New York, 1971, p. 1447.

84. Courville C, Mason VR: The heart in acromegaly. *Arch Intern Med* **61**:704, 1938.

85. Hejtmancik MR, Bradfield JY Jr, Herrmann GR: Acromegaly and the heart: A clinical and pathologic study. *Ann Intern Med* **34**:1445, 1951.

86. Pepine CJ, Aloia J: Heart muscle disease in acromegaly. *Am J Med* **48**:530, 1970.

87. McGuffin WL, Sherman BM, Roth J, et al: Acromegaly and cardiovascular disorders: A prospective study. *Ann Intern Med* **81**:11, 1974.

88. Ikkos D, Luft R, Sjögren B: Body water and sodium in patients with acromegaly. *J Clin Invest* **33**:989, 1954.

89. Strauch G, Vallotton MB, Touitou Y, et al: The renin-angiotensin-aldosterone system in normotensive and hypertensive patients with acromegaly. *N Engl J Med* **287**:795, 1972.

90. Cain JP, Williams GH, Dluhy RG: Plasma renin activity and aldosterone secretion in patients with acromegaly. *J Clin Endocrinol Metab* **34**:73, 1972.

91. Hamwi GJ, Skillman TG, Tufts KC Jr: Acromegaly. *Am J Med* **29**:690, 1960.

92. Jonas EA, Aloia JF, Lane FJ: Evidence of subclinical heart muscle dysfunction in acromegaly. *Chest* **67**:190, 1975.

93. Savage DD, Henry WL, Eastman RC, et al: Echocardiographic assessment of cardiac anatomy and function in acromegalic patients. *Am J Med* **67**:823, 1979.

94. Smallridge RC, Rajfer S, Davia J, et al: Acromegaly and the heart: An echocardiographic study. *Am J Med* **66**:22, 1979.

95. Savage DD, Eastman RC, Henry WL, et al: Echocardiographic evaluation of acromegalic patients (abstract). *Circulation* **54**(suppl II):22, 1976.
96. Martins JB, Kerber RE, Sherman BM, et al: Cardiac size and function in acromegaly. *Circulation* **56**:863, 1977.
97. O'Duffy JD, Randall RV, MacCarty CS: Median neuropathy (carpal-tunnel syndrome) in acromegaly. A sign of endocrine overactivity. *Ann Intern Med* **78**:379, 1973.
98. Klijn JGM, Lamberts SWJ, de Jong FH, et al: Interrelation between tumour size, age, plasma growth hormone and incidence of extrasellar extension in acromegalic patients. *Acta Endocrinol (Copenh)* **95**:289, 1980.
99. Epstein N, Whelan M, Benjamin V: Acromegaly and spinal stenosis: Case report. *J Neurosurg* **56**:145, 1982.
100. Mastaglia FL, Barwick DD, Hall R: Myopathy in acromegaly. *Lancet* **2**:907, 1970.
101. Lundberg PO, Osterman PO, Stalberg E: Neuromuscular signs and symptoms in acromegaly. In Walton JN, Canal N, Scarlata G, et al (eds): *First International Congress on Muscle Disease, Milan, Italy*. Excerpta Medica, Amsterdam, 1970, p. 531.
102. Pickett JBE III, Layzer RB, Levin SR, et al: Neuromuscular complications of acromegaly. *Neurology (NY)* **25**:638, 1975.
103. Southwick JP, Katz J: Unusual airway difficulty in the acromegalic patient—Indications for tracheostomy. *Anesthesiology* **51**:72, 1979.
104. Ovassapian A, Doka JC, Ramsa DE: Acromegaly—Use of fiberoptic laryngoscopy to avoid tracheostomy. *Anesthesiology* **54**:429, 1981.
105. Perks WH, Horrocks PM, Cooper RA, et al: Sleep apnoea in acromegaly. *Br Med J* **280**:894, 1980.
106. Hart TB, Radow SK, Blackard WG, et al: Sleep apnea in active acromegaly. *Arch Intern Med* **145**:865, 1985.
107. Cadieux RJ, Kales A, Santen RJ, et al: Endoscopic findings in sleep apnea associated with acromegaly. *J Clin Endocrinol Metab* **55**:18, 1982.
108. Sober AJ, Gorden P, Roth J, et al: Visceromegaly in acromegaly. *Arch Intern Med* **134**:415, 1974.
109. Thomson JA, McCrossan J, Mason DK: Salivary gland enlargement in acromegaly. *Clin Endocrinol (Oxf)* **3**:1, 1974.
110. Jadresic A, Banks LM, Child DF, et al: The acromegaly syndrome. Relation between clinical features, growth hormone values and radiological characteristics of the pituitary tumours. *Q J Med* **202**:189, 1982.
111. Gordon DA, Hill FM, Ezrin C: Acromegaly: A review of 100 cases. *Can Med Assoc J* **87**:1106, 1962.
112. Lawrence JH, Tobias CA, Linfoot JA, et al: Successful treatment of acromegaly: Metabolic clinical studies in 145 patients. *J Clin Endocrinol Metab* **31**:180, 1970.
113. Tyrrell JB, Wilson CB: Pituitary syndromes. In Friesen SR (ed): *Surgical Endocrinology. Clinical Syndromes*. JB Lippincott, Philadelphia, 1978, p. 304.
114. De Pablo F, Eastman RC, Roth J, et al: Plasma prolactin in acromegaly before and after treatment. *J Clin Endocrinol Metab* **53**:344, 1981.
115. Ezrin C, Kovacs K, Horvath E: Hyperprolactinemia: morphologic and clinical considerations. *Med Clin North Am* **62**:393, 1978.
116. Tolis G, Bertrand G, Carpenter S, et al: Acromegaly and galactorrhea-amenorrhea with two pituitary adenomas secreting growth hormone or prolactin. *Ann Intern Med* **89**:345, 1978.
117. Kanie N, Kageyama N, Kuwayama A, et al: Pituitary adenomas in acromegalic patients: An immunohistochemical and endocrinological study with special reference to prolactin-secreting adenoma. *J Clin Endocrinol Metab* **57**:1093, 1983.
118. Spitz IM, Barzilai D, Luboshitzky R: Diminished prolactin reserve in acromegaly. *Metabolism* **29**:880, 1980.
119. Tolis G, Kovacs L, Friesen H, et al: Dynamic evaluation of growth hormone (GH) and prolactin (hPRL) secretion in active acromegaly with high and low GH output. *Acta Endocrinol (Copenh)* **78**:251, 1975.

120. Tourniaire J, Trouillas J, Chalendar D, et al: Somatotropic adenoma manifested by galactorrhea without acromegaly. *J Clin Endocrinol Metab* **61**:451, 1985.

121. De Pablo F, Eastman RC, Roth J, et al: Plasma prolactin in acromegaly before and after treatment. *J Clin Endocrinol Metab* **53**:344, 1981.

120. Tsuchiya H, Onishi T, Takamoto S, et al: Prolactin secretion in acromegalic patients before and after selective adenomectomy. *J Clin Endocrinol Metab* **61**:104, 1985.

123. Cobb WE, Reichlin S, Jackson IMD: Growth hormone secretory status is a determinant of the thyrotropin response to thyrotropin-releasing hormone in euthyroid patients with hypothalamic-pituitary disease. *J Clin Endocrinol Metab* **52**:324, 1981.

124. Root AW, Snyder PJ, Rezvani I, et al: Inhibition of thyrotropin-releasing hormone-mediated secretion of thyrotropin by human growth hormone. *J Clin Endocrinol Metab* **36**:103, 1973.

125. Lamberg BA, Pelkonen R, Aro A, et al: Thyroid function in acromegaly before and after transsphenoidal hypophysectomy followed by cryoapplication. *Acta Endocrinol (Copenh)* **82**:254, 1976.

126. Lamberg BA, Ripatti J, Gordin A, et al: Chromophobe pituitary adenoma with acromegaly and TSH-induced hyperthyroidism associated with parathyroid adenoma: Acromegaly and parathyroid adenoma. *Acta Endocrinol (Copenh)* **60**:157, 1969.

127. Kovacs K, Horvath E, Ezrin C, et al: Adenoma of the human pituitary producing growth hormone and thyrotropin. A histologic, immunocytologic and fine structural study. *Virchows Arch Pathol Anat* **394**:59, 1982.

128. Hamilton CR Jr, Maloof F: Acromegaly and toxic goiter. Cure of the hyperthyroidism and acromegaly by proton-beam partial hypophysectomy. *J Clin Endocrinol Metab* **35**:659, 1972.

129. Horn K, Erhardt F, Fahlbusch R, et al: Recurrent goiter, hyperthyroidism, galactorrhea and amenorrhea due to a thyrotropin and prolactin-producing pituitary tumor. *J Clin Endocrinol Metab* **43**:137, 1976.

130. Waldhausl W, Bratusch-Marrain P, Nowotny P, et al: Secondary hyperthyroidism due to thyrotropin hyperthyroidism due to throtropin hypersecretion: study of pituitary tumor morphology and thyrotropin chemistry and release. *J Clin Endocrinol Metab* **49**:879, 1979.

131. Goldfine ID, Lawrence AM: Hypopituitarism in acromegaly. *Arch Intern Med* **130**:720, 1972.

132. Roth J, Glick SM, Cuatrecasas P, et al: Acromegaly and other disorders of growth hormone secretion: Combined clinical staff conference at the National Institutes of Health. *Ann Intern Med* **66**:760, 1967.

133. Inada M, Sterline K: Thyroxine turnover and transport in active acromegaly. *J Clin Endocrinol* **27**:1019, 1967.

134. Cerasi E, Luft R: Insulin response to glucose loading in acromegaly. *Lancet* **2**:769, 1964.

135. Sönksen PH, Greenwood FC, Ellis JP, et al: Changes of carbohydrate tolerance in acromegaly with progress of the disease and in response to treatment. *J Clin Endocrinol* **27**:1418, 1967.

136. Linfoot JA, Chong CY, Lawrence JH, et al: Acromegaly. In Li CH (ed): *Hormonal Proteins and Peptides*. Vol. 111. Academic Press, New York, 1975, p. 191.

137. Wass JAH, Cudworth AG, Bottazzo GF, et al: An assessment of glucose intolerance in acromegaly and its response to medical treatment. *Clin Endocrinol (Oxf)* **12**:53, 1980.

138. Muggeo M, Bar RS, Roth J, et al: The insulin resistance of acromegaly: Evidence for two alterations in the insulin receptor on circulating monocytes. *J Clin Endocrinol Metab* **48**:17, 1979.

139. Muggeo M, Saviolakis GA, Businaro V, et al: Insulin receptor on monocytes from patients with acromegaly and fasting hyperglycemia. *J Clin Endocrinol Metab* **56**:733, 1983.

140. Hitman GA, Katz J, Lytras N, et al: Are there genetic determinants for the glucose intolerance of acromegaly? *Clin Endocrinol (Oxf)* **23**:817, 1985.

141. Nadarajah A, Hartog M, Redfern B, et al: Calcium metabolism in acromegaly. *Br Med J* **4**:797, 1968.

142. Sigurdsson G, Nunziata V, Reiner M, et al: Calcium absorption and excretion in the gut in acromegaly. *Clin Endocrinol (Oxf)* **2**:187, 1973.

143. Spanos E, Barrett D, MacIntyre I, et al: Effect of growth hormone on vitamin D metabolism. *Nature (Lond)* **273**:246, 1978.

144. Spencer EM, Tobiassen O: The mechanism of the action of growth hormone on vitamin D metabolism in the rat. *Endocrinology* **108**:1064, 1981.

145. Brown DJ, Spanos E, MacIntyre I: Role of pituitary hormones in regulating renal vitamin D metabolism in man. *Br Med J* **280**:277, 1980.

146. Burnstein S, Chen IW, Tsang RC: Effects of growth hormone replacement therapy on 1,25-dihydroxyvitamin D and calcium metabolism. *J Clin Endocrinol Metab* **56**:1246, 1983.

147. Chipman JJ, Zerwekh J, Nicar M, et al: Effect of growth hormone administration: reciprocal changes in serum 1,25-dihydroxyvitamin D and intestinal calcium absorption. *J Clin Endocrinol Metab* **51**:321, 1980.

148. Gertner JM, Tamborlane WV, Hintz RL, et al: The effects on mineral metabolism of overnight growth hormone infusion in growth hormone deficiency. *J Clin Endocrinol Metab* **53**:818, 1981.

149. Camanni F, Massara F, Losanna O, et al: Increased renal tubular reabsorption of phosphate in acromegaly. *J Clin Endocrinol Metab* **28**:999, 1968.

150. Takamoto S, Tsuchiya H, Onishi T, et al: Changes in calcium homeostasis in acromegaly treated by pituitary adenomectomy. *J Clin Endocrinol Metab* **61**:7, 1985.

151. Aloia JF, Petrak Z, Ellis K, et al: Body composition and skeletal metabolism following pituitary irradiation in acromegaly. *Am J Med* **61**:59, 1976.

152. Wermer P: Genetic aspects of adenomatosis endocrine glands. *Am J Med* **16**:363, 1954.

153. Farhi F, Dikman SH, Lawson W: Paragangliomatosis associated with multiple endocrine adenomas. *Arch Pathol Lab Med* **100**:495, 1976.

154. German WJ, Flanigan S: Pituitary adenomas: a follow-up study of the Cushing series. *Clin Neurosurg* **10**:72, 1964.

155. Kadowaki S, Baba Y, Kakita T, et al: A case of acromegaly associated with pheochromocytoma. [In Japanese.] *Saishin-Igaku* **31**:1402, 1976.

156. Kahn MT, Mullon DA: Pheochromocytoma without hypertension: report of a patient with acromegaly. *JAMA* **188**:74, 1964.

157. Miller GL, Wynn J: Acromegaly, pheochromocytoma, toxic goiter, diabetes mellitus, and endometriosis. *Arch Intern Med* **127**:299, 1971.

158. Anderson RJ, Lufkin EG, Sizemore GW, et al: Acromegaly and pituitary adenoma with phaeochromocytoma: A variant of multiple endocrine neoplasia. *Clin Endocrinol (Oxf)* **14**:605, 1981.

159. Tolis G, Koutsilieris M, Bertrand G: Endocrine diagnosis of growth hormone-secreting pituitary tumors. *Prog Endocrinol Res Ther* **1**:145, 1984.

160. Daggett PR, Nabarro JDN: Measurement of the 24 hour integrated plasma concentration of growth hormone, in assessing the response of acromegalic patients to treatment. *Clin Endocrinol (Oxf)* **7**:437, 1977.

161. Wass JAH, Thorner MO, Morris DV, et al: Long-term treatment of acromegaly with bromocriptine. *Br Med J* **1**:875, 1977.

162. Job JC, Sizonenko PC: Serum growth hormone (GH) in oral glucose test (OGT) in children. Comparison with insulin and arginine stimulation tests. In *Second International Symposium on Growth Hormone, Milano, Italy.* Excerpta Medica, Amsterdam, 1972, p. 46. (Abst.)

163. Lawrence AM, Wilber JF, Hagen TC: The pituitary and primary hypothyroidism. Enlargement and unusual growth hormone secretory responses. *Arch Intern Med* **132**:327, 1973.

164. Clemmons DR, Van Wyk JJ, Ridgway EC, et al: Evaluation of acromegaly by radioimmunoassay of somatomedin-C. *N Engl J Med* **301**:1138, 1979.

165. Mims RB, Bethune JE: Acromegaly with normal fasting growth hormone concentrations but abnormal growth hormone regulation. *Ann Intern Med* **81**:781, 1974.

166. Daughaday WH, Salmon WD, Alexander F: Sulfation factor activity of sera from patients with pituitary disorders. *J Clin Endocrinol Metab* **19**:743, 1959.

167. Furlanetto RW, Underwood LE, Van Wyk JJ, et al: Estimation of somatomedin-C levels in normals and patients with pituitary disease by radioimmunoassay. *J Clin Invest* **60**:648, 1977.

168. Cohen KL, Nissley SP: The serum half-life of somatomedins: Evidence for growth hormone dependence. *Acta Endocrinol (Copenh)* **83**:243, 1976.

169. Rieu M, Girard F, Bricaire H, et al: The importance of insulin-like growth factor (soma-tomedin) measurements in the diagnosis and surveillance of acromegaly. *J Clin Endocrinol Metab* **55**:147, 1982.

170. Wass JAH, Clemmons DR, Underwood LE, et al: Changes in circulating somatomedin-C levels in bromocriptine-treated acromegaly. *Clin Endocrinol (Oxf)* **17**:369, 1982.

171. Irie M, Tsushima T: Increase in serum GH after TRH injection in patients with acromegaly and gigantism. *J Clin Endocrinol Metab* **35**:97, 1972.

172. Jaquet P, Codaccioni JL, Oliver C, et al: *Fourth International Congress of Endocrinology, Wash-ington D.C., June 18–24, 1972*. Excerpta Medica International Congress Series No. 256. (Abst. 221.)

173. Schalch DS, Gonzalez-Barcena D, Kastin AJ, et al: Abnormalities in the release of TSH in response to thyrotropin-releasing hormone (TRH) in patients with disorders of the pituitary, hypothalamus and basal ganglia. *J Clin Endocrinol Metab* **35**:609, 1972.

174. Faglia G, Beck-Peccoz P, Ferrari P, et al: Plasma growth hormone response to thyrotropin-releasing hormone in patients with active acromegaly. *J Clin Endocrinol Metab* **36**:1259, 1973.

175. Rubin AL, Levin SR, Berstein RI, et al: Stimulation of growth hormone by luteinizing hormone-releasing hormone in active acromegaly. *J Clin Endocrinol Metab* **37**:160, 1973.

176. Hulting AL, Werner S, Wersäll J, et al: Normal growth hormone secretion is rare after microsurgical normalization of growth hormone levels in acromegaly. *Acta Med Scand* **212**:401, 1982.

177. Liuzzi A, Chiodini PG, Botalla L, et al: Growth hormone (GH)-releasing activity of TRH and GH-lowering effect of dopaminergic drugs in acromegaly: Homogeneity in the two responses. *J Clin Endocrinol Metab* **39**:871, 1974.

178. Chiodini PG, Liuzzi A, Botalla L, et al: Stable reduction of plasma growth hormone (HGH) levels during chronic administration of 2-Br-α-ergocryptine (CB-154) in acromegalic pa-tients. *J Clin Endocrinol Metab* **40**:705, 1975.

179. Jackson IMD: Diagnostic tests for the evaluation of pituitary tumors. In Post KD, Jackson IMD, Reichlin S (eds): *The Pituitary Adenoma* Plenum, New York, 1980, p. 219.

180. Evain-Brion D, Garnier P, Schimpff RM, et al: Growth hormone response to thyrotropin-releasing hormone and oral glucose-loading tests in tall children and adolescents. *J Clin Endocrinol Metab* **56**:429, 1983.

181. Maeda K, Kata Y, Ohgo S, et al: Growth hormone and prolactin release after injection of thyrotropin-releasing hormone in patients with depression. *J Clin Endocrinol Metab* **40**:501, 1975.

182. Maeda M, Kuzuya N, Masuyama Y, et al: Changes in serum triiodothyronine, thyroxine and thyrotropin during treatment with thyroxine in severe primary hypothyroidism. *J Clin Endocrinol Metab* **43**:10, 1976.

183. Gild AD I, Dickerman Z, Weizman R: Abnormal GH response to LH-RH and TRH in adolescent schizophrenic boys. *Am J Psychiatry* **138**:357, 1981.

184. Collu R, Leboeuf G, Letarte J, et al: Increase in plasma growth hormone levels following thyrotropin-releasing hormone injection in children with primary hypothyroidism. *J Clin Endocrinol Metab* **44**:743, 1977.

185. Faggiano M, Criscuolo T, Graziani M, et al: Persistent TRH-induced growth hormone release after short-term and long-term L-thyroxine replacement therapy in primary con-genital hypothyroidism. *Clin Endocrinol (Oxf)* **23**:61, 1985.

186. Dunn PJ, Donald RA, Espiner EA: Bromocriptine suppression of plasma growth hormone in acromegaly. *Clin Endocrinol (Oxf)* **7**:273, 1977.

187. Tolis G, McKenzie JM, Martin JB, et al: Growth hormone secreting pituitary tumors: Clinical presentation, diagnostic tests and therapeutic modalities. In Tolis G, Labrie F, Martin JB, Naftolin F, (eds): *Clinical Neuroendocrinology: A Pathophysiologic Approach*. Raven Press, New York, 1979, p. 437.

188. Losa M, Schopohl J, Stalla GK, et al: Growth hormone releasing factor test in acromegaly: Comparison with other dynamic tests. *Clin Endocrinol (Oxf)* **23**:99, 1985.

189. Golde DW, Herschman HR, Lusis AJ, et al: Growth factors. *Ann Intern Med* **92**:650, 1980.

190. Ashcraft MW, Hartzband PI, Van Herle AJ, et al: A unique growth factor in patients with acromegaloidism. *J Clin Endocrinol Metab* **57**:272, 1983.

191. Cardoso ER, Peterson EW: Pituitary apoplexy: A review. *Neurosurgery* **14**:363, 1984.
192. Rigolosi RS, Schwartz E, Glick SM: Occurrence of growth-hormone deficiency in acromegaly as a result of pituitary apoplexy. *N Engl J Med* **279**:362, 1968.
193. Lawrence AM: Hypothalamic hypopituitarism after pituitary apoplexy in acromegaly. *Arch Intern Med* **137**:1134, 1977.
194. Werner PL, Shah JH, Kukreja SC, et al: Recurrence of acromegaly after pituitary apoplexy. *JAMA* **247**:2816, 1982.
195. Wright AD, Hill DM, Lowy C, et al: Mortality in acromegaly. *Q J Med* **39**:1, 1970.
196. Brennan MD, Jackson IT, Keller EE, et al: Multidisciplinary management of acromegaly and its deformities. *JAMA* **253**:682, 1985.
197. Whitehead EM, Shalet SM, Davies D, et al: Pituitary gigantism: A disabling condition. *Clin Endocrinol (Oxf)* **17**:271, 1982.
198. Klein I, Parveen G, Gavaler JS, et al: Colonic polyps in patients with acromegaly. *Ann Intern Med* **97**:27, 1982.
199. Ituarte EA, Petrini J, Hershman JM, et al: Acromegaly and colon cancer. *Ann Intern Med* **101**:627, 1984.
200. Baskin DS, Boggan JE, Wilson CB: Transsphenoidal microsurgical removal of growth hormone-secreting pituitary adenomas. A review of 137 cases. *J Neurosurg* **56**:634, 1982.
201. Richards SH, Thomas JP: Treatment of acromegaly by transethmoidal hypophysectomy. *Q J Med* **193**:21, 1980.
202. Tucker HG, Grubb SR, Wigand JP, et al: The treatment of acromegaly by transsphenoidal surgery. *Arch Intern Med* **140**:795, 1980.
203. Besser GM, Wass JAH: The medical management of acromegaly. Secretory tumors of the pituitary gland. In Black PM, Zervas NT, Ridgway EC, et al. (eds): *Progress in Endocrine Research and Therapy*, Vol. 1. Raven Press, New York, 1984, p. 155.
204. Stonesifer BD, Jordan RM, Kohler PO: Somatomedin C in treated acromegaly: Poor correlation with growth hormone and clinical response. *J Clin Endocrinol Metab* **53**:931, 1981.
205. Moses AC, Molitch ME, Sawin CT, et al: Bromocriptine therapy in acromegaly: Use in patients resistant to conventional therapy and effect on serum levels of somatomedin C. *J Clin Endocrinol Metab* **53**:752, 1981.
206. Nortier JWR, Croughs RJM, Thijssen JHH, et al: Bromocriptine therapy in acromegaly: Effects on plasma GH Levels, somatomedin-C levels and clinical activity. *Clin Endocrinol (Oxf)* **22**:209, 1985.
207. Laws ER JR: The neurosurgical management of acromegaly. Secretory tumors of the pituitary gland. In Black PM, Zervas NT, Ridgway EC, et al (eds), *Progress in Endocrine Research and Therapy*, Vol. 1. Raven Press, New York, 1984, p. 169.
208. Hardy J, Somma M: Acromegaly: Surgical treatment by transsphenoidal microsurgical removal of the pituitary adenoma. In Tindall GT, Collins WF (eds): *Clinical Management of Pituitary Disorders*. Raven Press, New York, 1979, p. 209.
209. Tindall GT, Tindall SC: Transsphenoidal surgery for acromegaly: Long-term results in 50 patients. Secretory tumors of the pituitary gland. In Balck PM, Zervas NT, Ridgway EC, et al (eds): *Progress in Endocrine Research and Therapy*, Vol. 1. Raven Press, New York, 1984, p. 175.
210. Schaison G, Couzinet B, Moatti N, et al: Critical study of the growth hormone response to dynamic tests and the insulin growth factor assay in acromegaly after microsurgery. *Clin Endocrinol (Oxf)* **18**:541, 1983.
211. Hardy J, Somma M, Vezina JL: Treatment of acromegaly: Radiation or surgery? In Morley JP (ed): *Current Controversies in Neurosurgery*. WB Saunders, Philadelphia, 1976, p. 377.
212. Laws ER, Piepgras DG, Randall RV, et al: Neurosurgical management of acromegaly. *J Neurosurg* **50**:454, 1979.
213. Roth J, Gorden P, Brace K: Efficacy of conventional pituitary irradiation in acromegaly. *N Engl J Med* **282**:1385, 1970.
214. Lawrence AM, Pinsky SM, Goldine ID: Conventional radiation therapy in acromegaly. *Arch Intern Med* **128**:369, 1971.
215. Eastman RC, Gorden P, Roth J: Conventional supervoltage irradiation is an effective treatment for acromegaly. *J Clin Endocrinol Metab* **48**:931, 1979.

216. Bloom B, Kramer S: Conventional radiation therapy in the management of acromegaly. Secretory tumors of the pituitary gland. In Black PM, Zervas NT, Ridgway EC, et al. (eds): *Progress in Endocrine Research and Therapy,* Vol. 1. Raven Press, New York, 1984, p. 179.

217. Aloia JF, Archambeau JO: Hypopituitarism following pituitary irradiation for acromegaly. *Horm Res* **9:**201, 1977.

218. Jenkins JS, Ash S, Bloom HJG: Endocrine function after external pituitary irradiation in patients with secreting and nonsecreting pituitary tumours. *Q J Med* **41:**57, 1972.

219. Kanis JA, Gillingham FJ, Harris P, et al: Clinical and laboratory study of acromegaly. Assessment before and year after treatment. *Q J Med* **43:**409, 1974.

220. Aristizabal S, Caldwell WL, Avila J: The relationship of time-dose fractionation factors to complications in the treatment of pituitary tumors by irradiation. *Int J Radiat Oncol Biol Phys* **2:**667, 1977.

221. Harris JR, Levene MB: Visual complications following irradiation for pituitary adenomas and craniopharyngiomas. *Radiology* **120:**167, 1976.

222. Kjellberg RN, Shintani A, Frantz AG, et al: Proton beam therapy in acromegaly. *N Engl J Med* **278:**689, 1968.

223. Kliman B, Kjellberg RN, Swisher B, et al: Proton beam therapy of acromegaly: A 20-year experience. Secretory tumors of the pituitary gland. In Black PM, Zervas NT, Ridgway EC, et al. (eds): *Progress in Endocrine Research and Therapy,* Raven Press, New York, 1984, p. 191.

224. Kjellberg RN, Kliman B: LIfetime effectiveness—A system of therapy for pituitary adenomas, emphasizing Bragg peak proton hypophysectomy. In *Recent Advances in the Diagnosis and Treatment of Pituitary Tumors.* Linfoot JA (ed): Raven Press, New York, 1979, p. 269.

225. Linfoot JA: Heavy ion therapy: Alpha particle therapy of pituitary tumors. In Linfoot JA (ed): *Recent Advances in the Diagnosis and Treatment of Pituitary Tumors.* Raven Press, New York, 1979.

226. Linfoot JA, Garcia JF, Hoye SA, et al: Treatment of acromegaly. *Proc R Soc Med* **63:**219, 1969.

227. Liuzzi A, Chiodini PG, Botalle L, et al: Decreased plasma growth hormone (GH) levels in acromegalics following CB 154 (2-Br-α-ergocryptine) administration. *J Clin Endocrinol Metab* **38:**910, 1974.

228. Parkes D: Bromocriptine. *N Engl J Med* **301:**873, 1979.

229. Besser GM, Wass JAH, Thorner MO: Acromegaly—results of long term treatment with bromocriptine. *Acta Endocrinol (Copenh) (Suppl)* **216:**187, 1978.

230. Benker G, Sandman K, Tharandt L, et al: Gel filtration studies of serum growth hormone in acromegaly following bromocriptine administration. *Horm Res* **11:**151, 1979.

231. Besser GM, Thorner MO, Wass JAH, et al: Bromocriptine treatment of acromegaly. *Q J Med* **45:**695, 1976.

232. Lindholm J, Riishede J, Vestergaard S, et al: No effect of bromocriptine in acromegaly. A controlled trial. *N Engl J Med* **304:**1450, 1981.

233. Cassar J, Mashiter K, Joplin GF: Bromocriptine treatment of acromegaly. *Metabolism* **26:**539, 1977.

234. Wass JAH, Clemmons DR, Underwood LE, et al: Changes in circulating somatomedin-C levels in bromocriptine-treated acromegaly. *Clin Endocrinol (Oxf)* **17:**369, 1982.

235. Chiba T, Chihara K, Minamitani N, et al: Effect of long term bromocriptine treatment on glucose intolerance in acromegaly. *Horm Metab Res* **14:**57, 1982.

236. Eskildsen PC, Svendsen PAA, Vang L, et al: Long-term treatment of acromegaly with bromocriptine. *Acta Endocrinol (Copenh)* **87:**687, 1978.

237. Thorner MO, Besser GM, Wass JAH, et al: Bromocriptine in acromegaly. *N Engl J Med* **305:**1092, 1981.

238. Wass JAH, Moult PJA, Thorner MO, et al: Reduction of pituitary tumour size in patients with prolactinomas and acromegaly treated with bromocriptine with or without radiotherapy. *Lancet* **2:**66, 1979.

239. Wass JAH, Williams J, Charlesworth M, et al: Bromocriptine in the management of large pituitary tumours. *Br Med J* **284:**1908, 1982.

240. Kendall-Taylor P, Upstill-Goddard G, Cook D: Longterm pergolide treatment of acromegaly. *Clin Endocrinol (Oxf)* **19:**711, 1983.

241. Eskildsen PC, Hommel E, Buchhave J: The effect of a new ergoline derivative, CU 32-085, in the treatment of acromegaly. A controlled study. *Clin Endocrinol (Oxf)* **22**:189, 1985.
242. Besser GM, Mortimer CH, Carr D, et al: Growth hormone release inhibiting hormone in acromegaly. *Br Med J* **1**:352, 1974.
243. Yen SSC, Siler TM, Devane GW: Effect of somatostatin in patients with acromegaly. *N Engl J Med* **290**:935, 1974.
244. Lamberts SWJ, Oosterom R, Neufeld M, et al: The somatostatin analog SMS 201-995 induces long-acting inhibition of growth hormone secretion without rebound hypersecretion in acromegalic patients. *J Clin Endocrinol Metab* **60**:1161, 1985.
245. Lamberts SWJ, Uitterlinden P, Verschoor L, et al: Long-term treatment of acromegaly with the somatostatin analogue SMS 201-995. *N Engl J Med* **313**:1576, 1985.
246. Adelman LS: The Pathology of Pituitary Adenomas. In Post KD, Jackson IM, Reichlin S (eds): *The Pituitary Adenoma,* Plenum Press, New York, 1980, pp. 47–62.
247. Reichlin S:Etiology of Pituitary Adenomas. In Post KD, Jackson IM, Reichlin S (eds): *The Pituitary Adenoma,* Plenum Press, New York, 1980, pp. 29–46.

Hypopituitary Dwarfism

Introduction

The pediatric definition of the term dwarfism is applied to children whose height is 4 standard deviations or more ($\geq$4 SD) below the mean of their coevals. Primary disturbances in growth regulation can be caused by a myriad of disease states. Intrinsic abnormalities of bone or cartilage as well as numerous systemic diseases affect growth (Table 26). The focus in this chapter

TABLE 26
Causes of Short Stature

Etiology	Comment
Constitutional	Often familial and corrects itself at puberty
Systemic disease	
Cardiac	Congenital heart disease
Renal	Chronic renal disease; Fanconi's syndrome
Gastrointestinal	Malabsorption syndromes
Pulmonary	Severe chronic lung disease
Hematological	Sickle cell; thalassemia major
Endocrine	
Pituitary dwarf	Monotropic hyposomatotropism
Laron dwarf	Failure to generate somatomedin
Cushing's syndrome	Antigrowth effects of steroids
Hypothyroidism	Juvenile or cretinism
Precocious puberty	Epiphyseal closure triggered by sex steroids
A-G syndrome	
Bartter's syndrome	Loss of K^+
Turner's syndrome	Genetically determined
Bone disease	
Rickets	Vitamin D deficiency (resistant variety)
Juvenile osteoporosis	
Achondroplasia	
Miscellaneous	
Emotional deprivation	
Metabolic diseases	

is on growth retardation caused by hyposomatotropism. Several terms used in the past, including primordial dwarfism, ateleotic dwarfism, genetic dwarfism, and constitutional dwarfism, have been mostly discarded currently, owing to the meaningless nature of such terms. It used to be believed that growth hormone (GH) deficiency represented a relatively uncommon (5–10%) cause of growth retardation. This belief was fostered in an era when GH deficiency was a relatively simple and straightforward entity (i.e., failure to grow was clearly a result of failure to generate GH by the anterior pituitary). In the short span of merely the past 10 years, GH-related growth disorders have come to be recognized as a heterogeneous group of disorders with diverse underlying mechanisms. Indeed, this entity is rapidly being elevated to the status of a fairly common etiology for growth retardation. Regardless of the frequency of occurrence, GH-related growth disorder may be the only group—besides hypothyroidism—in which some therapeutic rewards are attainable; a fact that is especially important, since delays in diagnosis can cost tremendous psychosocial maladjustments.

Pathophysiology

Several intricately linked factors play a role in controlling the rate of growth and the ultimate height of the child. In broad terms, the three factors that play an important role are hormonal factors, genetic factors, and the nutritional status (general health).

The main hormonal factor that regulates growth, obviously, is GH. The GH secretion by the somatotropes is under the control of hypothalamus, which exerts its influence by dual mechanisms. Growth hormone-releasing hormone (GHRH), or somatocrinin, is the physiological hypothalamic peptide that stimulates GH secretion and release by the somatotropes. The structure and function of GHRH are discussed in Chapter 2. The physiological suppressor of GH release is somatostatin, also secreted by the hypothalamus. The pulsatile nature of GH secretion is due to a positive drive from the hypothalamic GHRH, rather than to intermittent removal of somatostatin. This is reflected by the studies of Wehrenberg et al.,[1] who showed that the pulsatile secretion of GH can be abolished by the administration of monoclonal antibodies to the hypothalamic GHRH. Following release, GH induces the generation of a family of polypeptides, the somatomedins,[2] which mediate the growth promoting activities of GH. Notwithstanding the major role of somatomedins, the concept that GH may indeed have some direct role on cartilage growth is gaining acceptance.[3]

The role of genetics as a determinant of the ultimate height is well recognized. The role of genetic influences in GH deficiency (the isolated GH deficiency, type IA, or the Illig form) is unfolding with great clarity, owing to the availability of techniques using DNA probes. Finally, the role of nutrition and general health in growth promotion is crucial. Poor health and malnutrition impair somatomedin generation and/or somatomedin activity.[4–6] The interesting observation that fear of obesity and the resultant self-imposed caloric restriction can result in poor growth rate has brought the role of nutrition in growth to the forefront.[7,8]

Theoretically, GH-related growth problems can originate by several mechanisms:

1. Failure of the hypothalamus to secrete adequate GHRH (hypothalamic GHRH deficiency)
2. Excess secretion of somatostatin by the hypothalamus
3. Failure of the pituitary somatotropes to secrete GH (isolated GH deficiency)
4. Secretion of grossly inadequate amounts of GH by the somatotropes (partial GH deficiency)
5. Secretion of biologically inactive GH by the somatotropes (GH polymers, normal variant short stature)
6. Insensitivity to GH at the level of its receptors, with failure to generate somatomedin (Laron dwarfism, probably African pygmyism)
7. Insensitivity of cartilage to somatomedins

Etiology

GH-related growth defects can arise from several mechanisms. The remarkable widenings in the etiological perspectives of GH deficiency have

evolved as a consequence of three developments in the past decade: the availability of synthetic GHRH,[9] the emergence of accurate methodologies for
measurement of circulating somadomedins in plasma,[10-12] and the emergence
of genetic technology to study the GH gene.[13] This section focuses on the
well recognized, as well as the newer, etiologies underlying GH-dependent
growth disorders.

Hypothalamic GHRH Deficiency

In the 1970s, GH deficiency occurring as a result of hypothalamic GHRH
deficiency had been speculated based on the analogy that isolated gonadotropin failure and isolated thyroid-stimulating hormone (TSH) failure were
sometimes caused by deficiencies of their respective hypothalamic releasing
factors. This speculation eluded confirmation, since GHRH defied characterization, hence chemical synthesis. The long wait was over when two groups
of workers independently characterized and synthesized the human pancreatic growth hormone releasing factor (hpGHRH).[14,15] Immediately, and
often simultaneously, several workers[16-18] reported the rise of GH (and somatomedin C) in normal subjects following the intravenous administration of
hpGHRF. These observations were extended to patients diagnosed as idiopathic GH deficiency. Borges et al.[19] and Grossman et al.[20] from both sides
of the Atlantic, demonstrated that some patients presumed to have isolated
GH deficiency showed a hypothalamic pattern of response to GHRH administration. Since only some patients responded to GHRH, the possibility that
prolonged absence of GHRH may render the somatotropes dormant was
explored, again, by Borges et al.[21] These workers primed the somatotropes
by intermittent and pulsatile administration of GHRH and showed restoration
of somatotropin responsiveness to GHRH in five patients presumed to have
isolated GH deficiency. The situation is strikingly reminiscent of other hypothalamic releasing factor deficiencies, particularly GnRH. These studies
support the existence of a subset of patients with isolated GH deficiency in
whom the underlying problem is deficiency of GHRH.[9] This is also supported
by histological observations. Schecter et al.[22] examined, by immunocytochemical means, the pituitary gland of a patient with documented isolated GH
deficiency and demonstrated apparently normal-appearing somatotropes containing immunoreactive GH. This supports the notion that the pituitary somatotropes were intrinsically normal and that the problem originated at a
higher level.

The recognition of GHRH deficiency as a cause of GH deficiency has
significant impact on therapy.[23] Gelato et al.[24] documented that pulsatile
administration of GHRH increases short-term linear growth.

GHRH deficiency can also be acquired. The most common reasons for
acquired deficiency of GHRH are structural damage to the hypothalamus or
the stalk. Tumors, particularly the craniopharyngioma, and trauma are the
leading causes in children for such a phenomenon. Recently, Thorner et al.[25]
described two children with acquired GHRH deficiency secondary to an au-

tomobile accident, and a suprasellar cyst, respectively, who presented with arrested growth, low GH levels, and low somatomedin C, which significantly rose following GHRF administration. These workers treated the two children for 6 months with GHRH and found an impressive acceleration in growth rate throughout the treatment period.

Growth Hormone Deficiency from Excess Somatostatin

Although somatostatin-secreting tumors (somatostatinomas) can decrease GH secretion, this condition has not been described in children, nor does it cause growth retardation. In adults with somatostatin-secreting tumors, blunted GH response to nearly all provocative stimuli is a frequent observation.

Isolated Growth Hormone Deficiency

Isolated GH deficiency (IGHD) has long been considered the prototype of selective (monotropic) hormonal deficiency syndromes of the anterior pituitary. In the recent past, however, the heterogeneous nature of this entity is being increasingly recognized. Even if one excludes those cases of isolated GH deficiency secondary to hypothalamic GHRH deficiency, the entity is not a single disorder. In the classic sense, isolated GH deficiency occurs because the somatotropes are completely unable to synthesize GH. From a clinical perspective, several observations have indicated the existence of different subsets of patients with isolated GH deficiency. Three such observations merit mention. First, the measurement of GH levels—both under basal and stimulated conditions—reveals that not all patients with pituitary dwarfism caused by isolated GH deficiency are alike; some show no measurable levels at all in the serum, while others have measurable, albeit low, levels. Second, in response to exogenous GH therapy, some patients demonstrate a tendency to develop antibodies readily to exogenous GH, while others clearly do not; such a tendency is more often observed in patients with absolutely no measurable GH levels in the serum prior to treatment. Third, isolated GH deficiency is often inherited, with strikingly different inheritance patterns; thus, autosomal-recessive, autosomal-dominant, and even X-linked inheritance patterns have been noted.

These observations clearly support the tenet that isolated GH deficiency is not a single entity. While there is no universal classification that delineates the subtypes of isolated GH deficiency, the one proposed by Phillips[26] is outlined in Table 27. This classification is based on three variables: inheritance pattern, level of endogenous GH in the serum, and response to GH treatment, which is influenced by the generation of antibodies to the exogenously administered hormone. IGHD IA is characterized by an autosomal-recessive inheritance, absolute lack of GH, and the striking tendency to develop antibodies to the exogenous administration of GH. IGHD IB, II, and III are all characterized by low (but not absent) GH levels, as well as sustained response

TABLE 27
Isolated Growth Hormone Deficiency[a]

Type	Inheritance	Endogenous GH level	Response to GH Rx
IA	Autosomal recessive	Absent	Transient
IB	Autosomal recessive	Low	Present
II	Autosomal dominant	Decreased	Present
III	X-linked	Decreased	Present

[a] From Phillips.[26]

to GH therapy, presumably owing to the low titers of circulating antibodies against exogenous GH; the only differences are in the mode of inheritance.

Some insight into the mechanism of causation of IGHD, and its various subsets, has been gained in the past decade. The availability of newer techniques in genetic labeling has permitted the examination of the GH gene in normals and in patients with IGHD. One such technique is the DNA probe. The DNA probe is a piece of radiolabeled DNA that will hybridize with the fragments under investigation. Several techniques have been developed to yield authentic GH messenger RNA (mRNA) that can be used as a template. This template, with its specific reverse transcriptase, can be made to synthesize a strand of DNA complementary to it. The complementary DNA (cDNA) is in effect the GH gene, and therefore labeling this cDNA will result in the availability of a specific DNA probe that can seek out and examine the GH gene in the subject. The methodology for obtaining both the GH mRNA, as well as synthesizing DNA probes have greatly advanced in recent years. For a better understanding, the reader is referred to standard treatises on the subject.[27,28] The molecular basis for the development of IGHD may reside in abnormalities of the GH gene. The gene for GH is located in chromosome 17, in association with related genes and analogues. Four related analogues are located on the same chromosome in close proximity to the normal GH gene (GH-N); a silent chorionic somatomammotropin gene (CS-L), the expressed chorionic somatomammotropin gene (CS-A), its copy (CS-B), and an analogue of the GH gene (GH-V). There is significant homology in the base pairs of these analogues and the normal GH gene.

IGHD Type IA

At least one form of IGHD (type IA) can result from deletion of the GH gene. Normally during meiosis of the germ cells, there is separation of the duplicated chromosomes into daughter chromosomes. The process involves lining up of chromatids opposite each other, with each gene facing its counterpart. The mechanisms of exchanging information proceeds by transference of a particular gene from one chromatid to another, A to B, with an equivalent replacement of its counterpart transferred from B to A. The process by which gene exchange takes place heavily depends on proper and exact alignment of the chromatids, whereby identical strands of DNA recognize each other. In presence of malalignment, genetic information will be transferred from one chromosome to another, without replacement by its counterpart. As a consequence, one chromosome will be conferred with an extra gene, while its counterpart will have none of the particular gene in question. This phenomenon is referred to as gene deletion. When GH deletion occurs, the somatotropes will be deprived of the information required to produce GH, and therefore will never be able to secrete GH. Thus, patients with IGHD-IA, in whom deletion of the GH gene underlies the problem, GH is undetectable and, when GH treatment is instituted, antibodies are readily generated against the foreign GH. Performing genetic analysis in a family of patients with isolated GH deficiency, Phillips et al.[29] confirmed the fact that GH gene deletion, especially of the GH gene 1, was present in affected members.

IGHD Type IB

Patients with IGHD-IB suffer from a slightly different problem; they have low, but detectable, titers of GH and, probably because of this, do not generate large amounts of antibodies against exogenous GH. Examination of DNA samples from patients with the IGHD-IB type usually reveals a full complement of GH genes.[26] The assumption here is that although the genetic complement is normal, for some reason the genetic makeup precludes adequate synthesis of GH. Alternatively, it is possible that the genetic mutation may involve other aspects of GH physiology, such as production and release of GHRH from the hypothalamus.

IGHD Type II

The third type of deficiency, IGHD-II, resembles IGHD-IB, except that the inheritance pattern is autosomal dominant. The mechanism may involve inactivation of a normal GH gene by its abnormal counterpart. This can be theoretically explained by a mechanism that affects the presequence of the GH molecule. To understand this, one must view the entire process of GH secretion; normally, when GH is manufactured by the ribosomes, the first sequence of 26 amino acids do not form part of the ultimately secreted molecule. They are cleaved and serve to transport the hormone from the ribosomes, through the Golgi membrane to the secretory granules. The signal to terminate transcription comes when the first four of the 26 amino acids attach to a transfer particle. This particle subsequently binds to a site on the endoplasmic membrane, leading the way for the presequence to begin its travel to the secretory granules. Once this is attained, the transfer particle is released and transcription resumes. A defect in this intricate process, involving either the presequence, the transfer particle or the recognition site on the endoplasmic membrane could theoretically account for poor production of hormone and could be inherited from either parent through one abnormal gene.

IGHD Type III

Finally, IGHD-III resembles IGHD-IB and IGHD-II in that these patients possess measurable, but low, GH concentrations and respond to exogenous GH treatment without heavy antibody formation. The inheritance pattern is X-linked, and the GH genes are completely normal in this entity. Table 28 summarizes the presumed mechanisms underlying the types of familial IGHD.

Regardless of speculation, IGHD remains a singular example of an endocrine disorder in which the secretory signals for hormone production are absent, impaired, or abnormal. In addition to the idiopathic variety, GH deficiency can occur as a result of organic causes. Congenital maldevelopment of the pituitary, sequelae of inflammatory diseases such as meningitis, or encephalitis, trauma, radiation, (especially perinatal), and Hand-Schüller-Christian disease are important secondary etiologies of GH deficiency.

Secretion of Inadequate Growth Hormone

The variety of hyposomatotropism, (low secretory output) is characterized by a situation wherein GH secretion—especially in response to physiological stimuli—is completely inadequate. It is now clear that GH hyposecretion is not an all-or-none phenomenon. There appear to be a spectrum of GH abnormalities ranging from complete and absolute deficiency to intermittent, as well as suboptimal GH secretion. Spiliotis et al.[30] identified a group of short children with what they have termed neurosecretory dysfunction. These patients are capable of maintaining basal hormone secretion but do not secrete an adequate amount of GH during a 24-hr period. The 24-hr integrated profiles of these patients are characteristically devoid of the serrated, pulsatile pattern typical of normal controls. Since the pulsatile secretion of GH in normal humans is under neurotransmitter control, it is possible that a mal-

TABLE 28
Familial Idiopathic Growth
Hormone Deficiency

Type	Mechanism
IA	GH gene deletion
IB	Normal genetic complements
	? Mutation results in inadequate synthesis
	? Abnormal gene for GHRH
II	Normal genetic complement
	? Inactivation of normal GH gene by an abnormal counterpart
	? Involvement of "presequence" and transport of prohormone
III	Normal complement of GH gene mechanism unknown

function in neurotransmitter signal recognition may underlie the problem in some of these cases. The observation by Spiliotis et al.[30] that these patients responded rather gratifyingly to exogenous GH therapy—in some instances with doubling of growth velocity—clearly speaks in favor of overall deficient GH secretion. Recognition of this entity is important, as these short children will demonstrate normal results on GH stimulation testing; the low-output state can be recognized only by quantification of the 24-hr integrated or mean GH level. A low somatomedin C level is usually, but not invariably, seen in these children.

While recognition of partial defects in GH output have enhanced our understanding, it has raised concern over the type of patients who qualify for a trial of GH therapy. As we shall see later, the problem intensifies with the revelation that a normal GH status does not exclude GH-dependent growth failure.

Secretion of Bioinactive Growth Hormone

The concept that endocrine deficiency syndromes can occur as a result of secretion of an abnormal, biologically inactive molecule of hormone is not new. The classic example of such a phenomenon is represented by pseudoidiopathic hypoparathyroidism, a condition characterized by hypocalcemia, elevated levels of biologically inactive parathyroid hormone (PTH), and preservation of responsiveness of renal tubules to exogenously administered PTH. That such a circumstance may exist in relationship to GH was suggested by the report by Kowarski et al.[31] These workers described two 3-year-old boys with dwarfism and delayed bone age. The hormonal characteristics of these children included (1) normal GH responses after stimulation, (2) low somatomedin C levels that increased after GH administration, and (3) abnormally low radioreceptor activity of the circulating GH. These investigators hypothesized that the growth failure seen in these children was a consequence of secretion of GH which was immunoreactive but biologically inactive. The response of somatomedin C to exogenous GH separated these cases from the Laron type of dwarfism. The important aspect of the report was the remarkable growth rate (as much as 8–12 cm/year) attained following institution of GH therapy. Similar reports subsequently followed,[32,33] indicating the existence of patients with growth retardation, normal GH reserve by radioimmunoassay (RIA), low GH level in bioassay or radioreceptor systems, low somatomedin C, and an impressive response to exogenous GH therapy. In one study, Frazer et al.[33] employed a special radioreceptor assay involving the IM-9 lymphocyte as a source of human GH receptor. All five children in their report demonstrated a mean height 7.8 cm below the mean for age, with a mean RIA-GH exceeding 34.2 ± 3.5 ng/ml in response to stimulation and a significantly elevated ratio of RIA-GH/RRA-GH. The syndrome of secretion of bioinactive GH has now become an established entity and is considered an important cause of GH-dependent growth failure, since therapy with GH induces growth.

The second phase in our understanding of growth failure in the presence of normal immunoreactive GH was ushered in with the description of children with normal-variant short stature (NVSS), responding to treatment with GH. Rudman et al.[34] used the term normal-variant short stature to denote short children in whom no definite etiology could be found for growth retardation. On the basis of an elaborate 10-day evaluation of the metabolic and linear growth responses to GH, they identified four subgroups of patients: Patients in group 1 showed no anabolic or growth response; patients in subgroup 2 showed a weak anabolic response but no growth; and groups 3 and 4 showed both responses to GH, the magnitude of which was greater in subgroup 4.[35,36] They also pointed out that the response of somatomedin C to a 10-day trial of GH therapy clearly delineated the responsive from the nonresponsive groups. The important aspect of the studies reported by Rudman et al.[34–36] lies in the fact that children with NVSS constitute 30–50% of short children, a significant proportion of which could be GH responsive. If one projects the numbers of such patients in comparison to GH deficiency, the frequency of GH dependent NVSS far exceeds that of GH deficiency. The prevalence rate based on the figures published by Rudman et al.[35,36] has been questioned by others.[37] Despite the fact that patient selection may be conducive to overrepresentation, the implications of Rudman's work are threefold. First, those patients in subgroups 3 and 4 may not have been treated had the authors followed the hitherto strict conventional criteria for GH treatment; such a course would have been tragic, since these patients did respond to therapy. Second, the demonstration of normal GH responses to provocative testing does not exclude GH-dependent growth failure. Third, measurement of GH by radioreceptor assays and evaluation of somatomedin C response to exogenous GH administration must be given consideration in children with growth retardation.

The contention that the increment in somatomedin C level following the administration of GH for 10 days is predictive of the change in growth rate with long-term GH treatment has been questioned by some workers.[38] Similarly, the value of radioreceptor assays for GH has also come under controversy[39]; while these reservations are indeed valid and justifiable, the phenomenon of biological ineffective GH secretion indisputably exists.

The third phase in our understanding of growth failure in the presence of GH evolved with reports of pituitary dwarfism occurring in patients with abnormal GH polymers in their circulation. Valenta et al.[40] described a 14-year-old boy with growth failure (height below the fifth percentile for his age) and normal immunoreactive GH. Further studies showed decreased radioreceptor-assayable and bioassayable GH in the serum. The singular aspect of the case was that the plasma somatomedin levels were normal. When the patient's GH was analyzed by column chromatography, most of the IR-GH migrated to the 85,000- and 45,000-M_r range. These migratory positions are characteristic of "big-big" and "big" GH.[41] These studies highlighted and documented the abnormal structural characteristics of the GH secreted by the patient's somatotropes. Treatment with exogenous GH resulted in a rapid growth rate of 16 cm during the first year of therapy. A similar case was reported by Abucham-Filho et al.[42] who briefly reported a case of short stature with abnormal circulating GH, most likely a GH tetramer. The unique aspect of this case was the association of the GH polymer with hypothalamic deficiencies of TRH and possibly ACTH.

Growth Hormone Insensitivity

Laron Dwarfism

The term Laron dwarfism is applied to a syndrome characterized by inability of biologically normal GH to generate somatomedins. This form of familial dwarfism has been recognized in Israel and is characterized by all the clinical features of hypopituitary dwarfism, except that the immunoreactive GH levels are high.[43,44] While the prevalence of Laron dwarfism is clearly high in Jewish people of Asian descent, the condition can occur in other races. The characteristic abnormality in Laron dwarfism is failure to generate somatomedins by endogenous as well as exogenously administered GH.[45,46]

In the classic sense, Laron dwarfism represents target-organ resistance to practically all the actions of GH. The metabolic responses to GH administration are also impaired. In vitro, patients with Laron dwarfism demonstrate decreased binding of [^{125}I]-GH by the liver cell membrane. Furthermore, as an expression of resistance to all actions of GH, the erythrogenic precursors in the peripheral blood of patients with Laron dwarfism fail to be stimulated by GH in vitro. Finally, the fibroblasts of children with Laron dwarfism respond in vitro to IGF-II, indicating that the problem is proximal to somatomedin generation.

African Pygmyism

Whereas patients with Laron dwarfism demonstrate a defect in production of the entire family of somatomedins, African pygmyism is characterized by an isolated deficiency of insulinlike growth factor I. The early belief that the short stature of African pygmies is a result of somatomedin insensitivity[47–49] has been revised. Merimee et al.[50] studied the concentration of insulinlike growth factors IGF-I and IGF-II in 11 pygmies and found low levels of IGF-I in 10 of 11, while the levels of IGF-II were low in only one. Thus it appears that pygmies have a major defect in production of IGF-I, but not of IGF-II.

Insensitivity to Somatomedins

It is currently unclear whether target organ resistance to somatomedins plays any role in causing short stature. African pygmyism, which once belonged in this category, is no longer believed to result from target organ resistance to somatomedins.

Clinical Features

The five major facets of hypopituitary dwarfism are growth retardation, presence of associated somatic abnormalities, concomitant endocrinopathies, psychological changes, and abnormalities in glucose metabolism.

Growth Retardation

The slow growth rate in children with hypopituitary dwarfism usually becomes apparent between the first and third years of life. The growth retardation is manifested much later (8–10 years of age), when the hyposomatotropism is secondary to such tumors as craniopharyngioma. Nearly all hypopituitary dwarfs are 3–4 SD below the mean height appropriate for age. The patient's skeletal proportions are often symmetrical and childish. There is a striking contrast between the childish body proportions and the prematurely aged appearance from fine wrinkling of the face.

As with all other causes of short stature, the diagnosis of growth retardation cannot be made from a single abnormal height measurement. It is necessary to obtain a dynamic growth curve from serial height measurements.[51] Heights consistently below the fifth percentile, as well as a decelerated growth rate during the growing years, are a cause for concern. Such terms as constitutional growth delay and familial short stature should be used with caution while evaluating patients with short stature, since these diagnoses are retrospective. Constitutional growth delay is characterized by family history of shortness during childhood, delayed adolescence, and the attainment of a normal adult height ultimately. By contrast, familial short stature is used to describe patients whose ultimate adult height is short, with normal timing of puberty and a family history of short adult height.

Somatic Features

The facies of adults with hypopituitary dwarfism are often striking. Since the growth of the mandible, maxilla, and other facial bones is affected more than the cranial bones, the lower half of the face becomes disproportionately small. The underdeveloped nasal bridge, the small mouth, the small chin with micrognathism, and crowded teeth all impart characteristic facies to these patients. The integumental changes in patients with hypopituitary dwarfism consist of thinning of skin and hair, fine wrinkling and increased subcutaneous fat. Although the musculature is poorly developed, myopathy is not a feature of hypopituitary dwarfism. The hands and feet of these patients are impressively small.

When hypopituitary dwarfism is secondary to a congenital etiology, several somatic abnormalities may be present. These include cleft palate, absent septum pellucidum, optic nerve dysplasia, and iris-dental dysplasia (Reiger abnormality).

Concomitant Endocrinopathy

When hypopituitary dwarfism is secondary to tumors, the loss of other trophic hormones may be evident. When TSH deficiency is associated with GH deficiency, the growth retardation can be quite severe. Similarly, when gonadotropin deficiency is associated with GH failure, the retardation is accentuated by loss of the pubertal growth spurt. It is a well-noted observation that the other trophic hormone deficiencies appear much later than GH deficiency. The reason for this lag is unclear.

Diabetes insipidus secondary to loss of antidiuretic hormone (vasopressin) may be associated with GH deficiency when suprasellar lesions (e.g., craniopharyngioma, Hand-Schüller-Christian disease) are etiologically related.

Psychiatric Changes

The intelligence level of hypopituitary dwarfs is normal. The psychiatric changes seen in patients with hypopituitary dwarfism are those shared by patients with dwarfism from any etiology. These include dependency, isolation, and a desire to excel. The insecurities of a minority group, having to adjust to a culture in which the notion "bigger is better" prevails as an unspoken code, are understandable. The tendency to treat such patients as children because of their size is a contributing factor for the development of dependency tendencies. The emergence of support systems for "little people" has considerably aided in the attainment of mental well-being for many a patient with dwarfism.

Changes Related to Abnormal Glucose Handling

Patients with monotropic GH deficiency tend to demonstrate a tendency for fasting hypoglycemia. Defective gluconeogenesis (which is partly dependent on GH) may account for the tendency toward fasting hypoglycemia. These patients tend to be insulinopenic and demonstrate hypersensitivity to exogenous insulin.

By contrast, some hypopituitary dwarfs demonstrate an impaired glucose tolerance and even diabetes. Notably, the diabetes is mild, and characteristically, the incidence of microangiopathic complications is considerably lower. (Indeed, this observation prompted the recommendation of hypophysectomy as a form of treatment during the late 1960s to prevent progression of diabetic retinopathy.)

Diagnostic Studies

The stepwise approach to GH-related growth failure is outlined in Table 29.

TABLE 29
Diagnostic Evaluation of Growth Failure

Step 1:	Exclusion of non-GH endocrine disorders
	T$_4$, T$_3$, TSH, LH, FSH, cortisol (karyotype if indicated)
Step 2:	Determination of bone age
	Radiography of left hand and wrist
Step 3:	Assessment of GH secretory reserve
	GH response to two definitive stimuli (insulin, L-dopa)
Step 4:	Measurement of basal somatomedin-C level
Step 5:	Special endocrine studies
	For exclusion of bioinactive GH secretion
	Radioreceptor assay
	Somatomedin-C response to GH administration
	Chromatographic analysis of GH
	For exclusion of GH resistance (Laron type)
	Somatomedin-C response to GH administration
	For exclusion of sellar lesion
	Pituitary hormone reserve testing
	CT scans of sella and suprasellar area

Exclusion of Non-GH Endocrine Disorders

The initial step in the evaluation of patient with short stature (or a decreased growth velocity for age) is to exclude hypothyroidism and, in the case of phenotypic females, chromosomal anomalies. This can be readily done by measurement of thyroxine (T4), triiodothyronine (T3), TSH, luteinizing hormone (LH), follicle-stimulating hormone (FSH), and, when indicated, chromosomal analysis.

Determination of Bone Age

The next step is to determine bone age by obtaining radiographs of the left hand and wrist.[52,53] GH-dependent growth failure is characterized by a bone age growth rate that lags behind the chronological age. However, a delayed bone age is not unique for hypopituitarism and is often encountered in patients with hypothyroidism, hypogonadism, and constitutional delayed growth. The demonstration of a normal bone age in a child with a height below the third percentile is consistent with familial (or genetic) short stature.

Assessment of GH Reserve

In the child with short stature or impaired growth rate, delayed bone age, and euthyroidism, the next step is to evaluate the GH dynamics. It should be pointed out that the mere measure of random or basal GH levels is unsuited for screening to detect GH deficiency. When performing dynamic studies to detect underlying GH deficiency several basic facts bear emphasis:

1. Although physiological stimuli, such as exercise and sleep, have been employed as screening procedures to detect GH deficiency, the results

from these tests are not consistent enough to exclude or confirm it. In most instances, the patient will have to undergo provocative testing with pharmacological stimuli. The one exceptional instance is psychosocial dwarfism, in which characteristically GH response to physiological stimuli (sleep) is preserved, while responsiveness to pharmacological stimulation is temporarily impaired.

2. Several pharmacological stimuli can be employed to assess GH reserve-insulin hypoglycemia, L-dopa, arginine, glucagon, clonidine, and propranolol, to name a few. The dictum to remember is that the diagnosis of GH deficiency rests on demonstrating nonresponsiveness to two definitive stimuli. The normal peak of GH following insulin hypoglycemia, or L-dopa is equal to or above 10 ng/ml. Patients with severe GH deficiency never peak above 5 ng/ml, while less severe forms of GH deficiency are associated with peak GH levels of 6–9 ng/ml.

3. If concomitant hypothyroidism is present, this should be corrected before GH provocative testing is performed. This is because untreated hypothyroidism per se can affect GH responsiveness to provocative stimuli.

4. Rarely, the GH response to provocative stimuli can be preserved, but the 24-hr output of GH may be suboptimal. This entity can be suspected by the presence of low somatomedin C levels in plasma but can be documented only by performing 24-hr sampling of GH levels. The profile of the 24-hr secretory pattern in such patients is characterized by the absence of pulsatile bursts of GH secretion.

Measurement of Somatomedin C Level

In children older than 6 years of age, measurement of somatomedin C is a reasonably good screening test for GH-dependent growth problems. In children under age 6, there is considerable overlap between the somatomedin-C levels of normal and hypopituitary children. While a normal somatomedin-C level in a child above age 6 tends to exclude severe GH deficiency, it does not exclude partial deficiency states. On the false-positive side of the spectrum, there are several causes of lowered somatomedin activity (Table 30).

Special Endocrine Studies

Patients with short stature, delayed bone age, normal thyroid function, normal GH response to provocative stimuli, with low (or even normal) somatomedin C levels require additional tests to characterize the nature of the circulating GH. The concept that varying degrees of biologically inactive GH secretion may underlie the growth problem in such patients is gaining popularity. Additional studies required in this type of setting include the following:

1. Radioreceptor assay (RRA) for GH, to determine its biological activity[54]
2. Response of somatomedin C to a 10-day course of exogenous administration of GH (While a rise in somatomedin C following GH injections

TABLE 30
Decreased Somatomedin-C Activity[a]

Growth-hormone dependent
GH deficiency
Complete
Partial
Biologically inactive GH secretion
Laron dwarfism
Other endocrinopathies
Hypercortisolism
Hypothyroidism
Poorly controlled IDDM in children
General
Malnutrition
Renal failure
Cirrhosis with liver failure
Acute illness
Chronic debilitating illness

[a] GH, growth hormone; IDDM, insulin dependent diabetes mellitus.

is consistent with the secretion of bioinactive GH, it should be remembered that such a response is consistent with a normal state as well. The test will also help document GH resistance (Laron type), which is characterized by failure of somatomedin generation in response to exogenous administration of GH.)
3. Chromatographic elution of GH, to detect the presence of abnormal GH polymers (usually tetramers) in the circulation

When the responsiveness of GH to provocative testing is clearly below par, and this is coupled with a low circulating somatomedin C level, the diagnosis of GH deficiency is virtually certain. The only tests needed in such a circumstance are evaluation of the trophic hormone reserve of the other pituitary hormones and computed tomography (CT) scan of the sella to exclude intra- or perisellar lesions. The term isolated growth hormone deficiency is strictly used to denote unitropic loss of GH reserve with preservation of the other adenohypophyseal hormones. It is of interest to note that subtle defects in TSH and prolactin responsiveness has been noted to occur in patients with isolated GH deficiency.[55] Radiological evaluation of the sellar and perisellar region is mandatory to exclude craniopharyngioma. This is especially important because growth problems can be manifested before and after operative removal of craniopharyngiomas.[56] The algorithmic approach to growth retardation is outlined in Fig. 16.

Differential Diagnosis

Growth hormone deficiency accounts for no more than 10% of children with growth retardation. The causes of growth retardation are extensive: the

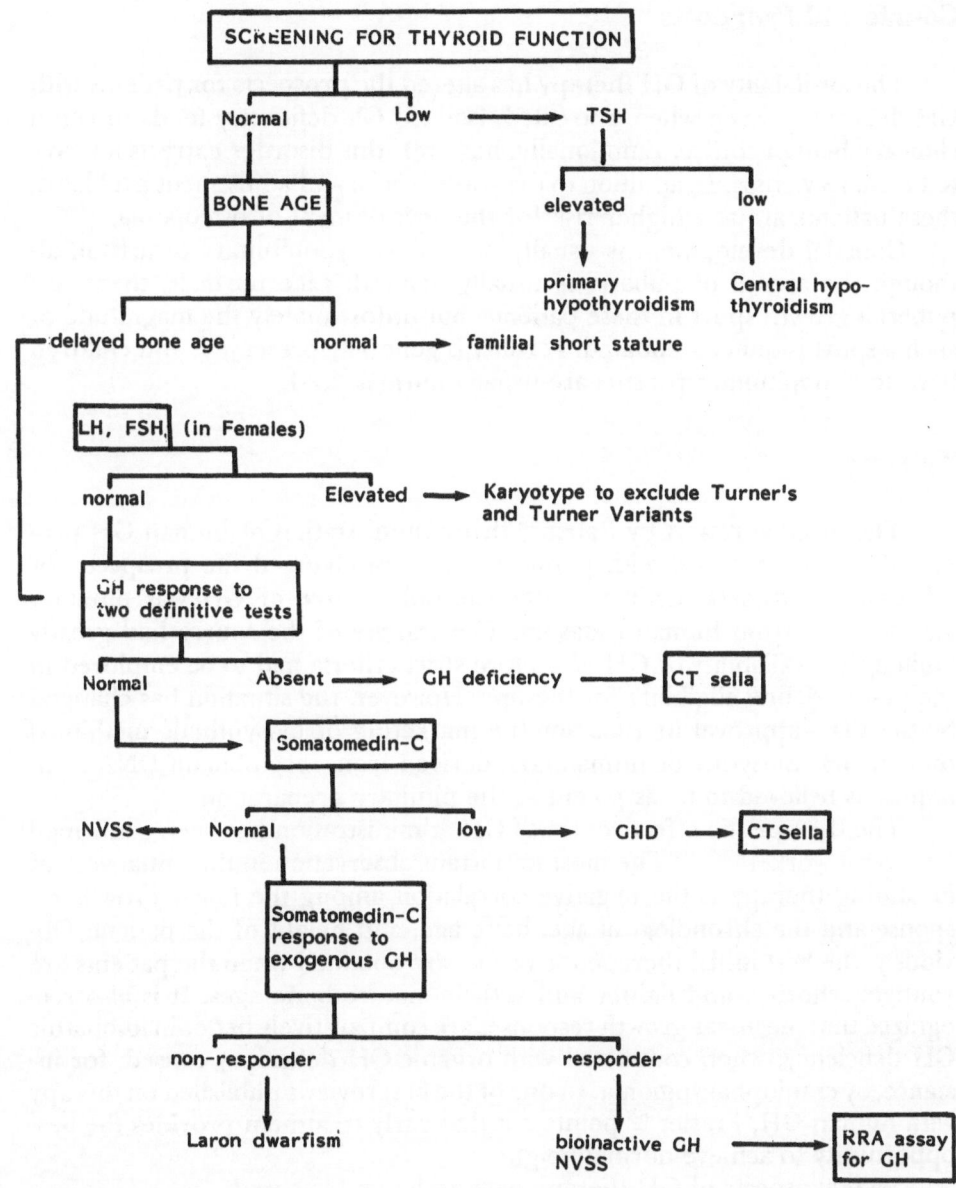

FIGURE 16. Algorithmic approach to short stature or subnormal growth velocity for age. NVSS, normal-variant short stature.

important ones are listed in Table 26. In clinical practice, the most frequent conditions that need to be differentiated from GH-dependent growth failure are hypothyroidism, gonadal dysgenesis, children with chronic ailments, and psychosocial dwarfism. The last one, also known as emotional deprivation syndrome, can mimic isolated GH deficiency to a remarkable extent.[57,58]

Course and Prognosis

The availability of GH therapy has altered the prospects for patients with GH deficiency. Even when untreated, isolated Gh deficiency tends to run a clinically benign course. Emotionally, however, this disorder extracts a heavy toll from its victims. In addition to the psychosocial and adjustment problems, these patients are at a higher risk for the development of osteopenia.

Gonadal development is usually normal in hypopituitary dwarfism, although the arrival of puberty is usually delayed. Like normals, there is a pubertal growth spurt in these patients, but unfortunately the magnitude of such a spurt is quite attenuated. Fertility is generally preserved, and children born to hypopituitary parents are usually normal sized.

Treatment

The original report by Raben[59] that administration of human GH promoted growth in a boy with pituitary dwarfism changed the prospects for GH-deficient dwarfs. Until recently, the only source of GH was pituitary glands taken from human cadavers. The scarcity of the source had greatly limited the availability of GH; therefore strict criteria had to be employed in the past to define eligibility for therapy. However, the situation has changed by the FDA approval in 1985 for the marketing of biosynthetic methionyl human GH. Biosynthetic human GH, derived from recombinant DNA techniques, is believed to be as potent as the pituitary preparation.

The therapeutic effectiveness of GH administration has been confirmed by several workers.[60-63] The most important observation in the initial year of instituting therapy is the negative correlation among the linear growth response and the chronological age, bone age, and height of the patient. Obviously, the best initial therapeutic results are obtained when the patients are younger, shorter, and lighter and with immature bone ages. It is also recognized that the linear growth responses are comparatively better in idiopathic GH deficiency, when contrasted with organic GH deficiency caused, for instance, by craniopharyngioma. In one of the best reviews published on therapy with human GH, Frasier [64] points out that early treatment provides the best opportunity to achieve normal height.

Several aspects of GH therapy need to be underscored.

1. The dosage of GH varies from patient to patient; calculating the dose based on body weight appears to be more efficacious than titrating a standard dose based on the response. The initial dose recommended is 0.06–0.1 units/kg IM three times per week. Therapy should be modified later, depending on the response.
2. Successful results have been attained by administration of GH on a weekly[65,66] twice-weekly,[67] or three-times-weekly schedule.[62]
3. When concomitant replacement endocrine therapy is administered, care should be exercised. The therapeutic response to GH gets blunted

when glucocorticoids are concomitantly administered, presumably because of peripheral antagonism. Should such therapy be indicated,the replacement maintenance dose should not exceed 10–15 mg hydrocortisone equivalent per square meter of body surface area (this is not applicable when the patient is under stress). Concomitant T4 replacement may also blunt the effectiveness of GH by causing epiphyseal fusion. Finally, administration of sex steroids, which also facilitate epiphyseal fusion, should be delayed until the maximal benefits of GH therapy have been attained.

4. Careful monitoring of bone age is needed, especially around puberty. Epiphyseal maturation would soon result in epiphyseal closure, a phenomenon that would render any further therapy with GH useless.

5. Therapy with GH should be continued during spontaneous puberty, which assists in the attainment of maximal height.

6. There is some controversy as to the predictive ability of rising somatomedin C concentrations following exogenous GH therapy. Dean et al.[68] noted considerable individual variations in the responses of somatomedin C level with GH therapy, minimizing its value as a tool to predict response. Similarly, Rosenfeld et al.[69] observed a lack of correlation between the growth rate following GH therapy and baseline, acute, or chronic somatomedin C levels. Their study found somatomedin C levels remaining constant despite duration of treatment and changes in growth rate.

7. After initial gratifying responses, the growth-promoting effect of GH therapy gradually diminishes. The reasons for a decline in growth velocity could be severalfold—a waning effect on somatomedin generation, a slowing of the catch-up phase of growth, and the development of antibodies to exogenously administered GH and all can contribute to the plateauing of response to therapy. Attempts to prolong or maintain the catch-up growth phase by intermittent therapy,[70] the use of long-acting hGH-gel preparation,[71] or the concomitant administration of anabolic steroids with hGH[72] have not met with consistent or uniform success.

8. The adverse effects of GH therapy are relatively few.
   a. Although the incidence of developing antibodies to exogenously administered GH can be as high as 30–60%,[73,74] clinically significant neutralization of hormone occurs in no more than 5–10% of patients receiving the hormone. It is generally believed that these antibodies nullify the effect of GH therapy, only when they are present in very high titers. The situation is analogous to the development of insulin antibodies in diabetics receiving insulin. Development of high-affinity antibodies, however, clearly can blunt the therapeutic efficacy of GH treatment. The method of extraction of GH from cadaver pituitaries probably plays a major role in generating anti-GH antibodies. It would be of interest to observe whether a similar phenomenon develops with the use of synthetic GH.

b. The second adverse effect of chronic GH therapy is the development of hypothyroidism. The frequency with which this phenomenon occurs following GH therapy has been hotly debated in the literature. Thus, some groups have reported a high incidence of hypothyroidism,[75-77] while others[78,79] have not found significant effects of Gh therapy on thyroid function. The consensus of opinion appears to be that thyroid function should be closely monitored during GH therapy, and the resulting hypothyroidism is more biochemical than clinical. The mechanisms for the development of hypothyroidism during Gh therapy are also speculative; the usual explanations offered are suppression of the TRH–TSH axis (supposedly mediated by a GH-induced somatostatin rise), accelerated conversion of T_4 to T_3, and the natural progression of the hypothalamic pituitary disease that caused the GH deficiency in the first place.

c. The discovery that Creutzfeldt-Jakob disease can be related to the administration of cadaveric human GH has provoked fear of a potential epidemic.[80] Several authenticated reports of this phenomenon[81-84] have served as grim reminder that human tissues can be sources of dangerous infectious disease. The sudden, explosive interest in synthetic hGH has been, in part, triggered by the fear of transmitting slow-growing viruses contaminating the cadaveric GH extracts.

The ready availability of synthetic hGH is bound to expand the indications for GH therapy in children with short stature. Although the recommendation for GH therapy has already extended to include children with biologically inactive GH secretion, as well as the subgroup of patients with normal variant short stature who respond to a 10-day trial of GH, these concepts have not been established in large series of patients. As these indications expand, the pressure on physicians to treat apparently normal, but slightly shorter children, will become inevitable. Clearly, there is a potential for abuse of synthetic hGH. Large therapeutic trials are required to define or redefine the patients who would receive additional benefit from GH therapy.

References

1. Wehrenberg WB, Brazeau P, Luben R, et al: Inhibition of the pulsatile secretion of growth hormone by monoclonal antibodies to hypothalamic growth hormone releasing factor. *Endocrinology* **111:**2147, 1982.
2. Phillips LS, Vassilopoulou-Sellin R: Somatomedins. *N Engl J Med* **302:**371, 1980.
3. Talwar GP, Pandian MR, Kumar N, et al: Mechanism of action of pituitary growth hormone. *Recent Prog Horm Res* **31:**141, 1975.
4. Phillips LS, Fusco AC, Unterman TG: Nutrition and somatomedin. XIV. Altered levels of somatomedins and somatomedin inhibitors in rats with streptozotocin-induced diabetes. *Metabolism* **34:**765, 1985.

5. Phillips LS, Orawski AT, Belosky DC: Somatomedin and nutrition. IV. Regulation of so-matomedin activity and growth cartilage activity by quantity and composition of diet in rats. *Endocrinology* **103**:121, 1978.

6. Phillips LS, Orawski AT: Nutrition and somatomedin. III. Diabetic control, somatomedin, and growth in rats. *Diabetes* **26**:864, 1977.

7. Pugliese MT, Lifshitz F, Grad G, et al: Fear of obesity: A cause of short stature and delayed puberty. *N Engl J Med* **309**:513, 1983.

8. Lucas AR: Undernutrition and growth. (Editorial.) *N Engl J Med* **309**:550, 1983.

9. Laron Z, Keret R, Bauman B, et al: Differential diagnosis between hypothalamic and pituitary hGH deficiency with the aid of synthetic GH-RH 1–44. *Clin Endocrinol (Oxf)* **21**:9, 1984.

10. Yalow RS, Hall K, Luft R: Radioimmunoassay of somatomedin B: Application to clinical and physiologic studies. *J Clin Invest* **55**:127, 1975.

11. Schalch DS, Heinrich UE, Koch JG, et al: Non-suppressible insulin-like activity (NSILA). I. Development of a new sensitive competitive protein-binding assay for determination of serum levels. *J Clin Endocrinol Metabl* **46**:664, 1978.

12. Furlanetto RW, Underwood LE, Van Wyk JJ, et al: Estimation of somatomedin-C levels in normals and patients with pituitary disease by radioimmunoassay. *J Clin Invest* **60**:648, 1977.

13. Miller WL, Eberhardt NL, Baxter JD: Growth hormone genes. In Black PM, Zervas NT, Ridgeway EC, et al (eds): *Secretory Tumors of the Pituitary Gland. Progress in Endocrine Research and Therapy.* Vol. 1. Raven, New York, 1984, p. 135.

14. Guillemin R, Brazeau P, Böhlen P, et al: Growth hormone-releasing factor from a human pancreatic tumor that caused acromegaly. *Science* **218**:585, 1982.

15. Rivier J, Spiess J, Thorner M, et al: Characterization of a growth hormone-releasing factor from a human pancreatic islet tumour. *Nature (Lond)* **300**:276, 1982.

16. Thorner MO, Rivier J, Spiess J, et al: Human pancreatic growth-hormone releasing factor selectively stimulates growth-hormone secretion in man. *Lancet* **1**:24, 1983.

17. Gelato MC, Pescovits O, Cassorla F, et al: Effects of a growth hormone releasing factor in man. *J Clin Endocrinol Metab* **57**:674, 1983.

18. Rosenthal SM, Schriock EA, Kaplan SL, et al: Synthetic human pancreatic growth hormone-releasing factor stimulates growth hormone secretion in normal man. *J Clin Endocrinol Metab* **57**:677, 1983.

19. Borges JLC, Blizzard RM, Gelato MC, et al: Effects of human pancreatic tumour growth hormone releasing factor on growth hormone and somatomedin-C levels in patients with idiopathic growth hormone deficiency. *Lancet* **2**:119, 1983.

20. Grossman A, Savage MO, Wass JAH: Growth-hormone-releasing factor in growth hormone deficiency: Demonstration of a hypothalamic defect in growth hormone release. *Lancet* **2**:137, 1983.

21. Borges JLC, Blizzard RM, Evans WS, et al: Stimulation of growth hormone (GH) and so-matomedin-C in idiopathic GH-deficient subjects by intermittent pulsatile administration of synthetic human pancreatic tumor GH-releasing factor. *J Clin Endocrinol Metab* **59**:1, 1984.

22. Schechter J, Kovacs K, Rimoin D: Isolated growth hormone deficiency: Immunocytochemistry. *J Clin Endocrinol Metab* **59**:798, 1984.

23. Rogol AD, Blizzard RM, Johanson AJ, et al: Growth hormone release in response to human pancreatic tumor growth hormone-releasing hormone-40 children with short stature. *J Clin Endocrinol Metab* **59**:580, 1984.

24. Gelato MC, Ross JL, Malozowski S, et al: Effects of pulsatile administration of growth hormone (GH)-releasing hormone on short term linear growth in children with GH deficiency. *J Clin Endocrinol Metab* **61**:444, 1985.

25. Thorner MO, Reschke J, Chitwood J, et al: Acceleration of growth in two children treated with human growth hormone-releasing factor. *N Engl J Med* **312**:4, 1985.

26. Phillips JA III: Genetic diagnosis: Differentiating growth disorders. Hosp Pract **20**:85, 1985.

27. Phillips JA III, Hjelle BL, Seeberg PH, et al: A molecular basis for familial isolated growth hormone deficiency. *Proc Natl Acad Sci USA* **78**:6372, 1981.

28. Phillips JA III: The molecular basis of familial growth hormone deficiency. In Messer A, Porter IH (eds): *Recombinant DNA and Medical Genetics.* Academic, New York, 1983, p. 9.

29. Phillips JA III, Parks JS, Hjelle BL, et al: Genetic analysis of familial isolated growth hormone deficiency Type I. *J Clin Invest* **70**:489, 1982.
30. Spiliotis BE, August GP, Hung W, et al: Growth hormone neurosecretory dysfunction: A treatable cause of short stature. *JAMA* **251**:2223, 1984.
31. Kowarski AA, Schneider J, Ben-Galim E, et al: Growth failure with normal serum RIA-GH and low somatomedin activity: Somatomedin restoration and growth acceleration after exogenous GH. *J Clin Endocrinol Metab* **47**:461, 1978.
32. Hayek A, Peake GT: Growth and somatomedin-C responses to growth hormone in dwarfed children. *J Pediatr* **99**:868, 1981.
33. Frazer T, Gavin JR, Daughaday WH, et al: Growth hormone-dependent growth failure. *J Pediatr* **101**:12, 1982.
34. Rudman D, Kutner MH, Blackston RD, et al: Normal variant short stature: Subclassification based on responses to exogenous human growth hormone. *J Clin Endocrinol Metab* **49**:92, 1979.
35. Rudman D, Kutner MH, Goldsmith MA, et al: Further observations on four subgroups of normal variant short stature. *J Clin Endocrinol Metab* **51**:1378, 1980.
36. Rudman D, Kutner MH, Blackston RD, et al: Children with normal-variant short stature: Treatment with human growth hormone for six months. *N Engl J Med* **305**:123, 1981.
37. Preece, MA: A new task for human growth hormone? *Br Med J* **283**:1145, 1981.
38. Van Vliet G, Styne DM, Kaplan SL, et al: Growth hormone treatment for short stature. *N Engl J Med* **309**:1016, 1983.
39. Friesen HG: Raben lecture: a tale of stature. *Endocr Rev* **1**:309, 1980.
40. Valenta LJ, Sigel MB, Lesniak MA, et al: Pituitary dwarfism in a patient with circulating abnormal growth hormone polymers. *N Engl J Med* **312**:214, 1985.
41. Gorden P, Hendricks CM, Roth J: Evidence for "big" and "little" components of human plasma and pituitary growth hormone. *J Clin Endocrinol Metab* **36**:178, 1973.
42. Abucham-Filho JZ, Czepielewski MA, Ribeiro SRR, et al: Abnormal growth hormone and dwarfism. *N Engl J Med* **313**:268, 1985.
43. Laron Z, Pertzelan A, Mannheimer S: Genetic pituitary dwarfism with high serum concentration of growth hormone—A new inborn error of metabolism? *Israel J Med Sci* **2**:152, 1966.
44. Laron Z, Pertzelan A, Karp M: Pituitary dwarfism with high serum levels of growth hormone. *Israel J Med Sci* **4**:883, 1968.
45. Laron Z, Pertzelan A, Karp M, et al: Administration of growth hormone to patients with familial dwarfism with high plasma immunoreactive growth hormone: Measurement of sulfation factor, metabolic and linear growth responses. *J Clin Endocrinol* **33**:332, 1971.
46. Laron Z, Kowadlo-Silbergeld A, Eshet R, et al: Growth hormone resistance. *Ann Clin Res* **12**:269, 1980.
47. Merimee TJ, Rimoin DL, Rabinowitz D, et al: Metabolic studies in the African pygmy. *Trans Assoc Am Physicians* **81**:221, 1968.
48. Merimee TJ, Rimoin DL, Cavalli-Sforza LL: Metabolic studies in the African pygmy. *J Clin Invest* **51**:395, 1972.
49. Rimoin DL, Merimee TJ, McKusick VA: Growth-hormone deficiency in man: An isolated, recessively inherited defect. *Science* **152**:1635, 1966.
50. Merimee TJ, Zapf J, Froesch ER: Dwarfism in the pygmy: An isolated deficiency of insulin-like growth Factor I. *N Engl J Med* **305**:965, 1981.
51. Rosenfeld RG, Hintz RL: Diagnosis and management of growth disorders. *Drug Therapy* **13**:61, 1983.
52. Rosenfeld RG: Evaluation of growth and maturation in adolescence. *Pediatr Rev* **4**:175, 1982.
53. Post EM, Richman RA: A condensed table for predicting adult stature. *J Pediatr* **98**:440, 1981.
54. Blethen SL, Chasalow FI: Use of a two-site immunoradiometric assay for growth hormone (GH) in identifying children with GH-dependent growth failure. *J Clin Endocrinol Metabl* **57**:1031, 1983.
55. Frisch H, Herkner K, Schober E, et al: Prolactin and thyrotrophin response to thyrotrophin-releasing hormone in growth hormone deficiency. *Arch Dis Child* **57**:769, 1982.

56. Bucher H, Zapf J, Torresani T, et al: Insulin-like growth factors I and II, prolactin, and insulin in 19 growth hormone-deficient children with excessive, normal, or decreased longitudinal growth after operation for craniopharyngioma. *N Engl J Med* **309**:1142, 1983.

57. Powell GF, Brasel JA, Blizzard RM: Emotional deprivation and growth retardation simulating idiopathic hypopituitarism. *N Engl J Med* **276**:1271, 1967.

58. Powell GF, Brasel JA, Raiti S, et al: Emotional deprivation and growth retardation simulating idiopathic hypopituitarism. II. Endocrinologic evaluation of the syndrome. *N Engl J Med* **276**:1279, 1967.

59. Raben MS: Treatment of a pituitary dwarf with human growth hormone. *J Clin Endocrinol Metab* **18**:901, 1958.

60. Tanner JM, Whitehouse RH, Hughes PCR, et al: Effect of human growth hormone treatment for 1 to 7 years on growth of 100 children, with growth hormone deficiency, low birthweight, inherited smallness, Turner's syndrome and other complaints. *Arch Dis Child* **46**:745, 1971.

61. Preece MA, Tanner JM, Whitehouse RH, et al: Dose dependence of growth response to human growth hormone in growth hormone deficiency. *J Clin Endocrinol Metab* **42**:477, 1976.

62. Milner RDG, Russell-Fraser T, Brook CGD, et al: Experience with human growth hormone in Great Britain: The report of the MRC working party. *Clin Endocrinol* **11**:15, 1979.

63. Aceto T, Frasier SD, Hayles AB, et al: Collaborative study of the effects of human growth hormone in growth hormone deficiency. I. First year of therapy. *J Clin Endocrinol Metab* **35**:483, 1972.

64. Frasier SD: Human pituitary growth hormone(hGH) therapy in growth hormone deficiency. *Endocr Rev* **4**:155, 1983.

65. Seip M, Trygstad O: Experiences with human growth hormone in pituitary dwarfism. *Acta Paediatr Scand* **55**:287, 1966.

66. Frasier SD, Aceto T, Hayles AB: Collaborative study of the effects of human growth hormone in growth hormone deficiency. V. Treatment with growth hormone administered once a week. *J Clin Endocrinol Metab* **47**:686, 1978.

67. Ferrandez A, Zachmann M, Prader A, et al: Isolated growth hormone deficiency in prepubertal children. Influence of human growth hormone on longitudinal growth, adipose tissue, bone mass and bone maturation. *Helv Paediatr Acta* **25**:566, 1970.

68. Dean HJ, Kellett JG, Bala RM, et al: The effect of growth hormone treatment on somatomedin levels in growth hormone-deficient children. *J Clin Endocrinol Metab* **55**:1167, 1982.

69. Rosenfeld RG, Kemp SF, Hintz RL: Constancy of somatomedin response to growth hormone treatment of hypopituitary dwarfism, and lack of correlation with growth rate. *J Clin Endocrinol Metab* **53**:611, 1981.

70. Preece MA, Tanner JM: Results of intermittent treatment of growth hormone deficiency with human growth hormone. *J Clin Endocrinol Metab* **45**:169, 1977.

71. Lippe B, Frasier SD, Kaplan SA: Use of growth hormone-gel. *Arch Dis Child* **54**:609, 1979.

72. MacGillivray MH, Kolotkin M, Munschauer RW: Enhanced linear growth responses in hypopituitary dwarfs treated with growth hormone plus androgen versus growth hormone alone. *Pediatr Res* **8**:103, 1974.

73. Parker ML: Antibody production in patients receiving human growth hormone and possible therapeutic implications. In: *Report of the Fifty-fourth Ross Conference on Pediatric Research, Baltimore, MD, March 1965*, p. 49.

74. Kaplan SL, Savage DCL, Suter S, et al: Antibodies to human growth hormone arising in patients treated with human growth hormone: Incidence, characteristics, and effects on growth. In: *Advances in Human Growth Hormone Research; A Symposium*. U.S. Department of Health, Education and Welfare, Publication No. NIH 74-612, 725, 1973.

75. Rosenbloom AL, Riley WJ, Silverstein JH, et al: Low dose single weekly injections of growth hormone: Response during first year of therapy of hypopituitarism. *Pediatrics* **66**:272, 1980.

76. Root AW, Bongiovanni AM, Eberlein WR: Diagnosis and management of growth retardation with special reference to the problem of hypopituitarism. *J Pediatr* **78**:737, 1971.

77. Lippe BM, Van Herle AJ, LaFranchi SH, et al: Reversible hypothyroidism in growth hormone-deficient children treated with human growth hormone. *J Clin Endocrinol Metab* **40:**612, 1975.
78. Rubio GR, Mellinger C, Zafar MS, et al: Evaluation of thyroid function during growth hormone therapy. *Metabolism 25:*15, 1976.
79. Cacciari E, Cicognani A, Pirazzoli P, et al: Effect of long-term GH administration on pituitary-thyroid function in idiopathic hypopituitarism. *Acta Paediatr Scand* **68:**405, 1979.
80. Brown P, Gajdusek DC, Gibbs CJ Jr, et al: Potential epidemic of Creutzfeldt-Jakob disease from human growth hormone therapy. *N Engl J Med* **313:**728, 1985.
81. Fatal degenerative neurologic disease in patients who received pituitary-derived human growth hormone. *MMWR* **34:**359, 1985.
82. Powell-Jackson J, Weller RO, Kennedy P, et al: Creutzfeld-Jakob disease following human growth hormone administration. *Lancet* **2:**244, 1985.
83. Koch TK, Berg BO, De Armond SJ, et al: Creutzfeldt-Jakob disease in a young adult with idiopathic hypopituitarism: Possible relation to the administration of cadaveric human growth hormone. *N Engl J Med* **313:**731, 1985.
84. Gibbs CJ Jr, Joy A, Heffner R, et al: Clinical and pathological features and laboratory confirmation of Creutzfeld-Jakob disease in a recipient of pituitary-derived human growth hormone. *N Engl J Med* **313:**734, 1985.

5

Thyrotropin and Thyrotropin-Releasing Hormone

Pituitary Thyrotropin (TSH)

Pituitary thyrotropin, or thyroid-stimulating hormone (TSH), as the name implies, is the principal regulator that governs the growth and function of the thyroid gland. The secretion, metabolism, regulatory control, and actions of this important glycoprotein form the basis of this chapter.

Structure and Chemistry of TSH

Thyrotropin is secreted by the pituitary thyrotropes. These cells represent 3–5% of the pituicyte population and can be recognized by their angular or polyhedral shape as well as the small size of their secretory granules. Thyrotropin is a glycoprotein with a molecular weight of 28,000 and consists of

TABLE 31
Subunits of TSH

Characteristic	α-Subunit	β-Subunit
Size	14,700 M_r	15,600 M_r
Biological activity	Inactive per se and in combination	Requires combination with α-subunit to exert bioactivity
Identity to other glycoproteins	Identical to α-subunits of LH, FSH, HCG	Nonidentical to the β-subunits of LH, FSH, HCG
Synthetic potential	Synthesized in molar excess	Free β-secretion minimal
Regulatory responses to T_3 or TRH	Concordant with TSH responses	Concordant with TSH responses

two subunits, α and β. These subunits are noncovalently linked.[1] The α-subunit is identical to the α-subunits of other glycoprotein hormones, i.e., luteinizing hormone (LH), follicle-stimulating hormone (FSH), and human chorionic gonadotropin (hCG). The β-subunit confers hormonal specificity. However, the β-chain per se does not display any biological activity, since it is essentially devoid of receptor binding, unless combined with the α-subunit. The α-subunit has a molecular weight of 14,700 and consists of 90 residues. The linear amino acid sequence of the α-subunit is identical to the α-subunits of LH, FSH, or hCG. The β-subunit of TSH consists of 110 amino acids with a molecular weight of 15,600. Three oligosaccharide moieties are attached to the peptide sequence of TSH. Studies using cell-free biosynthetic systems have suggested that TSH is synthesized as separate pre-α- and pre-β-subunit precursors.[2,3] These precursors are translocated to the rough endoplasmic reticulum, where the subunits are cleaved from their respective precursor peptides and processed into the complete hormone, TSH. This event involves the processing of the carbohydrate units linked to asparagine, a prerequisite for combination of the α- and β-subunits. The potential of TSH to synthesize the α-subunit is believed to be slightly greater than its potential to secrete the β-subunit. The relative proportion of the α-subunits secreted, as well as the ratio of the α-subunit to the intact TSH molecule assume significance when dealing with TSH-secreting tumors, as we shall see. The α- and β-subunits of TSH respond physiologically to regulatory influences such as thyroxine (T4) administration or TRH administration in a fashion concordant with that of intact TSH. This behavior too will assume significance in TSH-secreting tumors. The differences between the α- and β-subunits of TSH are summarized in Table 31.

Synthesis and Secretion of TSH

The stimulus for TSH secretion comes from the hypothalamic tripeptide thyrotropin-releasing hormone (TRH). TRH binds to high-affinity plasma membrane receptors on the thyrotrope. The precise mechanism by which TRH induces TSH secretion is somewhat unclear. The old theory of cyclic adenosine monophosphate (cAMP) dependency has given way to the notion that TRH-mediated TSH secretion (and release) is mediated by an increase in the intracellular calcium. (This theory of stimulus-secretion coupling is discussed under the action of TRH). TRH stimulation results in augmented incorporation of [^{14}C]alanine and [^{14}C]glucosamine. Subsequent to synthesis, the TSH is concentrated into condensed granules in the Golgi cysterni. Release of hormone occurs through a process of fusion with the plasma membrane and extrusion. The provocative effects of TRH on the secretion of TSH are powerfully modified by the circulating levels of thyroidal hormones. In vitro, the TRH-induced TSH secretion can be inhibited, even abolished, by addition of thyroid hormones or depletion of calcium in the medium.

Metabolism of TSH

Like all hormones, the ambient concentrations of TSH in the plasma reflect an equilibrium between the secretion on one hand and the clearance (by degradation or distribution) on the other. The plasma TSH level ranges between 0.5 and 8 µU/ml. Several observations pertain to the basal level of TSH measured in blood:

1. Most assays employed to measure TSH are not sensitive enough to detect levels below 0.5 or 1 µU/ml. This limits the use of the assay in differentiating normal from suppressed or deficient states.
2. In the basal state, TSH is relatively constant and nonpulsatile, in contrast to other pituitary hormones, such as growth hormone (GH), ACTH, and prolactin.
3. The TSH levels demonstrate a tendency to circadian periodicity, higher levels being noted toward the end of the working hours, near the onset of sleep. The low daytime levels and the gradual decline in the basal TSH levels over the morning are probably due to settling of the nocturnal elevation of TSH secretion.
4. The basal TSH levels bear a reciprocal relationship to the ambient concentrations of the circulating thyroid hormones (i.e., even a slight elevation of thyroidal hormone levels, while being well within the normal range decreases the TSH level in plasma and vice versa).
5. Several agents, in particular dopamine and glucocorticoids, can result in lowering of TSH levels in the plasma.
6. The common disorders characterized by an abnormal basal TSH level are outlined in Table 32.

The common clinical indications to obtain measurement of basal TSH level are as follows:

1. *To establish the etiology of hypothyroidism:* Primary hypothyroidism is associated with elevated TSH levels, whereas pituitary or hypothalamic hypothyroidism is associated with a lack of elevation of TSH. In the latter situation, even the demonstration of a normal TSH is indicative

TABLE 32
Disorders Characterized by Abnormal Basal TSH Levels

Elevated basal TSH	Decreased basal TSH
Primary hypothyroidism	Hyperthyroidism
Compensated euthyroidism	Hypopituitarism
Neonatal period (first day)	Depression
TSH-secreting tumors	Anorexia nervosa
Peripheral resistance to	Drugs (thyroid hormones,
thyroid hormones	L-dopa, dopamine)
Malnutrition	Glucocorticoids

of a breakdown in the normal reciprocal relationship between thyroid hormones and TSH.

2. *To demonstrate adequacy of therapy with L-thyroxine for replacement or suppressive purposes:* In patients receiving replacement therapy for primary hypothyroidism, the maintenance dose of L-thyroxine is one that normalizes the elevated basal TSH seen prior to initiation of therapy. When L-thyroxine is used for suppressive purposes, the aim is to achieve complete suppression of TSH below normal.

3. *To evaluate euthyroid goiters:* Compensated thyroid function can account for euthyroidism and a goiter at the expense of elevated TSH. Such is the situation in patients with compensated Hashimoto's thyroiditis, iodine-deficient goiters, defects in thyroid hormonogenesis, and certain instances of peripheral resistance to thyroidal hormones.

It should be noted that measurement of basal TSH is seldom indicated in the evaluation of the hyperthyroid patient unless unusual disorders are suspected (e.g., hyperthyroidism secondary to excessive secretion of pituitary TSH or hypothalamic TRH).

The TSH measured in the circulation by conventional RIA measures the intact molecule. The α-TSH represents 3–7% of the total circulating TSH and can be measured by separate subunit assays, with normal concentrations in the serum ranging between 0.5 and 9 ng/ml. β-TSH is generally unmeasurable in normal patients. In patients with primary hypothyroidism, all three forms of TSH, the intact hormone, the α- and β-subunit, are measurably elevated. The response of all three forms of TSH to TRH is augmented in primary hypothyroidism. By contrast, patients with TSH secreting tumors of the pituitary demonstrate elevation of only intact TSH and α-TSH, which do not respond to TRH.[4–6]

The secretion and release of TSH by the pituitary gland are controlled by two opposing forces, the negative effect of circulating thyroid hormones on the one hand and the stimulatory effects of the hypothalamic tripeptide thyrotropin-releasing hormone (TRH) on the other (Fig. 17). The roles of somatostatin and dopamine in the physiological regulation of TSH secretion are not entirely clear. Physiologically, in the presence of a normal hypotha-

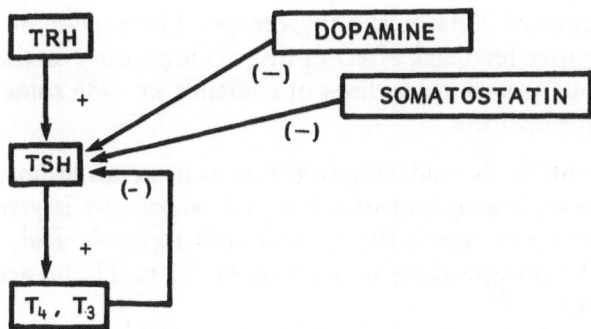

FIGURE 17. Regulation of TSH secretion.

lamic-pituitary-thyroid (HPT) axis, the plasma TSH bears a striking reciprocal relationship to the free thyroid hormone levels. The level of TSH in the plasma at any time is determined by the interaction between the negative feedback of thyroid hormones on the thyrotrope and the positive trophic effect of TRH on the thyrotrope. Thyrotropin-releasing hormone is a modified tripeptide (pyroglutamyl-histidyl-prolinamide) synthesized by the hypothalamus. It reaches the anterior pituitary via the hypophyseal portal blood vessels and binds to specific receptor sites on the thyrotrope membrane. It causes an increase in the synthesis and release of TSH by activating the cAMP systems within the thyrotrope or more likely by increasing the cytoplasmic free calcium within the thyrotrope. The importance of TRH in preserving the functional integrity of the thyrotrope is evidenced in tertiary hypothyroidism, a condition characterized by failure of the hypothalamus to generate TRH. In this disorder, hypothyroidism develops as a consequence of failure of the thyrotrope deprived of its hypothalamic drive.

Regulatory Control of TSH

Thyroid Hormones

Feedback regulation by thyroid hormones is the dominant force involved in control of TSH secretion. Administration of T4 or T3 to euthyroid patients is characteristically associated with two changes: first, a decrease in the resting level of TSH, and second, a blunting of the provocative effect of TRH on TSH. Conversely, a decrease in the circulating level of thyroid hormones is associated with a rise in the resting level of TSH as well as an exaggerated response of TSH to the administration of TRH. Thus, it appears that both the basal TSH level and the ability of the thyrotrope to respond to exogenous TRH are modified by the circulating level of thyroid hormones.[7-10]

The suppressive effects of thyroid hormones on TSH secretion and release are believed to be mediated primarily at the level of the pituitary gland rather than at the hypothalamic level. Whereas short-term administration of thyroid hormones decreases the release of TSH, long-term administration suppresses synthesis of TSH by the thyrotrope. The precise mechanism involved in the negative feedback effect of thyroid hormones on the thyrotrope is unclear, but four experimental lines of evidence provide some insight into the yet-unclear mechanisms.

1. Experiments in thyroidectomized rats indicate an important role for intrapituitary conversion of T4 to T3. When T4 is given to thyroidectomized rats previously treated with iopanoic acid, a drug that inhibits the intrapituitary conversion of T4 to T3, no acute decrease in TSH occurs.[11]

2. Several in vitro studies have demonstrated the ability of normal pituitary homogenates to convert T4 to T3. It is hypothesized that the binding of T3 to the nuclear receptors of the thyrotrope is important in initiating the acute suppression of TSH release following T4 or T3 administration.[12,13]

3. There is also in vitro experimental evidence to indicate that the administration of thyroid hormones is associated with a reduction of TRH receptors on the thyrotropes.[14,15]

4. Finally, there is the possibility that the nuclear-bound T3 induces the formation of a protein that interferes with the response of the thyroptrope to TRH.[16]

Whatever the mechanism, administration of thyroid hormones consistently lowers the TSH level and impairs the response of the thyrotrope to exogenous TRH. The relative influences of TRH and the thyroid hormones on the pituitary are balanced in a delicate fashion. Although the thyrotrope depends on TRH for stimulation and secretion of TSH, an excess of thyroid hormones can obliterate the effect of TRH on the secretion and release of TSH by the pituitary. This principle forms the basis of several dynamic tests using TRH.

TRH

The physiological control of TSH by TRH will be discussed at length in Section 2.3.

Hypothalamic Regulators Other Than TRH

Admittedly, TRH is the major hypothalamic influence that controls TSH secretion by the thyrotrope. There are, however, two other hypothalamic regulators of TSH secretion. Both hypothalamic dopamine and somatostatin possess an inhibitory effect over TSH secretion by the thyrotrope. It is controversial as to how important and physiological these influences are in the day-to-day regulation of TSH secretion and release.

Dopamine. The concept that a neurotransmitter such as dopamine can be an inhibitor of hormone secretion is hardly new. In fact, the major physiological regulation of prolactin is indeed mediated by a tonic inhibitory control exerted by hypothalamic dopamine (prolactin-inhibitory factor). Since the thyrotrope and the lactotrope share several functional similarities, and are linked by their tendency to respond to TRH, it seemed reasonable to suspect dopaminergic inhibition of the thyrotrope, analogous to the lactotrope. Several investigational lines of evidence suggest that dopamine exerts inhibitory control over thyrotrope function:

1. The administration of dopamine to healthy euthyroid subjects is associated with a significant drop in circulating TSH levels.[17,18] Similar results have been obtained with the use of dopamine agonist drugs, such as L-dopa.[19]

2. The use of dopamine receptor antagonists (such as metoclopramide) in primary hypothyroidism is associated with a rise in the circulating TSH level[20] and the use of dopamine receptor agonists (such as bromocriptine) is associated with a fall in circulating TSH level.[21] These findings are strikingly reminiscent of the response of circulating prolactin levels to administration of metoclopramide and bromocriptine.

3. These observations have been extended to normal euthyroid subjects as well; for instance, Scanlon et al.,[22] using a sensitive and precise radioimmunoassay for human TSH, demonstrated significant elevations in circulating TSH levels in euthyroid subjects following metoclopramide. As expected, the incremental rise was greater in females and inversely related to the basal TSH level. This latter finding suggests that dopamine may be one physiological determinant of the low daytime TSH level in the circulation.

4. The prior treatment with dopamine antagonists significantly enhances the TSH response to TRH, analogous to the effect of lowering the level of thyroidal hormones in circulation.

5. The greatly enhanced TSH response to TRH in patients with primary hypothyroidism may well be a reflection of dopaminergic depletion and the consequent release of the thyrotropes from the inhibitory clutches of dopamine. An identical phenomenon with prolactin is a well-accepted notion.

6. As with intact TSH, the free α- and β-subunits are also under inhibitory dopaminergic control. Subunit release after dopamine antagonism using metoclopramide was studied by Scanlon et al.,[23] who demonstrated that subunit release following metoclopramide is a phenomenon seen most clearly in subjects with primary hypothyroidism, where the dominant negative feedback of thyroid hormones is abolished. The more severe the hypothyroidism, the greater the subunit response to dopamine blockade. In this respect, the effect of dopamine blockade on subunit release in patients with primary hypothyroidism is analogous to the effect of TRH on subunit release.

The above observations, and the bulk of evidence irrefutably point to the mediation of dopamine in the inhibitory control of TSH secretion. The effect is of course, more noticable in patients with primary hypothyroidism. The physiological role of dopamine in the day-to-day regulation of TSH secretion is still unclear.

Somatostatin. Siler et al.[24] were among the earliest to demonstrate that concomitant administration of somatostatin and TRH resulted in significant blunting of the TRH-induced TSH release, without affecting the simultaneous release of prolactin. While the inhibitory effect of somatostatin on TRH-induced TSH release is unquestioned, again, the physiological significance of such an effect on the day-to-day regulation of TSH secretion is unclear. The clinical significance of such a phenomenon assumes importance in the following pathological circumstances:

1. The therapeutic administration of growth hormone (to treat GH deficiency) is often associated with blunting of TSH responsiveness to TRH.[25,26] Such a phenomenon may reflect a contemporaneous increase in somatostatin as a consequence of the positive feedback loop between GH and hypothalamic somatostatin.
2. For the same reason, a blunted (or even abolished) TSH response to TRH may be encountered in patients with acromegaly. The phenomenon is reversible with restoration of eusomatotropism.[27]
3. Reversible blunting of TRH-induced TSH release may be encountered in patients harboring somatostatinoma.

Action of TSH

Thyrotropin (TSH) stimulates thyroid function. It stimulates function and growth of the thyroid gland. The acute effects of TSH enhance and stimulate all facets of thyroid hormonogenesis as well as release. The effect of TSH on iodide transport is biphasic. In the early phases, there is an efflux of iodide out of the gland, but later there is a tremendous enhancement of iodide uptake by the gland. Thyrotropin greatly activates organification, iodide binding, coupling, and formation of thyroxine (T4) and triiodothyronine (T3). Histologically, colloid droplet formation is the hallmark of thyroid stimulation by TSH (or any other thyroid stimulator). The delayed effects of TSH on the thyroid gland relate to effects on growth, resulting in increased volume and number of cells as well as increased protein and nucleotide synthesis by the thyroid follicular cells.

The precise mechanisms involved in the thyroid regulation by TSH have been studied extensively. The predominant mechanism is by activation of the adenylate cyclase, cAMP system. In addition, the cyclic guanosine monophosphate (cGMP) system (coupled with calcium ions) is intimately involved in the mediation of TSH action. The role of iodide in modifying the action of TSH has also been well studied. It is recognized that in the presence of iodine deficiency there is enhancement of trapping (of iodine), a TSH-mediated effect, and in the presence of excess iodide there is inhibition of several effects of TSH on the thyroid gland. It is believed that following its trapping and oxidation, part of oxidized iodide is transformed into a compound called compound XI. This compound XI inhibits adenylate cyclase (or activates phosphodiesterase, which degrades cAMP) and thus negatively controls TSH-induced cAMP accumulation. In addition to iodine-mediated thyroregulation, the thyroid gland probably possesses alternate autoregulatory mechanisms, including changes in TSH receptors. For example, after maximal stimulation by TSH, the thyroid cells resist further stimulation by TSH but respond to other stimulators such as prostaglandin E_1 (PGE_1). Thus, it appears that although thyroid function is predominantly regulated by pituitary TSH, this can be modified at a local level by autoregulatory mechanisms that involve iodine and to a lesser extent other substances such as prostaglandins.

In summary, pituitary thyrotropin is controlled by dual mechanisms: negative feedback regulation by the circulating thyroid hormones and positive trophic regulation by hypothalamic TRH. Physiologically, TSH bears a reciprocal relationship to the level of thyroid hormone. The negative feedback exerted by thyroid hormones is a direct effect on the pituitary, probably dependent on the conversion of T4 to T3 within the pituitary gland. The trophic effect of TRH on the thyrotrope is blunted or lost in the presence of excess circulating thyroid hormones. The primary effect of TSH on the thyroid is to enhance function and promote growth under certain circumstances. Although the thyroid gland is driven mainly by TSH, there are built-in autoregulatory mechanisms that modify the effect of TSH on the thyroid.

Thyrotropin-Releasing Hormone

The pituitary thyrotrope is controlled by the elegant interplay between the negative feedback effect exerted by the thyroidal hormones, and the trophic effect exerted by TRH. A basic understanding of the physiology of TRH is essential for the understanding of the role of this peptide in altered states of thyrotrope function.

Chemistry and Distribution of TRH

Historically, TRH was the first hypothalamic factor to be isolated and characterized from ovine and porcine hypothalami.[28,29] TRH is a tripeptide and is often referred to by its chemical structure, pyroglutamyl histidylprolinamide.[30,31] The synthesis of TRH is carried out in several hypothalamic nuclei, from which it is secreted into the portal venous system to reach its target cells, the thyrotropes of the anterior pituitary. Distribution studies, using immunohistochemical techniques, have provided evidence in delineating the preoptic suprachiasmatic nucleus, and the periparaventricular area as the thyrotropic area in the rat hypothalamus.[32] Jackson and Reichlin[33] demonstrated that ablation of this area in the rat is accompanied by depletion of nearly all hypothalamic TRH, with the consequent development of hypothalamic hypothyroidism. The TRH, thus secreted, is transported via the

TABLE 33
Extrahypothalamic Distribution of TRH

Site of TRH distribution	Investigators
Posterior pituitary	Jackson and Reichlin[34]
Extrahypothalamic (motor nuclei of brain stem, motor neurons of the spinal cord)	Johansson et al.[32]
Medulla oblongata	Hokfelt et al.[35]
Ventral horn of spinal cord	Hokfelt et al.[35]
Thalamus, cerebral cortex	Koch and Okon[36]
Limbic system and cerebral cortex	Ogawa et al.[37]
Retina	Martino et al.[38]

stalk to the thyrotropes of the anterior pituitary. Although the pituitary thyrotrope can be regarded as the major target cell for the action of TRH, the distribution studies for demonstrating TRH have yielded intriguing results regarding the ubiquitous nature of this peptide.[32-38] Table 33 outlines several locations at which TRH or TRH-like material has been demonstrated.

The implications of these studies are threefold. First, the widespread nature of TRH in the CNS suggests that endogenous TRH must have physiological importance in the brain tissue of primates. In fact, the TRH content of extrahypothalamic neural tissue far exceeds the TRH content of the hypothalamus. Second, a role of TRH in the causation of such widely disparate entities such as endogenous depression,[39] Huntington's chorea,[40] and spinal shock[41] has been postulated but not unequivocally proven. Third, the ubiquitous nature of this peptide may be related to its ability to influence neurotransmitter action in the brain. For instance, TRH potentiates the excitatory effects of acetylcholine (ACh) on cortical neurons, and perhaps influences other neurotransmitters, such as norepinephrine, and dopamine, as well. Recent workers[42,34] have taken this concept even a step further, conferring the role of neurotransmitter to TRH. Hokfelt et al.[43] localized TRH, by immunocytochemical studies, to presynaptic nerve terminals, while Maeda and Frohman[44] convincingly demonstrated release of TRH from synaptosomal preparations in response to depolarizing stimuli. The consensus of opinions is that TRH functions in the brain as a neurotransmitter (in the classic sense) or as a neuromodulator in some instances.[45] In addition to neuronal tissue, TRH has been located in the GI tract and pancreas.

Metabolism of TRH

TRH is rapidly degraded in body fluids by enzymatic breakdown.[46,47] The enzymes responsible for metabolizing TRH are the TRH peptidases and the TRH amidases. The two major breakdown products resulting from metabolism of TRH are histidylproline-diketopiperazine, and deamido-TRH. It is of importance to recognize that the metabolites of TRH are not inert, and do possess biological activity. Table 34 outlines the reported actions of TRH metabolites. It is interesting to note that although none of the metabolites possesses TSH-releasing activity, the His-Pro-diketopiperazine derivative may demonstrate, in some areas, counteractivity to that of the parent compound.

Action of TRH

The actions of TRH can be viewed in terms of three perspectives: (1) effects on the pituitary gland, particularly the thyrotrope, and the lactotropes; (2) effects on the CNS; and (3) effects on the GI tract. Before evaluating these effects, it is necessary to review some new insights into the mechanism of TRH action. Since the primary target cells of TRH are the thyrotropes, it is not surprising that most of the in vitro work done to evaluate the mechanism of TRH action has employed pituitary tissue. Earlier postulates indicated that the TRH-mediated TSH release was consequent to the effect of TRH on directly stimulating the adenyl cyclase–cAMP system.[48] Recent work on the effects of TRH on TSH secretion and release would lead us to believe otherwise. The role of cytoplasmic free calcium in effecting the action of TRH on the thyrotrope is gaining acceptance as the mechanism of TRH action. The process of stimulus–secretion coupling has been extended to explain the action of TRH.[49] It is believed that TRH causes a rapid elevation of cytoplasmic free calcium within the thyrotropes.[50–53] This increase in cytoplasmic free calcium can be brought about by influx of extracellular calcium or by mobilization of intracellular calcium. The importance of the calcium-mediated stimulus–secretion coupling is brought to focus employing experiments that are aimed at inhibiting calcium influx. When verapamil, a calcium-channel blocker, is employed to block the influx of calcium into the anterior pituitary cells, there is virtual abolition of thyrotropin release and a partial inhibition of TRH-stimulated TSH release.[54]

Having established the importance of cytoplasmic free calcium as a major determinant of the TRH-mediated TSH release, the mechanism by which this is brought about merits reference. Recent data implicate an important role for an intracellular phospholipase C, which brings about the hydrolysis of phosphatidylinositol 4,5-biphosphate, resulting in the yield of inositol triphosphate and diacylglycerol.[55] Inositol triphosphate is believed to cause re-

TABLE 34
Actions of TRH Metabolites

His-Pro-Diketo piperazine
  Inhibits prolactin release
  Causes hypothermia
  Antagonizes ethanol narcosis
Deamido-TRH
  Causes symptoms similar to opiate withdrawal

lease of calcium from a nonmitochondrial pool. Martin[56] demonstrated that TRH rapidly activates the hydrolysis of polyphosphoinositides within the pituitary cells. The bulk of evidence points to the possibility that inositol triphosphate behaves like a second messenger coupling TRH stimulation to the secretion of TSH and prolactin. Within this framework, the clinical effects of TRH on the thyrotrope and the lactotrope are reviewed first.

The principal effect of TRH is to stimulate the secretion and release of TSH from the pituitary thyrotrope. The primary regulation of TSH secretion is modulated by the negative feedback exerted on the thyrotropes by the circulating thyroidal hormones. The role of TRH may be regarded as being the determinant of the set point of the interaction between TSH and the thyroid hormones.[57] The physiological response of the thyrotrope to an exogenous bolus of TRH is rather characteristic.[58] Following the intravenous administration of 200–500 μg of TRH, there is a brisk rise in the serum TSH reaching a peak within 15–30 min. The following general observations pertain to the TRH-induced TSH response.

1. The TRH-induced TSH response is dose dependent, to a point. The minimal dose of TRH required to elicit a TSH response can be as low as 10 μg IV. The magnitude of response is dose dependent, up to 400 μg.[59] Above this dose, there is no further increment in TSH release, probably reflecting maximal stimulation of the thyrotrope.
2. When the pituitary thyroid axis is intact, a rise in triiodothyronine can be seen within 90–150 min following intravenous TRH.
3. The magnitude of TSH response following TRH administration is directly proportional to the basal TSH concentrations, i.e., the higher the basal TSH, the greater the magnitude of response.
4. Sex and age influence the magnitude of the TSH response to TRH administration. The response in females is greater than in males, signifying an enhancing effect of estrogen on the thyrotrope. Also, in males there is a decline in the TRH-induced TSH response with age.[60]
5. The TSH response to TRH can be modified by nonthyroidal drugs, the most notable ones being dopamine and L-dopa. These drugs attenuate the TSH response to TRH.
6. Other hormones, in particular, glucocorticoids and GH, can blunt the TSH response to TRH. Thus, hypercortisolemic states and hypersomatotropism are associated with reversible impairment in the response of the thyrotrope to exogenously administered TRH.
7. Nonthyroidal illnesses in euthyroid subjects can affect the ability of the thyrotrope to release TSH in response to TRH. Three situations in particular are notable: endogenous depression,[61] renal failure, and starvation.

In addition to releasing TSH, the exogenous administration of TRH is associated with a brisk rise in prolactin. It is an interesting observation that the prolactin response to TRH may be encountered at doses too small to evoke a TSH response. It is generally regarded that TRH does not have a major role in the day-to-day regulation of prolactin secretion.[62] The TRH-induced prolactin release is also modulated by the level of circulating thyroidal hormones; thus, an augmented response is seen in patients with hypothyroidism, while an attenuated prolactin response is encountered in hyperthyroidism.[63]

In normal subjects, TRH administration has no effect on the release of GH, ACTH, or gonadotropins. In pathological states, however, administration of TRH is associated with a release of hormones other than TSH or prolactin. Thus, a paradoxical rise in GH level following TRH administration is seen in a significant proportion of acromegalics,[64] and a paradoxical rise in ACTH may be encountered in some patients with Cushing disease and Nelson syndrome.[65]

Diagnostic Applications of TRH Stimulation Test

The indications for performing a TRH study are outlined in Table 35. Each of these indicators deserve brief comment:

1. *To establish the diagnosis of hyperthyroidism in general and Graves' hyperthyroidism in particular:* When the clinical features of hyperthyroidism are mild and the laboratory data are borderline, the demonstration of an absent or blunted TSH response to TRH establishes the diagnosis of hyperthyroidism.[66,67] It must be remembered, however, that there are several conditions characterized by a blunted or absent TSH response to TRH (Table 36). The phenomenon of a blunted or even absent TSH response to TRH in the absence of hyperthyroidism (or hypopituitarism) is an intriguing one. Four examples of this phenomenon are briefly reviewed: euthyroid Graves' disease, acromegaly, hypercortisolism, and nonendocrine illness. The prototype of such a situation is euthyroid Graves' disease. This term signifies patients with clinical and biochemical euthyroidism presenting with ophthalmopathy or thyromegaly. In nearly 60–70% of such patients, the TSH response to TRH is blunted or absent,[68,69] despite circulating thyroid hormone levels in the perfectly normal range. This phenomenon reflects the exquisite sensitivity of the thyrotrope to minimal increments in the circulating thyroidal hormone levels,

TABLE 35
Indications for the TRH Study

To establish the diagnosis of hyperthyroidism
To evaluate TSH reserve in the presence of pituitary tumors
To establish the diagnosis of compensated primary
    hypothyroidism
To differentiate secondary from tertiary hypothyroidism
To ensure adequacy of suppressive therapy with L-thyroxine
To support the diagnosis of euthyroid Graves' disease
To evaluate acromegaly

TABLE 36
Conditions Characterized by Blunted or
Absent TSH Response to TRH

Hyperthyroidism
Euthyroid Graves' disease
Hypopituitarism
Hypercortisolemic states
Acromegaly, growth hormone therapy
Depression
Starvation
Chronic renal failure
Stress
Sick euthyroid syndrome
Advanced age (males)

although still remaining in the normal range. Alternately, it has been suggested that an inherent abnormality may be present in the thyrotrope of patients with euthyroid Graves' disease.[70]

The blunted TSH response to TRH in acromegalics is a well-recognized phenomenon.[64,71–73] Although the mechanism for such a phenomenon is at best speculative, one plausible explanation invokes the mediation of somatostatin. Experimental evidence in animals strongly favors the notion that GH exerts an ultrashort positive feedback control on hypothalamic somatostatin.[74,75] If so, the chronically elevated GH levels in acromegaly are conducive to elevating hypothalamic somatostatin, a well-known factor that attenuates the TSH response to TRH. This hypothesis is supported by the findings of Cobb et al.,[76] which indicate that the GH secretory status is a determinant of the TSH response to TRH in euthyroid subjects. Regardless of the mechanism, the message is clear: the demonstration of a blunted TSH response to TRH in an acromegalic need not always be indicative of structurally impaired thyrotrope reserve. The normalization of such a response upon restoration of eusomatotropism is further attestation to the functional nature of such a phenomenon.

The blunted TSH response to TRH in hypercortisolemic states is another example of one hormone (in this instance cortisol) affecting the response of another (TSH) to its physiological stimulus. Otsuki et al.[77] demonstrated that corticosteroids in large doses, over a protracted period, reduce the TSH response to TRH. Similarly, regardless of etiology, patients with Cushing's syndrome, demonstrate an impaired thyrotropin response to TRH. It is believed that excess cortisol may have direct suppressive effects on both the pituitary and the hypothalamus.

The blunted or absent TSH response to TRH in nonendocrine disorders constitutes an important entity, since its nonrecognition may result in establishing a mistaken diagnosis of endocrine disease. The major disorders that fall into this category are psychiatric illness (particularly endogenous depression), stress (particulary acute illness), chronic renal failure, and malnutrition.

The absent TSH response to TRH in endogenous depression is regarded by some workers as a marker of that disease.[61] Indeed, restoration of a normal TSH response to TRH may be seen with beneficial responses to treatment for depression.[78] This is not surprising given the fact that affective illness is associated with a plethora of reversible abnormalities of the hypothalamic pituitary axis. Maeda et al.[79] extended these observations to denote that in addition to a blunted TSH response to TRH, some patients with depression demonstrate a paradoxical GH response to TRH. The mechanism for the blunted TSH response to TRH is disputed. The hypercortisolemia seen in association with endogenous depression may, in part, be responsible for the reversible blunting of the TSH response to TRH seen in some patients with affective disorders. More intriguing is the observation by Spratt et al.[80] that of 645 patients admitted to an acute psychiatric disorders unit, 33% had elevated free T4 concentrations in their serum, several of these patients demonstrating a reversible attenuation of TSH response to TRH. It must be concluded that patients with psychiatric disorders represent a vastly heterogeneous group with a wide degree of variability in the behavior of the thyrotrope to TRH stimulation.

The blunted TSH response to TRH in critically ill patients is also a well recognized entity.[81–85] The mechanism(s) underlying such a phenomenon as well as its prognostic significance have not yet been completely elucidated. Since decreased renal function and the use of several drugs (particulary glucocorticoids and dopamine) that affect the TRH–TSH axis dominate the clinical setting of these patients, it is difficult to separate these effects from those of acute illness on the hypothalamic pituitary axis. Malnutrition and starvation are also important determinants that affect the TSH response to TRH.[86] Starvation is associated with a significant decrease in serum TSH.[87] Experiments in chronically starved rats indicate blunting of TSH response to TRH.[88]

2. *To evaluate TSH reserve.* In the presence of an intrasellar tumor, the demonstration of loss of TSH response to TRH is indicative of compromised thyrotrope reserve. Since TRH also stimulates release of prolactin, the adequacy of two hormones (TSH, prolactin) can be assessed by a single stimulus[89,90] (Fig. 18).

3. *To establish the diagnosis of primary hypothyroidism.* The TRH test has found application when hypothyroidism is mild or compensated with only marginal changes in the thyroxine or T3 values. In such instances, even with equivocal basal TSH levels, the response of TSH to TRH is exaggerated, establishing the diagnosis of subtle subclinical primary hypothyroidism.[91,92]

4. *To differentiate between pituitary (secondary) and hypothalmic (tertiary) hypothyroidism.* The TSH response to TRH in pituitary hypothyroidism is blunted or absent, whereas in hypothalamic hypothyroidism the response is delayed but sustained[93,94] (see Chapter 7).

5 *To ensure adequacy of suppressive therapy with L-T4.* This situation is especially important in patients with differentiated thyroid carcinoma on suppressive doses of L-T4. The adequacy of dosage can be determined by demonstrating abolition of the TSH response to intravenous TRH.

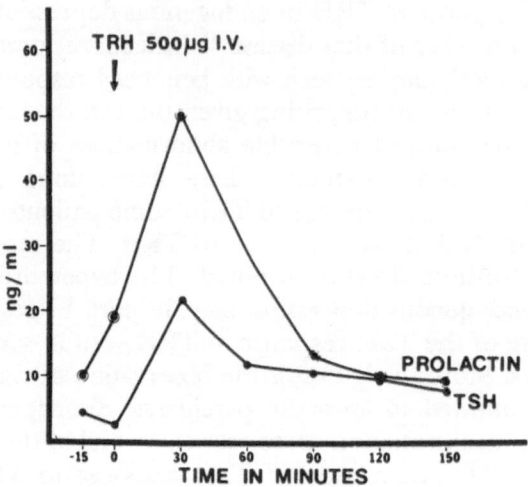

FIGURE 18. Effect of intravenous TRH administration on plasma levels of TSH and prolactin. (From Reichlin.[110])

6. *To establish the Gravesian nature of ophthalmopathy in euthyroid Graves' disease.* Patients with euthyroid Graves' disease (EGD) often demonstrate an abnormal TSH response to TRH. The usual abnormality is an impaired or blunted TSH response to TRH. This abnormality is encountered in approximately 60% of patients with euthyroid Graves' disease. Such a response, in the face of the euthyroid status, reflects the exquisite sensitivity of the thyrotrope to minimal increments in circulating thyroid hormone levels, albeit within the normal range.[60]

7. *To provide strong supportive evidence for hypersomatotropism.* The TRH test has diagnostic and prognostic value in acromegaly. Nearly 80–85% of acromegalics demonstrate an abnormal HGH response to TRH administration (normally TRH does not cause release of GH). There is often a dramatic and temporally related increase in the growth hormone levels of acromegalics 30–60 min following an intravenous bolus of TRH. When positive, the test strongly supports the diagnosis of acromegaly and can be used as a diagnostic tool and a parameter to predict the response to therapy. Restoration of eusomatotropism following surgery or radiation is associated with normalization of this abnormal response. Finally, acromegalics with the abnormal HGH response to TRH tend to respond more favorably to drug therapy with the ergot alkaloid bromocriptine. This paradoxical rise in GH in acromegalics following administration of TRH is thought to be a reflection of the nonspecific response of tumor cells to a variety of hypothalamic releasing factors. It is interesting to note that TRH can release growth hormone from rats in whom the pituitary is surgically separated from the hypothalamus and transplanted under the renal capsule.[95] A paradoxical rise in GH following TRH is not unique to acromegaly. A similar paradoxical response may be encountered in some patients with renal failure, depression, anorexia nervosa, and

chronic liver disease. The possibility of an "altered neurotransmitter milieu," as the underlying common factor for the paradoxical rise in GH seen in all these states is an attractive speculation.

Procedure for Performing the TRH Study

The sampling of blood for hormone determinations following TRH depends on the indication for which the test is performed. In its most simplistic form, when the test is done to document hyperthyroidism, all that are needed are a sample for TSH before and one 30 min after an intravenous bolus of TRH (short TRH test). If the test is performed to assess the pituitary reserve, samples are required at 30, 60, 90, and 120 min, and the samples are assayed for TSH as well as prolactin. If the test is performed to differentiate between pituitary and hypothalamic disease, the sampling may have to be extended to 180 min. The test is generally well tolerated. Nausea, desire to micturate, and lightheadedness are the usual side effects. Although TRH is effective orally, subcutaneously, or intramuscularly, the peak levels are achieved by the intravenous route following a bolus and is usually evident in the plasma 30 min following TRH. The normal response in our laboratory is a 8–20-μU/ml increment over the basal TSH value, or at least a threefold rise of TSH over the basal value. The magnitude of the response of TSH to TRH is logarithmic and dose related up to the 400-μg dose; most laboratories use 200–400 μg TRH. The magnitude of the TSH response is directly related to the basal concentration of TSH in plasma.

CNS Effects of TRH

Although TRH crosses the blood–brain barrier poorly, the effects of this peptide on the CNS are evident when administered intravenously. The CNS effects of TRH, it must be pointed out, are more impressive in animal experiments than in humans. The most notable effects on CNS are the following:

1. TRH appears to block some of the effects of endogenous opioid peptides.[96]
2. Since TRH appears to act in vivo as a partial opiate antagonist that spares the analgesic systems, its effect in facilitating recovery from spinal trauma had been studied by Faden et al.[41] These workers showed that the neurological recovery of cats that were subjected to cervical spinal trauma was significantly better after treatment with TRH than after saline or dexamethasone. This beneficial effect of TRH may find therapeutic application in patients with spinal trauma.
3. TRH also antagonizes the depressant effects of alcohol, and barbiturates.[97] The use of TRH in the treatment of endogenous depression remains controversial.
4. TRH affects the temperature regulation in rats, producing hyperthermia.[98] It is interesting to note that its dioketopiperazine metabolite does just the reverse, causing hypothermia.[99]

5. TRH, when given to animals, is known to possess a wide variety of behavioral and neurochemical effects such as motor activity, feeding, and drinking patterns, and even hibernation.[97]

The sum total of effects of TRH are highly suggestive of a role akin to that of a neurotransmitter or a neuromodulator.

Nonpituitary, Non-CNS Actions of TRH

The nonpituitary, non-CNS effects of TRH are limited to the GI tract. This is not surprising, since TRH is widely distributed within the GI tract, and since other hypothalamic peptides, particularly somatostatin, exert profound and pronounced effects on the GI tract. The most notable effect of TRH on the GI tract is its ability to inhibit gastric acid secretion. Dolva et al.[100] studied the effect of increasing doses of TRH on pentagastrin stimulated gastric secretion in normal subjects and demonstrated that TRH caused a stepwise inhibition of the volume, acid, and pepsin output of gastric juice. The same workers had demonstrated earlier[101] that intravenous TRH inhibits and/or retards the absorption of glucose and xylose from the gut. Both above-described effects are reminiscent of the effects of somatostatin on the GI tract.

The effects of TRH on the endocrine pancreas are less striking. Morley et al.[102] studied the effect of TRH on insulin and glucagon secretion by the endocrine pancreas, and concluded that TRH itself had no effect on insulin or glucagon release. However, they noted that TRH enhanced the arginine-induced glucagon release by the α cells. To a lesser extent, TRH enhanced the second phase of glucose-mediated insulin release. Since TRH is found in abundance within the pancreas, it is conceivable that local action of TRH may at least to a minor extent, influence islet cell physiology.

Potential Therapeutic Uses of TRH

Although TRH has found a secure place as a diagnostic tool for the evaluation of disorders of the hypothalamic-pituitary-thyroid axis, it has not found its niche as a therapeutic agent. The following represent its potential uses, some of which have indeed been tried and discarded.

1. *Depression:* Although the initial claims that TRH was a useful agent to treat endogenous depression were encouraging, this notion currently has been all but abandoned.[61,78]
2. *Motor disorders:* Although studies have suggested that TRH may alleviate the ataxia and the ocular movement abnormalities of patients with spinocerebellar degeneration,[103] such use is not widely accepted.
3. *Spinal trauma:* The observation that TRH is a partial physiological antagonist to certain opiate peptides, while sparing the analgesic system, has prompted work in evaluating its role in recovery of animals from experimentally induced spinal shock. While initial studies[41] clearly demonstrated that TRH improves neurological recovery after spinal trauma in cats, this is yet to be convincingly substantiated in humans.

4. *Differentiated thyroid carcinoma:* TRH has been used to enhance uptake or radioactive iodine by differentiated thyroid cancer,[104] but the results are rather unencouraging. It was thought that TRH may replace the need to administer bovine TSH (with allergic potential), for the purpose of augmenting uptake by the residual tissue in the thyroid bed or by functioning metastasis. Unfortunately this has not been borne out by subsequent studies.

Clinical Disorders Involving TSH and TRH

As with nearly all endocrine systems, hypofunction and hyperfunction dominate the spectrum of disorders that involve the TRH–TSH axis. The term central hypothyroidism is used to denote the hypothyroid state that results from inadequate TSH stimulation of an otherwise normal thyroid gland. Central hypothyroidism can result from loss of TSH reserve from any cause; or from deficiency of hypothalmic TRH. The terms secondary hypothyroidism and tertiary hypothyroidism are often used to denote, respectively, the pituitary and hypothalamic etiologies of hypothyroidism.

The most important cause of pituitary hypothyroidism is loss of thyrotrope reserve by tumor destruction. Less commonly, loss of TSH may occur as a consequence of Sheehan's postpartum necrosis, pituitary apoplexy, lymphocytic hypophysitis, granulomatous diseases, and suprasellar tumors, particularly craniopharyngiomas.

Hypothalamic hypothyroidism, characterized by defective TRH secretion, can be a consequence of several mechanisms. In its purest form, TRH deficiency occurs as a result of absent synthesis of TRH by the hypothalamic nuclei that normally secrete TRH. The triad that characterizes this form of hypothalamic hypothyroidism[93] is unitropic TSH deficiency secondary to TRH deprivation, absence of organic (space-occupying) lesions in the area of the hypothalamus, median eminence or the stalk, and a low TSH level that demonstrates responsiveness to the exogenous administration of TRH. A second variety of hypothalamic hypothyroidism results from destructive lesions in the hypothalamic region, usually tumors, or interruption of the hypothalamohypophyseal transport (stalk interruption), again as a consequence of tumors. A third and recent addition to hypothalamic hypothyroidism is the entity caused by the secretion of biologically inactive TSH secretion, correctible by protracted administration of exogenous TRH.[105]

The two disorders characterized by excess TSH secretion are the syndromes of inappropriate TSH secretion, and TSH-induced hyperthyroidism. Although both conditions are characterized by an elevation in circulating thyroid hormone levels, the clinical presentation and therapeutic implications are widely different in both the entities. The syndrome of inappropriate TSH secretion is usually a nontumorous condition, occurring as a consequence of peripheral or selective (pituitary) resistance to thyroidal hormones.[106] The patient is often euthyroid but can be hyper- or even hypothyroid. By contrast,

TABLE 37
Clinical Disorders of TSH and TRH

TSH hyposecretion
  Pituitary (secondary) hypothyroidism
    Pituitary tumors (compromised TSH reserve)
    Sheehan's syndrome
    Apoplexy
    Granulomatous disease
  Hypothalamic (tertiary) hypothyroidism
    Developmental (deficient TRH synthesis)
    Organic lesions (e.g., tumors, stalk section)
    Biologically ineffective TSH secretion
TSH hypersecretion
  Nonneoplastic inappropriate TSH secretion
    Peripheral resistance to thyroid hormone
    Pituitary resistance to thyroid hormones
  Neoplastic inappropriate TSH secretion
    TSH-secreting adenoma
      Without hypersecretion of other pituitary hormones
      With acromegaly, prolactin hypersecretion, etc.

TSH-secreting hyperthyroidism, as the name implies, results in clinical hyperthyroidism occurring from autonomous secretion by a thyrotrope adenoma.[107–109] These disorders are discussed in Chapters 6 and 7.

Table 37 outlines the disorders resulting from dysfunction of thyrotropin (TSH) and TRH.

References

1. Condliffe PG, Weintraub BD: Pituitary thyroid-stimulating hormone and other thyroid-stimulating substances. In Gray CH, James VHT (eds): *Hormones in Blood*. Academic, London, 1979, p. 499.
2. Chin WW, Habener JF, Kieffer JD, et al: Cell-free translation of the messenger RNA coding for the subunit of thyroid-stimulating hormone. *J Biol Chem* **153**:7985, 1978.
3. Kourides IA, Vamvakopoulos NC, Maniatis GM: mRNA-directed biosynthesis of α and β subunits of thyrotropin. *J Biol Chem* **254**:11106, 1979.
4. Kourides IA, Weintraub BD, Rosen SW, et al: Secretion of alpha subunit of glycoprotein hormones by pituitary adenomas. *J Clin Endocrinol Metab* **43**:97, 1976.
5. Lamberg BA, Pekonen F, Gordin A: TSH-induced hyperthyroidism and acromegaly due to pituitary tumour. *Ann Endocrinol* **39**:14A, 1978.
6. Smallridge RC, Wartofsky L, Dimond RC: Inappropriate secretion of thyrotropin: Discordance between the suppressive effects of corticosteriods and thyroid hormone. *J Clin Endocrinol Metab* **48**:700, 1979.
7. Utiger RD: Radioimmunoassay of human plasma thyrotropin. *J Clin Invest* **44**:1277, 1965.
8. Reichlin S, Utiger RD: Regulation of the pituitary thyroid axis in man: Relationship of TSH concentration to concentration of free and total thyroxine in plasma. *J Clin Endocrinol Metab* **27**:251, 1967.
9. Larsen PR, Frumess RD: Comparison of the biological effects of thyroxine and triiodothyronine in the rat. *Endocrinology* **100**:980, 1977.

10. Spira O, Birkenfeld A, Avni A, et al: TSH synthesis and release in the thyroidectomized rat: effect of T3. *Acta Endocrinol (Kbh)* **92**:502, 1979.

11. Silva JE, Larsen PR: Pituitary nuclear 3,5,3'-triiodothyronine and thyrotropin secretion: an explanation for the effect of thyroxine. *Science* **198**:617, 1977.

12. Larsen PR, Dick TE, Markowitz BP, et al: Inhibition of intrapituitary thyroxine to 3,5,3'-triiodothyronine conversion prevents the acute suppression of thyrotropin release by thyroxine in hypothyroid rats. *J Clin Invest* **64**:117, 1979.

13. Obregon MJ, Pascual A, Mallol J, et al: Evidence against a major role of L-thyroxine at the pituitary level: Studies in rats treated with iopanoic acid (Telepaue). *Endocrinology* **106**:1827, 1980.

14. De Lean A, Ferland L, Drouin J, et al: Modulation of pituitary thyrotropin releasing hormone receptor levels by estrogens and thyroid hormones. *Endocrinology* **100**:1496, 1977.

15. Gershengorn MC. Bihormonal regulation of the thyrotropin-releasing hormone receptor in mouse pituitary thyrotropic tumor cells in culture. *J Clin Invest* **62**:937, 1978.

16. Larsen PR: Thyroid-pituitary interaction: feedback regulation of thyrotropin secretion by thyroid hormones. *N Engl J Med* **306**:23, 1982.

17. Besses GS, Burrow GN, Spaulding SW, et al: Dopamine infusion acutely inhibits the TSH and prolactin response to TRH. *J Clin Endocrinol Metab* **41**:985, 1975.

18. Delitala G: Dopamine and TSH secretion in man. *Lancet* **2**:760, 1977.

19. Refetoff S, Fang VS, Rapoport B, et al: Interrelationships in the regulation of TSH and prolactin secretion in man. Effects of L-Dopa, TRH and thyroid hormone in various combinations. *J Clin Endocrinol Metab* **38**:450, 1974.

20. Scanlon MF, Weightman, DR, Mora B, et al: Evidence for dopaminergic control of thyrotrophin secretion in man. *Lancet* **2**:421, 1977.

21. Felt V, Nedvidkova J: Effect of bromocryptine on the secretion of thyrotropic hormone (TSH), prolactin (Pr), human growth hormone (HGH), thyroxine (T$_4$) and triiodothyronine (T$_3$) in hypothyroidism. *Horm Metab Res* **9**:274, 1977.

22. Scanlon MF, Weightman DR, Shale DJ, et al: Dopamine is a physiological regulation of thyrotrophin (TSH) secretion in normal man. *Clin Endocrinol (Oxf)* **10**:7, 1979.

23. Scanlon MF, Chan V, Heath M, et al: Dopaminergic control of thyrotropin, α-subunit, thyrotropin β-subunit, and prolactin in euthyroidism and hypothyroidism: Dissociated responses to dopamine receptor blockade with metoclopramide in hypothyroid subjects. *J Clin Endocrinol Metab* **53**:360, 1981.

24. Siler TM, Yen SSC, Vale W, et al: Inhibition by somatostatin on the release of TSH induced in man by thyrotropin-releasing factor. *J Clin Endocrinol Metab* **38**:742, 1974.

25. Porter BA, Refetoff S, Rosenfield RL, et al: Abnormal thyroxine metabolism in hyposomatotrophic dwarfism and inhibition of responsiveness to TRH during GH therapy. *Pediatrics* **51**:668, 1975.

26. Lippe BM, VanHerle AJ, LaFranchi SH, et al: Reversible hypothyroidism in growth hormone-deficient children treated with human growth hormone. *J Clin Endocrinol Metab* **40**:612, 1975.

27. McLaren EH, Hendriks S, Pimstone BL: Thyrotropin responses to intravenous thyrotropin-releasing hormone in patients with hypothalamic and pituitary disease. *Clin Endocrinol (Oxf)* **3**:113, 1974.

28. Guillemin R: Peptides in the brain: The new endocrinology of the neuron. *Science* **202**:390, 1978.

29. Schally AV: Aspects of hypothalamic regulation of the pituitary gland: Its implications for the control of reproductive processes. *Science* **202**:18, 1978.

30. Nair RMG, Barrett JF, Bowers CY et al: Structure of porcine thyrotropin releasing hormone. *Biochemistry* **9**:1103, 1970.

31. Burgus R, Dunn TF, Desiderio D, et al: Characterization of ovine hypothalamic hypophysiotropic TSH-releasing factor. *Nature (Lond)* **226**:321, 1970.

32. Johansson O, Hokfelt T: Thyrotropin releasing hormone, somatostatin, and enkephalin: distribution studies using immunohistochemical techniques. *J Histochem Cytochem* **28**:364, 1980.

33. Jackson IMD, Reichlin S: Brain thyrotropin-releasing hormone is independent of the hypothalamus. *Nature (Lond)* **267**:853, 1977.

34. Jackson IMD, Reichlin S: Distribution and biosynthesis of TRH in the nervous system. In Collu R, Barbeau A, Ducharme JR, Rochefort JG (eds): *Central Nervous System Effects of Hypothalamic Hormones and Other Peptides.* Raven Press, New York, 1979, p. 3.

35. Hokfelt T, Lundberg JM, Schultzberg M, et al: Co-existence of peptides and putative transmitters in neurons. In Costa E, Trabucchi M (eds): *Neural Peptides and Neuronal Communication.* Raven Press, New York, 1980, p. 1.

36. Koch Y, Okon E: Localization of releasing hormones in the human brain. *Int Rev Exp Pathol* **19**:45, 1979.

37. Ogawa N, Yamawaki Y, Kuroda H, et al: Discrete regional distribution of thyrotropin-releasing hormone (TRH) receptor binding in monkey central nervous system. *Brain Res* **205**:169, 1981.

38. Martino E, Nardi M, Vaudagna G, et al: Thyrotropin-releasing hormone-like material in human retina. *J Endocrinol Invest* **3**:267, 1980.

39. Kastin AJ, Ehrensing RH, Schalach DS, et al: Improvement in mental depression with decreased thyrotropin response after administration of thyrotropin-releasing hormone. *Lancet* **2**:740, 1972.

40. Spindel ER, Wurtman RJ, Bird ED: Increased TRH content of the basal ganglia in Huntington's disease. *N Engl J Med* **303**:1235, 1980.

41. Faden AI, Jacobs TP, Holaday JW: Thyrotropin-releasing hormone improves neurologic recovery after spinal trauma in cats. *N Engl J Med* **305**:1063, 1981.

42. Yarbrough GG: On the neuropharmacology of thyrotropin releasing hormone (TRH). *Prog Neurobiol* **12**:291, 1979.

43. Hokfelt T, Puxe K, Johansson O, et al: Thyrotropin releasing hormone (TRH)-containing nerve terminals in certain brain stem nuclei and in the spinal cord. *Neurosci Lett* **1**:133, 1975.

44. Maeda K, Frohman LA: Release of somatostatin and thyrotropin-releasing hormone from rat hypothalamic fragments *in vitro. Endocrinology* **106**:1837, 1980.

45. Cooper JR, Bloom FE, Roth RH (eds): *The Biochemical Basis of Neuropharmacology.* 3rd ed. Oxford University Press, New York, 1978, p. 259.

46. Matsui T, Prasad C, Peterkofsky A: Metabolism of thyrotropin releasing hormone in brain extracts. *J Biol Chem* **254**:2439, 1979.

47. Yanagisawa T, Prasad C, Peterkofsky A: The subcellular and organ distribution and natural form of histidylproline diketopiperazine in rat brain determined by a specific radioimmunoassay. *J Biol Chem* **255**:10290, 1980.

48. Gershengorn MC: Thyrotropin-releasing hormone. *Mol Cell Biochem* **45**:163, 1982.

49. Kolesnick RN, Gershengorn MC: Thyrotropin-releasing hormone and the pituitary. *Am J Med* **79**:729, 1985.

50. Albert PR, Tashjian AH Jr: Thyrotropin-releasing hormone-induced spike and plateau in cystosolic-free Ca^{2+} concentrations in pituitary cells. *J Biol Chem* **259**:5827, 1984.

51. Snowdowne KW, Borle AB: Changes in cystosolic ionized calcium induced by activators of secretion in GH_3 cells. *Am J Physiol* **246**:E198, 1984.

52. Kruskal BA, Keith CH, Maxfield FR: TRH-induced changes in intracellular $[Ca^{2+}]$ measured by microspectrofluorometry on individual Quin 2-loaded cells. *J Cell Biol* **99**:1167, 1984.

53. Gershengorn MC, Thaw C: Calcium influx is not required for TRH to elevate free cytoplasmic calcium in GH_3 cells. *Endocrinology* **113**:1522, 1983.

54. Geras E, Rebecchi MJ, Gershengorn MC: Evidence that stimulation of thyrotropin and prolactin secretion by thyrotropin-releasing hormone occur via different calcium-mediated mechanisms: studies with verapamil. *Endocrinology* **110**:901, 1982.

55. Berridge MJ: Inositol trisphosphate and diacyglycerol as second messengers. *Biochem J* **220**:345, 1984.

56. Martin TFJ: Thyrotropin-releasing hormone rapidly activates the phosphodiester hydrolysis of polyphosphoinositides in GH_3 pituitary cells. *J Biol Chem* **258**:14816, 1983.

57. Reichlin S, Martin JB, Jackson IMD: Regulation of thyroid-stimulating hormone (TSH) secretion. In Jeffcoate SL, Hutchinson JSM (eds): *The Endocrine Hypothalamus.* Academic Press, New York, 1978, p. 229.

58. Hershman JM: Clinical application of thyrotropin-releasing hormone. *N Engl J Med* **290**:886, 1974.

59. Snyder PJ, Utiger RD: Response to thyrotropin releasing hormone (TRH) in normal man. *J Clin Endocrinol Metab* **34**:380, 1972.

60. Burger HG, Patel YC: Thyrotropin releasing hormone-TSH. *Clin Endocrinol Metab* **6**:83, 1977.

61. Gold PW, Goodwin FK, Wehr T, et al: Pituitary thyrotropin response to thyrotropin-releasing hormone in affective illness: relationship to spinal fluid amine metabolites. *Am J Psychiatry* **134**:1028, 1978.

62. Harris AC, Christianson D, Smith MS, et al: The physiological role of thyrotropin-releasing hormone in the regulation of thyroid-stimulating hormone and prolactin secretion in the rat. *J Clin Invest* **61**:441, 1978.

63. Yamaji T: Modulation of prolactin release by altered levels of thyroid hormone. *Metabolism* **23**:745, 1974.

64. Irie M, Tsushima T: Increase of serum growth hormone concentration following thyrotropin-releasing hormone injection in patients with acromegaly or gigantism. *J Clin Endocrinol Metab* **35**:97, 1972.

65. Krieger DT, Condon EM: Cyproheptadine treatment of Nelson's syndrome: Restoration of plasma ACTH circadian periodicity and reversal of response to TRF. *J Clin Endocrinol Metab* **46**:349, 1978.

66. Hershman JM, Pittman JA Jr: Response to synthetic thyrotropin-releasing hormone in man. *J Clin Endocrinol Metab* **31**:457, 1970.

67. Haigler ED Jr., Pittman JA Jr., Hershman JM, et al: Direct evaluation of pituitary thyrotropin reserve utilizing synthetic thyrotropin releasing hormone. *J Clin Endocrinol Metab* **33**:573, 1971.

68. Lawton NF, Ekins RP, Nabarro JDN: Failure of pituitary response to thyrotropin-releasing hormone in euthyroid Graves' disease. *Lancet* **2**:14, 1971.

69. Chopra IJ, Chopra U, Orgiazzi J: Abnormalities of hypothalamo-hypophyseal thyroid axis in patients with Graves' ophthalmopathy. *J Clin Endocrinol Metab* **37**:955, 1973.

70. Chopra IJ, Chopra U, Vanderlaan WP, et al: Comparison of serum prolactin and thyrotropin responses to thyrotropin-releasing hormone in patients with Graves' ophthalmopathy. *J Clin Endocrinol Metab* **38**:683, 1974.

71. Hall R, Besser GM, Ormston BJ, et al: The thyrotrophin-releasing hormone test in disease of the pituitary and hypothalamus. *Lancet* **1**:759, 1972.

72. Brown J, Chopra IJ, Cornell JS: Thyroid physiology in health and disease. *Ann Intern Med* **81**:68, 1974.

73. Lamberg BA, Pelkomen R, Aro A, et al: Thyroid function in acromegaly before and after transsphenoidal hypophysectomy followed by cryoapplication. *Acta Endocrinol (Oxf)* **82**:254, 1976.

74. Hoffman DL, Baker BL: Effect of treatment with growth hormone on somatostatin in the median eminence of hypophysectomized rats. *Proc Soc Exp Biol Med* **156**:265, 1977.

75. Sheppard MC, Kronheim S, Pimstone BL: Stimulation by growth hormone of somatostatin release from the rat hypothalamus *in vitro. Clin Endocrinol (Oxf)* **9**:583, 1978.

76. Cobb WE, Reichlin S, Jackson I: Growth hormone secretory status is a determinant of the thyrotropin response to thyrotropin-releasing hormone in euthyroid patients with hypothalamic-pituitary disease. *J Clin Endocrinol Metab* **52**:324, 1981.

77. Otsuki M, Dakoda M, Baba S: Influence of glucocorticoids on TRH induced TSH response in man. *J Clin Endocrinol Metab* **36**:95, 1973.

78. Furlong FW, Brown GM, Beeching MF: Thyrotropin-releasing hormone: Differential antidepressant and endocrinological effects. *Am J Psychiatry* **133**:1187, 1976.

79. Maeda K, Kato Y, Ohgo S, et al: Growth hormone and prolactin release after injection of thyrotropin-releasing hormone in patients with depression. *J Clin Endocrinol Metab* **40**:501, 1975.

80. Spratt DI, Pont A, Miller MB, et al: Hyperthyroxinemia in patients with acute psychiatric disorders. *Am J Med* **73**:41, 1982.

81. Quint AR, Kaiser FE: Gonadotropin determinations and thyrotropin-releasing hormone and luteinizing hormone-releasing hormone testing in critically ill postmenopausal women with hypothyroxinemia. *J Clin Endocrinol Metab* **60**:464, 1985.

82. Kaptein EM, Grieb DA, Spencer CA, et al: Thyroxine metabolism in the low thyroxine state of critical nonthyroidal illnesses. *J Clin Endocrinol Metab* **53**:764, 1981.

83. Wood DG, Samols E: Impaired TSH secretion with decreased free T_4 in nonthyroidal illness (NTI). *Clin Res* **30**:494A, 1982. (Abs.)

84. Vierhapper H, Laggner A, Waldhausl W, et al: Impaired secretion of TSH in critically ill patients with "low T_4 syndrome." *Acta Endocrinol (Oxf)* **101**:542, 1982.

85. Wehmann RE, Gregerman RI, Burns WH, et al: Suppression of thyrotropin in the low-thyroxine state of severe nonthyroidal illness. *N Engl J Med* **312**:546, 1985.

86. Tibaldi JM, Surks MI: Animal models of nonthyroidal disease. *Endocr Rev* **6**:87, 1985.

87. Harris RC, Fang SL, Azizi F, et al: Effect of starvation on hypothalamic-pituitary function in the rat. *Metabolism* **27**:1074, 1978.

88. Campbell GA, Kurcz M, Marshall S, et al: Effects of starvation in rats on serum levels of follicle-stimulating hormone, luteinizing hormone, thyrotropin, growth hormone and prolactin; response to LH-releasing hormone and thyrotropin-releasing hormone. *Endocrinology* **100**:580, 1977.

89. Fleischer N, Lorente M, Kirkland J, et al: Synthetic thyrotropin-releasing factor as a test of pituitary thyrotropin reserve. *J Clin Endocrinol Metab* **34**:617, 1972.

90. Hall R, Ormston BJ, Besser GM, et al: The thyrotropin-releasing hormone test in diseases of the pituitary and hypothalamus. *Lancet* **1**:759, 1972.

91. Jackson, IMD: Thyrotropin-releasing hormone. *N Engl J Med* **306**:145, 1982.

92. Ormston BJ, Garry R, Cryer RJ, et al: Thyrotropin-releasing hormone as a thyroid-function test. *Lancet* **2**:10, 1971.

93. Pittman JA Jr, Haigler ED Jr, Hershman JM, et al: Hypothalamic hypothyroidism. *N Engl J Med* **285**:844, 1971.

94. Costom BH, Grumbach MM, Kaplan SL: Effect of thyrotropin-releasing factor on serum thyroid-stimulating hormone: An approach to distinguishing hypothalamic from pituitary forms of idiopathic hypopituitary dwarfism. *J Clin Invest* **50**:2219, 1971.

95. Muller EE, Salerno F, Cocchi D, et al: Interaction between the thyrotropin-releasing hormone-induced growth hormone rise and dopaminergic drugs: Studies in pathologic conditions of the animal and man. *Clin Endocrinol (Oxf)* **11**:645, 1979.

96. Morley JE: Extrahypothalamic thyrotropin releasing hormone (TRH)—Its distribution and its functions. *Life Sci* **25**:1539, 1979.

97. Prange AJ Jr, Nemeroff CB, Loosen PT, et al: Behavioral effects of thyrotropin-releasing hormone in animals and man: A review. In Collu R, Barbeau A, Ducharme JR, Rochefort J-G (eds): *Central Nervous System Effects of Hypothalamic Hormones and Other Peptides*. Raven Press, New York 1979, p. 75.

98. Brown M, Rivier J, Vale W: Actions of bombesin, thyrotropin releasing factor, prostaglandin E_2 and naloxone on thermoregulation in the rat. *Life Sci* **20**:1681, 1977.

99. Peterkofsky A, Battaini A: The biological activities of the neuropeptide histidylproline diketopiperazine. *Neuropeptides* **1**:105, 1980.

100. Dolva LO, Hanssen KF, Berstad A, et al: Thyrotropin-releasing hormone inhibits the pentagastrin stimulated gastric secretion in man—a dose response study. *Clin Endocrinol (Oxf)* **10**:281, 1979.

101. Dolva LO, Hanssen KF, Frey HMM: Actions of thyrotropin releasing hormone on gastrointestinal function in man. Inhibition of glucose and xylose from the gut. *Scand J Gastroenterol* **13**:599, 1978.

102. Morley JE, Levin SR, Pehlevanian M, et al: The effects of thyrotropin-releasing hormone on the endocrine pancreas. *Endocrinology* **104**:137, 1979.

103. Sobue I, Yamamoto H, Konagaya M, et al: Effect of thyrotropin-releasing hormone on ataxia of spinocerebellar degeneration. *Lancet* **1**:418, 1980.

104. Samaan NA, Beceiro JR, Stratton Hill C Jr, et al: Thyrotropin-releasing hormone (TRH) studies in patients with thyroid cancer. *J Clin Endocrinol Metab* **35**:438, 1972.
105. Beck-Peccoz P, Amr S, Menezes-Ferreira M, et al: Decreased receptor binding of biologically inactive thyrotropin in central hypothyroidism. *N Engl J Med* **312**:1085, 1985.
106. Weintraub BD, Gershengorn MC, Kourides IA, et al: Inappropriate secretion of thyroid-stimulating hormone. *Ann Intern Med* **95**:339, 1981.
107. Hamilton CR Jr, Maloof F: Acromegaly and toxic goiter: Cure of hyperthyroidism and acromegaly by proton-beam partial hypophysectomy. *J Clin Endocrinol Metab* **35**:659, 1972.
108. Kourides IA, Weintraub BD, Rosen SW, et al: Secretion of alpha subunit of glycoprotein hormones by pituitary adenomas. *J Clin Endocrinol Metab* **43**:97, 1976.
109. Lamberg BA, Pekonen F, Gordin A: TSH-induced hyperthyroidism and acromegaly due to pituitary tumor. *Ann Endocrinol* **39**:14A, 1978.
110. Reichlin S, Anatomical and Physiological Basis of Hypothalamic–Pituitary Regulation. In Post KD, Jackson IM, Reichlin S (eds): *The Pituitary Adenoma*, Plenum Press, New York 1980, pp. 3–28.

6

Inappropriate Pituitary Thyrotropin Secretion

Introduction

The term *inappropriate TSH secretion* is used to denote a diverse group of disorders characterized by an inappropriately elevated circulating level of thyroid-stimulating hormone (TSH), or thyrotropin, in the presence of elevated thyroidal hormones in the circulation. The increase in the thyroidal hormones is clearly a secondary phenomenon mediated by the biologically active thyrotropin. In the past decade, the syndrome of inappropriate TSH secretion is being recognized with increasing frequency, due in large part to awareness of this relatively new endocrine disorder. Before embarking on a discussion of the pathogenesis of this fascinating group of disorders, the following introductory remarks are pertinent:

1. Since the hallmark of this syndrome is an inappropriate TSH level, the recognition of this entity heavily rests on employing a highly sensitive radioimmunoassay for circulating TSH; an assay that is sensitive enough to differentiate suppressed TSH levels from inappropriately elevated levels, when viewed in light of circulating thyroidal hormone concentrations.

2. The clinical spectrum encountered in patients with the syndrome of inappropriate TSH secretion is impressively broad. Although woven with the same common etiological thread, the ultimate tapestry of clinical presentation can be highly variable in patients with this disorder; thus the clinician may encounter patients with euthyroidism, mild hypothyroidism or mild hyperthyroidism as expressions of inappropriate TSH secretion.

3. Anatomically and clinically, the two major forms of inappropriate TSH secretion are the nontumorous and tumorous varieties. Recent insights into the regulation of TSH secretion and the peripheral actions of thyroid hormones have provided methods to differentiate between the two.

4. The pathogenesis of the nontumorous varieties of inappropriate TSH secretion is intimately linked with resistance to thyroidal hormones at the peripheral level, the pituitary level, or both. The resultant inappropriate TSH secretion is relatively nonautonomous and may be viewed as TSH secretion reset at a higher threshold for negative feedback by thyroidal hormones.

5. By contrast, the tumorous variety of inappropriate TSH secretion reflects autonomous TSH secretion by a thyrotrope adenoma.

6. Some forms of inappropriate TSH secretion are familial.

Rare as these syndromes are, recognition of these disorders is crucial for dual purposes: to offer the right therapy as well as to avoid needless and often improper therapy. For instance, the mistaken diagnoses of hyperthyroid Graves' disease in many a case of inappropriate TSH secretion has resulted in unnecessary thyroidectomy, or radioactive iodine therapy. Such therapy is doubly tragic, since these procedures result in inadvertent removal of the only protective mechanism that keeps the patient with partial peripheral resistance from becoming hypothyroid. Therefore, an insight into the pathogenesis, presentation, and diagnostic features of these disorders merits attention.

Etiology and Pathogenesis

The pathogenesis of inappropriate TSH secretion can be best viewed in terms of the various syndromes that result from such a perturbation. The three major types (Table 38) that have emerged are nonneoplastic inappropriate TSH secretion resulting from predominantly pituitary resistance to thyroidal hormones; nonneoplastic inappropriate TSH secretion resulting from generalized—pituitary and peripheral—resistance to thyroidal hormones; and inappropriate TSH secretion by an autonomously functioning thyrotrope adenoma. This classification has the advantage of keeping in perspective the thyroidal status of the patient. For instance in the first type, which is represented by selective pituitary resistance, the clinical expression is mild hyperthyroidism; in the second type, characterized by both pituitary and peripheral resistance, the patient remains clinically euthyroid, or mildly hypothyroid; and the third type, typified by the thyrotrope adenoma, represents

TABLE 38
Spectrum of Inappropriate TSH Secretion

	Type	Clinical expression
I	Non-neoplastic inappropriate TSH secretion	
IA	From selective pituitary resistance	Mild hyperthyroidism
IB	From pituitary and peripheral resistance	Euthyroid or hypothyroid
II	Adenomatous inappropriate TSH secretion	
IIA	TSH-secreting adenoma	Hyperthyroid

the classic example of TSH-induced hyperthyroidism in the patient. The pathogenesis of each type deserves brief mention.

Nontumorous Inappropriate TSH Secretion Resulting from Selective Pituitary Resistance

In 1972, Emerson and Utiger[1] reported a patient with an unusual variety of hyperthyroidism characterized by elevated circulating thyroid hormones, clearly elevated TSH levels, slight suppression of TSH to the administration of thyroidal hormones, and a lack of TSH response to the exogenous administration of TRH. The patient had no evidence of pituitary tumor, and the authors suggested that the relatively autonomous TSH secretion may have been a result of chronic TSH stimulation by endogenous TRH. In 1975, Gershengorn and Weintraub[2] reported a patient with hyperthyroidism, elevated TSH levels in the presence of elevated circulating thyroidal hormone levels, partial TSH suppression to remarkably high doses of triiodothyronine (T3) administration, and a robust TSH response to exogenous TRH administration. The key observation was that the patient's hyperthyroidism was secondary to chronic thyroidal stimulation by TSH secreted by thyrotropes; these cells maintained responsiveness to TRH stimulation and thyroidal suppression, albeit in a quantitatively abnormal fashion. It was proposed that the threshold for negative feedback suppression by thyroidal hormones on the pituitary was reset at a higher level, accounting for the qualitatively normal, but quantitatively abnormal response to T3 administration. The basic defect, then, originates as a consequence of pituitary resistance to negative feedback suppression by thyroidal hormones. As a result, a cascade of events follow: sustained secretion of biologically active TSH by the thyrotrope, continual stimulation of an intrinsically normal thyroid gland, chronic hypersecretion of thyroxine (T4) and T3 by the thyroid, and the eventual expression of clinical hyperthyroidism. Obviously, the peripheral tissues respond to thyroidal hormones, since the peripheral effects of T4 and T3 are expressed as hyperthyroidism. Therefore, the resistance to thyroidal hormones, in this entity, is selective and limited to the thyrotrope.

The clinical features, hormonal studies, and dynamic data in the patient reported by Gershengorn and Weintraub[2] have emerged as prototypical for the nontumorous variety of inappropriate TSH secretion resulting from selective pituitary resistance to negative feedback by thyroidal hormones. Several case reports followed.[3–7] In most instances, the original five features described by Gershengorn and Weintraub were confirmed, that is, hyperthyroidism, inappropriate elevation of TSH, preservation of TRH responsiveness, suppressibility to extremely high doses of T3, and negative evidence for tumor in the pituitary. These phenomena are highly reminiscent of the hormonal features encountered in patients with pituitary dependent Cushing's disease. This disorder is also characterized by inappropriate ACTH secretion by corticotropes in the presence of hypercortisolemia, a higher-set threshold for negative feedback suppression to dexamethasone, preservation of ACTH response to exogenous administration of corticotropin-releasing-hormone, and in many instances a negative computed tomography (CT) study for tumors in the pituitary gland. Since it is believed that at least in some patients with pituitary dependent Cushing's disease a hypothalamic etiology may be operative, it is appropriate to consider the role of hypothalamic TRH in the causation of the selective pituitary resistance seen in these patients.

There are four possible hypotheses currently in vogue to explain the pathophysiological basis for the development of the pituitary resistance encountered in the nontumorous variety of inappropriate TSH secretion (Table 39).

1. An intrinsic abnormality in the thyrotrope, resulting in a decreased sensitivity to the inhibitory effects of thyroid hormone(s) is one possible explanation that has been advanced. It is not clear whether this intrinsic abnormality is a result of poor binding of thyroidal hormones to the thyrotrope or of a breakdown in the postreceptor binding events within thyrotropes. The fact that TSH suppression can be attained by administering megadoses of T3 suggests that the defect, regardless of mechanism, can be overcome.

2. A partial deficiency of pituitary T4-monodeiodinase is another possible mechanism that may account for the relative resistance of the thyrotrope to feedback inhibition by the thyroidal hormones. It is a well-accepted notion, based on studies in rats, that intrapituitary T3 content is the dominant factor that operates the feedback mechanism of thyroid hormones in the thyrotrope.[8–14] The enzyme T4-monode-

TABLE 39
Postulated Mechanisms for Selective
Pituitary Resistance

Decreased sensitivity of thyrotrope to negative feedback
　suppression
Partial deficiency of pituitary T4-monodeiodinase
Augmented TRH drive
Loss of hypothalamic inhibition exerted by dopamine or
　somatostatin

---

iodinase located within the pituitary gland is responsible for almost 50% of the intracellular T3 within the thyrotropes. The concept that partial deficiency of pituitary T4-monodeiodinase may account for the inappropriate TSH secretion seen in such patients is supported by the study of Rosler et al.[15] These workers studied six affected family members with the syndrome and showed that administration of T3, but not of T4, for a protracted period resulted in lowering of TSH level, and indeed causing a complete biochemical and clinical remission of the hyperthyroid state in three. The authors postulated that chronic administration of T3 (but not T4) results in raising the effective concentrations of T3 within the thyrotrope, leading to the suppression of TSH level. The lack of such a response to the long-term administration of T4 (although limited to only one patient) may have been attributable to the lack of intrapituitary enzymatic apparatus required to convert T4 to T3 in the pituitary. If confirmed, these workers' observation is highly significant, since therapy for this form of TSH-mediated hyperthyroidism would be, by necessity, the administration of T3.

3. An alternative explanation to account for the resistance of the thyrotrope to feedback suppression is that the disorder results from augmented TRH secretion; this would result in requiring higher levels of thyroid hormones to suppress the chronically stimulated thyrotrope mass. This hypothesis cannot be proved, unless TRH can be reliably measured in the hypophyseal portal blood. Also, the fact that TSH secretion in such patients was not totally inhibited when the circulating thyroid hormone levels were raised to a level that would normally abolish the TSH response to even large doses of exogenous TRH mitigates against a primary TRH hypersecretion. As of yet, the hypothesis that TRH hypersecretion may underlie the pituitary resistance seen in this disorder remains speculative.

4. Finally, the possibility that the pituitary resistance is due to loss of hypothalamic inhibitors that physiologically exert a tonic negative control over the thyrotropes is also speculative. There are at least two hypothalamic regulators that inhibit thyrotropin release, dopamine[16–19] and somatostatin.[20,21] It is theoretically possible, that removal of the dopaminergic inhibition may be conducive for the thyrotropes to develop resistance to negative suppression while retaining their ability to respond, even hyperrespond, to TRH. Plausible as this hypothesis is, defective suppression of TSH by dopamine or somatostatin has not been documented by any published case report.

Nontumorous Inappropriate TSH Secretion Resulting from Peripheral Target Organ Resistance

Another category of nontumoral hypersecretion of TSH occurs as a response to peripheral resistance (including the pituitary) to the action of thyroidal hormones. Such resistance is usually partial, since complete resistance of tissues to thyroidal hormones is an event not compatible with life. The partial resistance at the pituitary level results in failure of the thyrotrope to sense the circulating level of the T4 and T3, which, to begin with, are normal or mildly elevated, since these hormones are not fully internalized by the peripheral tissues. As a consequence two phenomena occur—the thyrotropes continue to secrete TSH, despite a normal or elevated T4 and T3; and the biologically active TSH continues to stimulate the thyroid gland, which responds by secreting more T4 and T3. This adaptive mechanism may be sufficient for overcoming the partial peripheral resistance, and keeping the patient euthyroid. However, with sustained TSH-mediated thyroidal stimulation, depending on the degree of the peripheral resistance, the tissues may respond, resulting in mild hyperthyroidism.

The most important aspect of the syndrome to be recognized is that the clinical expression depends on the degree of peripheral resistance and the degree of thyroidal hypersecretion. Thus, the patient may be mildly hypothyroid (when the peripheral resistance cannot be overcome), euthyroid (when adaptation has occurred) and mildly hyperthyroid (when the resistance is mild, and the thyroidal stimulation by TSH far exceeds the degree of peripheral resistance). The literature is replete with examples of each of the above-mentioned phenomena.[22] In 1967, Refetoff and associates[23] described the first reported case of peripheral target organ resistance; they reported a familial syndrome of deaf-mutism, stippled epiphysis, thyromegaly, elevated thyroid hormone levels, and measurable TSH levels despite the elevated thyroid hormone level. Although a euthyroid clinical state was encountered in three of the six siblings studied, certain studies aimed at evaluating the peripheral actions of thyroidal hormones provided results consistent with hypothyroidism (e.g., BMR, creatine excretion, ankle reflex time). In a follow-up report in 1972, Refetoff et al.[24] studied the metabolic responses of such patients to the administration of thyroid hormones and demonstrated varying degrees of attenuation of these responses. The net effect on growth, development, BMR, serum tyrosine, cholesterol, magnesium, and albumin turnover suggested compensated eumetabolism at the expense of thyroid overactivity and increased circulating thyroidal hormone levels. TRH testing, in the only sibling tested revealed an increase in TSH following TRH administration.[25] In 1973, Lamberg[26] reported on a 25-year-old woman who had had a goiter since birth and who had undergone two thyroidectomies in childhood, but who had remained entirely euthyroid. The notable features in her case included markedly elevated thyroidal hormones, increased daily turnover of thyroxine, normal TSH levels in plasma, and normal TSH responses to TRH, and ^{131}I uptake suppression to large doses of L-T4. Lamberg proposed that the peripheral tissues and the pituitary gland must have been in some way unresponsive to thyroid hormone levels. The thyromegaly and the recurrence of the goiter after surgery were adaptive mechanisms brought about by the sustained TSH secretion. In the same year, Bode et al.[27] also reported a similar syndrome of partial target organ resistance to thyroidal hormones.

In the past decade, resistance to thyroid hormones has become a well-accepted disease entity. Several workers have contributed to the gradually increasing awareness of this entity, both in children[28-30] and in adults.[31-35] The familial nature of the syndrome was also becoming established, suggesting an autosomal dominant pattern of inheritance. Brooks et al.[36] studied eight patients with thyroid hormone resistance in four generations of a kindred containing 19 members. The affected members were all clinically euthyroid, had goiters, demonstrated markedly increased T4, T3, and free hormones in the circulation, normal or slightly elevated TSH, and a normal or exaggerated TSH response to TRH suggestive of primary hypothyroidism, despite high concentrations of thyroidal hormones in their circulation. Bantle et al.[37] studied five patients from two unrelated families, presenting with goiter, elevated T4, T3, and measurable TSH levels, which responded to exogenous TRH administration. The message from all the above reports is abundantly clear. Patients with the syndrome of generalized resistance need the elevated thyroidal hormone concentrations to maintain a eumetabolic state. The thyromegaly as well as the increase in the secretion and turnover of the thyroid hormones are all clearly an adaptive response brought about by the hypersecreted TSH. Therapeutically, the removal of the thyroid would be most counterproductive and a likely event if the proper diagnosis is not made.

Having understood the basis for the development of the syndrome, the next aspect to focus is the reason for developing resistance at the level of target tissues. This requires a basic understanding of thyroid hormone action at the peripheral level. The free hormones, free T4 and free T3, enter their target cells either by passive diffusion or by uptake mediated by specific plasma membrane receptor sites. Once within the cell, these hormones bind to specific protein receptors in the nucleus. These proteins, nonhistone in character, demonstrate high affinity for T3.[38,39] Subsequently, the action of T3 is expressed by the production of mRNA, and new protein synthesis. In addition, both T4 and T3 specifically increase synthesis of Na^+-K^+ ATP-ase that is associated with the plasma membrane,[40] an effect directly responsible for the increased thermogenesis and oxygen consumption induced by thyroidal hormones. A third mechanism for hormonal expression resides in the ability of T4 and T3 to bind with high affinity sites in the mitochondria of target cells. Thus, nuclear activation, ATP-ase activation, and mitochondrial activation are the triple mechanisms that are involved in the expression of thyroidal hormonal action upon its target cells. The elucidation of the exact mechanism involved in patients with peripheral resistance to thyroid hormones is hampered by a major investigational hurdle; the tissues most sensitive to thyroidal hormones (heart, pituitary, kidney, liver) are not available for study. The only studies that can be performed are those that evaluate hormonal binding to nuclear sites on lymphocytes, or in some instances, cultured fibroblasts. Such studies have yielded conflicting results. Bernal et al.[41] studied a patient with peripheral resistance to thyroid hormones and demonstrated a marked decrease in the affinity of lymphocytic nuclei to T3. This finding was suggestive of the notion that the peripheral resistance observed in the patient may have been due to reduced receptor affinity for T3. Liewendahl et al.[42] reported

that the lymphocytic nuclei of three patients suffering from generalized resistance to thyroid hormones had normal K_a for T3, and a B_{max} at the lower
end of normal range. Elewaut et al.[43] studied T3 binding to lymphocytic
nuclei, as well as plasma cAMP response to intravenous glucagon in patients
with general resistance to thyroidal hormones and reported normal affinity.
Similar findings were reported by Bantle et al.,[37] who studied nuclear T3
receptors from cultured fibroblasts of a patient with peripheral resistance,
and showed a normal equilibrium association constant, and a maximal binding
capacity that was even greater than normal control values. These findings
suggest that the thyroid hormone resistance, at least in some patients, is not
due to a decrease in the number or affinity of nuclear T3 receptors. These
findings are in keeping with the data reported by Eil et al.[44] and Kanter et
al.[45] that support normal binding of thyroid hormones to receptors. In fact,
the study by Kanter et al.[45] demonstrated that fibroblasts from affected patients did not take up and degrade LDL normally in response to T3, indicating
a problem at the post-receptor level. In summary, the several studies that
have attempted to study the binding of T3 to nuclear receptors of patients
with peripheral resistance have yielded highly variable results, ranging from
greatly reduced,[41] mildly reduced,[42] normal,[43] and even slightly increased[37]
affinity. A postreceptor defect has also been postulated.

Clinical Features

The clinical expressions of the syndromes of inappropriate TSH secretion
are highly variable. Five major presentations are recognized. These include
hyperthyroidism, goiter, recurrence of goiter, or hyperthyroidism after surgery, requirement of large replacement doses of thyroxine, and space-occupying lesions of the sella.

Hyperthyroidism: This is usually mild and can be caused by both the nontumorous as well as the tumorous varieties of inappropriate TSH secretion.
The degree of thyroidal hormone elevation in the circulation far exceeds the
mild nature of the hyperthyroidism, and is felt by many workers to reflect
the presence of mild peripheral resistance in these patients. The hyperthyroidism is characteristically devoid of Gravesian stigmata such as proptosis
and related features of infiltrative ophthalmopathy. However, exceptionally,
proptosis has been reported in association with TSH-induced hyperthyroidism
caused by a pituitary neoplasm.[46]

Goiter: Thyromegaly is almost always present in patients with inappropriate TSH secretion and represents the effects of chronic stimulation by
TSH. The thyromegaly tends to be moderate, diffuse, and non-nodular.

Recurrence after thyroidectomy: This is a fairly frequent feature and represents the response of the remnant thyroid tissue to the stimulatory effects
of TSH an ongoing problem not corrected by surgery. This is somewhat
reminiscent of the situation in pituitary-dependent Cushing's disease, where
subtotal adrenalectomy for treatment results in recurrence of the disease

reflecting the response of the remnant adrenals to chronic continual hypersecretion of ACTH. The tendency for postsurgical recurrence is by no means limited to TSH-mediated hyperthyroidism, since other forms of hyperthyroidism particulary Graves' hyperthyroidism, can recur after subtotal thyroidectomy.

Requirement of large doses of replacement levothyroxine therapy following thyroidectomy or radioactive iodine ablation for either a goiter or hyperthyroidism: This represents a major clue for suspecting peripheral resistance to thyroid hormone. In the usual setting, the patient continues to complain of hypothyroid symptoms, despite supraphysiological replacement doses of levothyroxine, at times exceeding 500 μg/day. This observation, when coupled with marked elevation of T4 levels (consequent to exogenous administration) in association with an inappropriately normal or elevated TSH level virtually clinches the diagnosis.

TSH-secreting pituitary tumor: Adenomas originating from pituitary thyrotropes represent an important subgroup of TSH-mediated hyperthyroidism. Since the report by Hamilton et al.[47] of TSH hypersecretion by pituitary tumors, the syndrome is being recognized with increasing frequency.[48] Tumoral TSH secretion has been linked with secretion by chromophobe adenoma[47,49] as well as pure thyrotrope adenomas.[50] The TSH secretion can

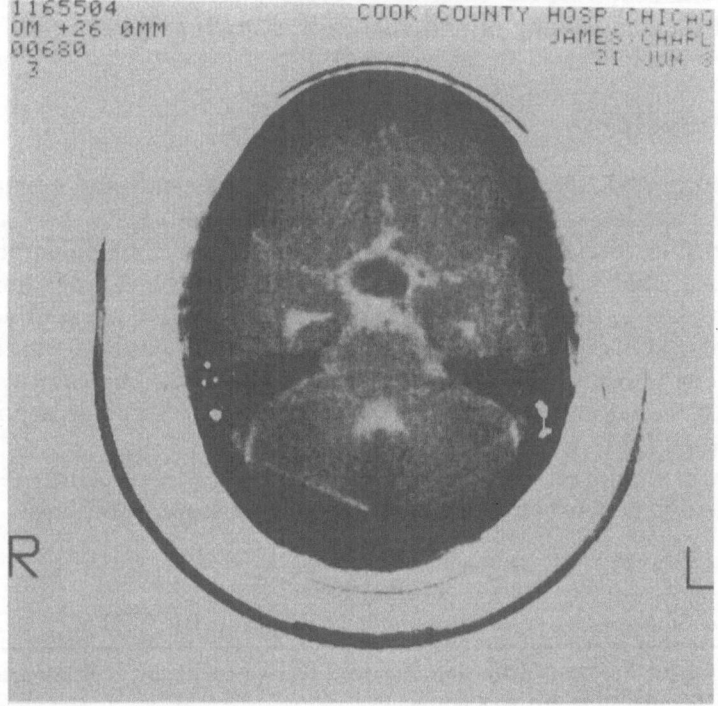

FIGURE 19. CT scan demonstrating large pituitary tumor with suprasellar extension. The excised tumor contained cells demonstrating strong immunopositivity for thyroid-stimulating hormone (TSH).

occur alone, or in combination with hypersecretion of other hormones. Thus, TSH induced hyperthyroidism can occur in conjunction with growth hormone hypersecretion[51,52] or prolactin hypersecretion.[50,53–55] These tumors can be small (microadenomas) or large enough to cause sellar enlargement[50] and even destruction of the sella with significant para- and suprasellar extension[46] (Fig. 19).

To summarize, the syndrome of inappropriate TSH secretion should be suspected under the following circumstances:

1. Thyromegaly, especially with elevated thyroidal hormones in a clinically euthyroid person
2. Thyromegaly, in association with hyperthyroidism especially when the hyperthyroidism is mild, chronic, and recurrent and when there is a discordance between the degree of elevation of thyroidal hormones and the mildness of the hyperthyroid state
3. Hyperthyroidism in association with radiological evidence of pituitary neoplasm
4. Hyperthyroidism in association with clinical or laboratory evidence of hypersecretion of growth hormone, prolactin, or ACTH
5. Thyromegaly in association with hypometabolic symptoms and elevated thyroidal hormone levels in the circulation
6. Patients requiring extremely large doses of replacement L-T4 therapy, and yet complaining of hypometabolic symptoms

Diagnostic Studies

Once suspected, the diagnostic approach to the syndrome of inappropriate TSH secretion is fairly straightforward (Table 40).

Step 1: Since the key factor in the diagnosis is the demonstration of inappropriate TSH levels viewed in correlation with the thyroid hormone levels in circulation, the crucial test is the TSH assay. While the routine measurement of TSH is by no means recommended in all patients with hyperthyroidism, its judicious use cannot be overemphasized. The circumstances under which measurement of TSH in hyperthyroidism assumes importance are outlined in Table 41.

The assay employed for measuring TSH should be sufficiently sensitive to make the distinction between normal levels and suppressed levels. When

TABLE 40
Diagnostic Approach to the Syndrome of Inappropriate TSH Secretion

Step I:	Obtaining TSH level to demonstrate the inappropriate nature of TSH relationship to the circulating thyroidal hormones
Step II:	TSH subunit assay, to demonstrate the ratio of α-TSH to molar TSH
Step III:	Dynamic tests—TRH infusion and T3 suppression
Step IV:	CT study of sella to exclude tumor

TABLE 41
Indication for TSH Measurement in Patients Who Are Hyperthyroid

1. Patients with biochemical hyperthyroidism, thyromegaly, and clinical euthyroidism or even mild hypothyroidism
2. Patients with chronic nonprogressive mild hyperthyroidism
3. Patients with recurrence of hyperthyroidism following surgery or radioactive iodine
4. Patients with mild hyperthyroidism with absent stigmata of Graves' disease (ophthalmopathy) or thyroidites
5. Patients with hyperthyroidism in conjunction with hypersomatotropism, or hyperprolactinemia
6. Patients with hyperthyroidism and a space-occupying lesion of the pituitary gland

such an assay is used, the demonstration of a TSH level in excess of 1.7 μU/ml in the presence of elevated circulating thyroid hormone levels is highly suspicious of inappropriate TSH secretion. It is important to recognize that when a sensitive assay is employed these patients may demonstrate only minimal elevations in TSH. With the availability of highly sensitive and specific radioimmunoassays for TSH, the separation of suppressed from normal basal TSH levels is becoming increasingly feasible, even to the extent of screening patients for suspected thyrotoxicosis.[56] The range of elevation of TSH can be as high as 400 μU/ml. The distinction between tumoral from non tumoral causes for the inappropriate TSH secretion cannot be made on the basis of basal TSH levels alone; this would require further tests, such as measurement of the subunits, as well as the response of TSH (and its subunits) to various dynamic studies.

Step 2: The measurement of the basal level of the free α subunit as well as its ratio to the molar TSH in the basal state provide clues in differentiating patients with inappropriate secretion of TSH secondary to tumors from those without tumors. Patients with TSH-secreting tumors consistently demonstrate a disproportionately elevated α subunit in the plasma relative to the intact (molar) TSH,[57] while patients with nontumoral inappropropriate TSH secretion (secondary to pituitary resistance to thyroidal hormones) demonstrate free α subunit levels in proportion to the molar TSH level.[58] The basal α-subunit levels in tumorous patients is generally in excess of 10 ng/ml. The level of the β-subunit of TSH, on the other hand, is normal or even low in comparison to the level of the intact TSH. Thus, measurement of the basal α-subunit level provides the first clue to the presence of an underlying tumoral etiology of TSH-mediated hyperthyroidism. While interpreting the α-subunit level, the other causes for an elevated α subunit should be kept in mind. These include the postmenopausal state, pregnancy, renal failure,[59] and α-subunit secretion by nonthyrotrope tumors of the pituitary,[60] as well as ectopic subunit secretion by malignant tumors elsewhere.[61]

Step 3: After demonstrating elevated TSH levels, and disproportionately elevated α level (elevated α-TSH/molar TSH ratio) in patients with the syndrome of inappropriate TSH secretion, the next step is to demonstrate autonomicity (or lack of it) by dynamic studies. The most important study is the measurement of all three forms of TSH (the α-subunit, molar TSH, and the β-subunit) in response to the intravenous administration of TRH. In general, patients with TSH-secreting tumors fail to increase the α-subunit, the β-subunit, and the molar TSH level in response to TRH,[62] while patients with nontumoral inappropriate TSH secretion demonstrate a brisk rise in the levels of all three forms of TSH.

The response of TSH and its subunits to T3 administration may also provide information in delineating the tumoral from the nontumoral varieties of inappropriate TSH secretion. When TSH-induced hyperthyroidism occurs as a consequence of pituitary resistance to thyroidal hormones, the negative feedback threshold is reset at a higher level. Therefore, the thyrotropes are not completely autonomous and retain some ability for suppression to exogenous thyroidal hormones, albeit at a much higher dose than normals, often requiring more protracted administration than normals; thus, α-TSH and β-TSH and the molar TSH demonstrate partial suppression to thyroidal hormone administration. In contrast, the TSH secretion of tumors is usually quite autonomous, and fails to show suppression of the subunits as well as the molar TSH in response to cytomel administration.

The response of TSH and its subunits to glucocorticoid administration does not separate the tumorous from the nontumorous variety of inappropriate TSH secretion.[1,27,62] The role of dopamine agonists, and antagonists in differentiating tumorous from nontumorous cases has not been clearly defined. In one report,[22] therapy with bromocriptine lowered the level of TSH and its α-subunit in a nontumorous case of TSH hypersecretion while having no effect on the TSH levels of a patient with TSH-secreting adenoma. In a similar vein, Sisson et al.[63] showed that patients with nonadenomatous inappropriate TSH secretion usually demonstrate a brisk and exaggerated TSH response to dopamine antagonism. The patient with TSH secreting pituitary adenoma described by Scanlon et al.[50] demonstrated no response to dopamine antagonism.

In summary, there are four hormonal criteria that assist in making a diagnosis of adenomatous inappropriate TSH secretion:

1. Inappropriately elevated TSH levels in the presence of increased circulating thyroidal hormone levels.
2. An elevated basal α-TSH subunit level.
3. An elevated basal α-TSH/molar TSH ratio.
4. Absence of responsiveness of α-TSH, β-TSH, and molar TSH to the administration of TRH and thyroidal hormones (T3).

Failure to respond to dopamine agonists (bromocriptine) as well as antagonists (metoclopramide) may be included in these criteria if confirmed in larger number of patients.

Step 4: Computed tomographic evaluation of the sella turcica is an integral part in the diagnostic evaluation of patients with the syndrome of inappropriate TSH secretion and complements the hormonal tests outlined above. Although data are scanty, it appears that most TSH-secreting adenomas are visualized by CT.[22]

Treatment

The therapy of inappropriate TSH secretion, with the limited data available, seems to depend on two major factors: the presence or absence of tumor in the pituitary gland, and the metabolic status of the patient. The presence of pituitary tumor can be determined on the basis of hormonal data and CT (Table 42).

The determination of the metabolic status can be relatively easy if symptoms and signs of hyperthyroidism are evident. In the absence of symptoms, the metabolic status may be difficult to ascertain, necessitating several studies such as BMR, ankle reflex–relaxation time, systolic time intervals, lipid levels, indices of muscle and skeletal protein catabolism, etc. For therapeutic purposes, patients with the syndrome of inappropriate TSH secretion can be divided into four major groups: patients with hypometabolism and inappropriate TSH secretion, patients with euthyroidism and inappropriate TSH secretion, those with nonadenomatous TSH-mediated hyperthyroidism, and those with adenomatous TSH-mediated hyperthyroidism.

1. Inappropriate TSH secretion with elevated thyroid hormone levels seen in conjunction with mild hypothyroidism usually represents target organ resistance to thyroidal hormones at both the peripheral level as well as the pituitary level. The original reports by Refetoff et al.[23–25] exemplify the nature of disorder seen in this group of patients. Thy-

TABLE 42
Differences between Adenomatous and Nonadenomatous TSH Hypersecretion

Feature	Nonadenomatous inappropriate TSH secretion	Adenomatous TSH secretion
Circulating thyroidal hormone level	Elevated	Elevated
TSH level	Elevated; occasionally inappropriately "normal"	↑ ↑
α-TSH subunit	Normal or minimally elevated	Clearly elevated
Ratio of α-TSH/molar TSH	<1	>1
Response of α-TSH and molar TSH to TRH	Normal, exaggerated	No response
Suppression to T3	Preserved, albeit at higher doses	Nonsuppressible
Response to dopamine agonists and agonists	Preserved	Nonresponsive
CT of sella	No tumor	Tumor present
Associated hypersecretory syndromes (HGH, prolactin)	Absent	May be present
Clinical progression	Slow, mild, and benign	Can be aggressive with compressive sequelae

romegaly, retardation of growth and development, and hearing disorders are the presenting features. Severe overt hypothyroidism is unusual, probably owing to the partial degree of the peripheral resistance encountered in such patients. Treatment is with supraphysiological doses of thyroxine to overcome the resistance. Clinical improvement as well as improvement in the metabolic parameters are observed, usually in a slow and gradual fashion.

2. Patients with euthyroidism and inappropriate TSH secretion represent the group with well-compensated generalized target organ resistance. Regarded as the most frequent manifestation of inappropriate TSH resistance, the clinical presentation of these patients is characterized by the triad of thyromegaly, clinical euthyroidism, and elevated thyroidal hormones in the circulation. This condition is often mistaken for hyperthyroid Graves' disease by the unwary. In order to avoid unnecessary therapeutic intervention, every patient with elevated thyroid hormone levels and clinical euthyroidism must undergo a TRH stimulation study. In contrast with patients with hyperthyroid Graves' disease, who show no increase in TSH following TRH, patients with generalized resistance to thyroidal hormones demonstrate a robust TSH response to TRH. No treatment is required for this group of patients; especially contraindicated are antithyroid drug therapy, which, by lowering thyroid hormone levels, will lead to the development of hypothyroidism.

3. Patients with nonadenomatous inappropriate TSH secretion with hyperthyroidism, pose a therapeutic problem. The basic defect in this disorder is excess TSH secretion by thyrotropes functioning at a higher threshold for negative feedback suppression by thyroidal hormone. It is therefore not surprising that treatment modalities aimed at partial or subtotal removal of the target gland (the thyroid) often lead to recurrence of hyperthyroidism. Treatment with antithyroid drugs, although seemingly reasonable in view of the hyperthyroid state results in further increment in goiter size and circulating TSH levels. Attempts to ablate the thyroid with radioactive iodine may result in further pituitary hyperplasia, and TSH increase—a situation quite analogous to the development of Nelson syndrome following bilateral total adrenalectomy for pituitary dependent Cushing's disease. In the animal model for the syndrome (the LAF_1 mouse), chronic hyperstimulation of thyrotropes results in the development of tertiary tumor.[64] The natural history of this entity in humans is not well known. Rosler et al.[15] described two patients in whom inappropriate TSH secretion had existed for 27 and 61 years, respectively, without resulting in tumor formation. In the absence of clear-cut information regarding the natural history, and given the fact that antithyroid drugs, surgery, and radioactive iodine are less than satisfactory choices, the only remaining option for these patients is therapy aimed at reducing TSH secretion by the thyrotropes. This can be done by the use of cautious exogenous administration of thyroidal hormones. The use of thyroid

hormones to treat hyperthyroidism may on the surface appear paradoxical. However, this is logical because the inappropriate TSH secretion in these patients is a consequence to selective pituitary resistance, which is capable of suppression to prolonged administration of thyroidal hormones. Regardless of the nature of the defect, the pituitary thyroid feedback mechanism operates normally in these patients, albeit at a higher level. TSH suppression following the administration of thyroidal hormones has been well documented in patients with this syndrome.[52,53,62] Rosler et al.[15] administered 50 μg T3 once a day to three patients with TSH-induced hyperthyroidism secondary to nonadenomatous inappropriate TSH secretion. The use of 50 μg T3 daily did not result in hypermetabolism but effectively suppressed TSH secretion. More interestingly, the TSH concentrations remained suppressed throughout the day, although the T3 was administered only once a day. The authors pointed out the need for protracted treatment for at least 2–3 months. Complete biochemical and clinical remission of the disease was seen as long as treatment was continued. The role of somatostatin or dopamine agonists in the treatment of such patients has not been completely elucidated. Sporadic case reports of responsiveness to bromocriptine therapy have appeared in the literature.[65]

4. The treatment for patients with TSH secreting adenomas is selective adenomectomy, or hypophysectomy often combined with postsurgical irradiation of the pituitary gland to prevent recurrence. Interim measures include the use of adequate antithyroid drug therapy and in same instances radioactive iodine therapy to the thyroid gland. Regardless of the forms of therapy directed on the thyroid gland, definitive treatment directed against the pituitary is clearly indicated.

References

1. Emerson CH, Utiger RD: Hyperthyroidism and excessive thyrotropin secretion. *N Engl J Med* **287**:328, 1972.
2. Gershengorn MC, Weintraub BD: Thyrotropin-induced hyperthyroidism caused by selective pituitary resistance to thyroid hormone: A new syndrome of "inappropriate secretion of TSH." *J Clin Invest* **56**:633, 1975.
3. Kourides IA, Pekonen F, Weintraub BD: Absence of thyroid-binding immunoglobulins in patients with thyrotropin-mediated hyperthyroidism. *J Clin Endocrinol Metab* **51**:271, 1980.
4. Kourides IA, Ridgway ED, Weintraub BD, et al: Thyrotropin-induced hyperthyroidism: Use of alpha and beta subunit levels to identify patients with pituitary tumors. *J Clin Endocrinol Metab* **45**:534, 1977.
5. Elewaut A, Mussche M, Vermeulen A: Familial partial target organ resistance to thyroid hormones. *J Clin Endocrinol Metab* **43**:575, 1976.
6. Sato T, Saida K, Suzuki Y, et al: A case of the syndrome of inappropriate secretion of TSH. *Endocrinol Jpn* **26**:623, 1979.
7. Novogroder M, Utiger R, Boyar R, et al: Juvenile hyperthyroidism with elevated thyrotropin (TSH) and normal 24 hour FSH, LH, GH and prolactin secretory patterns. *J Clin Endocrinol Metab* **45**:1053, 1977.

8. Schadlow AR, Surks MI, Schwartz HI, et al: Specific triiodothyronine binding sites in the anterior pituitary of the rat. *Science* **176**:1252, 1972.

9. Larsen PK, Dick TE, Markowitz BP, et al: Inhibition of intrapituitary thyroxine to 3,5,5'-triiodothyronine conversion prevents the acute suppression of thyrotropin release by thyroxine in hypothyroid rats. *J Clin Invest* **64**:117, 1979.

10. Silva JE, Larsen PR: Contributions of plasma triiodothyronine and local thyroxine monodeiodination to triiodothyronine to nuclear triiodothyronine receptor saturation in pituitary, liver and kidney of hypothyroid rats. *J Clin Invest* **61**:1247, 1978.

11. Gershengorn MC, Geras E, Marcus-Samuels BE, et al: Receptor affinity and biological potency of thyroid hormones in thyrotropic cells. *Am J Physiol* **237**:E142, 1979.

12. Larsen PR: Thyroid-pituitary interaction: Feedback regulation of thyrotropin secretion by thyroid hormones. Seminars in Medicine of Beth Israel Hospital. *N Engl J Med* **306**:23, 1982.

13. Obregon MJ, Pascual A, Mallol J, et al: Evidence against a major role of L-thyroxine at the pituitary levels: Studies in rats treated with iopanoic acid (Telepaque). *Endocrinology* **106**:1827, 1980.

14. Silva JE, Dick TE, Larsen PR: The contribution of local tissue thyroxine monodeiodination to the nuclear 3,5,3'-triiodothyronine in pituitary, liver and kidney of euthyroid rats. *Endocrinology* **103**:1196, 1978.

15. Rosler A, Litvin Y, Hage C, et al: Familial hyperthyroidism due to inappropriate thyrotropin secretion successfully treated with triiodothyronine. *J Clin Endocrinol Metab* **54**:76, 1982.

16. Besses GS, Burrow GN, Spaulding SW, et al: Dopamine infusion acutely inhibits the TSH and prolactin response to TRH. *J Clin Endocrinol Metab* **41**:985, 1975.

17. Delitala G: Dopamine and TSH secretion in man. *Lancet* **2**:760, 1977.

18. Refetoff S, Fang VS, Rapoport B, et al: Interrelationships in the regulation of TSH and prolactin secretion in man: Effects of L-dopa, TRH and thyroid hormone in various combinations. *J Clin Endocrinol Metab* **38**:450, 1974.

19. Scanlon MF, Weightman DR, Mora B, et al: Evidence for dopaminergic control of thyrotrophin secretion in man. *Lancet* **2**:421, 1977.

20. Siler TM, Yen SSC, Vale W, et al: Inhibition by somatostatin on the release of TSH induced in man by thyrotropin-releasing factor. *J Clin Endocrinol Metab* **38**:742, 1974.

21. Porter BA, Refetoff S, Rosenfield RL, et al: Abnormal thyroxine metabolism in hyposomatotrophic dwarfism and inhibition of responsiveness to TRH during GH therapy. *Pediatrics* **51**:668, 1975.

22. Weintraub BD, Gershengorn MC, Kourides IA, et al: Inappropriate secretion of thyroid-stimulating hormone. *Ann Intern Med* **95**:339, 1981.

23. Refetoff S, Dewind LT, DeGroot LJ: Familial syndrome combining deaf-mutism, stippled epiphyses, goiter and abnormally high PBI; possible target organ refractoriness to thyroid hormone. *J Clin Endocrinol Metab* **27**:279, 1967.

24. Refetoff S, DeGroot LJ, Benard B, et al: Studies of a sibship with apparent hereditary resistance to the intracellular action of thyroid hormone. *Metabolism* **21**:723, 1972.

25. Refetoff S, DeGroot LJ, Barsano CP: Defective thyroid hormone feedback regulation in the syndrome of peripheral resistance to thyroid hormone. *J Clin Endocrinol Metab* **51**:41, 1980.

26. Lamberg BA: Congenital euthyroid goiter and partial peripheral resistance to thyroid hormones. *Lancet* **1**:845, 1973.

27. Bode HH, Danon M, Weintraub BD, et al: Partial target organ resistance to thyroid hormone. *J Clin Invest* **52**:776, 1973.

28. Agerbaek H: Congenital goiter presumably resulting from tissue resistance to thyroid hormones. *Israel J Med Sci* **8**:1859, 1972.

29. Schneider G, Keiser HR, Bardin CW: Peripheral resistance to thyroxine: A cause of short stature in a boy without goitre. *Clin Endocrinol (Oxf)* **4**:111, 1975.

30. Maenpaa J, Liewendahl K: Peripheral insensitivity to thyroid hormones in a euthyroid girl with goitre. *Arch Dis Child* **55**:207, 1980.

31. Seif FJ, Scherbaum W, Klingler W: Syndrome of elevated thyroid hormone and TSH blood levels-a case report. *Acta Endocrinol (Copenh)* **87**(suppl 215):81, 1978.

32. Lamberg BA, Rosengard S, Leiwendhal K, et al: Familial partial peripheral resistance to thyroid hormones. *Acta Endocrinol (Copenh)* **87**:303, 1978.

33. Tamagna EI, Carlson HE, Hershman JM, et al: Pituitary and peripheral resistance to thyroid hormone. *Clin Endocrinol (Oxf)* **10**:431, 1979.

34. Mihailovic V, Feller MS, Kourides IA, et al: Hyperthyroidism due to excess thyrotropin secretion: follow-up studies. *J Clin Endocrinol Metab* **50**:1135, 1980.

35. Kaplan MM, Swartz SL, Larsen PR: Partial peripheral resistance to thyroid hormone. *Am J Med* **70**:1115, 1981.

36. Brooks MH, Barbato AL, Collins S, et al: Familial thyroid hormone resistance. *Am J Med* **71**:414, 1981.

37. Bantle JP, Seeling S, Mariash CN, et al: Resistance to thyroid hormones: A disorder frequently confused with Graves' disease. *Arch Intern Med* **142**:1867, 1982.

38. Oppenheimer JH, Koerner D, Schwartz HL, et al: Specific nuclear triiodothyronine binding sites in rat liver and kidney. *J Clin Endocrinol Metab* **35**:330, 1972.

39. Surks MI, Koerner D, Dillman W, et al: Limited capacity binding sites for L-triiodothyronine in rat liver nuclei. *J Biol Chem* **248**:7066, 1973.

40. Lo CS, Edelman IS: Effect of triiodothyronine on the synthesis and degradation of renal cortical (Na^+ + K^+)-adenosine triphosphate. *J Biol Chem* **251**:7834, 1976.

41. Bernal J, Refetoff S, DeGroot LJ: Abnormalities of triiodothyronine binding to lymphocyte and fibroblast nuclei from a patient with peripheral tissue resistance to thyroid hormone action. *J Clin Endocrinol Metab* **47**:1266, 1978.

42. Liewendahl K, Rosengard S, Lamberg BA: Nuclear binding of triiodothyronine and thyroxine in lymphocytes from subjects with hyperthyroidism, hypothyroidism and resistance to thyroid hormones. *Clin Chim Acta* **83**:41, 1978.

43. Elewaut A, De Baets M, Vermeulen A: Triiodothyronine binding to lymphocyte nuclei and plasma cyclic AMP response to intraveous glucagon in patients with general resistance to thyroid hormones. *Acta Endocrinol (Oxf)* **97**:54, 1981.

44. Eil C, Fein H, Smith T, et al: Nuclear binding of ^{125}I-T_3 in intact skin fibroblasts from patients with peripheral thyroid hormone resistance. In *Fifty-Seventh Meeting of the American Thyroid Association, Minneapolis, 1981* (abst. T2).

45. Kanter R, Chait A, Kenny MA, et al: Familial resistance to thyroid hormone in three generations: Studies of TSH and of responses to T_3 by cultured cells. *Fifty-Seventh Meeting of the American Thyroid Association, Minneapolis, 1981* (abst. T3).

46. Yovos JG, Falko JM, O'Dorisio TM, et al: Thyrotoxicosis and a thyrotropin-secreting pituitary tumor causing unilateral exophthalmos. *J Clin Endocrinol Metab* **53**:338, 1981.

47. Hamilton CR Jr, Adams LC, Maloof F: Hyperthyroidism due to thyrotropin-producing pituitary chromophobe adenoma. *N Engl J Med* **283**:1077, 1970.

48. Tolis G, Bird C, Bertrand G, et al: Pituitary hyperthyroidism: Case report and review of the literature. *Am J Med* **64**:177, 1978.

49. Baylis PH: Case of hyperthyroidism due to a chromophobe adenoma. *Clin Endocrinol (Oxf)* **5**:145, 1976.

50. Scanlon MF, Howells S, Peters JR, et al: Hyperprolactinaemia, amenorrhoea and galactorrhoea due to a pituitary thyrotroph adenoma. *Clin Endocrinol (Oxf)* **23**:35, 1985.

51. Hamilton CR Jr, Maloof F: Acromegaly and toxic goiter: cure of hyperthyroidism and acromegaly by proton-beam partial hypophysectomy. *J Clin Endocrinol Metab* **35**:659, 1972.

52. Lamberg BA, Pekonen F, Gordin A: TSH-induced hyperthyroidism and acromegaly due to pituitary tumour. *Ann Endocrinol* **39**:14A, 1978.

53. Horn K, Erhardt F, Fahlbusch R, et al: Recurrent goiter, hyperthyroidism, galactorrhea and amenorrhea due to a thyrotropin and prolactin-producing pituitary tumor. *J Clin Endocrinol Metab* **43**:137, 1976.

54. Duello TM, Halmi NS: Pituitary adenoma producing thyrotropin and prolactin. An immunocytochemical and electron microscopic study. *Virchows Arch [A]* **376**:255, 1977.

55. Benoit R, Pearson-Murphy BE, Robert F, et al: Hyperthyroidism due to a pituitary TSH secreting tumor with amenorrhea-galactorrhea. *Clin Endocrinol (Oxf)* **12**:11, 1980.

56. Evans M, Croxson MS, Wilson TM, et al: The screening of patients with suspected thyrotoxicosis using a sensitive TSH radioimmunoassay. *Clin Endocrinol (Oxf)* **22**:445, 1985.

57. Kourides IA, Weintraub BD, Rosen SW, et al: Secretion of alpha subunit of glycoprotein hormones by pituitary adenomas. *J Clin Endocrinol Metab* **43**:97, 1976.

58. Kourides IA, Ridgway EC, Weintraub BD, et al: Thyrotropin-induced hyperthyroidism: Use of alpha and beta subunit levels to identify patients with pituitary tumors. *J Clin Endocrinol Metab* **45**:534, 1977.
59. Blackman MR, Weintraub BD, Kourides IA, et al: Discordant elevation of the common alpha subunit of the glycoprotein hormones compared to beta subunits in serum of uremic patients. *J Clin Endocrinol Metab* **53**:39, 1981.
60. MacFarlane IA, Beardwell CG, Shalet SM, et al: Glycoprotein hormone alpha subunit secretion by pituitary adenomas: Influence of external irradiation. *Clin Endocrinol (Oxf)* **13**:215, 1980.
61. Rosen SW, Weintraub BD, Vaitukaitis JL, et al: Placental proteins and their subunits as tumor markers. *Ann Intern Med* **82**:71, 1975.
62. Smallridge RC, Wartofsky I, Dimond RC: Inappropriate secretion of thyrotropin: Discordance between the suppressive effects of corticosteroids and thyroid hormone. *J Clin Endocrinol Metab* **48**:700, 1979.
63. Sisson JC, Gniadek TC, Forbes H: Inappropriate secretion of thyrotrophin: A functional abnormality with a familial occurrence. *Clin Res* **29**:59A, 1981.
64. Furth J, Moy P, Hershman JM, et al: Thyrotropic tumor syndrome. *Arch Pathol Lab Invest* **96**:217, 1973.
65. Takamatsu J, Mozai T, Kuma K: Bromocriptine therapy for hyperthyroidism due to increased thyrotropin secretion. *J Clin Endocrinol Metab* **58**:934, 1984.

7

Central Hypothyroidism—
Trophoprivic Hypothyroidism

Definition

The term *central hypothyroidism* refers to the development of thyroid failure secondary to loss of TSH drive. The thyroid gland, which is intrinsically normal, gradually fails to function, since it is deprived of its tropic stimulus (hypothyrotropic hypothyroidism). The hypothyrotropic milieu can result from intrinsic pituitary disease (secondary hypothyroidism) or as a consequence of hypothalamic disease resulting in partial or complete loss of thyrotropin-releasing hormone (tertiary hypothyroidism).

Etiology and Pathogenesis

The causes of central hypothyroidism can be viewed in terms of four etiological perspectives: intrinsic pituitary diseases, deficient secretion of TRH, interruption of TRH transport from the hypothalamus to the pituitary, and secretion of an "abnormal" TSH that is biologically less active. The most common of these causes is TSH deficiency resulting from intrinsic pituitary disease.

Intrinsic Pituitary Disease

This entity represents the classic situation of target organ failure resulting from trophic hormone deficiency. The causes for this form of TSH deficiency are outlined in Table 43. The most frequent cause for intrinsic pituitary failure is tumorous destruction of the thyrotrope population, resulting in a compromised reserve. Tumor destruction results in loss of several hormones, often in a sequential fashion. Thus, loss of gonadotropins and loss of growth hormone (GH) are the earliest events followed by loss of TSH, ACTH, and prolactin. It is important to recognize that diminished TSH reserve demonstrated by reverse testing antedates the development of clinically overt hypothyroidism by months to even years. In most instances of TSH deficiency as a consequence of tumor encroachment, there are associated deficiencies of gonadotropins and/or growth hormone (GH). While chromophobe adenoma represents the most common reason for tumor-related TSH deficiency in adults, craniopharyngioma represents the leading cause for such a phenomenon in children.[1] The TSH deficiency seen in association with hypophysectomy, vascular disorders, and hypophysitis is generally part of multiple trophic hormone failure.[2-7]

Isolated TSH deficiency due to intrinsic pituitary (nonhypothalamic) disease is rare[8,9] (Table 44). When strict criteria are used to define the non-TRH-mediated variety of isolated TSH deficiency, this rare disorder appears to be limited to two varieties described in the literature. In 1971, Miyai et al.[10] described two sisters, offspring of a consanguineous marriage, presenting with nongoitrous cretinism. The diagnosis of isolated pituitary TSH deficiency

TABLE 43
TSH Deficiency

Pituitary tumors (chromophobe adenoma)
Craniopharyngioma
Vascular—
  Sheehan's syndrome
  Apoplexy
Granulomatous disease
Autoimmune
  Lymphocytic hypophysitis
Hypophysectomy

TABLE 44
Isolated TSH Deficiency (Nonhypothalamic)[a]

Familial isolated TSH deficiency[10]
Isolated TSH deficiency associated with
    pseudohypoparathyroidism[16-18]
Functional, reversible TSH deficiency
  Sick euthyroid syndrome[25]
  Children on GH therapy[27-30]

[a] GH, growth hormone; TSH, thyroid-stimulating hormone.

was based on undetectable TSH levels in the presence of marked hypothyroidism, integrity of other pituitary trophic hormones, failure of the TSH level in plasma to rise after administration of TRH, and an appreciable increase in thyroidal radioiodine uptake following the administration of exogenous TSH. The hereditary nature of this form of isolated TSH deficiency, as well as the background of consanguinity, is highly reminiscent of isolated familial GH deficiency, a disorder thought to be inherited in an autosomal-recessive pattern.[11,12]

Isolated TSH deficiency may also be encountered in association with pseudohypoparathyroidism, a prototypical disorder exemplifying target organ resistance to trophic hormone. It is now well recognized that deficient activity of receptor–cyclase coupling protein (guanine nucleotide regulatory protein) underlies the pathogenesis of pseudohypoparathyroidism. As a consequence, there is defective generation of cAMP, or a breakdown in the signals after cAMP generation.[13-15] Selective TSH deficiency has been reported in association with pseudohypoparathyroidism.[16-18] The basis for such an association may reside in a common mechanism, i.e., defective cAMP generation within the thyrotropes. Most cases of isolated TSH deficiency were described during the 1960s at a time when the physiology of TSH secretion was not entirely understood. Recent insights into the regulation of TSH secretion have shifted the emphasis away from cAMP mediation and have focused on stimulus–secretion coupling.[19,20] The new concept that an increase in the cytoplasmic free calcium is a major determinant for the secretion and release of TSH, and perhaps other pituitary hormones as well, should be taken into account while explaining the TSH deficiency in association with pseudohypoparathyroidism, a disease characterized by hypocalcemia and poor signal recognition. In fact, deficiency of prolactin secretion,[21] as well as deficiency in multiple hormones of the anterior pituitary,[22] have been described in association with pseudohypoparathyroidism. More recently, partial gonadotropin resistance[23] has been described in pseudohypoparathyroidism, conveying the global nature of tissue resistance to diverse hormones encountered in this fascinating disorder. Finally, it should be remembered that besides selective TSH deficiency, patients with pseudohypoparathyroidism may demonstrate an exaggerated TSH response to TRH.[24]

Functional, often reversible, isolated TSH deficiency can be encountered in patients with sick euthyroid syndrome. A subset of patients with the sick euthyroid syndrome demonstrate low T_3, low T_4, and low TSH levels that fail to increase following TRH administration.[25] It has been suggested that the low thyroxine status in this subset of patients with severe illness may in fact result from a breakdown in the normal negative feedback regulation between the thyroid and the pituitary[26] and that acute illness decreases TSH secretion. It is debatable if these sick patients should be termed euthyroid, because the hormonal milieu of these patients is analogous to that of thyrotropin deficiency.

The transient secondary hypothyroidism that occurs in children treated with GH is also a reflection of functional impairment of TSH secretion by the thyrotropes. Porter et al.[27] and Lippe et al.[28] simultaneously reported that children receiving GH therapy for GH deficiency can develop decrements in thyroxine levels and inhibition of TSH response to exogenous TRH as well as a reversible hypothyroid state. The clinical impact of these findings is significant, since an attenuation in growth response to therapy may occur as a result of the hypothyroid state. Similar findings were also reported by other workers.[29,30] The mechanism for such a phenomenon is speculative. Since GH and somatostatin are interregulated by a positive ultrashort feedback loop, it is possible that the elevated GH concentrations in the circulation (caused by therapy with exogeneous GH administration) may stimulate somatostatin release, which in turn can suppress the ability of the thyrotrope to respond to endogenous TRH. Although attractive, this hypothesis has not been substantiated.

Deficient TRH Secretion

In 1971, Pittman et al.[31] reported a 19-year-old man who had developed hypothyroidism, growth retardation, and diabetes insipidus following severe head injury. The remarkable feature in this patient was an extremely low TSH level (below $1.2 \mu U/ml$) in conjunction with a low serum thyroxine level. More importantly, the low TSH levels demonstrated a significant incremental response to the administration of exogenous synthetic TRH. The term "hypothalamic hypothyroidism" was used to underscore the fact that the normal TSH response to TRH suggested TRH deficiency as the cause for hypothyroidism. The term *tertiary* hypothyroidism is also in vogue to describe the entity. The development of selective unitropic pituitary hormone deficiencies occurring as a consequence of releasing factor deficiencies is a well-documented concept. Experimentally, in the rat, ablation of the thyrotropic area of the hypothalamus (the preoptic suprachiasmatic nucleus and the periparaventricular area) results in the depletion of nearly 70% of overall hypothalamic TRH content and induces severe hypothyroidism in the animal.[32,33] In humans, the syndrome of hypogonadotropic hypogonadism, Kallmann's syndrome,[34] is a classic example of spontaneous familial hypogonadism resulting from deficiency of gonadotropin-releasing hormone (GnRH). Also some forms of isolated growth hormone failure may develop as a result of deficiency of

TABLE 45
TSH Deficiency (Secondary) to Deficient
TRH Secretion

Spontaneous
  Familial
  Sporadic
    Isolated TRH deficiency
    Associated with deficiency of other releasing factors
Tumors
  Craniopharyngiomas
  Suprasellar tumor
Metastatic disease
  Breast
  GI tract
Sheehan's syndrome
Granulomatous disease
  Hand-Schüller-Christian disease
Trauma
  Basal skull fracture

---

growth hormone-releasing hormone (GHRH).[35] The hypothyroidism resulting from TRH deficiency belongs in the same clinical category as the above two disorders and is prototypical of the releasing factor deficiency syndromes.

TRH deficiency can be isolated or part of other releasing factoropathies. A definite familial pattern has not been established for hypothalamic hypothyroidism. The spontaneous variety of this disorder must be extremely rare, contrasting with the relative frequency of GnRH deficiency. Hypothalamic hypothyroidism is more frequently a result of trauma or destructive lesions in the suprasellar region, particularly tumors such as craniopharyngiomas. Basal skull fracture is a rare but important cause of hypothalamic hypothyroidism,[36] underscoring the need for a properly obtained past history. Rarely, hypothalamic hypothyroidism may result from Sheehan's syndrome[37] (Table 45).

Interruption of TRH Transport

The TRH secreted by the hypothalamus reaches the anterior pituitary via the median eminence and the pituitary stalk. The peptide is carried through the hypothalamic–hypophyseal portal circulation. Interruption of the stalk at any level results in depriving the thyrotropes of their trophic drive from the hypothalamus. This is not restricted to the thyrotropes alone, however, since the somatotropes, corticotropes, and the gonadotropes also depend on an intact avenue of transport for their respective hypothalamic releasing hormones. As a consequence, interruption of the stalk, exemplified most dramatically by pituitary stalk section, results in secondary hypopituitarism. This condition is characterized by decreased function of all populations of pituicytes, except the lactotropes; these cells not only survive but also thrive exceedingly well, since their inhibitory factor, hypothalamic dopamine, is no longer available to exert a tonic negative control. Thus, dual hormonal effects are expressed: low TSH, HGH, FSH, LH, and ACTH coupled with hyper-prolactinemia—the pituitary isolation syndrome.

Clinically, TSH deficiency caused by interruption of the pituitary stalk is almost always a result of tumors: craniopharyngiomas that impinge on the stalk from above and pituitary tumors such as chromophobe adenomas that impinge on the stalk when they extend superiorly. It is remarkable that hormonal evaluation can provide evidence of early stalk interruption, at a stage when CT may be equivocal.

Secretion of Biologically Inactive TSH

The observation that several patients presumed to have hypothalamic hypothyroidism demonstrated normal, or even elevated basal TSH levels, raised the possibility that the hypothyroidism in such patients may have been due to secretion of TSH with reduced biological activity. Several laboratories around the world reported conflicting data on the nature of the basal (and TRH stimulated) TSH levels in patients with hypothyroidism caused by hypothalamic-pituitary disorders.[38–40] In addition, Mitsuma et al.[41] reported that some patients with hypothalamic hypothyroidism demonstrated a diminished thyroidal response to TRH administration, suggesting that the TSH released by TRH administration did not cause optimal thyroidal stimulation. This observation was supported by Faglia et al.,[42] who studied 25 patients with central hypothyroidism with a normal or exaggerated TSH response to exogenous TRH; the T_3 response evoked by the endogenous TSH released after TRH, was compared with the T_3 response to TRH in 15 euthyroid subjects. Their results indicated a significantly decreased T_3 response to TRH administration in some patients with central hypothyroidism. The assumption was that although exogenous TRH brought about a robust increase in the release of endogenous TSH, the biological effectiveness of the latter was open to question. The observation that the β-subunit levels in some of these patients were selectively increased is also an intriguing one.[43] The concept that the pituitary gland of these patients secreted a material which is immunologically identical to the "standard" TSH, behaving normally to provocative and suppressive maneuvers, but with little or no biological activity, gained momentum with the application of cytochemical bioassays. Faglia et al.[44] studied the bioactivity of the basal and TRH-stimulated TSH of patients presumed to have hypothalamic hypothyroidism, and demonstrated a striking discordance between the immunoreactive level and bioassay level obtained by cytochemical assay. To define the mechanism of defective hormonal action further, Beck-Peccoz et al.[45] studied the adenylate cyclase-stimulating bioactivity and receptor-binding activity of the purified immunoreactive TSH obtained from patients with central hypothyroidism. As expected, the immunoreactive TSH from these patients showed a markedly decreased propensity to stimulate adenylate cyclase from, or bind to human thyroid membrane preparations.

The most remarkable aspect of the study was the restoration of normal bioactivity and binding following protracted TRH treatment. This fact lends support to the fact that the TRH regulates not only the quantity of TSH secreted but also the quality (molecular configuration) of that hormone. In all patients treated with TRH, thyroid hormone secretion was normalized.

The concept that target organ deficiency, i.e., thyroid failure, can occur as a result of biologically inactive trophic hormone, in this instance TSH is hardly new. Kreiger[46] reported a patient with multiple endocrine problems characterized by a suprasellar mass, clinical hypothyroidism, a low T4 level coupled with an elevated TSH level that possessed very little biological activity. More recently, Dickstein and Barzilai[47] described a patient with a chromophobe adenoma, panhypopituitarism, markedly elevated TSH levels in the presence of a low serum thyroxine, and a low thyroidal uptake that remarkably increased following exogenous TSH administration. Notably, this patient's hypothyroidism (caused by secretion of biologically inactive TSH by the pituitary tumor) completely resolved after removal of the tumor. The message to underscore is that hypothyroidism can come as a result of biologically inactive TSH secreted by tumors or by normal thyrotropes.[48] When such a phenomenon is noted in the absence of a demonstrable tumor, the possibility that the thyrotropes secrete an abnormal TSH can be verified in a methodical sequence by demonstrating that (1) the basal TSH levels are normal or elevated, (2) the immunoreactive TSH responds well to exogenous TRH, (3) the thyroidal (T_3) response to the TSH thus released is poor, and (4) the purified immunoreactive TSH binds poorly with human thyroid membranes, and inadequately stimulates adenylate cyclase in the bioassay system, and (5) a novel therapeutic concept that can be added to the above criteria is normalization of the defective action of TSH by treatment with TRH for a protracted period. This would indicate that the basic defect in this syndrome is the inability of endogenous TRH to stimulate the secretion of high-quality TSH by the thyrotropes. The major etiological categories of hypothyroidism caused by hypothalamic pituitary dysfunction are outlined in Table 46.

TABLE 46
Central Hypothyroidism

Intrinsic pituitary deficiency
    Spontaneous
    Secondary to organic lesions
Hypothalamic TRH deficiency
    Spontaneous
    Secondary to organic lesions
Interruption of TRH transport
    Tumors
Secretion of biologically inactive TSH
    Tumors
    Nontumorous

Clinical Features

General

The clinical features of hypothyroidism caused by pituitary–hypothalamic disorders are quite similar to those of primary hypothyroidism. Five general remarks are appropriate:

1. Although in general the manifestations of hypothyroidism tend to be milder than in primary hypothyroidism, occasionally myxedema, and even myxedema coma, can result from secondary hypothyroidism. The severity noted in some cases is probably a reflection of chronicity.
2. Secondary hypothyroidism must be excluded in any hypothyroid patient. The reason for this is the frequent association of secondary hypoadrenalism with secondary hypothyroidism. The therapeutic implication is crucial, because L-T4 administration to a hypoadrenal patient will result in catastrophic results by precipitating acute adrenal crisis.
3. Tumors of the pituitary and suprasellar region constitute the most important and frequent etiology of secondary hypothyroidism; a missed diagnosis of the latter can result in blindness from chiasmal compression.
4. The clinical distinction between primary and secondary hypothyroidism at the bedside can be extremely difficult The presence of thyromegaly, Graves' stigmata, or a past history of ablative therapy are the best clues to suggest a primary thyroidal etiology. Severe orthostatic hypotension, marked hypopigmentation, complete loss of axillary hair (in females), and visual-field defects strongly point to a pituitary etiology. Several findings have, at best, only marginal value in making the clinical differentiation: alopecia, menstrual irregularities, skin discoloration, mild hypotension, galactorrhea, growth retardation, and presence of serous effusions.
5. While the distinction between primary versus secondary hypothyroidism is an essential one, the distinction between secondary and tertiary hypothyroidism is not important as far as therapy is concerned.

TABLE 47
Symptoms of Hypothyroidism

Tiredness	Increased somnolence
Lack of energy	Muscle weakness
Bloated feeling	Constipation
Weight gain	Cold intolerance
Headaches	Chest discomfort
Muscle cramps	Dry skin
Decreased libido	Hair loss
Psychiatric changes (depression)	

TABLE 48
Signs of Hypothyroidism

Mild hypothyroidism
  Mostly symptoms
  Dry skin
Moderate hypothyroidism
  Dry, scaly, cold skin
  Sinus bradycardia
  Delay in the relaxation phase of DTR
  Periorbital puffiness
  Hair loss
Severe hypothyroidism
  All the above
  Multisystem involvement

Many symptoms of hypothyroidism are vague and notoriously nonspecific and can be encountered in euthyroid individuals with a spectrum of disorders such as anemia, depression, and plain obesity (Table 47). The early stage of the disease is characterized by a plethora of symptoms in the absence of any objective physical findings. It is therefore essential to consider screening for hypothyroidism in patients with any of the symptoms listed in Table 47. Since hypothyroidism occurs much more frequently in females, the tendency to minimize the importance of these symptoms is to be deplored.

The obviousness of the physical findings of hypothyroidism is directly related to the duration and severity of the hypothyroid state (Table 48). These findings can be subtle enough to be missed by the untrained eye or striking enough to permit instant recognition of myxedema. The term myxedema is indicative of severe chronic, untreated hypothyroidism characterized by the accumulation of mucoproteinaceous material in various body spaces, particularly in the subcutaneous tissue, resulting in edema. This is rare, however, in patients with central hypothyroidism.

Cutaneous

The early signs of hypothyroidism are reflected in the skin, which is characteristically dry, cold (vasoconstricted), and often scaly. The combination of hypothyroid symptoms and the skin changes is the most frequent presentation that is encountered. As the hypothyroidism progresses, the severity of symptoms increases, and other changes become obvious. The dry skin continues to become drier, thick, and coarse, and changes in the hair become evident. These consist of changes in the texture, resulting in coarse brittle hair that falls off easily on brushing. The alopecia is usually localized but rarely can become severe and more generalized. Loss of hair in the outer third of eyebrows can be striking in some patients. Varying degrees of periorbital puffiness is usually present. At this stage, most patients demonstrate mild sinus bradycardia and a clear delay in the relaxation phase of the deep tendon reflexes (DTR), particularly impressive over the Achilles tendon. Thus, the four important changes of hypothyroidism are to be found in the skin, hair, pulse rate, and DTRs. The presence of thyromegaly in the hypothyroid patient strongly indicates primary thyroid disease as the etiology, but an absence of such enlargement is consistent with both primary and pituitary disease.

Progressive severe hypothyroidism gradually evolves into myxedema. The changes in the skin are accentuated by desquamation and extreme dryness. A characteristic lemon-yellow tinge is imparted to the skin as a result of hypercarotenemia. The puffy face, the coarse, brittle hair, the hoarse, low-pitched voice, and the inappropriate affect (myxedema madness) are unmistakable. Dependent edema and accumulation of protein-rich transudates in the pleural, pericardial, and even the peritoneal cavity can result in mistaken diagnoses of pleuropulmonary, cardiac, or hepatic disorders. Myxedematous patients are at high risk of developing myxedema coma, a lethal complication.

Severe hypothyroidism affects practically every organ system, regardless of etiology. The effects of chronic hypothyroidism on the cardiovascular, respiratory, gastrointestinal, neuromuscular, reproductive, and hematopoietic systems highlight the importance of thyroid hormones in the normal functioning of these organ systems.

Cardiovascular[49-56]

The three facets of hypothyroid heart disease are cardiomyopathy, pericardial effusion, and coronary atherosclerosis. The cardiac muscle dysfunction in hypothyroidism is characterized by abnormal hemodynamic parameters, such as a decrease in cardiac output, stroke volume, and left ventricular ejection fraction. It is important to recognize that even asymptomatic patients with subclinical hypothyroidism may demonstrate abnormal myocardial contractile function. With protracted hypothyroidism, congestive heart failure may develop. Hypothyroidism should be suspected as the cause of congestive heart failure when the edema is disproprotional to dyspnea and pulmonary congestion, when there is lack of significant tachycardia, and in the presence of exquisite sensitivity to digintoxication. The characteristic electrocardiographic (ECG) changes include short-amplitude QRS complexes, sinus bradycardia, nonspecific ST, T changes, and intraventricular conduction defects. The addition of even small doses of L-T4 results in a salubrious diuretic response. In general, the cardiac size in patients with secondary hypothyroidism tends to be smaller than in patients with primary hypothyroidism.

Although pericardial effusion occurs more commonly in primary hypothyroidism and more frequently in severe disease and seldom results in hemodynamic compromise, these effusions have been described in patients with pituitary hypothyroidism or mild disease and can rarely result in cardiac tamponade. The pericardial effusion characteristically demonstrates slow resolution after initiation of L-T4 therapy.

The hypothyroid state strongly favors the development of coronary atherosclerosis as a consequence of the increased cholesterol and triglyceride levels as well as the accumulation of mucopolysaccharides in the intima of small blood vessels. The incidence of symptomatic coronary heart disease is low in relationship to the high prevalence of coronary atherosclerosis in autopsy studies and is probably a reflection of the decreased myocardial oxygen consumption seen in this disease. Unfortunately, when myocardial oxygen consumption is raised with L-T4 therapy, the symptoms appear, leading to angina or even acute myocardial infarction. It is generally regarded that the incidence of coronary atherosclerosis is much lower in secondary hypothyroidism.

Respiratory[57-60]

The most important effect of hypothyroidism is on the respiratory center, which is rendered relatively sluggish to hypoxia and hypercarbia. This sluggishness, in its extreme form, is exemplified in myxedema coma, where CO_2 retention as a result of respiratory hypoventilation dominates the picture. This central effect is aggravated by the decreased "bellows action" of the thorax in myxedematous patients.

Gastrointestinal[61-64]

Severe hypothyroidism affects intestinal motility, resulting in paralytic ileus and even intestinal obstruction. Myxedema should be recognized as a rare cause of adynamic ileus, since surgical intervention will prove catastrophic for these patients. Rarely, myxedema can result in malabsorption syndrome. Ascites is also a rare manifestation of myxedema and is almost always seen in primary myxedema.

Neuromuscular[65-69]

The two common neurological features of hypothyroidism are abnormal stretch reflexes and entrapment neuropathies. Less commonly, a cerebellar syndrome and myxedema coma can complicate the picture of hypothyroidism. The characteristic neurological sign of hypothyroidism is the pseudomyotonic reflex, i.e., the delayed relaxation phase. This finding, although quite frequently encountered in hypothyroidism, is not unique to this disease. Diverse disease states such as diabetic neuropathy, hypoalbuminemia, and pernicious anemia may be associated with a delay in the relaxation of DTRs. Carpal tunnel syndrome and entrapment neuropathies occur in approximately one-third of hypothyroid patients.

The cerebellar syndrome associated with myxedema is characterized by ataxia and other cerebellar signs, mimicking a cerebellar tumor. The condition responds favorably to L-T3 therapy. Myxedema coma is the penultimate manifestation of chronic untreated hypothyroidism. Obtundation, hypoventilation, hypothermia, and hypotension dominate the clinical picture.

Myxedema coma is often precipitated by infection or the use of sedatives. Even with aggressive therapy myxedema coma carries an extremely high mortality rate.

The muscle disease of hypothyroidism is of two forms: a biochemical syndrome and a clinical syndrome. The biochemical syndrome is extremely common in hypothyroid patients and consists of elevated muscle enzymes (CPK, aldolase) in the absence of any myopathic symptoms or signs. The clinical syndrome of hypothyroid myopathy is particularly common in children and young adults and is characterized by the bizarre combination of subjective and objective weakness in muscles that appear hypertrophied (Hoffmann syndrome). Cramps, stiffness, and weakness dominate the symptomatology of patients with the Hoffmann type of hypothyroid myopathy. Both the biochemical and clinical forms of hypothyroid myopathy respond excellently to L-thyroxine therapy. Again, musculoskeletal involvement is encountered less frequently in secondary hypothyroidism.

Hematopoietic

Anemia is an extremely common finding in all varieties of hypothyroidism. Microcytic anemia is often secondary to menorrhagia, and normocytic normochromic anemia results from poor erythropoiesis. In patients with severe disease, intestinal malabsorption of iron, folic acid and vitamin B_{12} further complicates the hematopoietic picture.

Table 49 illustrates the protean manifestations of chronic untreated hypothyroidism.

In addition to the well-recognized multisystem involvement of chronic hypothyroidism, there are three important metabolic alterations associated with this disorder.

1. The tendency to develop fasting hypoglycemia is particularly evident in patients with pituitary hypothyroidism, since deficiency of GH and ACTH is often associated with varying degrees of impaired gluconeogenesis and glycogenolysis.
2. The hyponatremia and hypotonicity of plasma with a less than maximally dilute urine can mimic the syndrome of inappropriate ADH secretion, as well as concomitant adrenocortical insufficiency (Chapter 21).
3. Finally, the perturbations of lipid metabolism in hypothyroidism are noteworthy. Practically every variety of dyslipoproteinemia has been described in association with the hypothyroid state. Although elevated cholesterol (and β-lipoproteins) is the most impressive abnormality, patients with hypothyroidism may demonstrate elevated triglycerides, reflecting increases in VLDL, chylomicrons, and even the intermediate LDL. These abnormalities are most striking in primary thyroid failure but can be seen in pituitary hypothyroidism, albeit to a lesser degree.

TABLE 49
Multisystem Involvement of Hypothyroidism

System	Manifestation
General constitutional	Fatigue, lethargy, memory lapses, headaches, muscle aches, cold intolerance, depression, exercise intolerance

Hematologic	Anemia
	    Microcytic: Fe deficiency from poor absorption or blood loss
	    Normocytic
	    Macrocytic: folate or B_{12} deficiency

Cardiovascular	"Hypothyroid heart," congestive heart failure Pericardial effusion Bradycardia, hypertension Coronary heart disease
Respiratory	Decreased ventilatory drive, sluggish respiratory center, impaired bellows action of lungs, tendency to retain CO_2
Gastrointestinal	Atrophic gastritis Malabsorption Constipation Myxedema ileus
Musculoskeletal	Asymptomatic CPK ↑ Hoffman's syndrome (triad of increased muscle stiffness and weakness and marked elevation of CPK)
Neurological	Pseudomyotonic stretch reflexes Mental changes Myxedema coma Cerebellar ataxia Carpal tunnel syndrome
Renal	Decreased free water clearance Tendency to water intoxication SIADH-like state with hyponatremia, hypotonic plasma, less than maximally dilute urine and natriusis
Joints	Arthralgias Hyperuricemia; attacks of gout in the genetically predisposed
Reproductive	Decreased libido Menorrhagia, oligomenorrhea, amenorrhea

Diagnostic Studies

The diagnostic approach to the patient with suspected central hypothyroidism should be followed in a systematic stepwise method:

Step 1: Measurement of basal TSH (to confirm a central etiology)

Step 2: Evaluation of the TSH response to TRH (in an attempt to delineate a pituitary versus hypothalamic origin)

Step 3: Evaluation of other pituitary hormones (to define presence of hypo- or hypersecretion of the other trophic hormones)

Step 4: Radiological evaluation of the sella and suprasellar region by computed tomography (to exclude space-occupying lesions, and if present to define extent)

Basal TSH Level

The first step in the diagnostic approach to a patient with central hypothyroidism is the demonstration of an inappropriately low TSH level in the presence of low thyroidal hormone concentrations in the circulation. In general, the basal levels of TSH are low in secondary hypothyroidism. For instance, the patients with secondary hypothyroidism studied by Snyder et al.[70] showed a mean basal TSH level of 3.6 ± 0.3 μm/ml. It should be noted that in an individual patient, there may be considerable overlap between the basal TSH level of normals and patients with central hypothyroidism.[71,72] The important principle to remember is that in the presence of low T_4 or T_3 levels in the circulation, the demonstration of even a normal basal TSH concentration in plasma is suggestive of TSH deficiency. The demonstration of a low to normal basal TSH levels in patients with hypothyroidism mitigates against the diagnosis of primary hypothyroidism, since the sine qua non for primary thyroid failure is a raised TSH concentration in plasma in conjunction with low circulating thyroid hormone levels.[73,74] Although several investigators have reported either undetectable or low basal levels in patients with hypothyroidism caused by diseases of the hypothalamus or the pituitary,[75–79] there have been reports of normal or even elevated basal TSH levels in patients with central hypothyroidism.[39] In the study reported by Patel and Burger,[39] the basal TSH level was elevated or normal in 19 of 21 patients with secondary hypothyroidism. Such a paradoxical finding could be explained by three hypotheses: assay insensitivity, quantitatively maximal response of the thyrotropes to the decline in thyroid hormone levels, or secretion of an abnormal TSH molecule that is biologically ineffective but detectable by radioimmunoassay (RIA). As far as assay insensitivity, most of the currently employed TSH assays are reasonably sensitive with normal ranges of 0.2–3.5 μU/ml. Regarding the second possibility, in the presence of a slowly evolving lesion of the hypothalamus or pituitary, the pituitary thyrotropes probably go through a phase of maximal stimulation by the low circulating thyroid hormone levels, accounting for the normal or slightly elevated values. As far as the third possibility is concerned, secondary hypothyroidism has been associated with the secretion of biologically inactive TSH by tumorous[46–47] as well as nontumorous[45] pituitary tissue. Regardless of the mechanism, the message seems to be that while a low basal TSH level in a hypothyroid patient points to central hypothyroidism, a normal or even slightly elevated basal level of TSH does not mitigate against it. It is in such cases that evaluating the TSH response to TRH becomes quite helpful.

TRH Stimulation Test

In the hypothyroid patient with low levels of both thyroid hormones as well as thyrotropin, the clear-cut implication is that the hypothyroidism has evolved secondary to some form of TSH deficiency. The purpose for further testing the TSH response to exogenous administration of TRH is threefold:

1. *To determine the underlying mechanism of TSH deficiency,i.e., to differentiate the pituitary (secondary) from the hypothalamic (tertiary) forms of hypothyroidism:* Such a distinction does convey important diagnostic and prognostic connotations. For example, if TRH testing revealed a "hypothalamic" response pattern, the TSH deficiency can be assumed to have resulted from deficient TRH secretion or interruption of its transport. In the background of an underlying pituitary tumor, such a finding should strongly suggest suprasellar extension, often with stalk interruption. The CT study in such a setting may have to be performed using higher cuts to demonstrate and determine the extent of suprasellar extension. Prognostically, TSH deficiency that occurs as a result of stalk compression may be completely reversible when the compression is removed. This contrasts with the TSH deficiency that occurs as a result of destruction of thyrotropes by tumor encroachment, which is usually irreversible.

2. *To determine the degree of the compromised TSH reserves:* Failure of thyrotropes, like all other hypofunctional states, is not an all-or-none phenomenon. Varying degrees of TSH deficiency can be associated with secondary thyroid failure.

3. *To obtain baseline information regarding TSH reserve:* This can be subsequently compared with data obtained after therapeutic intervention.

In the classic setting, the TRH study can differentiate between purely "pituitary" from "hypothalamic" varieties of central hypothyroidism.[40,70,75–78] The classic response of hypothyroid patients with intrinsic pituitary disease is a flat, or suboptimal rise following TRH.[74,80,81] The classic response of the hypothalamic variety of hypothyroidism is a normal but slightly delayed (60-min peak) TSH response following TRH.[76,77] The classic response in patients with primary thyroid failure is an exaggerated response of TSH following TRH. Figure 20 outlines the TSH response to TRH in three patients prototypical of primary, secondary, and tertiary forms of hypothyroidism. These responses are consistent with the classic concepts of regulatory mechanisms controlling the hypothalamic–pituitary–thyroid axis.

Unfortunately, anomalous responses occur frequently when the TRH study is employed in the evaluation of patients with hypothalamic pituitary diseases that cause hypothyroidism. These anomalies occur often enough to undermine the TSH response to TRH as the sole dynamic study for diagnosis. Three anomalies deserve comment:

1. *Intrinsic pituitary disease:* Although a "flat" or impaired response is the anticipated phenomenon when this entity disease underlies the TSH deficiency, a significant proportion of such patients may show a normal response, mimicking a hypothalamic pattern. The percentage of patients demonstrating such a response is variable in the studies reported.[39,70,82,83] The reasons for such a phenomenon are not clear, but several possibilities exist: reduced TSH degradation, undetected concomitant hypothalamic disease, or a final last-breath response of

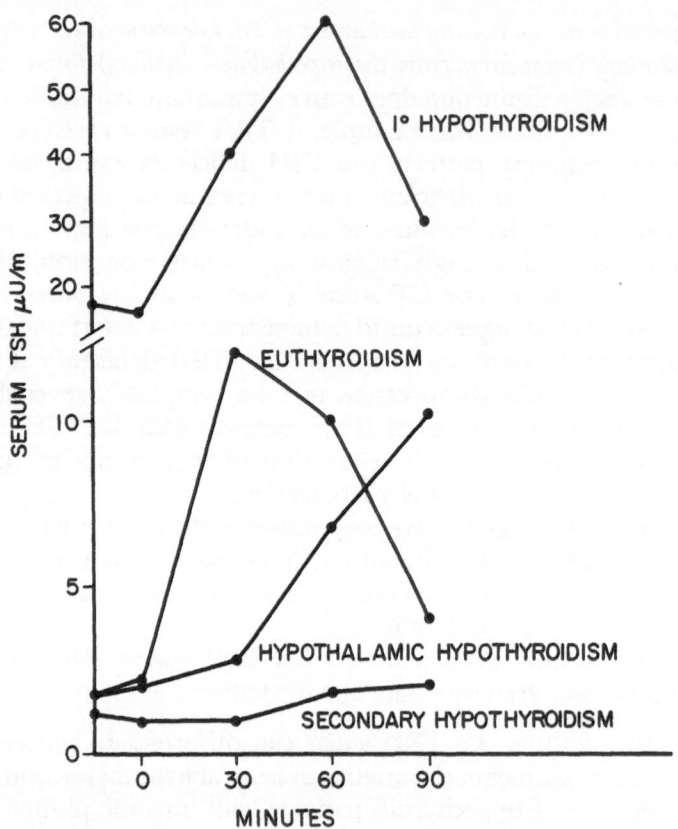

FIGURE 20. Plasma TSH response to intravenous thyroid-releasing hormone (TRH) in normals and patients with primary, secondary, and tertiary hypothyroidism.

the thyrotropes to a powerful physiological stimulus given in supraphysiological doses. One factor that is being increasingly recognized as an important determinant of the TSH response to TRH is the GH secretory status of the patient. Cobb and co-workers[84] reported that children with GH failure often demonstrated TSH responses to TRH that were greater than normal, and often prolonged. When TSH deficiency is associated with pituitary tumors, there is often concomitant GH deficiency as well. The latter phenomenon may be responsible for improving the TSH response to TRH in patients with TSH deficiency, especially when the deficiency is not total. Even more confounding is the description of prolonged, even exaggerated, TSH responses to TRH in patients with secondary hypothyroidism.[85] Such a phenomenon is particularly likely to be encountered in patients with secondary hypothyroidism resulting from secretion of biologically inactive TSH by tumorous[46,47] as well as nontumorous[45] pituitary tissue. In these instances, the TSH response to TRH mimics that seen in primary hypothyroidism. The distinction can be quite difficult and would require additional tests, such as evaluation of the response of thyroidal radioiodine uptake following injection of TSH (TSH stimulation test) or the measurement of T_3 release following the administration of TSH.

2. *Tertiary hypothyroidism:* Although a normal, but often delayed TSH response is the anticipated pattern when this entity underlies the TSH deficiency, occasionally the response of TSH following TRH may be impaired in a minority of patients with this disorder. While in some cases priming with TRH may be necessary, in other cases, the impairment may persist despite repeated exposure to TRH. In such instances it must be remembered that several factors can modify the TSH response to TRH. Some of these factors, such as depression, malnutrition, stress of acute illness, or glucocorticoid therapy, are often self-evident and may provide a ready explanation for the often reversible TSH impairment.

3. *Primary hypothyroidism:* Although an exaggerated TSH response to TRH is the hallmark of primary hypothyroidism, rarely one may encounter a normal basal TSH with a blunted rise after TRH. These are generally patients with severe hypothyroidism, often old, and often quite ill. The attenuated TSH responses in these patients probably reflects global slowing of all processes, including the ability of the thyrotropes to recognize the low thyroid hormones levels, or to respond after TRH administration. After treatment with thyroxine supplementation, there is restoration of the classic response pattern of primary hypothyroidism, i.e., elevated basal level with a brisk and exaggerated response following TRH.

Thus, it is apparent that while the use of the TRH study has importance in the evaluation of the hypothyroid patient, too much emphasis cannot be placed on this test alone.

Evaluation of Other Pituitary Hormones

The third step in the evaluation of patients with central hypothyroidism is evaluation of pituitary function with dual purposes in mind—to detect compromised reserve of the other pituitary hormones (GH, gonadotropins, corticotropin, and prolactin) as well as to detect hypersecretory syndromes involving growth hormone, corticotropin, prolactin and glycoprotein subunits. Since tumors of the pituitary and the suprasellar region are the leading cause of TSH deficiency, it is necessary to exclude compromised reserve of other trophic hormones. This can be done by measuring basal levels of gonadotropins, as well as employing provocative stimuli to evaluate the secretory reserve of growth hormone, ACTH, and prolactin.

The demonstration of hyperprolactinemia in conjunction with TSH deficiency suggests a possible hypothalamic etiology as the underlying mechanism. In the study by Snyder et al.,[70] thyrotropin reserve and prolactin secretion was evaluated in 100 patients with pituitary hypothalamic disease. Their data indicated that abnormal prolactin secretion was evident in the vast majority of hypothyroid patients with hypothalamic involvement. Besides their diagnostic value, studies of pituitary reserve testing serve as baseline data for post therapeutic comparisons.

Radiological Studies

The fourth, and final, step in the evaluation of patients with central hypothyroidism is computerized tomographic studies to exclude tumors in and around the pituitary gland. The resolution capacity of the currently used tomographic equipment not only correctly identifies tumors within the pituitary gland but is capable of accurately delineating supra-and parasellar extension of the tumor. This imaging technique is also excellent for defining suprasellar lesions, particularly craniopharyngioma, Rathke's pouch cysts, and germ cell tumors. An enlarged sella in the presence of hypothyroidism, is highly supportive of a pituitary tumor resulting in secondary hypothyroidism. However, the literature is replete with examples of an enlarged sella encountered in association with primary hypothyroidism.[86-95] This phenomenon represents secondary enlargment of the pituitary in response to target organ failure. The elevated basal TSH level as well as the exaggerated TSH response to TRH would clearly help in establishing the primary nature of the thyroid failure as well as the secondary nature of the pituitary enlargement.

Differential Diagnosis

Although the clinical expressions and treatment of thyroid failure are the same regardless of the etiology, it is important to establish the central nature of the hypothyroidism to avoid missing an intrasellar or suprasellar tumor. Table 50 outlines the differentiating features of the three varieties of hypothyroidism.

Treatment

The treatment of the hypothyroid state that results from deficiency of TSH secretion involves focus on three aspects: replacement therapy with thyroxine, prevention of adrenocortical insufficiency owing to an increase in the metabolic status, and treatment of the underlying etiology that led to the TSH deficiency.

Replacement with levothyroxine for TSH deficiency is identical to such therapy in patients with primary hypothyroidism.[96,97] Caution must be exercised in four groups of patients: the elderly, those with underlying coronary heart disease, those with severe hypothyroidism and those with underlying secondary adrenal insufficiency. The initial dosage of L-thyroxine is 0.025 mg/day. It should be emphasized that in patients with secondary hypothyroidism caused by TSH deficiency, levothyroxine therapy should not be in-

TABLE 50
Comparative Features of Various Forms of Hypothyroidism

Feature	Primary hypothyroidism	Secondary hypothyroidism	Tertiary hypothyroidism
Circulating thyroid hormone levels	Low	Low	Low
Basal TSH level	Elevated	Low; occasionally may be normal; may be elevated with biologically inactive TSH secretion	Low; may be normal.
TSH response to TRH	Exaggerated	Absent or blunted; occasionally may be normal	Normal, delayed and prolonged. Occasionally may be exaggerated or impaired
RAI uptake following TSH adminsitration	Absent response	Normal response	Normal response
T₃ release following TRH	Absent	Present; except in biologically inactive TSH secretion	Present, except in biologically inactive TSH secretion

stituted in the absence of glucocorticoid therapy. The only exception to this would be the rare patient with isolated TSH deficiency. The dose of L-thyroxine can be gradually and slowly increased to 0.1–0.15 mg/day, providing no adverse effects are encountered. Generally, patients tolerate levothyroxine quite well, with remarkable improvement within a month. Most patients note an improvement in the cold intolerance, the fatigue, and the malaise associated with the disease and describe a distinct sense of well-being. Some of this may be due to the concomitant use of glucocorticoid therapy. The follow-up parameters are clinical assessment of the thyroidal status, and the measurement of circulating thyroidal hormone levels. Unlike primary hypothyroidism, the basal plasma TSH has no place in foretelling the restoration of biochemical euthyroidism or in titrating the dose of administered levothyroxine.

The second aspect of treatment is the prevention of precipitating adrenal insufficiency as a result of increasing the metabolic demands secondary to the administration of L-thyroxine. In a sense, the hypothyroid state is protective to patients with hypoadrenalism, since they tolerate hypocortisolism quite well in the basal hypometabolic state. Thus, most patients with combined TSH and ACTH deficiency seemingly do well despite an appallingly low cortisol level. When the metabolic demands are raised by the administration of thyroxine alone, the adrenal glands, functioning at the brink of desperate exhaustion, give up and decompensate. In such a setting, within weeks of institution of L-thyroxine therapy, the patient collapses, often rather suddenly, without the prodromal symptoms of acute adrenocortical failure that are associated with addisonian crisis. Therefore, patients with secondary hypo-thyroidism should be replaced with glucocorticoids, before the metabolic demands are raised by levothyroxine. This can be done in one of two ways; cortisone acetate in physiological replacement doses (25 mg in the morning, and 12.5 mg in the evening) given for a week before L-thyroxine therapy, with the subsequent continuation of both; alternatively, cortisone and levo-thyroxine can be started simultaneously, but higher doses of cortisone acetate are employed in an attempt to build up the glucocorticoid status of the patient in anticipation of the peak thyroxine levels to come later. The choice between the two regimens depends on the urgency to treat the hypothyroid state. If this is mild, physiological glucocorticoid replacement can be started and lev-othyroxine can be safely initiated a week later at a dose of 0.05 to even 0.1 mg/day. When the hypothyroidism is moderate or severe, it is unwise to defer levothyroxine, especially since the drug reaches its peak rather slowly (3 weeks). In this setting, therapy can be initiated with both thyroxine and cortisone acetate. In patients with isolated TSH deficiency, (documented by demon-strating unequivocally normal reserve of ACTH) L-thyroxine can be admin-istered without concomitant glucocorticoid therapy.

The use of cortisone and L-thyroxine in patients with hypopituitarism can unmask previously unsuspected diabetes insipidus. In patients with su-prasellar lesions, particularly craniopharyngiomas, the development of ACTH (and TSH) deficiencies results in masking of the polyuria, because hypoad-renalism and hypothyroidism markedly decrease the free water clearance by the renal tubules. When these deficiencies are corrected by the use of L-thyroxine and cortisone acetate, the diabetes insipidus is allowed to manifest once again.

The third aspect to focus in the management of patients with TSH de-ficiency is treatment of the underlying tumors which are the leading cause of TSH deficiency; this should be appropriately managed by surgery, radiation, or both.

Finally, the chronic use of TRH has not found a role in the long-term management of tertiary hypothyroidism. The hypothyroid state, regardless of etiology, responds excellently to L-thyroxine. The effectiveness, the ease, and the cost of lifelong therapy with L-thyroxine are unparalleled. Thus, even when the deficiency of TSH, or TRH are the etiological mechanisms for thyroid failure administering TSH or TRH is hardly a consideration. Few diseases are as gratifying to treat as hypothyroidism; since replacement is lifelong, the patient should be offered the safest, surest, and least expensive remedy, which is orally administered levothyroxine. Since different commer-cial preparations of thyroxine have different bioavailabilities, it is advisable to keep the patient on the same preparation of thyroxine at all times.[98–100]

References

1. Jenkins JS, Gilbert CJ, Ang V, et al: Hypothalamic pituitary function in patients with craniopharyngioma. *J Clin Endocrinol Metab* **43**:394, 1976.

2. Sheehan LL, Murdoch R: Post-partum necrosis of the anterior pituitary: Pathological and clinical aspects. *J Obstet Gynocol Br Commonw* **45**:456, 1938.
3. Smith CW Jr, Palmer R, Howard RB: Variations in endocrine gland function in post-partum pituitary necrosis. *J Clin Endocrinol Metab* **19**:1420, 1959.
4. Asa SL, Bilbao JM, Kovacs K, et al: Lymphocytic hypophysitis of pregnancy resulting in hypopituitarism: A distinct clinicopathologic entity. *Ann Intern Med* **95**:166, 1981.
5. Mayfield RK, Levine JH, Gordon L, et al: Lymphoid adenohypophysitis presenting as a pituitary tumor. *Am J Med* **69**:619, 1980.
6. Cardoso ER, Peterson EW, et al: Pituitary apoplexy: A review. *Neurosurgery* **14**:363, 1984.
7. Mohr G, Hardy J: Hemorrhage, necrosis, and apoplexy in pituitary adenomas. *Surg Neurol* **18**:181, 1982.
8. Lohrenz FN, Fernandez R, Doe RP: Isolated thyrotropin deficiency: Review and report of three cases. *Ann Intern Med* **60**:990, 1964.
9. Sawin CT, McHugh JE: Isolated lack of thyrotropin in man. *J Clin Endocrinol Metab* **26**:955, 1966.
10. Miyai K, Azukizawa M, Kumahara Y: Familial isolated thyrotropin deficiency with cretinism. *N Engl J Med* **285**:1043, 1971.
11. Rimoin DL, Merimee TJ, McKusick VA: Growth-hormone deficiency in man: An isolated, recessively inherited defect. *Science* **152**:1635, 1966.
12. Kumahara Y Okada Y, Miyai K, et al: Typical cases of isolated growth hormone deficiency with autosomal recessive inheritance. *Acta Endocrinol (Kbh)* **63**:618, 1970.
13. Drezner M, Neelon FA, Lebovitz HE: Pseudohypoparathyroidism Type II: A possible defect in the reception of the cyclic AMP signal. *N Engl J Med* **289**:1056, 1973.
14. Levine MA, Downs RW Jr, Singer M, et al: Deficient activity of guanine nucleotide regulatory protein in erythrocytes from patients with pseudohypoparathyroidism. *Biochem Biophys Res Commun* **94**:1319, 1980.
15. Farfel Z, Abood ME, Brickman A, et al: Deficient activity of receptor–cyclase coupling protein in transformed lymphoblasts of patients with pseudohypoparathyroidism, Type I. *J Clin Endocrinol Metab* **55**:113, 1982.
16. Turner RW, Takamura T: Pseudohypoparathyroidism and hypothyroidism. *Ann Intern Med* **56**:276, 1962.
17. Winnacker JL, Becker KL, Moore CF: Pseudohypoparathyroidism and selective deficiency of thyrotropin: An interesting association. *Metabolism* **16**:644, 1967.
18. Zisman E, Lotz M, Jenkins ME, et al: Studies in pseudohypoparathyroidism: Two new cases with a probable selective deficiency of thyrotropin. *Am J Med* **46**:464, 1969.
19. Gershengorn MC: Thyrotropin-releasing hormone. *Moll Cell Biochem* **45**:163, 1982.
20. Kolesnick RN, Gershengorn MC: Thyrotropin-releasing hormone and the pituitary. *Am J Med* **79**:729, 1985.
21. Carlson HE, Brickman AS, Botazzo GF: Prolactin deficiency in pseudohypoparathyroidism. *N Engl J Med* **296**:140, 1977.
22. Shapiro MS, Bernheim J, Gutman A, et al: Multiple abnormalities of anterior pituitary hormone secretion in association with pseudohypoparathyroidism. *J Clin Endocrinol Metab* **51**:483, 1980.
23. Wolfsdorf JI, Rosenfield RL, Fang VS, et al: Partial gonadotrophin-resistance in pseudo-hypoparathyroidism. *Acta Endocrinol (Copenh)* **88**:321, 1978.
24. Werder EA, Illig R, Bernasconi S, et al: Excessive thyrotropin response to thyrotropin-releasing hormone in pseudohypoparathyroidism. *Pediatr Res* **9**:12, 1975.
25. Quint AR, Kaiser FE: Gonadotropin determinations and thyrotropin-releasing hormone and luteinizing hormone-releasing hormone testing in critically ill postmenopausal women with hypothyroxinemia. *J Clin Endocrinol Metab* **60**:464, 1985.
26. Wehmann, RE, Gregerman RI, Burns WH, et al: Suppression of thyrotropin in the low-thyroxine state of severe nonthyroidal illness. *N Engl J Med* **312**:546, 1985.
27. Porter BA, Refeteoff S, Rosenfield RL, et al: Abnormal thyroxine metabolism in hyposomatrotrophic dwarfism and inhibition of responsiveness to TRH during GH therapy. *Pediatrics* **51**:668, 1975.

28. Lippe BM, VanHerle AJ, LaFranchi SH, et al: Reversible hypothyroidism in growth hormone-deficient children treated with human growth hormone. *J Clin Endocrinol Metab* **40**:612, 1975.

29. Connors MH: Alteration of stimulated TSH and prolactin response in children treated with human growth hormone. *Life Sci* **21**:1505, 1977.

30. Lombardi G, Minozzi M, Faggiano M, et al: Plasma immunoreactive T_3 TSH and ACTH before and after provocative tests in idiopathic hypopituitary dwarfism. *J Clin Endocrinol Metab* **40**:143, 1975.

31. Pittman JA Jr, Haigler ED Jr, Hershman JM, et al: Hypothalamic hypothyroidism. *N Engl J Med* **285**:844, 1971.

32. Johansson O, Hökfelt T: Thyrotropin releasing hormone, somatostatin, and enkephalin: Distribution studies using immunohistomchemical techniques. *J Histochem Cytochem* **28**:364, 1980.

33. Jackson IMD, Reichlin S: Brain thyrotropin-releasing hormone is independent of the hypothalamus. *Nature (Lond.)* **267**:853, 1977.

34. Kallmann RJ, Schoenfeld WA, Barrera SE: The genetic aspects of primary eunuchoidism. *Am J Ment Defic* **48**:203, 1944.

35. Laron Z, Keret R, Bauman B, et al: Differential diagnosis between hypothalamic and pituitary hGH deficiency with the aid of synthetic GH-RH 1–44. *J Clin Endocrinol (Oxf)* **21**:9, 1984.

36. Woolf PD, Schalch D: Hypopituitarism secondary to hypothalamic deficiency. *Ann Intern Med* **78**:88, 1973.

37. Singer PA, Mestman JH, Manning PR, et al: Hypothalamic hypothyroidism secondary to Sheehan's syndrome. *West J Med* **120**:416, 1974.

38. Hall R: The immunoassay of thyroid-stimulating hormone and its clinical application. *Clin Endocrinol (Oxf)* **1**:115, 1972.

39. Patel YC, Burger HG: Serum thyrotropin (TSH) in pituitary and/or hypothalamic hypothyroidism: Normal or elevated basal level and paradoxical responses to thyrotropin-releasing hormone. *J Clin Endocrinol Metab* **37**:190, 1973.

40. Faglia G, Beck-Peccoz P, Ferrari C, et al: Plasma thyrotropin response to thyrotropin-releasing hormone in patients with pituitary and hypothalamic disorders. *J Clin Endocrinol Metab* **37**:595, 1973.

41. Mitsuma T, Shenkman L, Suphavai A, et al: Hypothalamic hypothyroidism: Diminished thyroidal response to thyrotropin-releasing hormone. *Am J Med Sci* **265**:315, 1973.

42. Faglia G, Ferrari C, Paracchi A, et al: Triiodothyronine response to thyrotropin-releasing hormone in patients with hypothalamic-pituitary disorders. *Clin Endocrinol (Oxf)* **4**:585, 1975.

43. Faglia G, Beck-Peccoz P, Ballabio M, et al: Excess of β-subunit of thyrotropin (TSH) in patients with idiopathic central hypothyroidism due to the secretion of TSH with reduced biological activity. *J Clin Endocrinol Metab* **56**:908, 1983.

44. Faglia G, Bitensky L, Pinchera A, et al: Thyrotropin secretion in patients with central hypothyroidism: Evidence for reduced biological activity of immunoreactive thyrotropin. *J Clin Endocrinol Metab* **48**:989, 1979.

45. Beck-Peccoz P, Amr S, Menezes-Ferreira M, et al: Decreased receptor binding of biologically inactive thyrotropin in central hypothyroidism: Effect of treatment with thyrotropin-releasing hormone. *N Engl J Med* **312**:1085, 1985.

46. Kreiger DT: Glandular end organ deficiency associated with secretion of biologically inactive pituitary peptides. *J Clin Endocrinol Metab* **38**:964, 1974.

47. Dickstein G, Barzilai D: Hypothyroidism secondary to biologically inactivethyroid-stimulating hormone secretion by a pituitary chromophobe adenoma: Recovery after removal of the tumor. *Arch Intern Med* **142**:1544, 1982.

48. Illig R, Krawczynska H, Torresani T, et al: Elevated plasma and hypothyroidism in children with hypothalamic hypopituitarism. *J Clin Endocrinol Metab* **41**:722, 1975.

49. Hall R, Scanlon MF: Hypothyroidism: Clinical features and complications. *Clin Endocrinol Metab* **8**:29, 1979.
50. Crowley WF Jr, Ridgway EC, Bough EW, et al: Noninvasive evaluation of cardiac function in hypothyroidism. *N Engl J Med* **296**:1, 1977.
51. Bough EW, Crowley WF, Ridgway EC, et al: Myocardial function in hypothyroidism. *Arch Intern Med* **138**:1476, 1978
52. Vora J, O'Malley BP, Petersen S, et al: Reversible abnormalities of myocardial relaxation in hypothyroidism. *J Clin Endocrinol Metab* **61**:269, 1985.
53. De Groot LJ: Thyroid and the heart. *Mayo Clin Proc* **47**:864, 1972.
54. Basteine PA: Hypothyroidism and coronary heart disease. *Acta Cardiol (Brux)* **37**:365, 1982.
55. Becker C: Hypothyroidism and atherosclerotic heart disease: Pathogenesis, medical management, and the role of coronary artery bypass surgery. *Endoc Rev* **6**:432, 1985.
56. Gupta MP, Kim S, Kang J, et al: Isolated TSH deficiency presenting as myxedema heart disease. *JAMA* **217**:205, 1971.
57. Wilson WR, Bedell GM: The pulmonary abnormalities in myxedema. *J Clin Invest* **39**:42, 1960.
58. Nordquist D, Stenderg K, Orndahl G: Myxedema coma and CO_2 retention. *Acta Med Scand* **166**:189, 1960.
59. Zwillich CW, Pierson DJ, Hofeldt FD, et al: Ventilatory control in myxedema and hypothyroidism. *N Engl J Med* **292**:662, 1975.
60. Orr WC, Males JL, Imes NK: Myxedema and obstructive sleep apnea. *Am J Med* **70**:1061, 1981.
61. Bastenie PA: Paralytic ileus in severe hypothyroidism. *Lancet* **1**:413, 1946.
62. Hohl RD, Nixon RK: Myxedema ileus. *Arch Intern Med* **115**:145, 1965.
63. Wells I, Smith B, Hinton M: Acute ileus in myxedema. *Br Med J* **1**:211, 1977.
64. Duret RL, Bastenie PA: Intestinal disorders in hypothyroidism: Clinical and manometric studies. *Dig Dis Sci* **16**:723, 1971.
65. Ramsey I: *Thyroid Disease and Muscle Dysfunction*. Heineman, London, p. 126, 1974.
66. Pearce J, Aziz H: The neuromyopathy of hypothyroidism. *J Neurol Sci* **9**:243, 1969.
67. Adams RD, Rosman NP: Hypothyroidism and the neuromuscular system. In Werner S, Ingbar S (eds). *The Thyroid*. Harper & Row, Hagerstown, MD, 1978, p. 901.
68. Wilson J, Walton JN: Some muscular manifestations of hypothyroidism. *J Neurol Neurosurg Psychiatry* **22**:320, 1959.
69. Klein I, Parker M, Shebert R, et al: Hypothyroidism presenting as muscle stiffness and pseudohypertrophy: Hoffman's syndrome. *Am J Med* **70**:891, 1981.
70. Snyder PJ, Jacobs LS, Rabello MM, et al: Diagnostic value of thyrotrophin-releasing hormone in pituitary and hypothalamic diseases. *Ann Intern Med* **81**:751, 1974.
71. Odell WD, Utiger RD, Wilber JF, et al: Estimation of the secretion rate of thyrotropin in man. *J Clin Invest* **46**:953, 1967.
72. Hall R, Ormston BJ, Besser GM, et al: The thyrotrophin-releasing hormone test in diseases of the pituitary and hypothalamus. *Lancet* **1**:759, 1972.
73. Utiger RD: Thyrotrophin radioimmunoassay: Another test of thyroid function. *Ann Intern Med* **74**:627, 1971.
74. Fleischer N, Burgus R, Vale W, et al: Preliminary observations on the effect of synthetic thyrotropin releasing factor on plasma thyrotropin levels in man. *J Clin Endocrinol Metab* **31**:107, 1970.
75. Anderson MS, Bowers CY, Kastin AJ, et al: Synthetic thyrotropin-releasing hormone. *N Engl J Med* **285**:1279, 1971.
76. Costom BH, Grumbach MM, Kaplan SL: Effect of thyrotropin-releasing factor on serum thyroid-stimulating hormone. *J Clin Invest* **50**:2219, 1971.
77. Foley TP Jr, Owings J, Hayford JT, et al: Serum thyrotropin responses to synthetic thyrotropin-releasing hormone in normal children and hypopituitary patients. *J Clin Invest* **51**:431, 1972.

78. Bowers CY, Schally AV, Schalch DS: Activity and specificity of synthetic thyrotropin-releasing hormone in man. *Biochem Biophys Res Commun* **39:**352, 1970.

79. Mayberry WE, Gharib H, Bilstad JM, et al: Radioimmunoassay for human thyrotrophin. Clinical value in patients with normal and abnormal thyroid function. *Ann Intern Med* **74:**47, 1971.

80. Haigler E, Hershman J, Pittman J: Response to synthetic thyrotropin-releasing hormone (TRH) in man. *Clin Res* **19:**373, 1971.

81. Hershman JM, Pittman JA Jr: Response to synthetic thyrotropin-releasing hormone in man. *J Clin Endocrinol Metab* **31:**457, 1970.

82. Jackson IMD: Thyrotropin-releasing hormone. *N Engl J Med* **306:**145, 1982.

83. McLaren EH, Hendricks S, Pimstone EL: Thyrotrophin responses to intravenous thyrotrophin-releasing hormone in patients with hypothalamic and pituitary disease. *Clin Endocrinol (Oxf)* **3:**113, 1974.

84. Cobb WE, Reichlin S, Jackson IMD: Growth hormone secretory status is a determinant of the thyrotropin response to thyrotropin-releasing hormone in euthyroid patients with hypothalamic-pituitary disease. *J Clin Endocrinol Metab* **52:**324, 1981.

85. Faglia G, Beck-Peccoz PB, Ambrosi B, et al: Prolonged and exaggerated elevations in plasma thyrotropin (HTSH) after thyrotropin releasing factor (TRF) in patients with pituitary tumors. *J Clin Endocrinol Metab* **33:**999, 1971.

86. Jackson IMD, Hall R: Pituitary enlargement resulting from primary thyroid disease. *Proc R Soc Med* **63:**578, 1970.

87. Lawrence AM, Wilber JF, Hagen TC: The pituitary and primary hypothyroidism. *Arch Intern Med* **132:**327, 1973.

88. Balsam A, Oppenheimer JH: Pituitary tumor with primary hypothyroidism. *NY State J Med* **75:**1737, 1975.

89. Vagenakis AG, Dole K, Cbaverman LE: Pituitary enlargement, pituitary failure, and primary hypothyroidism. *Ann Intern Med* **8:**195, 1976.

90. Katevuo K, Valimaki M, Ketonen L, et al: Computed tomography of the pituitary fossa in primary hypothyroidism. Effect of thyroxine treatment. *Clin Endocrinol (Oxf)* **22:**617, 1985.

91. Pita JC Jr, Shafey S, Pina R: Diminution of large pituitary tumor after replacement therapy for primary hypothyroidism. *Neurology (NY)* **29:**1169, 1979.

92. Silver BJ, Kyner JL, Dick AR, et al: Primary hypothyroidism. Suprasellar pituitary enlargement and regression on computed tomographic scanning. *JAMA* **246:**364, 1981.

93. Katz MS, Gregerman RI, Horvath E, et al: Thyrotroph cell adenoma of the human pituitary gland associated with primary hypothyroidism: Clinical and morphological features. *Acta Endocrinol (Copenh)* **95:**41, 1980.

94. Samaan NA, Osborne BM, Mackay B, et al: Endocrine and morphologic studies of pituitary adenomas secondary to primary hypothyroidism. *J Clin Endocrinol Metab* **45:**903, 1977.

95. Groff TR, Shulkin BL, Utiger RD, et al: Amenorrhea-galactorrhea and suprasellar pituitary enlargement as presenting features of primary hypothyroidism. *Obstet Gynecol (Suppl)* **63:**865, 1984.

96. Brennan MD: Clinical pharmacology series on pharmacology in practice—Thyroid hormones. *Mayo Clin Proc* **55:**33, 1980.

97. Stock JM, Surks MI, Oppenheimer JH: Replacement dosage of L-thyroxine in hypothyroidism. *N Engl J Med* **290:**529, 1974.

98. Ramos-Gabatin A, Jacobson JM, Young RL: In vivo comparison of levothyroxine preparations. *JAMA* **247;**203, 1982.

99. Ingbar JC, Borges M, Iflah S, et al: Elevated serum thyroxine concentration in patients receiving "replacement" doses of levothyoxine. *J Endocrinol Invest* **5:**77, 1982.

100. Hennessey JV, Burman KD, Wartofsky L: The equivalency of two L-thyroxine preparations. *Ann Intern Med* **102:**770, 1985.

Adrenocorticotropic Hormone

Chemistry and Synthesis

Human ACTH is a linear polypeptide containing 39 amino acids. ACTH is secreted by the corticotropes of the anterior pituitary. The secretion of ACTH is linked to that of β-lipotropin and β-endorphins, two closely related peptides. It is now apparent that several structurally related peptides are secreted by corticotropes.[1,2] These peptides, all possessing a common heptapeptide sequence (Met-Glu-His-Phe-Arg-Trp-Gly), originate from a single large-molecular-weight precursor, pro-opio-melano-cortin, POMC.[3,4] This glycoprotein has a molecular weight of 30,000–36,000. Although it was known that corticotropin, β-lipotropin, and β-endorphin in the rodent anterior pituitary are derived from the single precursor, pro-opio-melano-cortin,[5] it was unclear during the early 1980s if the human anterior pituitary processes the precursor in an identical fashion. For example, several studies had observed that the

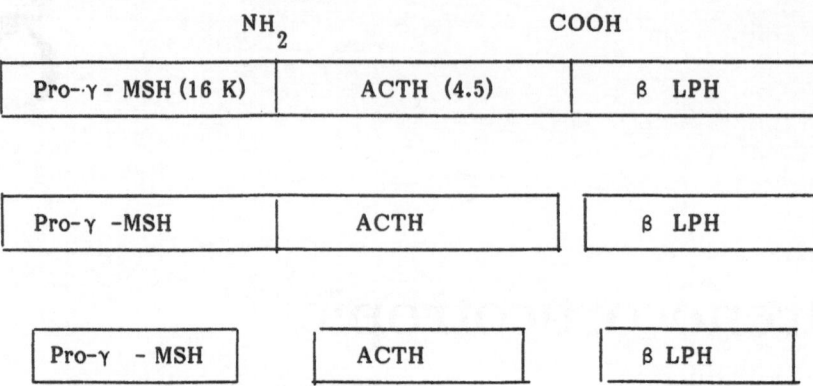

FIGURE 21. Initial processing of POMC in the anterior lobes.

13,000-M_r ACTH (which is found in the rodent adenohypophysis) is not present in the human.[6,7] The first detailed examination of the synthesis, processing, and release of ACTH and related peptides was reported by Allen et al.[8] who studied pituitary tissue removed from a patient with Nelson syndrome and maintained in monolayer culture. The important observations included the absence of 13,000-M_r ACTH (and 3500-M_r β-endorphin) and the presence of 24,000–26,000-M_r ACTH, and β-LPH containing intermediate proteins.

If one visualizes the proopiomelanocortin as a trisegmented molecule, the mid-region is occupied by ACTH, with a molecular weight of 4500 (termed 4500-M_r ACTH). The ACTH molecule is flanked by β LPH at the carboxy terminal, and by pro-γ-MSH at the amino-terminal (see Fig. 21). The actual sequence of processing of the pro-opiomelanocortin into the hormonally active subfragments begins in the anterior pituitary. The first cleavage results in formation of two fragments: β-LPH and the pro-γ-MSH-ACTH residue (or 21,000-M_r ACTH). Further cleavage of the latter results in the formation of two residues: pro-γ-MSH and ACTH, both of which are active. The subsequent processing takes place in the intermediate lobe of the pituitary gland. The first step is the cleavage of 4500-M_r ACTH to form α MSH, and CLIP (corticotropinlike intermediate lobe peptide), which represents the C-terminal portion of the ACTH (Fig. 22). The β-LPH subunit is further cleaved into

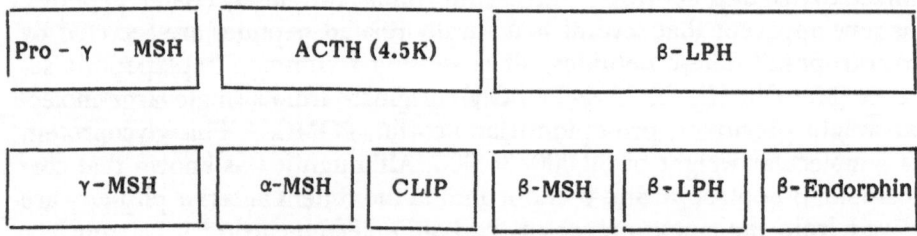

FIGURE 22. Intermediate lobe processing.

α- and β-endorphins. Pro-γ-MSH is broken down into smaller fragments of MSH in the intermediate lobe.[9,10]

The above schema highlights several points:

1. There are significant differences in the processing methods of the anterior lobe and the intermediate lobe of the pituitary.

2. The large peptide pro-opiomelanocortin, is the common precursor for ACTH, β-LPH, and endorphin. The terms big ACTH and big-big ACTH represent this precursor, which does not possess the biological activity of ACTH unless further cleaved.

3. The intermediate lobe possesses a unique ability in processing the three fragments derived from pro-opiomelanocortin (Pro-γ-MSH, ACTH, and β-LPH) into much smaller fragments. Thus, detection of these cleavage products, (γ-MSH, α-MSH, CLIP, α-LPH, and β-endorphins) from the "big three," indicates a pituitary origin. This is of relevance in cases of ectopic secretion of ACTH-like peptides by tumors. These tumors secrete the same type of precursors as the anterior pituitary (ACTH, or β-lipotropin) but do not possess the enzymatic machinery of the intermediate lobe, to cleave these big molecules into smaller cleavage products. Ratter et al.[11] characterized the molecular forms of ACTH that circulate in the plasma in normal and pathological states. They reported that the 1–39-ACTH is the predominant form of ACTH in the circulation of normals as well as those with pituitary-dependent Cushing's disease, Addison's disease, and Nelson's syndrome; while patients with the ectopic ACTH syndrome demonstrated a larger-molecular-size form of ACTH (22,000-M_r ACTH), in addition to the 1–39 fragment. The larger forms of ACTH can be identified by gel filtration and by their concanavalin A–Sepharose chromatographic affinity. Rarely, pituitary-dependent Cushing's disease may be associated with secretion of larger forms of ACTH precursors into the circulation.[12–14] These patients develop "atypical" Cushing's disease, characterized by highly invasive tumors, along with clinical and biochemical features of the ectopic ACTH syndrome.

4. The β-endorphin processed at the intermediate lobe is probably devoid of opiate activity, since it is immediately actylated. Importantly, β-endorphin is secreted stoichiometrically with ACTH.[15] It is believed that β-endorphin and ACTH are synthesized by the same cells, with dynamic patterns that parallel each other. Thus, corticotropin-releasing factor (CRF) stimulates the secretion of both, while dexamethasone suppresses both the peptides concomitantly.[16] The concordance of both peptides in responding to physiological stimuli such as stress[17] or running[18] or chemical stimuli such as metyrapone[19] has been well documented. It is therefore somewhat unexpected that infusion of β-endorphin consistently lowers ACTH and cortisol levels in normal human subjects. Taylor et al.[20] infused human synthetic endorphin in 10 normal subjects and measured ACTH and cortisol levels before

and every 30 min after the infusion for 210 min. They noted that the ACTH and cortisol levels decreased below the basal levels in all ten subjects. They postulated that the β-endorphin-mediated ACTH suppression was exerted at the hypothalamic level by short field back loop inhibition or by decreasing dopamine levels in the hypothalamus. A clinical correlate of this phenomenon is encountered in the endorphin-secreting adenoma, which results in ACTH deficiency.[21]

Control of ACTH Secretion

The regulation of ACTH control is multifaceted, with the predominant control exerted by the hypothalamus. In addition, the pituitary ACTH secretion is regulated directly by ambient cortisol levels in the circulation. The mechanisms of ACTH regulation may be viewed from the following four vantage points: hypothalamic CRF regulation of ACTH, negative feedback mechanisms, stress-mediated ACTH release, and biological rhythms and ACTH release.

Corticotropin-Releasing Factor

As early as 1948, it had been suggested by Harris[22] that the pituitary gland was controlled by a higher hypothalamic influence. In 1955, experimental support for this theory was provided by Guillemin and Rosenberg,[23] who proposed the humoral hypothalamic control of the pituitary gland by studying combined tissue cultures. The characterization of a 41-residue peptide from sheep hypothalami[24] and the sequential analysis of cDNA,[25] led to the isolation of the human CRF.[26] The homology between ovine and human CRF is so close that they differ by only seven amino acids. This fact, coupled with the observation that ovine CRF is a potent stimulator of ACTH release in humans,[24] has resulted in the use of ovine CRF for extensive investigation in humans. Although rat CRF is identical to human CRF,[27] it is not being used for clinical studies.

Human CRF contains 41 amino acids and is synthesized by neurons located in the paraventricular nucleus of the hypothalamus. Extensive studies in the rat brain have disclosed CRF immunoreactive neurons in the hindbrain and the limbic systems.[28,29] The significance of finding CRF receptors in these locations is unclear, but the proximity to the central sympathetic system suggests a relationship to the CRF release in response to stress. Following secretion, CRF is released into the hypothalamic portal circulation to reach the anterior pituitary where it binds to receptors on the corticotropes and stimulates the secretion and release of ACTH.

The availability of synthetic corticotropin-releasing factor (CRF-41) has permitted the evaluation of normal responses of corticotropes to the administration of CRF. Lytras et al.[30] evaluated the response of 20 normal subjects to the intravenous administration of 100 μg CRF-41. In all normal volunteers, serum cortisol rose following CRF-41 administration. The peak cortisol responses were seen either at 45 or 60 min after the injection of CRF, with a mean maximum increment of 276 ± 38 nmoles above the basal cortisol level. The two important additional observations were that the cortisol response to CRF negatively correlated with the basal cortisol level, and that the cortisol response to CRF could be abolished by prior administration of dexamethasone. This latter observation lends strong support to the theory that glucocorticoids may directly exert their negative feedback effect on the pituitary corticotropes. The ratio of the peak ACTH level following CRF to the basal ACTH level in normals ranged between 1.2 and 4.0, indicating that the magnitude of cortisol responses following CRF was more striking than the plasma ACTH response. The authors also noted that hypoglycemia evoked a greater cortisol response than the administration of CRF-41.

The response of ACTH, cortisol to CRF-41 is dose related. Nakahara et al.[31] administered 500 μg CRF-41 to normals and noted that the total hormone released, as well as the peak levels, were higher than with a 100-μg bolus. At this dose, ACTH rose 3.6 times the basal level, the cortisol peaks were 2.6 times the basal level, and even the aldosterone levels rose 1.6 times the basal level; most investigators use 1 μg CRF/kg body weight.

The corticotropin-stimulating effect of CRF is mediated by cAMP generation and accumulation. Giguere et al.[32] demonstrated a 2.5-fold stimulation of cAMP content in rat pituitary cells maintained in culture 60 sec after addition of 100 nM of synthetic ovine CRF. A maximal response of cAMP content at 400% above control was observed 2–30 min after addition of CRF. Synthetic CRF in this in vitro study led to a parallel stimulation of ACTH and cAMP release into the culture medium. The almost instantaneous increase in cellular cAMP suggests that accumulation of the cyclic nucleotide precedes ACTH secretion induced by CRF. Compartmentalization of cyclic nucleotide into different pools is suggested by their observation that CRF induces ACTH secretion at doses much lower than required for nucleotide accumulation. Pretreatment with dexamethasone demonstrated no effect on CRF induced cAMP accumulation, but resulted in marked inhibition of ACTH release induced by CRF as well as 8-bromoadenosine 3'5'-cyclic monophosphate. These data provide strong support to the notion that glucocorticoids directly and potently inhibit ACTH secretion at the pituitary level, and they do so at a site distal to that of cAMP formation.

CRF-mediated ACTH release is augmented in the presence of low cortisol and high vasopressin levels in circulation. Physiologically, the ACTH response to the CRF in normal subjects is inversely related to the basal cortisol level. DeCherney et al.[33] studied the effect of circadian rhythm on the response of plasma ACTH and cortisol to the intravenous administration of ovine CRF. After performing the study in different time slots, they concluded that although the time of administration of ovine CRF does not have a major influence on the CRF-mediated ACTH response, the cortisol response to CRF was much greater when CRF was given in the late afternoon or evening. The fact that the ACTH response to CRF is inversely related to basal cortisol concentrations is best exemplified in pathological states of hypocortisolism. Hermus et al.[34] studied eight patients with Addison's disease (primary adrenal failure) and demonstrated an augmented ACTH response to CRF, analogous to the TSH response to TRH in primary hypothyroidism. These workers were the first to demonstrate that in the absence of cortisol negative feedback, the absolute ACTH response to CRF is greatly augmented.

Another factor that can modulate and augment the ACTH response to CRF is vasopressin, the hypothalamic peptide secreted by paraventricular and supraoptic nuclei. In vitro studies by Gillies et al.[35] and Vale et al.[36] convincingly demonstrated that vasopressin augments the CRF mediated ACTH release from pituitary tissue maintained in tissue culture. In humans, Milsom et al.[37] studied the plasma ACTH response to CRF before and after hypertonic saline infusion, a powerful stimulus for endogenous vasopressin release. They were able to demonstrate an almost threefold increase in the ACTH response to CRF when endogenous plasma AVP was elevated by the hypertonic saline. Clearly, AVP potentiates the ACTH response to CRF. This has led to the possibility of combining the administration of CRF with hypertonic saline infusion (which provokes endogenous vasopressin) as a provocative test for ACTH reserve in lieu of insulin-induced hypoglycemia, a procedure that elicits a greater ACTH response in normals, than either CRF or vasopressin alone.

Although CRF is the dominant humoral mediator of ACTH release, other factors are also capable of serving as potent stimuli for ACTH release (Table 51); the most important one of which is vasopressin. Administration of intravenous vasopressin is followed by an increase in ACTH-mediated cortisol level. The consistency with which vasopressin elicits an ACTH response has resulted in the use of vasopressin as a diagnostic tool to study ACTH reserve. It was once thought that the provocative effect of vasopressin on ACTH release may, in part, be due to shared similarities in the chemical structure of CRF and vasopressin. However, it is now recognized that vasopressin-induced ACTH release is independent of any inherent CRF-like activities of vasopressin. The potentiation of CRF-mediated ACTH release by vasopressin has been already alluded to.

The profoundly provocative effect of hypoglycemia on ACTH and cortisol secretion is a well-recognized one.[38] Thus, the hypoglycemia test serves

TABLE 51
Factors That Stimulate ACTH
Release

Corticotropin-releasing factor (CRF)
Vasopressin
Insulin hypoglycemia
Angiotensin II
Catecholamines

as the "gold standard" for assessment of ACTH reserve. The ACTH–cortisol response to hypoglycemia is probably multifactorial, consisting of components mediated by CRF, components mediated by catecholamines, as well as by the poorly recognized hypothalamic glucoreceptors. The potential dangers of hypoglycemia in the patient with hypopituitarism has always been a concern with the use of hypoglycemic challenge. Perhaps in the future, the combined use of CRF administration with vasopressin may obviate the need for using the hypoglycemic challenge.

Angiotensin II and catecholamines are also important stimuli that release ACTH. It is controversial as to whether the ACTH response to these stimuli represents a direct effect or a CRF-mediated one.

Thyrotropin-releasing hormone (TRH) and luteinizing hormone-releasing hormone (LH-RH) do not lead to ACTH release in normal subjects. It is of interest that in animals with a well-developed intermediate lobe, TRH can elicit an ACTH response.[39] The only situation where physiologically TRH induces a paradoxical ACTH release is during pregnancy.[40] The reason for this phenomenon may well reside in the hypothesis that the intermediate lobe becomes functionally active in the pregnant woman.[41,42] The paradoxical rise of ACTH in response to TRH in some patients with pituitary-dependent Cushing's disease and Nelson's syndrome is well recognized.

In summary, the dominant regulator of ACTH secretion by the corticotropes is the hypothalamic peptide CRF; vasopressin, catecholamines, and angiotensin II play a significant, possibly, synergestic role, while other hypothalamic peptides such as TRH and LRH play no role in ACTH secretion in physiological states. The major determinant of CRF-mediated ACTH release is the ambient cortisol levels, which exert a negative feedback on the hypothalamic pituitary unit.

Negative Feedback Mechanisms

While it is generally accepted that glucocorticoids exert their negative feedback effect on both the hypothalamus and the pituitary, it is not completely clear as to which is more important. The receptors for the mechanism of negative feedback are believed to be present in both the pituitary and the hypothalamus. The bulk of evidence, based on studies that evaluate CRF-induced ACTH release, suggests that glucocorticoids exert a profound inhibitory effect on the pituitary. Hermus et al.[43] and Lytras et al.[30] clearly, and repeatedly, documented the negative correlation between CRF-induced ACTH secretion and basal cortisol levels; i.e., the higher the basal cortisol, the lower the ACTH response to ovine CRF. In its extreme form, this is manifested in patients with Cushing's syndrome caused by adrenal tumors, where characteristically there is a total and absolute absence of ACTH to the administration of CRF. In a less dramatic form, it is represented in some depressed patients with hypercortisolemia who also show a blunted ACTH response to CRF administration.[44] In vitro work by Giguere et al.[32] has also provided data supporting a predominantly pituitary locus for negative feedback suppression.

Using preparations of dispersed anterior pituitary cells maintained in primary culture, these workers showed that pretreatment with dexamethasone significantly inhibited the ACTH release induced by CRF as well as by a cAMP derivative. The failure of ACTH to respond to CRF in the presence of hypercortisolemia is strikingly reminiscent of the failure of TSH to respond to TRH in the presence of hyperthyroidism. The negative feedback effect of hypercortisolism on the hypothalamic pituitary axis forms the basis of the various dexamethasone suppression tests employed in the diagnostic evaluation of hypercortisolism.

There are, however, situations in which the negative feedback effect of glucocorticoids can be overridden; the most impressive example of such a phenomenon is during stress, a situation characterized by continuous ACTH secretion despite hypercortisolemia. It is believed that the stress-induced CRF elevation is of a magnitude that far surpasses the negative feedback exerted on the corticotropes by glucocorticoids. Support for such a hypothesis is provided by the work of Kendall et al.[45], which indicates that both the duration of prior corticosteroid administration as well as the intensity of the ACTH-releasing stimulus are important determinants of corticotrope activation. The stress-mediated release of ACTH is the most powerful stimulus for the corticotropes.

Stress-Mediated Release

The stress-mediated ACTH response is an adaptive process, mediated and integrated by the brain and other parts of the central nervous system. The two major participants of the stress response (general adaptation syndrome) are the hypothalamic–pituitary–adrenal axis and the sympathetic nervous system. Activation of the former results in secretion of cortisol, while activation of the latter, leads to release of catecholamines. CRF probably plays an important role in mediating the ACTH response, because stalk section impairs activation of this system. In addition, the administration of CRF into the cerebral ventricles of laboratory animals elicits cardiovascular and metabolic responses identical to the stress syndrome.[46] The demonstration of CRF (and its receptors) in the limbic system, and the hindbrain also provides suggestive evidence for CRF involvement in the stress response. The proximity of these structures to the central sympathetic system is also suggestive of possible catecholamine mediation of ACTH release, in a synergistic fashion. The afferent arc of the stress reflex is in the peripheral nervous system, because denervation abolishes the stress response secondary to limb injury. Various kinds of stress—burns, fever, trauma anesthesia, hypoglycemia, surgery—can result in identical elevation of cortisol. The clinical importance of the stress-triggered CRF–ACTH–cortisol response is enormous, since patients with untreated primary or secondary adrenal insufficiency may do well in a basal state but may profoundly and critically decompensate under the stress of infection or surgery; unless urgent steroid replacement is instituted, these patients succumb to the stress.

Biological Rhythms and ACTH Release

The circadian periodicity of cortisol levels are preceded by parallel changes in ACTH level. This nyctohemereal rhythm of ACTH is under hypothalamic control, which in turn is influenced by the mid-brain. The mid-brain is believed to exert a continuous inhibitory control over CRF release. The bursts of ACTH release are superimposed on the inherent nyctohemeral rhythm. Peak ACTH levels are seen at about the time of awakening, while the lowest ACTH levels are seen 1–2 hr after going to sleep. The cortisol levels closely parallel the ACTH peaks and nadirs. The circadian periodicity of CRF–ACTH–cortisol axis is lost in blind patients, as well as in those with Cushing's syndrome.

Action of ACTH

The predominant action of ACTH is stimulation of the growth and function of the adrenal cortex. Thus, both weight maintenance and steroidogenesis are dependent on ACTH. In the absence of ACTH, the adrenal glands undergo atrophy, a phenomenon that can be reversed by administering ACTH. In fact, the early bioassay for ACTH were based on the weight maintaining activity of the sera administered to hypophysectomized rats. On the functional end, ACTH promotes the synthesis and release of all the adrenocortical hormones, particularly the glucocorticoids. While all steps of adrenal steroidogenesis are activated, the major step influenced by ACTH is conversion of cholesterol into pregnenolone. It is believed that cAMP is the intracellular mediator of the steroidogenic action of ACTH. There is controversy as to whether low or high-affinity receptors in the adrenal cortex are linked to the activation of cAMP following ACTH stimulation.

While the secretion of glucocorticoids and androgens are entirely dependent on the trophic action on ACTH, the secretion of mineralocorticoids by the zona glomerulosa is not. The predominant stimulus for mineralocorticoid secretion is the renin–angiotensin–aldosterone system (RAAS). The zona glomerulosa has receptors for both angiotensin II as well as ACTH. Administration of pharmacological doses of ACTH is followed by a prompt increase in aldosterone levels.

Assay of ACTH

Adrenocorticotropin in the plasma may be measured by a sensitive radioimmunoassay. The blood should be collected in chilled tubes containing EDTA, and the plasma separated immediately. The preparations of tubes for the standard, the control, for the nonspecific binding and for the unknown plasma are conducted in the conventional manner; 200 μl [^{125}I]-ACTH is added to each tube, mixed, and allowed to incubate for 20–24 hr at 2–8°C. The next step consists of adding 500 μl of preprecipitated second antibody complex consists of normal rabbit serum preprecipitated with goat antirabbit serum and polyethylene glycol diluted in BSA-borate buffer with merthiolate. The mixture is incubated for 2 hr at 2–8°C, following which it is centrifuged for 20 min at room or refrigerator temperature. After decanting the supernatants, the radioactivity in the tube is counted in a gamma counter and the ACTH concentration is calculated.

The basal ACTH concentrations range from 30–100 pg/ml. Measurement of basal levels does not permit adequate separation of normals from ACTH-deficient persons. The clinical ability of measuring basal ACTH concentrations is discussed in a subsequent section of this chapter (ACTH dynamics in health and disease).

Disorders of ACTH Secretion

Several clinical disorders are characterized by altered states of ACTH secretion:

1. The prototype of ACTH hypersecretion is exemplified by pituitary-dependent Cushing's disease.[47] The pivotal abnormality in this entity is the inappropriate secretion of ACTH relative to the ambient cortisol levels. Although microadenoma of the adenohypophysis represents the most common etiology, Cushing's disease can result from corticotrope hyperplasia[48] or originate from the pars intermedia.[49] Cushing's disease is a heterogeneous disorder consisting of CRF-dependent as well as CRF-independent varieties.[50]

2. Nelson's syndrome also represents a disorder characterized by hypersecretion of ACTH. The rapid growth of a (previously small) corticotrope adenoma following bilateral adrenalectomy, generalized hyperpigmentation, and perisellar pressure effects are the triad of Nelson syndrome.[51,52] The hormonal dynamics of Nelson syndrome share several similarities with Cushing's disease.[53] It should be noted that the plasma ACTH levels in Nelson syndrome are astronomical (>1000 pg), while the levels in Cushing's disease are in the normal to mildly elevated range. This may be a reflection of the size of the tumor and the population of the corticotropes.

3. The third example of ACTH hypersecretion is illustrated by the ectopic ACTH-secreting syndromes. It is now recognized that the incidence of secretion of the ACTH precursors (POMC) by malignant tumors is extremely high,[54] but the clinical syndromes occur only when the enzymatic machinery necessary for cleaving the precursor peptides are present in the tumor. The characteristic hormonal features of ectopic ACTH-producing syndrome are the markedly elevated plasma levels of ACTH, often with a larger molecular weight (big ACTH), elevated level of β-LPH (and perhaps other POMC-related peptides), and marked elevation of cortisol levels. While it is recognized that ectopic ACTH-secreting syndrome can mimic Cushing's disease,[55–57] the reverse phenomenon can also occur. Hale et al.[14] described a

patient with pituitary-dependent Cushing's disease with all the characteristics of the ectopic ACTH syndrome. Thus, marked elevation of ACTH with a larger molecular weight, elevation of β-LPH, and other precursors derived from POMC, along with dexamethasone nonsuppressibility were noted. The ubiquitous presence of ACTH in several nonpituitary tissues is probably the reason for the high immunoreactive ACTH level seen in diverse malignant disorders.[58,59] Saito et al.[60] demonstrated in the rat that high-molecular-weight ACTH-like materials are widespread in normal extrapituitary tissues and probably represent a precursor form of the 4500-M_r ACTH.

4. On the hypofunctional end, *ACTH deficiency* may occur as an isolated event[61,62] (isolated ACTH deficiency) or as part of panhypopituitarism. Tumors represent the most important etiology for ACTH deficiency.

5. Functional deficiency of ACTH can occur by a variety of mechanisms. The most omnipotent example of such a phenomenon is the occurrence of ACTH deficiency following exogenous steroid administration. Corticotropic dormancy can also be encountered following selective microadenomectomy for Cushing's disease, albeit on a transient basis.[63] A novel mechanism for functional ACTH deficiency is the occurrence of secondary ACTH deficiency in conjunction with an endorphinoma. Endorphin hypersecretion has far reaching effects on the rest of the pituitary, resulting in decreased ACTH, LH and FSH, while increasing prolactin; such was the case in the patient with endorphin-secreting tumor reported by Trouillas et al.[21]

ACTH Dynamics in Health and Disease

There are several clinical situations in which the study of ACTH dynamics becomes necessary. The various tests that are employed for evaluating perturbations in ACTH dynamics can be viewed from three perspectives. First, the value of measuring basal concentrations of ACTH in the plasma is outlined; second, the dynamic studies that are provocative to ACTH release are discussed; these include the use of metyrapone, insulin hypoglycemia, vasopressin, CRF, and other lesser stimuli. Third, and finally, the dynamic studies that suppress ACTH release are reviewed. In this category belong the various dexamethasone-suppression tests.

Basal ACTH Concentrations in Plasma

Plasma ACTH can be measured by using a sensitive radioimmunoassay. The four facts that govern the interpretation of a single ACTH sample (customarily drawn at 8 AM) are the following:

1. The collection of blood for ACTH should be performed with meticulous care. Blood must be collected in chilled tubes containing EDTA on ice. Separation of the plasma should be carried out immediately.

2. The concentration of ACTH in the plasma is affected by the pulsatile nature of ACTH secretion. ACTH, like cortisol, is secreted in "bursts." If 24-hr profiles of ACTH in the plasma are plotted by sampling every 10 or 20 min, the pattern obtained is a "jagged" one with several secretory spikes being noted.[50] Therefore, the ACTH concentration in blood drawn randomly is unlikely to be representative of the overall prevailing concentrations of the hormone in circulation. Yet, calculating the integrated plasma ACTH levels by performing numerous timed samples is cumbersome and impractical.

3. The basal ACTH level in the plasma can be properly interpreted only when a simultaneously obtained cortisol level is available. The situation is analogous to the PTH : Cal concentrations in plasma. An elevated ACTH in conjunction with a low cortisol implies differently than a similarly elevated ACTH with a high serum cortisol.

4. When commercial laboratories are used, the different standards, and the quality of the assay used should be taken into consideration to avoid misinterpretation of ACTH values.

With the above generalizations considered, there are circumstances in which a single plasma ACTH level can be diagnostically revealing:

1. The combination of a normal to mildly elevated plasma ACTH in conjunction with hypercortisolemia is highly suggestive of pituitary-dependent Cushing's disease.

2. The demonstration of a markedly elevated ACTH level in a hyper-cortisolemic patient is almost diagnostic of Cushing's syndrome caused by ectopic ACTH-secreting tumor.

3. The demonstration of a suppressed (<20 pg) ACTH level in a patient with hypercortisolemia is indicative of an adrenal tumor causing Cushing's syndrome.

4. An elevated ACTH level in the presence of a low cortisol level is strongly suggestive of Addison's disease, while a low to normal ACTH in the presence of hypocortisolemia favors a hypothalamic-pituitary etiology, i.e., secondary hypoadrenalism.

5. A markedly elevated ACTH level is the hallmark of Nelson syndrome in the postadrenalectomized patient with Cushing's disease. Table 52 summarizes the diagnostic value of the basal plasma ACTH concentrations in various disorders.

Provocative Tests for ACTH Secretion

This section focuses on the use of insulin, vasopressin, metyrapone, and CRF in the evaluation of corticotrope function.

Insulin Tolerance Test[64-66]

Insulin-induced hypoglycemia has been the "gold standard" for testing the integrity of the entire hypothalamic pituitary adrenal axis. Following an

TABLE 52
Diagnostic Value of Basal ACTH Measurement

Condition	Basal ACTH (pg/ml)	Cortisol
Hypercortisolism		
Cushing's disease	↑ (80–150)	↑
Ectopic ACTH syndrome	↑ ↑ (>200)	↑ ↑
Adrenal neoplasm	↓ (<20)	↑
Iatrogenic Cushing's	↓	↑
Hypocortisolism		
Addison's disease	↑ ↑ (>250)	↓
ACTH deficiency	↓ (>100)	↓
Eucortisolism		
Nelson's syndrome	↑ ↑ ↑	Normal (on replacement)
Nonendocrine		
Stress	↑	↑
Anesthesia, surgery	↑	N or ↑
Depression	↑	N or ↑
Hypoglycemia	↑	↑

acute drop in the blood glucose levels there is a marked increased in cortisol levels. The cortisol response to hypoglycemia, evident within 45–60 min after the administration of 0.1 unit of regular insulin per kg body weight, is mediated through the hypothalamic pituitary axis, because stalk section abolishes the response. Moderate to marked increase in ACTH levels precede the rise in cortisol levels. The consistency with which the cortisol levels rise following hypoglycemia, and the magnitude of cortisol response that follows hypoglycemia have rendered it an invaluable tool for assessment of ACTH reserve. The mechanisms for the ACTH rise following hypoglycemia is probably multifactorial. It is believed that the response of the hypothalamic pituitary unit to hypoglycemia is analogous to the stress response. Intracellular glucoreceptors within the hypothalamus are highly sensitive to a drop in blood glucose, and release CRF in response to hypoglycemia. In addition, the catecholamine surge that follows hypoglycemia also serves as an additive stimulus for ACTH release by the pituitary gland. It is not clear whether vasopressin is also involved in the hypoglycemia-mediated release of CRF–ACTH. Whatever the mechanism, the rapid drop in blood glucose serves as an excellent provocative stimulus for the release of ACTH and cortisol. For test purposes, a drop in the blood glucose by 50% of the baseline is used as an indicator of adequate hypoglycemia. The normal response is an increase in the serum cortisol level greater than an absolute value of 25 μg, or a 10-μg increase above the baseline.

Excellent as the test is, two comments regarding the hypoglycemia test are appropriate. First, the test can result in precipitous and sustained hypoglycemia, especially in patients with hypopituitarism. It is mandatory that the test be performed under close supervision. Second, the absence of an ACTH–cortisol response to hypoglycemia does not differentiate between a pituitary versus a hypothalamic etiology. This is because the response to hypoglycemia is mediated by both components. Nevertheless, insulin hypoglycemia continues to remain the screening test for determining the functional integrity of the corticotrope reserve.

Vasopressin Stimulation Test[67,68]

The intravenous administration of vasopressin to normal subjects is associated with a rise in serum cortisol, usually occurring 60 min after the injection. Vasopressin is a powerful stimulus for ACTH release in vitro and in vivo. The mechanism by which the corticotropes respond to vasopressin has not been completely elucidated. It used to be thought that vasopressin might possess some structural similarity to CRF, thereby resulting in receptor acceptance by the corticotropes. It is not clear whether vasopressin consistently induces cAMP generation within the corticotropes prior to ACTH release. The fact that vasopressin directly stimulates the pituitary corticotrope was hailed as a distinct advantage of the vasopressin test since it could discriminate pituitary ACTH deficiency from hypothalamic CRF deficiency. However, the availability of CRF-41 has rendered this advantage superfluous. Furthermore, vasopressin administration is not without side effects, the most hazardous being precipitation of coronary spasm (angina) or even myocardial infarction.

CRF Stimulation Test

The availability of synthetic corticotropin-releasing factor (CRF-41) has added a new dimension to the evaluation of ACTH dynamics in various disorders of the hypothalamic-pituitary axis.[30,31,44] Following the intravenous administration of CRF-41, at a dose of 1 μg/kg, an increase in ACTH and cortisol is seen within 30–60 min after the injection. The magnitude of the ACTH-cortisol response is inversely proportional to the basal cortisol level.[31,34] The CRF test has found clinical application in three areas: (1) in the evaluation of hypercortisolism; (2) in conjunction with inferior petrosal sinus sampling, to improve the diagnostic yield of the procedure; and (3) in the etiological evaluation of hypoadrenalism. The first two areas are covered in Chapter 9, which deals with hypercortisolism. The value of the CRF study as a test to assess the reserve of ACTH is the focus in this section. The uniqueness of the CRF-test lies in the fact that it represents the only available stimulus that directly tests corticotrope reserve. This contrasts sharply with metyrapone and insulin, stimuli that provoke both the hypothalamus and pituitary. The availability of CRF can, at least in theory, separate hypothalamic from pituitary etiology of corticotropin deficiency. Hermus et al.[34] evaluated the ACTH response to 100 μg of ovine CRF in 11 patients with secondary adrenal insufficiency and were able to separate them into two categories: those having little or no ACTH and cortisol responses (pituitary deficiency); and a second group with a prolonged ACTH response with a biphasic pattern and a delayed second peak, followed in all patients by a marked cortisol increase (hypothalamic deficiency). These data demonstrate that the ovine CRF test can discriminate between hypothalamic and pituitary causes of secondary hypoad-renalism.

It should be pointed out that the standard insulin hypoglycemia test provides information regarding the integrity of the entire hypothalamic pi-tuitary adrenal axis, and is a more potent stimulus for ACTH release than CRF. Most authorities would agree that the insulin tolerance test should be used initially to screen for inadequacy of the entire HPA axis, and the CRF test should be subsequently employed to discriminate between pituitary and hypothalamic etiologies. It is not clear whether the ovine CRF augmented by vasopressin is superior to the insulin tolerance test.

Metyrapone Test

Metyrapone is a drug that inhibits the conversion of 11-deoxycortisol (compound S) to cortisol (compound F) by inhibiting the enzyme 11β-hy-droxylase within the adrenal cortex. As a result, there is a decrease in com-pound F level, which stimulates the hypothalamic-pituitary axis to secrete more ACTH. The increased amounts of ACTH stimulate the adrenal cortex with resultant activation of steroidogenesis. However, owing to the block in the final step, the steroidogenesis stops short of synthesizing increased amounts of compound S.

Therefore, the triple response of a normal person, i.e., one with normal adrenals and an intact hypothalamic pituitary axis, to the oral administration of metyrapone is as follows: a decrease in cortisol F, an increase in ACTH, and an increase in deoxycortisol S, which is the precursor product proximal to the block created by metyrapone. This is reflected in the normal person as an increase in the urinary-17 hydroxycorticosteroids, which measure com-pound S. (The 17-OHCS normally measures both compound F and compound S, but metyrapone precludes formation of compound F, and hence most of the 17-OHCS consists of the precursor compound S.)

The proper interpretation of the metyrapone test depends on three pre-requisites:

1. The completeness of the enzymatic block created by the drug—Unless the block is significant enough to lower the cortisol level, the subse-quent phenomena will not take place. Adequacy of the block should always be confirmed by demonstrating a significant lowering of cortisol level in plasma.
2. The integrity of the hypothalamic pituitary axis, since the ACTH response to declining cortisol levels is the key phenomenon
3. The integrity of the adrenal cortex to respond to the endogenous ACTH drive—The metyrapone test cannot be interpreted without knowing whether the adrenal glands are viable. For instance, a failure to increase urinary 17-OHCS following metyrapone can be indicative of either hypopituitarism (ACTH lack) or Addison's disease (primary adrenal disease). However, if the presence of adrenal responsiveness

has been established prior to performing the metyrapone test, the failure to increase the 17-OHCS following metyrapone can only mean hypopituitarism. It is therefore essential to perform an adrenal stimulation test by the administration of exogenous ACTH before doing the metyrapone test. If the adrenal glands are shown to respond to exogenous ACTH, one can assume that a similar response can be expected with increases in the endogenous ACTH, thus, attempts to stimulate the endogenous ACTH reserve by metyrapone are valid. If, one the other hand, the adrenals fail to respond to exogenous ACTH, the metyrapone test is not indicated, since lack of response to metyrapone can no longer be interpreted.

The test is performed by measuring basal levels of urinary 17-hydroxycorticosteroids, serum cortisol, and ACTH before, and following the oral administration of 750 mg metyrapone every 4 hr for six doses. A failure to increase the 17-OHCS in the urine collection the day after metyrapone is indicative of either inadequate block, ACTH lack, or primary adrenal insufficiency. Adequacy of block can be ensured by demonstrating a significant lowering of cortisol in the serum, while adequacy of adrenal function can be ensured by prior assessment of the adrenal response to exogenous ACTH. With these two prerequisites satisfied, a failure to increase the 17-OHCS following metyrapone is diagnostic of ACTH deficiency.

Patients receiving anticonvulsant therapy and those who are hypothyroid or depressed may also demonstrate blunted responses to metyrapone.

While metyrapone challenge is used primarily to diagnose ACTH deficiency, it can also be employed in the differential diagnosis of hypercortisolism (Cushing's disease). Patients with pituitary-dependent Cushing's disease increase their 17-OHCS following metyrapone, while those with suppressed ACTH secondary to an autonomous adrenal tumor demonstrate no response.

Suppression Tests

The suppression tests evaluate the ability of the hypothalamopituitary axis to suppress in response to orally administered dexamethasone, a potent glucocorticoid. Obviously, these tests assume importance only when hyperfunction is suspected.

1. *Overnight dexamethasone test:* This test evaluates the plasma cortisol level following 1 mg dexamethasone given at midnight (before sleep). The dexamethasone given before sleep abolishes the nyctohemeral release of ACTH, hence the cortisol level drawn at 8 AM will be below 5 µg%. Dexamethasone is potent at this dose but does not interfere with plasma measurements of cortisol. The value of the test lies in the fact that it is convenient and inexpensive and, if the patient suppresses below 5 µg, hypercortisolism is nearly completely excluded. Unfortunately, there are a wide variety of conditions characterized by nonsuppression:

obesity (30–45%), drug intake (estrogens, diphenylhydantoin), alcoholism, and endogenous depression. Thus, the only valuable information provided by the test is when the results are suppressible (i.e., a decline below 5 µg following oral dexamethasone.

2. *Low-dose dexamethasone test:* This test involves administration of 0.5 mg of dexamethasone four times per day orally for 2 days and evaluation of the 24-hr urinary 17-OHCS before and after; normal suppression is defined as a decrease in 17-OHCS below 3.5 mg after dexamethasone. The only group of patients who suppress to the low dose, in terms of those who showed nonsuppression to the overnight test, are obese patients and some patients on medication; since the urinary-free cortisol is a superior index for hypercortisolism, owing to the fact that it is not affected by obesity or medications, the low-dose test has nothing more to offer in these circumstances than the urinary-free cortisol does. Therefore, at many centers, the low-dose test is circumvented in favor of the 24-hr urinary-free cortisol.

3. *High-dose dexamethasone test:* This test evaluates the 17-OHCS before and after the administration of 2 mg oral dexamethasone four times per day for 2 days. Suppression here is defined as a drop in the 17-OHCS by at least 50% of the baseline. This test has some discriminatory value in separating the various etiologies of hypercortisolism in the following manner:

   a. The characteristic response of pituitary-dependent Cushing's disease is suppression to the high dose, while not suppressing to the low dose.

   b. The characteristic response of adrenal tumors is nonsuppression to high dose, since endogenous ACTH is already maximally suppressed, and '

   c. The characteristic response of an ectopic ACTH-secreting tumor is also nonsuppression to the high dose.

   Despite the above classic responses, it should be realized that there are several exceptions to the general rules:

   a. In 20–25% of patients with pituitary-dependent Cushing's disease, suppression may not occur with a high dose, necessitating a "high-high" dose, since they are functioning at a phenomenally higher threshold.

   b. A small (5%) number of patients with adrenal tumors (micronodular disease) may demonstrate suppression of ACTH to high-dose dexamethasone thus mimicking pituitary-dependent Cushing's disease.

   c. The rare bronchial carcinoid, ectopically secreting ACTH (or CRF), may also demonstrate preservation of suppression to the high-dose dexamethasone test.

   d. Endogenous depression is often associated with abnormal steroid dynamics, the most frequent one being nonsuppression to the low dose but preservation of suppression to the high dose mimicking pituitary disease. It is particularly important to keep this entity in

TABLE 53
Dynamic Testing in States of Hypercortisolism: General Principles

Condition	Overnight DXM (1 mg DXM at midnight PO)	Low-dose DXM (0.5 mg QID for 2 days)	High-dose DXM (2 mg QID for 2 days)	Ancillary tests ACTH stimulation	Metapyrone	Plasma ACTH	Other
Exogenous obesity	75% suppress, 25% may not	Suppression	Not indicated				
Endogenous depression and stress	Close to 40% may not show suppression	90% suppression	90% suppression				
Diphenylhydantoin, estrogens	No suppression	No suppression	Suppression				
Pituitary ACTH-dependent Cushing's	No suppression	No suppression	Suppression	Exaggerated response	Exaggerated response	"Normal" to modest ↑	
Adrenal adenoma	No suppression	No suppression	No suppression	50% show no response	May or may not respond	Suppressed	CT scan
Adrenal CA	No suppression	No suppression	No suppression	No response	No response	Suppressed	CT scan, 17-KS
Ectopic ACTH syndromes	No suppression	No suppression	No suppression	No response	No response	↑↑	
Definition of normal response	Post-DXM value under 5 µg is suppression	17-OHCS less than 3.5 mg in 24 hr is suppression	17-OHCS less than 50% of baseline is suppression				

mind, since the urinary free cortisol can also be elevated in depressed patients. The constellation of elevated urinary-free cortisol and abnormal suppression data in such patients may result in an erroneous diagnosis of pituitary-dependent Cushing's disease.

Table 53 outlines the hormonal dynamic tests involving suppression tests in diverse conditions. The differential diagnosis of a pituitary response (i.e., nonsuppression to low dose but preservation of suppression to a high dose of dexamethasone) includes the occasional adrenal adenoma, the rare bronchial carcinoid with ectopic ACTH secretion, the frequently encountered patient with endogenous depression, and the patient on anticonvulsants. The differential diagnosis of nonsuppression to the high-dose dexamethasone test includes adenoma or carcinoma of adrenal, ectopic ACTH-secreting neoplasms, as well as the minority of patients with pituitary-dependent Cushing's disease functioning at a markedly high setpoint for negative feedback suppression. These principles are dealt with in greater depth in Chapter 9.

References

1. Krieger DT, Liotta AS, Brownstein MJ, et al: ACTH, β lipotropin and related peptides in brain, pituitary and blood. *Recent Prog Horm Res* **36**:277, 1980.
2. Loh YP, Gainer H: The role of the carbohydrate in the stabilization, processing and packaging of the glycosylated adrenocroticotropin-endorphin common precursor in toad pituitaries. *Endocrinology* **105**:474, 1979.
3. Loh YP: Immunological evidence for two common precursors to corticotropins, endorphins and melanotropin in the neurointermediate lobe of the toad pituitary. *Proc Natl Acad Sci USA* **76**:796, 1979.
4. Crine P, Gossard G, Siedah NG et al: Concomitant synthesis of endorphin and melanotropin from two forms of proopimelanocortin in the rat pars intermedia. *Proc Natl Acad Sci USA* **76**:5085, 1979.
5. Eipper BA, Mains RE: Structure and biosynthesis of adreno-adrenocorticotropin/endorphin and related peptides. *Endocr Rev* **1**:1, 1980.
6. Allen RG, Orwoll E, Kendall JW, et al: The distribution of forms of ACTH and β-endorphin in normal, tumorous and autopsy pituitary tissue: Virtual absence of 13K ACTH. *J Clin Endocrinol Metab* **51**:375, 1980.
7. Miller WL, Johnson LK, Baxter JD, et al: Processing of the precursor to corticotropin and β lipotropin in humans. *Proc Natl Acad Sci USA* **77**:5211, 1980.
8. Allen R, Orwoll E, Kammer H, et al: Corticotropin/β-litropin biosynthesis, processing and release in Nelson's syndrome. *J Clin Endocrinol Metab* **53**:887, 1981.
9. Loh YP, Loriaux LL: Adrenocorticotropic hormone, β-lipotropin, and endorphin-related peptides in health and disease. *JAMA* **247**:1033, 1982.
10. Loh YP, Gainer H: In vitro evidence that glycosylation of pro-opiocortin and corticotropins influences their proteolysis by trypsin and blood. *Mol Cell Endocrinol* **20**:35, 1980.
11. Ratter SJ, Lowry PJ, Besser GM, et al: Chromatographic characterization of adrenocorticotrophin in human plasma. *J Endocrinol (Oxf)* **85**:359, 1980.
12. Fuller PJ, Lim ATW, Barlow JW, et al: A pituitary tumor producing high molecular adrenocorticotropin-related peptides: Clinical and cell culture studies. *J Clin Endocrinol Metab* **58**:134, 1984.
13. Ratter SJ, Gillies G, Hope J, et al: Pro-opiocortin related peptides in human pituitary and ectopic ACTH secreting tumours. *Clin Endocrinol (Ocf)* **18**:211, 1983.

14. Hale A, Millar JBG, Ratter SJ, et al: A case of pituitary dependent Cushing's disease with clinical and biochemical features of the ectopic ACTH syndrome. *Clin Endocrinol (Oxf)* **22:**479, 1985.
15. Guilleim R, Vargo T, Rossier J, et al: β-endorphin and adrenocorticotropins are secreted concomitantly by the pituitary gland. *Science* **197:**1367, 1977.
16. Vale W, Rivier C, Yang L, et al: Effects of purified hypothalamic corticotropin-releasing factor and other substances on the secretion of adrenocorticotropin and β-endorphin-like immunoactivities in vitro. *Endocrinology* **103:**1910, 1978.
17. Nakao K, Nakai Y, Jingami H, et al: Substantial rise of plasma beta-endorphin levels after insulin-induced hypoglycemia in human subjects. *J Clin Endocrinol Metab* **49:**838, 1979.
18. Carr D, Bullen BA, Skeinar GS, et al: Physical conditioning facilitates the exercise-induced secretion of beta-endorphin and beta-lipotropin in women. *N Engl J Med* **305:**560, 1981.
19. Nakao K, Nahai Y, Oki S, et al: Presence of immunoreactive beta-endorphin in normal human plasma. *J Clin Invest* **62:**1395, 1978.
20. Taylor T, Dluhy RG, Williams GH: β-endorphin suppresses adrenocorticotropin and cortisol levels in normal human subjects. *J Clin Endocrinol Metab* **57:**592, 1983.
21. Trouillas J, Girod C, Sassolas PA, et al: A human β-endorphin pituitary adenoma. *J Clin Endocrinol Metab* **58:**242, 1984.
22. Harris GW: Neural control of the pituitary gland. *Physiol Rev* **28:**134, 1948.
23. Guillemin R, and Rosenberg E: Humoral hypothalamic control of anterior pituitary: A study with combined tissue cultures. *Endocrinology* **57:**599, 1955.
24. Vale W, Spiess J, Rivier C, et al: Characterization of a 41-residue ovine hypothalamic peptide that stimulates secretion of corticotropin and beta-endorphin. *Science* **213:**1394, 1981.
25. Furutani Y, Morimoto Y, Shibahara S, et al: Cloning and sequence analysis of cDNA for ovine corticotropin-releasing factor precursor. *Nature (Lond)* **301:**537, 1983.
26. Shibahara S, Morimoto Y, Furutani Y, et al: Isolation and sequence analysis of the human corticotrophin-releasing factor precursor gene. *EMBO J* **2:**775, 1983.
27. Rivier J, Spiess J, Vale W: Characterization of rat hypothalamic corticotropin-releasing fact. *Proc Natl Acad Sci USA* **80:**4851, 1983.
28. Olschowka JA, O'Donohue TL, Mueller GP, et al: The distribution of corticotropin releasing factor-like immunoreactive neurons in rat brain. *Peptides* **3:**995, 1982.
29. De Souza EB, Perrin MH, Insel TR, et al: Corticotropin-releasing factor receptors in rat forebrain: autoradiographic identification. *Science* **224:**1449, 1984.
30. Lytras N, Grossman A, Perry L, et al: Corticotrophin releasing factor: Responses in normal subjects and patients with disorders of the hypothalamus and pituitary. *Clin Endocrinol (Oxf)* **20:**71, 1984.
31. Nakahara M, Shibasaki T, Shizume K, et al: Corticotropin-releasing factor test in normal subjects and patients with hypothalamic-pituitary adrenal disorders. *J Clin Endocrinol Metab* **57:**963, 1983.
32. Giguere V, Labrie F, Cote J, et al: Stimulation of cyclic AMP accumulation and corticotropin release by synthetic ovine corticotropin-releasing factor in rat anterior pituitary cells: Site of glucocorticoid action. *Proc Natl Acad Sci USA* **79:**3466, 1982.
33. DeCherney GS, DeBold CR, Jackson RV, et al: Diurnal variation in the response of plasma adrenocorticotropin and cortisol to intravenous ovine corticotropin-releasing hormone. *J Clin Endocrinol Metab* **61:**273, 1985.
34. Hermus ARMM, Pieters GFFM, Pesman GJ, et al: ACTH and cortisol responses to ovine corticotrophin-releasing factor in patients with primary and secondary adrenal failure. *Clin Endocrinol (Oxf)* **22:**761, 1985.
35. Gillies G, Linton E, Lowry PJ: Corticotrophin releasing activity of the new CRF is potentiated several times by vasopressin. *Nature (Lond)* **299:**355, 1982.
36. Vale W, Vaughen J, Smith M, et al: Effects of synthetic ovine corticotropin-releasing factor, glucocorticoids, catecholamines, neurohypophysial peptides and other substances on cultured corticotropin cells. *Endocrinology* **113:**1121, 1983.
37. Milsom SR, Conaglen JV, Donald RA, et al: Augmentation of the response to CRF in man: Relative contributions of endogenous angiotensin and vasopressin. *Clin Endocrinol (Oxf)* **22:**623, 1985.

38. Donald RA: Plasma immunoreactive corticotropin and cortisol response to insulin hypogly-
    cemia in normal subjects and patients with pituitary disease. *J Clin Endocrinol Metab* **32:**225,
    1971.
39. Leroux P, Tonon MC, Jegou S, et al: In vitro study of frog (Rana Ridibunda pallas): Neu-
    rointermediate lobe secretion by use of a simplified perifusion system. *Gen Comp Endocrinol*
    46:13, 1982.
40. Pieters GFFM, Smals AGH, Goverde HJM, et al: Paradoxical increase of ACTH and cortisol
    in response to TRH in pregnant women, evidence of intermediate lobe activity? *Neth J Med*
    25:56, 1982.
41. Clark D, Thody AJ, Shuster S, et al: Immunoreactive α MSH in human plasma in pregnancy.
    Nature (Lond) **273:**163, 1978.
42. Visser M, Swaab D: αMSH in the human pituitary. *Front Horm Res* **4:**42, 1977.
43. Hermus A, Pieters G, Smals A, et al: Plasma adrenocorticotropin, cortisol and aldosterone
    responses to corticotropin-releasing factor: Modulatory effect of basal cortisol levels. *J Clin
    Endocrinol Metab* **58:**187, 1984.
44. Chrousos GP, Schuermeyer TH, Doppman J, et al: Clinical applications of corticotropin-
    releasing factor. *NIH Conf Ann Intern Med* **102:**344, 1985.
45. Kendall JW, Egans ML, Stott AK, et al: The importance of stimulus intensity and duration
    of steroid administration in suppression of stressinduced ACTH secretion. *Endocrinology*
    90:525, 1972.
46. Sutton RE, Koob GF, Le Moal M, et al: Corticotropin releasing factor produces behavioral
    activation in rats. *Nature (Lond)* **297:**331, 1982.
47. Burch WM: Cushing's disease. *Arch Intern Med* **145:**1106, 1985.
48. Taylor HC, Velasco ME, Brodkey JS: Remission of pituitary-dependent Cushing's disease
    after removal of nonneoplastic pituitary gland. *Arch Intern Med* **140:**1366, 1980.
49. Lamberts SWJ, De Lange SA, Stefanko SZ: Adrenocorticotropin-secreting pituitary aden-
    omas originate from the anterior or the intermediate lobe in Cushing's disease: Differences
    in the regulation of hormone secretion. *J Clin Endocrinol Metab* **54:**286, 1982.
50. Van Cauter E, Refetoff S: Evidence for two subtypes of Cushing's disease based on the
    analysis of episodic cortisol secretion. *N Engl J Med* **312:**1343, 1985.
51. Nelson DH, Meakin JW, Thorn GW: ACTH-producing pituitary tumors following adre-
    nalectomy for Cushing's syndrome. *Ann Intern Med* **52:**560, 1960.
52. Moore TJ, Dluhy RG, Williams GH, et al: Nelson's syndrome: Frequency, prognosis, and
    effect of prior irradiation. *Ann Intern Med* **85:**731, 1976.
53. Krieger DT, Luria M: Plasma ACTH and cortisol responses to TRF, vasopressin or hypo-
    glycemia in Cushing's disease and Nelson's syndrome. *J Clin Endocrinol Metab* **44:**361,
    1977.
54. Gewirtz G, Yalow RS: Ectopic ACTH production in carcinoma of the lung. *J Clin Invest*
    53:1022, 1974.
55. Strott CA, Nugent CA, Tyler FH: Cushing's syndrome caused by bronchial adenomas. *Am
    J Med* **44:**97, 1968.
56. Mason AMS, Ratcliffe JG, Buckle RM, et al: ACTH secretion by bronchial carcinoid tumors.
    Clin Endocrinol (Oxf) **1:**3, 1972.
57. Northrup G, Baldwin D, Farber LP, et al: Dexamethasone suppression of urinary 17-hy-
    droxycorticoids in a patient with an ACTH-producing bronchial adenoma. *Presbyterian-St.
    Lukes Med Bull* **9:**43, 1970.
58. Ratcliffe JG, Knight RA, Besser GM, et al: Tumor and plasma ACTH concentrations in
    patients with and without the ectopic ACTH syndrome. *Clin Endocrinol (Oxf)* **1:**27, 1972.
59. Wolfsen AR, Odell WD: Pro ACTH: use of early detection of lung cancer *Am J Med* **66:**765,
    1979.
60. Saito E, Iwasa S, Odell WD: Widespread presence of large molecular weight adrenocorti-
    cotropin-like substances in normal rat extrapituitary tissues. *Endocrinology* **113:**1010,
    1983.
61. Stacpoole PW, Interlandi JW, Nicholson WE, et al: Isolated ACTH deficiency: A hetero-
    geneous disorder. *Medicine (Baltimore)* **61:**13, 1982.
62. Limjuco RA, Sherman L, Kolodny HD: Isolated ACTH deficiency. *NY State J Med* 439, 1976.

63. Fitzgerald PA, Aron DC, Findling JW, Brooks RM, et al: Cushing's disease: Transient secondary adrenal insufficiency after selective removal of pituitary microadenomas; evidence for a pituitary origin. *J Clin Endocrinol Metab* **54**:413, 1982.
64. Aizawa T, Greer MA: The primary site of action of hypoglycemia in stimulating ACTH secretion is in the medial basal hypothalamus. In *Sixty-second Annual Meeting of the Endocrine Society 1980*, p. 104.
65. Baylis PH, Heath DA: Plasma-arginine-vasopressin response to insulin induced hypoglycaemia. *Lancet* **2**:428, 1977.
66. Arky RA, Freinkel N: The response of plasma human growth hormone to insulin and ethanol-induced hypoglycemia in two patients with "isolated adrenocortical defect." *Metabolism* **13**:547, 1964.
67. Eddy RL: Aqueous vasopressin provocative test of anterior pituitary function. *J Clin Endocirnol* **28**:1836, 1968.
68. Gillies G, Lowry P: Is corticotropin releasing factor modulated vasopressin? *Nature (Lond)* **283**:698, 1980.

9

Cushing's Disease

Cushing's disease, or pituitary-dependent hypercortisolism, is an entity characterized by the development of hypercortisolism secondary to inappropriate ACTH secretion by the pituitary gland. Since the original description by Harvey Cushing in 1932,[1] there have been remarkable strides in our understanding of the pathophysiology, laboratory diagnosis and therapeutic modalities of this disease.

Etiology and Pathogenesis

Pituitary tumors are the most important and frequent anatomic etiology for Cushing's disease (CD). Thus, microadenomas (tumors under 10 mm), noninvasive macroadenomas (tumors larger than 10 mm, but confined to the sella), and invasive macroadenomas are all known to cause CD. Rare cases of CD secondary to hyperplasia of the corticotropes probably represent the early stages in the evolution of the disease. Microadenomas represent the most common etiology, accounting for nearly 80% of CD.

While the anatomical etiology of CD is relatively clear, the pathophysiological basis for the development of the disease is far from explicit. It is believed that an abnormality in the hypothalamic–pituitary–adrenal (HPA) axis sets the stage for the development of this disorder. The characteristic hallmark of a normal HPA axis is the prompt recognition of and brisk suppression to even minor increases in circulating cortisol levels. This unique sensitivity to suppress in response to negative feedback is blunted in pituitary-dependent CD. Consequently, the hypothalamic pituitary threshold for suppression is raised and the ACTH secretory rate is consistently inappropriate, relative to the circulating cortisol levels at any given time. The raised threshold to negative feedback is evidenced by the response of patients with CD to the standard dexamethasone suppression tests, that is, lack of physiological suppression to low dose dexamethasone but preservation of suppression to higher doses. The degree of abnormality is variable from patient to patient, some requiring a higher dose than that used in the conventional test. Nevertheless, they do suppress, denoting preservation of physiological cues, albeit operating at a higher threshold. This response pattern is classic for pituitary-dependent CD.

This simplistic concept of reset hypothalamic pituitary axis has been under rigorous scrutiny during the past decade. The controversy revolves around a very basic issue: Is Cushing's disease a "primary derangement of the pituitary gland" as characterized by Harvey Cushing, or is it a result of an overactive hypothalamic driving mechanism? The proponents of both theories have drawn on histological, biochemical, hormonal and pharmacological facts to support their respective premise. The ongoing polemic has great significance in terms of its impact on choosing the options for therapy; if the disease is a result of a primary pituitary tumor, a properly performed transsphenoidal microadenectomy should be curative, while if a hypothalamic overdrive were causing the problem, medical therapy to correct the overdrive would be the logical initial choice. The conflicting data in the literature can be sorted, if viewed in three perspectives: evidence that favors a primary *hypothalamic* defect, evidence that favors a primary pituitary defect, and evidence that points to the existence of anatomical subtypes encountered in seemingly identical corticotrope adenomas.

Primary Hypothalamic Etiology

There are several lines of evidence to suggest that CD is a result of corticotrope overactivity secondary to an accentuated hypothalamic drive. Anatomically, this is supported by the observation that a small but significant number of patients with CD fail to show any tumor in the pituitary gland despite the most scrutinizing exploration. Indeed, remission of pituitary-dependent CD after removal of nonneoplastic pituitary gland has been reported.[2] There is a large body of data derived from dynamic studies. Two earlier reports, one by Liddle[3] and another by Orth and Liddle,[4] documented that some patients with pituitary-dependent CD continue to demonstrate abnormal dexamethasone (and metyrapone) responses despite a clinical cure and normalization of urinary steroids excretion following treatment. The question that is obviously raised is whether ACTH-secreting adenomas arise de novo or occur as a consequence of stimulation from higher centers. The availability of assays for circulating ACTH level in the plasma has greatly facilitated study of ACTH dynamics in pituitary-dependent CD. The pituitary ACTH secretion in CD responds to a variety of seemingly unrelated agents. Thus, thyrotropin-releasing hormone (TRH),[5] vasopressin,[5,6] and luteinizing hormone-releasing hormone (LHRH)[7] are all capable of releasing ACTH from pituitary tumor that cause CD, denoting nonautonomy of the tumor, and partial preservation of responsiveness.

A major pharmacological link of evidence that points to a central nervous system site as the cause for pituitary-dependent CD comes from studies with cyproheptadine. This drug is a serotonin antagonist, which in some patients with CD effectively lowers cortisol and ACTH levels, restores normal suppressibility and can even induce a sustained remission of CD and the Nelson's syndrome.[5,8] While the action of this drug is not completely understood, it acts on the central nervous system (CNS) and is capable of blocking the release of corticotropin-releasing factor (CRF) from the hypothalamus of the rat.[9] More specifically, does cyproheptadine reverse the abnormality in ACTH feedback inherent to patients with CD? Lankford et al.[10] studied the two phases of cortisol feedback suppression of ACTH in nine patients who had been treated either by adrenalectomy or transsphenoidal microadenectomy. In normal humans, cortisol-induced ACTH suppression consists of an early rate-dependent phase and a delayed dose-dependent phase, the two phases being temporally and dynamically distinct. A characteristic abnormality of patients with CD, especially those who undergo adrenalectomy, is an initial paradoxical rise of ACTH in response to cortisol. Lankford et al.[10] demonstrated the persistence of cortisol ACTH feedback abnormality even after removal of the pituitary adenoma. They further showed that cyproheptadine reversed the abnormality in all patients suggesting that higher centers must have an important role in the pathophysiology of CD. Another pharmaco-

TABLE 54
Cushing's Disease: Primary Hypothalamic Disorder

Feature	Reference
Infrequent occurrence of radiologically evident pituitary tumors	Liddle[13]
Presence of abnormal EEG patterns	Krieger et al.[14]
Persistence of abnormal hypothalamic–pituitary axis despite apparent cure	Liddle[3] Orth and Liddle[4]
Failure to find any tumor in a small percentage of patients despite scrutinizing search at surgery	Boggan et al.[15] Hardy[16]
Response to cyproheptadine	Aronin et al.[8]
Response to CRF	Chrousos et al.[141]
Cushing's disease that be mimicked by metastic CRF-secreting neoplasm	Carey et al.[12]
Persistence or recurrence of hypercortisolism despite resection of adenomas	Lamberts et al.[17]

logical agent, bromocriptine, can also lower ACTH levels in patients with CD.[11] The characterized and isolation of CRF has perhaps provided the strongest support to the notion that hypothalamic (or even more central) influences underlie the pathophysiology of CD. The demonstration of brisk ACTH responses to the intravenous administration of ovine (CRF) clearly implies that the pituitary adenoma cells are not only nonautonomous but are possibly hypersensitive to the hypothalamic peptide. Finally, ACTH-dependent Cushing's syndrome can result from ectopic secretion of CRH. Carey et al.[12] described a patient with carcinoma of the prostate, metastatic to the median eminence of the hypothalamus. The patient manifested excessive ACTH secretion not suppressible by dexamethasone. The pituitary demonstrated hyperplasia of the corticotropes, while the extracts from the tumor and the metastases contained CRF (but not ACTH) measurable by both RIA and bioassay. This was the first documentation of CRF-induced hypercortisolism.

The development of Nelson's syndrome following bilateral adrenalectomy for CD is often cited as evidence to support a hypothalamic etiology for CD. Nelson's syndrome, characterized by markedly increased plasma ACTH levels, despite glucocorticoid replacement, shares several similarities in hormone dynamics with untreated CD. The response to TRH, cyproheptadine, bromocriptine, and corticotropin-releasing hormone (CRH) administration are strikingly identical in both diseases.

In summary, these lines of evidence form a compelling set of arguments to suggest a hypothalamic etiology in the development of CD (Table 54).

Primary Pituitary Etiology

The two major reasons for considering the pituitary gland as the primary seat of the disease are the high percentage of microadenomas found during surgery, and the increasing number of "cures" attained following successful microadenomectomy by transsphenoidal surgery (TPS). In the series of 72 patients reported by Hardy[16] and another series of 100 patients reported by Boggan et al.,[15] the incidence of detecting microadenomas is approximately 80%; and the overall cure rates of tumors confined to the sella turcica were 88% and 87%, respectively. Such figures strongly argue in favor of CD representing a problem arising primarily within the pituitary gland. If this is indeed so, the reason for the partial autonomy of the adenoma cells would require explanation. The ability to study the behavior of corticotrope adenoma cells in tissue culture has provided insight into this phenomenon. Ludecke et al.[18,19] showed that isolated corticotropes are only partially suppressible to dexamethasone, strikingly analogous to the partial suppressibility to dexamethasone seen in patients with CD. It is therefore conceivable that the classic response seen in pituitary-dependent CD is an inherent feature of the adenoma cells, independent of any resetting mechanism at the level of the hypothalamus.

Several recent in vitro studies have also evaluated the effect of various drugs on the isolated tumor cells grown in tissue culture. Suda et al.[20] demonstrated that both cyproheptadine and reserpine directly inhibit the release of ACTH and β-endorphin in vitro from adenomas removed from patients with pituitary-dependent CD. Similar observations have been extended to bromocriptine.[21,22] These powerful data have clearly dealt a blow to the hypothalamic theory of origin of CD, since these drugs obviously can directly act on the pituitary gland, circumventing all mediation by the hypothalamus.

Additional support to the primary pituitary origin of CD is derived from the observation that transient hypoadrenalism develops after successful extirpation of the ACTH-secreting microadenoma. Fitzgerald et al.[23] prospectively evaluated 12 patients who underwent pituitary microsurgery for CD; 11 of 12 developed postoperative hypoadrenalism with deficient adrenal responsiveness to exogenous ACTH administration. Such a response reflects suppression of normal corticotrope function secondary to prior hypercortisolism, reminiscent of the transient postoperative hypocalcemia following removal of a parathyroid adenoma. All patients eventually revealed normal corticotrope function within a time frame that was quite consistent with dormancy of the suppressed population of normal corticotropes. Such a sequence is consistent with a primary pituitary origin for the tumor.

A final argument in favor of the pituitary origin of the disorder is the corticotroph mass hypothesis proposed by Jeffecoate et al.[24] These workers reported two cases of pituitary-dependent CD; these patients underwent incomplete TPS and, despite persistence of the disease (and in the presence of residual tumor), demonstrated restoration of full suppression to dexamethasone. It is conceivable that, at least in these two cases, the apparent restoration of suppressibility may merely reflect a reduction in the mass of the tumor cells. Adenoma cells may behave similar to normal corticotropes, the partial suppressibility merely reflecting their total number. When that number was reduced, as with incomplete TPS, normal suppressibility was restored, despite active disease. Such a hypothesis obviates the need to invoke any hypothalamic mediation or even a resetting of the receptors at either level, the hypothalamus or the pituitary.

TABLE 55
Cushing's Disease: A Primary Pituitary Disorder

Feature	Reference
High incidence of microadenoma at surgery	Hardy[16]
	Boggan et al.[15]
The extremely high success rate with selective microadenomectomy	Hardy[16]
	Boggan et al.[15]
	Burch[25]
In vitro response of adenoma cells to dexamethasone	Ludecke et al.[18]
In vitro response of adenoma cells to cyproheptadine, reserpine, and bromocriptine	Suda et al.[20]
	Lamberts et al.[21]
Demonstration of dormancy of normal corticotrope population following successful removal of adenoma	Fitzgerald et al.[23]

It would appear that a formidable array of data can be lined up to support the notion that CD is a "primary derangement of the pituitary gland" as originally proposed by Harvey Cushing. Table 55 summarizes these data.

Subtypes of the Disease

While controversy continues to remain as to the pathophysiology of CD, evidence for the presence to both subtypes of CD has been presented by Van Cauter and Refetoff.[26] These workers studied the cortisol pulses, that is, variations in the circulating cortisol levels in the plasma, over a 24-hr period. The "cortisol profile" of normal subjects shows a characteristic jagged pattern with 7–10 pulsatile bursts of secretory activity. The cortisol spikes are due to concomitant ACTH spikes, in turn attributed to pulsatile release of corticotropin releasing factor by the hypothalamus. Van Cauter and Refetoff delineated two types of patterns in patients with pituitary-dependent CD. One group had a hypopulsatile pattern (normal number and absolute height of spikes, but the height relative to the preceding trough was lower than that of normal controls); the second group had a hyperpulsatile pattern (absolute and relative height of the spikes were greater than those in normal controls). The former group may represent patients with CRF-independent ACTH hypersecretion, and the latter group may represent patients with CRF dependent ACTH hypersecretion.

Anatomy and Histology

The most common anatomical lesion responsible for pituitary-dependent CD is a tumor in the anterior pituitary, usually under 10 mm (microadenoma). Macroadenomas, less commonly invasive macroadenomas and rarely diffuse hyperplasia of corticotropes constitute less frequent etiologies. It should be realized that in approximately 15–20% of patients with hormonally documented pituitary-dependent CD, no abnormalities may be found despite the most careful search during surgery. The location of the microadenoma is usually lateral. In the series reported by Boggan et al.,[15] the tumor arose laterally in 60 of 82 cases operated upon. Hardy[16] reported the gross anatomical findings in 72 patients subjected to transsphenoidal exploration for pituitary-dependent CD and found microadenomas in 52. The likelihood of finding adenomas that were located deeply in the central wedge of the pituitary was more in the case of the smallest adenomas.

While indeed most ACTH-secreting microadenomas arise from the anterior lobe, they can also arise from the intermediate lobe, an area of the pituitary that is anatomically not well delineated in the human. Yet, tumors arising from this region assume special importance in CD. Lamberts et al.[17] were the first to point out the differences in ACTH regulation between adenomas originating from the anterior lobe, and those arising from the intermediate lobe. The five characteristics of ACTH-secreting adenoma that arise from the intermediate lobe are (1) the presence of argyrophilic nerve fibers coursing in and around the tumor, suggesting a neural origin; (2) resistance to dexamethasone suppression, but responding to dopamine agonists such as bromoergocriptine; (3) frequent association with hyperprolactinemia; (4) poor visualization by CT of the pituitary; and (5) frequent failure to respond to transsphenoidal microadenectomy (i.e., persistence or recurrence of tumor despite the apparently complete removal of the tumor).

The relatively unsatisfactory response to surgery contrasts sharply to the highly favorable outcome following TPS for adenomas of the anterior pituitary and has tremendous impact when planning the surgical approach for such patients. The reasons for the unique differences that set these tumors of the intermediate lobe apart from the conventional microadenomas of the anterior pituitary are less clear.

Lamberts et al.[17] proposed that depletion of hypothalamic dopaminergic neural input may be the biochemical abnormality that underlies the development of hyperplasia of the corticotropes in the intermediate lobe followed by adenoma formation in some. An identical syndrome occurs in an strikingly parallel animal model in the horse (equine Cushing's disease). Orth et al.[27] described a horse and a pony suffering from myopathy, diabetes, and hirsutism. The animals manifested hypercortisolism with elevated plasma levels of ACTH, disproportionate elevation of α- and β-melanophore-stimulating hormones, and poor suppressibility to dexamethasone, with responsiveness to intravenous dopamine, bromocriptine, and pergolide.

Histopathologically, the typical pituitary corticotrope adenoma consists of cells that are well granulated and basophilic, staining positive with periodic acid-Schiff (PAS) stains. The normal acinar architecture of the pituitary is replaced by these cells that form nodular masses. A characteristic feature of these cells is the presence of microfilaments measuring 70 Å, considered to represent cytokeratin.[28] When the accumulation of these filaments is extensive, the appearance resembles the characteristic Crooke hyalinization.[29] This histological hallmark, best brought out by the staining with the Alcian blue–PAS–orange G stain is specific enough for corticotrope adenoma to obviate need for ultrastructural or immunocytochemical techniques. When in

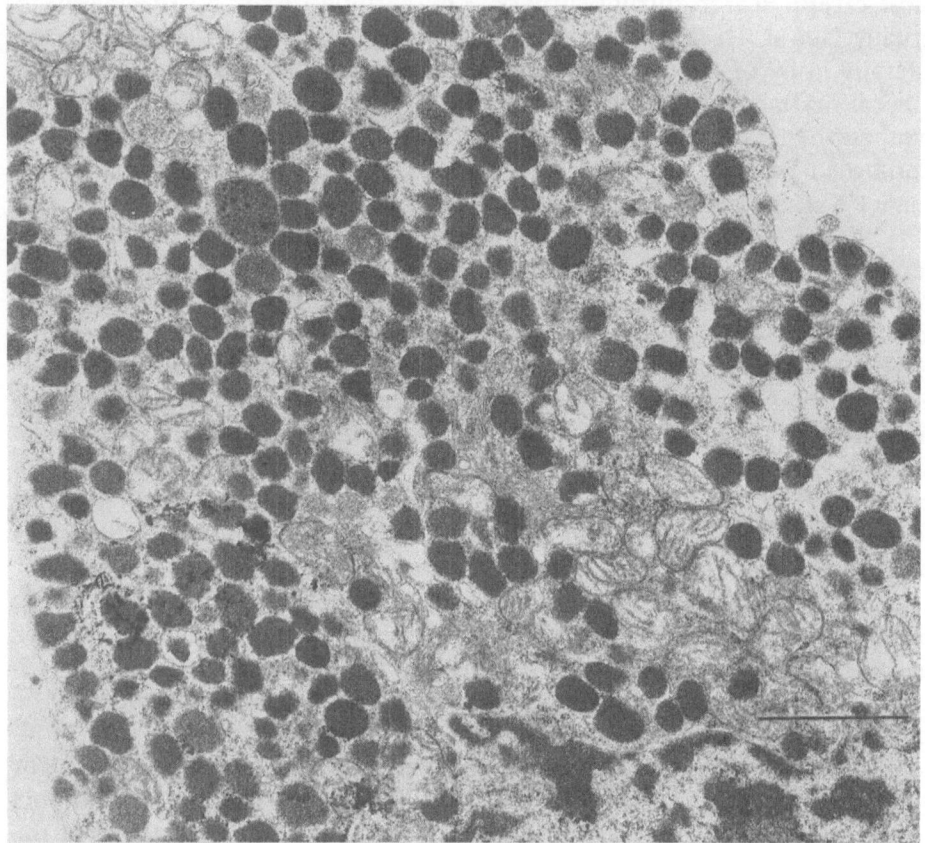

FIGURE 23.   Electron micrograph of a densely granulated corticotropic adenoma. (Scale bar: 1 μm.) (From Adelman.[198])

doubt, definitive diagnosis of corticotropin-secreting adenoma can be made by employing specific immunoperoxidase stains. When intermediate lobe tumors are suspected, silver stains need to be employed to demonstrate the argyrophilic strands that traverse the tumor. Figure 23 illustrates the electron-microscopic appearance of a densely granulated corticotrope adenoma.

Clinical Features

The clinical features of CD are a result of prolonged overproduction of cortisol. The features of hypercortisolism can be striking enough to be recognized at a glance, or subtle enough to be missed by the untrained eyes and mind. The classic textbook manifestations of fat deposition in the face, neck, supraclavicular area, and neck resulting in the familiar moon faces, buffalo hump, and truncal obesity with relative sparing of extremities are becoming less common. This is probably because sensitive tests have made it possible to diagnose the disease at an early stage. For descriptive purposes, the clinical features can be viewed from several vantage points: classic presentation, early manifestations, atypical manifestations, and clinical differences among the various types of hypercortisolism. Finally, the section will be concluded by correlating the physical findings to the well-known (and the not so well-known) effects of glucocorticoid hormones.

Classic Presentation

The detailed descriptions of the first 12 reported patients with CD by Harvey Cushing in 1932[1] rank among the best descriptions of a new syndrome in the archives of medical literature. The reiteration of the same description half a century (and several hundred patients) later by numerous reviewers, is attestation to that fact.

The evolution of pituitary-dependent CD is a slow process, gradually evolving into the plethoric features which have become strongly associated with that disease. Table 56 outlines the most frequently encountered features in patients with hypercortisolism.

Several series have analyzed the incidence of the numerous physical findings encountered in patients with hypercortisolism in general.[30-32] The recurring themes that emerge upon reviewing the literature are that (1) a high index of suspicion is required to make an early diagnosis, and (2) although several features of hypercortisolism can be caused by diverse disorders, when they occur collectively in the same patient CD is more likely to be present. The old adage that the presence of thin skin, thin muscle, and thin bones in a fat person should raise the possibility of Cushing's is an excellent one, since atrophy of the skin (84%) myopathy (90%), osteopenia (50%), and obesity, defined here as >115% IBW (80%), are very frequent features of chronic hypercortisolism. A review by Ross and Linch[33] analyzed the presenting features in 70 patients with documented CD seen in the past 30 years, with a view to identify the discriminatory features that would aid one in making an early diagnosis. These investigators noted that the three most discriminatory

TABLE 56
Frequency of Clinical Features in Cushing's Syndrome

Feature	Incidence (%)
Obesity	88–95
Plethora	60–90
Thin skin	80
Diastolic hypertension	76–87
Striae	50
Muscle weakness	60–90
Easy bruisability	42–65
Menstrual disorders	65–85
Hirsutism in women	64–80
Psychiatric problems	42–60
Backache (osteoporosis), edema	40–48

findings were the presence of ecchymoses, myopathy and hypertension. Three fourths of the patients had diastolic blood pressure in excess of 90 mm Hg: myopathy, evident objectively, was present in 56% of patients, and psychiatric disturbances figured prominently in two-thirds of their patients. While only 3% of patients were not obese, it was noted that the distribution of fat was not a particular discriminatory factor.

Hypertension, one of the most common features in patients with pituitary-dependent CD, is modest, often responding excellently to therapy. The mechanism(s) underlying the hypertension of CD are not well delineated. Biglieri et al.[34] studied adrenal secretion both in vivo and vitro in patients with CS and concluded that mineralocorticoid mediation was not a factor in the causation of hypertension in this disease. The two abnormalities that have been documented in patients with CD, that is, an increased plasma-renin activity[35] and an accentuated vascular response to pressor agents,[36] may be due to ACTH or glucocorticoids themselves or may represent causally unrelated factors.

The psychopathy of Cushing's syndrome deserves special emphasis, since it often highlights the clinical presentation. Mental changes are reported to occur in 50–60% of patients with Cushing's syndrome.[37,38] These include marked emotional lability, paranoid ideation, confusion, and disorientation. In severe cases, delusions, hallucinations, depression, and delerium may occur. Glaser[39] reported a 10% incidence of suicide attempts in patients with Cushing's syndrome. It is interesting to note that euphoria is an uncommon facet of Cushing's psychopathy, whereas patients placed on steroids are often euphoric. It is known that steroids may have a direct effect on neuronal tissue. It has been shown that steroids directly induce electrophysiological changes in the neurons, such as diminished synaptic delay and slowing of axonal conduction. Harvey Cushing recognized that emotional disturbances were prominent features of the disease; in fact, one of his patients was found in a mental hospital with depression, irritability and memory loss.[1] Recently, Starkman et al.[40] demonstrated a statistically significant correlation between the neuropsychiatric disability in Cushing's syndrome and the levels of cortisol and ACTH. Their study is suggestive of the fact that the incidence and severity of psychiatric manifestations are greater in patients with hypercortisolism due to elevated pituitary ACTH levels, i.e., central or pituitary-dependent CD. They observed that patients with adrenal adenomas (low ACTH levels) did not have as severe a psychiatric abnormality as the patients with central CD (high ACTH levels) in spite of having comparable cortisol levels. The obvious assumption is that ACTH contributes more to the psychopathy of Cushing's syndrome than do cortisol levels. The implication that β-endorphins may be related to the Cushing psychopathy has generated considerable interest in recent years. β-Endorphin is an opiatelike peptide extracted from the normal pituitary gland. β-Endorphin and ACTH are often secreted concomitantly to various stimuli that normally stimulate ACTH. Therefore, it seems plausible that when excessive ACTH secretion is present (central CD), a concomitant hypersecretion of β-endorphins may occur as well. A recent report demonstrating high endorphin levels in the CSF of psychotic patients has attempted

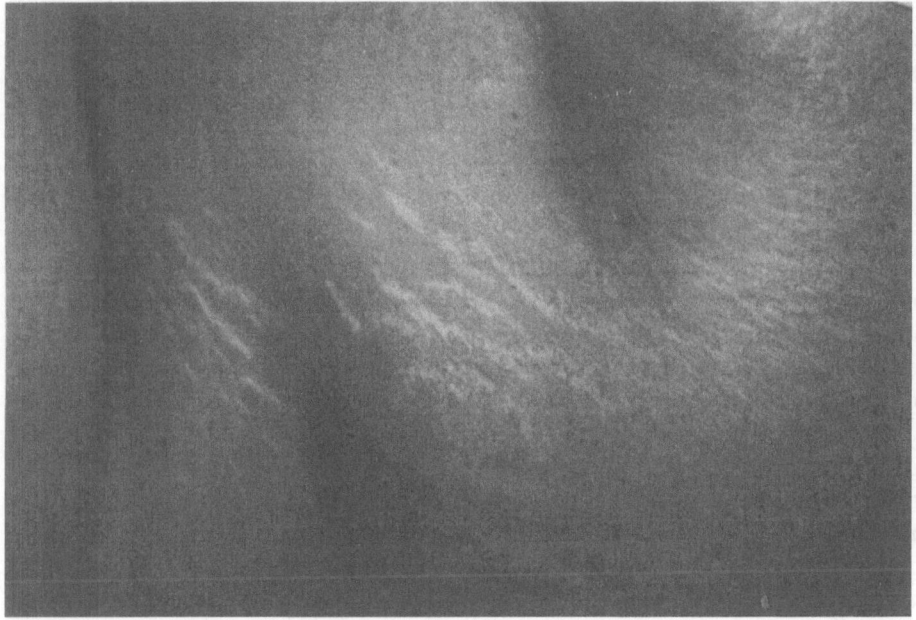

FIGURE 24. This 30-year-old patient with hypercortisolism demonstrates cutaneous striae with a violaceous hue.

to link mental disease with endorphins.[41] The significance of ACTH, endorphins, and the mental illness seen in CD awaits elucidation.

The myopathy associated with hypercortisolism is primarily due to protein catabolism caused by excess glucocorticoids. Like most endocrine myopathies, the proximal group of muscles are involved more commonly than the distal subjective features are more striking than objective evidence, and the biochemical abnormalities are even less striking. Rarely, overt severe wasting of the muscles may be encountered. Khaleeli et al.[42,43] recently characterized the clinical, biochemical, functional, and structural abnormalities in six patients with Cushing's syndrome, three of whom suffered from pituitary-dependent CD. Force measurements, using both the myometer and strain gauge techniques, demonstrated quadriceps weakness in every patient. The histological appearance by light microscopy was consistent with type II fiber atrophy in some, but not all, patients with endogenous hypercortisolism. Electromyographic evidence of myopathy was common to both endogenous and exogenous hypercortisolism. The skeletal muscle content of sodium was high, but that of creatine was low. Notably, the skeletal muscle content of potassium was normal. Several of these abnormalities were reversible upon restoration of eucortisolemia.

The thinning of the skin, in many an experienced endocrinologists's view, is a remarkably good clue to hypercortisolism, which is the only endocrine cause for measurable thinning of the skin.[44,45] The easy bruisability is not due to any abnormalities in coagulation, but merely represents increased fragility of the capillaries in the thinned-out skin. The striae, which can be quite impressive when classic, assume a violaceous hue, and are due to loss of collagen caused by glucocorticoid excess (Fig. 24).

Hyperpigmentation, a finding that was noted in five of the original 12 patients described by Cushing,[1] is due to the pigmentary effects of β-lipotropin. This finding is more frequent in patients with ectopic ACTH secretory syndrome.[45a]

Hirsutism, glucose intolerance, and hypertension is an impressive triad seen in a high percentage of patients with noncushingoid obesity as well. In many obese patients, these features may even be associated with subjective fatigue, mild depression, and menstrual irregularities, all secondary to obesity per se. Therefore it is easy to miss CD in a patient with obesity. The occurrence of easy bruisability, puffy facies, significant psychopathy, or edema should increase the index of suspicion in the detection of hypercortisolism. As pointed out by Gold,[46] diversity merits more emphasis, because the incidence of the occurrence of any single feature alone varies so widely among reported series that no single finding is a requisite for the diagnosis. On the other hand, the commonplace nature of several features of this syndrome are such that, if surveyed individually, half the population of United States would be encompassed.[46]

Early Manifestations

A high index of suspicion is required to recognize the disease in its early stages. Since pituitary-dependent CD evolves slowly and progresses gradually, the changes in appearance may be so insidious as to not concern the patient. In this regard, pituitary-dependent CD resembles acromegaly, a closely related disorder of the pituitary gland. The following clues may be helpful in suspecting the disease in its early phase.

1. Development of "puffiness (plethora) of face in a patient with hypertension of recent onset (Fig. 25)
2. Combination of edema of feet and recent-onset hypertension in an obese patient
3. Any patient with a history of recent weight gain, easy bruisability, and hypertension
4. Presence of spontaneous hypokalemia in patients with recent-onset hypertension
5. Combination of hirsutism and menstrual irregularities in a patient with recent weight gain
6. Obesity with significant muscle weakness
7. Unexplained psychiatric disturbances, especially depression, in association with any of the above
8. Growth retardation in an obese child

It should be noted that only a very small percentage of obese patients turn out to harbor CD; indeed, Cushing's syndrome seldom if ever causes morbid obesity. Nevertheless, CD, a disorder that can kill if unrecognized but

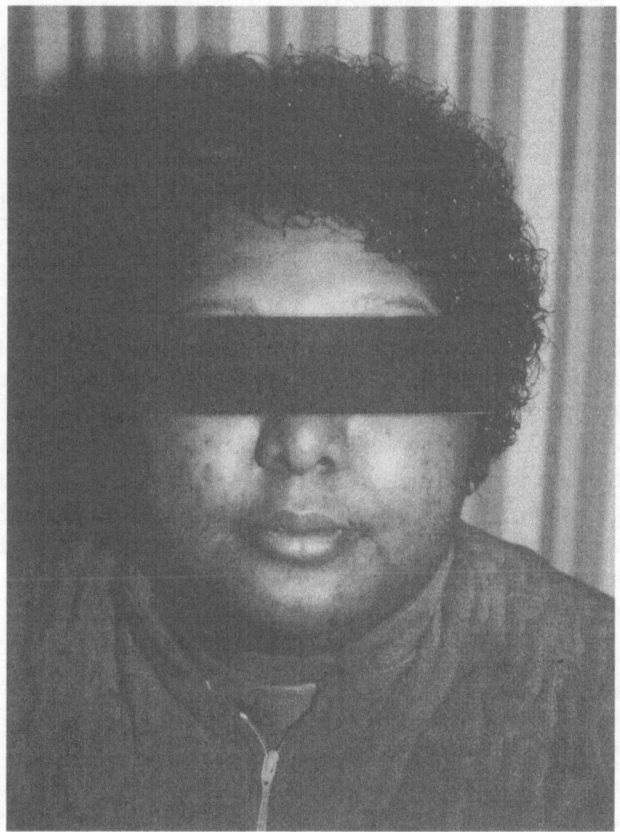

FIGURE 25. Same patient as in Fig. 24, demonstrating puffiness of face with acne and facial hirsutism.

that can be cured if detected, is an important etiology to be kept in mind when evaluating any patient with recent onset of obesity. If the question, Can this be due to hypercortisolism? is asked in every instance, the diagnosis is less likely to be missed. The five clues that may possess discriminatory value, in no particular order of importance, are easy bruisability (ecchymoses), puffiness of face, muscle weakness, hirsutism, and associated psychiatric problems.

Atypical Features

The term *atypical* may not be quite appropriate in discussing certain unusual findings in a disease that is as rare, as unpredictable, and as fascinating as Cushing's syndrome. Unusual deposition of fat in uncommon areas, lipomatosis, may be seen in this syndrome. The most well-known of this phenomenon is the mediastinal widening that occurs as a consequence of mediastinal lipomatosis.[47,48] Rarely, fat deposition may occur in the paracardiac space.[49] A more dangerous location for fat accumulation is the epidural space. Such epidural lipomatosis has been reported to be severe enough to cause compression of the spinal cord.[50–52] Very rarely, the rectum can be displaced anteriorly by the presacral deposition of fat,[53] or a fatty mass can be seen in the liver.[54] Although these phenomena occur more often with exogenous steroid administration, they are recognized features of endogenous hypercortisolism of any etiology.

Nephrolithiasis, once considered unusual in hypercortisolism, is being noted with increasing frequency. In the series of Ross and Linch,[33] renal calculi were present in 15% of patients with endogenous hypercortisolism. In the same series, loss of scalp hair was present in a small but significant percent (13%) of patients.

Since the vast majority of patients with pituitary-dependent CD harbor only small (micro) adenomas, headaches, visual-field cuts, and other parasellar phenomena are highly unusual. These features, when present, strongly point to the presence of invasive macroadenomas. It should be noted, however, that headaches are extremely common in patients with hypercortisolism regardless of the etiology, even in the absence of any increase in the sellar pressure. Other unusual but important features include the occurrence of cyclical edema,[55] increased tendency for developing superficial cutaneous fungal infections,[56] and thromboembolic phenomena.[57]

Etiological Differentiation

This may well nigh be impossible on clinical grounds alone. Although the syndrome evolves slowly when pituitary disease underlies the problem, and evolves rapidly when caused by adrenal adenomata, this may not be useful in a given case. The presence of galactorrhea and generalized hyperpigmentation may point to the pituitary as the site of CD. However, these findings are not consistently present. The occurrence of *virilization* in association with features of hypercortisolism is a strong sign of adrenocortical carcinoma.[58,59]

When Cushing's syndrome is caused by ectopic ACTH secretion, three distinct syndromes evolve: cachectic, pigmentary, and plethoric. The most frequent ones are cachectic and pigmentary. The cachectic form of ectopic ACTH-secreting syndrome, the prototype of which is lung cancer, is characterized by severe weight loss, darkening of skin, and metabolic changes of profound cortisol excess (hypokalemic alkalosis and hyperglycemia). In the pigmentary form, the clinical picture is dominated by increased skin pigmentation, with little or no clinical evidence of Cushing's syndrome. The reason for the paucity of plethoric findings in these circumstances is that the malignancy does not allow the patient enough time to develop cushingoid features. Occasionally, however, patients with ectopic ACTH secretion may indeed manifest the plethoric features of CD. The four classic examples of such a phenomenon are represented by ectopic ACTH secretion by carcinoids of the bronchus[60,61] carcinoid tumors of the pancreatic islets,[62] medullary carcinoma of the thyroid,[63,64] and pheochromocytoma.[65,66] The ectopic ACTH secreted by these tumors, which are benign or of low-grade malignancy, results in chronic stimulation of both adrenal glands. In all cases of ectopic ACTH-secreting syndrome, the disease results in stimulation of both adrenal glands. To this extent, the disorder resembles pituitary-dependent CD. The vast majority of tumors that secrete ACTH ectopically are indeed autonomous. A rare exception to this rule is bronchial carcinoid, which may demonstrate some degree of responsiveness to suppression tests. Whether these cases represent ectopic secretion of a CRF-like peptide is a controversial issue.

When Cushing's syndrome is secondary to iatrogenic causes, the most common form of Cushing's, the clinical features are again identical to other forms of plethoric Cushing's. There is a higher incidence of certain physical findings when Cushing's syndrome is caused by exogenous steroid administration. These include a higher incidence of ocular findings such as cataracts and papilledema, painful myopathy, ischemic necrosis of the femoral head, and pancreatitis. The reasons for these phenomena are not clear. It is also recognized that the incidence of fractures, and opportunistic infections are greater in iatrogenic Cushing's syndrome. Table 57 outlines the clinical presentations of the four forms of hypercortisolism.

Pathophysiological Correlations

The overproduction of cortisol in patients with Cushing's syndrome is clearly correlated with five expressions of such a phenomenon: catabolic effects on the skin, muscle, and bone; anti-insulin effects; CNS effects; anti-inflammatory effects; and the consequences on growth.

The catabolic effect on the skin and the muscle have been already alluded to. It is noteworthy that myopathy, a catabolic, protein-losing effect of glucocorticoids can be offset by the increase in anabolic androgens seen in association with adrenocortical carcinoma. The catabolic effects of glucocorticoids on bone metabolism deserve special mention, since this contributes to significant morbidity. Most patients with iatrogenic or spontaneous Cushing's syndrome suffer from a severe and significant form of bone loss. The changes in the skeletal system that occur in association with hypercortisolism have been extensively reviewed in the radiology literature.[67] In addition to osteopenia, hypercalciuria (urinary calcium excretion >250 mg/day) has been reported in approximately 45% of patients with hypercortisolism.[30] Considerable controversy revolves around the mechanisms that cause bone loss. While it is generally accepted that glucocorticoid excess suppress intestinal calcium absorption, the mechanism for such a phenomenon is far from clear. Most investigators have described a secondary increase in parathyroid hormone levels to counteract the decrease in intestinal absorption.[68] The effects of short-term administration of glucocorticoids on intestinal calcium absorption and its relationship to circulating PTH and vitamin D metabolite concentrations have been studied by Hahn and co-workers.[69] They concluded that the reduced intestinal calcium absorption following glucocorticoid administration could not be attributed to changes in immunoreactive PTH or circulating concentrations of the major known metabolites of vitamin D. To help resolve the difference cited in the literature, Findling et al.[70] studied the relationship

TABLE 57

Hypercortisolism: A Clinical Perspective[a]

Pituitary-dependent Cushing's	Adrenal adenoma	Adrenal carcinoma	Ectopic ACTH syndrome	Iatrogenic
Slow evolution	Slow evolution	Rapid evolution	Cachexia	History of steroid therapy
Plethoric features	Plethoric features	Virilization	Pigmentation	Cataracts
Headaches	Unilateral disease	Absence of myopathy	Metabolic ($\downarrow$ K, $\uparrow$ glucose)	Papilledema
Field cuts		Dissemination to lung, bone	Rarely cushingoid	Femoral head necrosis
Galactorrhea			(carcinoids of bronchus,	
Pigmentation			islet cell tumors, MCT,	
Both adrenals enlarged			tumors, MCT, pheo)	

[a] From Kannan.

of vitamin D metabolites and PTH to calcium and phosphorus homeostasis in seven patients with spontaneous Cushing's syndrome. Their data suggest that endogenous hypercortisolism decreases tubular reabsorption of phosphorus and increases PTH, which in turn results in an increase in 1,25-$(OH)_2$ D_3. These effects may contribute to the bone loss caused by the direct action of cortisol on bone. Histological and biochemical studies have demonstrated that glucocorticoids are capable of not only inhibiting bone formation but also can directly stimulate bone-resorbing cells.[71] The increased PTH secretion assumes importance as a contributory factor, especially in view of data that support the notion that parathyroidectomy abolishes the osteoclastic effect of steroids on bone in animals.[72] These facts may play an important role in the prevention and treatment of steroid related osteopenia.

The anti-insulin effect of hypercortisolemia is thought to be exerted at the postreceptor level. Nosadini et al.[73] studied five women with Cushing's syndrome and impaired oral glucose tolerance tests. By plotting the insulin-induced disposal dose–response curves obtained with the use of the euglycemic clamp procedure, these workers demonstrated a marked lowering of maximal glucose disposal (MGD). There were no significant differences in the insulin binding capacity of erythrocytes and monocytes between normals and patients with Cushing's syndrome. It is believed that glucocorticoid excess impairs tissue glucose disposal through a postreceptor mechanism.

The consequences of hypercortisolism on growth assume importance when the disease strikes youngsters.[74] Although the dynamics of growth hormone in adults with hypercortisolism may demonstrate a blunting in provoked secretion,[75] such is not the case in children; growth hormone dynamics in children with iatrogenic or spontaneous steroid excess is generally considered to be normal.[76,77] These observations, coupled with the fact that growth hormone therapy fails to induce linear growth in hypercortisolemic children,[78,79] are suggestive of the fact that glucocorticoids may interfere with somatomedin generation or action. Indeed, several studies have demonstrated low levels of net circulating somatomedin activity following glucocorticoid administration.[80–82] There is some controversy as to whether this reflects decreased production or indicates antagonism to the activity of these peptides. The results from studies using bioassay systems that test the direct effect of glucocorticoids on cartilage growth have yielded conflicting results.[83,84] A recent concept gaining favor is that hypercortisolemia can lower somatomedin activity by inducing changes in somatomedin inhibitors. These inhibitors of somatomedin activity are present in normal subjects.[85] Unterman and Phillips[86] demonstrated that glucocorticoid administration is followed by an increase in circulating somatomedin inhibitors, which can account for the fall in the net somatomedin activity in the sera of patients with hypercortisolemia. Regardless of the mechanism, hypercortisolemia results in a significant impairment of linear growth resulting in retarded skeletal maturation. The vertebral growth arrest lines (zones of increased density corresponding with the vertebral plates) may linger as residual radiological changes, years after cure, serving as a grim reminder of childhood Cushing's.[87]

Laboratory Diagnosis

The diagnostic studies involved in the evaluation of patients with Cushing's syndrome fall into three categories. The first phase of workup involves establishing the presence of true hypercortisolism. The second phase consists of tests that are aimed at establishing a primary pituitary etiology for the hypercortisolism in contrast to a primary adrenal or ectopic source. The third and final phase consists of performing the appropriate localization procedures as indicated. During all phases of the workup, the yield from the various diagnostic studies ranges from being conclusive and straight forward to frustratingly equivocal. Although the availability of newer tools such as assays for measurement of urinary free cortisol and plasma ACTH, high-resolution CT, and the CRH tests have all enhanced our diagnostic armamentarium, each phase of workup for this disease can be fraught with uncertainty. The following discussion will focus on the step-by-step evaluation of the patient suspected of endogenous hypercortisolism.

Phase I: Screening for True Hypercortisolism

A basic understanding of the synthesis, secretion, and circulation of glucocorticoids is essential to elucidate the best screening tests for hypercortisolism. Figure 26 outlines the pathways, and enzymes involved in adrenal steroidogenesis.

The major stimulus for glucocorticoid secretion is ACTH. Under the influence of this hormone, adrenal steroidogenesis takes place through all the necessary enzymatic steps. After secretion and release, the cortisol becomes

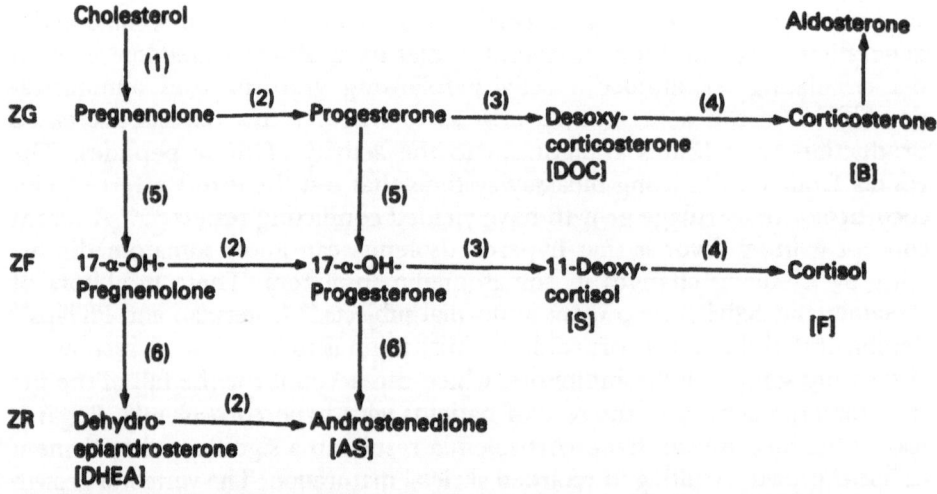

FIGURE 26. Adrenal steroidogenesis. Enzymes: (1) desmolase, (2) 3-β-OH-dehydrogenase, (3) 21-hydroxylase, (4) 11-hydroxylase, (5) 17α-hydroxylase, (6) 17,20-desmolase. ZG, zona glomerulosa; ZF, zona fasciculata; ZR, zona reticularis.

bound to a plasma globulin called transcortin. Transcortin has an extremely high affinity for cortisol (75% of all cortisol). Transcortin also binds to a variable extent with other steroids, such as progesterone, corticosterone, deoxycorticosterone, and deoxycortisol. The binding of cortisol to transcortin is reversible. Cortisol is also bound to albumin (15%). Approximately 10% of cortisol is free. It is the free fraction that moves into the cell to exert its action. The measurement of plasma cortisol (either by RIA or by fluorometric methods) measures only the cortisol bound to transcortin. The plasma cortisol determination by radioimmunoassay has replaced the fluorimetric methods for cortisol in plasma. Conditions that alter the transcortin would alter cortisol measurements. Thus, pregnancy, estrogen therapy, and obesity are associated with elevated cortisol levels due to elevated transcortin levels.

The glucocorticoids are mainly inactivated in the liver. By a series of reduction reactions, cortisol is converted to dihydro- and tetrahydrocortisol, which is then processed through the glucuronyl transferase system to become tetrahydrocortisol-glycuronide. Similarly, deoxycortisol is excreted as tetrahydrodeoxycortisol. These two metabolites are measured as 17-hydroxycorticosteroids, in the urine (Porter-Silber chromogens). Approximately 30% of cortisol secreted is metabolized to tetrahydrocortisol glucuronide. The collective measurement of 17-hydroxycorticosteroids in the urine in general reflects glucocorticoid activity, even though tetrahydrodeoxycortisol is not a biologically active compound (compound-S). If one wishes to extract or fractionate the amount of tetrahydrocortisol from the tetrahydrodeoxycortisol, this can be done by extractions using carbontetrachloride. In practice, however, this is not generally necessary, since the contribution of 11-deoxycortisol to the urinary 17-hydroxycorticoids is minimal. The 17-hydroxycorticoids can be affected by drugs such as spironolactone, dilantin, estrogens, and exogenous steroid administration.

In normal individuals, cortisol is secreted by the adrenal cortex in a series of bursts rather than in a steady continuous fashion.[88,89] If 24 profiles of serum cortisols are obtained by measuring the cortisol level in the plasma every 20 or 30 min, 7–10 secretory spikes can be identified. This phenomenon of pulsatile or episodic secretion should be taken into account when interpreting a sample of plasma cortisol drawn in an isolated period of time. In general, however, the amplitude of the spikes are greater in the early morning hours and lowest during the late night, with a noticeably protracted quiescent period around midnight. The cortisol spikes are a reflection of ACTH spikes that precede them.

When screening for a disorder that is potentially lethal, but that can generally be curable if diagnosed, it is desirable to have a screening test with the highest sensitivity; that is, the test should yield very low or negligible false negatives. If a false-positive diagnosis were made, the test could be repeated or other tests performed for confirmation. As screening tests, the combination of the overnight dexamethasone test, and the collection of urine for 24-hr free cortisol approach a reasonable standard of screening. Unfortunately, no single screening procedure for hypercortisolism is 100% free of false-negative results. Several tests are used: overnight dexamethasone suppression test, 24-hr urinary free cortisol (UFC), timed integrated concentration of cortisol (IC), 24-hr urinary 17-hydroxycorticoids, random plasma cortisol, evaluation of the circadian rhythm, and urinary steroid profiles.

Overnight Dexamethasone Suppression Test

This test is based on the principle that when 1 mg dexamethasone is administered by mouth to normal individuals at midnight, before sleep, it abolishes the nyctohemeral ACTH rise. As a consequence, the 8 AM cortisol level obtained the following morning will be suppressed below a defined limit of 5 μg%. It should be noted that the dexamethasone administered is potent enough to induce suppression of the hypothalamic pituitary axis, but does not interfere in the assay of plasma cortisol. Often 15–30 mg of flurazepam by mouth is given to ensure a rested night. The unique features of the overnight dexamethasone suppression test are the ease with which it can be performed, the inexpensiveness, and above all, the extremely low incidence of false-negative results. Crapo et al.[90] examined the data from several series and pointed out the following important observations. First, the number of patients with Cushing's syndrome who suppress normally is negligible; only 3 of 154 patients (1.9%) with Cushing's syndrome demonstrate normal suppression, one of whom on repeat testing demonstrated nonsuppression, the other two were not retested. Second, the incidence of false-positive results (i.e., failure to suppress in the absence of hypercortisolism) was very low in normal, nonobese, nonmedicated ambulatory patients (i.e., 5 of 466, or 1.1%). Third, the incidence of false-positive results was high in obese controls (13% of 173), in hospitalized patients and in those with chronic illness. Table 58 outlines the conditions characterized by nonsuppression to the overnight dexamethasone test.

As a screening test, when cost effectiveness is considered, the overnight dexamethasone test is highly recommended as first line screening.[91–94] Meikle[95] suggested that the discriminatory value of the test can be enhanced by the simultaneous measurement of plasma levels of dexamethasone along with plasma cortisol level at 8 AM. The demonstration of plasma dexamethasone levels high enough to exert suppression clearly aids in interpretation of the overnight dexamethasone suppression tests. The assay of dexamethasone in plasma will permit identification of those patients with inappropriately low plasma levels of the drug due to either delayed or altered drug metabolism. This is particularly the case in patients receiving anticonvulsant therapy. The clinical importance of concomitantly assaying the plasma dexamethasone level has been highlighted by two reports, one by Meikle et al.[96] and another by Caro et al.,[97] in which the diagnosis of Cushing's syndrome would have been difficult to establish without the information relating to the plasma level of dexamethasone or its metabolism. The assay, however, is not routinely available. Therefore, in clinical practice, when the overnight dexamethasone test is equivocal or abnormal, the next step in the approach to the problem is measurement of free cortisol in a 24-hr urine sample.

TABLE 58
Nonsuppression to Overnight Dexamethsone Test

Condition	Mechanism
All forms of Cushing's syndrome	
Pituitary dependent	Partial pituitary autonomy; reset hypothalamus
Adrenal adenoma or carcinoma	ACTH is already maximally suppressed by autonomous secretion of steroids by the adrenal tumor
Ectopic ACTH-producing neoplasma	Tumor elsewhere secretes ACTH autonomously
False-positive results	
Obesity	↑ transcortin (which binds the dexamethasone and decreases bioavailability)
Estrogen therapy (conjugated equine estrogen >1.25 mg)	↑ transcortin
Diphenylhydantoin	Metabolism of dexamethasone (increased hepatic mitochondrial inactivation to less polar metabolites)
Alcohol	? ↑ clearance
Malabsorption	Poor absorption of the administered drug
Depression and other psychiatric diseases	Abnormal hypothalamic–pituitary axis
Stress (e.g., chronic illness, hospitalization, surgery)	Activation of hypothalamic–pituitary axis

24-hr Urinary Free Cortisol

Regarded by many as the initial choice for screening, the measurement of free cortisol in the urine has several advantages.[98,99] First, and most importantly it measures the amount of cortisol that is free (i.e., unbound to transcortin) and metabolically active. The binding of cortisol to its globulin (transcortin or cortisol binding globulin) typically begins to reach saturation at levels of serum cortisol of approximately 25 μg.[100] When the concentrations of plasma cortisol exceed this level, there is a disproportionate increase in the level of the unbound or free cortisol. Since the bound cortisol is not filtered by the kidneys, measurement of free cortisol reflects the amount of cortisol above and beyond the saturation of transcortin; therefore, the urinary free cortisol measures the integrated glucocorticoid secretory activity in a 24-hr period; second, the urinary free cortisol is not affected by increases in transcortin, unlike plasma cortisol, which can be elevated whenever transcortin is increased (e.g., pregnancy, obesity). Third, and most importantly, under circumstances of hypersecretion, because of the saturation of transcortin, urinary free cortisol increases exponentially, while metabolite secretion (17-OHCS) increases only linearly. This is the reason why measurement of 24-hr urinary free cortisol is clearly superior to measurement of 17-hydroxycorticosteroid in the 24-hr urine, in screening for hypercortisolism. The value of urinary free cortisol in the diagnosis of Cushing's syndrome has been reviewed by Trecan et al.[101] These workers studied the sensitivity, specificity and predictive value of Cushing's syndrome based on measurement of basal and postdexamethasone levels of free cortisol in 24-hr urines. The subjects consisted of 26 patients proved to have Cushing's syndrome (20 with CD) and 93 normal subjects suspected of harboring hypercortisolism but who turned out not to have the syndrome. On the basis of their results, both the sensitivity and the specificity of urinary free cortisol measurement approached 96%. The value of the basal collection in predicting the presence of Cushing's syndrome using 95 μg as cutoff, was 80%. This increased of 100% when the postsuppression cut off number of 10 μg was employed. The predictive value for the absence of Cushing's syndrome, based on a basal level of ≤95 μg, was 99%. The authors also noted that measurement of 24-hr urinary 17-OHCS provided only partial separation of the two groups with considerable overlap.

Excellent as it may be, the 24-hr urine free cortisol suffers from the banal problems of completeness of urine collection. It is generally believed the urine free cortisol is not significantly affected by modest decreases in glomerular filtration rate (GFR). It should be noted that urinary free cortisol can be elevated in patients with psychiatric illness, particularly depression, and in stress. Most workers would agree that the urinary free cortisol serves as an excellent diagnostic tool in the obese patient suspected of having Cushing's syndrome, especially in the face of an equivocal overnight dexamethasone suppression test.

Timed Integrated Cortisol Concentration

The measurement of the integrated concentration (IC) of cortisol in the plasma is distinctly advantageous as compared with measurement of the hormone in a single sample drawn at an isolated point in time. This is because of the episodic or pulsatile nature of cortisol secretion by the adrenal cortex. The term *integrated* denotes the concentration of hormone in pooled blood collected at a constant rate over a period of time, the advantage being that it has a smaller variability and reflects the overall prevailing blood level than a single sample drawn at a given point in time. Several studies[88,102] have characterized the normal episodic pattern of cortisol secretory activity. Patients with CD are no exception to the fluctuating pattern,[103] a phenomenon that may greatly minimize the value of a single random cortisol level. Studies that determine the 24-hr integrated mean cortisol concentrations are highly representative of adrenal activity[104] but are impractical and cumbersome to employ as screening tests. Therefore, attempts have been made to evaluate the value of 3- or 6-hr integrated concentration of cortisol for diagnosing Cushing's syndrome. Zadik et al.[105] studied the clinical value of a shortened, practical version of the 24-hr integrated measurement, in its ability to detect Cushing's; 68 normal subjects and 13 patients with Cushing's syndrome constituted the study. When the integrated concentration of cortisol in different time slots were compared between the two groups, they noted that there was no overlap between in the 24-hr integrated plasma cortisols between the two groups. More importantly, the integrated concentration of cortisol drawn between 2000 and 0200 hr clearly separated controls from patients with Cushing's syndrome. The conclusion drawn was that the test, when performed between the hours of 2000 and 0200 (a period during which integrated cortisol levels in normal subjects are at a nadir), demonstrated a discriminatory value that was as good as more protracted 24-hr sampling. Halbreich et al.[106] studied an even shorter version of the test and evaluated the mean or integrated cortisol concentrations between the afternoon hours of 1300–1600 and concluded that it was a reliable and powerful indicator of cortisol hypersecretion. It must be emphasized that, at present, measurement of integrated cortisol concentration has not replaced the more conventional screening tests. At best they may be used when the clinical suspicion is strong and the other screening tests are equivocal.

24-hr Urinary 17-OHCS

Although the 24-hr urinary 17-OHCS measure the metabolites of glucocorticoids the amount of 17-OHCS excreted in a 24-hr period does not clearly separate normals from hypercortisolemic patients. This is not surprising since the Porter-Silber reaction (the color reaction between corticosteroids and phenylhydrizine) measures only 50–60% of extractable glucocorticoid metabolites in the urine. It is well known that a significant number (as high as 10%) of patients with hypercortisolism may show normal 24-hr 17-OHCS levels in the urine, while as many as 25% of noncushingoid obese subjects may demonstrate an elevated 24-hr urinary 17-OHCS.[90,107–109] While it is true that the discriminatory value of urinary 17-OHCS can be improved by expressing the 17-OHCS in terms of the urinary creatinine, the false-negative results are still too high for it to be a useful screening test. In general, most but not all patients with hypercortisolism demonstrate urinary 17-OHCS levels greater than 9 mg/day per urine creatinine.[110] Although the test is not suitable for screening, the measurement of 17-OHCS plays a dominant role in the standard dexamethasone suppression tests. This is because the standardization of this test is based on comparisons of 17-OHCS values before and after oral dexamethasone administration—the Liddle test.

Random Cortisol Level

Measurement of a random cortisol level is an extremely unreliable screening method for hypercortisolism. The plasma cortisol, measured by RIA, is affected by multiple variables, such as increase in the cortisol-binding globulin, drugs, and stress. The physiological range of cortisol in normal plasma is wide and variable and markedly fluctuates during the 24-hr cycle as a result of episodic secretion. The variability between the highest and lowest values in a single 24-hr period can be as much as a 10-fold difference. This magnitude of fluctuation undermines the meaningful interpretation of a single cortisol determination drawn at random. It is for this reason that spot cortisol determinations in the plasma are poor discriminators between normals and patients with hypercortisolism. Moreover, the rise in plasma cortisol associated with increased cortisol-binding globulin overlaps greatly with the levels seen in hypercortisolemic subjects.

Evaluation of Circadian Rhythm

A loss of circadian variation of the plasma cortisol, reported for the first time in 1960[111] has been regarded as a time-honored test for detecting hypercortisolism. While it is true that in normal subjects the ACTH-cortisol secretory pattern demonstrates a circadian rhythm, the precise demonstration of loss of rhythm would depend on performing multiple determinations for cortisol in plasma during crucial periods of the day. The fact that cortisol is secreted by the adrenal cortex in bursts with variable amplitudes may mask the phenomenon of circadian rhythm, merely because of the timing of blood sampling. Thus, even in normal subjects, depending on the precise moment at which blood is sampled, the circadian pattern of cortisol may not be readily evident. Furthermore, in contrast to the commonly held belief, some patients with pituitary-dependent CD may indeed demonstrate preservation of the circadian rhythm. Glass et al.[112] reported two such patients who underwent hourly blood sampling for 24 hr. Analysis of the cortisol patterns revealed statistically significant circadian rhythms in both. The amplitude of the rhythms fell within the range for the amplitude of the diurnal rhythm of cortisol seen in normal subjects. Similarly Van Cauter et al.[113] reported the preservation of diurnal pattern of cortisol rhythm in two patients with CD. Despite these interesting observations, it is still widely believed that the vast majority of patients with CD lack diurnal variation. The conflicting data may be representative of the fact the pituitary-dependent CD is a heterogeneous disorder characterized by CRH-independent as well as CRH-dependent forms.[26] If so, the lack of circadian rhythm would be expected to be a striking feature of the CRH-independent form, while preservation of the diurnal rhythm would be anticipated in the CRH-dependent variety of CD. Improved methods for interpreting hormone rhythms, such as using statistical analyses by periodograms, may permit better understanding of the phenomenon.

Urinary Steroid Profile

The metabolites of cortisol and androstenedione undergo a unique alteration in patients with hypercortisolism. Physiologically, both steroids are metabolized in the liver by enzymes containing 5α-reductase and 11β-hydroxysteroid dehydrogenase activity. These enzymes are highly sensitive to circulating T3 level. For instance, patients with hypothyroidism have tremendous impairment in the activity of these enzymes. Since hypercortisolism of any cause inhibits the peripheral deiodination (outer ring deiodination) of T4 to T3, an analogous situation develops. As a result, the hepatic metabolism of glucocorticoids and androgen shifts in a different direction, yielding a singular pattern, characterized by 5β- and 11β-hydroxysteroid metabolites in the urine (see Table 59). The only other condition characterized by such a shift in metabolism is hypothyroidism. Since these metabolites can be readily

TABLE 59
Neutral Steroid Profiles in Normal and Hypercortisolemic Patients

5α compounds	5β compounds	11-OH compounds	11-Keto compounds	Normal pattern	Cushing's
Androstenedione (AS)	Etiocholanolone (EC)	Tetrahydrocortisol (THF)	Tetrahydrocortisone (THE)	AS > EC	EC > AS
Allotetrahydrocortisol (allo-THF)	Tetrahydrocortisol (THF)			THE > THF	THF > THE
				allo-THF > THF	THF > allo-THF

measured by gas chromatography, several workers have evaluated the discriminative value of this neutral steroid profile in correctly identifying the presence of Cushing's syndrome.[114,115] Recently, Phillipou et al.[116] studied seven patients with Cushing's syndrome and noted that in every case there was a striking alteration in the hepatic metabolism of cortisol and androstenedione. Thus, the ratios of etiocholanalone/androstenedione, and THF/Allo-THF, were increased, while the ratio of THE/THF was decreased. There were no false negatives in any patient with Cushing's syndrome in all three series. The only condition that mimics the profile seen in hypercortisolemic patients is hypothyroidism; rarely, patients with 5α-reductase deficiency, or porphyria, may yield similar profiles. Until a larger number of patients are evaluated, the urinary steroid profile is not likely to replace the urinary free cortisol or the overnight dexamethasone test as the screening for Cushing's syndrome.

Phase II: Establishing the Etiology of Hypercortisolism

The second step in the approach to the patient with hypercortisolism is determining the source of the disorder, that is, whether it is originating from the pituitary, the adrenals, or an ectopic source. The three diagnostic procedures helpful in this exercise are the standard dexamethasone suppression test, the plasma ACTH level, and the metyrapone test. When results of all three tests point in the same direction, the etiological diagnosis can be established with a high confidence limit. If all three do not point to the same site, one or more need to be repeated, and other diagnostic avenues must be perused. The introduction of a fourth test—the CRF test—has added a new dimension to the diagnostic approach for hypercortisolism. It is too soon to speculate whether the new is going to replace the old conventional methods of testing. The principles, classic responses, atypical responses, diagnostic accuracy of these tests are as follows: the standard dexamethasone suppression test, the Plasma-ACTH assay, the metyrapone test, and the ovine CRF test.

Standard Dexamethasone-Suppression Test

The pioneering work by Liddle[117] in 1960 has laid the framework for what is still considered by many to be the "gold standard test" in the etiological diagnosis of hypercortisolism. The passage of a quarter century has seen the emergence (as well as the exit) of several "newer" tests claiming superior diagnostic accuracy. The standard dexamethasone test or more appropriately, Liddle's test, has withstood the test of time, mostly because of the solid principles on which it is founded. When Liddle reported his findings in 1960,[117] the thorough analysis of the test results that led to the criteria that he established for normals ranks among the best investigative work ever reported in the literature. A brief outline of these criteria is necessary for further understanding of the abnormal states. The standard dexamethasone test consists of a 2-mg (low-dose) and a 8-mg (high-dose) dexamethasone suppression test.

The test involves measuring the 24-hr urine for 17-OHCS before, during, and after the low and high doses of dexamethasone administered orally. The normal response to the low dose (2 mg/day) for 2 days is a decline in 24-hr urinary 17-OHCS to below an absolute value of 3.5 mg/g creatinine per day. All obese noncushingoid patients will normally suppress to the low dose, while regardless of the etiology, patients with Cushing's fail to do so. The high-dose dexamethasone, which extends the test for 2 more days, involves administering 8 mg/day for 2 days, and comparing the urinary 17-OHCS of the second day of high dose with the basal level. A decline in the 17-OHCS or the urinary free cortisol by 50% or more of the baseline is defined as suppression. Most patients with pituitary-dependent CD demonstrate suppression to the high dose, while patients with adrenal tumors and ectopic ACTH-secreting tumors will not. The discriminatory value of the high dose dexamethasone suppression test is based on the autonomy of these tumors as well as the preservation of partial responsiveness in pituitary-dependent CD. To reiterate, the classic patterns for the various types of hypercortisolemic states are (1) central Cushing's (pituitary-dependent), which demonstrates no suppression to the low dose, but adequate suppression to the high dose; in this entity, the servo-mechanism for feedback at the pituitary (or hypothalamic) level is set at a higher threshold; (2) adrenal Cushing's (adenoma or carcinoma), which fails to demonstrate suppression to a low or high dose; this type is classically autonomous, since the ACTH is already maximally suppressed by the increase in circulating glucocorticoids; and (3) Ectopic ACTH-secreting tumors, which again fail to demonstrate suppression to any dose of dexamethasone, since they are independent of pituitary ACTH. There are however, notable exceptions to the "classic" pattern in each type of Cushing's.

Central Cushing's: When there is a strong clinical or radiological suspicion of pituitary-dependent Cushing's, and the standard dexamethasone test reveals no suppression to a high dose, the following dynamic phenomena must be kept in mind and appropriate testing undertaken:

1. Pituitary ACTH-dependent Cushing's nonsuppressible to high-dose dexamethasone (As many as 15–30% of patients with pituitary dependent Cushing's may not show the classic suppression to high-dose dexamethasone.[118,119] These patients may require a high–high-dose dexamethasone suppression test (4–8 mg QID), since their hypothalamic-pituitary threshold for suppression is greatly elevated.[120,121])

2. Central Cushing's (pituitary ACTH dependent) nonsuppressible to dexamethasone, but suppressible to intravenous cortisone (This unusual but important phenomenon, in which the receptors at the hypothalamic-pituitary level do not recognize dexamethasone but do "see" cortisol, was first reported by Carey et al.[122])

3. Pituitary-dependent CD caused by intermediate lobe tumors (often resistant to suppression by high-dose dexamethasone, although they may suppress to bromocriptine.[17])

4. Central Cushing's (pituitary ACTH dependent) with a paradoxical increase in cortisol during the dexamethasone suppression test[123]

    5. Central Cushing's (pituitary ACTH dependent) with periodic
       hormonogenesis[124]

The latter two phenomena are recognized to be part of the same entity, in
which pituitary tumors secrete ACTH on a cyclical basis. Variable cycles have
been reported, i.e., once in 12 days, or as long as once in 80 days. Between
the "bursts", all tests may be normal, or slightly elevated, but during the
secretory "bursts" the ACTH and cortisol secretions are totally autonomous
occurring randomly and independently of the fortuitous administration of
dexamethasone. In these cases, it is often impossible to make the correct
diagnosis by the dexamethasone test. The patient should be retested at a later
date.

    Adrenal Cushing's: The phenomenon of adrenal tumors that retain ACTH
suppressibility to high-dose dexamethasone, especially an entity called macronodular hyperplasia of the adrenal glands, has been described.[125] These
are basically cases of long-standing hyperplasia of the adrenals, resulting in
a single or multiple "autonomous" adenoma(s) that still retain the memory to
suppress. More recently, Smalls et al.[126] systematically compared the biochemical and pathological features of 13 patients with macronodular hyperplasia; their study reconfirms the notion that the entity may be a result of
long-standing CD with varying degrees of pituitary dependence and adrenal
autonomy.

    Ectopic ACTH-secreting tumors: While the characteristic response of these
tumors is nonsuppression to any dose of dexamethasone, some such tumors
demonstrate suppression to high-dose dexamethasone. A notable example of
this phenomenon is the ACTH-secreting carcinoid tumors, usually of the
bronchus. These may mimic pituitary-dependent CD to a remarkable degree.[127–129]

    1. Nearly one-third of these tumors demonstrate some preservation of
       suppression to high-dose dexamethasone, yielding dynamic data iden-
       tical to pituitary-dependent Cushing's syndrome.
    2. The plasma ACTH level, in contrast to most cases of ectopic ACTH
       syndrome, is often not strikingly high, overlapping ranges that can be
       seen in pituitary-dependent Cushing's.
    3. The tumor can be clinically, radiologically, and biochemically silent,
       explaining the asymptomatic nature, negative chest radiographs, and
       normal serotonin metabolites encountered.
    4. Because of its benign nature, the patient may evolve into plethoric
       Cushing's, heightening the similarity to pituitary-dependent Cushing's
       syndrome.

It is therefore not surprising that the above features render the situation
conducive to missing the real cause of hypercortisolism in silent ACTH-secreting carcinoids. In some cases, the only definitive method for establishing
or excluding a pituitary source of ACTH is selective venous sampling of the
inferior petrosal sinus.[120]

    Finally, it must be realized that abnormal test results may occur in the absence of true hypercortisolism.[130] Patients with severe endogenous depression may show persistent abnormalities in ACTH dynamics, demonstrating failure to suppress with a low dose, but good suppression with high dose. In such patients even the diagnosis of Cushing's becomes dubious, much less the etiological separation.[131]

The above exceptions in no way mitigate against the overall value of the standard dexamethasone test. The results of the standard dexamethasone test assume greater significance when viewed in conjunction with the plasma ACTH level, the second test to be performed in the etiological diagnosis of hypercortisolemic patients.

Plasma ACTH Assay

Since the concentration of circulating ACTH levels in the plasma of normal humans is highly responsive to the ambient concentrations of cortisol in the plasma, measurement of ACTH should permit distinction between the various causes of hypercortisolism; and indeed it does so in a significant number of patients. Theoretically, patients with hypercortisolemia due to adrenal tumor should have very low, i.e., suppressed levels of ACTH, while patients with CD should demonstrate ACTH levels inappropriate to the level of circulating cortisol level i.e., a mildly elevated or even "normal" level. Patients with hypercortisolemia secondary to ectopic ACTH-secreting syndrome generally demonstrate markedly elevated levels of ACTH. There are, however, several important issues that may cloud the interpretation of the ACTH assay:

1. The assay for ACTH in plasma is a very delicate one. Blood has to be collected with meticulous care using chilled EDTA containing tubes on ice; the blood must be centrifuged and the plasma separated within 2 hr of collection. Failure to collect the sample properly could spuriously elevate or decrease the plasma ACTH level.
2. ACTH levels in plasma is subject to a great deal of fluctuation due to the inherent pulsatile nature of its secretion from the pituitary. Therefore, the level of hormone in a randomly drawn sample of blood is not reflective of the overall prevailing concentration of the hormone. The same diagnostic limitations that applied to the random serum cortisol are also applicable to a single ACTH level drawn at random.
3. The range of basal ACTH in the plasma of patients with pituitary dependent Cushing's is frustratingly wide (from as low as 30 pg to as high as 180 pg). Most, but not all, patients with hypercortisolism due to adrenal tumor demonstrate basal ACTH levels below 10 or 20 pg. The level of ACTH in ectopic ACTH-secreting bronchial carcinoids often overlaps the range seen with pituitary-dependent Cushing's.
4. Finally, the differences in assay systems employed by various reference laboratories make it impossible to interpret the ACTH assay by merely looking at a number.

TABLE 60
Interpretative Value of DXM Test Coupled with ACTH Assay

Suppression to high-dose dexamethasone	Plasma basal ACTH	Diagnosis
Suppressible	Normal or mildly elevated (50–100)	Cushing's disease
Nonsuppressible	Low (<10)	Cushing's syndrome secondary to adrenal tumor
Nonsuppressible	High (>200)	Cushing's syndrome secondary to ectopic ACTH
Suppressible	Low (<20)	Macronodular hyperplasia
Suppressible	Modest elevation (150–200)	Ectopic ACTH secondary to carcinoid (or) Cushing's disease

The plasma ACTH assay viewed in conjunction with the standard dexamethasone test provides additional information (Table 60).

Metyrapone Test

One of the main disadvantages of the standard dexamethasone suppression test is the inherent difficulties of ensuring a proper and complete 24-hr urine collection. Regardless of whether the parameter compared is free cortisol or the more conventional 17-OHCS, the collection for 7 days (2 basal, 4 for the low- and high-dose days, and one postcontrol) are difficult even under the best circumstances. For this reason, it would be desirable to have a dynamic test that does not involve urine collection and yet has a discriminatory value that equals (or excels) the standard dexamethasone test; the metyrapone test of the eighties seems to satisfy both these objectives.

Metyrapone is a chemical that blocks the 11β-hydroxylation of 11-deoxycortisol to cortisol. When administered at a dose of 750 mg by mouth every 4 hr for six doses, metyrapone almost completely blocks the 11-hydroxylation of 11-deoxycortisol. As a result, the concentration of cortisol rapidly declines, causing the hypothalamic pituitary unit to respond by releasing more ACTH. This in turn would lead to an increase in the level of 11-deoxycortisol, the immediate precursor steroid proximal to the block. The availability of a sensitive RIA for the measurement of 11-deoxycortisol in the plasma has greatly simplified the metyrapone test, obviating the nuisance of urine collection. A number of smaller series have evaluated the diagnostic utility of metyrapone.[132,133] The largest series that has prospectively compared the value of the metyrapone test vs. the standard high-dose dexamethasone test in 25 unselected patients with hypercortisolism is that of Sindler et al.[134] These workers compared the two tests in their ability to determine accurately the etiology of Cushing's syndrome. The results of their study indicated that the metyrapone test was more accurate than the standard high-dose dexamethasone test in differentiating pituitary-dependent CD from adrenal tumors causing hypercortisolism. All patients with CD demonstrated a post metyrapone 11-deoxycortisol level greater than 10 μg/dl, while all patients with adrenal adenomas had a suppressed 11-deoxycortisol level (<10 μg/dl) post-metyrapone. The reasons for these behavioral patterns are straightforward. The normal response to metyrapone administration is an increase in the plasma level of 11-deoxycortisol. This response is preserved in patients with pituitary-dependent Cushing's and is abolished in adrenal tumors causing Cushing's, since the ACTH is maximally suppressed. Metyrapone has more far reaching effects than just blocking 11β-hydroxylase. It is believed that metyrapone has a dual action on cortisol biosynthesis—a distal block on 11β-hydroxylase, and a proximal block early in the steroidogenesis that involves cholesterol cleavage.[135] The latter block is not significant in normal subjects since it can be overcome by ACTH, but in patients who have no ACTH in plasma (i.e., patients with adrenal tumors), metyrapone leads to significant lowering of 11-deoxycortisol. The response of ectopic ACTH secreting syndrome to metyrapone is mostly similar to adrenal tumors, resulting in very little or no increase in 11-deoxycortisol following metyrapone. The convenience and accuracy of the metyrapone test has made it an important dynamic test in determining the etiology of Cushing's. However, the predictive value of the metyrapone test is not exempt from the bizarre exceptions that pervade all diagnostic tests for Cushing's syndrome. There have been reported instances of unilateral adrenal neoplasms responding to metyrapone administration.[136]

CRF Stimulation Test

In a sense, all of the aforementioned three diagnostic studies—the dexamethasone suppression tests, the plasma ACTH level, and the metyrapone study—are focused on the responsiveness of the corticotrope. Therefore, when in the early 1980s, the structure of CRF was characterized, a new method to study aberrant corticotrope behavior appeared on the horizon. Vale et al.[137] characterized the 41-residue ovine hypothalamic peptide CRF and showed that it effectively stimulates the secretion of corticotropin and β-endorphin. The in vivo effectiveness of synthetic ovine CRF was established by Rivier et al.,[138] who demonstrated the selective effects of this peptide on human corticotropes. As expected, numerous studies from international centers began reporting the diagnostic use of ovine CRF in altered states of corticotrope function. Orth et al.[139] were among the earliest investigators to study the response pattern of ACTH to CRF stimulation. They reported two patients with mild CD who underwent testing with CRF shortly before, and 1 week after, successful transsphenoidal microadenectomy. When CRF was given in a dose of 1 μg/kg body weight, a brisk exaggerated ACTH response was seen in the first patient. This hyperresponsive ACTH pattern disappeared following removal of the microadenoma. Similar but less impressive results were seen in the second patient. The authors suggested that the response patterns in these two patients with ACTH secreting microadenomas was consistent with the fact that CRF may be involved in the pathogenesis of the CD seen in these two patients. Muller et al.[140] reported on the effects of intravenous CRF on ACTH concentrations in seven patients with pituitary-dependent CD, two patients with adrenal tumors, and one with the ectopic ACTH-secreting syndrome. All seven patients with pituitary-dependent CD demonstrated a brisk hyperresponsive ACTH pattern, despite the fact that four had normal baseline ACTH levels. By contrast, the ACTH response to CRF in the other two groups of patients remained unaltered.

Chrousos et al.[141] presented the hitherto largest series to date in 1984; they studied the response of ACTH to CRF in 22 patients with various forms of hypercortisolism. Again, all 13 patients with pituitary-dependent CD showed a robust increase in the already elevated basal ACTH level. They confirmed prior data that following microadenectomy there was restoration of the CRF-induced ACTH (and cortisol) responses to near-normalcy or normalcy. Once again there was confirmation of the fact that patients with ectopic ACTH syndrome (with very high basal ACTH levels) and those with adenal neoplasms (with very low basal ACTH levels) behaved identically with absolutely no increase in ACTH level when CRF was administered. The results of these studies clearly seemed to impress one fact—the failure of ACTH to respond to a bolus of CRF in the setting of Cushing's clearly excludes the pituitary as the source of the problem.

While the enthusiasm for the CRF test was gaining momentum, a few studies began to raise some doubt as to whether all patients with pituitary-dependent CD behave in a homogeneous fashion. Pieters et al.[142] studied the ACTH response to intravenous CRF administration in five patients with pituitary-dependent CD. While they noted that the absolute increments of ACTH (and cortisol) levels of these patients following CRF administration were significantly higher than in normals, the CRF responses in individual patients were variable. For instance, in three patients, the ACTH increase following CRF was within the mean ± 1 SD of the response seen in normal subjects. More interestingly, one patient with CD with bilateral micronodular hyperplasia demonstrated a negligible ACTH response to CRF. Pieters et al. concluded that the CRF–ACTH axis in patients with pituitary-dependent CD may reveal hyperresponsive, normoresponsive, and even hyporesponsive patterns when tested with exogenous CRF. When the behavior of pituitary adenomas removed surgically and maintained in culture were studied, somewhat unexpected data emerged. Suda et al.[143] evaluated the ACTH responsiveness in vitro by using superfusion of pituitary adenoma tissue removed from 16 patients with CD. Their data indicate that adenoma tissue, in some patients, demonstrated a strikingly low sensitivity to CRF. Similar data had been previously reported by Shibasaki et al.[144] who demonstrated that the response of pituitary adenomas removed from two patients with Cushing's disease to a variety of agents demonstrated great variability. In particular, CRF failed to stimulate the secretion of ACTH or β-endorphin from adenoma tissue of both patients, while vasopressin, 3-isobutyl methylxanthine and high concentrations of potassium brought about a prompt increase in the secretion of these POMC-derived peptides. These findings suggested that these corticotrope adenomas had either lost their receptors to CRF or, the postreceptor mechanisms were nonfunctional. The reason(s) for the dichotomy between the in vivo and the in vitro ACTH responses to CRF in some patients with pituitary-dependent CD is not clear. Perhaps the in vivo ACTH response to CRF is modulated by other factors, such as vasopressin or angiotensiogen.[143]

These conflicting data regarding the responsiveness of ACTH to CRF in pituitary-dependent CD clearly indicate the need for more studies.[146] It is unlikely that the CRF test will wholly replace the need for the conventional dynamic tests that have been in vogue for decades. In a resounding editorial article resonating with clarity, David Orth[145] has placed in perspective the new and the old in CD. Although the CRF test may separate most patients with pituitary dependent Cushing's disease from the ectopic ACTH syndrome, patients in the latter group with recent-onset hypercortisolemia and incomplete pituitary suppression may respond to CRF, mimicking the dynamics of CD.[145] The author's conclusion was that "the demonstration of relative versus absolute resistance to glucocorticoid inhibition, as shown by the dexamethasone-suppression test, will continue to remain the best single method to differentiate Cushing's disease from Cushing's syndrome due to adrenal tumor or ectopic ACTH secretion."

TABLE 61
Diagnostic Studies in Evaluation of Cushing's Syndrome

Test	Comment
Screening	
The overnight dexamethasone-suppression test	Limited by the high percentage of false-positive results
The 24-hr urinary free cortisol	Limited by the problems of complete urine collection
Etiological diagnosis	
The standard dexamethasone suppression tests	High-dose test distinguishes pituitary from nonpituitary dependency in 60–80% of cases
Plasma ACTH level	Serves as a valuable adjunct to other dynamic tests
Metyrapone test	Separates pituiatry dependent from nondependent Cushing's; discriminatory value equals, perhaps excels, the standard dexamethasone test
CRF test	New technique to view an old principle (i.e., ACTH responsiveness in hypercortisolism)
Localizing procedures	
CT scan of adrenals	Outlines abnormalities in nearly all cases of adrenal tumors
CT scan of sella	Abnormal in nearly 20–70% of pituitary-dependent Cushing's disease
Selective venous catheterization of inferior petrosal sinus for ACTH gradient	When performed properly, establishes or excludes the pituitary origin of ACTH excess

Table 61 outlines the important diagnostic tests in Cushing's syndrome.
In summary, the following diagnostic approach is followed at Cook County
Hospital in the evaluation of hypercortisolism. As a first step, the overnight
dexamethasone test is performed as an outpatient. If the post-dexamethasone
8 AM cortisol is below 5 μg, no further workup is indicated unless the clinical
findings are extremely strong. If the patient fails to suppress to the overnight
dose, a 24-hr urine is collected for measurement of free cortisol. If the values
exceed 150 μg, a high-dose dexamethasone suppression test is performed. If
the patient shows suppression (50% decline in basal 17-OHCS or urinary free
cortisol), pituitary-dependent disease is suspected; if the patient fails to show
suppression, an adrenal tumor or ectopic source of ACTH is suspected. Further hormonal confirmation of either entity is provided by performing a basal
ACTH level in plasma and the metyrapone test using 11-deoxycortisol level
as the parameter. If *all* tests point in the direction of the pituitary, the sella
turcica is evaluated by CT; if all tests, on the other hand, point towards an
adrenal source, both adrenals are evaluated by CT When the CT of sella is
normal, but the hormonal data point to the pituitary, the only definitive
method to document the source of the ACTH is by selective venous sampling
of the inferior petrosal sinus; when ectopic ACTH syndrome is suspected,
appropriate radiological studies are undertaken to detect the primary tumor.
The algorithm is outlined in Fig. 27. The value of an "overnight high-dose
dexamethasone suppression test is currently under investigation.[194]

Phase III: Anatomic Localization of the Etiology

The anatomical localization of the source of hyperticolism depends on
the information derived from the steroid dynamics. When all data point to a
pituitary source, the localizational procedures are aimed at imaging the pituitary by CT. When the hormonal data are indicative of an adrenal neoplasm,
the focus shifts to CT of the adrenal. When the data are inconclusive, computerized tomography of both the adrenals and the pituitary are indicated.
An additional procedure, selective venous sampling of the inferior petrosal
sinus, has added a new dimension for the diagnosis of difficult cases. The
advent of computerized tomography of the adrenal glands has obviated the
need for adrenal venography selective catheterization of adrenal veins, and
imaging procedures with iodocholesterol. These procedures are not discussed,
since their practical application in the diagnostic workup of hypercortisolism
has been superseded by newer and better diagnostic tests.

Computed Tomography of the Adrenals

The adrenal glands, located retroperitoneally, amid the abundant fat that
is always present in patients with Cushing's syndrome, lend themselves to
excellent anatomical depiction by CT. The reader is referred to standard
treatises in radiology[161] for the characteristics of the normal and abnormal
adrenal by computerized tomography. The limbs of the normal adrenal glands

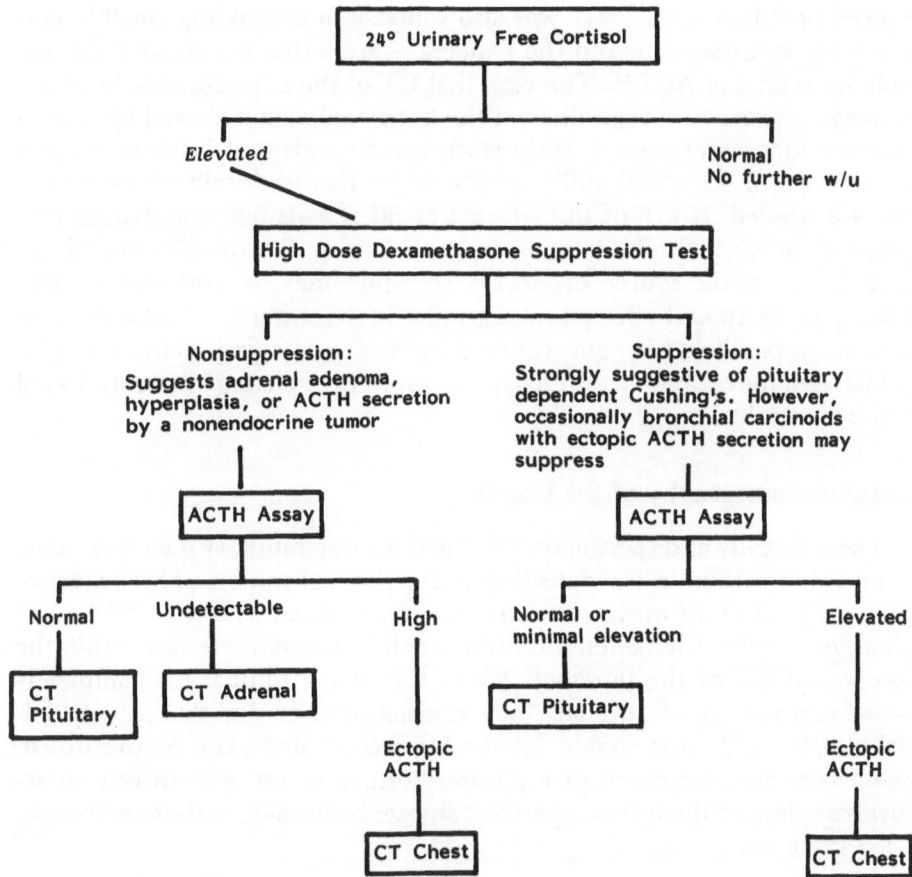

FIGURE 27. Algorithmic approach for evaluating hypercortisolism.

measure 3–4 mm in thickness, by CT. The outer margin of the adrenal limbs are invariably concave. The singular advantage of computerized tomography of the adrenals is that it is always abnormal when intrinsic adrenal pathology (adenoma or carcinoma) is causing the hypercortisolism. Adrenal tumors are almost always 2 cm or larger in diameter and are easily visualized by tomography. Since the cholesterol content of these adrenal tumors is abundant, they are 10–20 Hounsfield units lower in density as compared with the adjacent soft tissue. White et al.[162] noted that CT correctly identified all 15 patients in their series with adrenal tumors, distinguishing five carcinomas from the 10 adenomas. The appearance of both adrenals in pituitary-dependent Cushing's ranges from normal to slight enlargement bilaterally. The classic appearance, when present, is generalized thickening of the adrenal limbs bilaterally. In general, the bilateral increase in size of the adrenal glands tends to be greater when ectopic ACTH syndrome underlies the etiology of hypercortisolism. In the series of White et al.[162] CT was also valuable in identifying small lesions in the lung, mediastinum, and the pancreas—areas that are strongholds for neoplasms secreting ACTH. The view that CT of the adrenals should be the first imaging procedure regardless of the hormonal data is shared by several authorities for several reasons. If the study reveals unilateral, intrinsic adrenal disease (which it does with 100% accuracy), no further localizational procedures are needed. If CT of the adrenal gland reveals bilateral disease (enlargement) or normal adrenals, the appropriate studies to differentiate pituitary from ectopic source of ACTH secretion may be undertaken. The efficiency of computed adrenal tomography in detecting focal adrenal mass lesions is unparalleled by any other noninvasive imaging technique. The capabilities of the computed adrenal tomography in depicting bilateral adrenal nodules is variable.

Computed Tomography of the Pituitary

The sensitivity and specificity of CT in depicting pituitary microadenoma are much lower than in the detection of the adrenal tumors. The incidence of detecting pituitary microadenomas can range from as low as 20% to as high as 70%.[163,164] Microadenomas smaller than 1–2 mm are not within the resolution of CT of the pituitary. When CT of the pituitary is completely normal, and the CT of adrenals show normal-sized or slightly enlarged adrenal glands, an ACTH source for the hypercortisolism, can be presumed. Unequivocal documentation of a pituitary origin would have to rest on selective sampling of the inferior petrosal sinuses bilaterally, to demonstrate an ACTH gradient.

Selective Venous Sampling of the Inferior Petrosal Sinus

The venous anatomy of the pituitary is such that each half of the gland is drained by a venous plexus into either the ipsilateral inferior petrosal sinus or into the intercavernous sinus crossing the floor of the pituitary fosse.[165] Figure 28 diagrammatically outlines the venous drainage of the pituitary gland. The pituitary is drained via the cavernous sinuses posteriorly into the superior and inferior petrosal sinuses, and from there into the superior and inferior petrosal sinuses, and from there into the jugular bulb and vein. The blood from the transverse, sigmoid, and occipital sinuses also drains into the jugular bulb. Since there is increased admixture of blood in the jugular bulb and distal to it, sampling of the petrosal sinus is critical to determine the presence or absence of a hormone gradient. Although experimental data in humans are lacking, experiments in the rhesus monkey suggest that pituitary venous drainage is lateralized. Oldfield et al.[166] determined the extent of inter mixing of blood between the cavernous sinuses and between both inferior petrosal sinuses by injecting ^{99m}Tc colloidal sulfur into the superior orbital vein of rhesus monkeys. By comparisons of radioactivity in the interior petrosal sinuses bilaterally, these workers showed that the mean relative radioactivity was negligible in the contralateral inferior petrosal sinuses. Their

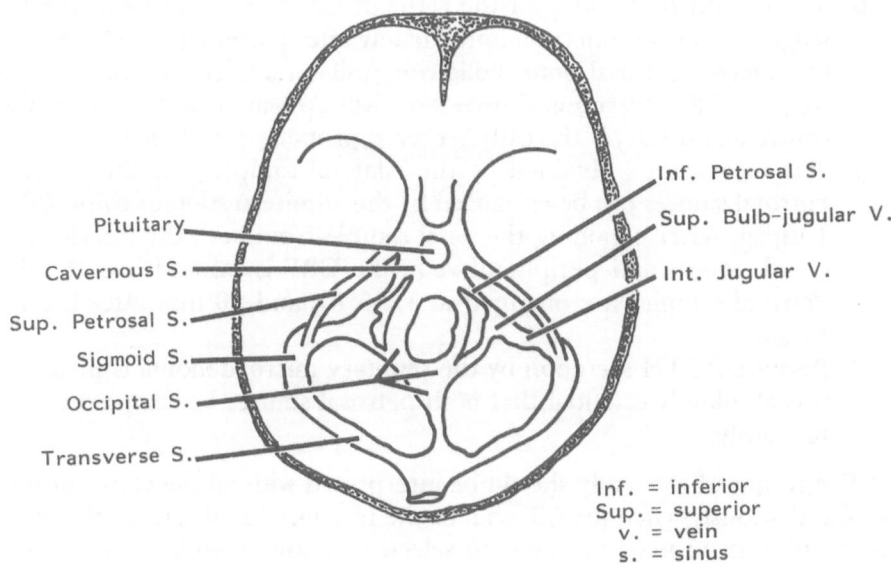

Pituitary

Cavernous S.

Sup. Petrosal S.

Sigmoid S.

Occipital S.

Transverse S.

Inf. Petrosal S.

Sup. Bulb-jugular V.

Int. Jugular V.

Inf. = inferior
Sup. = superior
v. = vein
s. = sinus

FIGURE 28.   Venous drainage of the pituitary gland.

findings suggest that mixture of blood between the cavernous sinuses and the inferior petrosal sinuses is insignificant. Thus, the bulk of evidence in humans favors the notion that pituitary venous drainage is lateralized. Therefore, a laterally placed microadenoma will result in increased ACTH levels in the ipsilateral petrosal sinus. This premise has been confirmed by several studies.[120,167–170] Accordingly, measurement of the ACTH gradient (the ratio of central to peripheral ACTH concentration) by catheterization of the inferior petrosal sinus is accurately representative of the pituitary origin of the tumor. Several important aspects of this procedure deserve emphasis.

1. The central-to-peripheral ACTH concentration must be interpreted cautiously, owing to the short half-life of ACTH and the intrinsically fluctuant nature of ACTH secretion.
2. Failure to collect and handle the sample properly for an assay that is difficult may result in misleading numbers.
3. The need for bilateral petrosal sinus sampling has not been sufficiently emphasized in the literature. Such a procedure will clearly demonstrate an ipsilateral gradient on the side of the microadenoma, if it were laterally situated. A midline microadenoma will demonstrate equal gradients higher than normal, information that is likely to help the neurosurgeon. If only one side is sampled, and if it were not the side at which the tumor was lateralized, the information yielded would not be clearly conclusive in localization of the source, or the side.
4. Lateralizing gradients are not consistently obtained with jugular venous sampling. It is therefore essential that the catheter be engaged in the inferior petrosal sinus.
5. The catheters should not be advanced to the cavernous sinus, owing to the risk of cavernous sinus thrombosis.

6. The correct positioning of the catheter can be verified by gentle retrograde contrast injection immediately after placement of catheter in the inferior petrosal sinus. Following unilateral injection, the contrast will fill both cavernous sinuses and will appear as reflux down the contralateral side, if the catheter were properly positioned.

7. The information obtained by the bilateral sampling of the inferior petrosal sinuses can be enhanced by the administration of ovine CRF, 1 μg/kg. After obtaining the basal sample from both the inferior petrosal sinuses and peripheral vein, the CRF is infused over 1 min. Petrosal samples are obtained at 1,3,5,10, and 30 min after the infusion.

8. Because ACTH secretion by the pituitary microadenoma is pulsatile, it is absolutely essential that both petrosal sinuses be sampled simultaneously.

The results of the study should be interpreted with all the above factors in mind. Obviously when the CT scan of the pituitary has disclosed microadenoma, there is no need to resort to selective venous sampling. But in the patient with hypercortisolism and a normal sella, with dynamic data that suggest a pituitary source, a properly performed selective venous sampling of the inferior petrosal sinus greatly helps in confirming or excluding a pituitary origin. Typical findings that characterize a laterally placed tumor are a large ACTH gradient on the ipsilateral petrosal sinus, a less impressive mildly increased ACTH gradient on the contralateral side, a prompt and dramatic rise in ACTH gradient on the ipsilateral side after administration of ovine CRF, and a less marked gradient on the contralateral side following CRF administration.

When performed correctly, the study is thought to be an accurate predictor of the site of the microadenoma in nearly every case. The information is especially helpful for the neurosurgeon if microsurgery fails to show a tumor. If the side of the ACTH secretion is known, a hemipituitectomy done on the side of gradient is curative.

Differential Diagnosis

Obesity

Obesity can be differentiated from true hypercortisolism by the measurement of urinary free cortisol levels in a 24-hr sample. Unless complicated by psychiatric illness, obesity is characterized by normal 24-hr urinary free cortisol. When free cortisol measurements are not available, the low-dose dexamethasone test can be performed, the results of which are nearly always normal in obesity.

Drug Interference

Various states of drug interference can cause confusing results in the dynamic tests employed in the evaluation of hypercortisolism; of particular importance is diphenyl hydantoin. Patients on this anticonvulsant drug often show failure to suppress to the overnight and low-dose dexamethasone tests. This is owing to accelerated clearance of dexamethasone occasioned by the effect of diphenylhydantoin on hepatic microsomal induction. An alternate method of studying steroid dynamics in patients receiving the drug is by administering hydrocortisone and measuring the plasma level of corticosterone.[147] Noncushingoid patients on diphenylhydantoin decrease their 8 AM plasma corticosterone level to below 270 ng/dl and below 50% of the baseline following 50 mg hydrocortisone at midnight.

Psychiatric Illness

Psychiatric illness, especially depression can mimic several facets of the abnormal steroid dynamics of Cushing's syndrome. Thus, increased cortisol secretion, more secretory spikes[148] and even increased urinary free cortisol levels[149] are often encountered in patients with depression. Absence of diurnal variation, and abnormalities in the overnight and even the low-dose dexamethasone suppression tests can occur in patients with depression. The combination of an elevated urinary free cortisol, abnormal suppression to low-dose dexamethasone and preservation of suppression to high-dose dexamethasone test closely resemble the dynamics of pituitary-dependent CD. The frequency with which abnormal steroid dynamics are encountered in endogenous depression, reverting to normal with treatment, has rendered the dexamethasone suppression test a parameter in the diagnosis and followup of patients with unipolar and bipolar depressive illness. Since depression is often a prominent aspect of Cushing's syndrome and shares several facets of hormone dynamics with that syndrome, it may be difficult to establish the presence of true hypercortisolism in patients who are depressed. The following may help in identifying Cushing's syndrome in a depressed patient.

1. The cortisol response to insulin-induced hypoglycemia is preserved in most cases of endogenous depression.[150]
2. The steroid dynamics revert to normal upon treatment of the depressive disorder, in contrast to Cushing's syndrome, where the abnormal suppression data persist despite adequate treatment with antidepressants
3. Recently, the CRF test has been used as a tool in the diagnostic evaluation of depression. The ACTH response to CRF in patients with depression is significantly blunted.[151,195] It is believed that such a response is appropriate for a pituitary gland exposed to the negative feedback of chronic hypercortisolemia, resulting in the restraining effect to CRF stimulation. Such a response separates the depressed patient from one with pituitary-dependent CD.

Alcoholic "Pseudo-Cushing's"

Chronic alcoholism shares several similarities in the clinical and hormonal features of hypercortisolism:

1. The clinical syndrome of pseudo-Cushing's caused by chronic alcoholism includes weakness, plethora, rounding of face, fatigue, peripheral muscle wasting, thin skin, and even purple striae. The picture may be further complicated by the occurrence of hypertension, glucose intolerance, and osteopenia in chronic alcoholism. In an individual patient, however, only a few of these features may be present.
2. The serum cortisol levels are often elevated, frequently demonstrating failure to suppression with the low-dose dexamethasone.[152]
3. Abnormal diurnal rhythms may be associated with the high cortisol levels.[152,153]
4. Rarely, hyperresponsiveness of the hypothalamic pituitary axis to metyrapone and an elevated secretory rate of cortisol may be encountered.[154] The similarity to pituitary-dependent Cushing's syndrome may be further accentuated by the abnormally blunted plasma cortisol response to hypoglycemia.[155]

The reasons for the abnormal steroid dynamics seen in chronic alcoholism could be a result of stress, alcohol withdrawal, intercurrent illness, or concomitant psychiatric disease. There is a paucity for studies that have evaluated the steroid dynamics in the nonimbibing phase of chronic alcoholics. It is generally believed that these abnormalities resolve during abstinence. Measurement of urinary free cortisol in the non imbibing phase is probably the best means of screening for hypercortisolism.

Adrenal Neoplasms

The distinction between the various types of Cushing's syndrome's can be attained by the combination of the high-dose dexamethasone test, the metyrapone test and the plasma ACTH immunoassay. Characteristically, patients with pituitary-dependent CD demonstrate preservation of suppression to high-dose dexamethasone and are responsive to metyrapone, both features being lost in patients with Cushing's syndrome caused by adrenal tumors. The level of the adrenal androgen dehydroepiandrosterone sulfate (DHEA-S) is depressed in patients with adrenal tumors causing Cushing's syndrome, often remaining low for as long as 2 years following surgical extirpation of tumor.[156] The DHEA-S levels in CD are normal or may be minimally elevated. A paradoxical response of ACTH to TRH and luteinizing hormone-releasing hormone (LH-RH) is seen in most patients with Cushing's disease, while such a response is absent when adrenal tumors underlie the etiology of Cushing's syndrome. Table 62 outlines the differences between Cushing's syndrome secondary to Cushing's disease and adrenal tumors.

Macronodular adrenocortical hyperplasia may be associated with conflicting dynamic data, since this entity is characterized by varying degrees of adrenal autonomy coupled with varying degrees of pituitary dependency, suppression tests, basal ACTH levels, and ACTH response to CRF may not always provide clear separation of this entity from CD and from unilateral tumors of the adrenal. Computed tomography may or may not reveal the macronodular disease, but always indicates bilateral disease.

TABLE 62
Differences between Pituitary-Dependent Cushing's Disease and Adrenal Neoplasms

Test	Pituitary-dependent Cushing's disease	Adrenal neoplasm
Response to 24-hr 17 OHCS or UFC to high dose dexamethasone	Preserved	Lost
Response of 11-deoxycortisol to metyrapone	Present	Absent
Plasma ACTH level	Normal to minimally elevated	Suppressed
Plasma DHEA-S level	Normal to minimally elevated	Suppressed
ACTH response to CRF	Preserved; may be exaggerated	Absent
Paradoxic ACTH response to TRH, LH-RH	Often present	Absent
CT scan of pituitary	May be abnormal	Normal
CT scan of adrenals	Both adrenal enlarged	Unilateral neoplasm

Adrenal Carcinoma

Adrenal carcinoma should be suspected when virilization occurs in association with hypercortisolism. This rare disease is a devastating disorder that affects younger patients, although it can occur at any age. The most common presentations of adrenal carcinomas are the development of Cushing's syndrome and virilization. In most instances these two syndromes occur together. Adrenal carcinomas causing virilization alone, without a concomitant increase in glucocorticoids, have been described, but the reverse has not, i.e., whenever an adrenal carcinoma causes Cushing's syndrome there is invariably an elevation of 17-ketosteroids. The hormonal hypersecretion by an adrenal carcinoma is a rather paradoxical phenomenon because when adrenal cells become malignant they lose many enzyme systems needed to produce hormones. These tumors are therefore regarded as rather inefficient hormone producers. The implications of this observation are twofold. First, when there is evidence of hormone hypersecretion at the time of diagnosis, it means that the tumors have been present for a long time and have grown large enough to produce hormone hypersecretion. Second, because of the enzyme deletions, several precursors are released in the circulation. Thus, DHEA, estrone, estradiol, and 11-deoxycortisol may be detected in the urine before virilization has occurred. The rapid development of Cushing's, the presence of virilizing signs, and a palpable mass would clearly indicate the clinical diagnosis of adrenocortical carcinoma. In addition to virilization adrenal carcinomas can result in precocious puberty in both sexes, and feminization of males.

Ectopic ACTH Syndrome

The recognition of ectopic ACTH in its characteristic presentation of weight loss, pigmentation, with metabolic phenomena (e.g., hypokalemia alkalosis, hyperglycemia) is relatively simple. In most patients the underlying malignancy is apparent and the plethoric features of Cushing's syndrome are absent. The cortisol levels are usually quite high (>50 μg) as is the plasma ACTH level (>250 pg). These patients are nonresponsive to the administration of dexamethasone, metyrapone, and CRF. The exceptional features of the ectopic secretion of ACTH by carcinoids has been already alluded to. The occurrence of plethoric Cushing's in the setting of suppressibility to high-dose dexamethasone with only modest increments in plasma ACTH levels closely resembles pituitary-dependent CD. The metyrapone test, or the CRF test have been evaluated in a large series of patients with dexamethasone-suppressible ectopic ACTH syndrome. The only definitive method of excluding a pituitary source of ACTH in such patients with an occult primary is by measuring the ACTH gradient by simultaneous bilateral inferior petrosal catheterization, preferably coupled with the administration of CRF just prior to sampling.[196,197]

Periodic Hormonogenesis

One of the most problematic tasks is establishing the diagnosis (much less the etiology) of hypercortisolism is when it occurs in the setting of periodic hormonogenesis. Apparently these represent examples of a tumor on a timer. The responses to dexamethasone administration are highly variable, ranging from no response to a paradoxical response. Several reports have highlighted this entity. The first documentation of such a phenomenon was by Bailey[157] in a patient with a malignant bronchial carcinoid that secreted ACTH cyclically with dramatic steroid cycles every 18 days. Brown et al.[124] described cycles of 11 days in a patient with a chromophobe adenoma causing CD. The cyclical changes in steroid levels in these cases was unaffected by dexamethasone, and the apparent paradoxical response to dexamethasone was fortuitous. The obvious difficulty is that one cannot interpret the steroid levels as an effect of drug administration and the increase or decrease in steroid levels is totally a reflection of periodic hormogenesis. The most remarkable example of such a phenomenon is the case reported by Lieberman et al.[123] These workers studied a patient with CD secondary to a chromophobe adenoma for 243 days and presented evidence for periodicity in cortisol production with cycles occurring every 85 days associated with laboratory remissions in between and paradoxical responses to dexamethasone administration. Thus, it appears that paradoxical responses to dexamethasone are best explained by the presence of an autonomous rhythm and periodic changes in hormone levels with nothing more than a coincidental temporal relation to the administration of dexamethasone.

Since it has been implied that cortisol secretion in such cases is independent of stimulation or feedback, it is interesting to note that Jordan et al.[158] studied a patient with cyclic Cushing's syndrome of pituitary origin, apparently set at a 35-day cycle. The fact that the pituitary did respond to feedback was evidenced by a decline in urinary free cortisol level with cyproheptadine and a dramatic rise in ACTH in response to the stress of anesthesia.

The phenomenon of periodic hormonogenesis (cyclical Cushing's syndrome) is not restricted to pituitary-dependent CD. Even though pituitary- and ACTH-secreting tumors predominate in the display of this astonishingly colorful spectacle of periodic hormonogenesis, adrenal tumors are not exempt.[159,160] The importance of recognizing the phenomenon of periodic hormonogenesis is threefold. First, the results of a high-dose dexamethasone test could be misinterpreted as nonsuppressible when, in fact, it is due to the periodic hormonogenesis; second, between cycles, the hormonal data may be normal, thereby establishing that the hormonal diagnosis could be difficult; and third, intermittent clinical effects of Cushing's syndrome such as cyclical edema,[55] intermittent hypokalemia, periodic mental changes, intermittent loss of diabetic control, and osteopenia can elude diagnosis if this entity is not kept in mind.

Treatment

The ideal treatment for pituitary-dependent CD should be one that will completely remove the source of ACTH hypersecretion, restore cortisol secretion to normal, avoid hypopituitarism, and be associated with negligible morbidity and mortality. These ideals, at least theoretically, are best fulfilled by selective microadenectomy by transsphenoidal pituitary surgery (TPS). Other forms of therapy—bilateral adrenalectomy, pituitary radiation, and therapy with drugs that lower ACTH-cortisol hypersecretion—fall short of these ideal requisites of therapy. The most satisfactory results for a "cure" are obtained when primary treatment is directed at the pituitary gland. Attempts at treating the hypothalamus with chemical therapy are fraught with unpredictable and transitory responses, while therapy aimed at the adrenals is drastic, with the potential for trading off two diseases (Addison's and Nelson's) for one disorder. With the increasing expertise that is emerging for pituitary microsurgery, this is clearly becoming the first, and in many cases the only choice. There are several options for treating patients with pituitary-dependent CD.

Transsphenoidal Pituitary Microsurgery

The procedure of transsphenoidal pituitary surgery has gained immense popularity in the past decade. Several workers have published their experience with this procedure in the treatment of pituitary-dependent CD, the two largest being that of Boggan et al.[15] from San Francisco, who treated 100 patients with CD between 1974 and 1981, and the series of 75 patients of Hardy[16] from Montreal. A smaller but equally good series by Burch[25] has also outlined the results of this procedure in 19 patients who underwent pituitary microsurgery for CD in a 6-yr period. The important aspects to focus on are the success rate of the procedure, the definition of "cure," the adverse effects, and the reasons for failure.

It is now generally accepted that the cure rate of transsphenoidal pituitary surgery (TPS) far exceeds every other available therapeutic modality for CD. The overall cure rate was noted to be 88% in the series by Hardy and 87%

in Boggan's series; these figures are applicable only to tumors confined to the sella turcica. The success rate drops significantly when the tumor is no longer confined to the sella (25 to 48%). It can be therefore emphatically stated that when the tumor is confined to the sella, TPS offers the patient the best chance of a cure. It is interesting to note that in all the three series, the incidence of finding no obvious tumor during surgery was approximately 20%. Before discussing the neurosurgeon's options in such a situation, it is necessary to understand the exploratory strategy practiced by most neurosurgeons who perform microsurgery.[171] The initial incision is in the midline carried into the neurohypophysis. If an adenoma is not visualized, both the lateral wings should be inspected to detect a resectable adenoma. If no adenoma is seen in either location, a final surgical sweep of the pars intermedia is made to detect a tumor in that region. If the exploration is completely negative, and the surgeon has data from selective inferior petrosal sinus catheterization that denote a unilateral source, then a hemipituiectomy on the side of the increased gradient is performed. If no catheterization data are available, the choice between a medial wedge resection, and a total hypophysectomy is a difficult one, often resting with the individual neurosurgeon's prior experience. Since the incidence of negative exploration is only about 20%, in the vast majority of patients an adenoma will be found (even with no abnormalities on CT) that can be successfully resected. It is in these situations that a cure rate greater than 80% can be offered to patients by the experienced and expert neurosurgeon.

The efficacy of TPS in children and adolescents, a group normally treated with conventional radiation, has been established by the report of Styne et al.[172] They treated 15 children and adolescents with CD with TPS and were able to achieve a cure in 14, with a negligible incidence of hypopituitarism.

The success of TPS can be evaluated within a few days after surgery. If the source of the pituitary ACTH was completely removed, the adrenals rapidly become hypofunctional. This is because the normal corticotropes have been rendered dormant owing to the suppressive effect of chronic hypercortisolism. It follows then, that secondary hypoadrenalism, (transient as it may be) is an anticipated reflection of successful resection of an ACTH-producing pituitary microadenoma.[23] In fact, if hypocortisolism does not develop immediately after TPS, it is unlikely that the patient is cured, despite normalization of urinary free cortisol and restoration of normal suppressibility to dexamethasone.[25] Such patients tend to show a high rate of recurrence of hypercortisolism.[173] In anticipation of transient secondary hypoadrenalism patients undergoing TPS are covered with hydrocortisone starting on the day of surgery. The transient secondary hypoadrenal state caused by prolonged suppression of normal corticotropes lasts for a variable period of time ranging from weeks to even months. It appears that hypocortisolism (with a low ACTH level) in the postoperative period is the best prognostic parameter for defining cure.[25]

The adverse effects of TPS are few. The morbidity associated with the procedure is very low. Transient diabetes insipidus is the most frequent complication. Less commonly, CSF leak, meningitis, and optic nerve damage may occur. The mortality, when it occurs, is more likely to be seen in patients with invasive tumors. It should be emphasized that the single most important factor that determines the outcome of surgery is the experience of the neurosurgeon. Such experience is invaluable in localizing the adenoma, and in properly exploring the entire pituitary. When the exploration is negative, the correctness of the further decision (medial wedge resection, versus selective partial central hypophysectomy, versus total hypophysectomy) is entirely a reflection of the neurosurgeon's prior experience. It is therefore essential that the patient be referred to a center that has an established track record.

There are several reasons for the persistence or recurrence of hypercortisolism following TPS. The three major reasons for such failure are when corticotrope hyperplasia underlies the disorder; when the tumor is invasive; and when tumors originating from the intermediate lobe cause CD. There is a fourth reason—erroneous diagnosis—the most notorious example being ectopic ACTH secretion caused by an occult relatively benign neoplasm mimicking the dynamics of pituitary-dependent CD.[196] Corticotrope hyperplasia can be the cause of pituitary-dependent CD in a small minority of patients. There are no definite preoperative methods for making such a diagnosis. The existence of such an entity has been clearly documented in the literature.[2,174–176] The fact that corticotrope hyperplasia can coexist in addition to a discrete adenoma of the adenohypophysis can further confound the issue.[44] Although corticotrope hyperplasia can occur in a diffusely scattered form, clusters of corticotrope nests are especially found in the central posterior area of the adenohypophysis. The therapeutic choice for corticotrope hyperplasia has not been well established, the recommended options ranging from total hypophysectomy[176] to selective central partial hypophysectomy.[16]

The poor response rate of intermediate lobe tumors to pituitary microsurgery has been recognized since Lamberts et al.[17] proposed the existence of two types of ACTH-secreting microadenomas—one from the adenohypophysis and one originating from the intermediate lobe. Tumors that originate from the intermediate lobe are probably of neural origin, and are histologically characterized by argyrophilic nerve fibers traversing the tumors, as well as by the hormonal triad of poor suppression to dexamethasone, responsiveness to bromocriptine, and hyperprolactinemia. If such patients could be identified preoperatively, the option for treatment with bromocriptine is a valid one to offer the patient, since surgical cures for this subset of patients is not gratifying. If the condition is identified during surgery by the characteristic neurofibrillary fibers within the adenoma, the choice between extensive surgery versus partial surgery with adjunctive bromocriptine therapy postoperatively is a difficult one to make with the current state of our knowledge.

The third reason for therapeutic failure with TPS is when the tumor is invasive. Lateral invasion of tumor is a well-recognized situation for noncure.[15] Under such circumstances, the choices are between complete hypophysectomy versus bilateral adrenalectomy. Each case should be individualized, since the price to be paid here for the cure of CD is the exchange for one or more deficiency states.

Selective adenomectomy still provides the best chance for a permanent cure. As our understanding of the disease progresses, and newer methods of identifying the poor responders become available, the success rate is likely to improve above the excellent chances of a cure for this disease.

Bilateral Adrenalectomy

This surgical method reverses hypercortisolism immediately, and for the most part permanently. The only instances of recurrence are when some adrenal tissue is inadvertently left behind; when the capsule is ruptured and clusters of cells are released into the retroperitoneal bed, where they germinate and gradually begin to hyperfunction (retroperitoneal adrenal remnants); or when adrenal rest cells in the gonads (usually testes) become functional under the chronic stimulatory influence of ACTH. The objections to bilateral total adrenalectomy as the first line of therapy for CD are as follows:

1. Removing the target organ(s), without removing the obvious seat of primary pathology is scientifically unsound strategy.
2. Permanent adrenal failure, with the need for lifelong replacement therapy is a high price to pay in an age when selective microadenectomy of the pituitary is an established procedure.
3. The morbidity and mortality of bilateral adrenalectomy is significant. Most centers acknowledge a 5% mortality with this procedure. The postoperative complications include infection, bleeding, thromboembolic phenomena, and pancreatitis.
4. The most important objection to bilateral total adrenalectomy is the possibility of developing Nelson's syndrome.[177] This entity is characterized by the development of generalized hyperpigmentation, progressive enlargement of the sella turcica, and the development of visual-field cuts. Hormonally, marked elevations of ACTH and β-lipotropin-endorphin peptides are seen. Nelson's syndrome represents growth of a pre-existing tumor, growth of which is enhanced by adrenalectomy. Cook et al.[178] demonstrated the restraining effect of glucocorticoids on ACTH during all stages of pituitary-dependent Cushing's. When the suppressive effect of glucocorticoids is eliminated by removal of the adrenal, ACTH secretion and tumor growth are enhanced. In some cases, tumor growth is considerably accelerated resulting in rapidly progressive, invasive tumors. The dynamic data in Nelson's syndrome are almost identical to pituitary dependent CD. The only notable difference is that the ACTH of patients with Nelson's syndrome is responsive to hypoglycemia.[5] The incidence of Nelson's syndrome following bilateral adrenalectomy for CD ranges from 8 to 35%, higher in children.[179] It is not clear whether Nelson's syndrome can be prevented by prior radiation to the pituitary. Once the syndrome has developed, pituitary surgery becomes mandatory.

The indications for bilateral adrenalectomy are as follows:

1. Hypercortisolism is life threatening, and a rapid 100% effective modality is imminently required to correct it.
2. Pituitary surgery has failed to cure the patient's disease.
3. Hypercortisolism recurs after a period of remission following pituitary surgery.
4. Macronodular hyperplasia is suspected on the basis of hormonal studies or computed tomography of adrenals.

Pituitary Radiation

Conventional radiation to the pituitary gland is a time-honored therapeutic modality for CD. About 5000 rad is delivered to the pituitary gland over a 4–5 week period. The main advantage of the procedure is that it is relatively free from major side effects. Conventional radiation has enjoyed the reputation of being an innocuous form of therapy, since it is devoid of complications such as hypopituitarism or optic nerve damage. The major drawback of conventional pituitary radiation, particularly in adults, is the relatively low cure rates attained. Orth et al.[14] evaluated the results of conventional radiation given to a large group of patients with pituitary-dependent CD. Of the 50 patients treated, complete cure was attained in only 10. An improvement was seen in another 13 patients. Results of therapy were seen from 6 to 18 months following radiation. The low cure rate and the long delay in attaining a remission have been compelling reasons to minimize the enthusiasm for this therapy. The results with conventional radiation to the pituitary are more encouraging in children and adolescents with the disease. In the report by Jennings et al.[180] a cure was seen in 13 of 15 children with CD treated by conventional radiation. The lack of adverse effects, particularly hypopituitarism, is especially important in this group of patients where preservation of growth and sexual maturation are vital concerns while planning therapy. Considering this aspect, and coupled with its high cure rate, conventional radiation is favored by many as the first option in children and adolescents with CD.

Proton-beam (or heavy-particle) radiation is a highly effective method of treating pituitary-dependent CD. The major limitations of this modality is the non availability of the procedure in most centers and the higher incidence of complications such as hypopituitarism and oculomotor nerve palsies. The value of prophylactic radiation to the pituitary gland (prior to bilateral adrenalectomy) in the prevention of Nelson's syndrome is controversial and unsettled.[4,177,181]

Neuropharmacotherapeutic Agents

The role of neuropharmacological agents in the treatment of CD is at best adjunctive. The basis for the use of drugs that modulate ACTH secretion rests on the ability of certain drugs to impair or deplete monoamine concentrations in the hypothalamus. Hypothalamic release of CRF can be affected by neurotransmitter concentration in the hypothalamus. Thus inhibitors of serotonin, (and in some instances dopamine agonists) are capable of lowering CRF and indirectly ACTH concentrations. The notion that these drugs exert their action solely on the hypothalamus must be reexamined in view of in vitro data that show a direct inhibitory effect of these drugs on the adenoma cells maintained in culture.[143] In addition to serotonin antagonists, and dopaminergic agonists, noreprinephrine depletors as well as agonists to gamma amino butyric acid have extended the neuropharmacological armamentarium. Several general observations are pertinent in this regard:

1. Although the basis for drug therapy is solidly based on the abnormal hormone dynamics that characterize pituitary-dependent CD, these data cannot be transposed to all patients with CD. The heterogeneity of this disease precludes predictability of response in an individual case.
2. Even when drug therapy works, as exemplified by cyproheptadine, the discontinuation of the drug is usually attended by a return of hypercortisolism. Recommendation of protracted, possibly lifelong, therapy is not justified when more definitive cures are available.
3. Side effects of the drugs employed should be considered when long-term therapy is contemplated.
4. Definitive therapy should not be deferred while waiting for a drug-induced remission that never may come.

Cyproheptadine is the most notable drug used to treat pituitary-dependent CD. Much of our knowledge regarding this drug is due to the pioneering studies of Krieger et al.[182,183] The administration of cyproheptadine is attended by lowering of cortisol and ACTH levels, restoration of suppressibility to dexamethasone and even a clinical remission in some patients with pituitary-dependent CD. The drug has to be administered in large doses (24 mg/day), with the expectation of no more than a 50% chance of remission, which lasts only as long as the drug is continued. There are no definite methods of prospectively identifying those patients who are unlikely to respond. The most serious, and undesirable effect of cyproheptadine therapy is weight gain.

The use of bromocriptine is likely to benefit patients with Cushing's disease originating from the intermediate lobe. Although the suppressive effect of dopamine agonists is well established in such patients,[17,21,184] the success rate of chronic dopamine agonist therapy has not been clearly established. Similarly, the use of reserpine[185] has not been well established.

Several newer drugs have entered the scene in the recent past. Sodium valproate, a γ-amino butyric acid (GABA) agonist has been shown to be effective in lowering the ACTH level in some patients with CD and Nelson's syndrome.[186,187] The effectiveness of treatment of CD with sodium valproate has been noted by Jones et al.[188] The drug has also been successfully employed to lower ACTH levels in patients with Nelson's syndrome.[189] The use of adrenolytic agents (o'p'-DDD aminoglutethemide, metyrapone, or trilostane) is not discussed, as they are not primary agents employed for the treatment of CD. The use of some of these agents in the preoperative preparation of the patient is resorted to when the clinical expression is severe. Most authorities who perform TPS do not employ these agents routinely in the preoperative management.[25] The prolonged use of ketoconozole in the management of CD has been described by Sonino et al.[190] to correct the hypercortisolism quickly, following TPS.

Appendix

The following case vignettes represent the spectrum of presentation seen in association with hypercortisolism.

Case 1

A 32-year-old woman presented with progressive weight gain of recent onset, puffiness of face, and mild hypertension. On close questioning, a history of easy bruisability was obtained. She had also noted a mild increase in facial hair, although her menstrual periods were normal. She did not complain of muscle weakness. The physical examination was negative for striae, truncal obesity, moon face, or buffalo humps. No galactorrhea or virilization was evident. Blood pressure was 140/100 mm Hg, and the random blood glucose was 220 mg%. The plasma cortisol level following 1 mg dexamethasone at midnight was 14 µg%. She was further evaluated for hypercortisolism with the following results:

Basal 24-hr urinary free cortisol	250 µg
Basal 24-hr urinary 17-OHCS	12 mg/g creatinine
24-hr urinary 17-OHCS post-high-dose dexamethasone	5 mg%
Basal plasma cortisol	38 µg
Basal plasma ACTH	90 pg
Plasma 11-deoxycortisol postdose metyrapone	14 µg%
CT scan of the sella turcica	No tumor

Comments: This patient's clinical presentation represents the mild expression of hypercortisolism that characterizes early disease. The combination of recent onset of weight gain, hypertension, hirsutism, facial puffiness, and glucose intolerance prompted screening for hypercortisolism. The ease of performing overnight dexamethasone test was the reason for choosing this procedure in the outpatient setting. The abnormal result (>5 µg cortisol postdose dexamethasone) prompted the urinary collection for free cortisol, which was clearly indicative of hypercortisolism. In this patient's case, the subsequent (phase II) tests all pointed to a pituitary etiology; the suppression to high-dose dexamethasone, the "normal" basal ACTH level in the presence of an elevated cortisol, and the brisk response of 11-deoxycortisol to metyrapone were uniformly indicative of pituitary-dependent CD. However, the CT of the sella failed to reveal a pituitary tumor. Although the plasma ACTH level and the response to metyrapone administration were not in favor of ectopic ACTH syndrome, exceptionally carcinoid tumors may mimic pituitary-dependent CD. This may be especially so in early cases, in which recent onset of hypercortisolemia may not have totally abolished pituitary responsiveness. Therefore, the patient underwent additional studies.

Computed tomography of adrenals revealed minimal enlargement of both adrenals, consistent with CD, and less frequently with ectopic ACTH syndrome. The chest radiograph, the 5-hydroxyindoleacetic acid (5-HIAA) level in the urine, and the CT scan of the chest were normal.

Selective venous catheterization of the inferior petrosal sinuses was recommended but declined by the patient. In view of the fact that a significant number of patients with CD show normal pituitaries by CT, and the dynamic data seen in this patient are only rarely, and exceptionally, encountered in ectopic ACTH syndrome, the working diagnosis was CD and the therapeutic recommendation transsphenoidal surgery.

On transsphenoidal exploration, the neurosurgeon found a 2-mm microadenoma in the right side. Following microadenomectomy, ACTH level declined to 20 pg and cortisol level fell to 6 μg, with no demonstrable response to exogenous ACTH administration. The patient was temporarily placed on hydrocortisone supplementation; 2 months after surgery, she was tapered off supplementation and has remained eucortisolemic ever since.

Case 2

A 40-year-old woman was evaluated for weight gain, plethoric facies, marked muscle weakness, and mental depression, of 1-yr duration. On physical examination, the blood pressure was 150/100. She appeared cushingoid with characteristic rounding of face and cervical fat accumulation. There were no signs of hirsutism, virilization acne, or striae. The nails revealed evidence of chronic superficial fungal infection, and there was galactorrhea upon pressure on the nipples. The laboratory data were as follows:

Basal 24-hr urianry free cortisol	620 μg
Basal 24-hr urinary 17-OHCS	20 mg
24-hr urinary 17-OHCS post-high-dose	18 mg
Basal 8 AM cortisol	40 μg
Basal 8 AM ACTH	50 pg/ml
Post-metyrapone	11 μg
CT scan of the pituitary	Normal
CT scan of the adrenals	4.5-cm mass in the left adrenal with a normal-sized right adrenal
Basal prolactin	80 ng
17 ketosteroids in 24-hr	8 mg

Comments. This patient, in contrast with the previous case, presented with unequivocal clinical evidence of overt cushingoid features, but the laboratory data are conflicting. The markedly elevated urinary free cortisol reflects severe hypersecretion of glucocoticoids by the adrenal(s). The high-dose Liddle test demonstrates no suppression, suggesting either a neoplasm of the adrenal or ectopic ACTH syndrome. As many as 30% of patients with pituitary-dependent CD may also demonstrate failure to suppress to high-dose dexamethasone.[108,117,118] Therefore, additional information was sought from the plasma ACTH assay, which revealed in a single 8 AM sample a concentration of 50 pg, a level that is not consistent with either the ectopic ACTH syndrome (where a very high level is anticipated) or adrenal neoplasm (where a suppressed ACTH level is expected). The ACTH level and the response of 11-deoxycortisol to metyrapone in this case are consistent with pituitary-dependent CD, while the CT scan clearly demonstrates a tumor in the left adrenal.

In such a circumstance, it should be recognized that the mere demonstration of a tumor in the adrenal gland does not unequivocally establish causality; this is especially so when ACTH levels are not completely suppressed. Therefore, the possibility of macronodular hyperplasia was a strong consideration in this patient. The fact that the contralateral gland was not atrophied further suggests, but does not confirm, macronodular disease. Most patients with adenoma would demonstrate a shrunken contralateral adrenal. Iodocholesterol scans could be employed to demonstrate bilateral disease when macronodular hyperplasia is suspected.[191,192]

In this patient, despite nonsuppression to high-dose dexamethasone, a negative CT scan of the pituitary, and an abnormal CT scan of the adrenal, primary pituitary-dependent CD was suspected on the following grounds: (1) a plasma ACTH level, that upon repetition multiple times during the day ranged from 25 to 90 pg; (2) the 11-deoxycortisol response to metyrapone administration; (3) the hyperprolactinemia; and (4) the normal contour of the contralateral adrenal gland.

However, definite proof of a pituitary origin rests on the demonstration that the source of ACTH was indeed from that gland. The patient underwent selective venous sampling of the inferior petrosal sinus for ACTH level. The ACTH concentration in the inferior petrosal sinus was 170 pg, in contrast to the peripheral vein concentration of 40 pg. The concentration in the venous effluent from the left adrenal was 45 pg, establishing the fact that her problem originated from the pituitary.

The patient underwent transsphenoidal surgery. No tumor was found upon exploration. Since her disease was severe and persistence or recurrence of disease was undesirable, a complete hypophysectomy was performed.

The pathophysiology of macronodular hyperplasia is unclear. The disorder should be kept in mind whenever discordant dynamic data are obtained, even when CT reveals unilateral disease. The 17-ketosteroid should be measured in every suspected case of macronodular hyperplasia, owing to the reported association of this disease with carcinoma of the adrenal.[193] The patient is being followed closely for recurrence, or growth of adrenal tumor.

Case 3

A 38-year-old woman was evaluated for severe depression, associated with recent-onset hypertension, ankle edema, fatigue, and menstrual irregularities. She had been treated with phenothiazines and a small dose of tricyclic antidepressant for the past 3 months. She was referred by her physician for exclusion of Cushing's syndrome as the underlying cause of her depression. Physical examination found her to be a withdrawn apathetic woman. No overt evidence of cushingoid features were evident. The positive findings included mild decrease in muscle strength in the proximal group of muscles, moderate hypertension (150/110 mm Hg) trace-pitting ankle edema, galactorrhea, and significant mental depression. The only biochemical abnormality was a postprandial glucose level of 210 mg%. The initial hormonal data included the following:

8 AM plasma cortisol	30 µg%
8 AM plasma cortisol after overnight DXM	10 µg%
Basal ACTH	110 pg
Baseline 24-hr urine free cortisol	140 µg
Baseline 24-hr urine 17-OHCS	12 mg
24-hr urine 17-OHCS post-high-dose dexamethasone	5 mg
Prolactin level	70 ng

Comments. The inherent difficulties in documenting Cushing's syndrome in a patient with endogenous depression are several. Psychiatric illness, particularly depression, can mimic several facets of the dynamics of CD. This patient's elevated plasma cortisol, the failure to suppress with overnight dexamethasone and even the elevated urinary 17-OHCS, free cortisol, and plasma ACTH level could all be secondary to her endogenous depression. The elevated prolactin level could be due to the phenothiazine therapy she was receiving. The suppression to high-dose dexamethasone, in the presence of an elevated urinary free cortisol is shared by both endogenous depression as well as pituitary-dependent Cushing's disease. The options in this case were as follows: (1) to evaluate the response of ACTH and cortisol to insulin-induced hypoglycemia, (2) to perform studies following adequate treatment of the depression, and (3) to perform CT of the sella and/or the adrenals.

The patient was studied with insulin-induced hypoglycemia. Her basal cortisol level of 28 µg rose to a maximum of 42 µg, when the blood glucose declined to 46 µg%. The concomitantly measured ACTH levels increased to 200 pg at the nadir of blood glucose concentrations. Such responses are more characteristic of depression than of CD, in which the response to hypoglycemia is blunted. It was decided to restudy the patient in 8 weeks, following appropriate psychiatric therapy. The repeat studies when she had improved demonstrated a baseline cortisol of 12.4 µg, which declined to 3.8 µg following 1 mg dexamethasone given at midnight. No further workup was deemed necessary. The possible role of CRF tests in depressive illness has already been discussed.[194]

Case 4

A 40-year-old man presented with a 1-yr history of fatigue, weight gain, muscle weakness, facial plethora, and intermittent edema. On physical examination, he appeared obviously cushingoid with mooning of the face, truncal obesity, and purple striae. The following data were obtained:

Basal 8 AM cortisol	42 μg%
8 AM cortisol post 1 mg dexamethasone	38 μg%
24-hr urine free cortisol	400 μg
24-hr urine 17-OHCS	38 μg%
24-hr urine 17-OHCS post-high-dose dexamethasone	20 μg%
Plasma ACTH level	176 pg
CT of the sella	Normal
CT of the adrenals	Bilaterally enlarged

This patient demonstrated marked hypercortisolism and preservation of suppression to high-dose dexamethasone. The ACTH level in the plasma was clearly, but not markedly, elevated. The CT of the adrenals, as well as the CT of the sella were consistent with pituitary-dependent Cushing's disease. Routine films of the chest and spine were unremarkable.

The patient underwent transsphenoidal pituitary surgery. On exploration, no tumor was evident. A near total hypophyectomy was performed by piecemeal removal. Several histological sections were evaluated for clumping of cells, which was not found. The final pathology report was read as corticotrope hyperplasia. Postoperatively, the patient continued to demonstrate hypercortisolism, ranging from 30 to 45 μg plasma cortisol. Prior to discharge, the pituitary reserve testing demonstrated absent reserve of TSH, FSH, LH, hGH, and prolactin. The ACTH level ranged between 90 and 150 pg and failed to respond to hypoglycemia, a finding interpreted as consistent with residual CD. The patient was placed on L-T4, 0–15 mg, and testosterone.

Six months later, the patient's condition had deteriorated. The serum cortisol averaged 50 μg (AM), the urinary free cortisol ranged between 400 and 500 μg and the plasma ACTH between 150 and 100 pg. The patient had gained more weight, looked more cushingoid than ever, and was troubled by severe weakness, hypertension, and depression. A CT scan of the sella showed a normal sella with very little pituitary tissue within the sella. The chest film was reported normal, but the lower thoracic spine had begun to show evidence of osteopenia. CT of the adrenal revealed bilateral enlargement, 7 mm in thickness. The patient was placed on medical therapy with the adrenolytic agent o,p'-DDD, to control hypercortisolism. He was also given 5000 rad to the pituitary gland in an attempt to ablate any residual pituitary tumor in that region secreting ACTH. Unfortunately, at that time selective venous sampling of the inferior petrosal sinus was not a readily available procedure.

The patient continued to be ill, partly from the adrenolytic therapy and partly from the hypercortisolism. Three months after adrenolytic therapy, cortisol levels had declined to 18 μg, but urinary 17-OHCS were still 11 mg in 24 hr. The chest radiograph was suspicious for a lesion in the right lower lobe. CT scan of the chest revealed a mass lesion in the corresponding area. The patient underwent a thoracotomy for removal of the mass lesion, which histologically proved to be a well-differentiated carcinoid of the bronchus. The tumor tissue was strongly positive for immunofluorecent staining with ACTH antibody. Following lobectomy, the plasma cortisol and plasma ACTH levels declined dramatically. Unfortunately the patient developed several postoperative complications—infection, pneumonia, pulmonary emboli, and aspiration, to name a few—and eventually succumbed.

The tragedy of errors in this case impressively illustrate the extent to which the mimicry of ectopic ACTH-secreting syndrome can be carried. Retrospectively, during the initial evaluation, two additional measures might have thrown doubt on the diagnosis of pituitary-dependent CD: (1) if a metyrapone test had been performed, a nonresponsive pattern might have been demonstrated, casting doubt on the diagnosis of Cushing's disease; and (2) the initial plasma ACTH level was slightly higher than expected for CD, and perhaps multiple samples might have demonstrated higher values (>200 pg), again casting a shadow on the diagnosis of pituitary-dependent CD. All the subsequent measures, including medical adrenalectomy and radiation of the pituitary, were options in desparation, since the real cause was not evident. The final irony of this "retrospectoscopy" was that the original radiograph of the chest, when viewed in light of the recent radiographic film of the chest, faintly showed a density in the right lower lung field. The lack of availability of selective venous catheterization of the inferior petrosal sinuses at that time compounded the problems in this nightmarish diagnostic quagmire. The availability of this procedure, coupled with dynamic testing using CRF, should certainly minimize the unfortunate events that occurred in this case.

References

1. Cushing H: The basophillic adenomas of the pituitary body and their clinical manifestations. *Bull Johns Hopkins* **50:**137, 1932.
2. Taylor HC, Velasco ME, Brodkey JS: Remission of pituitary dependent Cushing's disease after removal of nonneoplastic pituitary gland. *Arch Intern Med* **140:**1366, 1980.
3. Liddle GW: Tests of pituitary-adrenal suppressibility in the diagnosis of Cushing's syndrome. *J Clin Endocrinol Metab* **20:**1539, 1960.
4. Orth DN, Liddle GW: Results of treatment in 108 patients with Cushing's syndrome. *N Engl J Med* **285:**243, 1971.
5. Krieger DT, Luria M: Plasma ACTH and cortisol responses to TRF, vasopressin or hypoglycemia in Cushing's disease and Nelson's syndrome. *J Clin Endocrinol Metab* **44:**361, 1977.
6. Krieger DT: Medical treatment of Cushing's disease. In Tolis G (ed): *Clinical neuroendocrinology: A Pathophysiological Approach.* Raven, New York, 1979, p. 423.
7. Pieters GFFM, Smals AGH, Benraad TJ, et al: Plasma cortisol response to thyrotropin-releasing hormone and luteinizing hormone-releasing hormone in Cushing's disease. *J Clin Endocrinol Metab* **48:**874, 1979.
8. Aronin N, Krieger DT: Sustained remission of Nelson's syndrome after stopping cyproheptadine treatment. *N Engl J Med* **302:**453, 1980.
9. Jones MT, Birmingham M, Gillham B, et al: The effect of cyproheptadine on the release of corticotrophin releasing factor. *Clin Endocrinol (Oxf)* **10:**203, 1979.
10. Lankford HV, Tucker HSG, Blackard WG: A cyproheptadine-reversible defect in ACTH control persisting after removal of the pituitary tumor in Cushing's disease. *N Engl J Med* **305:**1244, 1981.
11. Lamberts SWJ, Klijn, JGM, de Quijada M, et al: The mechanism of the suppressive action of bromocriptine on adrenocorticotropin secretion in patients with Cushing's disease and Nelson's syndrome. *J Clin Endocrinol Metab* **51:**307, 1980.
12. Carey RM, Varma SK, Drake CR Jr, et al: Ectopic secretion of corticotropin releasing factor as a cause of Cushing's syndrome. *N Engl J Med* **311:**13, 1984.
13. Liddle GW: Pathogenesis of glucocorticoid disorders. *Am J Med* **53:**638, 1972.
14. Krieger DT, Glick SM: Sleep EEG stages and plasma growth hormone concentration in states of endogenous and exogenous hypercortisolemia or ACTH elevation. *J Clin Endocrinol Metab* **39:**986, 1974.
15. Boggan JE, Tyrrell, JB, Wilson CB: Transsphenoidal microsurgical management of Cushing's disease. *J Neurosurg* **59:**195, 1983.
16. Hardy J: Cushing's disease: 50 years later. *Can J Neurol Sci* **9:**375, 1982.
17. Lamberts SWJ, De Lange SA, Stefanko SZ: Adrenocorticotropin-secreting pituitary adenomas originate from the anterior or the intermediate lobe in Cushing's disease: Differences in the regulation of hormone secretion. *J Clin Endocrinol Metab* **54:**286, 1982.
18. Ludecke DK, Schabet M, Saeger W: In vitro secretion of adenoma and anterior lobe cells in two typical cases of Cushing's disease. *Neurosurgery* **12:**549, 1983.
19. Ludecke DK, Westphal M, Schabet M, et al: In vitro secretion of ACTH, β-endorphin and β-lipotropin in Cushing's disease and Nelson's syndrome. *Horm Res* **13:**259, 1980.
20. Suda T, Tozawa F, Mouri T, et al: Effects of cyproheptadine, reserpine, and synthetic corticotropin-releasing factor on pituitary glands from patients with Cushing's disease. *J Clin Endocrinol Metab* **56:**1094, 1983.

21. Lamberts SWJ, Klijn JGM, Quijada M, et al: The mechanism of the suppressive action of bromocriptine on adrenocorticotropin secretion in patients with Cushing's disease and Nelson's syndrome. *J Clin Endocrinol Metab* **51**:307, 1980.

22. Ishibashi M, Yamaji T: TRH stimulation and dopaminergic and antiserotonergic inhibition of ACTH release from cultured pituitary adenoma tissues of Nelson's syndrome. *Program of the Sixth International Congress of Endocrinology, Melbourne, Australia, 1980*, p. 407.

23. Fitzgerald PA, Aron DC, Findling JW: Cushing's disease: Transient secondary adrenal insufficiency after selective removal of pituitary microadenomas; evidence for a pituitary origin. *J Clin Endocrinol Metab* **54**:413, 1982.

24. Jeffcoate WJ, Dauncey S, Selby C: Restoration of dexamethasone suppression by incomplete adenomectomy in Cushing's disease. *Clin Endocrinol (Oxf)* **23**:193, 1985.

25. Burch WM: Cushing's disease. *Arch Intern Med* **145**:1108, 1985.

26. Van Cauter E, Refetoff S: Evidence for two subtypes of Cushing's disease based on the analysis of episodic cortisol secretion. *N Engl J Med* **312**:1343, 1985.

27. Orth DN, Holscher MA, Wilson MG, et al: Equine Cushing's disease: Plasma immunoreactive proopiolipomelanocortin peptide and cortisol levels basally and in response to diagnostic tests. *Endocrinology* **110**:1430, 1982.

28. Neumann PE, Horoupian DS, Goldman JE, et al: Cytoplasmic filaments of Crooke's hyaline change belong to the cytokeratin class. *Am J Pathol* **116**:214, 1984.

29. Felix IA, Horvath E, Kovacs K: Crook's hyalinization in corticotroph cell adenomas of the human pituitary: A histological, immunocytological and electron microscopic study of three cases. *Acta Neurochir* **58**:235, 1981.

30. Ross EJ, Marshall JP, Friedman M: Cushing's syndrome: Diagnostic criteria. *Q J Med* **35**:149, 1966.

31. Bertagna C, Orth DN: Clinical and laboratory findings and results of therapy in 58 patients with adrenocortical tumors admitted to a single medical center (1951–1978). *Am J Med* **71**:855, 1981.

32. Urbanic RC, George JM: Cushing's disease—18 years experience. *Medicine (Baltimore)* **60**:14, 1981.

33. Ross J, Linch DC: Cushing's syndrome—killing disease: Discriminatory value of signs and symptoms aiding early diagnosis. *Lancet* **2**:646, 1982.

34. Biglieri EG, Hane S, Slaton PE Jr, et al: In vivo and in vitro studies of adrenal secretions in Cushing's syndrome and primary aldosteronism. *J Clin Invest* **42**:516, 1963.

35. Krakoff L, Nicolis G, Amsel B: Pathogenesis of hypertension in Cushing's syndrome. *Am J Med* **58**:216, 1975.

36. Mendlowitz M, Gitlow S, Noftchi N: Work of digital vasoconstriction produced by infused norepinephrine in Cushing's syndrome. *J Appl Physiol* **13**:252, 1958.

37. Gabrilove JL: Neurologic and psychiatric manifestations of the classic endocrine syndromes. *Res Pub ARNMD* **43**:419, 1966.

38. Spillane JD: Nervous and mental disorders in Cushing's syndrome. *Brain* **74**:72, 1951.

39. Glaser GH: Psychotic reactions induced by corticotrophin (ACTH) and cortisone. *Psychosom Med* **15**:280, 1953.

40. Starkman M, Schteingart DE: Neuropsychiatric manifestations of patients with Cushing's syndrome. *Arch Intern Med* **141**:215, 1981.

41. Terenius L, Wahlstrom A, et al: Increased CSF levels of endorphins in chronic psychosis. *Neurosci Lett* **3**:157, 1976.

42. Khaleeli AA, Edwards RHT, Gohil K, et al: Corticosteroid myopathy: A clinical and pathological study. *Clin Endocrinol (Oxf)* **18**:155, 1983.

43. Khaleeli AA, Betteridge DJ, Edwards RHT, et al: Effect of treatment of Cushing's syndrome on skeletal muscle structure and function. *Clin Endocrinol (Oxf)* **19**:547, 1983.

44. Cryer P, Ludmerer KM, Kissane JM (eds): Bruising and thin skin in a 54 year old woman. *Am J Med* **79**:101, 1985.

45. Ferguson JK, Donald RA, Weston TS, et al: Skin thickness in patients with acromegaly and Cushing's syndrome and response to treatment. *Clin Endocrinol (Oxf)* **18**:347, 1983.

45A. Imura H, Matsukusa S, Yamamoto H, et al: Studies on ectopic ACTH-producing tumors II. Clinical, biochemical features of 30 cases. *Cancer* **35**:1430, 1975.

46. Gold EM: The Cushing's syndromes: Changing views of diagnosis and treatment. *Ann Intern Med* **90:**829, 1979.
47. Teates CD: Steroid-induced mediastinal lipomatosis. *Radiology* **96:**501, 1970.
48. Santini LC, Williams JL: Mediastinal widening (presumable lipomatosis) in Cushing's syndrome. *N Engl J Med* **284:**1357, 1971.
49. Van De Putte LB, Wagenaar JPM, San KH: Paracardiac lipomatosis in exogenous Cushing's syndrome. *Thorax* **28:**653, 1973.
50. Lipson SJ, Naheedy MII, Kaplan MM, et al: Spinal stenosis caused by epidural lipomatosis in Cushing's syndrome. *N Engl J Med* **302:**36, 1980.
51. Lee M, Lekias J, Gubby SS, et al: Spinal cord compression by extradural fat after renal transplantation. *Med J Aust* **1:**201, 1975.
52. George WE, Wilmott M, Greenhouse A, et al: Medical management of steroid-induced epidural lipomatosis. *N Engl J Med* **308:**316, 1983.
53. Sowerbutts JG: Some uses for presacral oxygen insufflation. *J Faculty Radiol* **10:**201, 1959.
54. Christian CD Jr, Schneider RP: Fatty tumor of the liver in a patient with Cushing's syndrome. *Arch Intern Med 143:*1605, 1983.
55. Chajek T, Romanoff H: Cushing's syndrome with cyclical edema and periodic secretion of corticosteroids. *Arch Intern Med 136:*441, 1976.
56. Findling JW, Tyrrell B, Aron DC, et al: Fungal infections in Cushing's syndrome. *Ann Intern Med* **95:392, 1981.**
57. Dal Bo Zanon R, Fornasiero L, Boscaro M, et al: Increased factor VIII associated activities in Cushing's syndrome: A probable hypercoaguable state. *Thromb Hemost* **47:**116, 1982.
58. Didolkar MS, Bescher AR, et al: Natural history of adrenal cortical carcinoma. *Cancer* **47:**2153, 1981.
59. Hutter AM, Kayhoe DE: Adrenal cortical carcinoma—clinical features of 138 patients. *Am J Med* **41:**572, 1966.
60. Fachnie JD, Zafar MS, Mellinger R, et al: Pituitary carcinoma mimics the ectopic adrenocorticotropin syndrome. *J Clin Endocrinol Metab* **50:**1062, 1980.
61. Liddle GW, Nicholson WE, Island DP, et al: Clinical and laboratory studies of ectopic humoral syndromes. *Recent Prog Horm Res* **25:**283, 1969.
62. Singer W, Kovacs K, Ryan N, et al: Ectopic ACTH syndrome: Clincopathological correlations. *J Clin Pathol* **31:**591, 1978.
63. Williams ED, Morales AM, et al: Thyroid carcinoma and Cushing's syndrome. *J Clin Pathol* **21:**129, 1968.
64. Birkenhager JC, Upton GV: Meduallary thyroid carcinoma. Ectopic production of peptides with ACTH-like, CRF-like and prolactin production stimulating activity. *Acta Endocrinol (Kbh)* **83:**280, 1976.
65. Melone CR, Tucci J, et al: Cushing's syndrome due to bilateral adrenocortical hyperplasia caused by a benign adrenal medullary tumor. *J Clin Endocrinol Metab* **26:**1192, 1966.
66. Spark RF, Connolly PB: ACTH secretion from a functioning pheochromocytoma. *N Engl J Med* **301:**416, 1979.
67. Howland WJ Jr, Pugh DG, Sprague RG: Roentgenologic changes of the skeletal system in Cushing's syndrome. *Radiology* **71:**69, 1958.
68. Hahn TJ: Corticosteroid-induced osteopenia. *Arch Intern Med 138:*882, 1978.
69. Hahn TJ, Halstead LR, Baran DT: Effects of short term glucocorticoid administration on intestinal calcium absorption and circulating vitamin D metabolite concentrations in man. *J Clin Endocrinol Metab* **52:**111, 1981.
70. Findling JW, Adams ND, Lemann J, et al: Vitamin D metabolites and parathyroid hormone in Cushing's syndrome: Relationship to calcium and phosphorus homeostasis. *J Clin Endocrinol Metab* **54:**1039, 1982.
71. Teitelbaum SL, Malong JD, Kahn AJ: Glucocorticoid enhancement of bone resorption by rat peritoneal macrophages in vitro. *Endocrinology* **108:**795, 1981.
72. Jee WSS, Park HZ, Roberts WE, et al: Corticosteroid and bone. *Am J Anat* **129:**477, 1970.
73. Nosadini R, Del Prato A, Valerio TA, et al: Insulin resistance in Cushing's syndrome. *J Clin Endocrinol Metab* **57:**529, 1983.

74. McArthur RG, Cloutier MD, Hayles AB, et al: Cushing's disease in children. *Mayo Clin Proc* **47:**318, 1972.

75. Frantz AG, Rabkin MT: Human growth hormone: Clinical measurement, response to hypoglycemia and suppression by corticosteroids. *N Engl J Med* **271:**1375, 1964.

76. Vazquez AM, Schutt-Aine JC, Kenny FM, et al: Effect of cortisone therapy on the diurnal pattern of growth hormone secretion in congenital adrenal hyperplasia. *J Pediatr* **80:**433, 1972.

77. Sturge RA, Beardwell C, Hartog M, et al: Cortisol and growth hormone secretion in relation to linear growth: Patients with Still's disease on different therapeutic regimens. *Br Med J* **3:**547, 1970.

78. Preece MA: The effect of administered corticosteroids on the growth of children. *Postgrad Med J* **52:**625, 1976.

79. Morris HG, Jorgensen JR, Elrick H, et al: Metabolic effects of human growth hormone in corticosteroid-treated children. *J Clin Invest* **47:**436, 1968.

80. Phillips LS, Belosky DC, Young HS, et al: Nutrition and somatomedin. VI. Somatomedin activity and somatomedin inhibitory activity in serum from normal and diabetic rats. *Endocrinology* **104:**1519, 1979.

81. Green OC, Winter RJ, Kawathara FS, et al: Pharmacokinetic studies of prednisolone in children: Plasma levels, half-life values, and correlation with physiologic assays for growth and immunity. *J Pediatr* **93:**299, 1978.

82. Elders MJ, Wingfield BS, McNatt ML, et al: Glucocorticoid therapy in children: effect on somatomedin secretion. *Am J Dis Child* **129:**1393, 1975.

83. Clark I, Umbreit W: Effect of cortisone and other steroids upon in vitro synthesis of chondrotin sulfate. *Proc Soc Exp Biol Med* **86:**558, 1954.

84. Phillips LS, Herington AC, Daughaday WH: Steroid hormone effects on somatomedin I. Somatomedin action in vitro. *Endocrinology* **97:**780, 1975.

85. Phillips LS, Fusco AC, Unterman TG, et al: Somatomedin inhibitor in uremia. *J Clin Endocrinol Metab* **59:**764, 1984.

86. Unterman TG, Phillips LS: Glucocorticoid effects on somatomedins and somatomedin inhibitors. *J Clin Endocrinol Metab* **61:**618, 1985.

87. Bessler W: Vertebral growth arrest lines after Cushing's syndrome. Case report. *Diag Imaging* **51:**311, 1982.

88. Krieger DT, Allen W, Rizzo F, et al: Characterization of the normal temporal pattern of plasma corticosteroid levels. *J Clin Endocrinol Metab* **32:**266, 1971.

89. Weitzman ED, Fukushima D, Nogeire C, et al: Twenty-four hour pattern of the episodic secretion of cortisol in normal subjects. *J Clin Endocrinol Metab* **33:**14, 1971.

90. Crapo L: Cushing's syndrome: A review of diagnostic tests. *Metabolism* **28:**955, 1979.

91. Nugent CA, Nichols T, Tyler FH: Diagnosis of Cushing's syndrome—single dose dexamethasone suppression test. *Arch Intern Med* **116:**172, 1965.

92. McHardy-Young S, Harris PWR, Lessof MH, et al: Single-dose dexamethasone suppression test for Cushing's syndrome. *Br Med J* **2:**740, 1967.

93. Tucci JR, Jagger PI, Lauler DP, et al: Rapid dexamethasone suppression tests for Cushing's syndrome. *JAMA* **199:**379, 1967.

94. Seidensticker JF, Folk RL, Wieland RG, et al: Screening test for Cushing's syndrome with plasma 11-hydroxycorticosteroids. *JAMA* **202:**87, 1967.

95. Meikle AW: Dexamethasone suppression tests: Usefulness of simultaneous measurement of plasma cortisol and dexamethasone. *Clin Endocrinol (Oxf)* **16:**401, 1982.

96. Meikle AW, Lagerquist LG, Tyler FH: Apparently normal pituitary-adrenal suppressibility in Cushing's syndrome: Dexamethasone metabolism and plasma levels. *J Lab Clin Med* **86:**472, 1975.

97. Caro JF, Meikle AW, Check JH, et al: Normal suppression to dexamethasone in Cushing's disease: An expression of decreased metabolic clearance for dexamethasone. *J Clin Endocrinol Metab* **47:**667, 1978.

98. Murphy BEP: Clinical evaluation of urinary cortisol determinations by competitive protein-binding radioassay. *J Clin Endocrinol Metab* **28:**343, 1968.

99. Eddy RL, Jones AL, Gilland PF, et al: Cushing's syndrome: A prospective study of diagnostic methods. *Am J Med* **55:**621, 1973.

100. Daughaday WH, Mariz IK: Corticosteroid-binding globulin: Its properties and quantitation. *Metabolism* **10**:936, 1961.
101. Trecan GV, Laudat MH, Thomopoulos JP, et al: Urinary free corticoids: Evaluation of their usefulness in diagnosis of Cushing's syndrome. *Acta Endocrinol (Copenh)* **103**:110, 1983.
102. Hellman L, Nakada F, Curti J, et al: Cortisol is secreted episodically by normal man. *J Clin Endocrinol Metab* **30**:411, 1970.
103. Hellman L, Weitzman ED, Roffwarg H, et al: Cortisol is secreted episodically in Cushing's syndrome. *J Clin Endocrinol Metab* **30**:686, 1970.
104. de Lacerda L, Kowarski AA, Migeon CJ: Integrated concentration and diurnal variation of plasma cortisol. *J Clin Endocrinol Metab* **36**:227, 1973.
105. Zadik Z, de Lacerda L, Kowarski A: Evaluation of the 6-hour integrated concentration of cortisol as a diagnostic procedure for Cushing's syndrome. *J Clin Endocrinol Metab* **54**:1072, 1982.
106. Halbreich U, Zumoff B, Kream J, et al: The mean 1300-1600 h plasma cortisol concentration as a diagnostic test for hypercortisolism. *J Clin Endocrinol Metab* **54**:1262, 1982.
107. Ernest I: Steroid excretion and plasma cortisol in 41 cases of Cushing's syndrome. *Acta Endocrinol (Copenh)* **51**:511, 1966.
108. Nichols T, Nugent CA, Tyler FH: Steroid laboratory tests in the diagnosis of Cushing's syndrome. *Am J Med* **45**:116, 1968.
109. Streeten DHP, Stevenson CT, Dalakos TG, et al: The diagnosis of hypercortisolism. Biochemical criteria differentiating patients from lean and obese normal subjects and from females on oral contraceptives. *J Clin Endocrinol Metab* **29**:1191, 1969.
110. Liddle GW: Cushing's syndrome. In Eisenstein AB (ed): *The Adrenal Cortex*. ed. 1. Little, Brown, Boston, 1967, p. 523.
111. Doe RP, Vennes JA, Flink EB: Diurnal variation of 17-hydroxycorticosteroids, sodium, potassium, magnesium and creatinine in normal subjects and in cases of treated adrenal insufficiency and Cushing's syndrome. *J Clin Endocrinol Metab* **20**:253, 1960.
112. Glass AR, Zavadil AP III, Halberg F, et al: Circadian rhythm of serum cortisol in Cushing's disease. *J Clin Endocrinol Metab* **59**:161, 1984.
113. Van Cauter E, LeClerq R, Van Haelst L, et al: Simultaneous study of cortisol and TSH daily variations in normal subjects and patients with hyperadrenalcorticism. *J Clin Endocrinol Metab* **39**:645, 1974.
114. Mizutani S, Sonoda T, Seki T, et al: Excretion patterns of urinary 17-KS and 17-OHCS in patients with Cushing's syndrome. *Urol Int* **29**:341, 1974.
115. Moolenaar AJ, Van Seters AP: Gas chromatographic determination of steroids in the urine of patients with Cushing's syndrome. *Acta Endocrinol* **67**:303, 1971.
116. Phillipou G: Investigation of urinary steroid profiles as a diagnostic method in Cushing's syndrome. *Clin Endocrinol (Oxf)* **16**:433, 1982.
117. Liddle GW: Tests of pituitary-adrenal suppressibility in the diagnosis of Cushing's syndrome. *J Clin Endocrinol Metab* **20**:1539, 1960.
118. Lamberts SWJ, de Jong FH, Birkenhager JC: Evaluation of diagnostic and differential diagnostic tests in Cushing's syndrome. *Neth J Med* **20**:267, 1977.
119. Nichols T, Nugent CA, Tyler FH: Steroid laboratory tests in the diagnosis of Cushing's syndrome. *Am J Med* **45**:116, 1968.
120. Findling JW, Aron DC, Tyrell JB, et al: Selective venous sampling for ACTH in Cushing's syndrome. Differentiation between Cushing's disease and ectopic ACTH syndrome. *Ann Intern Med* **94**:647, 1981.
121. Linn JE, Bowdoin B, Farmer A, et al: Observations and comments on failure of dexamethasone suppression. *N Engl J Med* **277**:403, 1967.
122. Carey RM: Suppression of ACTH by cortisol in dexamethasone-nonsuppressible Cushing's disease. *N Engl J Med* **302**:275, 1980.
123. Lieberman B, Wajchenberg BL, Tambascia MA, et al: Periodic remission in Cushing's disease with paradoxical dexamethasone response: An expression of periodic hormonogenesis. *J Clin Endocrinol Metab* **43**:913, 1976.
124. Brown RD, Van Loon GR, Orth DN, et al: Cushing's disease with periodic hormonogenesis. *J Clin Endocrinol Metab* **36**:445, 1973.

125. Aron DC, Findling JW, Fitzgerald PA, et al: Pituitary ACTH dependency of nodular hyperplasia in Cushing's syndrome. Report of two cases and review of the literature. *Am J Med* **71**:302, 1981.

126. Smalls AGH, Pieters GFFM, Van Haelst UJG, et al: Macronodular adrenocortical hyperplasia in long standing Cushing's disease. *J Clin Endocrinol Metab* **58**:25, 1984.

127. Mason AMS, Ratcliffe JG, Buckle RM, et al: ACTH secretion by bronchial carcinoid tumors. *Clin Endocrinol (Oxf)* **1**:3, 1972.

128. Northrop G, Baldwin D, Faber LP, et al: Dexamethasone suppression of urinary 17-hydroxycorticoids in a patient with an ACTH-producing bronchial adenoma. *Presbyterian-St. Lukes Med Bull* **9**:43, 1970.

129. Strott CA, Nugent CA, Tyler FH: Cushing's syndrome caused by bronchial adenomas. *Am J Med* **44**:97, 1968.

130. Aron DC, Tyrell JB, Fitzgerald PA, et al: Cushing's syndrome: Problems in diagnosis. *Medicine (Baltimore)* **60**:25, 1981.

131. Gruen PH: Endocrine changes in psychiatric diseases. *Med Clin North Am* **62**:285, 1978.

132. Meikle AW, Jubiz W, Hutchings M, et al: Simplified metyrapone test with determination of plasma 11-deoxycortisol (metyrapone test with plasma S). *J Clin Endocrinol Metab* **23**:985, 1969.

133. Spiger N, Jubiz W, Meikle AW, et al: Single dose metyrapone test. Review of a four-year experience. *Arch Intern Med* **135**:698, 1975.

134. Sindler BH, Griffing GT, Melby JC: The superiority of the metyrapone test versus the high-dose dexamethasone test in the differential diagnosis of Cushing's syndrome. *Am J Med* **74**:657, 1983.

135. Carballeira A, Fishman LM, Jacobi GD: Dual sites of inhibition by metyrapone of human adrenal steroidogenesis: Correlation of in vivo and in vitro studies *J Clin Endocrinol Metab* **42**:687, 1976.

136. Matthews JI, Fariss BB, Chertow BS, et al: Adrenal adenoma with variable response to dexamethasone. *J Clin Endocrinol Metab* **34**:902, 1972.

137. Vale W, Spiess J, Rivier C, et al: Characterization of a 41-residue ovine hypothalamic peptide that stimulates secretion of corticotropin and β-endorphin. *Science* **213**:1394, 1981.

138. Rivier C, Brownstein M, Spiess J, et al: In vivo corticotropin-releasing factor-induced secretion of adrenocorticotropin, β-endorphin, and corticosterone. *Endocrinology* **110**:272, 1982.

139. Orth DN, DeBold CR, DeCherney GS: Pituitary microadenomas causing Cushing's disease respond to corticotropin-releasing factor. *J Clin Endocrinol Metab* **55**:1017, 1982.

140. Muller OA, Stalla GK, Werder K: Corticotropin releasing factor: A new tool for the differential diagnosis of Cushing's syndrome. *J Clin Endocrinol Metab* **57**:227, 1983.

141. Chrousos GP, Schulte HM, Oldfield EH, et al: The corticotropin-releasing factor stimulation test—an aid in the evaluation of patients with Cushing's syndrome. *N Engl J Med* **310**:622, 1984.

142. Pieters FFM, Hermus ARMM, Smals AGH: Responsiveness of the hypophyseal–adrenocortical axis to corticotropin-releasing factor in pituitary dependent Cushing's disease. *J Clin Endocrinol Metab* **57**:513, 1983.

143. Suda T, Tomori N, Tozawa F: Effects of corticotropin-releasing factor and other materials on adrenocorticotropin secretion from pituitary glands of patients with Cushing's disease in vitro. *J Clin Endocrinol Metab* **59**:840, 1984.

144. Shibasaki T, Nakahara M, Shizume K, et al: Pituitary adenomas that caused Cushing's disease or Nelson's syndrome are not responsive to ovine corticotropin-releasing factor in vitro. *J Clin Endocrinol Metab* **56**:414, 1983.

145. Orth DN: The old and new in Cushing's syndrome. *N Engl J Med* **310**:649, 1984.

146. Lytras N, Grossman A, Tomlin PS, et al: Corticotrophin releasing factor: Responses in normal subjects and patients with disorders of the hypothalamus and pituitary. *Clin Endocrinol (Oxf)* **20**:71, 1984.

147. Meikle AW, Stanchfield JB, West CD, et al: Hydrocortisone suppression test for Cushing's syndrome. *Arch Intern Med* **134**:1068, 1974.

148. Sachar EJ, Hellman L, Roffwarg HP, et al: Disrupted 24-hour patterns of cortisol secretion in psychotic depression. *Arch Gen Psychiatry* **28**:19, 1973.

149. Carroll BJ, Curtis GC, Davies BM, et al: Urinary free cortisol excretion in depression. *Psychol Med* **6**:43, 1976.
150. Butler PWP, Besser GM: Pituitary adrenal function in severe depressive illness. *Lancet* **1**:1234, 1968.
151. Chrousos GP: Clinical application of corticotropin-releasing factor (NIH Conference). *Ann Intern Med* **102**:344, 1985.
152. Frajria R, Angeli A: Alcohol-induced pseudo-Cushing's syndrome. *Lancet* **1**:1050, 1977.
153. Jenkins RM, Page M McB: An atypical case of alcohol-induced cushingoid syndrome. *Br Med J* **282**:1117, 1981.
154. Lamberts SWJ, Klinjn JGM, de Jong FH, et al: Hormone secretion in alcohol-induced pseudo-Cushing's syndrome. *JAMA* **242**:1640, 1979.
155. James VHT, Landon J, Wynn U, et al: A fundamental defect of adrenocortical control in Cushing's disease. *J Endocrinol (Oxf)* **40**:15, 1968.
156. Yamaji T, Ishibashi M, Sekihara H, et al: Serum dehydroepiandrosterone Sulfate in Cushing's syndrome. *J Clin Endocrinol Metab* **59**:1164, 1984.
157. Bailey RE: Periodic hormonogenesis: A new phenomena; periodicity in function of a hormone producing tumor in man. *J Clin Endocrinol Metab* **32**:317, 1971.
158. Jordan RM, Ramos-Gabatin A, Kendall JW, et al: Dynamics of adrenocorticotropin (ACTH) secretion in cyclic Cushing's syndrome: Evidence for more than one abnormal ACTH biorhythm. *J Clin Endocrinol Metab* **55**:531, 1982.
159. Spark RF, Connolly PB: ACTH secretion from a functioning pheochromocytoma. *N Engl J Med* **301**:416, 1979.
160. Cook DM, Kendall JW: Cushing's syndrome—current concepts of diagnosis and therapy. *West J Med* **132**:111, 1980.
161. Weyman PJ, Glazer HS: The adrenals. In Lee JKT, Sagel SS, Stanley RJ (eds): *Computed Body Tomography*. Raven, New York, 1983, p. 379.
162. White FE, White MC, Drury PL, et al: Value of computed tomography of the abdomen and chest in investigation of Cushing's syndrome. *Br Med J* **284**:771, 1982.
163. Chambers EF, Turski PA, LaMasters D: Regions of low density in the contrast-enhanced pituitary gland: normal and pathologic processes. *Radiology* **144**:109, 1982.
164. Davis PC, Hoffman JC Jr, Tindall GT, et al: Prolactin-secreting pituitary microadenomas: inaccuracy of high presolution CT imaging. *AJNR* **5**:721, 1984.
165. Green HT: The venous drainage of the human hypophysis cerebri. *Am J Anat* **100**:435, 1957.
166. Oldfield EH, Girton ME, Doppman JL: Absence of intercavernous venous mixing: evidence supporting lateralization of pituitary microadenomas by venous sampling. *J Clin Endocrinol Metab* **61**:644, 1985.
167. Kley HK, Stolze T, Kruskemper HL: Jugular-vein sampling of ACTH. *N Engl J Med* **297**:731, 1977.
168. Corrigan DF, Schaaf M, Whaley RA, et al: Selective venous sampling to differentiate ectopic ACTH secretion from pituitary Cushing's syndrome. *N Engl J Med* **296**:861, 1977.
169. Oldfield EH, Chrousos GP, Schulte HM, et al: Preoperative localization of ACTH secreting microadenomas by bilateral and simultaneous inferior petrosal sinus sampling. *N Engl J Med* **312**:100, 1985.
170. Manni A, Latshaw RF, Page R, et al: Simultaneous bilateral venous sampling for adrenocorticotropin in pituitary-dependent Cushing's disease: evidence for lateralization of pituitary venous drainage. *J Clin Endocrinol Metab* **57**:1070, 1983.
171. Smyth HS: Pituitary basophilism. Microsurgical observations on the pathogenesis of Cushing's disease. In *Workshop on Pituitary Pathology. Second Meeting of the International Pituitary Pathology Club, Oaxtepec, Morelos, Mexico, June 19–22, 1984*, p. 3.
172. Styne DM, Grumbach MM, Kaplan SL, et al: Treatment of Cushing's disease in childhood and adolescence by transsphenoidal microadenomectomy. *N Engl J Med* **310**:889, 1984.
173. Pont A, Gutierrez-Hartman A: Cushing's disease: recurrence after a surgically induced remission. *Arch Intern Med* **139**:938, 1979.
174. McNichol AM: Patterns of corticotropic cells in the adult human pituitary in Cushing's disease. *Diag Histopathol* **4**:335, 1981.

175. Schnall AM, Kovacs K, Brodkey JS, et al: Pituitary Cushing's disease without adenoma. *Acta Endocrinol* **94**:297, 1980.
176. Ludecke D, Kautzky R, Saeger W, et al: Selective removal of hypersecreting pituitary adenomas. *Acta Neurochir* **35**:27, 1976.
177. Nelson DH, Meakin JW, Thorn GW: ACTH-producing pituitary tumors following adrenalectomy for Cushing's syndrome. *Ann Intern Med* **52**:560, 1960.
178. Cook DM, Kendall JW, Allen JP, et al: Nycothemeral variation and suppressibility of plasma ACTH in various stages of Cushing's disease. *Clin Endocrinol (Oxf)* **5**:303, 1976.
179. Moore TJ, Dluhy RG, Williams GH, et al: Nelson's syndrome: Frequency, prognosis, and effect of prior irradiation. *Ann Intern Med* **85**:731, 1976.
180. Jennings AS, Liddle GW, Orth DN: Results of treating childhood Cushing's disease with pituitary irradiation. *N ENgl J Med* **297**:957, 1977.
181. Barnett AH, Livesey JH, Friday K, et al: Comparison of preoperative and postoperative ACTH concentrations after bilateral adrenalectomy in Cushing's disease. *Clin Endocrinol (Oxf)* **18**:301, 1983.
182. Krieger DT, Amorosa L, Linick F: Cyproheptadine-induced remission of Cushing's disease. *N Engl J Med* **293**:893, 1975.
183. Krieger DT: Cyproheptadine: Drug therapy for Cushing's disease. In Muller EE (ed): *Neuroactive Drugs in Endocrinology.* Elsevier/North-Holland, Amsterdam, 1980, p. 361.
184. Lamberts SWJ, Birkenhager JC: Bromocriptine in Nelson's syndrome and Cushing's disease. *Lancet* **2**:811, 1976.
185. Miura K, Aida M, Mihara A, et al: Treatment of Cushing's disease with reserpine and pituitary radiation. *J Clin Endocrinol Metab* **41**:511, 1975.
186. Elias AN, Gwinup G, Valenta LJ: Effects of valproic acid, naloxone and hydrocortisone in Nelson's syndrome and Cushing's syndrome. *Clin Endocrinol (Oxf)* **15**:151, 1981.
187. Dornhorst A, Jenkins JS, Lamberts SWJ, et al: The evaluation of sodium valproate in the treatment of Nelson's syndrome. *J Clin Endocrinol Metab* **56**:985, 1983.
188. Jones MT, Gillham B, Beckford U: Effect of treatment with sodium valproate and diazepam on plasma corticotropin in Nelson's syndrome. *Lancet* **1**:1179, 1981.
189. Gomi M, Iida S, Itoh Y, et al: Unaltered stimulation of pituitary adrenocorticotrophin secretion by corticotrophin releasing factor following sodium valproate administration in a patient with Nelsón's syndrome. *Clin Endocrinol (Oxf)* **23**:123, 1985.
190. Sonino N, Boscaro M, Merola G, et al: Prolonged treatment of Cushing's disease by ketoconazole. *J Clin Endocrinol Metab* **61**:718, 1985.
191. Beierwaltes WH, Wieland DM, Yu T, et al: Adrenal imaging agents: Rationale, synthesis, formulation and metabolism. *Semin Nucl Med* **8**:5, 1978.
192. Thrall JH, Freitas JE, Beierwaltes WH: Adrenal scintigraphy. *Semin Nucl Med* **8**:23, 1978.
193. Anderson DC, Child DF, Sutcliffe CH, et al: Cushing's syndrome, nodular hyperplasia and virilizing carcinoma. *Clin Endocrinol (Oxf)* **9**:1, 1978.
194. Tyrrell JB, Findling JW, Aron DC, et al: An overnight high-dose dexamethasone suppression test for rapid differential diagnosis of Cushing's syndrome. *Ann Intern Med* **104**:180, 1986.
195. Gold PW, Loriaux DL, Roy A, et al: Responses to corticotropin-releasing hormone in the hypercortisolism of depression and Cushing's disease. *N Engl J Med* **314**:1329, 1986.
196. Findling JW, Tyrrell JB: Occult ectopic secretion of corticotropin. *Arch Intern Med* **146**:929, 1986.
197. Mellinger RC: The conundrum of Cushing's syndrome. *Arch Intern Med* **146**:858, 1986.
198. Adelman S: The Pathology of Pituitary Adenomas. In Post KD, Jackson IM, Reichlin S (eds): *The Pituitary Adenoma,* Plenum Press, New York, 1980, pp. 47–62.

10

Isolated ACTH Deficiency

Introduction

The term *isolated ACTH deficiency* refers to the occurrence of hypo-adrenalism secondary to defective synthesis and release of ACTH, with preservation of function of all other pituitary trophic hormones. It is an example of monotropic (unitropic) pituitary hormone deficiency. Before embarking on a discussion of this entity, a few general introductory remarks are appropriate:

1. Isolated ACTH deficiency is rare, comprising less than 50 well-documented cases in the world literature. This contrasts with isolated growth hormone (GH) failure and isolated gonadotropin deficiency, both of which are encountered with considerably greater frequency than isolated ACTH deficiency.
2. The heterogeneous nature of isolated ACTH deficiency may be the reason for marked differences—both in clinical presentations as well as hormonal data—encountered in patients with this deficiency. Apparently, selective ACTH deficiency is not an all-or-none disorder.
3. The lack of availability of sensitive assays for ACTH in the past coupled with the dearth of diagnostic studies delineating the hypothalamic from the pituitary varieties of ACTH deficiency have hampered our understanding of the real nature of this disorder. The availability of

synthetic corticotropin releasing factor (CRF) for diagnostic purposes may, at least in part, solve this problem.

4. Finally, the diagnosis of selective ACTH deficiency can be entirely missed owing to the nonspecific nature of the symptoms and the seemingly normal baseline studies obtained in many patients with this disorder. Selective ACTH deficiency falls in the category of diseases that are potentially fatal if missed, but completely and gratifyingly treatable if detected.

Incidence

Isolated ACTH deficiency is a rare disorder. Steinberg et al.[1] reported the first case of isolated ACTH deficiency in 1954. Their patient was a 50-year-old woman with a chronic history of periodic weakness, fatigue, and syncope whose illness had been punctuated by episodes of hypotension, hypoglycemia, and hyponatremia. The hormonal data consisted of extremely low urinary 17-OHCS and 17-KS excretion, which increased following ACTH administration. The thyroid function and urinary gonadotropic activity by bioassay were normal. Steinberg et al. hypothesized prophetically that the patient suffered from unitropic hormone deficiency. Until that time, pituitary ACTH deficiency was considered to occur as part of hypopituitarism, often following the sequential failure of gonadotropins GH and thyrotropin. The concept that a single hormone could be lost with preservation of the rest of the glandular function was a new concept in 1954. The patient described responded gratifyingly to glucocorticoids or chronic ACTH therapy. The one aspect of her illness that could not be explained was the addisonian type of pigmentation seen in the patient, which also resolved following treatment. Since Steinberg's original report, nearly 50 other cases have been reported in the world literature.

Isolated ACTH deficiency is a disease of adult life, the mean age of diagnosis being around 40 years. The disease is seen with almost equal frequency in both sexes. Selective ACTH deficiency has been diagnosed during pregnancy[2,3] and postpartum.[4] The temporal relationship to pregnancy and the role of postpartum hemorrhage in the causation of selective ACTH deficiency are discussed in Section 3.1. Occasionally, the disorder can be encountered in children.[5–7] In children reported with isolated ACTH deficiency, a history of complicated delivery may often be obtained. A possible relationship to perinatal trauma in these cases is speculative.

Etiology

The etiology of isolated ACTH deficiency is obscure. The rarity of this disease, the heterogeneous nature of its expression, and the lack of diagnostic tests to assess corticotropic reserve directly have all contributed to this obscurity. Clearly, tumors of the pituitary gland are not causally related to isolated ACTH deficiency. When secondary hypoadrenalism results from invasive macroadenomas of the pituitary, impairment of one or more of the other pituitary trophic hormones is usually evident on clinical or hormonal grounds. The etiology of isolated ACTH deficiency can be viewed from three perspectives: its relationship to post partum hemorrhage, its association with autoimmune disorders, and the role of the hypothalamus in the development of this disease.

Postpartum Hemorrhage

The development of selective corticotrope failure is an atypical manifestation of Sheehan's syndrome. This syndrome is usually characterized by severe—often pan—hypopituitarism, with clear loss of gonadotropin reserve. But, isolated ACTH deficiency may result from postpartum pituitary hemorrhage. Stacpoole et al.[8] reported a 20-year-old woman who developed galactorrhea, amenorrhea, and features of adrenal insufficiency 14 months following a complicated delivery. Hormonal evaluation of the patient documented extremely low plasma cortisols and 17-OHCS, which increased dramatically following intravenous ACTH; the plasma ACTH was low, in the presence of a markedly decreased plasma cortisol level; this finding, coupled with the urinary metabolite response to ACTH administration, was diagnostic of secondary hypoadrenalism. The patient demonstrated no ACTH or cortisol response to lysine vasopressin, indicating that an intrinsic pituitary problem, rather than a hypothalamic one, was the underlying cause of ACTH deficiency. Assessment of the remainder of pituitary hormones disclosed adequate reserve of thyroid-stimulating hormone, luteinizing hormone, follicle-stimulating hormone, and growth hormone (TSH, LH, FSH, and GH, respectively), coupled with hyperprolactinemia. Computed tomography (CT) of the sella revealed an empty sella. The temporal circumstance of this patient's illness, and the radiographically empty sella, led Stacpoole and co-workers to conclude that selective postpartum necrosis was the causative factor for the ACTH deficiency. Although Sheehan's syndrome may rarely manifest with isolated trophic hormone deficiencies,[9,10] their report[8] was the first to document unequivocally isolated ACTH deficiency in such a setting.

The development of selective ACTH deficiency during the puerperium need not necessarily be related to postpartum bleeding or shock. Richtsmeier et al.[4] reported a 31-year-old woman who manifested several impressive features of hypocortisolism during the puerperium following a normal delivery. She died from cardiac arrest, and the postpartum findings showed marked atrophy of both adrenal glands; the pituitary gland at autopsy was small, and approximately 5–10% of the anterior pituitary cells were infiltrated with lymphocytes. The striking finding was the demonstration of no apparent immunoreactive ACTH within the cells, while prolactin, TSH, GH, and LH-/FSH-secreting cells were readily recognized by the immunoperoxidase technique. The isolated ACTH deficiency in this case was caused by lympho-cytic hypophysitis, a disorder that is clearly related to pregnancy and puer-perium.[11–14] Since lymphocytic hypophysitis is an autoimmune disorder, it is appropriate to consider next the association of isolated ACTH deficiency to autoimmune disorders.

Autoimmunity

The case reported by Richtsmeier et al.[4] of isolated ACTH deficiency and lymphocytic hypophysitis strongly confers an autoimmune etiology in that particular instance. However, there is only scanty and indirect evidence in the literature to assign a major role to autoimmune disease in the causation of isolated ACTH deficiency. There are, however, several associations that merit mention. Primary hypothyroidism is seen with an impressive frequency in patients with selective ACTH deficiency. Hashimoto et al.[15] described a patient with primary hypothyroidism in whom isolated ACTH deficiency de-veloped 5 years later. It is noteworthy that circulating anti-ACTH antibodies were detected in her serum by radioimmunological methods, a finding anal-ogous to the detection of antibodies to prolactin in patients with lymphocytic hypophysitis. Hashimoto and colleagues also noted a hyperresponsive TSH response to TRH administration in 10 out of 22 Japanese patients with isolated ACTH deficiency. While this finding is suggestive of incipient or overt primary thyroid failure, it should be recognized that an exaggerated TSH response to TRH may be encountered in hypocortisolemic states per se.[16,17] The patient with lymphocytic hypophysitis and ACTH deficiency described by Richtsmeier et al.[4] also demonstrated histological evidence of lymphocytic infiltration in the thyroid gland at autopsy. Stephens et al.[18] have also observed the devel-opment of isolated ACTH deficiency in a male patient in whom hypothy-roidism had developed several years after radioactive iodine therapy for Graves hyperthyroidism. It is too soon to say whether the combination of isolated ACTH deficiency and autoimmune thyroid disease represent a form of plu-riglandular failure. Coexistent diabetes mellitus has been described in a patient with isolated ACTH deficiency.[19] This association is reminiscent of the well-known relationship between diabetes mellitus and primary adrenal insuffi-ciency, both of which share an autoimmune background.[20,21]

Hypothalamic Origin

The possibility that isolated ACTH deficiency may occur as a result of deficiency of corticotropin-releasing factor (CRF) has been suggested.[5–7] Such a premise is indeed tenable, since hypothalamic deficiencies resulting in TSH,[22–24] GH[25] and gonadotropin[26,27] failure are well-recognized entities.[28] It is con-ceivable that isolated failure of the corticotrope can result from hypothalamic deficiency of CRF. If such is the case, it is difficult to reconcile with the onset of such a phenomenon in adult life, in the absence of organic lesions of the hypothalamus or the stalk. In children, however, such a premise is more tenable. Martin and Martin[5-7] studied three children with isolated ACTH deficiency who demonstrated a cortisol response to lysine vasopressin administration. Since all three children were products of complicated labor, the assumption was that perinatal damage to the hypothalamic area resulted in decreased synthesis and release of ACTH.

Similarly, two out of four patients with isolated ACTH deficiency reported by Stacpoole et al.[29] demonstrated a significant increment in ACTH following lysine vasopressin. The availability of ovine CRF has added another dimension to the diagnostic evaluation of ACTH deficiency. Hermus et al.[30] characterized the response patterns of plasma ACTH levels to the administration of CRF-41 in patients with intrinsic pituitary disease and those with ACTH deficiency secondary to hypothalamic etiologies (usually tumors). As expected, patients with intrinsic pituitary disease showed a negligible rise in plasma ACTH following CRF-41 administration, while patients with ACTH deficiency caused by a hypothalamic etiology demonstrated a delayed biphasic response. The application of this test to patients with isolated ACTH deficiency awaits elucidation.

Although the etiology of isolated ACTH deficiency remains unsettled and speculative, the histopathological appearance of the disorder is even less certain. The paucity of autopsy material certainly is a factor. Perkoff et al.[31] were among the first to report the marked decrease in the basophils of the anterior pituitary in a patient with isolated ACTH deficiency. They described a 49-yr-old man with isolated ACTH deficiency who was successfully treated with glucocorticoid therapy but who subsequently died of esophageal malignancy. At autopsy, the most striking feature was the complete absence of basophils; no bioassayable ACTH could be extracted from the pituitary gland. The supraoptic and paraventricular nuclei of the hypothalamus were normal. The remainder of the anterior pituitary cells consisted of well-granulated eosinophils and chromophobes. The brain and the neurohypophysis were found to be normal.

Clinical Features

In a critical review, Stacpoole et al.[29] summarized the clinical features of 43 cases of isolated ACTH deficiency reported in the world literature, including four of their own. The two observations that emerge upon reviewing the literature are the nonspecific nature of symptoms with which these patients presented and the extreme variability in the expression of the syndrome. The symptoms of hypocortisolism due to ACTH deficiency are similar to those in Addison's disease. However, the tendency to develop acute adrenal crisis is less pronounced than in patients with Addison's disease. Fatigue, weakness, and weight loss were seen in one-fourth to one-third of patients.

A dominant feature that is present in more than 50% of patients with isolated ACTH deficiency is hypoglycemia. This was particularly evident in children and during pregnancy.[2] The attention to hypoglycemia as an important feature of isolated ACTH deficiency dates back to 1968, when Hung and Migeon[32] reported the occurrence of hypoglycemia in a 2-yr-old boy with probable isolated ACTH deficiency. They further noted that the patient had no adrenal medullary responsiveness to hypoglycemia initially; this phenomenon may be a reflection of the dependency of the enzyme phenylethanolamine-N-methyltransferase (PNMT) on glucocorticoids and ACTH. This enzyme catalyzes the conversion of norepinephrine to epinephrine. The patient's medullary responsiveness to insulin-induced hypoglycemia was restored to normal upon institution of glucocorticoid therapy. In both adults and children, ACTH deficiency should be sought as a cause of hypoglycemia.

Hypotension was encountered in 26% of patients and hyponatremia in 9% of patients with isolated ACTH deficiency reviewed by Stacpoole et al.[29] When compared with addisonian patients, severe hyponatremia is less common in patients with isolated ACTH deficiency. However, severe hyponatremia can and does occur in these patients. This is particularly so in the presence of sodium depletion. It is well recognized that patients with panhypopituitarism may develop hyponatremia either from blunted aldosterone responses to salt restriction or from excess retention of water.[33] Williams et al.[34] observed a blunted aldosterone response to sodium depletion in hypopituitary patients. It appears that the ability of the renin-angiotensin-aldosterone system to respond to sodium depletion is impaired in patients with ACTH-cortisol deficiency. Major et al.[35] reported a female patient with profound hyponatremia (Na level of 110 mEq), negative sodium balance, and excessive urinary losses of sodium. The plasma-renin activity and the basal aldosterone levels were low despite the profound hyponatremia. The aldosterone level responded adequately to the administration of exogenous ACTH and angiotensin II infusion. It was noteworthy that the natriuresis was reversed by glucocorticoid therapy but not with mineralocorticoid treatment. Furthermore, substitution doses of cortisol improved the disturbed renin dynamics as well. The authors interpreted these data to indicate that cortisol, in low doses, has a permissive effect on the renin-angiotensin-aldosterone system and that secondary hyporeninism may underlie the hyponatremia of ACTH deficiency states. Merriam and Baer[36] extended these observations in their report of a 36-yr-old woman with a chronic wasting illness associated with hyponatremia, hypotension, and ACTH deficiency. They too noted markedly reduced aldosterone excretion despite the presence of hyponatremia. More importantly, they noted that chronically adequate (6 months) glucocorticoid levels were needed to maintain a normal aldosterone response in their patient. The importance of these studies lies in recognizing the fact that although ACTH deficiency does not directly affect the zona glomerulosa structurally, the functional mechanisms that normally evoke an aldosterone response become attenuated, even lost, in the presence of hypocortisolism. The consequence, in some patients, is urinary sodium loss and hyponatremia analogous to that seen in primary adrenal failure.

Sparseness of sexual hair in females is also a feature of isolated ACTH deficiency, although less impressive than in patients with Addison's disease.

The incidence of acute adrenal crisis is also less in comparison with Addison's disease but is an important complication.[37] Death from adrenal crises in isolated ACTH deficiency has been reported.[38] The precipitating factors are usually infections, anesthesia, or surgical procedures.

Our finding that has defied explanation is the development of hyperpigmentation in some patients with isolated ACTH deficiency. In fact, the very first reported case of isolated ACTH deficiency, by Steinberg et al.,[1] had manifested the classic generalized hyperpigmentation characteristic of Addison's disease. A similar case was reported by Stolbach et al.[39] This finding is paradoxical, since both ACTH and the pigmentary hormones are derived from the same precursor, pro-opiomelanocortin, secreted by the adrenocorticomelanotropic cell.[40].

A rare manifestation of ACTH deficiency is the stiff-man syndrome.[41] This syndrome is characterized by progressive stiffness, and painful muscle spasms affecting the axial and limb musculature. George et al.[42] reported the case of a 42-yr-old woman in whom cramps and stiffness of her extremities developed following delivery. The clinical and electrophysiological criteria were consistent with the stiff man (woman) syndrome in her case. Hormonal evaluation demonstrated deficiencies of ACTH, GH, and prolactin, consistent with postpartum hypopituitarism. The remarkable aspect of the case was the complete resolution of the neuromuscular syndrome following replacement with glucocorticoids. The role of the GH and prolactin deficiencies is questionable, in light of the dramatic improvement with substitution glucocorticoid therapy.

Diagnostic Studies

The diagnostic approach to the patient with isolated ACTH deficiency should be taken in a stepwise fashion:

Step I: Establishing hypocortisolism (e.g., urinary 17-OCHS, cortisol)
Step II: Delineating primary from secondary hypoadrenalism (standard ACTH stimulation test)
Step III: Documenting ACTH deficiency (metyrapone)
Step IV: Differentiating pituitary from hypothalamic disease (vasopressin, O-CRF)
Step V: Documenting integrity of other pituitary hormones (reserve testing)
Step VI: Exclusion of tumors in sellar or suprasellar area (CT scan)

The diagnostic studies for the evaluation of isolated ACTH deficiency can be classified into three categories (see Table 63). Measurement of basal plasma cortisol or ACTH is not a good screening test when ACTH deficiency is suspected. Several patients with the disorder demonstrate misleadingly normal basal cortisol (and ACTH) levels in the plasma. Indeed, in the report by

TABLE 63
Diagnostic Studies in Evaluating Isolated ACTH Deficiency

Test	Comment
Basal studies	
Plasma cortisol	Usually low, may be normal
Plasma ACTH	Usually low, can be in normal range
24-hr urinary 17-OHCS	Very low, often below 1.5 mg/24 hr
Dynamic studies	
Rapid ACTH stimulation test	Cortisol response often blunted due to dormancy of adrenals, but aldosterone response preserved
Standard ACTH stimulation test	Classic "stepladder" pattern of increase in urinary 17-OHCS with 3–5 days of ACTH infusion
Insulin hypoglycemia test	No ACTH or cortisol response to hypoglycemia
Metyrapone test	No response in ACTH or 11-deoxycortisol levels following metyrapone
Lysine vasopressin test	Variable response; patients with hypothalamic CRF deficiency may show an increase in ACTH, cortisol following lysine vasopressin
CRF test	Same as above
Pituitary function tests	
TSH reserve T4, T3R, TSH, TRH test	To demonstrate adequacy of thyrotropin, growth hormone, prolactin, and gonadotropin reserve
Growth hormone reserve	Hypoglycemia challenge L-dopa
Prolactin reserve	TRH test
LH, FSH reserve	Basal LH, FSH, LH–Rh test
CT scan of the pituitary	To exclude an organic lesion in the pituitary or suprasellar area

Stacpoole et al.[29] all four patients with selective ACTH deficiency demonstrated measurable ACTH levels in the plasma, often in the low-normal range. By contrast, the 24-hr urinary 17 OHCS is nearly always low, generally below 1.5–2 mg/24 hr.

The cortisol response to a single bolus of IV ACTH does not differentiate between primary failure (addisonian) and secondary adrenal failure (ACTH deficiency). The lack of response in Addison's disease is due to compromised secretory reserve, while the sluggish inadequate cortisol response that typifies ACTH deficiency is due to dormancy of the zona glomerulosa chronically deprived of ACTH priming.[43] The aldosterone response to ACTH, however, will differentiate between primary adrenal failure (where no aldosterone response is seen) from ACTH deficiency, where a rise in aldosterone is observed following an IV bolus of ACTH.[44]

The "gold standard" tests for the documentation of ACTH deficiency are the standard ACTH stimulation test and the metyrapone test. The former evaluates the adrenocortical responsiveness under protracted ACTH stimulation, while the latter evaluates the integrity of the hypothalamic pituitary axis in terms of responding to declining cortisol levels.

The standard ACTH stimulation test is performed by infusing 40 units of ACTH in 8 hr for 3–5 days with concurrent daily measurements of 24-hr urinary 17-OHCS. The test can also be performed by administering ACTH gel 40 units twice a day IM for 3–5 days with measurements of 17-OHCS with continued ACTH administration. Table 64 illustrates this response pattern seen in one of our patients with hypoadrenalism caused by ACTH deficiency. The gradual increase in metabolite excretion is a reflection of the gradual awakening of a dormant unprimed adrenal cortex.

Although the standard ACTH provides excellent separation between patients with Addison's disease (primary) and those with ACTH deficiency (secondary), occasionally there may be some difficulty in differentiating patients with limited adrenal reserve from those with ACTH deficiency. Therefore, even if the response of urinary 17-OHCS demonstrates a pattern consistent with ACTH deficiency, the confirmation of the diagnosis rests on demonstrating ACTH failure. This can be achieved by performing the metyrapone test or the insulin hypoglycemia challenge test.

The metyrapone test evaluates the ability of the hypothalamic–pituitary axis to release ACTH in response to an acute lowering of basal cortisol levels. Metyrapone blocks the enzyme 11β-hydroxylase and impairs conversion of 11-deoxycortisol to cortisol, lowering the concentration of the latter. This decrement is sensed by the hypothalamic–pituitary axis, which is prompted to respond by releasing ACTH. The effect of ACTH release is reflected as an increase in plasma 11-deoxycortisol (the precursor proximal to the block) or its urinary metabolites, 17-OHCS.

The characteristic response of patients with ACTH deficiency is failure to increase the plasma 11-deoxycortisol (or 17-OHCS) after metyrapone. A similar response is to be expected in Addison's disease as well. However, performance of the ACTH-stimulation test prior to the metyrapone would have demonstrated a viable, responsive adrenal in patients with ACTH deficiency, while characterizing the nonresponse of addisonian patients.

TABLE 64
Standard ACTH Stimulation Test in a Patient with ACTH Deficiency

	Cortisol (μg/ml)	DHEA (ng/ml)	17-OH (mg/24 hr)	17-KS (mg/24 hr)
Baseline	5	300	1.8	4
Baseline	6	300	1.6	3.8
ACTH (40 U)	6.2	300	1.6	3.8
ACTH (40 U)	8	400	2.2	5
ACTH (40 U)	15	>1000	6	8
ACTH (40 U)	20	>1000	8	12
ACTH (40 U)	20	>1000	10	16

One concern in performing the metyrapone test in a patient with dangerously low cortisol levels is the further lowering with metyrapone. Indeed, when the serum cortisol is very low, metyrapone can drastically lower the cortisol to dangerous levels, with catastrophic consequences.[38] When in doubt, it is therefore advisable to administer ACTH gel 40 units twice a day, for 3 days, followed by 3 control days, and then to perform the metyrapone test by orally administering 750 mg every 4 hr for 6 doses.

The insulin hypoglycemia challenge directly stimulates the hypothalamic–pituitary axis with a resultant release of ACTH and cortisol. Considered the most potent stimulus for ACTH release, the insulin test should be performed under supervision. Patients with ACTH deficiency fail to demonstrate an increase in ACTH or cortisol despite severe hypoglycemia.

After having established ACTH deficiency as the cause of hypoadrenalism, the next step in the etiologic workup is to distinguish hypothalamic from intrinsic pituitary disease. This can be attained by the use of lysine vasopressin[45,46] or more recently, with the use of ovine CRF.[47,48] Lysine vasopressin directly stimulates the pituitary to release ACTH. Lysine vasopressin is usually administered intravenously at a rate of 5 U/hr for 2 hr. The normal response is a mean increase in cortisol by 20–25µg, and an ACTH increase by 90–100 pg/ml. The mechanism of its action is probably related to its stressful effects, rather than any inherent structural similarity to hypothalamic CRF. Some patients with isolated ACTH deficiency respond to lysine vasopressin, suggesting an underlying hypothalamic disorder. Lysine vasopressin should not be administered to patients with coronary heart disease because of its coronary-spastic properties. The recent availability of CRF is a more direct method to delineate pituitary, from hypothalamic varieties of ACTH deficiency. Hermus et al.[30] reported their experience with o-CRF in delineating the pure pituitary from hypothalamic forms of ACTH deficiency in patients with organic disease of the pituitary and hypothalamus. As expected, patients with intrinsic pituitary disease do not respond to o-CRF with an ACTH release, while patients with ACTH deficiency secondary to a hypothalamic origin demonstrated a delayed biphasic response.

The next step in the evaluation of patients with selective ACTH deficiency is the demonstration of integrity of the other pituitary trophic hormones. The integrity of TSH and prolactin can be assessed by the intravenous administration of TRH; the reserve of LH and FSH can be screened with the use of LH-RH; assessment of GH can be adequately performed by the use of hypoglycemia or L-dopa.

The final step in the workup is to exclude organic etiology in the sellar of suprasellar region. This can be effectively done by CT of the sella and the suprasellar region.

The criteria for establishing the diagnosis of selective ACTH deficiency have been outlined by Stacpoole et al.[29]

1. Low basal urinary 17-OHCS with or without low basal plasma cortisol
2. Low or normal basal ACTH

3. Stimulation of cortisol, 17-OHCS, or both during prolonged ACTH administration
4. Failure of urinary 17-OHCS (or plasma 11-deoxycortisol) to rise following metyrapone
5. Integrity of the other pituitary hormones

To these criteria may be added the exclusion of pathogy in the sellar or suprasellar area by CT.

Treatment

The treatment of isolated ACTH deficiency is replacement therapy with hydrocortisone. Aspects of glucocorticoid therapy are discussed in Chapter 15.

References

1. Steinberg A, Shechter FR, Segal HI: True pituitary unitropic deficiency. *J Clin Endocrinol* **14**:1519, 1954.
2. Satterfield RG, Williamson HO: Isolated ACTH deficiency and pregnancy. *Obstet Gynecol* **48**:693, 1976.
3. Kratz F, Graef V: Uber einen Fall einer Gravidität bei isolierten Ausfall von ACTH. *Klin Wochenschr* **51**:1062, 1973.
4. Richtsmeier AJ, Henry RA, Bloodworth JMB, et al: Lymphoid hypophysitis with selective adrenocorticotropic hormone deficiency. *Arch Intern Med* **140**:1243, 1980.
5. Martin MM, Martin ALA: Hypoglycemia due to isolated ACTH deficiency. *South Med J* **62**:1539, 1969.
6. Martin MM, Martin ALA: Corticotropin releasing factor (CRF) deficiency presenting as idiopathic hypoglycemia. *International Congress of Pediatrics, Vienna, Austria,* 1971, p. 251.
7. Martin MM, Martin ALA: Idiopathic hypoglycemia—A defect in hypothalamic ACTH-releasing factor secretion. *Pediatr Res* **5**:396, 1971.
8. Stacpoole PW, Kandell TW, Fisher WR: Primary empty sella, hyperprolactinemia, and isolated ACTH deficiency after postpartum hemorrhage. *Am J Med* **74**:905, 1983.
9. Sheehan HL: Atypical hypopituitarism. *Proc R. Soc Med* **54**:43, 1961.
10. Smith CW Jr, Palmer R, Howard RB: Variations in endocrine gland function in post-partum pituitary necrosis. *J Clin Endocrinol Metab* **19**:1420, 1959.
11. Asa, SL, Bilbao JM, Kovacs K, et al: Lymphocytic hypophysitis of pregnancy resulting in hypopituitarism: A distinct clinicopathologic entity. *Ann Intern Med* **95**:166, 1981.
12. Hume R, Roberts GH: Hypophysitis and hypopituitarism: report of a case. *Br Med J* **2**:548, 1967.
13. Lack EE: Lymphoid "hypophysitis" with end organ insufficiency. *Arch Pathol Lab Med* **99**:215, 1975.
14. Mazzone T, Kelly W, Ensinck J: Lymphocytic hypophysitis: Associated with antiparietal cell antibodies and vitamin B_{12} deficiency. *Arch Intern Med* **143**:1794, 1983.
15. Hashimoto K, Takahara J, Takaya Y, et al: Anaphylactic shock after synthetic adrenocorticotropin in a patient with isolated adrenocorticotropin and β-lipotropin deficiency. *J Clin Endocrinol Metab* **51**:1175, 1980.
16. Nicoloff JT, Fisher DA, Appleman MD Jr: The role of glucocorticoids in the regulation of thyroid function in man. *J Clin Invest* **49**:1922, 1970.

17. Stryker TD, Molitch ME: Reversible hyperthyrotropinemia, hyperthyroxinemia, and hyperprolactinemia due to adrenal insufficiency. *Am J Med* **79**:271, 1985.
18. Stephens WP, Goddard KJ, Laing I, et al: Isolated adrenocorticotrophin deficiency and empty sella associated with hypothyroidism. *Clin Endocrinol* **22**:771, 1985.
19. Abramson EA, Arky RA: Coexistent diabetes mellitus and isolated ACTH deficiency: Report of a case. *Metabolism* **17**:492, 1968.
20. Riley WJ, Maclaren NK, Neufeld M: Adrenal autoantibodies and Addison's disease in insulin-dependent diabetes mellitus. *J Pediatr* **97**:191, 1980.
21. Neufeld M, Maclaren NK, Blizzard RM: Two types of autoimmune Addison's disease associated with different polyglandular autoimmune (PGA) syndromes. *Medicine (Baltimore)* **60**:355, 1981.
22. Pittman JA Jr, Haigler ED Jr, Hershman JM, et al: Hypothalamic hypothyroidism. *N Engl. J Med* **285**:844, 1971.
23. Singer PA, Mestman JH, Manning PR, et al: Hypothalamic hypothyroidism secondary to Sheehan's syndrome. *West J Med* **120**:416, 1974.
24. Martin LG, Martul P, Connor TB, et al: Hypothalamic origin of idiopathic hypopituitarism. *Metabolism* **21**:143, 1972.
25. Laron Z, Keret R, Bauman B, et al: Differential diagnosis between hypothalamic and pituitary hGH deficiency with the aid of synthetic GH-RH 1-44 *Clin Endocrinol* **21**:9, 1984.
26. Kallmann RJ, Schoenfeld WA, Barrera SE: The genetic aspects of primary eunuchoidism. *Am J Ment Defic* **48**:203, 1944.
27. Woolf PD, Schalch D: Hypopituitarism secondary to hypothalamic deficiency. *Ann Intern Med* **78**:88, 1973.
28. Lieblich JM, Rogol AD, White BJ, et al: Syndrome of anosmia with hypogonadotropic hypogonadism (Kallmann syndrome): Clinical and laboratory studies in 23 cases. *Am J Med* **73**:506, 1982.
29. Stacpoole PW, Interlandi JW, Nicholson WE, et al: Isolated ACTH deficiency: A heterogeneous disorder: Critical review and report of four new cases. *Medicine (Baltimore)* **61**:13, 1982.
30. Hermus ARMM, Pieters GFFM, Pesman GJ, et al: ACTH and cortisol responses to ovine corticotrophin-releasing factor in patients with primary and secondary adrenal failure. *Clin Endocrinol* **22**:761, 1985.
31. Perkoff GT, Eik-nes K, Carnes WH, et al: Selective hypopituitarism with deficiency of anterior pituitary basophils: A case report. *J Clin Endocrinol* **20**:1269, 1980.
32. Hung W, Migeon CJ: Hypoglycemia in a two-year-old boy with adrenocorticotropic hormone (ACTH) deficiency (probably isolated) and adrenal medullary unresponsiveness to insulin-induced hypoglycemia. *j Clin Endocrinol* **28**:146, 1968.
33. Bethune JE, Nelson DH: Hyponatremia in hypopituitarism. *N Engl J Med* **272**:771, 1965.
34. Williams GH, Rose LI, Dluhy RG, et al: Aldosterone response to sodium restriction and ACTH stimulation in panhypopituitarism. *J Clin Endocrinol Metab* **32**:27, 1971.
35. Major P, Kuchel O, Boucher R, et al: Selective hypopituitarism with severe hyponatremia and secondary hyporeninism. *J Clin Endocrinol Metab* **46**:15, 1978.
36. Merriam GR, Baer L: Adrenocorticotropin deficiency: Correction of hyponatremia and hypoaldosteronism with chronic glucocorticoid therapy *J Clin Endocrinol Metab* **50**:10, 1980.
37. Nichols ML, Brown RD, Granville GE, et al: Isolated deficiency of adrenocorticotropin (ACTH) and lipotropins (LPHs). *J Clin Endocrinol Metab* **47**:84, 1978.
38. Odell WD, Green GM, Williams RH: Hypoadrenotropism: The isolated deficiency of adrenotropic hormone. *J Clin Endocrinol* **20**:1017, 1960.
39. Stolbach L, Eppes RB, Stockdale F, et al: Hypopituitarism presenting as hypoadrenotropism. *Ohio State Med J* **61**:721, 1965.
40. Phifer RF, Orth DN, Spicer SS: Specific demonstration of the human hypophyseal adrenocortico-melanotropic (ACTH/MSH) cell. *J Clin Endocrinol Metab* **39**:684, 1974.
41. Gordon EE, Januszko DM, Kaufman L: A critical survey of stiff-man syndrome. *Am J Med* **42**:582, 1967.

42. George TM, Burke JM, Sobotka PA, et al: Resolution of stiff-man syndrome with cortisol replacement in a patient with deficiencies of ACTH, growth hormone, and prolactin. *N Engl J Med* **310**:1511, 1984.
43. Speckart PF, Nicoloff JT, Bethune JE: Screening for adrenocortical insufficiency with cosyntropin (synthetic ACTH). *Arch Intern Med* **128**:761, 1971.
44. Dluhy RG, Himathongkam T, Greenfield M: Rapid ACTH test with plasma aldosterone levels: Improved diagnostic discrimination. *Ann Intern Med* **80**:693, 1974.
45. Eddy RL: Aqueous vasopressin provocative test of anterior pituitary function. *J Clin Endocrinol* **28**:1836, 1968.
46. Hedge GA, Yates MB, Marcus R, et al: Site of action of vasopressin in causing corticotropin release. *Endocrinology* **79**:328, 1966.
47. Lytras N, Grossman A, Perry L, et al: Corticotrophin releasing factor: Responses in normal subjects and patients with disorders of the hypothalamus and pituitary. *Clin Endocrinol* **20**:71, 1984.
48. Nakahara M, Shibasaki T, Shizume K, et al: Corticotropin-releasing factor test in normal subjects and patients with hypothalamic-pituitary-adrenal disorders. *J Clin Endocrinol Metab* **57**:963, 1983.

11

LH and FSH:
The Gonadotropins

LH and FSH

The gonadotropins, luteinizing hormone (LH) and follicle-stimulating hormone (FSH), are synthesized by the pituitary gonadotropes, which constitute 5–10% of the pituitary cell population. The pituitary gonadotropes are randomly distributed within the central wedge and the lateral wing of the gland. The same cell secretes both LH and FSH, since immunoperoxidase stains localize both LH and FSH to the same cell. The glycoprotein secretory granules measure 275–375 nm in diameter and appear basophilic by conventional staining techniques. These cells are periodic acid-Schiff (PAS) positive; they have been referred to as castration cells, because they enlarge with striking vacuolation following gonadectomy. Measurement of gonadotropin content within the pituitary has revealed extremely low hormonal content in prepubertal children; during the reproductive phase the glandular content of LH, FSH varies; approximately 700 IU of LH and 200 IU of FSH is contained within the pituitaries of menstruating women with a three fold increase in LH after menopause.

Chemistry

LH and FSH are glycoprotein hormones consisting of an α- and a β-subunit. The α-subunit is common to LH and FSH, as well as thyroid-stimulating hormone (TSH) and human chorionic gonadotropin (HCG).[2] The β-subunit is unique, conferring biological and immunological specificity. The β-subunits however, require combination with α-subunits to affect the biological actions of the hormone. The α-subunits contain 96 amino acids and possess two complex carbohydrate moieties at amino acid positions 56 and 82. The β-subunit of LH contains 115 amino acids with two asparagine-linked complex carbohydrate moieties at positions 13 and 30.[3] The β-subunit of FSH also contains 115 amino acids, but in contrast to LH, the two asparagine-linked carbohydrate moieties are located at positions 7 and 24.[4]

Secretion and Release

LH and FSH are released from the gonadotropes by a process of exocytosis, which involves fusion of the secretory granule with the cell membrane followed by extrusion into the extracellular fluid space. LH and FSH circulate in the plasma in a free state unbound to carrier proteins. The basal levels of LH and FSH in the circulation range between 5 and 20 mU/ml. A disproportionate rise in FSH is observed postmenopausally. In prepubertal children, the concentrations of both gonadotropins are low.

The plasma concentrations of LH and FSH demonstrate impressive fluctuation with 5–20-min pulsations, as the pituitary gonadotropes secrete gonadotropins in an episodic pulsatile fashion.[5–9] Typically, in the normal male LH is secreted as well-defined pulses recurring at approximately 2 hour intervals.[10] In normal ovulating females, the pulsatile secretory pattern of LH and FSH is complicated by the dynamic sequence of change in the hypothalamic pituitary ovarian axis. In addition to distinct peaks throughout the day, LH demonstrates a striking pulse related to sleep.[11,12] This phenomenon of sleep entrainment is absent in prepubertal children. The LH programming to sleep is an important event that heralds the development of puberty. Boyar et al.[13] demonstrated that pubertal LH programming is initiated at an earlier stage in children with precocious puberty and suggested that changes in adrenal androgen secretion during puberty may be the event that initiates pubertal LH programming.

Since the gonadotropin levels in the serum are subject to enormous variation a randomly drawn single determination of LH or FSH is difficult to interpret except when these levels are markedly elevated. Interpretation of single basal levels are even more difficult in the ovulating female owing to the effects of the hormonal changes involved in the normal menstrual cycle.

Control of Gonadotropin Secretion

The control and release of gonadotropin secretion are modulated by dual mechanisms: trophic influence from the hypothalamus and negative feedback by the sex steroids in the circulation. Before each of these dual aspects is discussed, three fundamental statements require emphasis: (1), in both sexes, both gonadotropins are controlled by a single hypothalamic peptide, gonadotropin-releasing hormone (GnRH); (2), in both sexes, the respective sex steroids (testosterone or 17β-estradiol) exert a negative feedback on the hypothalamic-pituitary unit; and (3), the hypothalamic–pituitary unit of females is uniquely capable of responding to positive feedback by estrogens, whereas males never do. Exposure of the fetus (or neonate) to aromatizable androgens such as testosterone or androstenedione forever is believed to abolish the ability of the hypothalamus to respond to positive feedback with estrogens. This phenomenon (lack of hypothalamic positive feedback to estrogen) is called defeminization of the hypothalamus. The ability of LH to surge in response to estrogen is absent in normal males and in patients with testicular feminization syndrome, but not in patients with XY gonadal dysgenesis.[14,15]

Hypothalamic control over gonadotropin secretion is mediated by GnRH. There are several impressive lines of evidence to support the important role of this hypothalamic peptide.

1. LH and FSH levels of prepubertal children are extremely low, and the pituitary gland does not begin to secrete significant quantities of gonadotropins until puberty. Although the hormonal events of puberty are far from clear, it is believed that the pulsatile release of GnRH by the hypothalamus and the establishment of gonadotrope sensitivity to GnRH represent the two major steps. The prepubertal pituitary gonadotropes characteristically fail to show an LH peak in response to a bolus of synthetic GnRH. However, at the time of puberty, the gonadotropes become sensitive and "tuned in" to respond to GnRH. This may be a consequence of improved receptor sensitivity of the gonadotropes or may be a response to the pulsatile release of GnRH by the hypothalamus.
2. Failure of the hypothalamus to generate GnRH results in classic Kallmann syndrome with hypogonadotropic hypogonadism.
3. The exogenous administration of synthetic GnRH to a normal adult results in a four- to tenfold rise in LH and a two- to fourfold rise in FSH over the basal level, indicating the provocative influence of this hypothalamic peptide.

The negative feedback of the sex steroids on gonadotropin secretion is exerted at the hypothalamic and pituitary levels. In the male, LH secretion is negatively controlled by the serum testosterone level; i.e., a high testosterone level suppresses and a low testosterone level stimulates LH, the hormone responsible for testosterone production. The FSH secretion by the gonadotrope is negatively controlled by the protein inhibin. This substance, secreted by the Sertoli cells of the testes, correlates with spermatogenesis. A low sperm count is associated with a low level of inhibin, which stimulates the synthesis and release of FSH and spermatogenesis (Fig. 29). Although the feedback mechanism is compartmentalized in the above manner, persistent administration of testosterone will suppress the hypothalamus (GnRH), resulting in a lowering of both LH and FSH.

The negative feedback of sex steroids on the control of gonadotropin secretion in the female is more complex. Three aspects are to be underscored.

1. In contrast to males, there are no separate feedback mechanisms for LH and FSH.
2. Estrogens exert their negative feedback effect on the pituitary as well as the hypothalamus. The suppressive effect of estrogens on the pituitary is evidenced by the fact that the administration of ethinyl estradiol is associated with a lowering of the basal concentrations of FSH and LH as well as a blunted response of these two hormones to ex-

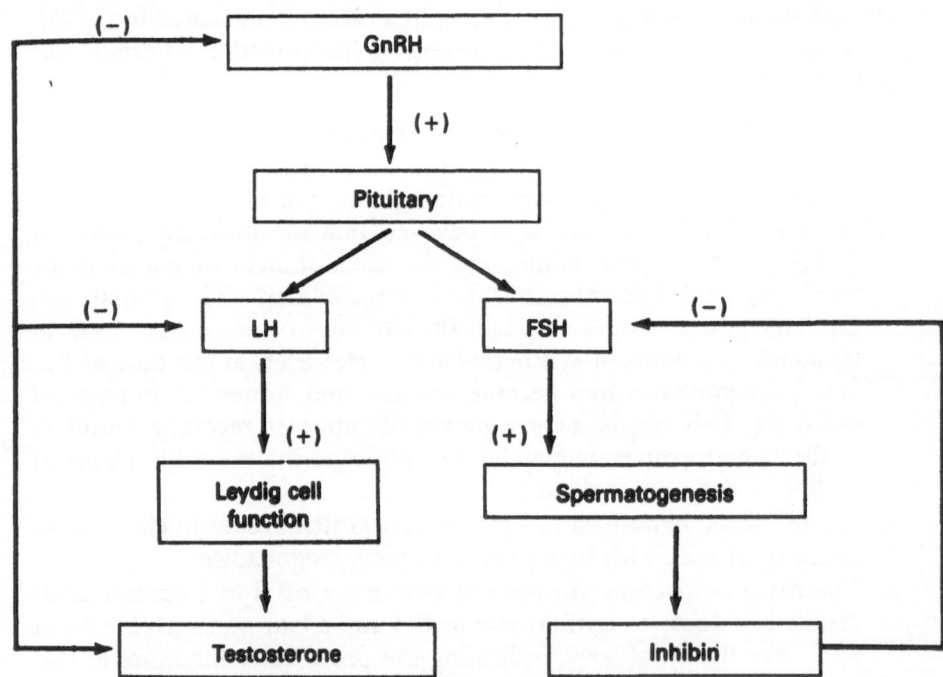

FIGURE 29. Feedback regulation of gonadotropins in the male.

ogenous LHRH. (The situation is analogous to the response of the thyrotrope to increasing thyroid hormone levels.)

3. The effects of estrogen on the hypothalamus are rather paradoxical. The hypothalamus can be visualized as consisting of a tonic portion and a cyclic portion. The tonic hypothalamus responds to negative feedback from estrogens, whereas the cyclic part responds to positive feedback from estrogens. This concept helps in explaining the hormonal events during a normal menstrual cycle. At the beginning of the cycle, the low estrogen level stimulates the tonic hypothalamus to secrete GnRH, with consequent increases in LH and FSH. During the mid-cycle, the high estrogen level stimulates the cyclic portion to release GnRH, which in turn results in an abrupt surge of LH. This preovulatory surge of LH is responsible for rupture of the mature Graafian follicle, resulting in ovulation.

Actions of Gonadotropins

In Males

The two functions of the testis are secretion of testosterone and spermatogenesis. The gonadotropins are essential for initiating and maintaining both functions. Testosterone is secreted by the Leydig cells and has three important actions:

1. In the fetus, testosterone, by virtue of conversion to dihydrotestosterone, is exclusively responsible for virilizing the male fetus. In the absense of this hormone (or when complete resistance to its action occurs), the male fetus will not virilize, resulting in formation of female external genitalia (male pseudohermaphroditism). The secretion of testosterone by the fetal testis is under the regulatory control of placental gonadotropins (hCG), and not the pituitary LH.

2. Testosterone and dihydrotestosterone are responsible for pubertal virilization of boys, resulting in the development of secondary sexual characteristics. Thus, the pubertal hair growth in the face, the deepening of voice, the pubertal growth spurt, and the increase in muscle mass are stimulated by these androgens. Development of scrotal rugosity, scrotal pigmentation, and sexual hair on the scrotum, on the undersurface of the penis, and medial surface of the thighs is also stimulated by these hormones. The phallic growth to adult proportions, the development of male pubic hair pattern, and dense hair growth in the axilla, chest, and back are impressively striking aspects of testosterone effects. Growth of the prostate during puberty is also mediated by testosterone. Thus, testosterone and dihydrotestosterone are the virilizing hormones of male puberty. Pubertal activation of Leydig cells is entirely under the control of pituitary LH.

3. Testosterone also plays a permissive role in the maintenance of normal

spermatogenesis. It appears that for spermatogenesis to proceed normally, high intratesticular concentrations of the hormone are required.

Thus, testosterone and dihydrotestosterone are responsible for the formation of male genitalia, the development of maleness during puberty, and for proper spermatogenesis. The role of these hormones in the development of aggressive behavior, in sexual arousal, and in preserving libido is less clearly defined. The bulk of evidence in subhuman primates indicates a role for the androgen in maintenance of aggressive and sexual behavior. The data in humans are strongly supportive but not conclusive.

Regulation of Testosterone Secretion.   Testosterone, the most potent androgen, is secreted by the Leydig cells in the interstitium of the testes. The secretory activity of the Leydig cells assumes crucial importance during two periods of life. First, in utero, the fetal testes of the male fetus demonstrate intense secretory activity and secrete testosterone, the prohormone required to virilize the fetus into a male. Failure of this mechanism results in the birth of an infant with female external genitalia.

The second burst of secretory activity occurs during puberty, resulting in virilization of the boy into a man in the androgenic sense of that term. Failure of this mechanism results in sexual immaturity. During both crucial periods, the testes are stimulated by trophic hormones: in the fetal period, the stimulus is chorionic gonadotropin, and during puberty, the stimulus comes from the pituitary gonadrotropin, luteinizing hormone (LH). The focus of discussion in this section is limited to testicular function during postnatal life.

The secretory activity of the Leydig cells is in abeyance until the time of puberty. The reason for such dormancy is lack of adequate gonadotropins in the circulation, which are required to induce the receptor activity on the surface of Leydig cells. Puberty is preceded by a heightened sensitivity of the Leydig cell receptors to the action of pituitary LH. Under the trophic stimulus of LH, the secretory activity of the Leydig cells starts and, in the normal male, continues for the rest of life, showing a tendency to decelerate during old age (male climacteric). The precise mechanisms by which LH stimulates testosterone synthesis are not clear, but binding of LH to the receptors on the Leydig cell surface and increased generation of cyclic AMP are important steps in mediation of the action. Mendelson et al.[16] showed that maximal biosynthesis of testosterone occurs at LH concentrations that result in only 10% of the maximal cAMP generation. With continued stimulation by LH, there is down regulation of receptors.

The biochemical steps involved in the synthesis of testosterone are strikingly similar to adrenal androgen synthesis. The starting point for synthesis is cholesterol. The first step, the formation of pregnenolone, is crucial and is believed to be the step that is intensely activated in response to stimulation by LH. The pregnenolone is directed into two pathways: the main pathway

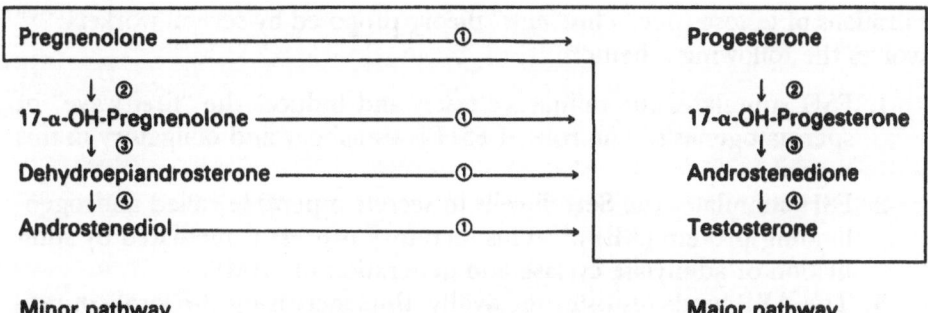

FIGURE 30. Pathways for testosterone synthesis by the testes. Enzymes: (1)-3-β-hydroxysteroid dehydrogenase; (2) 17-α-hydroxylase; (3) 17,20-desmolase; (4) 17-β-hydroxysteroid dehydrogenase.

is conversion to progesterone, 17α-hydroxyprogesterone, androstenedione, and testosterone; a minor pathway involves conversion to 17α-hydroxypregnenolone, DHEA, and androstenediol, which is eventually converted into testosterone. This "minor" pathway in the testes becomes the "major" pathway in the adrenal, and vice versa (Fig. 30).

The second function of the testes is spermatogensis. Sperm production takes place in the seminiferous tubules, which constitute the bulk of testicular tissue. The process of spermatogenesis takes place in the germinal epithelium of the seminiferous tubules. Spermatozoa are derived from the immature spermatogonia, which divide to form spermatocytes. These spermatocytes immediately undergo meiotic (reduction) division to form the haploid cells called spermatids. Through a series of metamorphoses, the spermatids eventually become the motile, flagellated spermatozoa.

In the past it was believed that the dual functions of the testes were strictly compartmentalized and that each gonadotropin independently controlled each function of the testes. According to the older theory, FSH is responsible for stimulating the immature prepubertal testes and initiating spermatogenesis as well as maintaining it. The negative feedback between spermatogenesis and FSH is mediated by a protein called inhibin, secreted by the Sertoli cells. When spermatogenesis fails, inhibin levels decline, signaling FSH to be released in an effort to stimulate spermatogenesis. Thus, the standard theory is fairly simple: FSH stimulates spermatogenesis, and LH stimulates testosterone, which is required, in a permissive fashion, for FSH to stimulate spermatogenesis.

Many recent studies[17,18] have modified this theory to some extent. These studies in animal models suggest that although the induction and initiation of spermatogenesis (the first wave) are clearly dependent on FSH, subsequent spermatogenesis can be maintained by testosterone alone in the absence of FSH. In other words, the immature prepubertal seminiferous tubules require FSH to initiate the process of spermatogenesis, but once this becomes established, the key factor that continues the process is high intratesticular concentrations of testosterone. This "new" theory proposed by several workers[17–20] involves the following scheme:

1. FSH stimulates the immature testes and induces the "first wave" of spermatogenesis. The role of FSH is transitory and obligatory in this regard.
2. FSH stimulates the Sertoli cells to secrete a peptide called androgen-binding protein (ABP).[21] This secretory process is mediated by stimulation of adenylate cyclase and generation of cAMP.
3. The ABP binds testosterone avidly, thus increasing the local concentrations of testosterone to a remarkably high degree.
4. The ABP–testosterone complex passively diffuses into the germinal epithelium and reaches the nucleus, where testosterone probably initiates transcription and formation of new messenger RNA (mRNA)—a documented effect of the androgen in other target tissues.
5. The newly formed mRNA initiates spermatogenesis by effecting conversion of spermatogonia.
6. In addition to its role in initiating spermatogenesis, FSH may also influence the sensitivity of the Leydig cells to LH by regulating the number of LH receptors on these cells.[22]

The current theory attaches importance to the concentrations of testosterone in the interstitium in maintenance of spermatogenesis, once it has been initiated by FSH. Attractive as this theory is, it does not explain the indisputable finding that spermatogenesis declines when testosterone is administered to the normal male. Thus, the exact mechanisms involved in hormonal regulation of spermatogenesis are not completely clear.

In Females

The role of gonadotropins in the maintenance of ovarian function extends in three directions: the first is estradiol synthesis, the second is in stimulating follicular maturation and inducing ovulation, and the third is in maintenance of secretory function of the corpus luteum. Current evidence suggests that both hormones, LH and FSH, are required for all three functions.[23] In the past, follicle maturation was thought to be mediated by FSH and ovulation by LH. Although this may still be predominantly so, studies using individual preparations of LH and FSH indicate that both gonadotropins work synergistically.

The secretion of estradiol by the granulosa cells clearly requires FSH and, to a lesser extent, LH. The mid-cyclical surge of LH (and, to a lesser extent, FSH) is crucial to ovulation. The maintenance of the corpus luteum is predominantly a function of LH.

Follicular Maturation and Ovulation.[24] Follicular maturation begins only after puberty. Prior to puberty, the ovary is filled with immature primary follicles containing oogonia that are not capable of cell division. The process of maturation is initiated by gonadotropins, in particular FSH. (The situation is analogous to spermatogenesis, a process requiring FSH for initiation.) The maturation of the graafian follicle not only depends on the availability of FSH but is also quite dependent on the intraovarian content of estradiol secreted by the granulosa cells of the follicle. Each month, one ovum is selected to ovulate, although several follicles grow and mature. The mechanisms involved in such a highly selective process, whereby only a single ovum undergoes ovulation while the rest undergo atresia, are unknown. There are several established concepts in follicular maturation:

1. Pituitary gonadotropins are absolutely necessary for follicular maturation, since administration of estrogen alone to hypophysectomized women does not result in proper maturation. The crucial role of gonadotropins in stimulating follicular growth has been amply demonstrated in hypophysectomized female rats. In such a setting, both FSH and LH are required for proper preovulatory growth of the follicle.

2. Estradiol secreted by the granulosa cells is also crucial for follicular maturation. While FSH is the dominant gonadotropin involved in promoting follicular growth, this action is dependent on steroid hormones secreted by the granulosa cells (Fig. 31). Estradiol synthesis is regulated by both FSH and LH. It is believed that estrogens stimulate granulosa cell proliferation and inhibit atresia.[25,26] Synthesis of androgens is also regulated by LH and FSH. In contrast to estrogens, the androgens secreted by the ovary inhibit granulosa cell proliferation while stimulating follicular atresia.[27] Estrogens are also responsible, in conjunction with FSH, for induction of LH receptors in the preovulatory follicle for purposes of ovulation.

3. The increase in the size of the follicle and the changes in cells that

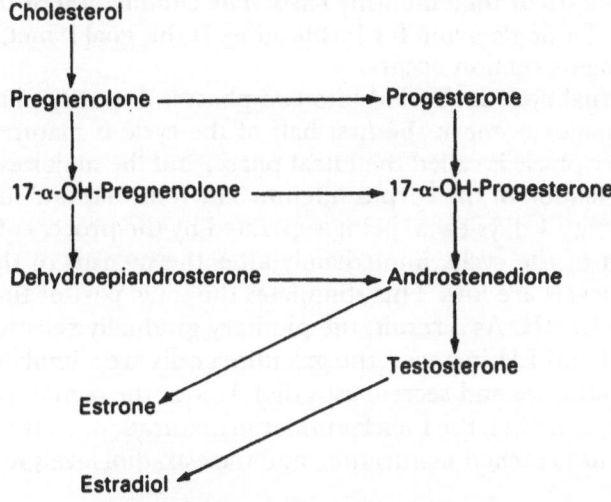

FIGURE 31. Ovarian steroidogenesis.

surround the follicle are dependent on estradiol concentrations. The follicle, which originally measures 50 μm enlarges about 400 times in size. The oocyte within the follicle also enlarges, and the surrounding granulosa cells proliferate and begin to secrete increasing concentrations of estradiol under the intense stimulation by FSH and LH. The stromal cells surrounding the follicles proliferate and become arranged in concentric layers around the follicle. The stromal cells close to the follicle are called theca interna cells, whereas the theca externa cells merge with the adjacent stroma.

4. By the time the follicle matures, the pituitary releases a surge of LH (and to a lesser extent FSH); 24 hr after this surge, ovulation occurs, and the ovum is extruded from the follicle. The follicle emptied of the ovum becomes a sealed new structure called the corpus luteum. Following ovulation, the theca interna cells, the granulosa cells, the theca externa cells undergo extreme mitosis with increased vascularity and fat content. This process, called "leutinization" of the granulosa cells and the theca interna cells, is mediated by LH. The corpus luteum secretes progesterone and to a lesser extent estradiol.

Thus, one set of actions of gonadotropins, i.e., stimulation of sex steroid synthesis, clearly affects another set of actions of gonadotropins, i.e., follicular growth, follicular atresia, and ovulation.

The hormonal highlights of ovulation are the preovulatory surge of LH and FSH and the postovulatory secretion of progesterone by the corpus luteum. The sequence of events that occurs before and after ovulation can be viewed in terms of the hormonal milieu of the normal menstrual cycle in females.

The Menstrual Cycle. The menstrual cycle is a constellation of cyclic hormonal events that occur on a monthly basis. The culmination of the events is the production of a single ovum for fertilization. If this goal is met, pregnancy ensues; if not, menstruation occurs.

The menstrual cycle is divided into two phases: the early phase is follicular, since the major event in the first half of the cycle is maturation of the follicle; the latter phase is called the luteal phase, and the major event during this phase is formation of the corpus luteum. The follicular and luteal phases are approximately 14 days each, being separated by the process of ovulation.

At the start of the cycle, immediately after the mensus of the previous cycle, estradiol levels are low. This stimulates the tonic part of the hypothalamus to secrete GnRH. As a result, the pituitary gradually releases FSH and LH. As the FSH and LH increase, the granulosa cells are stimulated by these hormones to synthesize and secrete estradiol. Under the combined influence of estradiol, LH, and FSH, the follicles undergo maturation. By the fourteenth day, the follicle has reached maturation, and the estradiol levels have reached their peak.

During this late follicular phase, the high concentrations of estradiol stimulate the cyclic hypothalamus. This is mediated by positive feedback in contrast to the tonic hypothalamus, which can be stimulated only by negative feedback. The cyclic hypothalamus, stimulated by estradiol, secretes GnRH a second time, resulting in a predominantly LH surge. This surge, 24 hr later, causes rupture of the mature follicle.

After ovulation, the estradiol concentrations decline gradually. The luteal phase is characterized by sustained secretion of progesterone. Toward the late luteal phase, estradiol is also secreted by the corpus luteum. If fertilization occurs, the corpus luteum persists; if not, it undergoes atrophy. The atrophy of the corpus luteum is accompanied by a drop in progesterone and estradiol levels. This flux in hormonal levels results in withdrawal bleeding by the uterus.

Thus, a normal ovulatory menstrual cycle would seem to require the close interlocked relationship between the gonadotropins themselves, as well as between the gondotropins and the steroid hormones secreted by the ovaries.

Clinical Value of LH, FSH Measurement

The pituitary gonadotropins, FSH and LH, are often evaluated together. Like GH and ACTH, the pituitary gonadotropins are pulsatile. Especially in the female, the pituitary gonadotropins demonstrate significant fluctuation depending on the phase of the menstrual cycle, rendering the interpretation of an isolated sample quite difficult and often meaningless. The clinical indications for obtaining basal FSH and LH levels and for performing dynamic studies of these hormones are mostly in the evaluation of hypogonadal patients.

Basal Studies

1. *Male and female hypogonadism*: In both sexes, measurement of basal LH and FSH with concomitant measurement of the respective sex steroids constitutes the first step in the hormonal evaluation of hypogonadism. Three possible combinations are encountered, depending on the etiology of hypogonadism (Table 65). It should be pointed out that in

TABLE 65
Gonadotropin Levels in Hypogonadism

LH, FSH	Testosterone or 17β-estradiol	Interpretation	Condition
↑	↓	Hypergonadotropic Hypogonadism	Primary gonadal failure
↓	↓	Hypogonadotropic Hypogonadism	Hypothalamic or pituitary disease
Normal	↓	Normogonadotropic	Hypothalamic or pituitary disease

patients with marked lowering of sex steroids, even the demonstration of normal LH and FSH should be considered as a poor suboptimal pituitary response indicating a blunted output of FSH and LH to hypotestosteronemia or hypoestrogenemia.

2. *Secondary amenorrhea*: In patients with secondary amenorrhea, the basal measurements of LH and FSH are likely to help only if these gonadotropins are elevated (which indicates a primary gonadal disorder). It is difficult to interpret low or normal LH and FSH in a single sample, since the ranges seen in normal and hypopituitary patients do overlap considerably.

3. *Hirsutism*: In patients with hirsutism, the demonstration of an elevated basal LH: FSH ratio is suggestive of polycystic ovarian disease. However, this abnormality may not be evident unless multiple samplings are done, preferably on different phases of the menstrual cycle.

Dynamic Studies

The gonadotropin dynamics can be evaluated by studying the LH and FSH response to the administration of exogenous synthetic LH-RH. Theoretically, the two indications for doing such a study are the evaluation of gonadotrope reserve and the differentiation between pituitary and hypothalamic hypogonadism.

1. Evaluation of the gonadotropin reserve by the use of LH-RH. Adults respond to the intravenous administration of LH-RH by a four- to tenfold increase in LH and a two- to fourfold increase in FSH. The vast degree of variability in the normal response does pose difficulties in interpreting marginal responses. A clearly flat response of both LH and FSH to LH-RH is indicative of compromised gonadotropin reserve.

2. The differentiation between hypothalamic and hypopituitary hypogonadism with the use of LH-RH has not proven to be as effective as originally thought. Theoretically, hypopituitary patients do not respond to LHRH, whereas those with hypothalamic hypogonadism demonstrate a delayed, but normal response to LHRH. In practice, however, such clear-cut patterns of separation have not emerged, minimizing the application of a single bolus of LHRH as a diagnostic test for this purpose.

GnRH, LHRH

Synthesis

The hypothalamic control over the secretion of LH and FSH by the pituitary gonadotropes is mediated by GnRH. This peptide is synthesized by hypothalamic neurons located primarily in two regions of the hypothalamus: the preoptic region and the arcuate region. The GnRH neurons at the preoptic region possess long axons that project into the median eminence; the GnRH

neurons in the arcuate region are located more caudally and possess shorter
axons that also project into the median eminence. Experiments with female
rats have implied that GnRH neurons in the preoptic regions are stimulated
by estrogens to induce preovulatory surges of GnRH (positive feedback) while
those neurons at the arcuate region are suppressed by estrogens (negative
feedback). In addition to these two major sites GnRH is also found in the
lamina terminalis, a specialized structure located at the rostral tip of the third
ventricle.

Chemistry

Biochemically, GnRH is a decapeptide. The structure and synthesis of
GnRH were originally proposed by Matsuo et al.[28] and later confirmed by
Burgus et al.[29] Modification of the structure can yield numerous compounds
with increased potency and longer duration of action; indeed, numerous
analogues can be developed by altering the structure of GnRH to synthesize
peptides with agonist and antagonist properties.[30] The ability to produce
super-LHRH agonists by altering the molecular structure of the peptide has
revealed that substitutions at positions 6 and 10 of the decapeptide enhance
biological potency.

Transport

After synthesis by the hypothalamus GnRH is transported by the hypophysial portal vessels to the pituitary. The pituitary stalk serves as the avenue
of transport for the peptide. The existence of an alternate pathway for GnRH
transport has been suggested by McCann.[31] Accordingly, LHRH released
from the neurons at the organum vasculosum of the lamina terminalis directly
exudes into the CSF of the third ventricle from here, it is picked up by
specialized ependymal cells whose processes reach out to the median eminence. The importance of this pathway, if any, is unclear.

Control of GnRH Secretion

The major factor that modulates secretion of GnRH by the hypothalamus
is the level of sex steroids in the circulation. In addition to this predominant
regulator, two other factors play a role in control of GnRH secretion: autoregulation of GnRH secretion by GnRH receptors in the pituitary, and the
putative role of synaptic neurotransmitters.

Circulating Sex Steroids

Measurement of GnRH in the peripheral blood is quite difficult, mostly
because of its extremely short half-life in the circulation. In addition, GnRH
is largely limited to the hypophyseal portal circulation. Therefore, attempts
to separate the direct effects of gonadal steroids on the pituitary from the
direct effects of these steroids on the hypothalamus have met with consid-

erable difficulties. Several animal studies have demonstrated that sex steroids can modulate GnRH secretion and that the hypothalamus is an important site for the gonadal steroids to exert feedback effects.[32] The ability to measure LHRH in the portal blood of rats and rhesus monkeys has facilitated evaluation of GnRH responses to estradiol administration. Thus, several workers[33-35] have demonstrated that gonadal steroids feedback directly into the hypothalamus. However, there is some controversy with regard to whether the dominant effect of sex steroids is on the hypothalamus or the pituitary. The phenomenal work from the laboratory of Knobil[36] has confirmed that patterns of estrogen secretion by the ovary influenced the time courses of LH and FSH in the circulation throughout the menstrual cycle in female rhesus monkeys. The relative importance of the pituitary and hypothalamus in the estrogen-induced effects have been somewhat unclear. Experimental evidence does suggest that microinjection of estrogens into the hypothalamus abolishes the LH rise following castration.[37] However, studies employing measurement of LHRH in portal blood have provided conflicting results.[33] In a remarkable study using the Knobil primate model, Nakai et al.[38] conducted elaborate studies on oophorectomized rhesus monkeys. They induced hypothalamic lesions in such monkeys, abolishing all influenzes from endogenous GnRH production. Tonic gonadotropin secretion was maintained by administering LHRH by pump; the effects of estradiol administration were studied in this setting. The results indicated that despite exogenous administration of LHRH, estradiol administration causes a decline in LH and FSH levels, followed by a rise in these hormones. The results were remarkably similar to those obtained in otherwise intact oophrectomized animals. The conclusion therefore, was that in the rhesus monkey, estradiol exerted both negative and positive feedback effects on gonadotropin secretion at the level of the pituitary. These results are in keeping with the observation that administration of estrogens to normal females blunts the LH-FSH response to exogenous administration of LHRH.[39-41]

The consensus of opinion is that gonadal sex steroids exert their negative feedback on both the pituitary as well as on the hypothalamus, but the dominant negative feedback effect is on the pituitary. The positive feedback effect of estrogens—a phenomenon limited to normal females—is probably heavily dependent on mediation by the hypothalamus. The anatomical demonstration of two discrete populations of GnRH neurons—at the preoptic area and the arcuate region—has led to the belief that tonic and cyclic hypothamic GnRH exist. The tonic portion responds to negative feedback, while the cyclic hypothalamus responds to positive feedback by estrogens. Although anatomically present in the male, the latter mechanism is developed only in the female.

Autoregulation of GnRH Receptors

It has been established that GnRH binds to high-affinity receptors in the pituitary gonadotropes. The recent developments of techniques for assessing GnRH receptors employing radioiodinated analogues of GnRH has permitted analysis of the relationship between GnRH and its receptors. The GnRH receptors are at their peak following castration, and following injections of GnRH.[42–44] Pieper et al.[45] suggested that the rise in GnRH receptors following castration of male rats is a result of the increased GnRH secretion. Further support to the notion that the number of GnRH receptors reflects changes in GnRH secretion by the hypothalamus was provided by Clayton et al.[46] These workers reported animals studies involving the induction of hypothalamic lesions and the use of GnRH anatagonists, to demonstrate that GnRH receptors in the pituitary are regulated in a positive manner by hypothalamic GnRH. Thus, as in other hormonal systems, the GnRH–GnRH receptor interaction tends to autoregulate hypothalamic secretion of GnRH. The receptors are upregulated when GnRH levels are low and vice versa.[47] It is believed that receptor regulation plays, at best, a permissive role in the regulation of GnRH secretion by the hypothalamus.

Putative Role of Neurotransmitters

Like other hypothalamic peptides, secretion of LHRH is influenced by neurotransmitters. Although dopaminergic, serotonergic, and histaminergic tracts exist in the GnRH neurons, the most significant neurotransmitters involved are noradrenergic.[48] Thus, norepinephrine appears to stimulate GnRH, an effect that can be abolished by the use of α-adrenergic blockade. Neurotransmitter mediation may play a role in the positive feedback of estrogen on GnRH release. It is believed that the positive feedback of estrogens is exerted on the preoptic area rich in LHRH neurons, which contain facilitatory noradrenergic synapses. This influenze is, at best, putative because the use of adrenergic receptor blocking drugs cannot completely abolish LHRH surges seen during the preovulatory phase.

Actions

GnRH exerts its provocative effects on the gonadotropes of the anterior pituitary by first binding to the high affinity surface receptors. The increase in the LH release, seen as early as 1 min following IV injection of synthetic GnRH, is brought about by stimulation of the adenylate cyclase–cAMP system. The release of LH and FSH from within the gonadotropes is brought about by a process of exocytosis. Like all other pituitary hormones, this involves fusion of the granule with the cell membrane, followed by transfer of hormone into the extracellular space. Also analogous to other anterior pituitary hormones, gonadotropin release is a calcium-dependent process. GnRH also promotes de novo synthesis of LH and FSH by the gonadotropes.

The crucial fact that deserves emphasis is that the physiological regulation of gonadotropin secretion and release depends on the pulsatile secretion of GnRH by the hypothalamus.[49] The endogenous LHRH pulse frequency is reflected in the episodic and pulsatile nature of LH secretion—a phenomenon that becomes established after puberty. Studies employing bolus doses of

LHRH have demonstrated that when administered for prolonged periods of time, or when given continuously, there is a dramatic decline in the LH, FSH response to LHRH, resulting in pituitary desensitization.[50–53] Such densensitization occurs independent of gonadal steroids, since the phenomenon is observed in postmenopausal females as well.[54] The mechanism for densensitization may be related to receptor blockage (downregulation) of the GnRH receptors located in the pituitary gonadotropes.[55] The concept that exogenous GnRH therapy can stimulate or inhibit gonadotropin release and secretion has had enormous impact on therapy with GnRH and its analogues.[56,57]

In addition to its effects on pituitary gonadotropes, GnRH has several extrapituitary effects, notably on the gonads,[58] but these actions are limited to nonprimate species.

Degradation

GnRH is degraded by specific endopeptidases located within the hypothalamus and pituitary.[59] The development of potent long-acting analogues is dependent on their resistance to cleavage by these endopeptidases. The half-life of exogenously administered GnRH is short. Studies of the pharmacokinetics of GnRH administered by the IV and SC routes have shown considerable differences, implying decreased bioavailability of GnRH by the SC route.[60]

In summary, the hormonal troika of GnRH, gonadotropins, and gonadal steroids works in a synchronized fashion to bring about the intricate processes of sexual maturation and fertility. Failure to generate these central signals results in the development of hypogonadotropic hypogonadism, as we shall see in Chapter 12.

References

1. McKerns KW (ed): *Structure and Function of the Gonadotropins.* Plenum, New York, 1978.
2. Shome B, Parlow AF: Human follicle stimulating hormone (hFSH): First proposal for the amino acid sequence of the alpha-subunit (hFSH-α) and first demonstration of its identity with the alpha-subunit of human luteinizing hormone (hLH-α). *J Clin Endocrinol Metab* **39:**199, 1974.
3. Shome B, Parlow AF: The primary structure of the hormone-specific beta subunit of human pituitary luteinizing hormone (hLH). *J Clin Endocrinol Metab* **36:**618, 1973.
4. Shome B, Parlow AF: Human follicle stimulating hormone: First proposal for the amino acid sequence of the hormone-specific, β-subunit (hFSH-β). *J Clin Endocrinol Metab* **39:**203, 1974.
5. Nankin HR, Troen P: Repetitive luteinizing hormone elevations in serum of normal men. *J Clin Endocrinol Metab* **33:**558, 1971.
6. Rubin RT, Kales A, Adler R, et al: Gonadotropin secretion during sleep in normal men. *Science* **175:**196, 1972.
7. Boyar RM, Perlow M, Hellman L, et al: Twenty-four hour pattern of luteinizing hormone secretion in normal men with sleep stage recording. *J Clin Endocrinol Metab* **35:**73, 1972.
8. Yea SSC, Tsai CC, Naftolin F, et al: Pulsatile patterns of gonadotropin release in subjects with and without ovarian function. *J Clin Endocrinol Metab* **34:**671, 1972.

9. Kapen S, Boyar RM, Hellman L, et al: Episodic release of luteinizing hormone at mid-menstrual cycle in normal adult females. *J Clin Endocrinol Metab* **36**:724, 1973.

10. Crowley Jr WF, Filicori M, Spratt D, et al: The physiology of gonadotropin-releasing hormone (GnRH) secretion in men and women. *Recent Prog Horm Res* **41**:473, 1985.

11. Boyar RM, Finkelstein J, Roffwarg H, et al: Identification of puberty by synchronization of luteinizing hormone release with sleep. *J Clin Invest* **51**:13a, 1972.

12. Boyar RM, Finkelstein J, Roffwarg H, et al: Synchronization of augmented luteinizing hormone secretion with sleep during puberty. *N Engl J Med* **287**:582, 1972.

13. Boyar RM, Finkelstein JW, David R, et al: Twenty-four hour patterns of plasma luteinizing hormone and follicle-stimulating hormone in sexual precocity. *N Engl J Med* **289**:282, 1973.

14. Van Look PFA, Hunter WM, Corder CC, et al: Failure of positive feedback in normal men and subjects with testicular feminization. *Clin Endocrinol (Oxf)* **7**:353, 1977.

15. Oshima H, Troen P: Endocrine and environmental influences on sexual roles. *Am J Med* **70**:1, 1981.

16. Mendelson C, Dufau ML, Catt KJ: Gonadotropin binding and stimulation of cyclic adenosine $3'5'$-monophosphate and testosterone production in isolated Leydig cells. *J Biol Chem* **250**:8818, 1975.

17. Setchell BP: Regulation of spermatogenesis and possible sites for contraceptive action. In Jeffcoate SL, Sandler M (eds): *Progress towards a Male Contraceptive*. Wiley, New York, 1982, p. 1.

18. Steinberger E, Steinberger A, Sanborn B: Endocrine control of spermatogenesis. *Basic Life Sci* **4**:163, 1974.

19. Hansson W, Weddington SC, McLean WS, et al: Regulation of seminiferous tubular function by FSH and androgen. *J Reprod Fertil* **44**:363, 1975.

20. Bremner WJ, Matsumoto AM, Sussman AM, et al: Follicle-stimulating hormone and human spermatogenesis. *J Clin Invest* **68**:1044, 1981.

21. Tindall DJ, Miller DA, Means AR: Characterization of androgen receptor in Sertoli cell-enriched testis. *Endocrinology* **101**:13, 1977.

22. Ketelslegers JM, Hetzel WD, Sherins RJ, et al: Developmental changes in testicular gonadotropin receptors: Plasma gonadotropins and plasma testosterone in the rat. *Endocrinology* **103**:212, 1978.

23. Ross GT, Lipsett MB: Hormonal correlates of normal and abnormal follicle growth after puberty in humans and other primates. *Clin Endocrinol Metab* **7**:561, 1978.

24. McNatty KP, Hillier SG, Van Den Boogaard AMJ, et al: Follicular development during the luteal phase of the human menstrual cycle. *J Clin Endocrinol Metab* **56**:1022, 1983.

25. Goldenberg RL, Reiter EO, Ross GT: Follicle response to exogenous gonadotropins: An estrogen mediated phenomenon. *Fertil Steril* **24**:121, 1972.

26. Harman SM, Louvet J-P, Ross GT: Interaction of estrogens and gonadotropins on follicular atresia. *Endocrinology* **96**:1145, 1975.

27. Louvet J-P, Harman SM, Schreiber JR, et al: Evidence for a role of androgens in follicular maturation. *Endocrinology* **97**:366, 1975.

28. Matsuo H, Baba Y, Nair RMG, et al: Structure of the porcine LH-and FSH-releasing hormone. I. The proposed amino acid sequence. *Biochem Biophys Res Commun* **43**:1334, 1971.

29. Burgus R, Butcher M, Amoss M, et al: Primary structure of the ovine hypothalamic luteinizing hormone-releasing factor (LRF). *Proc Natl Acad Sci USA* **69**:278, 1972.

30. Schally AV, Arimura A, Coy DH: Recent approaches to fertility control based on derivatives of LH-RH. *Vitamins Hormones* **38**:257, 1981.

31. McCann SM: Luteinizing-hormone-releasing hormone. *N Engl J Med* **296**:797, 1977.

32. Sarkar DK, Chiappa SA, Fink G, et al: Gonadotrophin-releasing hormone surge in pro-estrous rats. *Nature (Lond)* **264**:461, 1976.

33. Carmel PW, Araki S, Ferin M: Pituitary stalk portal blood collection in Rhesus monkeys: Evidence for pusatile release of gonadotropin-releasing hormone (LHRH). *Endocrinology* **99**:243, 1976.

34. Neill JD, Patton JM, Dailey RA, et al: Luteinizing hormone-releasing hormone (LHRH) in pituitary stalk blood of Rhesus monkeys: Relationship to level of LH release. *Endocrinology* **101**:430, 1977.

35. Sarkar DK, Fink G: Effects of gonadal steroids on output of luteinizing hormone releasing factor into pituitary stalk blood in female rats. *J Endocrinol* **80**:303, 1979.

36. Knobil E: On the control of gonadotropin secretion in the rhesus monkey. *Recent Progr Horm Res* **30**:1, 1974.

37. Ferin MPW, Carmel EA, Zimmerman M, et al: Location of intrahypothalamic estrogen-responsive sites influencing LH secretion in the female rhesus monkey. *Endocrinology* **95**:1059, 1974.

38. Nakai Y, Plant TM, Hess DL, et al: On the sites of the negative and positive feedback action of estradiol in the control of gonadotropin secretion in the rhesus monkey. *Endocrinology* **102**:1008, 1978.

39. Keye Jr WR, Jaffe RB: Modulation of pituitary gonadotropin-releasing hormone by estradiol. *J Clin Endocrinol Metab* **38**:805, 1974.

40. Yen SS, Lasley BL, Wang CF, et al: The operating characteristics of the hypothalamic-pituitary system during the menstrual cycle and observations of biological action of somatostatin. *Recent Progr Horm Res* **31**:321, 1975.

41. Thompson IE, Patton WC, Taymor ML: Modulation of pituitary responses to synthetic LH-RH by gonadal steroids in women with secondary amenorrhoea. *Acta Endocrinol (Kbh)* **82**:238, 1976.

42. Frager MS, Pieper DR, Tonetta S, et al: Pituitary gonadotropin-releasing hormone (GnRH) receptors: Effects of castration, steroid replacement, and the role of GnRH in modulating receptors in the rat. *J Clin Invest* **67**:615, 1981.

43. Clayton RN, Catt KJ: Regulation of pituitary gonadotropin-releasing hormone receptors by gonadal hormones. *Endocrinology* **108**:887, 1981.

44. Badger TM, Wilcox CE, Meyer ER, et al: Simultaneous changes in tissue and serum levels of luteinizing hormone, follicle-stimulating hormone and luteinizing hormone/follicle-stimulating hormone releasing factor after castration in the male rat. *Endocrinology* **102**:136, 1978.

45. Pieper DR, Gala RR, Regiani SR, et al: Dependence of pituitary gonadotropin-releasing hormone (GnRH) receptors on GnRH secretion from the hypothalamus. *Endocrinology* **110**:749, 1982.

46. Clayton RN, Channabasavaiah K, Stewart JM, et al: Hypothalamic regulation of pituitary gonadotropin-releasing hormone receptors: Effects of hypothalamic lesions and a gonadotropin-releasing hormone antagonist. *Endocrinology* **110**:1108, 1982.

47. Clayton RN: Gonadotropin-releasing hormone modulation of its own pituitary receptors: Evidence for biphasic regulation. *Endocrinology* **111**:152, 1982.

48. McCann SM, Moss RL: Putative neurotransmitters involved in discharging gonadotropin-releasing neurohormones and the action of LH-releasing hormone on the CNS. *Life Sci* **16**:833, 1975.

49. Knobil E: The neuroendocrine control of the menstrual cycle. *Recent Prog Horm Res* **36**:53, 1980.

50. Bremner WJ, Paulson CA: Two pools of luteinizing hormone in the human pituitary: Evidence from constant administration of luteinizing hormone-releasing hormone. *J Clin Endocrinol Metab* **39**:811, 1974.

51. Happ J, Neubauer M, Szamak F, et al: LH, FSH, and testosterone responses to prolonged infusion of GnRH in normal human males. *Horm Metab Res* **8**:249, 1976.

52. Rabin D, McNeil LW: Long-term therapy with luteinizing hormone-releasing hormone in isolated gonadotropin deficiency: Failure of therapeutic response. *J Clin Endocrinol Metab* **52**:557, 1981.

53. Rabin D, McNeill LW: Pituitary and gonadal desensitization after continuous luteinizing hormone-releasing hormone infusion in normal females. *J Clin Endocrinol Metab* **51**:873, 1980.

54. Heber D, Swerdloff RS: Down-regulation of pituitary gonadotropin secretion in postmenopausal females by continuous gonadotropin-releasing hormone administration. *J Clin Endocrinol Metab* **52**:171, 1981.

55. Clayton RN, Catt KJ: Gonadotropin-releasing hormone receptors: Characterization, physiological regulation, and relationship to reproductive function. *Endocr Rev* **2**:186, 1982.

56. Hammond CB, Ory SJ: Diagnostic and therapeutic uses of gonadotropin-releasing hormone. *Arch Intern Med* **145**:1690, 1985.
57. Sandow J: Clinical applications of LHRH and its analogues. *Clin Endocrinol* **18**:571, 1983.
58. Hsueh AJW, Jones PBC: Extrapituitary actions of GnRH. *Endocr Rev* **2**:437, 1981.
59. Kochman R, Kerdlehue B, Zor U, et al: Studies of enzymatic degradation of luteinizing hormone-releasing hormone by different tissues. *FEBS Lett* **50**:190, 1975.
60. Handelsman DJ, Jansen RPS, Boylan LM, et al: Pharmacokinetics of gonadotropin-releasing hormone: Comparison of subcutaneous and intravenous routes. *J Clin Endocrinol Metab* **59**:739, 1984.

12

Hypogonadotropic Hypogonadism

The term hypogonadotropic hypogonadism (HH) refers to the development of gonadal failure as a consequence of partial or complete absence of gonadotropin secretion. In the purest sense, failure to secrete luteinizing hormone (LH) and follicle-stimulating hormone (FSH) can result from intrinsic pituitary or hypothalamic disorders. In clinical practice and the literature, however, the term has come to signify a syndrome characterized by deficient secretion of gonadotropin-releasing hormone (GnRH) by the hypothalamus. When such a situation is associated with olfactory dysfunction, the term *Kallmann's syndrome*[1] is used to describe the entity. When olfactory function is intact, and no other intracranial pathology is evident, the term *idiopathic hypogonadotropic hypo-*

gonadism (IHH) is employed to define the syndrome. The term *isolated gonadotropin deficiency* is merely a generic term that emphasizes the integrity of the other adenohypophyseal hormones, whereas LH and FSH secretion are deficient.

Some general aspects of HH are comment-worthy:

1. The syndrome is characterized by extreme heterogeneity. Thus, the clinical and hormonal features of HH are impressively diverse. Consequently a spectrum exists in the clinical and endocrine profile of patients with HH. At one end of the spectrum are patients who are severely affected and who manifest complete hypogonadism, while at the other end are patients who present with postpubertal gonadal failure with regression of reproductive function. Genetic heterogeneity or variable expression of a single gene for hypogonadism and anosmia may account for the heterogeneous clinical and hormonal expressions.

2. The inheritance pattern of HH is also impressively variable. The family clustering of patients with HH was originally noted by Kallmann.[1] X-linked[2] as well as autosomal[3,4] modes of inheritance patterns have been described for HH.

3. In addition to central defects (in GnRH release), HH may be complicated by additional peripheral defects as well. Thus, as suggested by Bardin et al.,[5] in some male patients with Kallmann's syndrome both ends of the pituitary axis are abnormal. This leads to resistance of the Leydig cells alone, or of both the Leydig cells and germinal epithelium, to gonadotropins. Rarely, as has been described by Check et al.,[6] Leydig cell response to gonadotropins may be preserved, while germinal cell resistance may be encountered. The reason for such phenomena seen in association with Kallmann's syndrome is unclear.

4. Finally, the advent of extensive studies of gonadotropin secretory patterns in HH has revealed a variety of abnormal patterns.[7-9] These data, coupled with the physiological pulsatile nature of GnRH secretion,[10-12] have had an enormous impact on therapy for patients with HH.

Etiology

The etiology of HH—failure to generate GnRH in a proper fashion—can be viewed from three vantage points: genetic, anatomical, and hormonal.

Genetic Basis

Clearly, some forms of HH demonstrate a tendency to be inherited. In one extensive study of six kindreds of HH involving 30 affected patients, a positive family history was obtained in 35% of siblings.[3] Male-to-male transmission was also observed. It is currently believed that HH is inherited by an autosomal-dominant mode of inheritance with incomplete expression of the gene. The genes for anosmia and hypogonadism are variably expressed. Anosmia tends to be more common in males with Kallmann's syndrome than in females. The existence of anosmia alone in eugonadal relatives of patients with Kallmann's syndrome is well recognized. This contrasts with the rarity of hypogonadism with normal olfaction in relatives of patients with Kallmann's syndrome.[13] The inheritance pattern of HH has also been studied by Lieblich et al.,[14] whose data are in favor of an autosomal-recessive pattern of inheritance.

Anatomical Basis

It is believed that the gonadotropin failure seen in HH is a consequence of inadequate generation of GnRH by the hypothalamus. The association between hypothalamic dysfunction and olfactory disturbances is intriguing. The syndrome that goes by Kallmann's name was originally recognized in 1856 by the Spanish pathologist Maestre de San Juan,[15] who noted at autopsy, that the olfactory lobes were completely absent in a man with underdeveloped genitalia. Other workers, particularly Weidenreich[16] and Altman,[17] also described abnormal postmortem findings in the olfactory lobes of some patients with hypogonadism. The connection between the hypothalamic origin of the gonadal problem and the olfactory dysfunction was pointed out by the French physician De Morsier,[18,19] whose name is also sometimes used to denote the syndrome of olfactogenital dysplsia. More recently, Males et al.[20] demonstrated agenesis of the olfactory bulb in a patient with Kallmann's syndrome who underwent craniotomy. Attempts to connect the olfactory lobe agenesis and hypothalamic GnRH deficiency by a single embryological mechanism have been unconvincing. Similarly, there appear to be no physiological mechanisms to explain why GnRH deficiency coexists with olfactory lobe agenesis. The genes for olfactory lobe development and that for GnRH secretion must be identical, analogous, or in close proximity to each other on the same chromosome. The presence of such a genetic locus has not yet been fully established.

Hormonal Basis

The hormonal basis of HH lies at the hypothalamic level. Yet, hormonal data obtained in the early 1970s had provided contradictory, even puzzling results. If HH was a consequence of inadequate or absent secretion of GnRH, the administration of synthetic LHRH should be able to stimulate the intrinsically normal gonadotropes. While this is certainly true of some cases, the heterogeneity in the LH-FSH responses to a bolus of synthetic GnRH in patients with HH has been striking. Thus, the following combinations of responses to GnRH have been reported[7,9,21–23]: (1) responses of both LH and FSH to a single bolus of GnRH; (2) initial absence of LH, FSH response to

GnRH, which is restored after prolonged priming of gonadotropes with GnRH; (3) absent responses in LH-FSH before and after priming; (4) LH response alone without any response in FSH to GnRH administration; and (5) FSH response alone without a contemporaneous rise in LH following a single bolus of exogenous GnRH. Clearly, then, HH is not a single entity, and the patterns of gonadotropin response to a single bolus of LRH can resemble the hypothalamic or the pituitary pattern.

Two other observations have also been rather puzzling: first, the apparent paradox of gonadal failure in presence of normal or only minimally decreased levels of circulating gonadotropins[24]; and second, the apparent normalcy of LH, FSH responses to a bolus of (supraphysiological dose of) exogenous GnRH, in conjunction with clinically severe hypogonadism. Currently, five lines of observations suggest that the intrinsic problem in HH—heterogeneous as this disorder is—is a hypothalamic abnormality:

Aberrant secretion of GnRH in HH. Direct measurement of GnRH in the peripheral blood of patients has been impossible, since this peptide is mostly limited to the hypophyseal portal blood supply. Indirectly, however, the dynamics of GnRH can be assessed by carefully viewing the 24-hr secretory patterns of LH and FSH. The functional integrity of the hypothalamic pulse generator, a term used by Lincoln,[25] can be accurately estimated by 24-hr sampling of LH, FSH in the circulation. In animals, it has been shown that the pulses of LH and FSH are preceded temporally by bursts in GnRH secretion.[26,27] Therefore, it is reasonable to assume that a similar situation exists in humans. The normal secretory pattern of gonadotropins is characterized by the presence of well-defined LH pulses occurring at approximately 2-hr intervals, with intensification by sleep (sleep entrainment). In an elegantly performed study, Santoro et al.[24] reported the existence of several abnormal gonadotropin-secretory patterns in males with HH:

1. The overwhelming majority failed to show any detectable LH pulsations throughout the day. These patients were severely hypogonadal.
2. Some patients showed only sleep-related LH pulses, similar to the pubertal pattern. These patients had minimal pubertal development.
3. Some patients exhibited an abnormally low amplitude of LH pulsations in comparisons to normals.
4. Some patients showed a decrease in the frequency of LH pulses, demonstrating a slowing in the LH pulse frequency, often half that of their normal siblings. These patients, interestingly but as expected, had some degree of pubertal virilization and responded to the administration of clomiphene citrate. The use of this drug was attended with an augmentation effect, which elevated the GnRH pulse frequency of these patients to normals.
5. Rarely, patients with HH may demonstrate absolute normal-looking gonadotropin secretory patterns—both in terms of the amplitude and frequency—but the bioactivity of the secreted LH was demonstrably low. These studies indicate that a spectrum of abnormalities may be encountered in males with HH, ranging from absolute deficiency of

TABLE 66
Spectrum of Gonadotropin Secretory Patterns in Males with Hypogonadotropic Hypogonadism

1. Absent LH pulses throughout the day
2. Absent LH pulses through most of the day, with preservation of only the sleep-entrained LH rise
3. Preservation of pulsatile pattern but with peaks of abnormally low amplitude
4. Preservation of pulsatile LH pattern but with peaks of abnormally slow frequency
5. Preservation of pulsatile LH secretion of normal amplitude and frequency, but the secreted LH is biologically inactive

secretion to qualitative abnormalities in the LH-pulse generation (Table 66).

Santoro et al.[24] also noted a wide spectrum of secretory abnormalities of gonadotropins in females with HH. The most severe of these abnormalities is represented by the apulsatile pattern. This pattern is characterized by total lack of evidence for endogenous gonadotropin pulsation. Almost all patients with such a pattern had primary amenorrhea and demonstrated very little secretion of estradiol. A second subset of patients, like their male counterparts, demonstrated a pulsatile pattern, but the amplitude of the LH pulses was subnormal. A third group consisted of patients with a pulsatile pattern but with low-frequency pulse generation. This was the most frequent pattern in women with hypothalamic amenorrhea. The authors also described a fourth subset of patients with unclassified perturbations in which spontaneous recovery with restoration of normalcy was intermittently observed.

Thus, a very large body of evidence points to confirm the vast heterogeneity in gonatropin secretory activity among patients with HH. The more severe the hypogonadism, the more likely is one to encounter the apulsatile pattern with low to negligible LH (and FSH) levels throughout the day. The reversal of abnormal secretory patterns to a more normal pattern can be induced by pulsatile administration of GnRH exogenously. This observation forms the second major argument for a predominantly hypothalamic etiology in the development of HH.

Normalization of the abnormal pattern of gonadotropin secretion by pulsatile administration of GnRH. The administration of physiological small doses of synthetic GnRH in a pulsatile fashion by the use of portable peristaltic pumps has provided yet another unique opportunity to explore the basic etiology of HH. One of the accepted modes of delivery of GnRH is by the use of GnRH in doses of 25 µg/kg, administered as 2-hr pulses.[28] In most patients with HH, such therapy is associated with the appearance of pulsatile pattern of LH secretion in patients who previously demonstrated apulsatile patterns of gonadotropin secretion. Pulsatile administration of GnRH also corrects abnormal pulsatile patterns that were characterized by low amplitude or frequency. Barkan et al.[29] demonstrated that the gonadotropin responses to exogenous GnRH depended on the magnitude of the endogenous GnRH secretory defect. These workers differentiated those with partial defects from those with no evidence of GnRH secretion on the basis of presence or absence of spontaneous LH secretory pulses. They showed that continuation of pulsatile therapy for a 4-day period resulted in significant augmentation of LH secretion in patients with partial defects, while causing only minimal augmentation in patients with no endogenous GnRH secretion. It has also been noted that the frequency of pulse administration of GnRH can influence the 24-hr secretory pattern of gonadotropins in patients with HH. In one study,[30] complete normalization of hormone dynamics was attained when the GnRH was administered in pulses 45 min apart. The heterogeneity in gonadotropin responses to LHRH may be a reflection of differences in endogenous GnRH deficiency, as originally proposed by Boyar et al.[7]

General lack of response to clomiphene citrate in LH. Clomiphene citrate stimulates LH and FSH by stimulating endogenous GnRH. It has been repeatedly illustrated in the literature that patients with HH are usually unresponsive to the administration of clomiphene.[5,31-34] Hamilton et al.[35] indicated that while most patients with anosmia and HH are unresponsive to clomiphene, some patients with normal or subnormal (but present) olfactory sense may respond to a 4-week course of clomiphene. These patients probably represent those with some endogenous GnRH secretion, which permitted augmentation with clomiphene. Santoro et al.[24] also noted that a subset of patients with HH— those with qualitatively abnormal pulsatile secretion of gonadotropins—may respond to clomiphene. Notwithstanding the rare exceptions,[36] most patients with apulsatile gonadotropin secretion fail to respond to clomiphene, indicating a primary hypothalamic defect.

The animal model for HH (hpg mouse). GnRH deficiency is known to cause hypogonadism in certain mutant mice.[37] In these mice, GnRH deficiency has been confirmed by direct measurements of GnRH. The similarities in the hormonal dynamics, as well as in the histology of gonadotropes[38,39] between humans and this strain of affected mice leads one to believe that the disease is identical in both species. Charlton et al.[40] showed that exogenous administration of GnRH results in correction of the hypogonadism in this strain of mice with GnRH deficiency. Studies in the hpg mouse have supported the tenet that HH in humans is probably also a consequence of deficient GnRH secretion by the hypothalamus.

Therapeutic response to GnRH administration. Finally, the demonstration that some patients with HH can be "cured" by replacement therapy with exogenous GnRH (LRH) establishes the missing link in the etiology of the disorder. The impact of such therapy will be discussed under treatment.

In addition to deficient secretion of GnRH by the hypothalamus, hypogonadotropic hypogonadism can result from intrinsic pituitary disease. Three variations on such a theme deserve emphasis—isolated gonadotropin deficiency resulting from tumors, isolated FSH deficiency alone, and isolated gonadotrope failure occurring in association with the autoimmune pluriglandular syndrome.

Loss of gonadotropin reserve is often an early feature of tumorous hypopituitarism. The importance of excluding tumorous etiology in all patients with hypogonadotropic hypogonadism cannot be overemphasized. This is usually attainable by computed tomography (CT) of the sella and suprasellar region.

The rare syndrome of isolated FSH (but not LH) deficiency has been recognized in the literature.[41-43] These patients present with primary amenorrhea, low FSH levels unresponsive to exogenous GnRH administration, and a normal or elevated LH level. The latter is seen as a consequence of the integrity in the GnRH—LH axis, which responds to the low estradiol levels. The basic defect in these patients resides at the level of the pituitary gonadotropies, which are able to synthesize LH but not FSH. Since administration of GnRH to such patients is capable of eliciting a rise in the free α-subunit level, it is presumed that the intrinsic defect represents an inability to synthesize the β-subunit of FSH.[42,43] The reverse phenomenon, that is, the ability to synthesize only FSH but not LH, theoretically can lead to the rare syndrome of isolated LH deficiency. When such a syndrome occurs in males, the term *fertile eunuch* has been used to denote the resultant hypogonadal state. The fertility potential of a markedly undervirilized male is open to question, especially in light of the observation that adequate spermatogenesis depends on the availability of at least some testosterone. Isolated unihormonal gonadotropin deficiency represents errors in the enzymatic machinery within the gonadotropes, resulting in an inability to synthesize the intact FSH or LH.

The third, and latest, variant of isolated gonadotropin failure of pituitary etiology is gonadotrope involvement by autoimmune processes. The usual gonadal failure seen in association with pluriglandular autoimmune syndromes involves primary gonadal disease caused by autoantibodies. However, Barkan et al.[44] recently described two males with polyglandular autoimmune syndrome with isolated gonadotropin deficiency acquired after the onset of puberty. The endocrine profile of these patients was characterized by low LH and FSH levels, subresponsive to bolus as well pulsatile administration of GnRH, with preservation of the trophic hormone reserve of TSH, ACTH, GH, and prolactin. The authors suggested that selective gonadotropin deficiency can occur as a part of autoimmune hypophysitis.

To summarize, the etiology of HH, in most cases, resides at the level of the hypothalamus. The etiology is possibly multifactorial; the precise characterization of the basic defect in HH is currently unclear. The two notable features of HH are the remarkable heterogeneity of expression and the tendency for inheriting the disorder. When a pituitary etiology underlies the HH, tumors dominate the causative mechanism. Rare and recent examples of additional mechanisms include impaired synthetic machinery of the gonadotropes and autoimmune disease. Regardless of the mechanism, the clinical expression of the syndrome is hypogonadism.

Clinical Features

The clinical features of HH can be viewed from two perspectives: the hypogonadism and the associated somatic abnormalities seen in the syndrome.

Hypogonadism

The hypogonadal state that characterized HH is highly variable. In the prototypical variety of HH, which represents the complete expression, the clinical features are those of prepubertal hypogonadism.[45] The situation is a classic instance of puberty that never arrives. These patients are usually evaluated for delayed puberty, but much to the patient's (and parents') concern, puberty never arrives. The physical examination is characterized by marked hypogonadism. In males, the appearance is boyish, often with eunuchoidal skeletal proportions and small, soft prepubertal testes. In females, the constellation of sexual infantilism is represented by primary amenorrhea, poor or absent breast development, and prepubertal external genitalia often with eunuchoidal skeletal proportions. Thus, in both sexes severe retardation of sexual development is the presenting feature, when the expression is complete.

It is now well recognized that considerable heterogeneity exists in the clinical expression of HH. Santoro et al.[24] pointed out that females with HH may present with normal pubertal feminization followed by regression of reproductive function. It is not clear whether the entity called hypothalamic amenorrhea represents the milder end in the spectrum of hypogonadotropic hypogonadism. It has also been suggested that the spontaneous waxing and waning of hypothalamic dysfunction seen in females with mild HH in fact represents periodic reactivation of the hypothalamic program of endogenous GnRH secretion.[24]

Milder degrees of GnRH deficiency can also be encountered in males with HH. However, in contrast to females with HH, complete pubertal virilization is uncommon in males with HH, most of these patients demonstrating suboptimal pubertal virilization. Cryptorchidism often complicates HH.

Somatic Anomalies

In addition to anosmia, several somatic anomalies have been described in association with HH in both sexes. The presence of anosmia in a hypogonadal patient has enormous diagnostic impact, pointing towards HH as the etiology of hypogonadism. The anosmia in these patients may be easily missed since they do not complain about this. A more characteristic statement is that "food tastes flat." While testing for olfactory function, care should be taken not to use irritants, which are sensed by the trigeminal nerve. Testing with phenylethyl methyl ethyl carbinol (PEMEC) is the recommended stimulus for testing olfactory nerve function in all patients suspected of having Kallmann's syndrome.[46]

Midline craniofacial defects are often seen in patients with Kallmann's syndrome. The most frequent of these are cleft lip and cleft palate, which are seen in these patients with a much greater frequency than in the general population. Other somatic anomalies[1,5,14,18,47–53] involving the palate, hands, kidneys, the cardiovascular system, hearing, and vision are seen less frequently (see Table 67).

TABLE 67
Somatic Anomalies in Kallmann's Syndrome

Anosmia[1,18]
Cleft lip, cleft palate[14,20,47]
High-arched palate[14]
Abnormal metacarpals[14]
  Brachydactyly
  Camptodactyly
  Clinodactyly
Renal agenesis[48]
Cardiovascular
  Atrial septal defect[49]
  Epstein's anomaly[50]
  Conduction disturbances[51,52]
  Anomalous right aortic arch[53]
Deafness[5]
Color blindness[5]

Laboratory Diagnosis

The methodical approach in the diagnostic evaluation of patients with HH consists of five steps.

Measurement of Gonadotropin Levels

This initial step is to document the hypogonadotropic nature of the hypogonadism. Basal levels of LH and FSH are usually low, occasionally undetectable, in patients with HH. The inadequacy of the gonadotropin levels would become more apparent when viewed in the context of the low levels of circulating testosterone or 17β-estradiol in the plasma of these patients. Rarely, the basal gonadotropin level is normal. These are patients with milder expression of the disease. In most cases, however, measurement of basal LH, FSH levels is generally quite helpful in excluding primary gonadal failure and pointing in the direction of the hypothalamic–pituitary axis. 24 hour sampling of blood to obtain the secretory pattern of gonadotropins is impractical as is the determination of the sleep-entrained LH release. These procedures are limited to specialized centers, and while such studies have had an enormous impact on the understanding of HH, they are cumbersome and difficult to perform.

Evaluating the Response of LH, FSH to Exogenous Administration of GnRH

The gonadotropin response to exogenous administration of GnRH is impressively heterogenous. As early as 1973, Bell et al.[9] recognized this, and reported four types of responses to GnRH in patients with HH: an increase in both LH, and FSH, absent responses of both LH and FSH, a rise in LH only or a rise in FSH only. In patients who fail to show a rise in gonadotropins with exogenous GnRH, priming the gonadotropes with multiple doses of GnRH results in restoration of responsiveness to bolus doses of GnRH administration.[54,55] The duration of priming required is highly variable. For instance, Zarate et al.[56] studied the response to LH,FSH to synthetic LHRH in two brothers with HH, cryptorchism, and anosmia. Although the rise in LH, FSH was significant, it was subnormal. The magnitude of response after 5 days of intramuscular administration of LH-RH was no different from the initial response. The authors postulated that the responses may have been different had the LHRH been given more frequently, in larger doses or for longer periods. The latter was achieved in the studies by Yoshimoto et al.[55] They studied the LH FSH responses to 400-μg boluses of synthetic LH-RH before and after priming for 23 days. Of the nine patients with HH studied, restoration of the initially impaired responses to normal was achieved in five. The authors noted that at least 5 days of priming was necessary for optimizing the gonadotropin responses to synthetic GnRH.

A novel addition for evaluating the responses of LH, FSH to bolus GnRH involves priming by pulsatile administration of GnRH. This involves administration of GnRH by a pump as 90–120-min pulses for variable lengths of time and evaluating the gonadotropin response to bolus GnRH after such priming.

It is important to emphasize that the LH, FSH response to a single bolus of synthetic GnRH does not permit separation between the hypothalamic and pituitary varieties of disease. When synthetic GnRH (LH-RH) became available in the early 1970s, it was hoped that this diagnostic tool would help delineate between the hypothalamic and pituitary etiologies of hypogonadotropism. Unfortunately, this has not proved to be true. Several studies have clearly shown that the plasma LH responses to a single bolus of LH-RH in patients with HH are either absent or at best only minimal in nearly 90% of cases.[9,55–61] Furthermore, patients with hypopituitarism caused by Sheehan's syndrome have been reported as being capable of significantly (but subnormally) raising their LH levels after a bolus of synthetic LHRH.[55] Thus, a single bolus of GnRH administration does not provide separation between patients with hypothalamic versus pituitary disease; nor does it separate patients with constitutional delayed puberty, an important concern in the differential diagnosis.

Exclusion of Coexistent Endocrine Deficiencies

In the strictest sense, patients with HH exhibit an isolated or selective deficiency of gonadotropin secretion. Pituitary and hypothalamic functions are otherwise intact.[31,62] However, there have been reports of associated perturbations in other anterior pituitary and neurohypophyseal hormones as well. Three in particular deserve mention: prolactin, TSH, and AVP.

Yamaji et al.[63] studied the prolactin and TSH response to 500 μg IV

TRH in 15 males with HH and observed a significant blunting of prolactin response in eight and an attenuation of TSH response to TRH in two.

More recently, Spitz et al.[64] reported that the maximal responses of prolactin to TRH was impaired in all but one patient with HH. Those workers even suggested that the prolactin responses to TRH may serve as a useful parameter to distinguish between patients with HH and constitutional delayed puberty. As we shall see, this has been disputed by others. Rarely, the prolactin response to chlorpromazine may also be blunted.[14] Turksoy[65] described a female patient with Kallmann's syndrome who demonstrated a dissociation between the prolactin responsiveness to TRH and chlorpromazine.

Rarely, there have been reports of blunted TSH reserve,[63,14] GH reserve,[23,66] and impaired ACTH responses to metyrapone administration.[20] Males et al.[20] noted a subnormal rise in urinary 17-OHCS levels following metyrapone in two of six patients with Kallmann's syndrome. It is difficult to separate the effects of the low sex steroids on pituitary hormone reserve while testing and interpreting the latter. Low testosterone or estradiol levels can considerably affect the responses of GH, prolactin and TSH to their respective provocative stimuli. Only in the case of prolactin has it been shown conclusively that abnormal prolactin dynamics persist even after treating such patients with sex steroids.

Finally, since Kallmann's syndrome is believed to originate at the hypothalamic level, it is inevitable that abnormal osmoreceptor function would be described sooner or later in association with the syndrome. Hochberg et al.[67] reported osmoreceptor dysfunction and abnormal thirst regulation in two of six patients with Kallmann's syndrome; while these patients were able to respond to water deprivation, the threshold of plasma osmolality needed to release AVP was higher than normal. Abnormal thirst regulation was also encountered in some patients with Kallmann's syndrome.

Thus, in the classic sense, HH may not be as isolated as was once presumed. However, the associated hormonal perturbations are more biochemical than clinical.

Exclusion of Tumorous Lesions in the Sella or Suprasellar Region

Since tumors in and above the pituitary represent an important cause for acquired hypogonadotropic hypogonadism, these have to be excluded by CT radiology of the sella and suprasellar region. In children with delayed or absent puberty, exclusion of craniopharyngioma is mandatory.

Evaluation of Gonadal Responsiveness to hCG

Rarely, patients with HH may exhibit dual defects: one central and one peripheral. This problem is particularly seen in males with HH. The testicular unresponsiveness to gonadotropins may involve Leydig cell unresponsiveness, germinal cell unresponsiveness, or both. In a review by Lieblich et al.[14] of 23 patients with Kallmann's syndrome, no patient with gonadotropin unresponsiveness was encountered. Rare as the phenomenon may be, it is important to exclude this entity, since in its presence therapy with gonadotropins (or GnRH) will not be successful in stimulating testosterone production. The sensitivity of the gonads to gonadotropins can be evaluated by measuring testosterone responses to short-term hCG administration; 5000 units of hCG IM for 4 days. Check et al.[6] pointed out that demonstration of Leydig cell responsiveness to hCG administration does not automatically establish sensitivity of the germinal epithelium to FSH (or human menopausal gonadotropins). These workers described a hCG responsive patient with Kallmann's syndrome who was able to attain sexual maturation with hCG therapy, but remained persistently azoospermic despite hMG therapy. Testicular biopsy in this patient revealed Leydig cell hyperplasia but very little spermatogenesis. Thus, hCG sensitivity does not always ensure fertility.

Differential Diagnosis

Hypogonadotropic hypogonadism must be differentiated from constitutional delayed puberty and form other hypogonadal states characterized by sexual retardation.

Constitutional Delay of Puberty

Constitutional delay of puberty is caused by retarded activation of the hypothalamo-pituitary-gonadal axis beyond the age of 14. In normal children, the pubertal phenomena are set in motion by the age of 14 or 15 yrs. In contrast, patients with constitutional delay of puberty experience an unduly prolonged prepubertal stage. The arrival of normal puberty is heralded by a number of endocrine events. The five events that characterize the hormonal milieu of normal puberty are (1) the release of the hypothalamus from its exquisitive sensistivity to suppression by circulating sex steroids; (2) pulsatile release of GnRH from the hypothalamus[68]; (3) pulsatile gonadotropin release during sleep[69] (sleep entrainment); (4) elevation of serum gonadotropins and restoration of gonadotrope sensitivity to exogenous administration of GnRH[70,71]; and (5) elevation of serum levels of gonadal steroids and restoration of gonadal sensitivity to gonadotropins.

It is clear that the earliest hormonal change that serves as a harbinger of the soon-to-develop puberty is the pulsatile secretion of GnRH by the hypothalamus. For reasons unclear, the maturation of the hypothalamic signals that cause pulsatile release of GnRH are delayed in subjects with constitutional delay of puberty. Of course, in patients with HH, the hypothalamus partially or completely fails to secrete GnRH on a more permanent basis. Thus, at a given point in time, when a teenager presents with no signs of pubertal development, it is difficult on clinical grounds, to distinguish between a simply delayed puberty versus HH. When olfactory disturbances are present, or a family history of HH is obtained, the diagnosis of Kallmann's syndrome is easily predictable. In the absence of such features the distinction between these two entities; that is, HH versus constitutional delay of puberty can be extremely difficult, sometimes impossible.

Several tests have been suggested, and discarded, in the differentiation between HH and constitutional delay of puberty. Measurement of basal levels of LH and FSH does not differentiate between the two. The responses of LH, FSH to a bolus of GnRH are highly variable in both entities and do not permit clear differentiation.[72,73] Similarly, the gonadotropin response to clomiphene also fails to consistently differentiate patients with constitutional delay of puberty from HH. In 1983, Spitz et al.[64] suggested that the prolactin response to TRH is nearly always attenuated in patients with HH and can serve to distinguish these patients from those with constitutional delay of puberty, who show normal prolactin responsiveness to TRH. This notion has been seriously contested in the literature. Moshang et al.[74] studied the prolactin responses to TRH in 13 teenage boys with HH and compared these with the results obtained in 14 teenage boys with constitutional delay of puberty. They observed considerable overlap in the prolactin responses to TRH between the two groups. Similarly, Hawthorne et al.[75] were unable to differentiate between the two groups on the basis of their prolactin responses to TRH.

A new dimension has been added to the difficult problem of differentiating between these two conditions in a study reported by Partsch et al.[76] These workers presented data that enable one to distinguish between HH and constitutional delay of puberty by first priming the pituitary with GnRH and then evaluating the gonadotropin response to a bolus of GnRH. The key factor in their study was to administer GnRH in a pulsatile fashion (5 μg IV every 90 min by a portable pump) for a short period (36 hr); the incremental response of LH to a bolus of GnRH, administered after priming, clearly separated subjects with delayed puberty (who showed significant rises in LH) from patients with HH who showed little or no response to the bolus GnRH post priming. Their results are indicative of the fact that the pituitary LH reserve is lower in patients with HH after short-term priming by pulsatile GnRH administration. If confirmed in larger studies, the evaluation of the LH response to a bolus of GnRH following short-term (36-hr) priming with pulsatile GnRH administration may prove to be a superior test to differentiate patients with HH from those with constitutionally delayed puberty. Despite all the available tests, constitutional delay of puberty is often diagnosed by prolonged observation; if puberty arrives, the diagnosis of constitutional delay in puberty can be made retrospectively. Usually, delayed puberty is not delayed beyond 20 years of age.

Differentiation from Other Hypogonadal States

Hypogonadotropic hypogonadism in males needs to be distinguished from Klinefelter's syndrome and other syndromes caused by primary gonadal failure. The distinction can be readily made on clinical and laboratory grounds.

Gynecomastia is less pronounced in HH, as is mental retardation. When anosmia accompanies the sexual infantilism, the diagnosis of Kallmann's syndrome is unmistakable. Measurement of basal LH, FSH levels clearly differentiates HH from primary gonadal failure, where the gonadotropin levels are elevated (hypergonadotropic hypogonadism).

When cryptorchism complicates HH, the differential diagnosis revolves around the vanishing testes syndrome, also known as functional prepubertal castrate. The two differentiating features are the low gonadotropin levels and the preservation of gonadal responsiveness to exogenous hCG in HH. By contrast, patients with vanishing testes syndrome demonstrate high basal gonadotropin levels, and no gonadal response to hCG; both findings reflect the anorchic state.

In females with HH, the differential diagnosis revolves around other disorders presenting with sexual infantilism. Thus, the Turner variants with normal stature (those with isochromosome of short arm and deletion of the long arm), as well as phenotypic females with gonadal dysgenesis (46 XY or 46 XX), are the two important groups of patients that need to be distinguished from females with HH. The basal gonadotropin levels are elevated in these patients with primary gonadal failure. This, coupled when necessary with chromosomal studies, usually helps differentiate Turner variants and gonadal dysgenesis from HH.

Treatment

The treatment of hypogonadotropic hypogonadism is aimed at inducing pubertal virilization or feminization, as the case may be and restoring fertility. These dual goals are attainable by the administration of gonadotropins or by the pulsatile delivery of synthethic GnRH. Obviously, both forms of therapy are protracted. The use of androgens or estrogens, although most effective in inducing pubertal changes, will not restore fertility.

Gonadotropin Therapy

For the past two decades, the availability of two hormonal drugs—hCG and human menopausal gonadotrophin (hMG)—have provided excellent means for the treatment of hypogonadotropic hypogonadism. hCG simulates the biological effects of LH, while hMG mimics the biological effects of FSH. The use of therapy with hCG alone[77,78] or in conjunction with hMG[79–81] has resulted in successful virilization and fertility in males with HH. HCG therapy has held a track record for successful induction of virilization in males and feminization of females with HH.[82–85] In the past it was accepted that addition of hMG was deemed absolutely necessary for the initiation and maintenance of spermatogenesis. However, this concept has changed in recent years. Finkel et al.[86] evaluated the efficacy of gonadotropin therapy in stimulating spermatogenesis in 21 males with hypogonadotropic hypogonadism. The interesting observation that emerged from their studies were that the need for addition of human menopausal gonadotropin to hCG depended on the time of onset of hypogonadism; when the hypogonadism had its onset postpubertally administration of hCG alone was adequate to induce spermatogenesis, while patients with prepubertal onset of hypogonadism required addition of hMG to hCG for induction of spermatogenesis. This is in keeping with the concept that FSH is required mostly for initiating the first wave of spermatogenesis, following which LH alone can maintain adequate function of the seminiferous tubules. In this study,[86] two other observations were made. First, the pretreatment level of FSH did not predict the need for requiring hMG to initiate spermatogenesis and second, that patients with cryptorchism usually failed to have adequate spermatogenesis despite the use of both hCG and hMG.

Since hCG therapy needs to be given for protracted periods of time to effectively induce pubertal virilization, patients must be advised regarding the long wait that is entailed in such therapy. Ley et al.[87] showed that prior use of androgens does not adversely affect the success rate of subsequent treatment with hCG, or hMG–hCG combination in inducing spermatogenesis. The commonly held impression that prepubertal androgen therapy to males with HH forever damages fertility potential may have to be reconsidered in light of recent data. It has been recommended that such men can be safely virilized by the use of testosterone esters until they desire fertility. When fertility is desired, the testosterone can be stopped, and hCG is given alone for 1 year or so, until testicular volume reaches a plateau; at this stage, hMG is added to the regimen.[87] The advantages to this regimen are the low cost and the dramatic and quick virilization attained by the use of testosterone esters.

Gonadotropin therapy is expensive, and has to be administered three times per week for optimal results for a period ranging from 3 to 24 months. While the individual dose is variable, 2000 to 4000 IU of hCG are usually required for adequate virilization.

GnRH Therapy

The characterization and synthesis of GnRH (LH-RH) in the early 1970s had fostered hope for a new form of replacement therapy for patients with hypothalamic hypogonadism. Unfortunately, this hope failed to materialize in the early years following the availability of synthetic GnRH. The major reason for failure was not realizing that endogenous GnRH was secreted by the hypothalamus in a pulsatile fashion and that exogenous administration of GnRH on a continuous basis was counterproductive because it "desensitized" the gonadotropes. Belchetz et al.[88] demonstrated that administration of LHRH on an intermittent pulsatile fashion caused sustained secretion of LH and FSH, while continuous intravenous administration of the peptide inhibited the secretion of both the gonadotropins, rendering the pituitary refractory. The implication, therefore, to the clinician are that depending on the method of administration synthetic GnRH can either stimulate or inhibit gonadotropin secretion by the anterior pituitary.

In 1980, Knobil[89] further confirmed the pulsatile nature of GnRH secretion. Using a primate model he showed that replacement of GnRH by the pulsatile administration of the peptide activated the physiological episodic secretion of LH and FSH in the rhesus monkey. Shortly thereafter, several workers reported the successful use of pulsatile administration of LHRH to patients with hypogonadotropic hypogonadism. Valk et al.[28] demonstrated that the infusion of small amounts of LHRH every 2 hr to patients with HH for 5 days restored pulsatile gonadotropin secretion. It became evident that properly used, GnRH treatment for patients with HH was an effective means to achieve the triple objectives: induction of puberty, induction and maintenance of spermatogenesis, and induction of ovulation.

Successful induction of puberty by the pulsatile administration of GnRH has been reported by Hoffman and Crowley.[90] These workers treated six males with HH by using long-term low-dose subcutaneous GnRH in an episodic fashion by a portable infusion pump. The therapeutic outcome was clinically characterized by the occurrence of spontaneous erections, increase in testicular size, nocturnal emission, breast tenderness and the development of pubertal changes. The hormonal changes included elevations in the gonadotropin and testosterone concentrations, which were noticeable as early as 14 days after initiation of therapy. Spermatogenesis was achieved in three patients by the forty-third week of therapy. These observations were extended to eight more patients successfully treated and reported by the same investigators.[91] It is noteworthy that discontinuation of therapy resulted in return of the pre-treatment state with expression of the clinical and biochemical features of hypogonadotropic hypogonadism. Several investigators have reported successful induction of puberty by the pulsatile administration of GnRH to males with HH.[92-96] Successful induction of spermatogenesis is an inherent part of the therapeutic outcome following GnRH therapy. This has been aptly documented from the reports by Hoffman and Crowley[90] as well as others.[97]

In females with HH, pulsatile GnRH therapy has been used to induce pubertal changes and ovulation successfully.[98-100] Crowley and McArthur[98] demonstrated the occurrence of ovulation in a patient with Kallmann's syndrome who was treated with LHRH administered as 1.5-μg pulses every 120 min SC with a portable pump. The precise doses required to induce ovulation differ considerably among individual patients. For instance, Leyendecker et al.[99] noted that doses that were highly successful in inducing ovulation in patients with hypothalamic amenorrhea were not adequate to induce follicular maturation in females with Kallmann's syndrome. It appears that the doses employed range from as low as 1–5 μg per pulse[101] to as high as 20–100 μg per pulse. As in males, the treatment period is protracted, requiring several months to initiate pubertal changes and ovulation. Nevertheless, the results of pulsatile administration of LHRH to women with HH are highly encouraging, sometimes even culminating in pregnancy.[101] However, the results last only as long as the therapy is continued. When pulsatile GnRH therapy is discontinued there is prompt regression of the clinical and hormonal status to the pretreatment state.

The efficacy of GnRH therapy in the treatment of hypogonadotropic hypogonadism is now a well established fact. Pulsatile GnRH therapy to patients with HH has served as a fascinating model for the study of pubertal physiology. In a remarkable account, Seibel et al.[102] outlined the events surrounding the induction of puberty in a female with Kallman's syndrome treated for 230 days by the pulsatile administration of GnRH via the subcutaneous route. These workers showed that pulsatile administration of GnRH to a patient who presented with total lack of sexual development resulted in growth of pubic hair, breast development and cyclic changes in the ovarian diameter (by vetrasonographic monitoring) and doubling in the length of the uterine fundus concurrent with these changes, the authors noted cyclic changes in LH and estradiol along with the establishment of sleep entrainment for prolactin, but not for LH. It was also noted that widening of the pulse interval affected the LH (but not FSH) levels. The anatomic, physiologic and hormonal changes seen in the patient antedated the occurrence of menarche.

The use of pulsatile GnRH therapy is not without limitations. The practical problems of using a programmable pump are self evident. While the route of administration can be either intravenous or subcutaneous, most workers opt for the subcutaneous route. The computerized pump has to be carried throughout the treatment period, which can be as long as 2 years. Thus, patient compliance is a cardinal factor in the efficacy of GnRH therapy. Ultra frequent monitoring of clinical and hormonal responses necessitate numerous physician visits. Inevitably, the cost of GnRH pump therapy is a highly expensive one. The two adverse effects of therapy are the formation of antibodies and the development of allergic reactions. Van Loon and Brown[103] reported the occurrence of secondary drug failure during chronic therapy with LHRH, due to the formation of antibodies against the peptide. The potential for allergic reaction to GnRH is illustrated by the patient reported by Seibel et al.,[102] in whom the treatment, which had worked extremely well, had to be terminated due to the development of allergic reactions. The advantages and disadvantages of chronic GnRH therapy are outlined in Table 68. There have been no systematic studies that have compared the cost effectiveness of gonadotropin therapy with pulsatile GnRH therapy. It is generally felt that pulsatile GnRH therapy is more successful than gonadotropin therapy for the induction of spermatogenesis. It is also believed that induction of puberty can be achieved faster with hCG than with pulsatile GnRH therapy.

Some of the problems encountered in the use of GnRH pump therapy have been circumvented by combining GnRH therapy with other modalities. Klingmuller et al.[104] proposed initiating therapy with hCG for a year or so and following this with the pulsatile subcutaneous administration of GnRH for 3–4 months to induce spermatogenesis. Once spermatogenesis has occurred, intranasal GnRH may be substituted instead of pulsatile pump delivery. These workers were able to show that intranasal administration of GnRH (200 μg every 2 hr for 8 doses per day) effectively maintained normal sper-

TABLE 68
Pulsatile GnRH Therapy for Hypogonadotropic Hypogonadism

Advantages	Disadvantages
Represents the most physiological form of replacement therapy for HH	Requires use of a programmable pump throughout treatment
When administered in a pulsatile fashion, consistently stimulates gonadotropin secretion	Dosage and intervals of pulses need to be individually determined
	Cost often prohibitive
High success rate of inducing puberty, ovulation, and spermatogenesis	Development of pubertal changes a slow process, especially in females
	Development of antibodies to GnRH may limit response
	Allergic reactions a potential problem
	Superspecialist familiar with GnRH pump therapy required

matogenesis in the absence of any other treatment. This novel form of therapy may be used to maintain the effects of the GnRH without the need to use a programmable pump. It should be pointed out that initiation of puberty and stimulation of Leydig cells cannot be induced by nasal LHRH.[105] The use of intranasal LHRH as maintenance therapy, attractive as the notion may be, has not been substantiated in larger studies.

Thus, the clinician faced with treating the patient with HH has three options to consider:

1. To initiate puberty with gonadotropins: Treatment is started with hCG, which per se can stimulate spermatogenesis in some patients. When needed, hCG can be combined with hMG. This treatment is protracted and the success rate of inducing ovulation and spermatogenesis is not universal.

2. To initiate therapy with pulsatile subcutaneous GnRH administration by a programmable pump: This treatment requires the services of specialized medical personnel with experience in GnRH pump therapy. The success rate of inducing puberty, ovulation and spermatogenesis are quite good, even excellent, but the process is slow with the limitations of carrying a pump around. The therapy is also costly.

3. To start replacement with sex steroids (androgens or estrogens) for rapid and dramatic attainment of secondary sexual characteristics: When fertility is desired, the treatment can be terminated and the patient can either be placed on gonadotropin therapy or GnRH therapy. The concept that initiating therapy with androgens or estrogens does not cause irreversible harm to the reproductive potential is becoming an established one. Many patients are likely to accept this choice for two reasons: first, the attainment of secondary sexual characteristics within a reasonable period of time is virtually guaranteed with testosterone or estrogens; and second, the patient has the option

to try the more complicated and costly drugs when fertility is desired at a subsequent date.

The choice of therapy depends on individual preference, the modalities available to the physician, age of the patient (and consequently the urgency with which induction of puberty is required), compliance and cost factors, as well as the experience of the physician. Obviously, several factors need to be taken into consideration when planning therapy for patients with hypogonadotropic hypogonadism.

References

1. Kallman FJ, Schoenfeld WA, Barrera SE: The genetic aspects of primary eunuchoidism. *Am J Ment Defic* **48**:203, 1944.
2. Turner RC, Bobrow M, Bobrow LG, et al: Cryptorchidism in a family with Kallmann's syndrome. *Proc R Soc Med* **67**:33, 1974.
3. Merriam GR, Beitins IZ, Bode HH: Father-to-son transmission of hypogonadism with anosmia. *Am J Dis Child* **131**:1216, 1977.
4. Santen RJ, Paulsen CA: Hypogonadotropic eunuchoidism. I. Clinical study of the mode of inheritance. *J Clin Endocrinol Metab* **36**:47, 1973.
5. Bardin CW, Rose GT, Rifkind DB, et al: Studies of the pituitary–Leydig cell axis in young men with hypogonadotropic hypogonadism and hyposmia. Comparison with normal men, prepubertal boys and hypopituitary patients. *J Clin Invest* **48**:2046, 1969.
6. Check JH, Caro JF, Criden L, et al: Leydig cell responsiveness with germinal cell resistance to gonadotropin therapy in Kallman's syndrome. *Am J Med* **67**:495, 1979.
7. Boyar RM, Wu RHK, Kapen S, et al: Clinical and laboratory heterogeneity in idiopathic hypogonadotropic hypogonadism. *J Clin Endocrinol Metab* **43**:1268, 1976.
8. Kletzky OA, Davajan V, Nakamura RM, et al: Classification of secondary amenorrhea based on distinct hormonal patterns. *J Clin Endocrinol Metab* **41**:660, 1975.
9. Bell J, Spitz I, Slonim A, et al: Heterogeneity of gonadotropin response to LHRH in hypogonadotropic hypogonadism. *J Clin Endocrinol Metab* **36**:791, 1973.
10. Santen RJ, Bardin CW: Episodic luteinizing hormone secretion in man. Pulse analysis, clinical interpretation, physiologic mechanisms. *J Clin Invest* **52**:2617, 1973.
11. Merriam GR, Wachter KW: Algorithms for the study of episodic hormone secretion. *Am J Physiol* **243**:E310, 1982.
12. Crowley WF Jr, Filicori M, Spratt D, et al: The physiology of gonadotropin-releasing hormone (GnRH) secretion in men and women. *Recent Prog Horm Res* **41**:473, 1985.
13. Dornan J, Barnard JM, Faird NR: Lack of close linkage of hypogonadotropic hypogonadism with HLA. *Tissue Antigens* **15**:510, 1980.
14. Lieblich JM, Rogol AD, White BJ, et al: Syndrome of anosmia with hypogonadotropic hypogonadism (Kallmann syndrome). *Am J Med* **73**:506, 1982.
15. Maestre de San Juan A: Teratologia: falta total de los nervios olfactorios con anosmia en un individuo en quien existia un atrofia congenita de los testiculos y miembro virl. *El Siglo Medico* **3**:211, 1856.
16. Weidenreich F: Uber partiellen Riechlappendefect und Eunuchoidismus beim Menschen. *Z Morphol Antropol* **18**:157, 1914.
17. Altmann F: Uber Eunuchoidismus. *Virchows Arch [Pathol Anat]* **276**:455, 1930.
18. De Morsier G: Etudes sur les dysraphies cranio-encéphaliques. 1. Agenesie des lobes olfactifs (telencephaloschizis lateral) et des commissures calleuse et anterieure (telencephaloschizis median): la dysplasie olfacto-génitale. *Schweiz Arch Neurol Neurochir Psychiatr* **74**:309, 1954.
19. De Morsier G, Gauthier G: La dysplasie olfacto-genitale. *Pathol Biol (Paris)* **11**:1267, 1963.
20. Males JL, Townsend JL, Schneider RA: Hypogonadotropic hypogonadism with anosmia— Kallmann's syndrome. *Arch Intern Med* **131**:501, 1973.

21. Naftolin F, Harris GW, Bobrow M: Effect of purified luteinizing hormone releasing factor on normal and hypogonadotropic anosmic men. *Nature (Lond)* **232**:496, 1971.

22. Mortimer CH, Besser GM, McNeilly AS, et al: Luteinizing hormone and follicle stimulating hormone-releasing hormone test in patients with hypothalamic-pituitary gonadal dysfunction. *Br Med J* **4**:73, 1973.

23. Boyar RM, Finkelstein JW, Witkin M, et al: Studies of endocrine function in "isolated" gonadotropin deficiency. *J Clin Endocrinol Metab* **36**:64, 1973.

24. Santoro N, Filicori M, Crowley WF: Hypogonadotropic disorders in men and women: diagnosis and therapy with pulsatile gonadotropin-releasing hormone. *Endocrine Rev* **7**:11, 1986.

25. Lincoln D: Hypothalamic pulse generators. *Recent Prog Horm Res* **41**:369, 1985.

26. Clarke IJ: The relationships between GnRH and LH secretion. Program of the *Seventh International Congress of Endocrinology, Quebec City, Quebec, Canada* 1984, p. 80. (Abst. 66.)

27. Levine JE, Pau K-YF, Ramirez VD, et al: Simultaneous measurement of luteinizing hormone-releasing hormone release in unanesthetized, ovariectomized sheep. *Endocrinology* **111**:1449, 1982.

28. Valk TW, Corley KP, Kelch RP, et al: Hypogonadotropic hypogonadism: hormonal responses to low dose pulsatile administration of gonadotropin-releasing hormone. *J Clin Endocrinol Metab* **51**:730, 1980.

29. Barkan AL, Reame NE, Kelch RP, et al: Idiopathic hypogonadotropic hypogonadism in men: Dependence of the hormone responses to gonadotropin-releasing hormone (GnRH) on the magnitude of the endogenous GnRH secretory defect. *J Clin Endocrinol* **61**:118, 1985.

30. Delemarre-van de Waal HA, Schoemaker J: Prolonged pulsatile LRH treatment in different frequency schedules in a hypogonadotropic male. *Upsala J Med Sci* **89**:67, 1984.

31. Antaki A, Somma M, Wyman H, et al: Hypothalamic–pituitary function in the olfactogenital syndrome. *J Clin Endocrinol Metab* **38**:1083, 1974.

32. Tagatz G, Fialkow PJ, Smith D, et al: Hypogonadotropic hypogonadism associated with anosmia in the female. *N Engl J Med* **283**:1326, 1970.

33. Weinstein RL, Reitz RE: Pituitary–testicular responsiveness in male hypogonadotropic hypogonadism. *J Clin Invest* **53**:408, 1974.

34. Schroffner WG, Furth ED: Hypogonadotropic hypogonadism with anosmia (Kallmann's syndrome) unresponsive to clomiphene citrate. *J Clin Endocrinol Metab* **31**:267, 1970.

35. Hamilton CR, Henkin RI, Weir G, et al: Olfactory status and response to clomiphene in male gonadotrophin deficiency. *Ann Intern Med* **78**:47, 1973.

36. Boyar RM: The effect of clomiphene citrate in anosmic hypogonadotrophism. *Ann Intern Med* **71**:1127, 1969.

37. Cattanach BM, Iddon CA, Charlton HM, et al: Gonadotrophin-releasing hormone deficiency in a mutant mouse with hypogonadism. *Nature (Lond)* **269**:338, 1977.

38. McDowell IFW, Morris JF, Charlton HM: Characterization of the pituitary gonadotropin cells of hypogonadal (hpg) male mice: Comparison with normal mice. *J Endocrinol* **95**:321, 1982.

39. Kovacs K, Sheehand HL: Pituitary changes in Kallmann's syndrome: A histologic, immunocytologic, ultrastructural and immunoelectron microscopic study. *Fertil Steril* **37**:83, 1982.

40. Charlton HM, Halpin DMG, Iddon C, et al: The effects of daily administration of single and multiple injections of gonadotropin-releasing hormone on pituitary and gonadal function in the hypogonadal (hpg) mouse. *Endocrinology* **113**:535, 1983.

41. Rabin D, Spitz I, Bercovici B, et al: Isolated deficiency of follicle-stimulating hormone: Clinical and laboratory features. *N Engl J Med* **287**:1313, 1972.

42. Rabinowitz D, Spitz IM: Isolated gonadotropin deficiency and related disorders. *Isr J Med Sci* **11**:1011, 1975.

43. Rabinowitz D, Benveniste R, Lindner J, et al: Isolated follicle-stimulating hormone deficiency revisited. *N Engl J Med* **300**:126, 1979.

44. Barkan AL, Kelch RP, Marshall JC. Isolated gonadotrope failure in the polyglandular autoimmune syndrome. *N Engl J Med* **312**:1535, 1985.

45. Espiner EA, Donald RA: Pituitary and gonadal function in hypogonadotrophic hypogonadism. *Acta Endocrinol (Kbh)* **73**:209, 1973.

46. Rosen SW, Gann P, Rogol AD: Congenital anosmia: Detection thresholds for seven odorant classes in hypogonadal and eugonadal patients. *Ann Otol Rhinol Laryngol* **88**:288, 1979.

47. Rosen SW: The syndrome of hypogonadism, anosmia and midline cranial anomalies. *Proceedings of the Forty-Seventh Meeting of the Endocrine Society, 1965*, p. 123. (Abst.)

48. Nowakowski H, Lenz W: Genetic aspects in male hypogonadism. *Recent Prog Horm Res* **17**:53, 1961.

49. Dimitrovski C, Plaseski A, Bogoev M, et al: Kallmann's syndrome associated with atrial septal defect. *JAMA* **248**:1358, 1982.

50. Gauthier G: La dysplasie olfactor-génitale. *Acta Neuroveg* **21**:345, 1960.

51. Gould L, Reddy R: Cardiac abnormalities in a female patient with hypogonadotrophic hypogonadism and anosmia. *J Electrocardiol* **10**:279, 1977.

52. Kemman E, Conrad P, Jones J: Cardiac abnormalities in female hypogonadotrophic hypogonadism with anosmia. *Am J Obstet Gynecol* **126**:964, 1980.

53. Rosenberg MS, Riddick HD: Dynamic pituitary testing in a female with Kallmann's syndrome and associated cardiac anomaly. *Obstet Gynecol* **48**:230, 1976.

54. Colle ML, Asch RH, Greenblatt RB: Kallman's syndrome: effect of repeated stimulation of pituitary-gonadal axis with LH-RH. *J Reprod Med* **18**:31, 1977.

55. Yashimoto Y, Moridera K, Imura H: Restoration of normal pituitary gonadotropin reserve by administration of luteinizing hormone-releasing hormone in patients with hypogonadotropic hypogonadism. *N Engl J Med* **292**:242, 1975.

56. Zarate A, Kastin AJ, Soria J, et al: Effect of synthetic luteinizing hormone-releasing hormone (LH-RH) in two brothers with hypogonadotropic hypogonadism and anosmia. *J Clin Endocrinol Metab* **36**:612, 1973.

57. Naftolin F, Harris GW, Bobrow M: Effect of purified luteinizing hormone releasing factor on normal and hypogonadotropic anosmic men. *Nature (Lond)* **232**:496, 1971.

58. Roth JC, Kelch RP, Kaplan SL, et al: FSH and LH response to luteinizing hormone-releasing factor in prepubertal and pubertal children, adult males and patients with hypogonadotropic and hypergonadotropic hypogonadism. *J Clin Endocrinol Metab* **35**:926, 1972.

59. Schneider HP, Dahlen HG: Studies with synthetic LH-releasing hormone in the human. II. Evaluation of anterior pituitary gonadotropic function in various endocrine states. *Neuroendocrinology* **11**:328, 1973.

60. Yoshimoto Y, Moridera K, Imura H: Plasma luteinizing hormone and follicle stimulating hormone responses to luteinizing hormone-releasing hormone in patients with hypogonadism. *Endocrinol Jpn* **21**:191, 1974.

61. Spitz IM, Diamant Y, Rosen E, et al: Isolated gonadotropin deficiency: A heterogeneous syndrome. *N Engl J Med* **290**:10, 1974.

62. Winters SS, Mencklenburg RS, Sherins RJ: Hypothalamic function in men with hypogonadotrophic hypogonadism. *Clin Endocrinol* **8**:417, 1978.

63. Yamaji T, Shimamoto K, Kosaka K, et al: Heterogeneity of Prolactin and TSH response to TRH in hypogonadotropic hypogonadism. *J Clin Endocrinol Metab* **45**:319, 1977.

64. Spitz IM, Hirsch HJ, Trestian S: The prolactin response to thyrotropin-releasing hormone differentiates isolated gonadotropin deficiency from delayed puberty. *N Engl J Med* **308**:575, 1983.

65. Turksoy RN: Dissociation of prolactin responsiveness to thyrotropin-releasing hormone and chlorpromazine in a female with Kallmann's syndrome. *Fertil Steril* **32**:288, 1979.

66. Tanaka T, Matsumoto J, Kuwubara T, et al: A case of olfacto-genital syndrome with growth hormone deficiency. *Nippon Naika Gakkai Zasshi* **68**:189, 1979.

67. Hochberg Z, Moses AM, Miller M, et al: Altered osmotic threshold for vasopressin release and impaired thirst sensation: Additional abnormalities in Kallmann's syndrome. *J Clin Endocrinol Metab* **55**:779, 1982.

68. Corley KP, Valk TW, Kelch RP, et al: Estimation of GnRH pulse amplitude during puberal development. *Pediatr Res* **15**:157, 1981.

69. Boyar RM, Wu RHK, Kapen S, et al: Synchronization of augmented luteinizing hormone secretion with sleep during puberty. *N Engl J Med* **287**:582, 1972.

70. Job JC, Garnier PE, Chaussain JL, et al: Elevation of serum gonadotropins (LH and FSH) after LHRH injection in normal children and in patients with disorders of puberty. *J Clin Endocrinol Metab* **35**:473, 1972.

71. Roth JC, Grumbach MM, Kaplan SL: Effect of synthetic luteinizing hormone-releasing factor on serum testosterone and gonadotropins in prepubertal, pubertal and adult males. *J Clin Endocrinol Metab* **37**:680, 1973.

72. Sizonenko PC: Preadolescent and adolescent endocrinology: Physiology and pathophysiology. II. Hormonal changes during abnormal pubertal development. *Am J Dis Child* **132**:797, 1978.

73. Kelch RP, Hopwood NJ, Marshall JC: Diagnosis of gonadotropin deficiency in adolescents: limited usefulness of a standard gonadotropin-releasing hormone test in obese boys. *J Pediatr* **97**:820, 1980.

74. Moshang T Jr, Marx BS, Cara JF, et al: The prolactin response to thyrotropin-releasing hormone does not distinguish teenaged males with hypogonadotrophic hypogonadism from those with constitutional delay of growth and development. *J Clin Endocrinol Metab* **61**:1211, 1985.

75. Hawthorne G, Sheridan B, Traub AI, et al: Prolactin responses to thyrotrophin releasing hormone do not distinguish patients with isolated gonadotrophin deficiency from normals. *Clin Endocrinol* **23**:55, 1985.

76. Partsch CJ, Hermanussen M, Sippell WG: Differentiation of male hypogonadotropic hypogonadism and constitutional delay of puberty by pulsatile administration of gonadotropin-releasing hormone. *J Clin Endocrinol Metab* **60**:1196, 1985.

77. Luboshitzky R, Rosen E, Trestian S, et al: Hyperprolactinaemia and hypogonadism in men: Response to exogenous gonadotrophins. *Clin Endocrinol* **11**:217, 1979.

78. Levalle O, Bokser L, Pacenza N, et al: Restoration and maintenance of spermatogenesis by HCG therapy in patients with hypothalmo-hypophyseal damage. *Andrologia* **16**:303, 1984.

79. Lytton B, Kase N: Effects of human menopausal gonadotrophin on a eunuchoidal male. *N Engl J Med* **274**:1061, 1966.

80. Mancini RE, Seiguer AC, Lloret AP: Effect of gonadotropins on the recovery of spermatogenesis in hypophysectomized patients. *J Clin Endocrinol Metab* **29**:467, 1969.

81. Granville GE: Successful gonadotropin therapy of infertility in a hypopituitary man. *Arch Intern Med* **125**:1041, 1970.

82. Crooke AC, Davis AG, Morris R: Treatment of eunuchoidal men with human chorionic gonadotropin and follicle-stimulating hormone. *J Endocrinol* **42**:441, 1968.

83. Lunenfeld B, Mor A, Mani M: Treatment of male infertility. *Fertil Steril* **18**:583, 1967.

84. Paulsen CA, Espeland DH, Michals EL: Effects of HCG, HMG, hLH and hGH administration of testicular function. *Adv Exp Med Biol* **10**:547, 1970.

85. Sherins RJ, Winters SJ, Wachslicht H: Studies of the role of hCG and low dose FSH in initiating spermatogenesis in hypogonadotropic men. *Fifty-Ninth Annual Meeting of the Endocrine Society, Chicago, IL*, 1977, p. 212. (Abst. 312.)

86. Finkel DM, Phillips JL, Snyder PJ: Stimulation of spermatogenesis by gonadotropins in men with hypogonadotropic hypogonadism. *N Engl J Med* **313**:651, 1985.

87. Ley SB, Leonard JM: Male hypogonadotropic hypogonadism: Factors influencing response to human chorionic gonadotropin and human menopausal gonadotropin, including prior exogenous androgens. *J Clin Endocrinol Metab* **61**:746, 1985.

88. Belchetz PE, Plant TM, Nakai Y, et al: Hypophysial responses to continuous and intermittent delivery of hypothalamic gonadotropin-releasing hormone. *Science* **202**:631, 1978.

89. Knobil E: The neuroendocrine control of the menstrual cycle. *Recent Prog Horm Res* **36**:53, 1980.

90. Hoffman AR, Crowley WF Jr: Induction of puberty in men by long-term pulsatile administration of low-dose gonadotropin-releasing hormone. *N Engl J Med* **307**:1237, 1982.

91. Cutler GB Jr, Hoffman AR, Swerdloff RS, et al: Therapeutic applications of luteinizing-hormone releasing hormone and its analogs. *Ann Intern Med* **102**:643, 1985.

92. Skarin G, Nillius SJ, Wibell L, et al: Chronic pulsatile low dose GnRH therapy for induction of testosterone production and spermatogenesis in a man with secondary hypogonadotropic hypogonadism. *J Clin Endocrinol Metab* **55**:723, 1982.

93. Crowley WF Jr, McArthur JW: Induction of puberty in hypogonadotropic males: Use of low-dose pulsatile luteinizing hormone-releasing hormone (LH-RH) administration. *Endocrinology* **106**(Suppl) 260, 1980. (Abst. 743.)

94. Donald RA, Wheeler M, Sonksen PH, et al: Hypogonadotrophic hypogonadism resistant to hCG and responsive to LHRH: report of a case. *Clin Endocrinol* **18**:385, 1982.

95. Delemarre-van de Waal HA, Schoemaker J: Induction of puberty by prolonged pulsatile LRH administration. *Acta Endocrinol (Copenhagen)* **102**:603, 1983.

96. Mortimer GH, McNeilly AS, Fisher RA, et al: Gonadotropin-releasing hormone therapy in hypogonadal males with hypothalamic or pituitary dysfunction. *Br Med J* **4**:617, 1974.

97. Wheeler MJ, Sonksen FH, Jones RA, et al: The treatment of infertility in a male with hypogonadotrophic hypogonadism, with low dose pulsatile luteinizing hormone-releasing hormone (LHRH). *First Joint Meeting of the British Endocrine Societies*, 1982. (Abst. 119.)

98. Crowley WF, McArthur JW: Stimulation of the normal menstrual cycle in Kallmann's syndrome by pulsatile administration of luteinizing hormone-releasing hormone (LH-RH). *J Clin Endocrinol Metab* **51**:173, 1980.

99. Leyendecker G, Wildt L, Hansmann M: Pregnancies following chronic intermittent (pulsatile) administration of GnRH by means of a portable pump (Zyklomat): A new approach in the treatment of infertility in hypothalamic amenorrhea. *J Clin Endocrinol Metab* **51**:1214, 1980.

100. Schoemaker J, Simons AHM, von Osnabrugge GJC, et al: Pregnancy after prolonged pulsatile administration of luteinizing hormone-releasing hormone in a patient with clomiphene-resistant secondary amenorrhea. *J Clin Endocrinol Metab* **52**:882, 1981.

101. Miller DS, Reid RR, Cetel NS, et al: Pulsatile administration of low-dose gonadotropin-releasing hormone: Ovulation and pregnancy in women with hypothalamic amenorrhea. *JAMA* **250**:2937, 1983.

102. Seibel MM, Claman P, Oskowitz SP, et al: Events surrounding the initiation of puberty with long term subcutaneous pulsatile gonadotropin-releasing hormone in a female patient with Kallmann's syndrome. *J Clin Endocrinol Metab* **61**:575, 1985.

103. Van Loon GR, Brown GM: Secondary drug failure occurring during chronic treatment with LHRH: appearance of an antibody. *J Clin Endocrinol Metab* **41**:640, 1975.

104. Klingmuller D, Schweikert HU: Maintenance of spermatogenesis by intranasal administration of gonadotropin-releasing hormone in patients with hypothalamic hypogonadism. *J Clin Endocrinol Metab* **61**:868, 1985.

105. Happ J, Neubauer M, Egri A, et al: GnRH therapy in males with hypogonadotropic hypogonadism. *Horm Metab Res* **7**:526, 1975.

13

Prolactin

Introduction

Knowledge regarding the existence of a lactogenic principle in extracts of the anterior pituitary of cows dates back to 1928. However, the existence of prolactin as an independent hormone was not established until 1970, when Frantz and Kleinberg[1] concluded from bioassay data that prolactin was distinct from GH. The development of a specific and sensitive RIA in 1971 allowed the measurement of prolactin in health and in diverse disease states.[2,3] Within the short span of a decade, the causative role of this hormone in patients with secondary amenorrhea, infertility, galactorrhea, impotence, pituitary tumors, and hypothalamic disease has rendered prolactin-related problems the most common hormonal perturbation of the anterior pituitary. In fact, prolactin measurement in the serum has been referred to as the "sedimentation rate in endocrinology of the pituitary." To understand the role played by this hormone in various disease states it is essential to understand the physiological principles that regulate prolactin secretion in health. This understanding is largely based on interpretation of prolactin measured by RIA. The assumption that immunoreactive prolactin represents bioactive hormone has governed the established concepts regarding prolactin secretion. The recently developed bioassay for prolactin based on its ability to stimulate proliferation of a lymphoma-cell line (NB$_2$ cell line) has provided a means of evaluating the relationship between radioassayable and bioassayable prolactin[4]; while there may be discordance between these two facts of prolactin in the rat[5] it appears that in humans, a close correlation exists between the two.[6] The structure, regulation, and action of the hormone as well its impact in disordered lactotrope function are discussed in this chapter.

Structure and Chemistry

Prolactin (PRL) is a 198-amino acid peptide hormone synthesized by the lactotropes of the anterior pituitary. Prolactin synthesis may also take place in the placenta, in other areas of the brain and probably in neoplastic tissue. In recent years there has been considerable evidence that favors the heterogeneity of the prolactin molecule. By the use of immunoperoxidase electrophoresis, at least three variants of prolactin can be demonstrated in the plasma and the pituitaries of normal humans. Based on the work of Cowden et al.,[7] Suh and Frantz,[8] and Fang et al.,[9] three major immunoreactive peaks of prolactin can be recognized, with molecular weights of 23,000, 48,000, and 170,000. All these fractions have been shown to possess biological activity, as determined by the Nb$_2$ rat lymphoma cell line, but with differing biological potencies. The role of the larger prolactin molecules ("big" and "big-big" prolactins) in the various hyperprolactinemic states has not been clearly characterized. The secretion of these variants in patients harboring prolactinomas,[10] as well as those with idiopathic (or functional) hyperprolactinemia, is a well-recognized event. Recently, Jackson et al.[11] showed that patients with idiopathic hyperprolactinemia and normal menses demonstrated a high- (150,000–160,000) molecular-weight prolactin with diminished bioactivity in comparison with the 22,000-M_r species. These findings support the notion that the secretion of high-molecular-weight prolactin is a phenomenon not restricted to tumorous hyperprolactinemia.

Heterogeneity in the prolactin molecule is compounded by heterogeneity in the lactotropes as well. Immunocytochemical studies of normal lactotropes suggest the presence of two types based on size and location within the pituitary. It is not clear whether the heterogeneity of the prolactin molecule is related to morphological and anatomical differences in the lactotropes population. Dynamic studies of prolactin secretion, using normal pituitary cells grown in culture, have revealed functional heterogeneity in the response of lactotropes to provocative stimulation.[12] Thus, while some stimuli result in release of stored prolactin, others provoke the release of newly synthesized prolactin. Changes in prolactin synthesis are mediated by changes in the mRNA levels in the prolactin gene (prolactin gene transcription). In rats, the prolactin gene is rather large and consists of at least five coding sequences and four intervening sequences. The considerable homology between the human prolactin gene, and the human hGH gene suggests a common origin of these hormone genes from a single ancestral gene.[13] Molecular cloning techniques comparing the cDNA of GH and prolactin genes have highlighted the similarities in the nucleotide sequences between hGH, human prolactin, and human chorionic somatomammotropin. Niall et al.[14] postulated that the genes for all three hormones have resulted from duplication of a single ancestral hormone gene.

The similarities in the genes for GH and prolactin are especially interesting in light of the embryological origin of the somatotropes and the lactotropes. Both types of cells are derived from an acidophil stem cell line. This common origin contributes to the combined secretion of prolactin and GH by certain pituitary tumors. Such plurihormonal secretory phenomena are particularly prevalent in patients with acidophil stem cell adenomas and mammosomatotropic adenomas. The observation by Frawley and Neill[15] that 39% of prolactin cells are capable of also secreting GH is intriguing. By the use of reverse hemolytic placque assays, these workers were able to identify a pituitary cell type that secretes both GH and prolactin. It is not clear whether such cells represent variants of the primitive stem cells or whether they contribute significantly to the secretory output of both hormones under physiological conditions. It is generally believed that the glandular content of GH is 100-fold greater than the glandular content of prolactin.

Synthesis and Release

Prolactin is synthesized as a larger precursor on the rough endoplasmic reticulum (RER). The precursor hormone is rapidly cleaved and stored in the granules. Two pools of prolactin are known to exist within the lactotropes—a readily releasable older, stored prolactin pool, and a newly synthesized pool. Some provocative stimuli such as TRH are effective in causing release of the older stored pool of prolactin, whereas others (metoclopramide) cause release of a different, probably newer, pool of prolactin.

The lactotropes within the anterior pituitary are predominantly found in the posterior aspect of the lateral wings. However, clusters of lactotropes can be seen in the central mucoid wedge, intermingled with other cell types, particularly the gonadotropes. There is a growing body of in vitro evidence to suggest that the lactotropes in close proximity to the gonadotropes may release prolactin by paracrine control exerted on these lactotropes by their neighboring gonadotropes. Denef and Andries[16] presented evidence that suggests a paracrine interaction between gonadotropes and lactotropes in pituitary cell aggregates. When lactotropes and gonadotropes are cultured together, prolactin secretion by the lactotropes can be induced by the perfusion of GnRH. Such an effect is not observed when lactotropes are cultured alone. These findings support the clinical observations by Cetel et al.[17] that prolactin and LH pulses occur concomitantly and synchronously in women with primary ovarian failure. It is unlikely, however, that GnRH plays any significant role in the release of prolactin under physiological conditions.

Gene Transcription and cAMP Mediation

Prolactin synthesis and secretion are brought about by gene transcription. Changes in prolactin synthesis are intimately releated to changes in PRL mRNA levels. The regulation of prolactin gene transcription can be studied by the use of assays that quantitate the transcripts containing newly synthesized prolactin sequences. On the basis of several studies,[18–21] it appears that dopamine and dopaminergic agonists inhibit prolactin gene transcription, while estrogens stimulate prolactin synthesis by bringing about increases in PRL mRNA. The role of cAMP in mediating these changes is most impressive in the case of dopamine (and its agonists) and their effect on prolactin secretion. The effect of dopaminergic agents on the adenylate cyclase and cAMP levels of normal lactotropes maintained in culture has been the subject of several studies.[22–25] These agents cause inhibition of adenylate cyclase activity, which results in a decrease of intracellular cAMP levels. These observations have been extended to prolactinoma cells as well.[26,27] The decrease in cAMP levels is followed by a decrease in the PRL mRNA levels, imparting an important role to cAMP in regulating prolactin gene transcription. Maurer et al.[25] showed that the inhibitory effect of dopamine agonists on prolactin synthesis can be blocked and overcome by treatment of cultured pituitary cells with cAMP derivatives or phosphodiesterase inhibitors. It is indeed possible that the decrease in PRL mRNA seen following dopaminergic drugs is a direct result of decrease in the intracellular cAMP levels preceding it.

Estrogens and PRL Synthesis

The stimulatory effect of estradiol on prolactin synthesis has also been studied extensively. In contrast to dopamine and dopamine agonists, estradiol stimulates prolactin synthesis. This provocative effect is associated with an increase in PRL mRNA synthesis.[28] Stone et al.[29] demonstrated a four-to-five-fold increase in the PRL mRNA content after 24 hr of estradiol treatment. This magnitude of elevation is adequate to explain the well known stimulatory effect of estrogens on the de novo synthesis of prolactin.[30] The stimulatory effects of estrogens on prolactin synthesis are believed to represent a direct effect on the pituitary without any hypothalamic mediation.[31] These in vitro observations strongly support the notion that estradiol increases intracellular cAMP, and alters PRL gene transcription; as a result, there is increased PRL mRNA production followed by increased synthesis of the hormone. Nearly all the studies that support the above tenet involve the use of estradiol in extremely high concentrations. Physiologically, such a magnitude of elevation in estradiol level is seldom obtained in humans, with the possible exception of the pregnant state. The well-known effect of pregnancy in causing lacto-trope hyperplasia, hyperprolactinemia, and enlargement of macroprolacti-nomas is clearly estrogen related.

Calcium and PRL Secretion

It is well recognized that Ca^{2+} ions play a central role in the phenomenon of stimulus–secretion coupling. Like other endocrine cells, the lactotropes contain some amount of bound calcium within; the transmission of specific signals across the cell membrane can trigger intracellular mechanisms that lead to accumulation of free calcium ions within the cells. The calcium me-diation of TRH-induced PRL release has been the subject of intense study. Gershengorn[32] proposed the following sequence of events to explain one pathway by which TRH may stimulate PRL secretion:

1. TRH interacts with receptors on the lactotrope cell membrane.
2. TRH activates hydrolysis of phosphatidylinositol biphosphate to yield inositol triphosphate.
3. Inositol triphosphate interacts with nonmitochondrial pools of calcium to release Ca^{2+}.
4. There is an increase in cytoplasmic free Ca^{2+}.
5. Cytoplasmic free calcium ions activate the process of exocytosis, re-sulting in PRL secretion.

Martin[33] presented evidence to support dual intracellular signals, calcium and lipid, in the mediation of the TRH-induced PRL release; the lipid involved may be intracellular diacyl glycerol, which in concert with calcium ions has been shown to stimulate prolactin secretion in permeabilized pituitary cells.

The clinical significance of the role played by calcium in PRL secretion is unclear. However, there are two clinical situations that may be related to such a role: the association between pseudohypoparathyroidism and isolated prolactin deficiency and the association between the use of verapamil, a cal-cium channel blocker, and the development of hyperprolactinemia.

Regulation of PRL Secretion

Like other pituitary hormones, prolactin is controlled by the hypothal-amus, but unlike the other hormones of the anterior pituitary, the predom-inant control or prolactin is by tonic inhibition from the hypothalamus. In addition to this predominant negative control, the hypothalamus also exerts trophic control on the lactotropes. However, the impact of the prolactin-releasing factors is hardly a match for the powerful prolactin-inhibitory influence of the hypothalamus. The regulation of prolactin secretion is viewed from three perspectives. First, the inhibitory control exerted by the hypothalamus on the regulation of prolactin is examined. The major—even exclusive—role played by dopamine dominates this perspective. Second, the trophic effect of the hypothalamus is outlined. The supporting cast in this role is played by TRH, serotonin, histamine, and other peptides of lesser importance. The third perspective in the regulation of prolactin secretion focuses on the phenomenon of autoregulation between the pituitary lactotropes and the hypothalamus, particularly the median eminence.

Prolactin-Inhibitory Factor

Several lines of evidence strongly support the fact that the hypothalamus exerts a predominantly inhibitory effect upon prolactin secretion. Experimental evidence for such a phenomenon abounds; when the hypothalamic pituitary connection is disrupted by surgery or disease, this is accompanied by an elevation in the serum prolactin, with a concomitant decrease in the concentrations of the other adenohypophyseal hormones. This phenomenon, referred to as the pituitary isolation syndrome, underscores the importance of the hypothalamic portal connection in the maintenance of normal prolactin secretion. Thus, stalk section[34] or pituitary transplantation below the renal capsule[35] have become classic anatomic experiments to demonstrate "lactotrope escape" from hypothalamic inhibition when the pituitary stalk is disrupted. Clinically, when the hypothalamic prolactin inhibitory factor (PIF) is destroyed by disease of the hypothalamus, lack of PIF results in unopposed prolactin hypersecretion resulting in hyperprolactinemia. A similar situation develops when there is interruption in the transport of PIF from the hypothalamus to the pituitary. Also, crude extracts of hypothalamic tissue cause profound inhibition of prolactin from intact pituitary tissue maintained in monolayer culture. All the above experimental evidence confers a strong role for the PIF of the hypothalamus in the regulation of prolactin secretion.

Having established that PIF exerts a tonic inhibitory effect on the lactotropes, the next phase in prolactin physiology focused on characterization of the nature of PIF. It has now been established that PIF is in fact dopamine. The anatomical, experimental, and clinical lines of evidence that support this notion are as follows:

1. Dopamine,[36] as well as its agonists such as L-dopa, bromocriptine, and apomorphine, are potent inhibitors of prolactin secretion in vitro and in vivo.
2. Dopamine receptors have been demonstrated on the membrane of lactotropes.[37] Specific binding of these receptors with dopamine or its agonists is the event that initiates the prolactin-lowering effect of these drugs.

3. The prolactin lowering effect of dopamine and its agonists can be blocked and abolished by dopamine antagonist drugs such as meto-clopramide.[38]
4. The depletion of hypothalamic (or pituitary) dopamine by drugs such as chlorpromazine, and tricyclic antidepressants is associated with elevation of prolactin levels.
5. Anatomically, the median eminence is extremely rich in dopaminergic neurons and dopaminergic pathways.[39] These tuberoinfundibular neurons are primarily responsible for providing dopaminergic input to the pituitary gland.
6. The final criterion that establishes dopamine as the hypothalamic PIF is the convincing demonstration that dopamine is present in the portal blood in concentrations that are sufficient to inhibit prolactin release.[40]

The demonstration that dopamine is the physiological PIF has had an enormous impact on the therapy of patients with hyperprolactinemia. The role of dopamine agonist drugs, typified by bromocriptine, has extended to hyperprolactinemic patients without tumors, as well as those with small tumors, large tumors, and even invasive tumors.

Prolactin-Releasing Factors

The presence of prolactin-releasing activity in the hypothalamus has been suggested by the observation that partially purified hypothalamic extracts are capable of provoking prolactin release. These prolactin-releasing factors (PRF) are probably involved in the release of prolactin in response to various stimuli. The brisk release of prolactin in response to suckling, stress, sleep, activity, pain, pleasure (orgasm in females) anesthesia, hypoglycemia, and dehydration are mediated by these prolactin-releasing factors. The precise nature of PRF in unclear, but several contenders for the role exist. Table 69 outlines the

TABLE 69
Prolactin-Releasing Factors

Peptides
    Thyrotropin-releasing hormone (TRH)
    Endorphins (opioid peptides)
    Enkephalins
    Vasoactive intestinal polypeptide (VIP)
    Angiotensin II
    Gonadotropin-releasing hormone (GnRH)
    Oxytocin
    Gut hormones (cholecystokinin, bombesin, substance P)
Steroids
    Estradiol
Biogenic amine
    Serotonin
    Histamine
    Acetylcholine

numerous factors that have been implicated at some time or another as the physiological PRF. It is believed that other parts of the brain, such as the cerebral cortex, and the limbic system may be involved in the prolactin release associated with stress, sleep, and so forth.

The precise nature of the PRF involved in the suckling response is unclear. Suckling—or stimulation of the nipple during the postpartum period—represents a most powerful provocative stimulus for the release of prolactin. Stimulation of the nipple in the nonpregnant state, or in males is also associated with prolactin release, but to a lesser degree. The afferent arc for this suckling reflex originates in the axons or the nerve terminals in the nipple and areola. The cortex probably plays a role as relay center, signaling the hypothalamus to release a substance or substances with PRF activity. The efferent limb of the reflex represents the released prolactin and the receptors for prolactin located in the mammary tissue.

Although the precise nature of the PRF is unknown, three important substances with prolactin-releasing properties merit mention: thyrotropin-releasing hormone (TRH), biogenic amines (serotonin and histamine), and the endogenous opioid peptides.

Thyrotropin-Releasing Hormone

Although TRH is a powerful stimulus for the release (and even secretion) of prolactin, this peptide is probably not the physiological PRF. Shortly after the characterization of TRH, it became apparent that exogenous administration of TRH was associated with a brisk release of both TSH and prolactin.[41] It is of interest to note that the prolactin response to TRH can be seen with doses that are too small to provoke TSH response. The prolactin response to TRH is dose dependent to a point. The role of TRH as direct stimulator of prolactin release from the lactotropes is a well-established one and has become an important diagnostic tool to evaluate prolactin reserve. However, the role of TRH as a physiological regulator of prolactin secretion is a highly unsettled one. For example, in humans the prolactin response to suckling is not associated with a simultaneous release of TSH, a phenomenon that would have been expected, had the prolactin response been mediated by TRH.[42] The role of TRH-mediated prolactin increase in pathological states is even more controversial. While it is true that primary hypothyroidism is associated with elevations in serum prolactin levels,[43-46] this phenomenon, in contrast to the TSH elevation in primary hypothyroidism, is not universal. Furthermore, the prolactin elevation in patients with primary hypothyroidism on replacement lingers even after the TSH is normalized, implying different mechanism for the prolactin elevation. Indeed, dopamine depletion has been proposed as an important mechanism for the hyperprolactinemia of primary hypothyroidism.[47-49] Also, the occurrence of hyperprolactinemia and even galactorrhea in patients with thyrotoxicosis[50] would be difficult to explain on the basis of TRH mediation.

Direct stimulation of prolactin release by the exogenous administration of TRH has been employed to study prolactin dynamics in various disease states. Impairment in the prolactin response to TRH is classically encountered in patients with hypopituitarism, particularly resulting from Sheehan's syndrome. However, a similar pattern may be encountered in patients with prolactin-secreting tumors[51-55] and hypogonadotropic hypogonadism[56,57] and in hyperthyroidism.[58]

Biogenic Amines

In contrast to dopamine that lowers prolactin levels in the circulation, serotonin and histamine are important biogenic amines involved in stimulating prolactin release. The prolactin release following nursing, sleep and possibly stress may be mediated by serotonergic mechanisms. Ferrari et al.[59] showed that the use of serotonin antagonists cause prolactin suppression. However, the inherent dopaminergic properites of certain serotonin antagonists such as methysergide, should be taken into account when interpreting the prolactin suppression caused by these drugs. Despite convincing experimental evidence relating to the prolactin lowering effect of serotonin antagonists, these drugs have found no place in the treatment of hyperprolactinemia.

Histamine, another biogenic amine, may have a dual role in the regulation of prolactin secretion. When the H_1-receptors are stimulated by histamine this results in stimulating prolactin release, whereas stimulation of H_2-receptors leads to inhibition of prolactin release. This may be the mechanism for the hyperprolactinemia sometimes seen in association with the use of cimetidine, an H_2-receptor blocker.

Endogenous Opiates

Two observations have lent support to the endogenous opiate mediation in the release of prolactin; the observation that stimulation of opioid receptors results in hyperprolactinemia[60] and the finding that endorphins stimulate prolactin secretion.[61] Rivier et al.[62] and Foley et al.[63] demonstrated that the exogenous administration of β-endorphins is associated with stimulation of prolactin secretion. A clinical correlate of this phenomenon may be the hyperprolactinemia seen in association with endorphin-secreting pituitary adenomas. Trouillas et al.[64] described a 43-yr-old man who presented with a large pituitary tumor associated with mild hyperprolactinemia, hypocortisolemia, and hypogonadotropic hypogonadism. The surgically excised tumor was prolactin-immunonegative, and the hyperprolactinemia returned to normal 2 days following surgery. Since no suprasellar extension of the tumor was evident, the authors hypothesized a causal relationship between the reversible hyperprolactinemia and the β-endorphin hypersecretion by the tumor.

The role of endogenous opiates in the causation of hyperprolactinemia is controversial; while such a role has been perceived in the development of the hyperprolactinemia of chronic renal failure, this has not been conclusively established. Studies using naloxone, an opiate antagonist, to reverse the hyperprolactinemia of chronic renal failure or hypothalamic amenorrhea have not yielded consistent results. Thus, the role of endogenous opiates in the causation or perpetuation of hyperprolactinemia has not been elucidated.

Autoregulation of Prolactin Secretion

The servoregulatory mechanism involved in the autoregulation of prolactin is mediated between the pituitary lactotropes and the hypothalamic neurotransmitters, particularly within the median eminence. Prolactin secretion by the lactotropes is autoregulated by an ultrashort positive feedback loop between prolactin and hypothalamic dopamine. Thus, increased prolactin concentrations lead to an increase in the turnover of dopamine in the median eminence, which in turn decreases prolactin secretion by the lactotropes.[65] It is possible that alterations in this feedback system of dopaminergic regulation may underlie the development of the idiopathic or even the tumorous varieties of hyperprolactinemia. Indeed, patients with idiopathic hyperprolactinemia demonstrate convincing evidence for increased resistance to dopaminergic suppression. Serri et al.[66] demonstrated the differential effects of low-dose dopamine infusion on prolactin secretion in hyperprolactinemic patients as compared with normals.

In addition to hypothalamic dopamine, another substance—γ-aminobutyric acid (GABA)—has been isolated in the median eminence and has been shown to exhibit prolactin inhibiting activity.[67] Several workers have demonstrated the prolactin-suppressing activity of GABA both in vitro and in vivo.[68–70] Yet considerable variation and conflict exist in the reports that describe the effects of GABAergic drugs on prolactin secretion in animals.[71,72] Sodium valproate, an anticonvulsant with GABAergic properties, results in a significant decrease in prolactin levels of normals and hyperprolactinemic patients within 180 min after ingestion of the drug.[73] Although the median eminence contains plenty of GABAergic neurons, and pituitary cells contain receptors for GABA on their membrane, it is uncertain whether this neurotransmitter plays a significant role in the physiological autoregulation of prolactin secretion.

Most investigators favor the concept that autoregulation of prolactin occurs at the hypothalamic level, i,e an elevated prolactin increases dopamine turnover by the tuberoinfundibular neurons of the medial basal hypothalamus.[65,74,75] However, there has been some evidence in the literature to suggest that prolactin may directly inhibit its own secretion.[76,77] Assuming that such a mechanism does exist, it probably plays a minor role in the autoregulation of prolactin secretion.

In summary, the regulation of prolactin secretion is predominantly carried out by the tonic negative inhibitory influence of hypothalamic dopamine.[78] Although the hypothalamus clearly possesses prolactin-releasing activity, no single substance has qualified as the PRH. These substances mediate the release of prolactin in response to such diverse stimuli as stress, suckling, and sleep. The lactotropes effectively autoregulate prolactin secretion and release by an ultrashort positive feedback loop between prolactin and dopamine in the median eminence.

Actions of Prolactin

The main action of prolactin in humans is lactogenic. In lower vertebrates prolactin has diverse effects on metabolic and behavioral phenomena. In higher mammalian species, however, there is very little evidence to impart a significant metabolic role for this hormone. In female rats prolactin is essential for the normal development of breast tissue, an effect that cannot be induced by estrogens (or progesterone) in the absence of pituitary prolactin. Clearly, such is not the case in humans. Similarly, the effect of prolactin on the parental behavior of birds is a well-accepted notion. In this class, prolactin provides the parental orientation to facilitate a protectional and caring attitude towards the young. In many birds, courting and mating behavior is under control of gonadotropins, while the parenting behavior after parturition and hatching is controlled by prolactin, both phases being somewhat inhibitory to each other. Thus, during the parenting phase in this class, the reproductive phase (e.g., courting, mating) is in abeyance.

The major action of prolactin in all species is its lactogenic action. Prolactin prepares the breast tissue to provide milk in the puerperium. The preparative actions of prolactin begin as early as the first trimester of pregnancy. Under the combined influence of prolactin, estrogen, and progesterone, breast development takes place during pregnancy. However, galactorrhea does not occur, because the high estrogen levels antagonize the peripheral actions of prolactin. As the pregnancy proceeds, placental lactogen (chorionic somato-mammotropin) contributes to the lactogenic effects of the pituitary prolactin. The role assumed by glucocorticoids and insulin in the process of breast development during pregnancy is a permissive one. Breast enlargement reaches its peak during the last trimester. After delivery, when estrogen and progesterone levels decline precipitously, the prolactin, now at is peak, induces milk secretion. The continuing stimulus of suckling perpetuates prolactin secretion on a continual basis. Thus, the prolactin cycle during pregnancy and post-partum can be viewed as triphasic: the first phase is characterized by estrogen-induced prolactin secretion and preparation of the breast; the second phase, after delivery, is characterized by the estrogen withdrawal, which facilitates the successful interaction between hormone (prolactin) and a ripened target organ (mammary tissue); and the third phase is the continuous secretion of prolactin from the hyperplastic primed lactotropes in response to nursing the infant.

The metabolic effects of prolactin parallel those of GH, but with a markedly attenuated potency. Thus, prolactin is a weak generator of somatomedins; however, in the absence of GH (as in children operated on for craniophar-yngioma) prolactin can assume an important role in promoting growth in the absence of GH. The other metabolic effects of prolactin (e.g., lipolytic, mitogenic) are too weak to be physiologically significant in humans. The mitogenic effects of prolactin have permitted the development of a bioassay using a special line of cells—the Nb_2 lymphoma cell line. The ability of prolactin to induce proliferation of these cells is thought to indicate the bioactivity of prolactin. The effects of prolactin on water metabolism (conservation of water) and on sodium handling by the tubules are actions seen in lower vertebrates. The physiological significance of the demonstration of prolactin receptors in the liver, prostate, testes, ovaries, and adrenals has remained unclear.

Prolactin Dynamics in Health and Disease

The measurement of prolactin in the serum has become an integral part in the evaluation of patients with disorders of the hypothalamus, pituitary, gonads thyroid, and even the adrenals. Interpretation of prolactin levels requires an understanding of the factors that affect the basal prolactin levels both under physiological and pathological states.

Basal Levels in Health

Prolactin is measured by a sensitive and highly specific radioimmunoassay. In most laboratories, the basal prolactin ranges between 10–20 ng/ml for females. The mean levels in males is at least 3–5 ng/ml below females. Three physiological phenomena can affect the basal prolactin level: sleep,[81,82] spontaneous fluctuation,[83] and the luteal phase in females.[84] Prolactin, like GH, demonstrates a sleep-related (nyctohemeral) rise, without displaying a circadian periodicity. This sleep entrainment of prolactin usually begins at or around the time of puberty, reminiscent of the sleep entrainment of LH. The augmentation of prolactin by sleep is lost in some patients with prolactin-secreting tumors.

Basal prolactin levels can fluctuate, sometimes to an impressive degree. The stress of venipuncture may also add to the fluctuation. The pulsatile release of prolactin may occasionally be conducive to missing milder degrees

TABLE 70
**Physiological Factors That Cause a
Rise in Basal Prolactin**

Stress
Sleep
Exercise
Pregnancy
Nursing
Spontaneous fluctuation
Neonatal period
Luteal phase of menstrual cycles

of hyperprolactinemia. This can be obviated by obtaining three samples—20 min apart—and estimating the prolactin content in a single pooled sample. Regarding alterations of basal prolactin in relation to the menstrual cycle, a rise in the basal level can be encountered during the luteal phase. For the most part, however, prolactin levels remain reasonably stable throughout the normal menstrual cycle. Several factors can result in physiological elevation of prolactin (Table 70). The interpretation of the basal prolactin level should take these into account.

Basal Levels in Disease States

Hyperprolactinemia is an important marker of several clinical conditions. Approximately one third of patients seeking advice for secondary amenorrhea, or infertility may have an underlying prolactin problem. In males, approximately 5–10% of organic impotence may be related to an underlying prolactin disorder, usually macroprolactinomas. Nearly two-thirds of patients with pituitary tumors will demonstrate basal hyperprolactinemia. The above statistics clearly underscore the need to obtain prolactin measurements in females and males with gonadal disorders. The following settings warrant measurement of basal prolactin levels in the serum:

1. Females with menstrual irregularities (oligomenorrhea, secondary amenorrhea)
2. Males with impotence and erectile dysfunction
3. Galactorrhea in either sex
4. All patients with hypopituitarism
5. All patients with clinical or radiological evidence of a pituitary tumor
6. Patients with primary amenorrhea
7. Patients with clinical evidence of suprasellar disease—diabetes insipidus or visual-field cuts
8. Patients with unexplained premature osteopenia

The degree of hyperprolactinemia correlates with some clinical aspects, while showing no correlation with certain other facets. For instance, it is well known that the degree of hyperprolactinemia clearly bears no correlation to the presence or severity of galactorrhea. Obviously, the expression of prolactin action on the breast tissue will depend on the degree of estrogen priming as well as the degree of hypoestrogenemia. Thus, in the extreme setting of male prolactinomas, galactorrhea is unusual even when the basal level of prolactin is astronomical (> 2000 ng/ml). At the other extreme are women with copious galactorrhea with normal or only mildly elevated basal prolactin levels. The possibility that normoprolactinemic galactorrhea represents a disorder characterized by heightened end-organ sensitivity to normal circulating levels of prolactin is one hypothesis to explain such a situation. Alternatively, a discordance between bioassayable and radioassayable prolactin as the basis for normoprolactinemic galactorrhea remains to be proven.

While galactorrhea may not correlate with the presence or the degree of basal hyperprolactinemia, several other clinical facets do correlate with the degree of basal hyperprolactinemia. The degree of prolactin elevation has etiological significance. When the prolactin levels are between 50 and 100 ng/ml, the cause of hyperprolactinemia revolves around idiopathic (functional) hyperprolactinemia, drug-induced hyperprolactinemia, hypothalamic-stalk lesions, and chronic renal failure. When the basal prolactin levels exceed 200 ng/ml, prolactin-secreting pituitary tumors are usually responsible for the hyperprolactinemia. There is a close correlation between the degree of hyperprolactinemia and the size of the tumor; when the basal levels exceed 400–500 ng/ml, macroadenomas are usually present. The only exception are patients with secretion of macroprolactins (large-molecular-weight prolactin). Jackson et al.[85] described a 35-year-old woman with a 3-year history of documented hyperprolactinemia in excess of 350–400 ng/ml, without evidence of a pituitary tumor. Analysis of the prolactin in the patients' serum by chromatographic methods revealed that 85% of the circulating immunoreactive prolactin had a molecular weight greater than 100,000 (peak I or macroprolactin). This is in contrast to the other hyperprolactinemic states, where the predominant prolactin in the circulation is the $22,000\text{-}M_r$ hormone. Jackson and co-workers termed this condition macroprolactinemia. The same investigators also characterized peak I prolactin as the predominant fraction in women with idiopathic hyperprolactinemia and normal menses.[11] The reduced bioactivity of this fragment has been pointed out by other reports in the literature.[86,87] Despite these minor exceptions, the demonstration of serum prolactin levels above 200–250 ng/ml is virtually diagnostic of a prolactin-secreting pituitary tumor.

Another clinical facet that correlates with the degree of hyperprolactinemia is the resultant hypogonadal state. In generally, amenorrhea and hypoestrogenism are more common in patients with chronic and more severe hyperprolactinemia.

In most clinical situations, the measurement of basal prolactin levels in the serum is adequate for diagnostic purposes, obviating the need for performing dynamic studies of prolactin secretion. However, the literature is replete with dynamic studies of prolactin secretion is an attempt to unravel the pathophysiology of prolactin dysregulation.

Dynamic Studies

Several diagnostic agents with provocative and suppressive effects on prolactin secretion have been used to evaluate the differences in health and in various disorders of prolactin secretion, particularly tumors. It is best to view these diverse agents in terms of five categories, based on their action. First, there are agents that affect prolactin by dopaminergic mechanisms; to this class belong L-dopa, dopamine, and dopamine agonist drugs such as bromocriptine and apomorphine. All these agents, with varying degrees of potency, lower prolactin levels in normals as well as in patients with hyperprolactinemia. The second category of pharmacological agents are those that evaluate prolactin response to dopamine (DA) receptor blockade. The agents that belong to this category are domperidone and metoclopramide. DA receptor blockade in normals is attended with a brisk rise in prolactin levels. Patients with tumors often do show an attenuated prolactin response to dopamine antagonism (or DA receptor blockade); however, a similar attenuation may be seen in patients with stalk section and idiopathic hyperprolactinemia. The third category of pharmacotherapeutic agents used for the evaluation of prolactin dynamics are agents that affect dopamine metabolism within the CNS; the two agents prototypical of this group are the antidepressant nomifensine, which decreases synaptic reuptake of dopamine by the hypothalamus, and carbidopa, which inhibits the decarboxylation of dopamine in the periphery and the pituitary, but not in the hypothalamus, since it does not cross the blood-brain barrier. The fourth category of agents are those that directly stimulate prolactin release from the pituitary lactotropes; the prototype of this group is thyrotropin-releasing hormone (TRH). The fifth category of provocative agents are those that cause release of prolactin via hypothalamic substances with PRF activity, the release of prolactin in response to insulin hypoglycemia, dehydration, etc. Table 71 outlines the various agents used to study prolactin dynamics. Although not used very often in clinical practice, these tests have contributed enormously to the understanding of the intricate mechanisms involved in the regulation of prolactin. A brief sketch of these tests follows.

TABLE 71
Dynamic Testing of Prolactin Secretion

Group	Type	Agent	Effect on serum PRL
I	Dopaminergic	L-Dopa Dopamine Apomorphine Bromocriptine	Lowers prolactin
II	Dopamine antagonists (DA receptor blockade)	Metoclopramide Domperidone	Increases prolactin
III	Drugs that affect dopamine Metabolism	Carbidopa Nomifensine	No effect Lowers prolactin
IV	Drugs that directly stimulate prolactin release	TRH	Increases prolactin
V	Factors that stimulate prolactin release via hypothalamus	Stress Sleep Insulin hypoglycemia Exercise Dehydration	Increase prolactin

Dopaminergic Agents and PRL Secretion

Dopaminergic agents cause a prompt suppression of prolactin levels even when administered acutely. This prolactin-lowering effect of dopaminergic agents extends to hyperprolactinemic patients regardless of etiology, hence cannot be used to distinguish idiopathic hyperprolactinemia from CT negative microadenomas. The powerful suppressive effect of dopamine on prolactin secretion can be evidenced even when dopamine is infused at subphysiological (low-dose) levels. Connell et al.[88] showed that prior administration of low dose dopamine is capable of attenuating or even abolishing the prolactin response to TRH administration. The effect of estrogen priming on the inhibitory effects of dopamine (DA) on prolactin secretion is intriguing. It is believed that in the normal female, prolactin sensitivity to the inhibitory effects of dopamine infusion can be increased by estrogen priming.[89-91] This observation, when viewed with the fat that estrogen priming accentuates the TRH (or metoclopramide) mediated prolactin release, appears paradoxical. The effect of estrogen priming in primates seems to be one that sensitizes the lactotrope to DA inhibition and TRH stimulation; that is, it primes the lactotrope to respond better to physiological cues.

Dopamine Antagonists and PRL Secretion

Metoclopramide and domperidone are DA receptor blockers, which consequently elevate serum prolactin levels by abolishing the negative effect exerted by dopamine.[92,93] Both drugs cause a prompt release of prolactin. While domperidone does not cross the blood brain barrier (and hence its actions are limited to the pituitary), metoclopramide probably inhibits dopamine receptors at both the pituitary and hypothalamic level. It is believed that the magnitude of prolactin release seen after the administration of domperidone or metoclopramide is as good as, or better than, the magnitude of prolactin release following 200–400 μg IV TRH. It has also been postulated that TRH and DA blockade cause release of different pools of PRL within the lactotrope. In one recent study, Ho et al.[94] compared the prolactin release following TRH with the domperidone-induced PRL release and showed that the PRL rise seen after the combined administration of TRH and domperidone was significantly greater than the PRL rise encountered with either TRH or domperidone alone. The observation that TRH was able to induce a significant rise in PRL after maximal DA receptor blockade had been attained as well as the reverse observation that domperidone was able to still induce a significant rise in PRL despite maximal TRH stimulation provides powerful evidence for the existence of two pools of stored prolactin within the lactotrope.

The diagnostic use of domperidone in delineating patients with pituitary tumors has also been studied extensively. Cowden et al.[95] proposed that a reduced prolactin response to acute DA receptor blockade in a hyperprolactinemic patient is diagnostic of the presence of a prolactinoma. While it is true that nearly all normal subjects respond to DA blockade with a brisk release of prolactin, and nearly all patients with prolactinoma show a blunting, the response patterns in patients with stalk disease and idiopathic hyperprolactinemia may be normal or blunted, mimicking microadenoma. Thus, the diagnostic value of evaluating the prolactin response to acute DA blockade (with domperidone or metoclopramide) has the same advantages and limitations of evaluating the prolactin response to TRH administration. It is interesting to note that patients with puerperal hyperprolactinemia respond to acute DA blockade with a normal brisk rise in prolactin.[96] The combined evaluation of the TSH and prolactin responses to acute DA blockade with domperidone has been evaluated in normal and hyperprolactinemic patients. Scanlon et al.[97] and Massara et al.[96] shown that patients with pathological hyperprolactinemia, particularly of tumor etiology, demonstrate dual phenomena in response to DA receptor blockade; a blunted, even absent, prolactin response contrasted with a concomitant pronounced TSH response to domperidone. It has been suggested that patients with prolactin secreting microadenomas have an enhancement of the dopaminergic inhibitory effect on the normal thyrotropes surrounding the adenoma. When this dopaminergic tone is removed by acute DA blockade, the TSH levels rise. The simultaneous presence of PRL hyporesponsiveness and TSH hyperresponsiveness may be a reflection of a single phenomenon, an altered relationship between the tubero-infundibular dopaminergic area (TIDA) and the pituitary. Normally this system is inhibitory to both PRL and TSH. For reasons that are not entirely clear, the relationship is altered in patients with prolactin secreting microadenomas resulting in a markedly depressed lactotrope inhibition and an enhanced thyrotrope inhibition. The specific reasons for these contrasting effects are unknown.

Drugs That Affect DA Metabolism[98–102]

The L-dopa–carbidopa combination test and the nomifensine test are the two studies in the category. The principle, the interpretation and the diagnostic value of these two tests in the etiological evaluation of hyperprolactinemic patients are discussed in the next chapter.

Drugs That Directly Stimulate PRL

Prototypical of this category, TRH is a potent stimulatory of PRL release in normals. This response is blunted, even absent, in most patients with prolactin secreting tumors.[51–58] It is believed that the reason for such an impairment is the depletion of the TRH-releasable pool of prolactin stored within the lactotropes. A perfectly normal TRH induced prolactin response diminishes the likelihood of tumor, but a blunted PRL response to TRH can be encountered in both tumorous and nontumorous hyperprolactinemia.

Factors That Indirectly Stimulate PRF

These factors stimulate PRL via hypothalamic PRF. The prolactin response to sleep, stress, hypoglycemia etc are no longer performed, since they add very little to the diagnostic yield.

In summary, measurement of basal prolactin levels is the single—and probably the only—hormonal index required in the evaluation of patients with a prolactin-related clinical problem. The dynamic studies mentioned above, with the possible exception of the TRH study, are seldom performed on a routine basis. However, familiarity with the various factors mentioned above has clarified several facets in the ever-intriguing mechanisms that regulate prolactin secretion.

Disorders of Prolactin Secretion

Hypofunction of lactotropes has practically no consequences, except in the setting of post partum necrosis of the pituitary—Sheehan's syndrome. By contrast, hyperprolactinemia is the hormonal marker of disordered lactotrope function encountered in clinical practice. It should be emphasized that pregnancy and nursing are the only physiological causes of hyperprolactinemia. With these two exceptions, the demonstration of elevated prolactin levels in the circulation is always due to disturbed lactotrope function. Several clinical situations are characterized by hyperprolactinemia. there are five major mechanisms that can result in hyperprolactinemia.

1. *Loss of dopaminergic tone:* This is the mechanism that underlies some of the most common causes of hyperprolactinemia. Thus, drugs that block dopamine, pathological lesions that destroy the dopamine content of the hypothalamus or the median eminence, and stalk section lead to hyperprolactinemia by decreasing the dopamine (physiological PIF) that reaches the pituitary lactotropes. Idiopathic hyperprolactinemia is thought to arise from a raised threshold of lactotropes to dopaminergic suppression.

2. *"Autonomous" hypersecretion of prolactin:* This is the basis of the hyperprolactinemia associated with pituitary tumors that secrete prolactin, and represents the most common cause of hyperprolactinemia (60–70%). The broad spectrum of prolactinomas consists of microprolactinomas, macroprolactinomas, and invasive macroprolactinomas.

3. *Decreased clearance of prolactin:* This is the mechanism of hyperprolactinemia in patients with chronic renal failure, and perhaps in some cases of hypothyroidism.

4. *Chest wall diseases:* Several diseases of the chest wall can cause hyperprolactinemia by facilitating the afferent arc of the limb that stimulates hypothalamic prolactin-releasing factor(s).

5. *Ectopic secretion of prolactin:* This mechanism, doubted by some, is a rare cause of hyperprolactinemia and can be seen in association with certain neoplastic diseases.

TABLE 72
Causes of Hyperprolactinemia

Drugs
Hypothalamic-stalk disease
  Suprasellar tumors (e.g., craniopharyngiomas, germ cell tumors)
  Granulomatous disease
  Hypophysitis
  Metastatic disease
  Pituitary tumors with suprasellar invasion (pseudoprolactinomas)
  Stalk section
Prolactin-secreting tumors
Idiopathic hyperprolactinemia
Chest wall disease
  Herpes zoster
  Thoracotomy
  Burns
Decreased clearance
  Chronic renal failure
  ? Primary hypothyroidism
Thyroid disease
  Primary hypothyroidism
  Thyrotoxicosis
Ectopic secretion of prolactin
Other associated disorders
  Anorexia nervosa
  Polycystic ovaries
  Primary ovarian failure
  Empty sella syndrome
  Adrenal insufficiency

Table 72 outlines the numerous causes for hyperprolactinemia. Each category deserves brief mention.

Drug-Induced Hyperprolactinemia[103–109]

Drug-induced hyperprolactinemia can result from several mechanisms (Table 73) The two most important ones are the use of drugs that block DA receptors in the hypothalamus or the pituitary and drugs that deplete dopamine at either level. Drugs that block DA receptors include phenothiazines (chlorpromazine, fluphenazine, and perphenazine), butyrophenones such as heloperidol, tricyclic antidepressants (particularly amoxapine), and metoclopramide. Dopamine-depleting drugs are exemplified by reserpine and α-methyldopa. A third mechanism for hyperprolactinemia is represented by drugs that directly stimulate the lactotropes; estrogens are prototypical of this category. Of lesser importance is the hyperprolactinemia secondary to cimetidine, verapamil, and narcotics.

The three characteristics of drug-induced hyperprolactinemia are the mild to modest degree of prolactin elevation (usually in the 30–100-ng/ml

TABLE 73
Drug-Induced Hyperprolactinemia

Dopamine receptor-blocking drugs
  Phenothiazines (chlorpromazine)
  Tricyclic antidepressants
  Butyrephenones
  Metoclopramide
Dopamine-depleting drugs
  Reserpine
  Methyldopa
H_2-Receptor blocking drugs
  Cimetidine
Direct stimulation of lactotropes
  Estrogen
  Thyrotropin-releasing hormone (TRH)
Miscellaneous
  Verapamil
  Morphine

range), preservation of prolactin responsiveness to TRH administration, and prompt normalization upon withdrawal of the drug. Occasionally, however, the prolactin levels in drug-induced hyperprolactinemia can be impressive and may be coupled with impairment of prolactin response to TRH. These two findings may resemble the hyperprolactinemic state associated with tumors. Lankford et al.[110] performed TRH and metoclopramide stimulation tests in 10 women taking phenothiazines and showed that the prolactin response to TRH was exaggerated in most, while the prolactin response to metoclopramide was blunted in most.

The hyperprolactinemia associated with the use of oral contraceptive agents is more likely due to the fact that estrogens may have unmasked a prolactinoma than caused it. Although it is true that in vitro, estrogens increase the cAMP concentrations of the lactotrope and increase prolactin mRNA content, resulting in prolactin hypersecretion, the prolactin levels of patients on birth control pills are not impressively higher or different as compared with controls.[111] As for the postpill amenorrhea galactorrhea syndrome, this may represent the subset of patients who may have had menstrual irregularities and perturbations of lactotrope function prior to institution of treatment with oral contraceptive agents.

Hypothalamic-Stalk Disease[112]

Several clinical situations can impair the synthesis or transport of dopamine to the pituitary lactotropes. In its complete and classic form, this is represented by stalk section, which leads to hyperprolactinemia of the pituitary isolation syndrome. The resultant lactotrope escape from dopaminergic suppression leads to unrestrained prolactin secretion. The three main etiologies in clinical practice that cause hypothalamic-stalk disease leading to hyperprolactinemia are suprasellar tumors (particularly craniopharyngiomas), metastatic or granulomatous disease, and pituitary tumors that extend above and compress the stalk or the infundibulum. The term *pseudoprolactinoma* is applied to the situation where hyperprolactinemia results from interruption of the stalk by a superiorly enlarging nonsecretory, functionless pituitary tumor. The three characteristic features of pseudoprolactinomas are a prolactin level disproportionately low in comparison with the size of tumor seen by CT, decreased adenophyseal hormone reserve, and lack of prolactin immunopositivity in the tumor. Although the prolactin levels in such patients can be effectively lowered by dopamine agonist drugs, the pseudoprolactinomas will not shrink with such therapy because these tumors, in contrast to true prolactinomas, lack dopamine receptors.

Prolactin-Secreting Tumors

These tumors represent the most common cause of hyperprolactinemia. The clinical presentation, diagnostic studies, and therapeutic approach of these tumors are the subject of Chapter 14.

Idiopathic Hyperprolactinemia[113]

The term idiopathic hyperprolactinemia refers to the occurrence of hyperprolactinemia in patients without evidence of tumor in the pituitary gland (by high-resolution CT), or in the hypothalamic stalk region, and without any other recognizable cause for prolactin elevation. Although a small percentage of such patients may harbor microadenomas that are too diminutive to be visualized by the CT, most patients with idiopathic hyperprolactinemia diagnosed by current state-of-the-art CT equipment do not have microadenomas.

Chest Wall Disease[114-116]

Chronic tactile stimulation of the nipple–areolar region, or lesions in the chest wall such as burns, herpes zoster, or postsurgical scars, can cause hyperprolactinemia. Even though the lesion may be unilateral, the ensuing galactorrhea is often bilateral. Inspection of the chest wall is an integral part in the evaluation of galactorrhea.

Ectopic Prolactin Secretion

Ectopic secretion of prolactin is relatively rare in comparison with ectopic secretion of other hormones by neoplastic tissue. The rarity of this phenomenon is illustrated in the study conducted by Molitch et al.[117] These workers measured prolactin levels in 215 patients with a variety of malignancies and found that only two had modestly elevated prolactin in the absence of other recognized causes of hyperprolactinemia. Since these two patients suffered from breast cancer and lung cancer, stimulation of afferent nerves in the chest wall could have been a mechanism for the hyperprolactinemia. Although prolactin has been extracted from cancerous tissue of the bronchus and breast,[118,119] the frequency of finding hyperprolactinemia in these cancers is distinctly rare.[120,121]

Thyroid Disease

TRH possesses PRF-like properties; the mediation of this hypothalamic peptide was presumed to be the mechanism of the hyperprolactinemia seen in association with primary hypothyroidism. While the presence of such a mechanism has neither been proved nor disproved, other mechanisms seem to prevail in the causation of the hyperprolactinemia; these include decreased clearance of prolactin in the hypothyroid state as well as the possibility of chronic dopamine depletion in the hypothalamus and the median eminence. The hyperprolactinemia persists even after restoration of normal TSH levels with treatment and resolves quite slowly.

Rare Causes and Associations

Rarely, hyperprolactinemia is associated with diverse disorders such as anorexia nervosa, polycystic ovaries, primary ovarian failure, adrenocortical insufficiency, and primary empty sella syndrome. The mechanisms underlying the hyperprolactinemia in these instances are far from clear.

Regardless of the etiology or the mechanism of hyperprolactinemia, the resulting clinical features are often the same. In females, hyperprolactinemia can manifest in one of four ways: menstrual irregularities (oligomenorrhea, or amenorrhea), infertility with or without normal cycles, galactorrhea, and symptoms of hypoestrogenism such as decreased libido and dyspareunia. In males, the cardinal manifestation of hyperprolactinemia is impotence, and to a lesser extent infertility. Galactorrhea is relatively rare in males with hyperprolactinemia. Males with hyperprolactinemia are more likely to have large pituitary tumors with invasive characteristics. In both sexes, pituitary tumors are the most frequent cause of hyperprolactinemia.[122,123] One of the most important revelations of employing the prolactin assay inpatients with pituitary tumors is the observation that a high proportion of pituitary tumors secrete prolactin even when galactorrhea or amenorrhea are not present. One of the earlier studies that addressed this issue was that of Antunes et al.,[123] which shows that hyperprolactinemia was present in 65% of the 69 patients with documented pituitary tumors. Thus, determination of basal prolactin levels has become an integral part in the evaluation of patients with pituitary tumors.

The indications for obtaining prolactin levels are outlined in Table 74. The most common underlying cause for any or all the indications outlined is the prolactinoma. These tumors are discussed in Chapter 14.

TABLE 74
Indications for Obtaining Prolactin Level

Patients with menstrual irregularities
  Oligomenorrhea
  Secondary amenorrhea
  Primary amenorrhea (rare)
Women with infertility
  Menstrual irregularities
  Normal cycles, but decreased luteal phase
Men with impotence
Patients with galactorrhea
Patients with pituitary tumors
Patients with hypothalamic-stalk disease
Women with premature osteopenia

References

1. Frantz AG, Kleinberg DL: Prolactin: Evidence that it is separate from growth hormone in human blood. *Science* **170**:745, 1970.
2. Hwang P, Guyda H, Friesen H: A radioimmunoassay for human prolactin. *Proc Natl Acad Sci USA* **68**:1902, 1971.
3. Friesen H, Hwang P, Guyda H, et al: A radioimmunoassay for human prolactin. In: Boyns AR, Griffiths K (eds): *Prolactin and Carcinogenesis Proceedings of the Fourth Tenovus Workshop.* Alpha Omga Alpha, Cardiff, Wales, 1972, p. 64.
4. Tanaka T, Shiu RPC, Gout PW, et al: A new sensitive and specific bioassay for lactogenic hormones: Measurement of prolactin and growth hormone in human serum. *J Clin Endocrinol Metab* **51**:1058, 1980.
5. Klindt J, Robertson MC, Friesen HG: Episodic secretory patterns of rat prolactin determined by bioassay and radioimmunoassay. *Endocrinology* **111**:350, 1982.
6. Rowe RC, Cowden EA, Faiman C, et al: Corelation of Nb2 bioassay and radioimmunoassay values for human serum prolactin. *J Clin Endocrinol Metab* **57**:942, 1983.
7. Cowden EA, Friesen HG, Gout PW: Biologically active circulating prolactin in uremia. *Clin Res* **28**:695A, 1980.
8. Suh HK, Frantz AG: Size heterogeneity of human prolactin in plasma and pituitary extracts. *J Clin Endocrinol Metab* **39**:928, 1974.
9. Fang VS, Refetoff S: Heterogenous human prolactin from a giant pituitary tumor in a patient with panhypopituitarism. *J Clin Endocrinol Metab* **47**:780, 1978.
10. Rogol AD, Rosen SW: Prolactin of apparent large molecular size: The major immunoreactive prolactin component in plasma of a patient with pituitary tumor. *J Clin Endocrinol Metab* **38**:714, 1974.
11. Jackson RD:, Wortsman J, Malarkey WB: Characterization of a large molecular weight prolactin in women with idiopathic hyperprolactinemia and normal menses. *J Clin Endocrinol Metab* **61**:258, 1985.
12. Walker AM, Farquhar MG: Preferential release of newly synthesized prolactin granules is the result of functional heterogeneity among mammotrophs. *Endocrinology* **107**:1095, 1980.
13. Miller WL, Eberhardt NL, Baxter JD: Growth hormones genes. In Black PM, Zervas NT, Ridgway EC, et al (eds): *Secretory Tumors of the Pituitary gland. Progress in Endocrine Research and Therapy.* Vol. 1. Raven, New York, 1984, p. 135.
14. Niall HD, Hogan ML, Sayer R, et al: Sequences of pituitary and placental lactogenic and growth hormones: Evolution from a primordial paptide by gene duplication. *Proc Natl Acad Sci USA* **68**:866, 1971.

15. Frawley LS, Neill JD: Identification of a pituitary cell type that secretes both growth hormone and prolactin: Detection by reverse hemolytic plaque assays. In *Sixty-fifth Annual Meeting of the Endocrine Society*, 1983. (Abst. 918.)

16. Denef C, Andries M: Evidence for paracrine interaction between gonadotrophs and lactrophs in pituitary cell aggregates. *Endocrinology* **112**:813, 1983.

17. Cetel NS, Yen SSC: Concomitant pulsatile release of prolactin and lutenizing hormone in hypogonadal women. *J Clin Endocrinol Metab* **56**:1313, 1983.

18. Maurer RA: Dopaminergic inhibition of prolactin synthesis and prolactin messenger RNA accumulation in cultured pituitary cells. *J Biol Chem* **255**:8092, 1980.

19. Maurer RA: Transcriptional regulation of the prolactin gene by ergocryptine and cyclic AMP. *Nature (Lond)* **294**:94, 1981.

20. Maurer RA: Estradiol regulates the transcription of the prolactin gene. *J Biol Chem* **257**:2133, 1982.

21. Maurer RA, Gorski J: Effects of estradiol-17β and pimozide on prolactin synthesis in male and female rats. *Endocrinology* **101**:76, 1977.

22. Barnes GD, Brown BL, Gard TG, et al: Effect of TRH and dopamine on cyclic AMP levels in enriched mammotroph and thyrotroph cells. *Mol Cell Endocrinol* **12**:273, 1978.

23. Swennen L, Denef C: Physiological concentrations of dopamine decrease adenosine 3′,5′-monophosphate levels in cultured rat anterior pituitary cells and enriched populations of lactotrophs: Evidence for a causal relationship to inhibition of prolactin release. *Endocrinology* **111**:398, 1982.

24. Giannattasio G, De Ferrari ME, Spada A: Dopamine-inhibited adenylate cyclase in female rat adenohypophysis. *Life Sci* **28**:1605, 1981.

25. Maurer RA: Adenosine 3′,5′-monophosphate derivatives increase prolactin synthesis and prolactin messenger ribonucleic acid levels in ergocryptine-treated pituitary cells. *Endocrinology* **110**:1957, 1982.

26. De Camilli P, Macconi D, Spada A: Dopamine inhibits adenylate cyclase in human prolactin-secreting pituitary adenomas. *Nature (Lond)* **278**:252, 1979.

27. Spada A, Nicosia S, Cortelazzi L, et al: In vitro studies on prolactin release and adenylate cyclase activity in human prolactin-secreting pituitary adenomas. Different sensitivity of macro- and microadenomas to dopamine and vasoactive intestinal polypeptide. *J Clin Endocrinol Metab* **56**:1,1983.

28. Ryan R, Shupnik MA, Gorski J: Effect of estrogen on preprolactin messenger ribonucleic acid sequences. *Biochemistry* **18**:2044, 1979.

29. Stone RT, Maurer RA, Gorski J: Effect of estradiol-17β on preprolactin messenger ribonucleic acid activity in the rat pituitary gland. *Biochemistry* **16**:4914, 1977.

30. MacLeod RM, Abad A, Eidson LL: In vivo effect of sex hormones on the in vitro synthesis of prolactin and growth hormone in normal and pituitary tumor-bearing rats. *Endocrinology* **84**:1475, 1969.

31. Vician L, Shupnik MA, Gordki J: Effects of estrogen on primary ovine pituitary cell cultures: Stimulation of prolactin secretion, synthesis and preprolactin messenger ribonucleic acid activity. *Endocrinology* **104**:736, 1979.

32. Gershengorn MC: Intracellular mechanisms of calcium-mediated stimulation of prolactin secretion by thyrotropin-releasing hormone. In MacLeod RM, Thorner MO, Scapagnini U. (eds): *Prolactin. Basic and Clinical correlates*. Liviana, Padova, 1985, p. 155.

33. Martin TFJ: Dual intracellular signaling by Ca^{2+} and lipids mediates the actions of TRH. In MacLeod RM, Thorner MO, Scapagnini U. (eds): *Prolactin. Basic and Clinical correlates*. Liviana, Padova, 1985, p. 165.

34. Kanematsu S, Sawyer CH: Elevation of plasma prolactin after hypophysial stalk section in the rat. *Endocrinology* **93**:238, 1973.

35. Everett JW: Luteotrophic function of autografts of the rat hypophysis. *Endocrinology* **54**:685, 1954.

36. Birge CA, Jacobs LS, Hammer CT, et al: Catecholamine inhibition of prolactin secretion by isolated rat adenohypophysis. *Endocrinology* **86**:120, 1970.

37. Goldsmith PC, Cronin MJ, Weiner RI: Dopamine receptor sites in the anterior pituitary. *J Histochem Cytochem* **27**:1205, 1979.

38. Scanlon MF, Pourmond M: Some current aspects of clinical and experimental neuroendocrinology with particular reference to growth hormone, thyrotropin and prolactin. *J Endocrinol Invest* **2**:307, 1979.

39. Palkovits M: Topography of chemically identified neurons in the central nervous system: Progress in 1977–1979. *Med Biol* **58**:188, 1980.

40. Gibbs DM, Neill JD: Dopamine leels in hypophysial stalk blood in the rat are sufficient to inhibit prolactin secretion in vivo. *Endocrinology* **102**:1895, 1978.

41. Tashjian AH Jr, Barowsky NJ, Jensen DK: Thyrotropin releasing hormone: Direct evidence for stimulation of prolactin production by pituitary cells in culture. *Biochem Biophys Res Commun* **43**:516, 1971.

42. Gautvik KM, Weintraub BD, Graeber CT, et al: Serum prolactin and TSH: Effects of nursing and pyro-Glu-His-ProNH$_2$ administration in postpartum women. *J Clin Endocrinol Metab* **37**:135, 1973.

43. Edwards CRW, Forsyth IA, Besser GM: Amenorrhea, galactorrhea and primary hypothyroidism with high circulating levels of prolactin. *Br Med J* **3**:462, 1971.

44. Keye WR, Ho Yuen B, Knopf RF, et al: Amenorrhea, hyperprolactinemia and pituitary enlargement secondary to primary hypothyroidism. Successful treatment with thyroid replacement. *Obstet Gynecol* **48**:697, 1976.

45. Honbo KS, Van Herle AJ, Kellet KA: Serum prolactin in untreated primary hypothyroidism. *Am J Med* **64**:782, 1978.

46. Ross F, Nusynowitz ML: A syndrome of primary hypothyroidism, amenorrhea and galactorrhea. *J Clin Endocrinol Metab* **28**:591, 1968.

47. Feek CM, Sawers JSA, Brown NS, et al: Influence of thyroid status on dopaminergic inhibition of thyrotropin and prolactin secretion: Evidence for an additional feedback mechanism in the control of thyroid hormone secretion. *J Clin Endocrinol Metab* **51**:585, 1980.

48. Scanlon MF, Mora B, Shale DJ, et al: Evidence for dopaminergic control of thyrotropin (TSH) secretion in man. *Lancet* **2**:421, 1977.

49. Contreras P, Generini G, Michelsen H, et al: Hyperprolactinemia and galactorrhea: Spontaneous versus iatrogenic hypothyroidism. *J Clin Endocrinol Metab* **53**:1036, 1981.

50. Kapcala LP: Galactorrhea and thyrotoxicosis. *Arch Intern Med* **144**:2349, 1984.

51. Jaquet P, Grisoli F, Guibout M, et al: Prolactin secreting tumors: Endocrine status before and after surgery in 33 women. *J Clin Endocrinol Metab* **46**:459, 1978.

52. Healy DL, Pepperell RJ, Stockdale J, et al: Pituitary autonomy in hyperprolactinemic secondary amenorrhea: Results of hypothalamic-pituitary testing. *J Clin Endocrinol Metab* **44**:809, 1977.

53. Jeske W: The effect of metoclopramide, TRH and L-dopa on prolactin secretion in pituitary adenoma and in "functional" galactorrhoea syndrome. *Acta Endocrinol (Copenh)* **91**:385, 1979.

54. Lamberts SWJ, Birkenhäger JC, Kwa HG: Basal and TRH-stimulated prolactin in patients with pituitary tumours. *Clin Endocrinol (Oxf)* **5**:709, 1976.

55. Schlechte JA, Sherman BM: Abnormal regulation of prolactin secretion after successful surgery for prolactin-secreting pituitary tumours. *Clin Endocrinol (Oxf)* **15**:165, 1981.

56. Yamaji T, Shimamoto K, Kosaka K, et al: Heterogeneity of prolactin and TSH response to TRH in hypogonadotropic hypogonadism. *J Clin Endocrinol Metab* **45**:319, 1977.

57. Spitz IM, Hirsch HJ, Trestian S: The prolactin response to thyrotropin-releasing hormone differentiates isolated gonadotropin deficiency from delayed puberty. *N Engl J Med* **308**:575, 1983.

58. Yamaji T: Modulation of prolactin release by altered levels of thyroid hormones. *Metabolism* **23**:745, 1974.

59. Ferrari C, Caldara R, Romussi M, et al: Prolactin suppression by serotonin antagonists in man: Further evidence for serotonegic control of prolactin secretion. *Neuroendocrinology* **25**:319, 1978.

60. Morley JE: The endocrinology of the opiates and opiod peptides. *Metabolism* **30**:195, 1981.
61. Wardlaw SL, Wehrenberg WB, Ferin M, et al: High levels of β-endorphin in hypophyseal portal blood. *Endocrinology* **106**:1323, 1980.
62. Rivier C, Vale W, Ling N, et al: Stimulation in vivo of the secretion of prolactin and growth hormone by β-endorphin. *Endocrinology* **100**:238, 1977.
63. Foley KM, Kourides IA, Inturrisi CE: β-Endorphin analgesic and hormonal effects in humans. *Proc Natl Acad Sci USA* **76**:5377, 1979.
64. Trouillas J, Girod C, Sassolas G, et al: A human β-endorphin pituitary adenoma. *J Clin Endocrinol Metab* **58**:242, 1984.
65. Hokfelt T, Fuxe K: Effects of prolactin and ergot alkaloids on the tuberoinfundibular dopamine neurons. *Neuroendocrinology* **9**:100, 1972.
66. Serri O, Kuchel O, Buu NT, et al: Differential effect of low dose dopamine infusion on prolactin secretion in normal and hyperprolactinemic subjects. *J Clin Endocrinol Metab* **56**:255, 1983.
67. Schally AV, Redding TW, Arimura A, et al: Isolation of gamma-aminobutyric acid from pig hypothalami and demonstration of its prolactin release-inhibiting (PIF) activity in vivo and in vitro. *Endocrinology* **100**:681, 1977.
68. Nistri A, Costanti A: Pharmacological characterization of different types of GABA and glutamate receptors in vertebrates and invertebrates. *Prog Neurobiol* **13**:177, 1979.
69. Vijayan E, McCann SM: Effects of intraventricular injection of γ-aminobutyric acid (GABA) on plasma growth hormone and thyrotropin in conscious ovariectomized rats. *Endocrinology* **103**:1888, 1978.
70. Vijayan E, McCann SM: The effects of intraventricular injection of γ-aminobutyric acid (GABA) on prolactin and gonadotropin release in conscious female rats. *Brain Res* **155**:35, 1978.
71. Locatelli V, Cocchi D, Frigerio C, et al: Dual γ-aminobutyric acid control of prolactin secretion in the rat. *Endocrinology* **105**:778, 1979.
72. Takahara J, Yunoki S, Yakushiji W, et al: Stimulatory effects of gamma-hydroxy butyric acid on growth hormone and prolactin release in humans. *J Clin Endocrinol Metab* **44**:1014, 1977.
73. Melis GB, Paoletti AM, Mais V, et al: The effects of the gabaergic drug, sodium valproate, on prolactin secretion in normal and hyperprolactinemic subjects. *J Clin Endocrinol Metab* **54**:485, 1982.
74. Nicoll CS: Aspects of the neural control of prolactin secretion. In Martini L, Ganong WF (eds): *Frontiers in Neuroendocrinology.* Oxford University Press, New York, 1971, p. 291.
75. Nicholson G, Greeley GH Jr, Humm J, et al: Prolactin in cerebralspinal fluid: a probable site of prolactin autoregulation. *Brain Res* **190**:477, 1980.
76. Herbert DC, Ishikawa H, Rennels EG: Evidence for the autoregulation of hormone secretion by prolactin. *Endocrinology* **104**:97, 1979.
77. Melmed S, Carlson HE, Briggs J, et al: Autofeedback of prolactin in cultured prolactin-secreting pituitary cells. *Horm Res* **12**:340, 1980.
78. Ben-Jonathan N: Dopamine: A prolactin-inhibiting hormone. *Endocrine Rev* **6**:564, 1985.
79. Frantz AG: Prolactin. *N Engl J Med* **298**:201, 1978.
80. Thorner MO: Prolactin. *Clin Endocrinol Metab* **6**:201, 1977.
81. Nokin J, Vekemans M, l'Hermite M, et al: Circadian periodicity of serum prolactin concentration in man. *Br Med J* **3**:561, 1972.
82. Sassin JF, FRantz AG, Weitzman ED, et al: Human prolactin: 24 hour pattern with increased release during sleep. *Science* **177**:1205, 1972.
83. Parker DC, Rossman LG, VanderLaan EF: Sleep-related nyctohemeral and briefly episodic variation in plasma prolactin concentrations. *J Clin Endocrinol Metab* **36**:1119, 1973.
84. Robyn C, Delvoye P, Nokin J, et al: Prolactin and human reproduction. In Pasteels JL, and Robyn C (eds): *Human Prolactin.* International Congress Series, No. 308. Excerpta Medica, Amsterdam, 1973, p. 167.
85. Jackson RD, Wortsman J, Malarkey WB: Macroprolactinemia presenting like a pituitary tumor. *Am J Med* **78**:346, 1985.

86. Andersen AN, Pederawn H, Djursing H, et al: Bioactivity of prolactin in a woman with an excess of large molecular size prolactin, persistent hyperprolactinemia and spontaneous conception. *Fertil Steril* **38**:625, 1982.

87. Whittaker PG, Wilcox T, Lind T: Maintained fertility in a patient with hyperprolactinemia due to big, big prolactin. *J Clin Endocrinol Metab* **53**:863, 1981.

88. Connell JMC, Ball SG, Balmforth AJ, et al: Effect of low-dose dopamine infusion on basal and stimulated TSH and prolactin concentrations in man. *Clin Endocrinol (Oxf)* **23**:185, 1985.

89. Neill JD, Frawley LS, Plotsky PM, et al: Dopamine in hypophyseal stalk blood of the rhesus monkey and its role in regulating prolactin secretion. *Endocrinology* **108**:489, 1981.

90. Judd SJ, Rigg LA, Yen SSC: The effects of ovariectomy and estrogen treatment on the dopamine inhibition of gonadotropin and prolactin release. *J Clin Endocrinol Metab* **49**:182, 1979.

91. Valcavi R, Harris PE, Foord SM, et al: The influence of estrogens on the sensitivity of PRL, TSH and LH to the inhibitory actions of dopamine in hyperprolactinaemic patients. *Clin Endocrinol (Oxf)* **23**:139, 1985.

92. Ghigo E, Goffi S, Molinatti GM, et al: Prolactin and TSH responses to both domperidone and TRH in normal and hyperprolactinaemic women after dopamine synthesis blockade. *Clin Endocrinol (Oxf)* **23**:155, 1985.

93. Camanni F, Genazzani AR, Massara F, et al: Prolactin-releasing effect of domperidone in normoprolactinemic subjects. *Neuroendocrinology* **30**:2, 1980.

94. Ho KY, Smythe A, Lazarus L: The interaction of TRH and dopaminergic mechanisms in the regulation of stimulated prolactin release in man. *Clin Endocrinol (Oxf)* **23**:7, 1985.

95. Cowden EA, Ratcliffe JG, Thomson JA, et al: Tests of prolactin secretion in the diagnosis of prolactinoma. *Lancet* **1**:1155, 1979.

96. Massara F, Camanni F, Martra M, et al: REciprocal pattern of the TSH and PRL responses to dopamine receptor blockade in women with physiological or pathological hyperprolactinaemia. *Clin Endocrinol (Oxf)* **18**:103, 1983.

97. Scanlon MF, Rodriguez-Arnao MD, McGregor AM, et al: Altered dopaminergic regulation of thyrotrophin release in patient with prolactinoma: Comparison with other tests of hypothalamic pituitary function. *Clin Endocrinol (Oxf)* **14**:133, 1981.

98. Genazzani AR, De Leo V, Murru S, et al: Dynamic tests of prolactin secretion in hyperprolactinemic states: Carbidopa-L-dopa and indirectly acting dopamine agonists. *J Clin Endocrinol Metab* **54**:429, 1982.

99. Fine SA, Frohman LA: Loss of central nervous system component of dopaminergic inhibition of prolactin secretion in patients with prolactin-secreting pituitary tumors. *J Clin Invest* **61**:973, 1978.

100. Crosignani PG, Ferrari C, Malinverni A, et al: Effect of central nervous system dopaminergic activation on prolactin secretion in man: Evidence for a common central defect in hyperprolactinemic patients with and without radiological signs of pituitary tumors. *J Clin Endocrinol Metab* **51**:1068, 1980.

101. Muller EE, Genazzani AR, Murru S: Nomifensine: Diagnostic test in hyperprolactinemic states. *J Clin Endocrinol Metab* **47**:1352, 1978.

102. Ferrari C, Crosignani PG, Caldara R., et al: Failure of nomifensine administration to discriminate between tumorous and nontumorous hyperprolactinemia. *J Clin Endocrinol Metab* **50**:23, 1979.

103. Robinson B: Breast changes in the male and female with chlorpromazine or reserpine therapy. *Med J Aust* **2**:239, 1957.

104. Langer R, Ferin M, Sachar EJ: Effect of haloperidol and L-dopa on plasma prolactin in stalk-sectioned and intact monkeys. *Endocrinology* **102**:367, 1978.

105. Cooper DS, Gelenberg AJ, Wojcik JC, et al: The effect of amoxapine and imipramine on serum prolactin levels. *Arch Intern Med* **141**:1023, 1981.

106. Refetoff S, Frank PH, Roudebush C, et al: Evaluation of pituitary function. In DeGroot LJ, Cahill GF, Martini L, et al (eds): *Endocrinology.* Grune & Stratton, New York, 1979, p. 175.

107. Gelenberg AJ, Cooper DS, Doller JC, et al: Galactorrhea and hyperprolactinemia associated with amoxapine therapy. *JAMA* **242**:1900, 1979.
108. Gluskin LE, STrasberg B, Shah JH: Verapamil-induced hyperprolactinemia and galactorrhea. *Ann Intern Med* **95**:66, 1981.
109. Röjdmark S, Andersson DEH: Cimetidine effect on dopaminergic modulation of prolactin release in healthy women. *Metabolism* **31**:1042, 1982.
110. Lankford HV, Blackard WG, Gardner DF, et al: Effects of thyrotropin-releasing hormone and metoclopramide in patients with phenothiazine-induced hyperprolactinemia *J Clin Endocrinol Metab* **53**:109, 1981.
111. Davis JRE, Selby C, Jeffcoate WJ: Oral contraceptive agents do not affect serum prolactin in normal women. *Clin Endocrinol (Oxf)* **20**:427, 1984.
112. Molitch ME, Reichlin S: Hypothalamic hyperprolactinemia: Neuroendocrine regulation of prolactin secretion in patients with lesions of the hypothalamus and pituitary stalk. In MacLeod, RM, Thorner MO, Scapagnini U. (eds): *Prolactin. Basic and Clinical Correlates*. Liviana Press, Padova, 1985, p. 709.
113. Malarkey WB, Martin TL, Kim M: Patients with idiopathic hyperprolactinemia infrequently develop pituitary tumors. In MacLeod RM, Thorner MO, Scapagnini U. (eds): *Prolactin. Basic and Clinical Correlates*. Liviana Press, Padova, 1985, p. 705.
114. Morley JE, Dawson M. Hodgkinson H, et al: Galactorrhea and hyperprolactinemia associated with chest wall injury. *J Clin Endocrinol Metab* **45**:931, 1977.
115. Berger RL, Joison J, Braverman L, et al: Lactation after incision of the thoracic cage. *N Engl J Med* **274**:1493, 1966.
116. Weir JH: Post thoracotomy galactorrhea successfully treated with clomiphene citrate. *Am J Obstet Gynecol* **111**:106, 1971.
117. Molitch ME, Schwartz S, Mukherji B: Is prolactin secreted ectopically? *Am J Med* **70**:803, 1981.
118. Podmore J, Wilson B, Cowden EA, et al: Multiple hormones in human tumors. In Lehman GF (ed): *Carcino-embryonic proteins*. Vol. 1. Elsevier/North-Holland, New York, 1979, p. 457.
119. Rees LH, Bloomfield GA, Rees GM, et al: Multiple hormones in a bronchial tumor. *J Clin Endocrinol Metab* **28**:1090, 1974.
120. Turkington RW: Ectopic production of prolactin. *N Engl J Med* **285**:1455, 1971.
121. Davis S, Proper S, May PB, et al: Elevated prolactin levels in bronchogenic carcinoma. *Cancer* **44**:676, 1979.
122. Kleinberg DL, Noel GL, Frantz AG: Galactorrhea: A study of 235 cases, including 48 with pituitary tumors. *N Engl J Med* **296**:589, 1977.
123. Antunes JL, Housepian EM, Frantz AG, et al: Prolactin-secreting pituitary tumors. *Ann Neurol* **2**:148, 1977.

14

Prolactin-Secreting Tumors

Introduction

Prolactin-secreting tumors are the most frequently encountered pituitary disorders in clinical practice. The realization that nearly 40% of so-called nonsecretory and functionless pituitary tumors actually contain prolactin was an exciting one.[1] This realization in conjunction with the development of a sensitive immunoassay for prolactin, the emergence of high-quality computed tomography (CT), and the discovery of dopamine agonist drugs has resulted

TABLE 75
Spectrum of Hyperprolactinemia

Condition	Serum prolactin (ng/ml)	Comment
Idiopathic hyperprolactinemia	↑ mild (<100)	CT negative; no lesions in the pituitary or hypothalamus; benign, often self-limited course; does not progress into tumors
Microprolactinomas	↑ ↑ (100–300)	Tumors <1 cm by CT. No impairment in pituitary function; does not usually progress to microadenomas; negligible chances of enlargement during pregnancy
Macroprolactinomas	↑ ↑ ↑ >300	Tumors >1 cm by CT. May impair pituitary function and often causes chiasmal compression
Invasive prolactinomas	↑ ↑ ↑ (500–2000)	Tumor >1 cm with extension. Often impairs pituitary function and often causes chiasmal compression
Pseudoprolactinoma	Variably ↑	Large nonsecretory tumor (negative staining for prolactin), with pressure on hypothalamic-pituitary stalk. Interruption of dopamine transport elevates PRL

in a remarkably rapid growth in the understanding of prolactin-related disorders.

Prolactin-secreting tumors can be classified into micro- and macroprolactinomas. Such a distinction is important for practical reasons, since there are significant differences in sex incidence, presentation, natural history, and the prognosis between the two types. There is good reason to believe that only a small percentage of microadenomas go on to become macroadenomas. It is also important to recognize that not all pituitary tumors associated with elevated serum prolactin levels represent true prolactinomas. The term pseudoprolactinoma, originally used by Dr. Randall of Mayo Clinic, refers to the hyperprolactinemia that results from a large pituitary tumor (nonsecretory) compressing the stalk and interrupting dopamine transport. The diagnosis of a true prolactinoma rests on the demonstration of immunopositivity for prolactin (PRL) in the tumor cells. The spectrum of hyperprolactinemia involves five distinct anatomically- and perhaps physiologically—different entities; idiopathic hyperprolactinemia, microprolactinoma, macroprolactinoma, invasive macroprolactinoma, and pseudoprolactinoma (Table 75). The clinical presentations, course, and therapy of each entity are also widely divergent.

Etiology

The etiology of prolactinomas is largely unknown. The common physiological thread woven through all the disorders in the spectrum of hyperprolactinemia is a central defect in hypothalamic prolactin-inhibitory dopaminergic regulation. This is based on the remarkable response of all types of prolactin-secreting tumors to dopamine agonist drugs. However, the reasons that trigger, or perpetuate the process which lead to tumor formation are unclear. One obvious explanation is the hypothesis that lactotrope hyperplasia is a result of escape from the tonic inhibitory control of hypothalamic dopamine exerted on these cells. Experimental evidence is also suggestive of the fact that dopamine is involved in the control of cell division of the lactotropes. Studies by Lloyd et al.[2] demonstrated that when dopamine receptors of the lactotropes are blocked by a dopamine antagonists, this is followed by a striking increase in mitotic activity and DNA content of the lactotropes. Although it is known that prolactin-secreting tumors possess dopamine receptors, it is not known whether tumor cells are relatively resistant to endogenous (hypothalamic) dopamine. Such resistance can result in "lactotrope escape," followed by lactotrope hyperplasia and eventual adenomatous transformation. In the background of this perturbation in dopaminergic regulation of lactotrope function, the role of two other putative factors in the causation of prolactinomas merit consideration; the role of estrogens, and the role of regional vascularization of the anterior pituitary.

The role of estrogens in the causation of prolactinomas is speculative and is based primarily on experimental—and some clinical—observations. In rats, chronic treatment with diethylstilbestrol (DES) can induce prolactinomas that show functional and morphological similarities to human prolactinomas. Using DES-treated rat models, Phelps and Bartke[3] showed that the hyperprolactinemia effected by continued DES administration failed to increase the tuberoinfundibular dopamine neurons. Based on normal feedback principles, hyperprolactinemia normally increases dopamine content of these hypothalamic dopaminergic neurons, which in turn will decrease PRL level. If DES is withdrawn, in the rat model, dopamine fluorescence can be shown to rebound to normal levels, indicating that continued DES treatment somehow affected hypothalamic inhibition of PRL regulation. Clinically, the well-known effect of pituitary enlargement during pregnancy is mediated by estrogens. The increase in the lactotrope population seen during pregnancy largely contributes to the glandular enlargement. Upon estrogen withdrawal following delivery, the lactotrope population returns to normal, provided the infant is not breast fed. Also, the well-known effect of pregnancy on growth of macroprolactinomas is mediated by estrogens. Despite these observations, it is difficult to ascribe an etiological role to estrogens in the causation of prolactin-secreting tumors. There is little evidence that the low dose of estrogens present in currently used oral contraceptive agents causes microprolactinomas.[4] Davis et al.[5] surveyed 230 healthy women on oral contraceptive agents and found no differences in the distribution profile of serum PRL levels as compared with an identical group of women not using oral contraceptives. To the extent that preexisting macroprolactinomas can enlarge in the milieu of high concentrations of estrogens (as in pregnancy), the role of estrogens is relevant; the rest is speculative.

The role of a vascular etiology in the development of prolactinomas is also under investigation. Weiner et al.[6] suggested a vascular etiology to explain the focal nature of these tumors and their escape from dopamine inhibition. These workers believe that direct arterial blood supply to a region of the anterior pituitary may be a fairly common anomaly. Such an occurrence would be likely to decrease the dopamine concentration in that particular region due to dilution with systemic blood. Speculative as it is, this may predispose to the development of tumorigenesis when stimulated by factors such as estrogens. Data from experiments in rats are suggestive of such a phenomenon.

Histology

The hallmark of prolactin-secreting tumors is the immunopositivity with prolactin antisera. As with the somatotropinomas, a densely granulated and a sparsely granulated variety have been recognized, the latter variant being

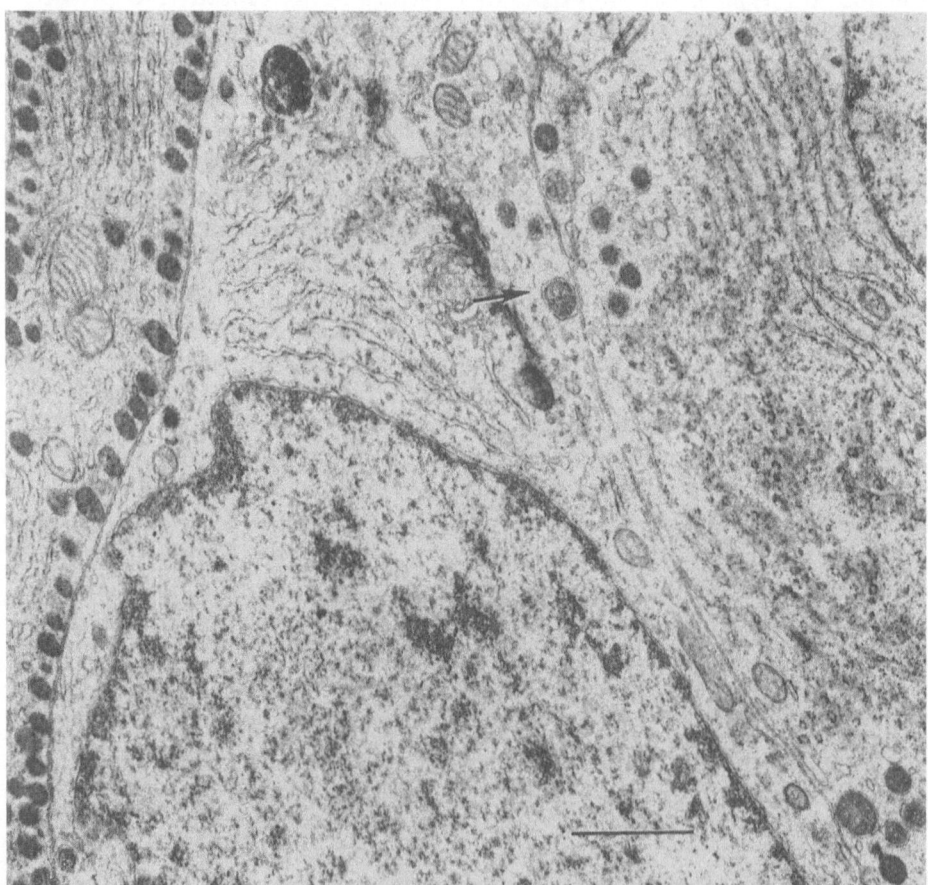

FIGURE 32.   Electron micrograph of a sparsely granulated prolactinoma demonstrating rich endoplasmic reticulin and misplaced exocytosis (arrow). Scale bar: 1 μm. (From Adelman.[174])

more common.[7,8] The appearance of the tumor cells in the densely granulated variety resembles normal lactotropes, with a prominent rough endoplasmic reticulum and Golgi apparatus. The granules of the densely granulated prolactinoma cells are impressively large, ranging from 600 nm to as large as 1200 nm. The densely granulated prolactinoma stains eosinophilic by conventional stains. The sparsely granulated prolactinoma is the more common variety and consists of smaller granules (200–300 nm) and more prominent endoplasmic reticulum and Golgi apparatus. The endoplasmic reticulum can be impressively prominent, often arranged in whorls referred to as "Nebenkerns." Another feature when prominent, permits recognition of the sparsely granulated prolactinoma; this is the presence of extrusions of granules into the extracellular space (misplaced exocytosis) (Fig. 32). The sparsely granulated variety may not uniformly show immunopositivity with prolactin antisera. This variety stains chromophobic by conventional stains.

Clinical Features

The clinical syndromes that result from hyperprolactinemia are hardly new. In 1954, at a time when PRL was a yet-uncharacterized hormone, Forbes et al.[9] described the remarkable triad of amenorrhea, galactorrhea and low urinary FSH levels. With prophetic accuracy, Dr. Anne Forbes hypothesized that the syndrome had to be somehow linked with prolactin. Their report was by no means the first to remark on the galactorrhea syndrome, since Frommel[10] had written about it 72 years earlier; however, it was left to Forbes et al.[9] to characterize systematically the seven major points associated with that syndrome. These constellation of features, distilled at a time when laboratory proof was nearly nonexistent, reverberate with clarity of thought. Three decades later, the seven features described by these workers continue to characterize the syndrome with the added testimony of having been proved by time and present-day techniques. The seven points made by Forbes et al.[9] were as follows:

1. The disease was distinct from acromegaly.
2. Nonpuerperal galactorrhea was the prominent clinical effect in the patients described.
3. Amenorrhea figured prominently in the 15 patients described.
4. Low urinary FSH levels were demonstrated by bioassay.
5. There was evidence of mild hyperfunction of the adrenal glands.
6. Underlying pituitary tumor was proposed as a cause for the amenorrhea-galactorrhea syndrome.
7. The presumption was that PRL hypersecretion represented the key hormonal abnormality in the syndrome described by these workers.

Prolactin-secreting microadenomas of the pituitary are the most common pituitary disease seen in clinical practice. The availability of a sensitive radioimmunoassay for PRL as well as the emergence of high-resolution computed tomography (CT) to detect microadenomas have contributed to the detection of high numbers of patients with prolactin-secreting tumors. Equally important is the awareness that hyperprolactinemia constitutes an important etiology for amenorrhea, oligomenorrhea, infertility, and impotence. An estimated 25–30% of cases of secondary amenorrhea are causally related to hyperprolactinemia.[11–14] Prolactin-related problems rank as the second most common cause for secondary amenorrhea, mandating screening for hyperprolactinemia in all women with secondary amenorrhea. A similar percentage (25–30%) of cases of infertility are caused by prolactin hypersecretion. In males, hyperprolactinemia represents a small (5%) but significant cause of impotence.

The presentation of prolactinomas in females is impressively different as compared with that of male prolactinomas. In females, prolactinomas are usually small, and display a rather benign course, often remaining static and seldom showing an aggressive behavior. In males, the presentation is characterized by large tumors, often displaying an aggressive behavior with neighborhood compression syndromes and impairment in pituitary reserve. The symptoms in females with hyperprolactinemia are secondary amenorrhea, oligomenorrhea, galactorrhea, and infertility. In addition, symptoms of estrogen deficiency (dyspareunia and decreased libido) are frequently encountered. The main symptom of hyperprolactinemia in males is decreased libido. Less frequently, infertility and galactorrhea are encountered in males with hyperprolactinemia. Symptoms such as headache and visual-field cuts result from invasive prolactinomas in both sexes but are encountered more frequently in males.

Prolactinomas in Women

Menstrual Irregularities

Secondary amenorrhea or oligomenorrhea is the classic symptom of prolactinomas in women, and is seen in approximately 65–70% of patients. Amenorrhea is more likely to be seen in the presence of a radiographically detectable tumor, and especially when seen in conjunction with galactorrhea. It should be emphasized that patients with prolactinomas can, and often do, present with regular cycles, but with anovulation, infertility, or a decreased luteal phase. In an infertile patient with intact regular cycles, PRL problems cannot be excluded just on the basis of a "regular" cycle. In general, hyperprolactinemic amenorrhea correlates with duration of hyperprolactinemia, and presence of tumor. In a large series of 144 women with secondary amenorrhea, Haesslein and Lamb[15] noted that when the amenorrhea was of 2 years duration, the incidence of detecting pituitary tumors was 17%, a percentage that increased to 25% when the amenorrhea was longer than 5 years. In the presence of secondary amenorrhea alone, i.e., in the absence of galactorrhea, the incidence of detecting hyperprolactinemia is approximately 22%; when galactorrhea and amenorrhea occur together, the incidence of detecting hyperprolactinemia is about 50%. Biller et al.[16] noted in their series of 70 patients that 25% of patients developed symptoms after discontinuing oral contraceptives. The mechanism of estrogen deficiency and the secondary amenorrhea that results from hyperprolactinemia is multifactorial. Three mechanisms may underlie the hypogonadism of hyperprolactinemia: hypothalamic, gonadal, and pituitary:

Hypothalamic Defect. The basal and mean LH and FSH levels in patients with prolactinomas are usually normal and similar to levels seen in normal women in the follicular phase of their menstrual cycle. The gonadotropin responses to exogenous GnRH administration are well preserved in most patients with hyperprolactinemia, unless gonadotropin reserve is compromised by a large tumor. The estrogen-induced positive feedback of gonadotropin release may be absent in patients with hyperprolactinemia.[17] It has been suggested that increased opioid inhibition of LH secretion may underlie the amenorrhea seen in the hyperprolactinemia caused by microadenoma.[18] Grossman et al.[19] infused high doses of naloxone into five patients with hyperprolactinemic amenorrhea and demonstrated a marked rise in LH and FSH levels in all five, following administration of naloxone. These investigators concluded that opiate-mediated tonic inhibition of hypothalamic GnRH plays an important role in the amenorrhea associated with hyperprolactinemia. The three major abnormalities variably seen in hyperprolactinemic patients, i.e., apulsatile gonadotropin secretion, absent LH surge, and low estradiol levels, are all restored to normal upon instituting dopamine agonist therapy, in some cases even before normalization of the prolactin level.

Gonadal Defect. High PRL levels have an inhibitory influence on ovarian steriodogenesis and follicular growth. It is believed that hyperprolactinemia induces a state of resistance to gonadotropins at the ovarian level. In contrast to the hypothalamic defect (in GnRH release), the defective ovarian response to gonadotropin stimulation may persist for weeks after normoprolactinemia has been achieved.

Pituitary Defect. In patients with macroprolactinomas, gonadotropin reserve can be impaired. In some cases, the impairment of reserve can be a result of compression of normal tissues by the large tumor. When tumor size is reduced by bromocriptine, the gonadotropin reserve improves in some patients, possibly since the compressive effect has been removed.

Infertility

Hyperprolactinemia is an important cause of infertility in 20–30% of patients with that problem. The lack of ovulation and the defective luteal phase are the two important reasons for infertility. Infertility caused by hyperprolactinemia is correctible in most patients when the PRL level is normalized. The only exceptions are patients with associated ovarian disorders and those who have impaired gonadotropin reserve due to destruction of the normal gonadotropes by the encroaching tumor.

Galactorrhea

Galactorrhea is encountered in 30–80% of patients, with hyperprolactinemia, depending on the zeal with which it is sought.[20,21] The usually quoted incidence of galactorrhea in hyperprolactinemic women is 30%. Jacobs et al.[22] noted galactorrhea in 37% of their patients with hyperprolactinemia. Molitch and Reichlin[23] summarized the collective data from seven series and noted an overall incidence of galactorrhea in 12.8%. Several clinical observations are pertinent with regard to hyperprolactinemia and galactorrhea: (1) there is no correlation between the presence of galactorrhea and the degree of hyperprolactinemia; (2) menstrual irregularities often precede the detection of galactorrhea by months to years; (3) when galactorrhea and amenorrhea coexist with hyperprolactinemia, the chances of radiological detection of pituitary tumors are as high as 70–80%; (4) galactorrhea is an extremely unusual manifestation of male prolactinomas, since the male breast tissue is unprimed by estrogens and progesterone; and (5) the galactorrhea can be unilateral.[24,25]

Prolactinomas in Men

The age difference in the presentation between men and women harboring prolactinomas is noteworthy. In women, hyperprolactinemia is a condition seen mostly between the ages of 20 and 35, although the disorder can be encountered at any age. In most series, the mean age is 28–30. By contrast, males with prolactinomas come to attention in their fifth or sixth decade. In one series of 55 men with prolactinoma reported by Hardy,[26] the mean age for males with prolactinomas was younger—39 years of age. Morphologically, the tumor is much larger in males, often causing significant compressive effects. Thus, the presentation of prolactinoma in males is dominated by the pressure effects caused by the tumor. Impotence is the leading symptom in men with prolactin excess.[27] The incidence of hyperprolactinemia as a cause of male impotence has been reported to range from 5 to 25%.[27–31] This may reflect an overestimation of the problem owing to the highly selected population of patients. Boyd et al.[32] noted that of 100 males screened for impotence, only three had elevated PRL levels. In addition to impotence, males with hyperprolactinemia may present with infertility[31,32] galactorrhea or with signs of decreased virilization such as reduced beard growth. It should be noted that galactorrhea is present at the most, in only 20–30% of males with hyperprolactinemia.[27,31,33] In addition to the effects of hyperprolactinemia, the most important symptoms that bring the patient to the physician are those related to the mass effect of tumor (e.g., headaches, chiasmal compression, deficient pituitary hormone reserve).

Several other clinical features are encountered in patients with hyperprolactinemia. Dyspareunia and diminished libido are a direct effect of estrogen deficiency.[34] Mild hirsutism is sometimes encountered in patients with hyperprolactinemia and may be related to the effect of PRL stimulating androgen production by the adrenal cortex. Bassi et al.[35] found elevated levels of dehydroepiandrosterone sulfate in the plasma of 10 patients with hyperprolactinemic amenorrhea. Headaches are often experienced by hyperprolactinemic women, even in the absence of macroprolactinomas.

Diagnostic Tests

The two studies required in the setting of a possible prolactin secreting tumor are the measurement of serum prolactin levels and CT study of the sella and suprasellar region. Exclusion of primary hypothyroidism, renal failure, and a properly obtained drug history to detect drug-induced hyperprolactinemia are basic steps that need to be performed in all patients with a possible PRL problem. In addition to the serum PRL levels and the CT study, most authorities would recommend baseline evaluation of pituitary reserve, particularly when macroadenomas underlie the etiology, and when ablative therapy is contemplated. In the author's experience, the diagnostic approach to patients with PRL-related problems can be viewed as four different groups:

1. When PRL levels are clearly elevated and the CT study shows a microadenoma, pituitary reserve testing yields very little information since hypopituitarism is extremely unlikely with a microadenoma. Basal LH and FSH levels are obtained in this situation, with the aim of excluding coexistent primary ovarian failure.
2. When the serum PRL levels are clearly elevated and the CT study reveals a macroadenoma (with or without suprasellar extension) the patient must undergo a complete pituitary reserve testing using TRH, insulin hypoglycemia, and LRH as provocative stimuli.
3. When the serum PRL levels are clearly elevated, but the CT study of the sella and suprasellar region is entirely normal, additional studies of PRL dynamics may yield information regarding the presence of a small CT-negative microadenoma. The four studies in this regard are evaluation of the PRL response to TRH, metoclopramide, L-dopa–carbidopa combination, and nomifensine. These tests are arguably redundant, since therapy for idiopathic disease and CT-negative microadenoma is essentially the same—dopamine agonist therapy.
4. When the serum PRL level is normal in the patient with galactorrhea associated with amenorrhea, oligomenorrhea, infertility, or a decreased luteal phase, multiple sampling of prolactin levels is indicated. This may be done by sampling every 20 min for four consecutive samples. If the mean PRLs are clearly in the normal range, the diagnosis of normoprolactinemic galactorrhea can be made. In these circumstances further studies of the PRL by bioassay or chromatography may be indicated.

TABLE 76
Diagnostic Approach to Hyperprolactinemia

Group	Feature	Recommended W/U	Additional tests
I	Elevated PRL CT reveals microadenoma	Basal LH, FSH (to exclude primary ovarian disease) T4, T3, TSH (to exclude primary hypothyroidism) Basal GH (to screen for acromegaly)	Complete pituitary reserve testing
II	Elevated PRL, CT reveals macroadenoma	Complete pituitary workup	
III	Elevated PRL levels with normal CT	TRH study	L-Dopa-carbidopa test Nomifensine study Metoclopramide study
IV	Normal PRL	Multiple sampling for PRL CT study	Bioassay Chromatographic studies

Table 76 summarizes the diagnostic approach in the four groups of patients with PRL-related problems. The diagnostic approach for the evaluation of the patient with galactorrhea/irregular cycles should be pursued in a systematic fashion. The first step is to obtain serum PRL levels. The second step is exclusion of drug history, primary hypothyroidism, chronic renal failure, and, rarely, acromegaly. The third step is radiological evaluation of the sella turcica and the suprasellar area by high-resolution CT. In most cases, this is the extent of the workup. Additional testing of pituitary reserve is indicated in some patients. The role of ancillary studies such as TRH study, L-dopa–carbidopa combination study, the nomifensine study, and metoclopramide testing is rather limited and seldom needed.

Prolactin Levels

Measurement of basal PRL levels constitutes the first step in the evaluation of the patient who presents with symptoms that could be related to a PRL problem. The normal range of PRL for females in our laboratory is 5–20 ng/ml, and for males 5–15 ng/ml. Occasionally, because of the pulsatile nature of PRL secretion, mild hyperprolactinemia can be missed unless multiple sampling is done at 10–15-min intervals to obtain a mean level. The degree of hyperprolactinemia has diagnostic and prognostic value. In general, when hyperprolactinemia is caused by drugs, idiopathic (functional) galactorrhea, and hypothalamic-stalk lesions the degree of PRL elevation tends to be mild to modest usually below 100 ng/ml. Molitch and Reichlin[36] evaluated six patients with lesions of the hypothalamus-stalk region, and 8 with large "nonfunctioning" pituitary adenomas with moderate to severe extrasellar extension and found the PRL levels to be modestly elevated (<100 ng/ml) in all but one. Occasionally, nonprolactin-secreting tumors can cause significant hyperprolactinemia (pseudoprolactinoma). This is unusual, however. Patients with tumorous hyperprolactinemia may show a wide range of elevation from 100 ng to as high as 3000 ng/ml. There seems to be a close correlation between the degree of hyperprolactinemia and tumor size. Patients harboring microadenomas demonstrate basal PRL levels of 150–300 ng/ml. Prolactin levels of >500 ng/ml are usually indicative of macroadenomas, with or without suprasellar extension. Males with PRL-secreting tumors have much higher basal prolactin levels than do females; the magnitude of hyperprolactinemia is impressive in male prolactinomas and can be as high as 4500 ng/ml.

The degree of hyperprolactinemia in patients with microprolactinomas does have some impact on the initial therapeutic recommendation; patients with microprolactinomas who have basal levels of >200 or 250 ng/ml do not respond as gratifyingly to microadenomectomy as those with levels below 200 ng/ml. Such is not the case with dopamine agonist therapy, which lowers PRL level regardless of the pretreatment basal level.

When the PRL level is completely normal despite repeated sampling, in the patient with galactorrhea with or without irregular menstrual cycles, the entity of normoprolactinemic galactorrhea is said to exist. The galactorrhea in such patients is thought to reflect increased tissue sensitivity to normal levels of PRL in the circulation. The possibility that the PRL in the serum was bioactive, but not measured by the immunoassay, has been one consideration to explain the galactorrhea in such instances. Johnston et al.[37] evaluated PRL secretion and biological activity in 20 females with normoprolactinemic galactorrhea; the basal bioassayable PRL, measured by its ability to stimulate proliferation of Nb_2 node rat lymphoma cells, was found to be normal in all patients. The presence of mild metabolic abnormalities, and galactorrhea in conjunction with normal immunoassayable and bioassayable PRL concentrations led these workers to conclude that increased tissue sensitivity to the lactogenic and metabolic actions of PRL underlied the clinical expression in these patients, with relative sparing of ovarian cyclical function.

Radiological Examination of the Sella and Suprasellar Regions

After having demonstrated hyperprolactinemia, the next step is to exclude some common, and not so common, causes of hyperprolactinemia. The three relevant causes to exclude are drug-induced hyperprolactinemia, primary hypothyroidism, and chronic renal failure. Drugs to be excluded are reserpine, methyldopa, antipsychotic agents, tranquilizers, metoclopramide, and so on. Hypothyroidism and chronic renal failure are readily excluded by measurement of T4, T3R, and TSH and by the BUN and creatinine. The occurrence of hyperprolactinemia in patients with primary hypothyroidism is well recognized.[38] However, the exact prevalence of this phenomenon is less certain; a highly variable incidence of hyperprolactinemia (5–39%) has been reported in untreated primary hypothyroidism.[39] The incidence of galactorrhea is even lower (less than 5%) in untreated primary hypothyroidism. In one study,[40] the incidence of hyperprolactinemia and galactorrhea correlated well with the duration and severity of hypothyroidism and was seen more commonly in premenopausal women. In general, PRL elevations occur later in the course of untreated primary hypothyroidism, and rarely, if ever, in males or postmenopausal females. The exact mechanism is unclear but is presumed to be due either to the effect of increased TRH levels on the lactotrope or to depleted dopaminergic content of the hypothalamus. The latter would explain the prevalence of hyperprolactinemia in chronic untreated cases of primary hypothyroidism, since it would take time for the hypothalamic dopamine to become depleted. The PRL abnormalities reverse with T4 therapy, albeit at a slower rate than the restoration of TSH levels to normal. The observation that hyperprolactinemia persists even after normalization of the TSH levels in patients with primary hypopthyroidism on treatment has cast doubt on the role of TRH in the causation of the hyperprolactinemia of hypothyroidism. Regardless of the mechanism, primary hypothyroidism must be excluded in all patients with hyperprolactinemia. The occasional development of hyperprolactinemia in hyperthyroidism is a well-noted and interesting observation.[41] Rarely, galactorrhea can be an early manifestation of acromegaly, at a stage when the somatic features of that disease have not yet been manifested.[42]

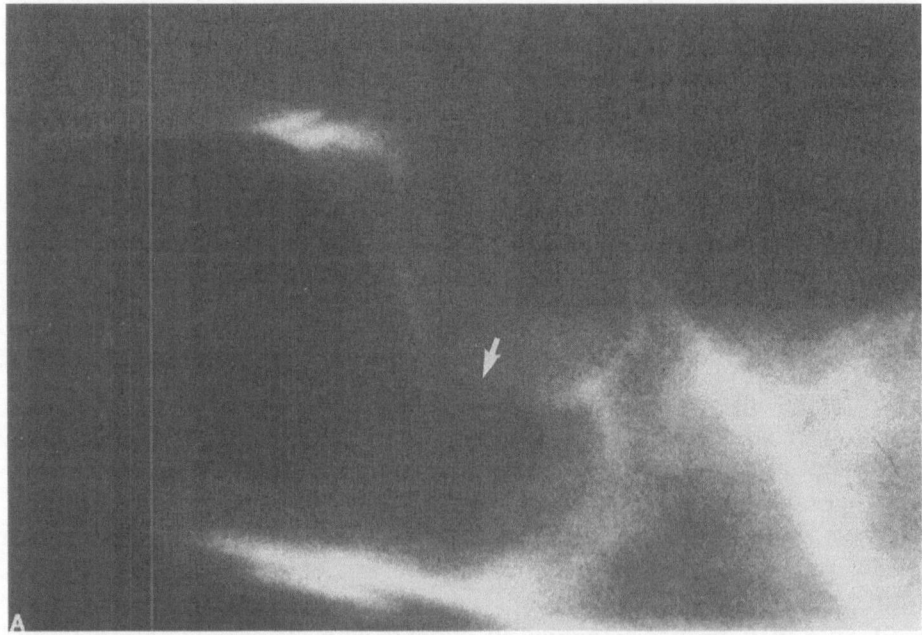

FIGURE 33.   (A) Lateral tomogram demonstrating bulging of the sella inferiorly with a break in the lamina dura (arrow). (B) Frontal tomogram demonstrating bulging of the sella inferiorly (arrows). (C) CT demonstrating a low-density area (arrows) within the sella. (From Wolpert.[175])

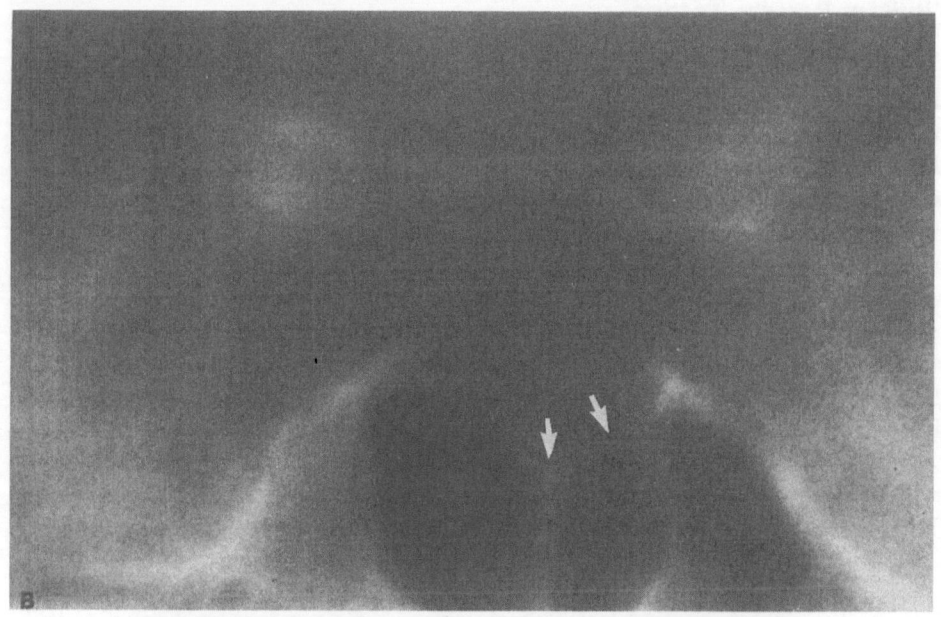

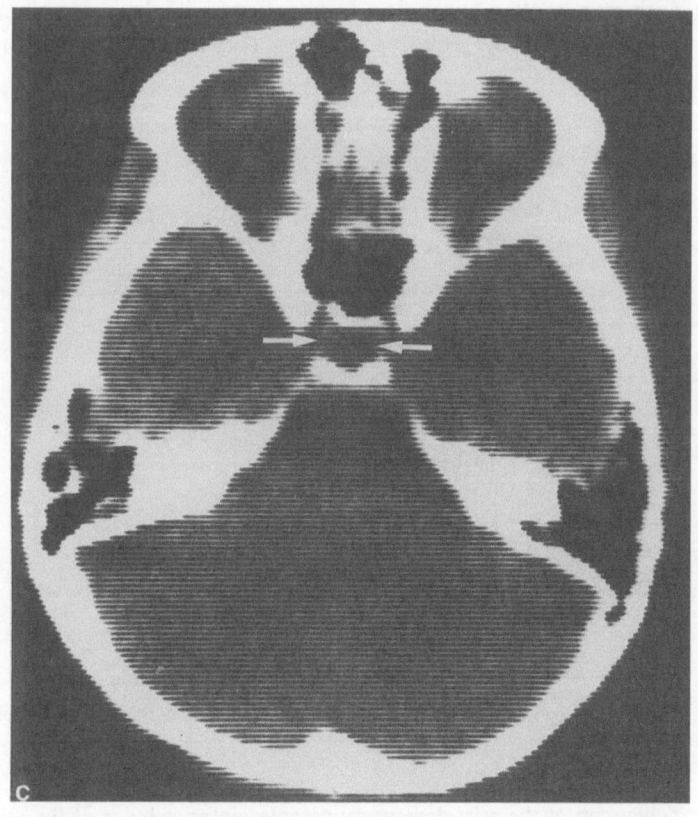

FIGURE 33.   *(Continued)*

Pituitary Reserve Testing

The third step in the approach to the patient with hyperprolactinemia is the radiological evaluation of the sella and suprasellar areas. This is best done by CT. While polytomography continues to be used as a method for evaluating the sella at some institutions, most centers have supplanted this technique with the CT study (Fig. 33). The CT study, in the hyperprolactinemic patient can be normal, may show a microadenoma, may demonstrate a macroadenoma with or without suprasellar extension, or rarely may demonstrate the empty sella syndrome. Small prolactinomas appear as a low-density (or even cystic) lesion by direct coronal, high-detail CT scanning. It is recommended that the scan be viewed with both soft tissue and bone techniques. The presence of associated changes (Fig. 34) in the sellar floor, such as thinning and depression are strong confirmatory signs, especially when contiguous to the soft tissue changes.[43,44] Occasionally, prolactinomas may appear hyperdense in the CT study.[45] The most important differential diagnoses of microprolactinomas by the CT study are cysts and incidental adenomas within the pituitary gland.

The CT study is invaluable in demonstrating the presence of suprasellar extension (Fig. 35). Nonsecretory pituitary adenomas can cause hyperprolactinemia by suprasellar extension. Therefore, when the CT study reveals a large tumor with suprasellar extension in a patient with hyperprolactinemia, the diagnostic considerations revolve around the macroprolactinoma with

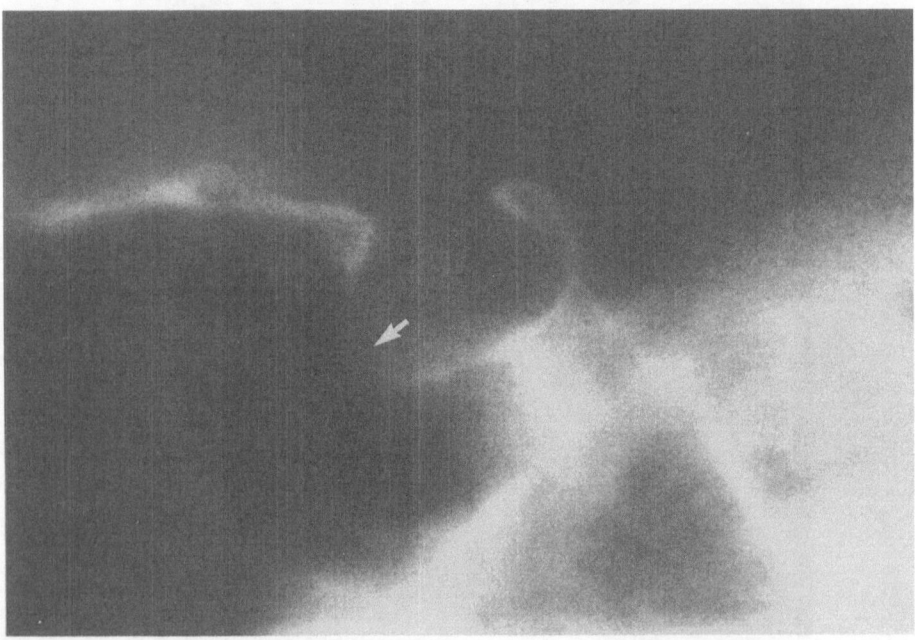

FIGURE 34. Tomogram of the sella demonstrating anteroinferior bulge of the sella. (From Wolpert.[175])

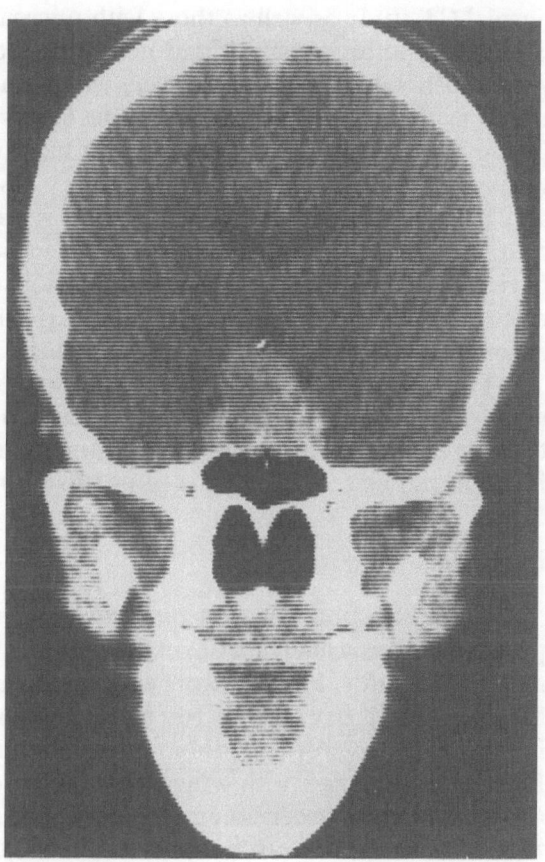

FIGURE 35. Coronal CT scan demonstrating a large enhancing pituitary tumor with suprasellar extension, indenting the third ventricle. (From Post.[176])

suprasellar invasion versus a pseudoprolactinoma. The serum prolactin level is usually markedly elevated in the invasive macroprolactinoma, while being only modestly elevated in the pseudoprolactinoma. This, however, may not always apply, since some patients with pseudoprolactinomas may demonstrate prolactin levels that are quite high.[17]

The association between hyperprolactinemia and the empty sella syndrome is well recognized[34,46–54] (see Chapter 18). Briefly, the development of hyperprolactinemia in patients with the empty sella syndrome could be on the basis of a coexistent microadenoma in the compressed, flattened pituitary gland or could occur as a result of compression of the infundibulum, which is kinked due to the anatomic deformity in the diaphragma sella.

Serum PRL measurement and CT are often the only two definitive tests required in the patient with galactorrhea and menstrual irregularities.

Ancillary Studies

The next step in the evaluation of the patient with hyperprolactinemia is testing the integrity of the other pituitary hormones. In patients with a normal CT study as well as those with microadenomas, the pituitary reserve is not likely to be impaired, and hence the total reserve testing is unlikely to yield much information. However, when surgery (or radiation) is contemplated as the initial therapy, a baseline reserve testing is clearly essential for future comparison. Also, patients with macroadenomas, especially when invasive, require evaluation of the pituitary reserve. While evaluating the LH and FSH response to LH-RH, the wide variability of response patterns encountered in hyperprolactinemic patients must be kept in mind. Most workers have reported normal gonadotropin secretory responses to LH-RH.[55–59] However, blunted[60–62] or even exaggerated LH and FSH responses[56,60,63] may be encountered in hyperprolactinemic patients.

The demonstration of impaired pituitary reserve in patients with invasive macroadenomas does not necessarily imply permanent damage or irreversibility of trophic hormone reserve. In many instances, restoration of normal function can be seen when the tumor size is reduced by bromocriptine.

Prolactin Dynamics

The distinction between tumorous and nontumorous causes of hyperprolactinemia is usually made on the basis of two studies: the serum PRL level and the CT study. Most patients with tumorous hyperprolactinemia will show basal PRL levels in excess of 100 ng/ml and an abnormal CT study. However, in some cases, microprolactinomas may present with lesser elevations in basal PRL levels, coupled with a negative CT study. In such circumstances, the differentiation between idiopathic hyperprolactinemia and a small CT-negative microadenoma can be quite difficult. Attempts to separate the two by dynamic studies have resulted in the evolution of several tests that evaluate PRL dynamics. Four of these merit mention: PRL response to TRH, L-dopa–carbidopa combination, nomifensine, and metoclopramide. The principle behind these studies is briefly outlined below, recognizing the fact that the diagnostic value of any or all of these tests is considerably limited by the heterogeneity of response seen in hyperprolactinemic patients.

TRH Study

In normal individuals, following the administration of exogenous TRH, there is a prompt increase in the serum PRL levels. Patients with prolactinomas often demonstrate a distinct impairment in the TRH-mediated PRL response.[62,64–69] A blunted PRL response to TRH is encountered in at least 65–70% of patients, depending on the definition of "blunting." Frantz[70] reviewed the results of 12 groups of investigators encompassing 212 patients with PRL-secreting tumors and found an impaired PRL response to TRH in 87% of patients studied. The limitation of the test lies in the fact that a similar blunting can be encountered in nearly two-thirds of patients with hyperprolactinemia resulting from nontumorous causes. Thus, a perfectly normal PRL response to TRH decreases the likelihood of a tumorous etiology, while an impaired PRL response to TRH lacks diagnostic specificity.

L-Dopa–Carbidopa Study

L-Dopa crosses the blood-brain barrier to reach the hypothalamus and must be decarboxylated by the enzyme dopa–decarboxylase to become dopamine. The increased hypothalamic dopamine is transmitted via the neurons of the median eminence and the pituitary stalk to the pituitary, where it causes prompt suppression of the lactotropes. L-Dopa also has a direct action on the pituitary, where it must undergo similar decarboxylation to become dopamine, directly suppressing the lactotropes. Carbidopa is a drug that blocks the conversion of L-dopa to dopamine, and hence prevents formation of dopamine. This drug is unique in that it does not cross the blood–brain barrier. Therefore, its actions are limited to the periphery and the pituitary. The administration of this drug affords an excellent method of separating the hypothalamic component from the pituitary component of L-dopa-induced PRL suppression. A normal person pretreated with carbidopa and then given L-dopa will still retain the L-dopa-induced PRL suppression because the hypothalamic component of PRL suppression by L-dopa is not affected by carbidopa (since the latter does not cross the blood–brain barrier). By contrast, in patients with tumorous hyperprolactinemia, even though L-dopa causes PRL suppression, this effect can be diminished or even abolished by prior treatment with carbidopa.[71] It was assumed that patients with microadenoma or tumors had an intrinsic loss of the hypothalamic component of dopaminergic suppression. The effects of L-dopa in such patients would have to be explained on the basis of pituitary suppression, an effect that was demonstrably abolished by prior treatment with carbidopa. This exciting study held promise as a useful pharmacological tool in separating tumorous from functional hyperprolactinemia. Unfortunately, similar patterns of response have been described in patients with functional hyperprolactinemia.[72] It is quite possible that there is a common defect in hypothalamic dopaminergic inhibition of PRL secretion in tumorous as well as nontumorous hyperprolactinemia. This is supported by the observations of other investigators,[73,74] who have noted a considerable overlap in the degree of inhibition following L-dopa–carbidopa combination in tumorous as well as nontumorous hyperprolactinemia. Thus, in a given situation, the L-dopa–carbidopa combination test would not assist in differentiating idiopathic hyperprolactinemia from a CT-negative microadenoma.

Nomifensine Study

Nomifensine, an antidepressant, prevents the synaptic reuptake of dopamine by the hypothalamus, thereby increasing the hypothalamic dopaminergic tone. The drug crosses the blood barrier but has no effect on the pituitary dopamine receptors. Nomifensine causes a lowering of PRL levels in normal subjects. In patients with functional hyperprolactinemia, there is suppression of PRL levels following administration of nomifensine. Muller et al.[75] reported that patients with microadenomas failed to lower PRL levels following the administration of nomifensine and proposed that nomifensine testing may be a valid tool for discriminating between microadenomas and functional hyperprolactinemia. As with all other tests in this class, subsequent reports show that the drug fails to discriminate patients with tumors from those without.[76–79] While it is true that no more than 10% of patients with tumors show normal PRL suppression following nomifensine administration, only one-third of patients with nontumorous hyperprolactinemia demonstrate normal suppression. The matter is further confounded by the fact that a significant proportion of normal individuals fail to show normal suppression to the drug. Thus, failure to suppress the PRL levels after nomifensine in a hyperprolactinemic patient may be difficult to interpret.

Metoclopramide Study

Metoclopramide is a dopamine antagonist, which when given to normal subjects, is a potent stimulator of PRL release. Attempts to separate tumorous from nontumorous etiologies of hyperprolactinemia by acute or chronic administration of metoclopramide have not been consistently helpful. Healy et al.[67] and Barbarino et al.[65] demonstrated that nearly all patients with PRL-secreting tumors demonstrate a blunted PRL rise following metoclopramide. The problem, as with the PRL response to TRH, lies in the fact that several patients with idiopathic hyperprolactinemia and postpartum lactational galactorrhea (when the mechanism is clearly different from tumors) demonstrate blunting of PRL responses to metoclopramide. Thus, the most that can be said for the study is that a clearly normal and brisk PRL response to metoclopramide decreases the likelihood of tumor.

Differential Diagnosis

Prolactin-secreting tumors need to be differentiated from "idiopathic hyperprolactinemia," from "pseudoprolactinomas" and from nontumorous hypothalamic-stalk lesions that cause hyperprolactinemia.

Idiopathic Hyperprolactinemia

Idiopathic or functional hyperprolactinemia is the term applied to patients with hyperprolactinemia in the absence of demonstrable lesions in the pituitary or the CNS system or of any other recognizable cause of hyperprolactinemia. The clinical features encountered in idiopathic hyperprolactinemia are analogous to those seen in patients with PRL-secreting microadenomas. Thus, menstrual irregularities and infertility dominate the clinical spectrum. In fact, some patients with "idiopathic hyperprolactinemia" may harbor microadenomas that are not visualized by computed tomography. However, it is not correct to assume that patients with idiopathic hyperprolactinemia will naturally evolve into microadenomas. Martin et al.[80] followed 41 patients with symptomatic idiopathic hyperprolactinemia for a mean period of 5.5 years, and showed that in the vast majority, the serum PRL levels stayed the same, or even decreased, and only one patient developed a pituitary tumor. Malarkey et al.[81] also noted that patients with idiopathic hyperprolactinemia seldom develop a pituitary tumor. The natural history of idiopathic hyperprolactinemia is so benign that it raises questions regarding any form of ablative therapy (or even chronic medical therapy) for a spontaneously self-limiting disorder.

The distinction between a CT-negative microadenoma and idiopathic hyperprolactinemia can be difficult and rests on the PRL levels, and the response of PRL to various dynamic tests, particularly TRH. Basal PRL levels in idiopathic hyperprolactinemia are usually below 100 ng/ml, while patients with microprolactinomas usually demonstrate basal values exceeding 100 ng/ml. When the response of PRL to TRH and metoclopramide are clearly preserved, the likelihood of tumor is less. While most patients with tumorous hyperprolactinemia demonstrate blunting of PRL responses to provocative stimuli, a significant proportion of patients with idiopathic hyperprolactinemia also demonstrate impairment, diminishing the diagnostic value of these dynamic studies.

Bioassay and gel chromatographic studies of circulating PRL have indicated that some patients with idiopathic hyperprolactinemia and normal menses demonstrate circulating PRL with higher molecular weight and decreased bioactivity. Jackson et al.[82] studied five women with idiopathic hyperprolactinemia and normal menses and characterized a 150,000–160,000-M_r PRL as the predominant circulating form. This finding contrasted with the predominance of the 22,000-M_r hormone in other hyperprolactinemic states. The bioactivity of this large-molecular-weight PRL, measured by using the Nb_2 rat lymphoma line, was found to be diminished in comparison with the 22,000-M_r PRL. These interesting studies raise speculations regarding the type of PRL secreted by the nontumorous, but hyperplastic lactotropes.

Pseudoprolactinomas

Pseudoprolactinomas are tumors of the pituitary or suprasellar region that do not actually secrete PRL but that cause hyperprolactinemia as a result of interruption in the synthesis or transport of dopamine. Such a phenomenon can result from a true chromophobe (null-cell) adenoma, craniopharyngioma germ cell tumors, or meningioma.[83] By distorting the pituitary stalk or the median eminence, these tumors can lead to dopamine deficiency, which in turn results in lactotrope escape and consequently hyperprolactinemia. In general, patients with pseudoprolactinomas have much lower basal PRL levels than those with true prolactinomas. However, Besser et al.[17] described a group of 34 patients who had surgery for pituitary macroadenomas that were immunonegative for PRL, in whom the preoperative PRL level in the serum ranged from 150 to 5600 mU/liter (normal 360 mU/liter). These patients with pseudoprolactinomas can manifest the same clinical features (e.g., galactorrhea, oligomenorrhea) as patients with true prolactinomas. The PRL response to bromocriptine does not distinguish between the true and pseudoprolactinomas. The normal lactotropes of patients with pseudoprolactinomas respond to dopaminergic agonism in the same way as tumor cells do. Since the reason for hyperprolactinemia in patients with psuedoprolactinoma is dopamine deficiency, the use of dopamine agonist drugs in this setting can be viewed as replacement therapy that will provide dopaminergic input and lower PRL hypersecretion by the otherwise normal lactotropes. However, in contrast to the true prolactinoma, the tumor cells of the pseudoprolactinoma do not possess dopamine receptors. Therefore, the pseudoprolactinoma will not shrink with dopamine agonist therapy. This is in contrast to true macroprolactinoms, two-thirds of which show appreciable size reduction with dopamine agonist therapy.

Attempts to recognize pseudoprolactinomas preoperatively by dynamic studies of PRL secretion have yielded inconsistent results. The most reliable indicator for the presence of pseudoprolactinoma is the basal level of PRL. Since these tumors are usually large, and since large prolactinomas are usually associated with an impressive degree of hyperprolactinemia, the demonstration of mild (or even modest) hyperprolactinemia in a patient harboring a large pituitary tumor is suggestive of pseudoprolactinoma. This, however, is not absolute.

Nontumorous Hypothalamic Stalk Lesions

In addition to tumors in the suprasellar region, several nontumorous lesions can result in hyperprolactinemia. The PRL level has been referred to as by Dr. Doughaday as the sedimentation rate of pituitary endocrinology. The five most important groups of nontumorous disorders that can result in hyperprolactinemia are granulomatous disease (particularly sarcoidosis),[84,85] hypophysitis (lymphocytic[86,87] and granulomatous[88,89]), metastatic disease[90-93] (particularly from the breast, lung, and colon), vascular disease (rarely, Sheehan's syndrome),[94-96] and metabolic disorders (such as Hand-Schüller-Christian disease). These lesions are discussed in Chapter 15.

Natural History

While several authorities contend that patients with galactorrhea demonstrate transitional tendencies, with progression of microadenomas to macroadenomas,[55,97,98] the current opinions strongly favor the notion that most microadenomas remain static. Long-term observations regarding the natural history of microadenomas strongly suggest that these small tumors seldom show progressive growth. In fact, resolution of abnormalities in the CT study can and do occur in patients with microadenomas. Furthermore, amelioration of symptoms such as oligomenorrhea can occur spontaneously in patients who harbor microadenomas. It is not unusual for PRL levels to decrease spontaneously in such patients without any therapy. Studies that support the notion that microadenomas essentially display a nonprogressive natural history are overviewed in the section dealing with treatment of microprolactinomas. Indeed, one of the major realizations in the 1980s has been the observation that a significant proportion of microprolactinomas remain static over extended periods of time. This fact undoubtedly must be considered when planning therapy for microprolactinoma.

The natural history of idiopathic hyperprolactinemia appears to be characterized by significant spontaneous resolution. The development of tumors in patients with idiopathic hyperprolactinemia is a rare enough event to preclude ablative choices of therapy.

Treatment

The therapeutic options for the patient with hyperprolactinemia have considerably widened in the past decade. The collaborative efforts of endocrinologists, neurosurgeons, radiologists, pathologists and basic scientists have vastly influenced the constant flux which seems to characterize the therapeutic decisions for hyperprolactinemic patients. Opinions regarding the correct choice for therapy have evolved through several phases; in the early 1970s, the rediscovery of transsphenoidal microsurgery, buttressed by the sophisticated technology of optic magnification seemed to promise more than other therapies. The second phase in the evolution of therapy for hyperprolactinemia dates back to the late 1970s. This phase was characterized by burgeoning information that elevated dopamine agonist drug therapy to the status of a miracle drug. The accumulation of a formidable amount of data—from both sides of the Atlantic—clearly indicated that bromocriptine was extremely effective and impressively safe, impressing even those who had been most critical of the drug in the earlier years. The third phase in the therapeutic choices for hyperprolactinemia unfolded in the early 1980s. This phase witnessed the arrival of data that tempered the initial glow of the results from microadenectomy, as well as data that reinforced the tumor-shrinking effect of dopamine agonists. Also, an understanding of the natural history of microprolactinomas has been instrumental in a slant toward conservative therapies. The current phase, in this dynamic cycle, is witnessing the birth of newer dopaminergic agonists—pergolide, lisuride terguride, mesulergin, to name a few—as well as a resurgence of interest in radiotherapy as an important option for therapy. At the time of writing, medical therapy with dopamine agonist drugs has transcended other modalities as the initial therapy for patients with hyperprolactinemia resulting from idiopathic hyperprolactinemia, microprolactinoma, noninvasive macroadenoma and even locally invasive macroadenoma. The safety and simplicity of bromocriptine therapy are such that to many this treatment for hyperprolactinemia (caused by pituitary tumors) represents a one-step move that ends the whole situation.

The goals of therapy—any therapy—for the treatment of hyperprolactinemia are (1) normalization or near-normalization of serum prolactin levels, (2) shrinkage or disappearance of tumor, (3) restoration of normal gonadal function, and (4) avoidance of hypopituitarism. Both bromocriptine and microsurgery satisfy several of these goals. However, there are several advantages that have rendered bromocriptine therapy a superior choice. First, in terms of ease and compliance, when instructed properly most patients can tolerate the drug extremely well. Second, in terms of efficacy, the successful outcome with bromocriptine therapy is as good as (or even better than) that of transsphenoidal pituitary surgery (TPS). Third, hypopituitarism is never a sequel of bromocriptine therapy, while such a guarantee can not be always offered by TPS. Fourth, the "curative" nature of TPS must be tempered by the observations that the recurrence of hyperprolactinemia within 5 years of TPS is significant, and can be as high as 30–50%. Fifth, progression or growth of prolactinoma while on bromocriptine is virtually unheard of and does not occur. Finally, even massive prolactinomas with suprasellar extension often respond dramatically—within days or weeks—rendering such therapy an excellent method for medical debulking.

The only disadvantage of bromocriptine therapy is that when discontinued the hyperprolactinemia returns, often with a resumption of tumor growth. In this sense, bromocriptine therapy cannot be viewed as a permanent tumorolytic treatment.

The approach to treatment of hyperprolactinemia involves a clear understanding of what can and cannot be achieved by the treatment, regardless of whether bromocriptine or TPS is chosen as the initial modality. The discussion of therapy for hyperprolactinemia can be viewed from the following three perspectives—the clinical setting, the treatment modality chosen, and the underlying pathology that may pose special problems and restrictions.

General Considerations: The Clinical Setting

Several important clinical factors should be considered before deciding on the best therapeutic option for the hyperprolactinemic patient. These include the patient's objectives, the presence and size of the tumor, the PRL level, the presence of neighborhood-compression syndromes, the compliance factor, and the available neurosurgical expertise.

1. *Objectives of the patient:* The patient's desire for fertility and or restoration of normal menses constitutes a major factor in recommendation of therapy. Microprolactinomas do not always necessitate therapeutic intervention. Data on the natural history of microprolactinomas strongly suggest that these tumors are usually static, seldom displaying an aggressive tendency or

rapid growth, and occasionally regress without any treatment. Several studies have reaffirmed the benign natural history of PRL-secreting microadenomas. Microprolactinomas do not naturally progress to macroadenomas. Koppelman et al.[99] studied the course of untreated hyperprolactinemia, mostly caused by microadenomas, in 25 women. The mean interval from onset of symptoms to reevaluation was 11.3 years. No patient was judged to have worsened clinically and indeed may have improved by clinical as well as hormonal criteria. Progression of the radiological abnormality was noted in only one patient. Menses had resumed in an impressive number (32%) of patients without any form of therapy. Koppelman and co-workers also noted that the greatest degree of clinical improvement tended to occur in patients whose symptoms were estrogen related. These data indicate that most patients with microadenomas, demonstrate a stable, and clearly benign course. Similar data were presented by Weiss et al.,[100] who followed 27 patients with microprolactinomas for 6 years; their data suggest that only 3 of their 27 patients showed any significant growth of tumor. In a 3–20-year longitudinal evaluation of 43 symptomatic patients with prolactinomas, March et al.[101] found progressive tumor growth in only two cases. These studies have fostered a growing notion that indeed one option that may be considered is to do nothing and merely follow the patient. Such a choice is entirely tenable in the patient with microadenoma or idiopathic hyperprolactinemia who does not wish to become pregnant, or to resume menstruation, and is not bothered by the galactorrhea. The long-term adverse effects of mild to moderate hyperprolactinemia on bone density have not been conclusively established; thus, the argument that therapeutic intervention of hyperprolactinemia is mandatory to prevent osteopenia is not based on solid grounds. Indeed, a PRL-secreting microadenoma is not a disorder requiring intervention when the possible benefits of such intervention are of no concern to the patient.

2. *Presence of a tumor in the pituitary:* This is an important concern for two obvious reasons. First, the presence of hyperprolactinemia in the absence of any abnormalities in the CT study of sellar and suprasellar region is indicative of idiopathic hyperprolactinemia. While a very small percentage of such patients may demonstrate microadenomas during surgery, the overall prevailing opinion is that surgical intervention is contraindicated in these patients. Recently, Malarkey et al.[81] followed 41 patients with idiopathic hyperprolactinemia and noted that if a tumor is not present during the initial evaluation, it was highly unlikely that a tumor would develop within 5–6 years. They also noted that spontaneous normalization of PRL occurred when the basal levels were below 40 ng/ml. Most importantly, their study underscored the fact that PRL levels remained stable in this group of patients even without any treatment. When treated, successful induction of pregnancy with dopamine agonist therapy was achievable in a very high percentage of these patients.

The second important reason for evaluating the pituitary hypothalamic region is to exclude pseudoprolactinomas (i.e., pituitary tumors that do not actively secrete prolactin but that cause hyperprolactinemia by interruption of dopamine transport). The cells of such tumors do not possess dopamine receptors and therefore cannot be expected to respond to dopamine agonist drug therapy. Several tumors in the suprasellar region—craniopharyngioma, in particular—can result in pseudoprolactinomas. Exclusion of such entities by careful CT studies of the infundibulum and the suprasellar region is mandatory before discussing therapeutic options.

3. *Size of the tumor:* The size of the tumor matters, only in terms of the surgical expectations. It has been amply documented that surgical cures for macroprolactinomas (especially those exceeding 3 cm) are generally unsatisfactory. A frequently quoted figure is a 30% of cure rate, with a 5–18% risk of developing trophic hormone deficiency. By contrast, the size of the tumor has no bearing on responsiveness to dopamine agonist therapy. In fact, some of the most dramatic responses are those seen when bromocriptine is administered to patients with macroprolactinomas.

4. *Degree of PRL elevation:* As with size of the tumor, surgical cures with TPS for PRL-secreting microadenoma are greatest when serum PRL levels are below 250 ng/ml. The prospects of immediate postoperative lowering of PRL diminish when the basal PRL exceeds 250–300 ng/ml. Furthermore, even in those who demonstrate postoperative normalization of PRL level, the frequency of late recurrence (within 5 years) is significant, approaching 17–50%.

5. *Presence of suprasellar extension:* The presence of neighborhood compression phenomena in association with large, invasive macroprolactinomas poses a therapeutic dilemma. Such patients do poorly with surgery, with little guarantee that the PRL level will normalize; furthermore, the risks of hypopituitarism as well as the possibility of recurrent hyperprolactinemia are quite real. Such patients should receive very guarded and realistic information regarding the neurosurgical outcome versus initial therapy with bromocriptine. The realization that shrinkage of the extrasellar extension with resultant disappearance of compression can be attained by dopamine agonist therapy has made the argument for bromocriptine a powerful choice in the initial management in such patients. However, follow-up management of such patients has to be quite rigorous, often with visual-field testing performed once, or even twice, a week.

6. *Compliance factors:* This becomes a major issue when dopamine agonist drugs are used. Since the tumor growth resumes when the drug is stopped, patients should be warned against careless attitudes. Often the drug has to be administered three times a day and must be taken with meals. Therefore, patients who miss meals and who suffer side effects of the drug taken on an empty stomach are candidates who are likely to miss their dosage sooner or later. Patients who are educated about bromocriptine usually understand the importance of complying with the regimen.

7. *Neurosurgical expertise:* Obviously, the degree of neurosurgical expertise available will have an overriding influence on the choice of therapy. It is believed that a surgeon qualified to perform TPS is one who performs 20–25 such procedures annually. The track record of the surgical team is an important factor that merits consideration in referring a patient.

Choice of Therapy

With the above generalizations, the role of dopamine agonist therapy, transsphenoidal surgery, and radiation therapy can be viewed.

Dopamine Agonist Therapy

The introduction of bromocriptine for the treatment of PRL disorders is based on the physiological principles governing normal regulation of prolactin secretion. Prolactin release by the normal lactotropes is under the tonic inhibitory control exerted by hypothalamic dopamine. Specific dopamine receptors are located in the lactotropes; when occupied by dopamine or its agonists, this results in marked inhibition of PRL release and secretion.

Bromocriptine has been in use for treatment of hyperprolactinemia since the early 1970s, and has been in use worldwide. Since the early report by Besser et al.[102] in 1972 that bromocriptine lowered PRL levels of patients with hyperprolactinemia, a plethora of reports from all around the world have confirmed the efficacy of the drug.[27,103–116] The unequivocal efficacy of bromocriptine can be viewed in terms of its effect on lowering PRL, its effects in causing clinical improvement, and its role in shrinking tumors that are large, even invasive. The price exacted for such unique and almost guaranteed benefits is the necessity to take the drug on a chronic basis, even indefinitely. The psychological price exacted from both patient and physician is the knowledge that bromocriptine, while controlling the disorder exceedingly well, cures nothing.

Effects on Serum PRL Level. Bromocriptine lowers, and normalizes, PRL levels in more than 90% of patients with hyperprolactinemia regardless of the etiology. This effect can be achieved with the use of the conventional dose of 2.5 mg tid, in most patients with hyperprolactinemia. In some cases, higher doses, even up to 40 mg, may be required, but this is exceptional. Even in those patients in whom complete normalization of PRL level is not attained, clinical improvement may be seen. This is owing to the fact that bromocriptine may lower the biologically active PRL preferentially, an effect that may go unnoted when immunoreactive PRL is measured to monitor response. It should be noted that bromocriptine is not 100% effective. Obviously, when the prolactinoma does not contain sufficient dopamine receptors, the dopamine agonist drug cannot be expected to work. The PRL lowering effect of bromocriptine is rapid, and long acting. The effectiveness of the drug is so remarkable, that PRL levels are lowered regardless of the degree of elevation in pretreatment levels.

Effects on Clinical Status. The clinical effects of bromocriptine are also impressively rapid. Within 2–6 months of instituting therapy there is return of regular menstruation, ovulation, and adequacy of luteal phase in 60–80% of patients with microprolactinomas. The clinical benefits occur regardless of the duration of the amenorrhea or infertility before treatment. Patients with histories of 10–15 years of oligomenorrhea/amenorrhea have been known to resume menses, and ovulation, months after instituting bromocriptine therapy for hyperprolactinemia. Similarly, in men potency returns to normal within 6 months of starting bromocriptine. The slightly longer duration in males may be due to the fact that men with hyperprolactinemia harbor larger tumors with much higher pretreatment PRL levels.

The success of bromocriptine in the treatment of hyperprolactinemic females is now well accepted, even by its worst critics. When the drug is used to treat idiopathic hyperprolalctinemia the success rate for restoring normal menstruation and ovulation is 80%. The response rate is slightly lower for patients with macroprolactinomas, at 60–70%. Approximately 10% of patients with macroadenomas may show no clinical improvement with the use of bromocriptine. In such cases, the the possibility of impaired gonadotropin reserve may have to be considered.[117,118] Once regular menstruation has resumed, the conception rate is approximately 50% within 2 months. Comparable results have been experienced by investigators around the world.

Effects on Tumor Size.   Recently it has become evident that in addition to its effect on lowering serum prolactin levels, bromocriptine reduces the tumor mass of the prolactinoma, often dramatically. Although reports of a tumor-shrinking effect of the drug had sporadically and anecdotally appeared in the literature in the mid-1970s, the first large review on the subject appeared in 1979. Wass et al.[119] reported that of 69 patients treated with bromocriptine, followed up to 6.5 years, there was evidence of tumor shrinkage in 23%. Indeed, these data, if anything, were underrepresentative of the tumor-shrinking effect of the drug, since the parameter used was comparisons in the size of the pituitary fossa by routine radiology. By the use of comparative CT, recent evidence indicates that when patients with large prolactinomas are treated with bromocriptine, clearly the tumor mass shrinks in at least two thirds of such patients.[120] The rapidity with which the tumor begins to shrink is often clinically mirrored by the quick and dramatic improvement noted in headache, vision, and the neighborhood compression syndromes that were manifest before therapy.[121] Objective evidence for shrinkage of the tumor by CT can be seen within 3–4 weeks of initiating therapy, although signs of improvement are noticeable as early as 10 days. Even with massive tumors that have extended above the sella, bromocriptine has been instrumental in shrinking the tumor to the point where it falls back into the sella and becomes contained within. In some cases, the tumor-shrinking effect on massive prolactinomas can be so impressive, and so complete, that the patient may develop CSF rhinorrhea as well as a spontaneous pneumoencephalogram. The responsiveness to bromocriptine can be, in sensitive patients, so dramatic that the analogy to "pulling the plug out of the bottle" has been used to denote the extreme degree of shrinkage caused by the drug. In addition to the shrinkage of tumor size, the phenomenon is associated with restoration of previously impaired pituitary function to normal. The hypopituitarism in these instances was caused by compression of the normal gland, which upon reduction of tumor size, springs back into normal action since the pressure is relieved by bromocriptine. It has been the impression of many workers that the large prolactinomas, with serum PRL levels greater than 500 ng/dl, appear to be more sensitive and show a greater reduction in size than smaller adenomas. It is not known whether larger prolactinomas possess more dopamine receptors, accounting for the incredible response encountered with the use of bromocriptine. The tumor-shrinking effect of bromocriptine has been now confirmed by several workers around the world and has become an established fact.[27,122-132] It has to be reemphasized that once the bromocriptine is discontinued, there is return of tumor regrowth with development of possible visual impairment.

Dopamine agonists work exceedingly well and bring about lowering of PRL, marked improvement in the clinical status, and reduction in tumor size to an impressive degree.[133] The mechanism of action of bromocriptine is believed to be exerted through its occupancy of dopamine receptors in the tumor cells. The drug replaces dopamine and provides the cells with a physiological inhibitor of PRL secretion[134,135] (dopamine agonism). The effects of bromocriptine can be inhibited, even abolished, by the use of selective dopamine antagonist drugs. Bromocriptine administration to rats with prolactinomas results in tremendous changes in the lysosomal pattern of these tumors, indicating a virtual blowout of the lysosomes. The drug probably increases internal cytolytic activity within tumor cells. The rapidity with which some tumors shrink and respond to bromocriptine therapy cannot be explained by its effects on cell replication alone. It is not known why these tumors shrink so quickly. A possible effect on vascularity is suspected but not proven.

The unequivocal therapeutic efficacy of bromocriptine has to be tempered by other concerns. In spite of its proven and accepted prowess as a treatment modality for hyperprolactinemia, it has not become a requiem for surgery. The major reason for this is the fact that bromocriptine has to be administered for indefinite and protracted periods of time. Although the drug has been used for 10 years, skepticism regarding "long-term" effects exists, and even flourishes, mostly in neurosurgical circles. Clearly bromocriptine therapy is not tumoricidal and can be regarded, at best, as a wonderful palliative treatment; this, the neurosurgeons point out, is second-rate treatment when microadenomas can be extirpated and "cured" by TPS. Recent studies[136,137] suggest that a course of dopamine agonist therapy may have a prolonged suppressive effect on PRL suppression. Although PRL levels are expected to rise promptly following discontinuation of the drug, some patients may show only a minimal rise after 18 months of therapy. Clearly, more studies are needed to delineate the group of patients who would show prolonged suppression of PRL after discontinuation of long-term dopamine agonist therapy.

Side Effects of Therapy. The two main side effects of bromocriptine therapy are nausea and postural hypotension. Both can be minimized, even abolished, if certain principles are followed. The drug should be started as a low dose (1.25 mg), and initially should be taken in the middle of meals rather than after. Initiating therapy at night would help decrease the possibility of postural hypotension, a complication that is mostly encountered in the early days of therapy. After 3–4 days, the dose is increased to 2.5 mg, and moved to an earlier schedule, in the middle of the evening meal. At 3-day intervals, the dose is increased to 2.5 tid, a dose that can be tolerated by most.

Other adverse effects, including constipation, vasospasm, and psychiatric symptoms, are less common and may be seen when much higher dosage is employed.

The cost of therapy with bromocriptine for 1 year approximates $750–1000 in the United States.

Several newer dopamine agonist drugs have entered the experimental scene. The three notable second-generation dopamine agonists are pergolide, lisuride, and mesulergin. At the time of writing, none of these drugs are approved by the FDA for their use in the United States. These three compounds, collectively referred to as ergoline, lack the tripeptide moiety present in bromocriptine. Lisuride is unique in that it demonstrates antagonism to serotonin in addition to its inherent dopamine agonist action. The newer dopamine agonist drugs are more potent than bromocriptine on a weight-to-weight basis, and are longer acting than bromocriptine, an advantage that may permit once or twice a day administration. Occasionally patients who show no response to bromocriptine may benefit from these drugs. While several studies[138–141] have reported the usefulness of the second generation of dopamine agonist drugs, it is unlikely that bromocriptine will be replaced by any of these.

Transsphenoidal Microadenomectomy

Transsphenoidal microsurgery offers a safe and effective method of removing microadenomas without impairing pituitary function. Although the procedure is by no means a new one, the pioneering work of Dr. Jules Hardy of Montreal, Canada, has resulted in a resurgence of interest in transsphenoidal pituitary surgery. Reports in the early seventies clearly fostered the belief that microadenomectomy is the initial mode of therapy for PRL-secreting microadenomas. In Hardy's series,[142] the restoration of normoprolactinemia and resumption of menses was 74% and 63%, respectively, with a subsequent fertility rate of 36%. Several reports soon followed from Hardy and co-workers,[143–147] as well as from others.[148–151] The success rate in terms of restoring euprolactinemia and resumption of menses was variable, ranging from 48 to as high as 70%.[151–155] The variability in results was partly due to the heterogeneity in surgical expertise. The restoration of euprolactinemia after surgery in the series reported by Chang et al.,[148] Hardy et al.,[142] Tindall et al.,[152] Domingue et al.,[154] and Post et al.[151] was 48%, 74%, 67%, 62%, and 70%, respectively, underscoring these variabilities. In a recent review, Hardy et al.[156] reported the collective experience with 300 patients operated on by his group. In 78% of cases, postoperative normalization (PRL <20 ng/dl) was attained.

The correlation between attainment of a cure and the preoperative PRL level (as well as the tumor size) has been well emphasized in surgical literature. When the preoperative PRL levels exceeded 200 to 250 ng/ml, the rate of successful normalization of PRL declines sharply. This has been illustrated in the large series reported by Faria et al.[155] In their series of 100 patients, 72 had a preoperative PRL level below 200 ng/ml and 28 had levels of >200 ng/ml. In the first group 78% demonstrated normalization of PRL level after surgery, with resumption of menses. This contrasted with the 39% of success rate in the group with preoperative PRL levels greater than 200 ng/ml. In the former group, 81% had tumors less than 1 cm in diameter. By contrast, 86% of patients in the group with a basal PRL level of >200 ng/ml had tumors larger than 1 cm. Thus, it appears that patients with prolactinomas smaller than 1 cm and basal PRL lower than 200 ng/ml had the best chances (approximately 75%) of attaining a cure.

Even when the PRL levels are successfully normalized postoperatively, the possibility of recurrence of hyperprolactinemia is quite real. In an earlier report, Tucker et al.[150] followed 45 women operated for prolactinoma for a period of 3.1 years and concluded that a normal PRL level 6 months after surgery probably indicated a definitive cure. However, a recent report from Notre Dame Hospital Canada, by Serri et al.[157] has thrown a shadow on the initial glow of TPS for prolactinoma. These workers followed 44 patients for 6.2 years, and reported that in the 24 patients who were presumed to have been cured (based on normalization of PRL level and resumption of menses), 12 patients returned with recurrence that was delayed up to as long as 5 years. The possibility of recurrence was higher in patients who had higher PRL levels during the immediate postoperative period; it was unrelated to intercurrent pregnancies. Other workers[158] have reported lower but impressive (17%) recurrence rates of hyperprolactinemia with 5 years of relatively successful microadenectomy. Thus, the efficacy of transsphenoidal microadenomectomy has to be reevaluated in light of these findings.

Finally, the prospects of cure are disappointing when the tumor is large. For example, the cure rates following surgery for macroprolactinomas are approximately 30–35%. Furthermore, the chance of developing hypopituitarism (often involving LH) is 5–18%.

Protagonists for neurosurgery as the initial therapy for prolactinomas claim that microadenectomy is safe and effective and also rapidly and often permanently cures hormonal hypersecretion. The major objective cited by neurosurgeons to the use of dopamine agonist therapy is their contention the the long-term effects of bromocriptine are largely unknown. The two factors that are beginning to undermine microadenectomy as the initial choice are the revelation that recurrence after apparent surgical cures is significant and the realization that 15 years of bromocriptine have failed to show any real problems with chronic use of this drug.

In summary, transsphenoidal microadenomectomy, as a primary choice of therapy for patients with hyperprolactinemia must be viewed from the following therapeutic perspectives:

1. The best chances for a "cure" (i.e., postoperative normalization of normal menses) with surgery are 75%.
2. The candidates who enjoy such a cure following microsurgery are those with tumors less than 1 cm in diameter, with preoperative serum PRL levels below 200 or 250 ng/ml.
3. The risk of developing hypopituitarism, usually involving one axis, is small but significant. The resultant deficiency state is usually transitory but occasionally can become permanent.
4. Even in those patients who are successfully managed by transsphenoidal microadenomectomy, the possibility of recurrence of hyperprolactinemia in 5 years can be as high as 50%.
5. The success rate in normalizing PRL and restoring gonadal function by surgery in patients with macroprolactinomas and invasive macroprolactinomas is disappointing and is no greater than 30–33%.

The above perspective must be discussed with the patient before deciding on the appropriate therapeutic options. Neurosurgeons believed that placing patients with pituitary prolactinomas on bromocriptine for indefinite, even life-long, periods is unjustified when a safe and effective modality for removing such tumors is available. This tenet must be reevaluated in light of the recurrence rate as well as the safety record held by bromocriptine treatment. Whereas most neurosurgeons are clearly impressed by the effects of bromocriptine therapy on both function and size of prolactinomas, there is reluctance on the surgeon's part to consider bromocriptine as primary therapy. The role of the drug, as viewed by neurosurgeons, is adjunctive, either for preoperative preparation, or for postoperative use, when surgery fails to normalize the PRL levels. This view, obviously, is at variance with endocrine opinion that the role of bromocriptine far surpasses these adjunctive indications.

Radiation Therapy

The role of radiation therapy in the management of PRL-secreting tumors is constantly being redefined. Once thought to have little or no place in the management of prolactinomas, radiation therapy is now being reevaluated as a viable alternative for patients with prolactinoma. The role of radiation therapy in the management of PRL-secreting tumors can be applicable in the following settings:

1. As adjunctive therapy postoperatively for macroadenomas that have been removed (to prevent recurrence, which occurs in 1 of 10 patients operated on for macroadenomas)
2. As adjunctive therapy to treat residual tumor tissue when less than gross total removal of tumor is achieved by neurosurgical means

3. In patients with macroadenomas who wish to become pregnant but who opt to be treated with bromocriptine (a course of radiation therapy *prior* to bromocriptine therapy is often recommended, done prophylactically to prevent expansion of microadenomas, should pregnancy occur)

4. An emerging role of radiation therapy as a primary mode of therapy for patients with macroprolactinomas, regardless of the issue of pregnancy

Since surgical excision for macroadenomas are disappointing, and treatment with bromocriptine for such patients represents an indefinite, perhaps life-long, period of therapy, radiation therapy has become one consideration in the treatment of patients with large PRL-secreting tumors. The early data available were constituted by a small number of patients with inadequate follow-up evaluation. The series reported by Kleinberg et al.,[34] Antunes et al.,[159] and Gomez et al.,[55] consisting of 8, 6, and 8 patients, respectively, who received radiation treatment showed that, while PRL levels declined by 75–90% compared with the baseline following radiation, none showed normalization. The lack of systematic study of patients radiated for prolactinomas was coupled with lack of uniformity in technique as well as follow-up. The combination of surgery with postoperative conventional radiation therapy was systematically evaluated by Sheline et al.[160] These workers treated 14 patients with grade III or IV PRL-secreting adenomas. Their data indicate that improvement in PRL levels (which can be as impressive as a 90% reduction) can be seen in most patients, but it may require several months, even years, before this effect becomes manifest. However, normalization of PRL level occurred in only a very small number of patients.

The study that has generated a great deal of interest is the experience from St. Bartholomew's Hospital, London, on the effect of radiation therapy in the treatment of small prolactinomas. Grossman et al.[161] studied 36 women who presented with oligomenorrhea or amenorrhea, infertility, galactorrhea, and abnormal findings on polytomography without significant suprasellar extension. All patients were treated with 4500 rad to the pituitary, and placed on dopamine agonist therapy. The followup extended from 2–10 years, after which they were evaluated after discontinuation of bromocriptine. Of the 28 patients who were evaluable, a progressive fall of PRL level was observed in 25, with complete normalization in 8. There was no instance of tumor expansion. The incidence of hypopituitarism, which was always a delayed feature, involved mostly GH. The incidence of clinically significant, or relevant, hypopituitarism occurred in only two patients and involved the gonadotropins. Such data, when confirmed by others, are likely to result in resorting to the combination of radiation with interim dopamine agonist therapy as the initial treatment for microadenomas, instead of surgery.

In summary, given the current flux in therapeutic attitudes in the treatment of hyperprolactinemia, only broad outlines can be sketched.[162,163] The choice of therapy is still largely individual. The following recommendations are based on the prevailing attitude among most endocrinologists, neurosurgeons, and radiotherapists. The treatment and recommendations can be viewed in terms of the four groups of patients with hyperprolactinemia: idiopathic hyperprolactinemia, microadenomas, macroadenomas, and invasive macroadenomas.

Idiopathic hyperprolactinemia: The therapeutic decision is relatively straightforward in this group of patients. Clearly, the one choice that is not a recommendation is surgery. If the patient's problem is infertility, and she wishes to conceive, the recommended therapy is bromocriptine. If the patient desires symptomatic relief for oligomenorrhea, amenorrhea, or estrogen deficiency, again, the therapeutic recommendation is bromocriptine. If the patient wishes no intervention, follow-up evaluation is the only required recommendation. Prognostically, if the fourth-generation CT is clearly normal at the time of evaluation, the changes of developing a tumor within the next 5–6 years are extremely small, even negligible.

Patients with microadenoma: First, in women with microprolactinomas with a basal PRL level below 200 ng/ml therapy becomes necessary only when infertility and menstrual irregularities require correction. Three choices are currently available to those patients who desire therapeutic intervention: (1) therapy with bromocriptine alone; (2) the performance of transsphenoidal selective microadenomectomy; and (3) the combination of megavoltage radiation coupled with long-term interim bromocriptine therapy. If the first choice is followed and bromocriptine therapy is chosen, the duration of the therapy should be clearly discussed with, and understood by, the patient. If pregnancy ensues bromocriptine should be discontinued, although there are no teratogenic effects on the fetus. Turkalj et al.[164] reviewed the outcome of 1410 pregnancies in 1335 women treated with bromocriptine in early pregnancy and found no ill effects on the fetus. However, since bromocriptine does pass through the placenta, it is advisable to discontinue the drug, in order to avoid possible late effects in the offspring (analogous to the DES situation). In women who do conceive, the effects of pregnancy on causing growth of the microadenoma is negligible, less than 0.5%.[165–167] The risk of such an occurrence is modest with macroadenomas, the often quoted incidence being 20%. In the woman who does not desire fertility, bromocriptine must be administered for long periods of time, even indefinitely. Periodic interruptions of therapy with measurement of PRL is an acceptable practice, given the benign and static course of microprolactinomas.

If neurosurgical removal of the tumor is chosen as the primary treatment of microadenomas, the patient must be familiarized with both the rate of success (75%) as well as the rate of recurrences (30–50%). The small but significant (5%) incidence of transient or permanent hypopituitarism cannot be ignored. Many endocrinologists today recommend the use of bromocriptine as primary therapy for microprolactinomas, reserving surgery when bromocriptine fails to normalize prolactin, a phenomenon that may be seen in a very small (5–10%) number of patients. On the contrary, most neurosurgeons would recommend microadenomectomy as the initial therapy for patients with microprolactinomas (especially when the prolactin level is below 200 ng/ml) reserving bromocriptine when the surgical procedure has failed to normalize the PRL level, a phenomenon that may be seen in a modest number (20%) of patients with microadenomas.

The third option of combining radiotherapy with bromocriptine has generated more enthusiasm in Great Britain than in the United States, where experience with this combination for treatment of microprolactinomas is limited.

Second, when the serum PRL is in excess of 200–250 ng/ml, the results of cure with surgery are discouraging enough not to recommend this option. Bromocriptine therapy (with or without radiation) becomes the mainstay of treatment for patients with prolactinomas with PRL levels greater than 200–250 ng/ml.

Patients with macroadenomas: The disappointing rate of cure (approximately 30–35%) with surgery for macroprolactinomas has resulted in evaluating bromocriptine in the treatment of macroprolactinomas. The effectiveness of bromocriptine in reducing the size of PRL-secreting tumors has been already alluded to. In a multicenter trial, Molitch et al.[168] systematically evaluated the effectiveness of bromocriptine in reducing tumor size. Thirty-nine patients with macroprolactinomas from nine centers, were treated using the same protocol. Normalization of prolactin was attained in 18; in 46% of patients, the reduction of tumor size was >50%, in 18% the size reduction was approximately 50%, and in 36% the reduction of tumor size was only 10–25%. The duration of treatment required for attainment of maximum reduction in tumor size ranged from 6 weeks to 6 months. Similarly, in another study, Warfield and colleagues[169] evaluated the response to bromocriptine in six males with large prolactinomas. During treatment, marked reduction in tumor size was noted in four, with striking improvement in visual fields and pituitary hormone reserve. The beneficial role of bromocriptine therapy in reducing the size of macroprolactinomas has been repeatedly confirmed.[170,171,122–133]

The ideal treatment for the patient with macroprolactinoma who wishes to conceive has to be formulated with several facts in mind. First, further enlargement of macroadenomas during pregnancy is a cause for concern. The risk of symptomatic tumor enlargement with pregnancy is estimated to be approximately 15.5% for patients with macroprolactinomas.[172] This contrasts sharply with the risk of pregnancy-related tumor enlargement in patients with microadenomas (below 1%). Second, the recommendation of prior surgery to prevent tumor growth during pregnancy is valid in principle; this, however, is tempered by the possibility that the patient may not be able to conceive either because surgery failed to normalize PRL or because the procedure resulted in inadvertent loss of gonadotropin reserve. Third, prior radiation clearly reduces the risk of pregnancy-related tumor growth of macroprolactinomas.[172] Fourth, the possibility of continuation of bromocriptine during pregnancy in such patients has not been fully investigated. Therefore, the choices for the patient with macroprolactinomas who wish to become pregnant are (1) to undergo surgery first, and if the procedure fails therapy with bromocriptine may be instituted; (2) to initiate therapy with bromocriptine, in conjunction with radiation therapy; and (3) to initiate therapy with bromocriptine, and defer pregnancy till PRL levels have completely normalized. Should the patient become pregnant, the drug may be reinstituted if the tumor shows radiographic evidence of enlargement. There is no single option that is acceptable as the ideal treatment in the difficult circumstance of the patient with macroprolactinoma who wishes conception.

The choice of therapy for patients with macroprolactinomas who do not wish to become pregnant is fairly simple; the choice rests between long-term (indefinite) bromocriptine therapy versus, the combination of radiation therapy with bromocriptine. The disappointing cure rates with surgery have swayed the therapeutic decision away from surgery as the initial therapeutic choice for such patients.

In the group of patients with macroprolactinomas with extrasellar extension (with field cuts, etc.), a group in whom a decade ago emergency pituitary decompression would have been considered, bromocriptine therapy is becoming established as a valid initial choice. Obviously these patients must be very closely monitored. In at least a third of these patients bromocriptine therapy results in a rapid response, with shrinkage of tumor improvement in field cuts and even in pituitary hormone reserve. Reversal of hypopituitarism in these patients reflects improvement in the compression effects exerted by the tumor on normal pituitary tissue. In many such patients, therapy with bromocriptine permits the tumor to retract back into the sella turcica rendering it more amenable to surgical debulking, if deemed necessary. If bromocriptine therapy fails to cause radiological improvement within 4 weeks, surgical decompression should be considered in these patients with suprasellar extension.

In summary, the therapeutic arena for prolactinomas is dominated by dopamine agonist therapy, represented by bromocriptine. It is remarkable that only 10 years ago the universally accepted indication for the drug was in treatment of idiopathic (nontumorous) hyperprolactinemia. The extension of the use of bromocriptine to involve hyperprolactinemic patients with microadenomas, macroadenomas, and even invasive macroadenomas represents a most remarkable therapeutic evolution.[173]

References

1. Peake GT, McKeel DW, Jarett L, et al: Ultrastructural, histologic and hormonal characterisation of a prolactin rich human pituitary tumour. *J Clin Endocrinol Metab* **29**:1383, 1969.
2. Lloyd HM, Meares JD, Jacobi J: Effects of oestrogen and bromocriptine on in vivo secretion and mitosis in prolactin cells. *Nature (Lond)* **255**:497, 1975.
3. Phelps C, Bartke A: Effect of chronic hyperprolactinemia on tuberoinfundibular dopaminergic neurons: Histofluorescence in male rats. In MacLeod RM, Thorner MO, Scapagnini U (eds): *Prolactin. Basic and Clinical Correlates*. Liviana, Padova, 1985, p. 615.
4. Franks S: Regulation of prolactin secretion by oestrogens: Physiological and pathological significance. *Clin Sci* **65**:457, 1983.

5. Davis JRE, Selby C, Jeffcoate WJ: Oral contraceptive agents do not affect serum prolactin in normal women. *Clin Endocrinol (Oxf)* **20**:427, 1984.

6. Weiner RI, Elias KA, Monnet F: The role of vascular changes in the etiology of prolactin secreting anterior pituitary tumors. In MacLeod RM, Thorner MO, Scapagnini U (eds): *Prolactin: Basic and Clinical Correlates.* Livana, Padova, 1985, p. 641.

7. Kovacs K, Korvath E, Cornblum B, et al: Pituitary chromophobe adenomas consisting of prolactin cells: A histologic, immunocytological and electron microscopic study. *Virchows Arch [A]* **366**:113, 1975.

8. Kovacs K: Morphology of prolactin producing adenomas. *Clin Endocrinol (Oxf)* **6**:71s, 1977.

9. Forbes AP, Henneman PH, Griswold GC, et al: Syndrome characterized by galactorrhea, amenorrhea and low urinary FSH: Comparison with acromegaly and normal lactation. *J Clin Endocrinol Metab* **14**:265, 1954.

10. Frommel R: Uber puerpale Atrophie des Uterus. *Z Geburtsh Gynakol* **7**:305, 1882.

11. Franks S, Murray MAF, Jequier AM, et al: Incidence and significance of hyperprolactinemia in women with amenorrhea. *Clin Endocrinol (Oxf)* **4**:597, 1975.

12. Jacobs HW, Hull MGR, Murray MAF, et al: Therapy-oriented diagnosis of secondary amenorrhea. *Horm Res* **6**:268, 1975.

13. Frantz AG, Kleinberg DL, Noel GL: Studies on prolactin in man. *Recent Prog Horm Res* **28**:527, 1972.

14. Franks S, Nabarro JDN, Jacobs HS: Prevalence and presentation of hyperprolactinemia in patients with "functionless" pituitary tumors. *Lancet* **1**:778, 1977.

15. Haesslein HC, Lamb EJ: Pituitary tumors in patients with secondary amenorrhea. *Am J Obstet Gynecol* **125**:759, 1956.

16. Biller BJ, Boyd AE III, Moltich M, et al: Clinical features of pituitary tumors: Galactorrhea-syndromes. In Post KD, Jackson IMD, Reichlin S (eds): *Pituitary Adenoma Update.* Plenum, New York, 1981, p. 65.

17. Besser GM, Wass JAH, Grossman A, et al: Clinical and therapeutic aspects of hyperprolactinemia. In MacLeod RM, Thorner MO, Scapagnini U (eds): *Prolactin. Basic and Clinical Correlates.* Liviana, Padova, 1985, p. 833.

18. Quigley ME, Sheehan KL, Casper RF, et al: Evidence for an increased opioid inhibition of luteinizing hormone secretion in hyperprolactinaemic patients with pituitary microadenoma. *J Clin Endocrinol Metab* **50**:427, 1980.

19. Grossman A, Moult PJA, McIntyre H, et al: Opiate mediation of amenorrhoea in hyperprolactinaemia and in weight-loss related amenorrhoea. *Clin Endocrinol (Oxf)* **17**:379, 1982.

20. Thorner MO, Besser GM: Bromocriptine treatment of hyperprolactinemic hypogonadism, *Acta Endocrinol* (Suppl 216) (Copenh) **88**:131, 1978.

21. Thorner MO: Prolactin. In Besser GM (ed): *Clinics in Endocrinology and Metabolism.* Vol. 6. WB Saunders, Philadelphia, 1977, p. 201.

22. Jacobs HS, Franks S, Murray MAF, et al: Clinical and endocrine features of hyperprolactinemic amenorrhea. *Clin Endocrinol (Oxf)* **5**:439, 1976.

23. Molitch ME, Reichlin S: The amenorrhea, galactorrhea and hyperprolactinemia syndromes. In Stollerman GH (ed): *Advances in Internal Medicine.* Year Book Medical, Chicago, 1980, pp. 37–60.

24. Lavric MV: Breast secretion in nulligravid women. *Am J Obstet Gynecol* **112**:1139, 1972.

25. Shevach AB, Spellacy WN: Galactorrhea and contraceptive practices. *Obstet Gynecol* **38**:286, 1971.

26. Hardy J, Beauregard H, Robert F: Prolactin-secreting pituitary adenomas: Transsphenoidal microsurgical treatment. In Robyn C, Harter M (eds): *Progress in Prolactin Physiology and Pathology.* Elsevier, New York, 1978, p. 361.

27. Carter JN, Tyson JE, Tolis G, et al: Prolactin-secreting tumors and hypogonadism in 22 men. *N Engl J Med* **299**:847, 1978.

28. Ambrosi B, Rosella B, Travaglini P, et al: Study of the effect of bromocriptine on sexual impotence. *Clin Endocrinol (Oxf)* **7**:417, 1977.

29. Skrabanek P, McDonald D, deValera E, et al: Plasma prolactin in amenorrhea, infertility and other disorders: A retrospective study of 608 patients. *Ir J Med Sci* **149**:236, 1980.

30. Spark RF, White RA, Connolly PB: Impotence is not always psychogenic. *JAMA* **243:**750, 1980.

31. Segal S, Yaffee H, Laufer N, et al: Male hyperprolactinemia: Effects on fertility. *Fertil Steril* **32:**556, 1979.

32. Boyd AE III, Hamilton D, Murray BG, et al: Medical management of prolactinomas. II. In Black PM, Zervas NT, Ridgway EC, et al (eds): *Secretory Tumors of the Pituitary Gland. Progress in Endocrine Research and Therapy.* Vol 1. Raven, New York, 1984, p. 65.

33. Thorner MO, Edwards CRW, Hanker JP, et al: Prolactin and gonadotropin interaction in the male. In Troen P, Nankin H (eds): *The Testis in Normal and Infertile Men.* Raven, New York, 1977, pp. 351–366.

34. Kleinberg DL, Noel GL, Frantz AG: Galactorrhea: 235 cases including 48 with pituitary tumors. *N Engl J Med* **296:**589, 1977.

35. Bassi F, Guisti G, Borsi L, et al: Plasma androgens in women with hyperprolactinemic amenorrhea. *Clin Endocrinol (Oxf)* **5:**61, 1977.

36. Molitch ME, Reichlin S: Hypothalamic hyperprolactinemia: Neuroendocrine regulation of prolactin secretion in patients with lesions of the hypothalamus and pituitary stalk. In MacLeod RM, Thorner MO, Scapagnini U (eds): *Prolactin. Basic and Clinical Correlates.* Liviana, Padova, 1985, p. 709.

37. Johnston DG, Haigh J, Prescott RWG, et al: Prolactin secretion and biological activity in females with galactorrhoea and normal circulating prolactin concentrations at rest. *Clin Endocrinol (Oxf)* **22:**661, 1985.

38. VanWyk JJ, Grumbach MM: Syndrome of precocious menstuation and galactorrhea in juvenile hypothyroidism. An example of hormonal overlap in pituitary feedback. *J Pediatr* **57:**416, 1960.

39. Honbo KS, Van herle AJ, Kellett KA, et al: Serum prolactin levels in untreated primary hypothyroidism. *Am J Med* **64:**782, 1978.

40. Contreras P, Generini G, Michelsen H, et al: Hyperprolactinemia and galactorrhea: Spontaneous versus iatrogenic hypothyroidism. *J Clin Endocrinol Metab* **53:**1036, 1981.

41. Kapcala LP: Galactorrhea and thyrotoxicosis. *Arch Intern Med* **144:**2349, 1984.

42. Tourniaire J, Trouillas J, Chalendar D, et al: Somatotropic adenoma manifested by galactorrhea without acromegaly. *J Clin Endocrinol Metab* **61:**451, 1985.

43. Taylor S: High resolution computed tomography of the sella. *Radiol Clin North Am* **20:**207, 1982.

44. Wolpert SM: The radiology of pituitary adenomas. *Semin Roentgenol* **19:**53, 1984.

45. Gardeur D, Naidich TP, Metzger J: CT analysis of intrasellar pituitary adenomas with emphasis on patterns of contrast enhancement. *Neuroradiology* **20:**241, 1981.

46. Hsu T-H, Shapiro JR, Tyson JE, et al: Hyperprolactinemia associated with empty sella syndrome. *JAMA* **235:**2002, 1976.

47. Badawy SZA, Nusbaum ML, Omar M: Hypothalamic-pituitary evaluation in patients with galactorrhea-amenorrhea and hyperprolactinemia. *Obstet Gynecol* **55:**1, 1980.

48. Gharib H, Frey HM, Laws ER Jr, et al: Coexistent primary empty sella syndrome and hyperprolactinemia: Report of 11 cases. *Arch Intern Med* **143:**1383, 1983.

49. Haney AF, Kramer RS, Wiebe RH, et al: Hypothalamic-pituitary function and radiographic evaluation of women with hyperprolactinemia and an empty sella turcica. *Am J Obstet Gynecol* **134:**917, 1979.

50. Futterweit W: Galactorrhea, amenorrhea, hyperprolactinemia and pseudotumor cerebri in a patient with primary empty sella syndrome: Case report with review of the literature. *Mt Sinai J Med* **49:**514, 1982.

51. Dominque JN, Wing SD, Wilson CB: Coexisting pituitary adenomas and partially empty sellas. *J Neurosurg* **48:**23, 1978.

52. Swanson JA, Sherman BM, Van Gilder JC, et al: Coexistent empty sella and prolactin-secreting microadenoma. *Obstet Gynecol* **53:**258, 1979.

53. Archer DF, Maroon JC, DuBois PJ: Galactorrhea, amenorrhea, hyperprolactinemia, and an empty sella. *Obstet Gynecol* **52**(Suppl):23, 1978.

54. Jones JR, DeHempel PAC, Kemmann E, et al: Galactorrhea and amenorrhea in a patient with an empty sella. *Obstet Gynecol* **49**:S9, 1977.

55. Gomez F, Reyes F, Faiman C: Nonpuerperal galactorrhea and hyperprolactinemia. *Am J Med* **62**:648, 1977.

56. Archer DF, Spring JW, Nankin HR, et al: Pituitary and gonadotrophin response in women with idiopathic hyperprolactinemia. *Fertil Steril* **27**:1158, 1976.

57. Aono T, Myake T, Shiroji T, et al: Impaired LH release following exogenous estrogen administration in patients with amenorrhea-galactorrhea syndrome. *J Clin Endocrinol Metab* **42**:696, 1976.

58. Mortimer CH, Besser GH, McNeilly AD, et al: Luteinizing hormone and follicle stimulating hormone releasing hormone test in patients with hypothalamic pituitary gonadal dysfunction. *Br Med J* **4**:73, 1973.

59. Wiebe RH, Hammond CB, Borchert LG: Diagnosis of prolactin-secreting pituitary microadenoma. *Am J Obstet Gynecol* **126**:993, 1976.

60. Lackelin GCL, Abu-Fadil S, Yen SSC: Functional delineation of hyperprolactinemic amenorrhea. *J Clin Endocrinol Metab* **44**:1163, 1977.

61. Van Campenhout J, Papas S, Blanchet P: Pituitary responses to synthetic luteinizing hormone-releasing hormone in thirty-four cases of amenorrhea or oligo-amenorrhea associated with galactorrhea. *Am J Obstet Gynecol* **127**:723, 1977.

62. Zarate Z, Jacobs LS, Canales ES: Functional evaluation of pituitary reserve in patients with the amenorrhea-galactorrhea syndrome utilizing luteinizing releasing hormone (LH-RH), L-dopa and chlorpromazine. *J Clin Endocrinol Metab* **37**:855, 1973.

63. Chang RJ, Keye WR, Young JR, et al: Detection, evaluation and treatment of pituitary microadenomas in patients with galactorrhea and amenorrhea. *Am J Obstet Gynecol* **128**:357, 1977.

64. Lamberts SWJ, Birkenhäger JC, Kwa HG: Basal and TRH-stimulated prolactin in patients with pituitary tumours. *Clin Endocrinol (Oxf)* **5**:709, 1976.

65. Barbarino A, de Marinis L, Menini E, et al: Prolactin-secreting pituitary adenomas: Prolactin dynamics before and after transsphenoidal surgery. *Acta Endocrinol (Kbh)* **91**:397, 1979.

66. Cowden EA, Ratcliffe JG, Thomson JA, et al: Tests of prolactin secretion in diagnosis of prolactinomas. *Lancet* **1**:1155, 1979.

67. Healy DL, Pepperell RJ, Stockdale J, et al: Pituitary autonomy in hyperprolactinemic secondary amenorrhea: Results of hypothalamic-pituitary testing. *J Clin Endocrinol Metab* **44**:809, 1977.

68. Jeske W: The effect of metoclopramide, TRH and L-dopa on prolactin secretion in pituitary adenoma and in "functional" galactorrheoea syndrome. *Acta Endocrinol (Kbh)* **91**:385, 1979.

69. Schlechte JA, Sherman BM: Abnormal regulation of prolactin secretion after successful surgery for prolactin-secreting pituitary tumours. *Clin Endocrinol (Oxf)* **15**:165, 1981.

70. Frantz AG: Endocrine diagnosis of prolactin-secreting pituitary tumors. In Black PM, Zervas NT, Ridgway EC, et al. (eds): *Secretory Tumors of the Pituitary Gland. Progress in Endocrine Research and Therapy.* Vol 1. Raven, New York, 1984, p. 45.

71. Fine SA, Frohman LA: Loss of central nervous system component of dopaminergic inhibition of prolactin secretion in patients with prolactin secreting pituitary tumors. *J Clin Invest* **61**:973, 1978.

72. Crosignani PG, Ferrari C, Malinverni A, et al: Effect of central nervous system dopaminergic activation on prolactin secretion in man: Evidence for a common central defect in hyperprolactinemic patients with and without radiological signs of pituitary tumors. *J Clin Endocrinol Metab* **51**:1068, 1980.

73. Genazzani ZR, De Leo V, Murru S, et al: Dynamic tests of prolactin secretion in hyperprolactinemic states: Carbidopa-L-dopa and indirectly acting dopamine agonists. *J Clin Endocrinol Metab* **54**:429, 1982.

74. Moriondo P, Travaglini P, Nisim M, et al: Evaluation of two inhibitory tests (nomifensine and L-dopa + carbidopa) for the diagnosis of hyperprolactinaemic states. *Clin Endicrinol (Oxf)* **13**:525, 1980.

75. Muller EE, Genazzani AR, Murru S: Nomifensine: diagnostic test in hyperprolactinemic states. *J Clin Endocrinol Metab* **47**:1352, 1978.

76. Ferrari C, Crosignani PG, Caldara MC, et al: Failure of nomifensine administration to discriminate between tumorous and nontumorous hyperprolactinemia. *J Clin Endocrinol Metab* **50**:23, 1979.

77. Kamoi K, Tchuchida I, Sato H, et al: Comparison of the responses in the nomifensine test with hyperprolactinemia due to prolactin-secreting pituitary tumors and nonprolactin-secreting hypothalamic tumors. *J Clin Endocrinol Metab* **53**:1285, 1981.

78. Camanni F, Genazzani AR, Massara F, et al: Prolactin responsiveness to nomifensine in patients with hyperprolactinemia of tumorous or uncertain etiology. *J Clin Endocrinol Metab* **51**:650, 1980.

79. Dallabonzana D, Spelta B, Botalla L, et al: Effects of nomifensine on growth hormone and prolactin secretion in normal subjects and in pathological hyperprolactinemia. *J Clin Endocrinol Metab* **54**:1125, 1982.

80. Martin TL, Kim M, Malarkey WB: The natural history of idiopathic hyperprolactinemia. *J Clin Endocrinol Metab* **60**:855, 1985.

81. Malarkey WB, Martin TL, Kim M: Patients with idiopathic hyperprolactinemia infrequently develop pituitary tumors. In MacLeod RM, Throner MO, Scapagnini U (eds): *Prolactin. Basic and Clinical Correlates.* Liviana, Padova, 1985, p. 705.

82. Jackson RD, Wortsman J, Malarkey WB: Characterization of a large molecular weight prolactin in women with idiopathic hyperprolactinemia and normal menses. *J Clin Endocrinol Metab* **61**:258, 1985.

83. Post KD, Kasdon DL: Sellar and parasellar lesions mimicking adenoma. In Post KD, Jackson I, Reichlin S, et al. (eds): *The Pituitary Adenoma.* Plenum, New York, 1980, p. 159.

84. Stuart CA, Neelon FA, Lebovitz HE: Hypothalamic insufficiency: The cause of hypopituitarism in sarcoidosis. *Ann Intern Med* **88**:589, 1978.

85. Vesely DL, Maldonodo A, Levey GS: Partial hypopituitarism and possible hypothalamic involvement in sarcoidosis: Report of a case and review of the literature. *Am J Med* **62**:425, 1977.

86. Asa SL, Bilbao JM, Kovacs K, et al: Lymphocytic hypophysitis of pregnancy resulting in hypopituitarism: A distinct clinicopathologic entity. *Ann Intern Med* **95**:166, 1981.

87. Mayfield RK, Levine JH, Gordon L, et al: Lymphoid adenohypophysitis presenting as a pituitary tumor. *Am J Med* **69**:619, 1980.

88. Rickards AG, Harvey PW: Giant cell granuloma and the other pituitary granulomata. *Q J Med* **23**:425, 1954.

89. Holck S, Laursen H: Prolactinoma coexistent with granulomatous hypophysitis. *Acta Neuropathol (Berl)* **61**:253, 1983.

90. Abrams HL, Spiro R, Goldstein N: Metastases in carcinoma: Analysis of 1000 autopsied cases. *Cancer* **3**:74, 1950.

91. Hagerstrand I, Schonebeck J: Metastases to the pituitary gland. *Acta Pathol Microbiol Scand* **75**:64, 1969.

92. Kovacs K: Metastatic cancer of the pituitary gland. *Oncology* **27**:533, 1973.

93. Teears RJ, Silverman EM: Clinicopathologic review of 88 cases of carcinoma metastatic to the pituitary gland. *Cancer* **36**:216, 1975.

94. Stacpoole PW, Kandell TW, Fisher WR: Primary empty sella, hyperprolactinemia and isolated ACTH deficiency after postpartum hemorrhage. *Am J Med* **74**:905, 1983.

95. Merker E, Futterweit W: Postpartum amenorrhea, diabetes insipidus and galactorrhea. *Am J Med* **56**:554, 1974.

96. Dadey SL, Hurxthal LM: Abnormal lactation: Report of a case with amenorrhea and diabetes insipidus. *Lahey Clin Bull* **10**:166, 1957.

97. Boyd AE III, Reichlin S, Turksoy RN: Galactorrhea-amenorrhea syndrome: Diagnosis and therapy. *Ann Intern Med* **87**:165, 1977.

98. Maas JM: Amenorrhea-galactorrhea syndrome: Before, during and after pregnancy. *Fertil Steril* **18**:857, 1967.

99. Koppelman MCS, Jaffe MJ, Rieth KG, et al: Hyperprolactinemia, amenorrhea, and galactorrhea: A retrospective assessment of twenty-five cases. *Ann Intern Med* **100:**115, 1984.

100. Weiss MH, Teal J, Gott P, et al: Natural history of microprolactinomas: Six-year follow-up. *Neurosurgery* **12:**180, 1983.

101. March CM, Kletzky OA, Davajan V, et al: Longitudinal evaluation of patients with untreated prolactin-secreting pituitary adenomas. *Am J Obstet Gynecol* **139:**835, 1981.

102. Besser GM, Parke L, Edwards CRW, et al: Galactorrhoea: Successful treatment with reduction of plasma prolactin levels by bromergocryptine. *Br Med J* **3:**669, 1972.

103. Yuen BH: Bromocriptine, pituitary tumours and pregnancy. *Lancet* **2:**1314, 1978.

104. Thorner MO, Evans WS, MacLeod RM, et al: Hyperprolactinemia: Current concepts of management including medical therapy with bromocriptine. *Adv Biochem Psychopharmacol* **23:**165, 1980.

105. Parkes D: Drug therapy: Bromocryptine. *N Engl J Med* **301:**873, 1979.

106. Nagulesparen M, Ang V, Jenkins JS: Bromocriptine treatment of males with pituitary tumors, hyperprolactinaemia, and hypogonadism. *Clin Endocrinol (Oxf)* **9:**73, 1978.

107. Mroueh AM, Siler-Khodr TM: Bromocriptine therapy in cases of amenorrhea-galactorrhea. *Am J Obstet Gynecol* **127:**291, 1977.

108. Fluckiger E: Pharmacology of prolactin secretion. In Falbusch R, et al (eds): *Treatment of Pituitary Adenomas.* Little, Brown, Boston, 1978, p. 351.

109. Aronoff SL, Daughaday WH, Laws ER Jr: Bromocriptine treatment of prolactinomas. (Letter.) *N Engl J Med* **300:**1391, 1979.

110. Besser GM, Mouk PJA: Prolactinomas and their management. In Jacobs HS (ed): *Advances in Gynecological Endocrinology.* 6th ed. London Royal College of Obstetricians and Gynecologists, London, 1978, p. 234.

111. Barrow DL, Tindall GT, Kovacs K, et al: Clinical and pathological effects of bromocriptine on prolactin-secreting and other pituitary tumors. *J Neurosurg* **60:**1, 1984.

112. Bergh T, Nillius SJ, Wide L: Bromocriptine treatment of seven women with primary amenorrhea and prolactin-secreting pituitary tumors. *Clin Endocrinol (Oxf)* **10:**145, 1979.

113. Vance ML, Evans WS, Thorner MO: Drugs five years later: Bromocriptine. *Ann Intern Med* **100:**78, 1983.

114. Velentzas C, Carras D, Vassilouthis J: Regression of pituitary prolactinoma with bromocriptine administration. *JAMA* **245:**1149, 1981.

115. Tindall GT, Kovacs K, Horvath E, et al: Human prolactin-producing adenomas and bromocriptine: A histological, immunocytochemical, ultrastructural, and morphometric study. *J Clin Endocrinol Metab* **55:**1178, 1982.

116. Thorner MO, Schran HF, Evans WS, et al: A broad spectrum of prolactin suppression by bromocriptine in hyperprolactinemic women: A study of serum prolactin and bromocriptine levels after acute and chronic administraiton of bromocriptine. *J Clin Endocrinol Metab* **50:**1026, 1980.

117. Thorner MO, Edwards CRW, Charlesworth M, et al: Pregnancy in patients presenting with hyperprolactinaemia. *Br Med J* **2:**771, 1979.

118. Thorner MO, Besser GM: Bromocriptine treatment of hyperprolactinaemic hypogonadism. *Acta Endocrinol. (Copenh)* [Suppl 216] **88:**131, 1978.

119. Wass JAH, Moult PJA, Thorner MO, et al: Reduction of pituitary tumor size in patients with prolactinomas and acromegaly treated with bromocriptine with or without radiotherapy. *Lancet* **2:**66, 1979.

120. Wass JAH, Williams J, Charlesworth M, et al: Bromocriptine in management of large pituitary tumors. *Br Med J* **284:**1908, 1982.

121. Clayton RN, Webb J, Heath DA, et al: Dramatic and rapid shrinkage of a massive invasive prolactinoma with bromocriptine: A case report. *Clin Endocrinol (Oxf)* **22:**573, 1985.

122. Chiodini P, Liuzzi A, Cozzi R, et al: Size reduction of macroprolactinomas by bromocriptine or lisguride treatment. *J Clin Endocrinol Metab* **53:**737, 1981.

123. George SR, Burrow GN, Zinman B, et al: Regression of pituitary tumors, a possible effect of bromergocryptine. *Am J Med* **66:**697, 1979.

124. Grisoli F, Vincentelli F, Jaquet P, et al: Prolactin secreting adenomas in 22 men. *Surg Neurol* **13**:241, 1980.
125. McGregor AM, Scanlon MF, Hall K, et al: Reduction in size of a pituitary tumor by bromocriptine therapy. *N Engl J Med* **300**:291, 1979.
126. Vaidya RA, Aloorkar SD, Rege NR: Normalization of visual fields following bromocriptine treatment in hyperprolactinemic patients with visual field constriction. *Fertil Steril* **29**:632, 1978.
127. Thorner MO, Martin WH, Rogol AD, et al: Rapid regression of pituitary prolactinomas during bromocriptine treatment. *J Clin Endocrinol Metab* **51**:438, 1980.
128. Thorner MO, Perryman RL, Rogol AD, et al: Rapid changes of prolactinoma volume after withdrawal and reinstitution of bromocriptine. *J Clin Endocrinol Metab* **153**:480, 1981.
129. Wollesen FO, Andersen T, Karle A: Size reduction of extrasellar pituitary tumors during bromcriptine treatment: Quantitation of effect on different types of tumors. *Ann Intern Med* **96**:281, 1982.
130. Nillius SJ, Bergh T, Lundberg PO, et al: Regression of a prolactin-secreting pituitary tumor during long-term treatment with bromocriptine. *Fertil Steril* **30**:710, 1978.
131. Landolt AM, Wutrich R, Fellmann H: Regression of pituitary prolactinoma after treatment with bromocriptine. *Lancet* **1**:1082, 1979.
132. Corenblum B, Webster BR, Mortimer CB, et al: Possible antitumor effects of 2-bromo-ergocryptine (CB-154 Sandoz) in 2 patients with large prolactin-secreting pituitary adenomas. *Clin Res* **23**:614A, 1975.
133. Nabarro JDN: Pituitary prolactinomas. *Clin Endocrinol (Oxf)* **17**:129, 1982.
134. MacLeod RM, Lehmeyer JE: Studies on the mechanism of the dopamine-mediated inhibition of prolactin secretion. *Endocrinology* **94**:1077, 1974.
135. Mashiter K, Adams E, Bear M, et al: Bromocriptine inhibits prolactin and growth-hormone release by human pituitary tumours in culture. *Lancet* **2**:197, 1977.
136. Sobrinho LG, Nunes MC, Calhaz-Jorge C, et al: Effect of treatment with bromocriptine on the size and activity of prolactin-producing pituitary tumors. *Acta Endocrinol (Copenh)* **96**:24, 1981.
137. Eversmann T, Fahlbusch R, Rjosk HK, et al: Persisting suppression of prolactin secretion after long-term treatment with bromocriptine in patients with prolactinomas. *Acta Endocrinol (Copenh)* **92**:413, 1979.
138. Grossman A, Bouloux P-MG, Loneragan R, et al: Comparison of the clinical activity of mesulergine and pergolide in the treatment of hyperprolactinaemia. *Clin Endocrinol (Oxf)* **22**:611, 1985.
139. Kleinberg DL, Lieberman A, Todd J, et al: Pergolide mesylate: A potent day-long inhibitor of prolactin in rhesus monkeys and patients with Parkinson's disease. *J Clin Endocrinol Metab* **51**:152, 1980.
140. Franks S, Horrocks PM, Lynch SS, et al: Effectiveness of pergolide mesylate in long term treatment of hyperprolactinaemia. *Br Med J* **286**:1177, 1983.
141. Kleinberg DL, Boyd AE III, Wardlaw S, et al: Treatment of prolactin and growth hormone secreting pituitary tumors with pergolide. *N Engl J Med* **309**:704, 1983.
142. Hardy J, Beauregard H, Robert F: Prolactin-secreting adenomas: Transsphenoidal microsurgical treatment. In Robyn C, et al (eds): *Progress in Prolactin Physiology and Pathology*. Elsevier/North-Holland, New York, 1978, p. 361.
143. Hardy J: Transsphenoidal surgery of hypersecreting pituitary tumors. In Kohler PO, Ross GT (eds): *Diagnosis and Treatment of Pituitary Tumors*. Excerpta Medica, Amsterdam, 1973, p. 179.
144. Hardy J: Transsphenoidal microsurgical removal of pituitary microadenoma. In Krayen-bühl H, Maspes PE, Sweet WH (eds): *Progress in Neurological Surgery*. S. Karger, Basel, 1975, p. 200.
145. Hardy J, Vezina JL: Transsphenoidal neurosurgery of intracranial neoplasm. In Thompson RA, Greene JR (eds): *Advances in Neurology*. Raven Press, New York, 1976, p. 261.

146. Hardy J: Microsurgical exploration of a normal sella turcia for a microadenoma. In Derome JP, Jedynak CP, Peillon F (eds): *Pituitary Adenomas: Biology, Physiopathology and Treatment.* Asclepios, France, 1980, p. 195.
147. Hardy J (ed): Prolactinoma. *Neurochirurgie* **27**(Suppl 1), 1981.
148. Chang RJ, Keye WR Jr, Young JR: Detection, evaluation, and treatment of pituitary microadenomas in patients with galactorrhea and amenorrhea. *Am J Obstet Gynecol* **128**:356, 1977.
149. Jacques P, Grisoli F, Guibout M, et al: Prolactin secreting tumors: Endocrine status before and after surgery in 33 women. *J Clin Endocrinol Metab* **46**:459, 1978.
150. Tucker H St G, Grubb SR, Wigand JP, et al: Galactorrhea-amenorrhea syndrome: Follow-up of forty five patients after pituitary tumor removal. *Ann Intern Med* **94**:302, 1981.
151. Post KD, Biller BJ, Adelman LS, et al: Selective transsphenoidal adenomectomy in women with galactorrhea-amenorrhea. *JAMA* **242**:158, 1979.
152. Tindall GT, McLanahan S, Christy JH: Transsphenoidal microsurgery for pituitary tumors associated with hyperprolactinemia. *J Neurosurg* **48**:849, 1978.
153. Wilson Ch.B, Dempsey SC: Transsphenoidal microsurgical removal of 250 pituitary adenomas. *J Neurosurg* **48**:13, 1978.
154. Domingue JN, Richmond IL, Wilson CB: Results of surgery in 114 patients with prolactin-secreting pituitary adenomas. *Am J Obstet Gynecol* **137**:102, 1980.
155. Faria MA Jr, Tindall GT: Transsphenoidal microsurgery for prolactinomas. *In* Givens Jr (ed): *Hormone Secreting Pituitary Tumors.* Year Book, Chicago, 1982, p. 275.
156. Hardy J: Transsphenoidal microsurgery of prolactinomas. In Black P McL, et al. (eds): *Progress in Endocrine Research and Therapy.* Vol. 1. Raven Press, New York, 1984, p. 73.
157. Serri O, Rasio E, Beauregard H, et al: Recurrence of hyperprolactinemia after selective transsphenoidal adenomectomy in women with prolactinoma. *N Engl J Med* **309**:280, 1983.
158. Rodman EF, Molitch ME, Post KD, et al: Long-term follow-up of transsphenoidal selective adenomectomy for prolactinoma. *JAMA* **252**:921, 1984.
159. Antunes JL, Housepian EM, Frantz AG, et al: Prolactin-secreting pituitary tumors. *Ann Neurol* **2**:148, 1977.
160. Sheline GE, Grossman A, Jones AE, et al: Radiation therapy for prolactinomas. In Black P McL, et al. (eds): *Secretory Tumors of the Pituitary Gland. Progress in Endocrine Research and Therapy.* Vol. 1. Raven Press, New York, 1984, p. 93.
161. Grossman A, Cohen BL, Charlesworth M, et al: The circulating prolactin response to radiotherapy in patients with prolactinoma. *Br Med J* **288**:1105, 1984.
162. Grossman A, Besser GM: Prolactinomas. *Br Med J* **290**:182, 1985.
163. Robinson AG, Nelson PB: Prolactinomas in women: Current therapies. *Ann Intern Med* **99**:115, 1983.
164. Turkalj I, Braun P, Krupp P: Surveillance of bromocriptine in pregnancy. *JAMA* **247**:1589, 1982.
165. Lamberts SWJ, Klijn JGM, deLange SA, et al: The incidence of complications during pregnancy after treatment of hyperprolactinemia with bromocriptine in patients with radiologically evident pituitary tumors. *Fertil Steril* **31**:614, 1979.
166. Griffith RW, Turkalj I, Braun P: Pituitary tumors during pregnancy in mothers treated with bromocriptine. *Br J Clin Pharmacol* **1**:393, 1979.
167. Gemzell C, Wang CF: Outcome of pregnancy in women with pituitary adenoma. *Fertil Steril* **31**:363, 1979.
168. Molitch ME, Elton RL, Blackwell RE, et al: Bromocriptine as primary therapy for prolactin-secreting macroadenomas: Results of a prospective multicenter study. *J Clin Endocrinol Metab* **60**:698, 1985.
169. Warfield A, Finkel DM, Schatz NJ, et al: Bromocriptine treatment of prolactin-secreting pituitary adenomas may restore pituitary function. *Ann Intern Med* **101**:783, 1984.
170. Johnston DG, Prescott RWG, Kendall-Taylor P, et al: Hyperprolactinemia. Long-term effects of bromocriptine. *Am J Med* **75**:868, 1983.

171. Spark RF, Baker R, Bienfang DC, et al: Bromocriptine reduces pituitary tumor size and hypersecretion. *JAMA* **247**:311, 1982.
172. Molitch ME: Pregnancy and the hyperprolactinemic woman. *N Engl J Med* **312**:1364, 1985.
173. Barbieri RL, Ryan KJ: Bromocriptine: Endocrine pharmacology and therapeutic applications. *Fertil Steril* **39**:727, 1983.
174. Adelman LS: The Pathology of Pituitary Adenomas. In Post KD, Jackson IM, Reichlin S (eds): *The Pituitary Adenoma*, Plenum Press, New York, 1980, pp. 47–62.
175. Wolpert SM: The Radiology of Pituitary Adenomas. In Post KD, Jackson IM, Reichlin S (eds): *The Pituitary Adenoma*, Plenum Press, New York, 1980, pp. 297–320.
176. Post KD, Kasdon DL: Sellar and Parasellar Lesions Mimicking Adenoma. In Post KD, Jackson IM, Reichlin S (eds): *The Pituitary Adenoma*, Plenum Press, New York, 1980, pp. 159–216.

15

Hypopituitarism

Introduction

Hypopituitarism occurs as a result of deficient secretion of one or more trophic hormones secreted by the pituitary gland.[1] Panhypopituitarism or total deficiencies of all the hormones is a rare disorder. More often, pituitary failure is characterized by selective deficiency of one or more trophic hormones. The pituitary gland demonstrates an amazing ability to withstand the destructive effects of infiltrative or neoplastic processes; thus, clinically significant hypopituitarism does not ensue until more than 85% of the reserve of the gland has been compromised. The clinical implications of this fact are threefold: (1) the evolution of the clinical syndrome can be quite a slow process, indeed often taking years to evolve; (2) partial deficiency of the trophic hormone(s) can be detected by sophisticated testing even in the absence of symptoms, and (3) even when only a single hormone is evidently affected (unitrophic) at the time of presentation, multiple trophic hormone deficiencies may develop later. This underscores the need for extended, even lifelong, follow-up of such patients. The traditional concept that gonadotropins and growth hormone (GH) are the earliest pituitary hormones to be lost followed sequentially by thyroid-stimulating hormone (TSH), adrenocorticotrophic hormone (ACTH), and prolactin (PRL), is observed more frequently when the pituitary reserve is compromised by tumor encroachment. However, the loss of hormone reserve can occur in any order.

Etiology

Basically two major mechanisms can cause pituitary hypofunction: loss of hormone reserve as a result of intrinsic pathology of the pituicytes (primary hypopituitarism) and loss of pituitary hormone reserve because of lack of the hypothalamic-releasing factors that drive the gland to secrete (secondary hypopituitarism). The most common etiology of primary hypopituitarism is destruction of the pituicytes by tumor, vascular insufficiency, or infiltrative disorders; the most common cause of secondary hypopituitarism is interruption

TABLE 77
Etiological Spectrum of Hypopituitarism

Etiology	Condition
Pituitary tumors	Usually chromophobe adenomas
Suprasellar tumors	Craniopharyngioma, meningioma germinoma, chordoma infundibuloma, hamartoma cholesteatoma, glioma
Vascular disease	Apoplexy
	Sheehan's syndrome
	Aneurysm of internal carotid artery
Autoimmune	Lymphocytic hypophysitis
Granulomatous disease	Granulomatous hypophysitis
	Sarcoidosis
	Hand-Schüller-Christian disease
Infiltrative disease	Metastatic disease
	Hemochromatosis
Cysts	Arachnoid cyst
	Developmental cysts
	Dermoid cysts
	Abscess
Developmental	Congenital; usually affects growth; associated with optic nerve abnormalities
Idiopathic	Idiopathic deficiency of hypothalamic-releasing factors (GHRH and GNRH) deficiency
Miscellaneous	Postradiation; post-basal skull fracture; emotional deprivation syndrome

of the hypothalamo–hypophyseal communication by pituitary tumors that extend above the sella or by suprasellar tumors that invade the stalk from above. Table 77 outlines the etiological spectrum of hypopituitarism.

It is essential that the multitude of causes of hypopituitarism be placed in proper perspective. In adults of both sexes, the most common cause for hypopituitarism is tumor of the pituitary gland, usually a chromophobe adenoma. In females, Sheehan's syndrome and lymphocytic hypophysitis bear a temporal relationship to pregnancy. In children, the most common etiologies of hypopituitarism are developmental causes and craniopharyngioma. The rapid development of hypopituitarism (particularly ACTH deficiency) in patients of either sex, regardless of age, should always alert the physician to the possibility of pituitary apoplexy or rupture of an aneurysm of the cavernous portion of the internal carotid artery—catastrophic events characterized by extremely high mortality unless emergency decompression of the sella is carried out.

Since the etiology of hypopituitarism is multifactorial, the important etiologies are briefly and individually discussed. These include the tumorous, vascular, autoimmune, granulomatous, and infiltrative etiologies of hypopituitarism.

Tumorous

Tumors in the sellar, suprasellar, and parasellar region represent the most important and most frequent cause of hypopituitarism. Chromophobe adenomas arising from within the pituitary gland are prototypical of intrasellar tumors. The frequency of detecting subtle abnormalities in pituitary hormone reserve is greater than the development of clinically overt hypopituitarism. Nelson et al.[2] evaluated pituitary function in 20 patients with chromophobe adenomas and showed impaired reserve of one or more trophic hormones in 15 of 20 cases (75%). The most consistent hormone deficiency encountered in these patients was GH deficiency, followed by deficiencies of gonadotropins, TSH, and ACTH. It is noteworthy that very few patients in this study manifested clinical features of hypopituitarism, underscoring the observation that clinically overt hypopituitarism is delayed in pituitary tumors.

There are four mechanisms for the development of trophic hormone deficiency in association with pituitary tumors:

1. Destruction of normal pituicytes by tumor encroachment, a feature of large and invasive tumors such as the chromophobe adenoma
2. Suprasellar expansion of the tumor resulting in impingement and compression of the hypothalamus, median, eminence, the stalk, or the vital neural connections with the hypothalamo–hypophyseal system
3. Loss of pituitary function secondary to apoplexy (bleeding) within the tumor
4. Functional trophic hormone deficiency secondary to the inhibitory effects of the products secreted by the tumor. [A classic example of

such an instance is the endorphin-secreting tumor (endorphinoma)]. Hypersecretion of endorphins by the tumor cells results in suppression of ACTH, follicle-stimulating hormone (FSH), and luteinizing hormone (LH), resulting in deficiencies of these hormones. Trouillas et al.[3] reported a patient with clinical and hormonal evidence of hypopituitarism (deficiencies of gonadotrophins and ACTH), mild hyperprolactinemia, and an enlarged sella secondary to pituitary tumor. Immunocytochemical staining of the tumor tissue revealed a very high content of endorphin-containing granules. The endocrine deficiency in this patient was reversed following removal of the endorphin-secreting tumor indicating the suppressive effects of endorphin hypersection on ACTH, FSH, and LH release.

Several suprasellar tumors can lead to hypopituitarism by destroying the mediobasal hypothalamus (where several releasing factors are generated), the median eminence, or the stalk. The triad of features characterizing such a phenomenon are hypopituitarism, diabetes insipidus, and hyperprolactinemia. Three tumors lead the list of suprasellar lesions: craniopharyngioma, meningioma, and dysgerminoma. All three occur in younger patients and can result in varying degrees of hypothalamic-pituitary insufficiency. Craniopharyngiomas are tumors that originate form the remnants of Rathke's pouch, the embryological structure from which the anterior pituitary is derived. Although perceived as a childhood tumor, the incidence in adults is impressive enough to consider craniopharyngioma in any patient with a suprasellar tumor.[4–7] Banna[8] reviewed 160 cases of craniopharyngioma and found that 13% of these tumors occurred in the third decade of life. The two characteristics of craniophryngiomas are the cystic nature of these tumors (50–75%) and the presence of calcification. The latter, however, may not be seen in 50% of adults with craniopharyngioma. Abnormalities in pituitary function are encountered in most patients with craniopharyngioma.[9]

Meningiomas[10] arise from the meninges overlying the bony continuation of the sella, the planum sphenoidale, the tuberculum sella, and the diaphragm sella. Parasellar meningiomas tend to be homogeneously solid and often cause hyperostosis. These tumors, although benign, can be quite adherent to the structures surrounding it, particularly the hypothalamus or the stalk. Dysgerminomas are malignant tumors that originate in the primitive germ cells. These tumors are characterized by their midline location, the ability to secrete β-HCG, and their radiosensitivity. The preoperative diagnosis of craniopharyngioma, meningioma, and dysgerminoma can be facilitated by computed tomography (CT).[11–13] Less frequent tumors in the region of the suprasellar area include chordomas, gliomas, hamartomas, ependymoma, and infundibuloma. These invasive tumors can be diagnosed only during surgery.

Vascular

Next to tumors of the sellar and suprasellar region, vascular diseases constitute an important etiology of hypopituitarism. The two vascular disorders that can result in hypopituitarism are pituitary apoplexy and Sheehan's syndrome. In 1937 Sheehan[14] described postpartum necrosis of the pituitary resulting in postpartum amenorrhea and failure of lactation. The pathophysiology of this syndrome was further characterized by Sheehan and Murdoch,[15,16] who demonstrated severe thrombosis of the pituitary sinusoids in patients suffering fatal pituitary necrosis during obstetrical shock. The vulnerability of the pituitary gland to ischemic necrosis is attributable to the several changes it undergoes during pregnancy. The pituitary gland enlarges during pregnancy, predominantly due to hyperplasia of the lactotrope population. The enlargement is greater with each successive pregnancy, with maximal enlargements occurring during the third trimester.

Estrogens are believed to play a dominant role in stimulating lactotrope hyperplasia. The enlarged gestational adenohypophysis is exquisitely vulnerable to vascular damage. According to Sheehan's original postulate, severe hemorrhage and the resultant hypotension lead to severe arteriolar spasm of the portal vessels that supply the anterior pituitary. The resulting ischemia can be severe enough to cause complete pituitary destruction. The total wipeout of pituitary function seen in some cases of Sheehan's syndrome is strikingly reminiscent of pituitary apoplexy, another vascular catastrophe. It has been suggested that Sheehan's syndrome may be the result of intravascular coagulation involving the hypothalamic portal vessels.[17]

The classic presentation of Sheehan's syndrome consists of the development of postpartum amenorrhea, lack of lactation, and the gradual development of panhypopituitarism. Several variations of this classic theme need to be recognized.

1. There may not be a classic history of bleeding or obstetric shock in all instances of Sheehan's syndrome. Absence of bleeding or shock has been observed in 5% of patients with postpartum necrosis of the pituitary.[18]
2. The hypopituitarism that follows postpartum necrosis can manifest during the first year after delivery or can be delayed as long as 15–20 years.
3. The resultant hypopituitarism can be complete or partial, even unitropic. Varying degrees of impaired trophic hormone deficiency, including isolated unitropic deficiencies, have been reported to follow postpartum necrosis.[19–22] Thus, panhypopituitarism is not an inevitable sequel.
4. The causative pregnancy need not always be the last one. When partial deficiency follows Sheehan's syndrome, subsequent pregnancy is possible. This underscores the importance a detailed history regarding all the pregnancies in the past, not merely the last one.
5. Although rare, the hypothalamus can be involved in Sheehan's syndrome. Autopsy studies by Sheehan and Summers[23] and by Whitehead[24] have shown considerable atrophy of the supraoptic and paraventricular nuclei, along with loss of neurons in the stalk and atrophy of the neurohypophysis in some patients dying from Sheehan's syndrome.

Thus, diabetes insipidus has been reported to occur in association with Sheehan's syndrome.[18,25–30] In some cases galactorrhea was also noted.[26,27] The combination of hypopituitarism with hyperprolactinemia and diabetes insipidus is the classic triad of hypothalamic hypopituitarism. Rarely, hypothalamic hypothyroidism has been reported in association with Sheehan's syndrome.[31]

Lymphocytic Hypophysitis

Lymphocytic hypophysitis has emerged as a distinct clinicopathological entity during the past decade. The first report of its existence dates back to 1962, when Goudie and Pinkerton[32] described the autopsy findings in a 22-year-old woman who had developed amenorrhea 2 months postpartum. The young woman, in addition to showing lymphocytic infiltration of the anterior pituitary, also had histological features of Hashimoto's thyroiditis. The notion that endocrine failure may be associated with, or even caused by, lymphocytic hypophysitis was supported by the case reported by Richtsmeier et al.,[33] who described ACTH deficiency in a woman who died following a complicated postpartum illness. The autopsy, again, revealed coexistent chronic thyroiditis, along with lymphocytic infiltration of the anterior pituitary. The clinical spectrum of lymphocytic hypophysitis began to unfold when Mayfield et al.[34] reported a 23-year-old woman, who presented with increased intracranial pressure, weight loss, and hypoglycemic episodes 7 months postpartum. X-ray films of the skull revealed an enlarged sella, and hormonal studies demonstrated endocrine deficiencies involving GH, ACTH, TSH, and prolactin, with preservation of gonadotropin reserve. At transsphenoidal surgery, a granular mass was removed that demonstrated a pituitary gland intensely infiltrated by lymphocytes. This report represented the first instance of lymphocytic hypophysitis diagnosed antemortem. Shortly after this report, Asa and colleagues[35] described two patients whose presentation was characterized by the constellation of a sellar mass with mild suprasellar extension, visual-field cuts, endocrine deficiency, mild hyperprolactinemia, and a biopsy appearance characteristic of lymphocytic hypophysitis. Thus, it had become clear that lymphocytic hypophysitis is characterized by five features: a temporal relationship to pregnancy or postpartum, enlargement of the sella (mimicking a tumor), endocrine deficiency, lymphocytic infiltration, and association with other autoimmune diseases.

The etiology of lymphocytic hypophysitis is autoimmune in nature. This is supported on the basis of experimental, clinical, immunological, and histological grounds:

1. Experimentally, Levine et al.[36] demonstrated that injection of pituitary tissue and adjuvant into the footpads of mice can result in hypophysitis, the gland showing a dense lymphocytic infiltrate. The severity of the reaction was much greater in pregnant rats.
2. Clinically, the association between lymphocytic hypophysitis and other autoimmune endocrinopathies is well established. The most common

autoimmunopathy seen in conjunction with lymphocytic hypophysitis is Hashimoto's thyroiditis. Lack et al.[37] demonstrated findings of parathyroiditis and mild adrenalitis in the autopsy of a patient with lymphocytic hypophysitis. The first report of a clinically significant autoantibody to an extra pituitary tissue in a living patient with lymphocytic hypophysitis was by Mazzone et al.[38] They demonstrated low vitamin B_{12} levels, anemia, and positive titers for antiparietal cell antibodies in a 37-year-old woman with lymphocytic hypophysitis presenting as a sellar tumor with significant hyperprolactinemia.

3. Immunological data supporting an autoimmune basis for lymphocytic hypophysitis are provided by the demonstration of autoantibodies against prolactin cells. Bottazzo et al.[39,40] reported the presence of increased titers of antibodies to pituitary tissue in patients with pluriglandular failure and hypoparathyroidism. These studies, however, are not routinely available to make an impact on the diagnosis of lymphocytic hypophysitis.

4. Finally, the histological appearance of the pituitary in lymphocytic hypophysitis is strikingly similar to Hashimoto's thyroiditis: a bona fide autoimmune disorder. The biopsy appearance of lymphocytic hypophysitis is characteristic: islands of adenohypophyseal cells separated by mononuclear inflammatory cells, usually lymphocytes and plasma cells. The infiltration with lymphocytes can be dense enough to form germinal centers.

In summary, lymphocytic hypophysitis must be considered in the differential diagnosis of Sheehan's syndrome (since the endocrine deficiency is temporally related to pregnancy), pituitary tumors (since the CT appearance is that of a mass lesion), and hyperprolactinemia (since suprasellar involvement may be seen in hypophysitis).

Granulomatous Hypophysitis

Granulomatous hypophysitis, like lymphocytic hypophysitis, is a rare cause of hypopituitarism, the diagnosis being made at autopsy or during surgery. Rickards and Harvey[41] offered the first systematic description of this entity in 1954. The histological criteria that characterize granulomatous hypophysitis are as follows:

1. Infiltration of the pituitary by giant cells, plasma cells, and to a lesser extent, lymphocytes
2. Granuloma formation, with multinucleated giant cellas and epithelioid cells at the periphery of the granuloma
3. Absence of caseation

The diagnosis of granulomatous hypophysitis rests on the exclusion of tuberculosis and syphilis, which are extremely rare, and sarcoidosis, which is quite common. The distinction between granulomatous hypophysitis and sarcoidosis can be difficult. The granulomas of sarcoid are larger and better delineated, with more prominent epithelioid cells.

The clinical presentation of granulomatous hypophysitis shares numerous similarities with both lymphocytic hypophysitis as well as sarcoidosis of the pituitary: an expanding sellar mass with suprasellar extension and hypopituitarism.[42–44] Panhypopituitarism is more common with giant cell granuloma of the pituitary than with sarcoidosis. Granulomatous hypophysitis, while occurring more commonly in females, bears no temporal relationship to pregnancy. The etiology of granulomatous hypophysitis is unknown.

Sarcoidosis

Sarcoidosis can affect the hypothalamus, the infundibulum, or the pituitary gland. However, it is an extremely rare cause of hypopituitarism. Stuart et al.[45] reported 10 patients with generalized sarcoidosis and hypopituitarism. The two notable observations were that the low levels of trophic hormones responded to exogenous administration of releasing factors and the high prevalence of optic nerve involvement by sarcoidosis. The preservation of response to releasing factors indicated that the cause of hypopituitarism was hypothalamic. A hypothalamic origin of disease is further strengthened by the observation that hyperprolactinemia is encountered in nearly one-third of patients with extrapulmonary sarcoidosis. In addition, diabetes insipidus dominates as the most frequent manifestation of CNS sarcoidosis. Pennell et al.[46] reviewed 54 cases of CNS sarcoidosis and found evidence of diabetes insipidus in 35%.

Sarcoidosis involving primarily the pituitary gland is extremely rare. Vesely et al.[47] reported a 20-year-old patient with generalized sarcoidosis with arrested growth and sexual maturation since age 16. The pituitary function testing demonstrated marked impairment in the pituitary reserve with little or no response to administration of releasing factors. They reviewed the literature of the previously documented 28 cases of sarcoidosis involving the pituitary. The general characteristics of sarcoidosis of the pituitary are the presence of granulomas within the anterior pituitary, normal or slightly enlarged sella turcica, high prevalence of concomitant hypothalamic involvement, infrequency of lytic bone lesions in the skull, and the favorable outcome following glucocorticoid therapy.

Hand-Schüller-Christian Disease

Hand-Schüller-Christian disease (HSCD) is a granulomatous disease that can affect the function of both the anterior and posterior pituitary.[48] In the chronic, disseminated form of HSCD, two endocrine disturbances dominate the clinical picture: growth retardation, which is seen in nearly 50% of the patients, and diabetes insipidus, encountered in more than 60%. The other clinical features of HSCD include the presence of exophthalmos, granulomatous ottitis media, pulmonary interstitial involvement, and osteolytic bone lesions. Braunstein and Kohler[49] pointed out the frequency with which GH deficiency accompanies vasopression deficiency (diabetes insipidus) in patients with HSCD. The implication of this observation is that a single hypothalamic lesion could account for both endocrine expressions of HSCD.

Metastatic Disease

Although metastatic disease to the hypothalamic pituitary region was described as early as 1913 by Harvey Cushing,[50] this entity has only recently received attention.[51-55,57-60] The breast and lung are the most frequent primary lesions, followed by malignant neoplasms of the gastrointestinal (GI) tract. In autopsy studies, the incidence of pituitary metastases has been reported to vary from 0.14 to 26%.[51-54] Since hypothalamic pituitary involvement is a reflection of hematogenous dissemination, most patients with pituitary metastases would demonstrate evidence of hepatic, cerebral, or skeletal metastases. There are, however, exceptions to this rule. Pituitary metastasis has been described as a solitary lesion in a patient with adenocarcinoma of the rectum in the absence of hepatic metastases.[56] James et al.[57] described a patient whose hypernephroma reappeared as an isolated pituitary metastasis 9 years after nephrectomy. Rarely, multiple myeloma[58] and thyroid carcinoma[59] have presented as a pituitary mass, reflecting metastatic phenomena.

The three major presenting manifestations of metastatic disease to the hypothalamic pituitary region are as follows:

1. As diabetes insipidus, a particularly emphasized manifestation of breast cancer[60]
2. As an enhancing mass in the sellar or suprasellar region, mimicking a pituitary adenoma[13,55]
3. As hypopituitarism with varying degrees of compromised pituitary reserve[61]

Although rare, metastatic disease to the pituitary should be considered an important etiology when evaluating pituitary lesions, in order to spare the patient unnecessary surgery. The CT appearance of pituitary metastases can be similar—almost indistinguishable—from that of an adenoma.[12] The past medical history of malignancy, the presence of overt focus of a primary neoplasm usually with evidence of dissemination, and erosions in the sellar wall are suggestive of metastatic disease to the pituitary.

Hemochromatosis

Hemochromatosis is characterized by the deposition of iron in various tissues. The most common expression of pituitary involvement by hemochromatosis is hypogonadotropic hypogonadism, which occurs in 19–75% of men with hereditary hemochromatosis.[62] To a lesser extent, impairment of GH, ACTH, TSH, and PRL has been reported to occur.[63-67] Kelly et al.[62] evaluated gonadal function in 41 males with homozygous hemochromatosis and demonstrated hypogonadotropic hypogonadism in only four. A notable observation in this study was that reversal of hypogonadism occurred with iron depletion therapy by aggressive phlebotomy.

Hemochromatosis can cause hypothalamic dysfunction in rare instances, leading to pituitary insufficiency. Williams et al.[68] reported the case of a patient with hemochromatosis secondary to sideroblastic anemia and repeated blood transfusions, who developed hypothyroidism and hypogonadism. Endocrine testing with hypothalamic-releasing factor stimulation tests revealed that the pituitary failure was secondary to hypothalamic involvement. This study raises the possibility that some of the earlier reports of pituitary hypogonadism may in fact be due to hypothalamic dysfunction, since repetitive infusions of gonadotropin-releasing hormone were not used to study such patients. Finally, hemochromatosis can directly affect the testes and cause primary testicular failure, with intact hypothalamic–pituitary function.[69,70]

Clinical Features

The clinical features of hypopituitarism reflect the effects of the loss of various trophic hormones.

Constitutional Symptoms

The most common constitutional symptoms of hypopituitarism are tiredness, easy fatiguability, and a general sense of ill health. These symptoms, which are invariably present when hypopituitarism becomes clinically significant, result from either ACTH deficiency (secondary hypoadrenalism) or TSH deficiency (secondary hypothyroidism). Changes in weight in either direction may occur, depending on the presence and relative degree of ACTH or TSH deficiency. When ACTH deficiency is significant, weight loss predominates; when TSH deficiency coexists, no change in weight or even a slight weight gain may be evident. However, since anorexia is present in most patients with hypopituitarism, weight loss is more frequently observed than weight gain. In extreme cases, the weight loss can be severe (pituitary cachexia or Simmond's syndrome). Depression and apathy are noted in more than one-third of patients.

ACTH Deficiency[71–74]

These cases are the most impressive and can be responsible for a strikingly dramatic presentation when adrenal crisis is precipitated by stress. In addition to easy fatiguability, constant tiredness, anorexia, weight loss, and apathy, these patients are at constant risk of developing adrenal crisis. When hypothyroidism is associated with ACTH deficiency, the former, in a sense, is protective, i.e., the metabolic demands are so lowered by the presence of concomitant hypothyroidism (secondary to TSH deficiency) that these patients manage to get by at the basal state despite very little cortisol production. But when the system is stressed by infection or trauma, rapid decompensation occurs and adrenal crisis develops. The presentation, severity, and clinical findings of adrenal crisis consequent to hypopituitarism are no different from those associated with primary adrenocortical failure. Although theoretically mineralocorticoid deficiency should not be a feature of chronic ACTH deficiency, it may occur, since glucocorticoids have a permissive role in the release of aldosterone by the zona glomerulosa in response to renin-mediated stimuli[75,76]

Several physical findings may be evident in the patient with ACTH deficiency. Low blood pressure and postural hypotension can be encountered in secondary adrenal insufficiency, although they are less common and less severe than in primary adrenocortical insufficiency. Lack of axillary hair, especially in females, is a striking indicator of loss of adrenal androgens; since ACTH and β-lipotropin share a common precursor peptide, ACTH deficiency is often associated with loss of the pigmentary hormone, β-lipotropin. This may result in some generalized hypopigmentation. Occasionally, the loss of β-lipotropin can be so impressive as to give the skin an alabaster tone.

In summary, the effects of chronic ACTH deprivation include nonspecific but extremely important constitutional symptoms, hypotension, decreased skin pigmentation, loss of axillary hair (especially in females), and a heightened proclivity to decompensate into adrenal crisis during stress. This last aspect is the most crucial to recognize, since untreated adrenal crisis is often fatal.

TSH Deficiency

The hypothyroidism that develops as a consequence of TSH deficiency is identical to primary thyroid failure. Characteristically, the thyroid gland is shrunken because of lack of TSH drive, hence is not palpable. Hypothyroidism coupled with pallor (in the absence of significant anemia) may be a helpful clue to suggest hypopituitarism. It should be pointed out that lack of sexual hair, decreased libido, amenorrhea, and even hypotension can all be encountered in patients with primary hypothyroidism. Therefore, these features, although often present in secondary hypothyroidism (hypopituitarism with TSH deficiency), are by no means pathognomonic for this entity. Pericardial effusions do occur in patients with pituitary hypothyroidism, but ascites is rare. The cutaneous, integumental, cardiorespiratory, hematologic, and metabolic consequences of hypothyroidism are the same regardless of whether the etiology is primary thyroidal or pituitary in origin (Chapter 7).

Gonadotropin Deficiency

Deficiencies of LH or FSH are often the earliest markers to indicate the presence of hypopituitarism. Decreased libido in males and oligomenorrhea or infertility in females are the most important symptoms. Atrophy of breasts or testes may be evident in longstanding cases.

Prolactin Deficiency

This becomes important only in the setting of the nursing mother in the postpartum period. The classic history of inability to lactate with amenorrhea following delivery is pathognomonic for postpartum pituitary necrosis of the Sheehan type.

Growth Hormone Deficiency

Deficiency of GH is a rare cause of growth retardation in children (Chapter 4). In both adults and children, hypoglycemia is an important consequence of GH failure. Characteristically, the hypoglycemia occurs during fasting and is particularly severe in the presence of concomitant ACTH deficiency.

Associated Features

Depending on the etiology of hypopituitarism, additional physical findings may be evident.

1. Headaches and visual-field cuts in a patient with hypopituitarism are virtually diagnostic of pituitary tumor. Rarely, hypophysitis, granulomatous disorders and metabolic disease can underlie this triad.
2. Diabetes insipidus (polyuria, polydipsia) in a patient with hypopituitarism signifies suprasellar disease caused by tumor (e.g., craniopharyngioma, dysgerminoma), by metastatic disease (e.g., from breast, colon), or by granulomatous disease (histiocytosis, Hand-Schüller-Christian disease, or sarcoidosis). The symptoms of diabetes insipidus are often masked by the concomitant presence of adrenocortical insufficiency. In the absence of glucocorticoids there is a marked reduction in the free water clearance by the renal tubule, imposing a nullifying effect on the vasopressin deficiency. When the glucocorticoid deficiency is corrected by the administration of cortisol, the diabetes insipidus will become manifest, as the renal tubules can now excrete free water.
3. Galactorrhea, paradoxical as it may seem, may occur in association with hypopituitarism and is seen when hypopituitarism is secondary to stalk inhibition. When the pituitary stalk is interrupted (by a tumor), the result is loss of several hypothalamic factors that are trophic to secretion of ACTH, TSH, FSH, LH, and GH. Therefore, with progressive loss of hypothalamic drive, deficiencies of any of these hormones may develop. However, *pari passu* with loss of the trophic releasing factors, inhibition of the stalk also results in loss of the hypothalamic prolactin-inhibitory factor (PIF). This leads to escape of the lactotropes from the tonic negative influence of PIF, resulting in hyperprolactinemia and even galactorrhea. Rarely, galactorrhea can

TABLE 78
Clinical Consequences of Hypopituitarism

Hormone lost	Clinical consequence
ACTH	Constitutional symptoms: weight loss, tiredness, apathy; susceptible to adrenal crisis; postural hypotension; loss of axillary hair (esp. females)
β-Lipotropin	Generalized hypopigmentation
TSH	Secondary hypothyroidism, atrophied thyroid gland
Gonadotropin	Decreased libido, oligomenorrhea, amenorrhea, infertility
PRL	Inability to lactate
GH	Growth retardation; fasting hypoglycemia
Vasopressin	Polyuria, polydipsia

occur in association with postpartum necrosis, when hypothalamic insult is coexistent with partial pituitary infarction.

The clinical features of hypopituitarism are summarized in Table 78.

Diagnostic Studies

Hypopituitarism should be suspected in any patient presenting with symptoms referable to failure of target glands, i.e., the thyroid, adrenal, or gonads or growth. There are three steps in the diagnostic evaluation of hypopituitarism: (1) documentation of hypopituitarism by hormonal studies, (2) delineating whether the pituitary failure is "primary" (intrinsic pituitary disease) or "secondary" (to impaired synthesis, secretion, or transport of hypothalamic releasing factors), and (3) defining the etiology of hypopituitarism, with particular reference to anatomic causes.

Hormonal Studies

Hormonal studies can be documented in one of two ways[77]: (1) by simultaneous evaluation of the relationship between the target gland hormone and the trophic hormone of the pituitary, or (2) by evaluating the reserve of each trophic hormone, employing appropriate provocative stimuli. The former is based on the negative feedback principle that when the target organ fails, the normal pituitary responds by secreting more of its trophic hormone. In essence, the demonstration of a low target-gland hormone (e.g., T_4) and a nonelevated trophic hormone (pituitary TSH) implies impairment of TSH reserve. In the second method of testing, regardless of levels of basal hormone, the capacity of the pituitary gland to respond to trophic stimuli (thyrotropin-releasing hormone, hypoglycemia, or metyrapone) is evaluated. Both methods possess advantages as well as disadvantages.

Certain general principles are applicable to the hormonal evaluation of pituitary function: measurement of a single basal sample of any pituitary hormone (without measuring the target gland hormone simultaneously) is absolutely useless for interpretation because (1) the "normal" ranges for all the pituitary hormones at the basal state considerably overlap with the ranges seen in patients with hypopituitarism, and (2) the pulsatile nature of pituitary hormones is impressive enough to preclude making diagnostic conclusions based on a single sample drawn at an isolated period in time. When both the target-gland hormone and the trophic hormone are clearly low, the diagnosis of hypopituitarism can readily be made. When the target-gland hormone is low but the trophic hormone level is normal, pituitary dysfunction can be suspected but cannot be established unless more detailed testing is undertaken. Finally, when the target-gland hormone is low and the trophic hormone is elevated, pituitary dysfunction has been excluded.

Clearly, in the presence of a low target-gland hormone, the simultaneous demonstration of a low level of trophic hormone strongly supports hypopituitarism, whereas a clearly elevated level of trophic hormone excludes the diagnosis. Equivocal numbers mandate formal pituitary reserve testing with provocative stimuli to exclude or establish pituitary dysfunction.

Formal pituitary reserve testing involves evaluation of the response of each pituitary hormone to a specific provocative stimulus. Thus, the ability of TSH and PRL to rise in response to intravenous TRH, the ability of ACTH and GH to peak in response to hypoglycemia, and the ability of LH and FSH to respond to intravenous luteinizing hormone-releasing hormone (LH-RH) are indicative of the pituitary reserve. Although these stimuli can be administered simultaneously, the test involves multiple samplings of six hormones at 30 intervals for 2–3 hr, rendering the test expensive. The drawback is that physiologically, the definition of a normal response can be highly variable from person to person, and interpretation should take into consideration the multitude of factors that affect these responses in normal persons. Individual testing of each trophic hormone is discussed in Chapters 2, 5, 8, 11, and 20.

Distinguishing Primary from Secondary Hypopituitarism

The second step in the evaluation of hypopituitarism established by hormonal tests is to determine whether the problem is primary or secondary. In general, a distinction between primary and secondary hypopituitarism can be made by evaluating the response of the pituitary hormones to respective trophic stimuli, such as TRH,[78,79] LH-RH,[80–82] GHRH[83,84] and ovine corticotropin-releasing hormone.[85] Patients with intrinsic pituitary disease show minimal or no response, whereas those with secondary hypopituitarism demonstrate a normal but delayed response. Testing hormone dynamics with releasing factors does not always consistently separate pituitary from hypothalamic disease. Patients with secondary hypopituitarism often tend to demonstrate hyperprolactinemia.

The third step is to determine the etiology of hypopituitarism. Since tumors represent the most important and most common etiology, the crucial test is imaging of the pituitary gland. Conventional X-ray films of the sella as well as polytomography have been replaced by the use of CT of the pituitary. However, conventional skull radiographs sometimes provide clues to the underlying etiology of hypopituitarism. Thus, the double-floor sign (tumor), erosion of the clinoids (tumor), suprasellar calcification(craniopharyngioma), and asymmetrical erosion by a calcified lesion (aneurysm) are important signs that can be detected by routine lateral skull films. Since hypopituitarism is caused only by the invasive macroadenomas, and these are invariably visualized by CT imaging, a normal CT scan virtually excludes tumor as the etiology of hypopituitarism. Metastatic lesions as well as suprahypophyseal lesions also lend themselves well to the resolution of CT. The possibility of a vascular etiology (postpartum necrosis) should be considered when the history is suggestive. Lymphocytic hypophysitis and granulomatous diseases are rare and difficult to document. In essence, since treatment of all nontumorous varieties of hypopituitarism is identical (i.e., replacement therapy), the major concern is to exclude tumors, which can be done effectively by CT scans.

Treatment

The treatment of hypopituitarism should focus on the dual aspects of replacing the lost hormone(s) as well as correcting the etiologic factor such as tumor. When hypopituitarism is caused by a pituitary or suprasellar tumor, surgery is mandatory to prevent further pituitary destruction as well as to avoid serious complications such as visual loss and increased intracranial tension.

When hypopituitarism is caused by nontumorous etiologies, the focus of therapy is on replacement of the deficient hormone(s). For example, TSH deficiency and the resultant secondary hypothyroidism readily can be corrected by the use of oral L-T4, 0.15 mg/day. Secondary adrenal failure from ACTH deficiency is treated by the use of oral hydrocortisone, 20 mg in the morning and 10 mg in the evening. With regard to replacement therapy with thyroxine and hydrocortisone, two facts should be stressed. First, when adrenal insufficiency coexists with hypothyroidism, thyroid replacement should be attempted only after building up the adrenal status. Vigorous thyroid hormone replacement in a patient with undertreated adrenal insufficiency imposes the unacceptable risk of precipitating adrenal crisis. The impact of such an error can be catastrophic. Second, patients with hypoadrenalism should be advised to double their replacement dose in the event of stress. Most importantly, these patients should wear identification tags to indicate the diagnosis and the need for intravenous steroids should they be found unconscious.

Gonadotropin deficiency and the resultant hypogonadism are best treated with testosterone or estrogen replacement therapy. The only exception to this is when fertility is desired, in which case therapy with human chorionic and human menopausal gonadotropins may be considered. Growth hormone deficiency in the child should be treated with injections of GH given once or twice a week under close supervision to monitor the response. Replacement therapy with glucocorticoids and L-T4 for hypopituitarism is lifelong. Therefore, when lifelong replacement therapy is being contemplated, establishing the correct diagnosis is crucial. The results of such therapy are extremely gratifying.

References

1. Randall RV, Scheithauer BW, Abboud CF: The anterior pituitary. In Volpe R (ed): *Clinical Medicine*. Vol. 8. Harper & Row, Philadelphia, 1983, p. 1.
2. Nelson JC, Kollar DJ, Lewis JE: Growth hormone secretion in pituitary disease. *Arch Intern Med* **133:**459, 1974.
3. Trouillas J, Girod C, Sassolas G, et al: A human β-endorphin pituitary adenoma. *J Clin Endocrinol Metab* **58:**242, 1984.
4. Kjellberg RN: Craniopharyngiomas. In Tindall GT, Collins WF (eds): *Clinical Management of Pituitary Disorders*. Raven, New York, 1979, p. 373.
5. Bartlett JR: Craniopharyngioma: A summary of 85 cases. *J Neurol Neurosurg Psychiatry* **34:**37, 1971.
6. Khan EA, Gosch HH, Seeger JF, et al: Forty-five years experience with the craniopharyngiomas. *Surg Neurol* **1:**5, 1973.
7. Hoff JT, Patterson RH: Craniopharyngioma in children and adults. *J Neurosurg* **36:**299, 1972.
8. Banna M: Craniopharyngioma: Based on 160 cases. *Br J Radiol* **49:**206, 1976.
9. Jenkins JS, Gilbert CJ, Ang V: Hypothalamic pituitary function in patients with craniopharyngioma. *J Clin Endocrinol Metab* **43:**394, 1976.
10. Shah RP, Leavens ME, Samaan NA: Galactorrhea, amenorrhea, and hyperprolactinemia as manifestations of parasellar meningioma. *Arch Intern Med* **140:**1608, 1980.
11. Fitz CR, Wortzman G, Harwood-Nash DC, et al: Computed tomography in craniopharyngiomas. *Radiology* **127:**687, 1978.
12. Daniels DL, Williams AL, Thornton RS, et al: Differential diagnosis of intrasellar tumors by computed tomography. *Radiology* **141:**697, 1981.
13. Naidich TP, Pinto RS, Kushner MJ, et al: Evaluation of sellar and parasellar masses by computed tomography. *Radiology* **120:**91, 1976.
14. Sheehan HL: Postpartum necrosis of anterior pituitary. *J Pathol* **45:**189, 1937.
15. Sheehan HL: Simmond's disease due to postpartum necrosis of the anterior pituitary. *Q J Med* **8:**277, 1939.
16. Sheehan HL, Murdoch R: Postpartum necrosis of the anterior pituitary: Pathological and clinical aspects. *J Obstet Gynecol Br Commonw* **45:**456, 1938.
17. McKay DG: *Pituitary-Disseminated Intravascular Coagulation*. Hoeber, New York, 1965, p. 405.
18. Aquilo F, Vega LA, Haddock L, et al: Diabetes insipidus syndrome in hypopituitarism of pregnancy. *Acta Endocrinol (Kbh) [Suppl]***60:**137H, 1969.
19. Smith CW Jr, Palmer R, Howard RB: Variations in endocrine gland function in postpartum pituitary necrosis. *J Clin Endocrinol Metab* **19:**1420, 1959.
20. Sheehan HL: Atypical hypopituitarism. *Proc R Soc Med* **54:**43, 1961.
21. Haddock L, Vega LA, Aguilo F, et al: Adrenocortical thyroidal and human growth hormone reserve in Sheehan's syndrome. *John Hopkins Med J* **131:**80, 1972.
22. Stacpoole PW, Kandell TW, Fisher WR: Primary empty sella, hyperprolactinemia and isolated ACTH deficiency after postpartum hemorrhage. *Am J Med* **74:**905, 1983.
23. Sheehan HL, Summer VK: The syndrome of hypopituitarism. *Q J Med* **18:**319, 1949.
24. Whitehead R: The hypothalamus in post-partum hypopituitarism. *J Pathol* **86:**55, 1963.
25. Spain AW, Geoghehan F: Diabetes insipidus in association with post-partum pituitary necrosis. *J Obstet Gynaecol Br Commonw* **53:**223, 1946.

26. Merker E, Futterweit W: Postpartum amenorrhea, diabetes insipidus and galactorrhea. *Am J Med* **56**:554, 1974.
27. Dadey SL, Hurxthal LM: Abnormal lactation: Report of a case with amenorrhea and diabetes insipidus. *Lahey Clin Bull* **10**:166, 1957.
28. Doxiades T, Tiliakos M: Diabetes insipidus in association with post-partum hypopituitarism. *Br Med J* **1**:23, 1956.
29. Evans HW: Sheehan's syndrome with diabetes insipidus. *Am J Med* **28**:648, 1960.
30. Ahn CS, Kim DS: Sheehan's syndrome associated with diabetes insipidus. *Lancet* **2**:1045, 1964.
31. Singer PA, Mestman JH, Manning PR, et al: Hypothalamic hypothyroidism secondary to Sheehan's syndrome. *West J Med* **120**:416, 1974.
32. Goudie RB, Pinkerton PH: Anterior hypophysitis and Hashimoto's disease in a young woman. *J Pathol* **83**:584, 1962.
33. Richtsmeier AJ, Henry RA, Bloodworth JMB Jr, et al: Lymphoid hypophysitis with selective adrenocorticotropic hormone deficiency. *Arch Intern Med* **140**:1243, 1980.
34. Mayfield RK, Levine JH, Gordon L, et al: Lymphoid adenohypophysitis presenting as a pituitary tumor. *Am J Med* **69**:619, 1980.
35. Asa SL, Bilbao JM, Kovacs K, et al: Lymphocytic hypophysitis of pregnancy resulting in hypopituitarism: A distinct clinicopathologic entity. *Ann Intern Med* **95**:166, 1981.
36. Levine S: Allergic adenohypophysitis: new experimental disease of the pituitary gland. *Science* **158**:1190, 1967.
37. Lack EE. Lymphoid "hypophysitis" with end organ insufficiency. *Arch Pathol Lab Med* **99**:215, 1975.
38. Mazzone T, Kelly W, Ensinck J. Lymphocytic hypophysitis associated with antiparietal cell antibodies and vitamin B_{12} deficiency. *Arch Intern Med* **143**:1794, 1983.
39. Bottazzo GF, Vandeli C, Mirakian R, et al: Autoantibodies to single endocrine cells in pituitary, pancreas and gut. In: Cumming IA, Funder JW, Mendelssohn FAO (eds): *Endocrinology 1980*. Australian Academy of Science, Canberra, 1980, pp. 95–99.
40. Bottazzo GF, Pouplard A, Florin-Christensen A, et al: Autoantibodies to prolactin-secreting cells of human pituitary. *Lancet* **2**:97, 1975.
41. Rickards AG, Harvey PW: Giant cell granuloma and the other pituitary granulomata. *Q J Med* **23**:425, 1954.
42. Holck S, Laursen H: Prolactinoma coexistent with granulomatous hypophysitis. *Acta Neuropathol (Berl)* **61**:253, 1983.
43. del Pozo JM, Roda JE, Montoya JG, et al: Intrasellar granuloma: Case report. *J Neurosurg* **53**:717, 1980.
44. Case records of the Massachusetts General Hospital: Case 5—1985. *N Engl J Med* **312**:297, 1985.
45. Stuart CA, Neelon FA, Lebovitz HE: Hypothalamic insufficiency: The cause of hypopituitarism in sarcoidosis. *Ann Intern Med* **88**:589, 1978.
46. Pennel WH: Boeck's sarcoid with involvement of the central nervous system. *Arch Neurol* **66**:728, 1951.
47. Vesely DL, Maldonodo A, Levey GS: Partial hypopituitarism and possible hypothalamic involvement in sarcoidosis: Report of a case and review of the literature. *Am J Med* **62**:425, 1977.
48. Rowland RS. Christian's syndrome and lipoid cell hyperplasias of the reticulo-endothelial system. *Ann Intern Med* **2**:1277, 1929.
49. Braunstein GD, Kohler PO. Pituitary function in Hand-Schüller-Christian disease. *N Engl J Med* **286**:1225, 1972.
50. Cushing H. Concerning diabetes insipidus and the polyurias of hypophyseal origin. *Boston Med Surg J* **168**:901, 1913.
51. Abrams HL, Spiro R, Goldstein N. Metastases in carcinoma: Analysis of 1000 autopside cases. **3**:74, 1950.
52. Hagerstrand I, Schonebeck J: Metastases to the pituitary gland. *Acta Pathol Microbiol Scand* **75**:64, 1969.

53. Kovacs K. Metastatic cancer of the pituitary gland. *Oncology* **27**:533, 1973.
54. Teears RJ, Silverman EM: Clinicopathologic review of 88 cases of carcinoma metastatic to the pituitary gland. *Cancer* **36**:216, 1975.
55. Max MB, Deck MDF, Rottenberg DA: Pituitary metastasis: incidence in cancer patients and clinical differentiation from pituitary adenoma. *Neurology (NY)* **31**:998, 1981.
56. Case Records of the Massachusetts General Hospital (Case 33–1983): *N Engl J Med* **309**: 418, 1983.
57. James RL, Arsenis G, Stoler M: Hypophyseal metastatic renal cell carcinoma and pituitary adenoma. *Am J Med* **76**:337, 1984.
58. Poon MC, Prchal JT, Murad TM: Multiple myeloma masquerading as chromophobe adenoma. *cancer* **43**:1513, 1979.
59. Johnson PM, Atkins HL: Functioning metastasis of thyroid carcinoma in the sella turcica. *J Clin Endorcinol Metab* **25**:1126, 1965.
60. Yap HY, Tashima CK, Blumenschein MD, et al: Diabetes insipidus and breast cancer. *Arch Intern Med* **139**:1009, 1979.
61. Modhi G, Bauman W, Nicolis G: Adrenal failure associated with hypothalamic and adrenal metastases: A case report and review of the literature. *Cancer* **47**:2098, 1981.
62. Kelly TM, Edwards CQ, Meikle AW, et al: Hypogonadism in hemochromatosis: reversal with iron depletion. *Ann Intern Med* **101**:629, 1984.
63. Walsh CH, Wright AD, Williams JW, et al: A study of pituitary function in patients with idiopathic hemochromatosis. *J Clin Endocrinol Metab* **43**:866, 1976.
64. McNeil LW, McKee LC, Lorber D, et al: The endocrine manifestations of hemochromatosis. *Am J Med Sci* **285**:7, 1983.
65. Stocks AE, Powell LW: Pituitary function in idiopathic haemochromatosis and cirrhosis of the liver. *Lancet* **2**:298, 1972.
66. Stocks AE, Martin FIR. Pituitary function in haemochromatosis. *Am J Med* **45**:839, 1968.
67. Charbonnel B, Chupin M, Legrand A, et al. Pituitary function in idiopathic hemochromatosis: hormonal study in 36 male patients. *Acta Endocrinol (Kbh)* **98**:178, 1981.
68. Williams TC, Frohman LA: Hypothalamic dysfunction associated with hemochromatosis. *Ann Intern Med* **103**:550, 1985.
69. Simon M, Franchimont P, Murie N, et al: Study of somatotropic and gonadotropic pituitary function in idiopathic hemochromatosis (31 cases). *Eur J Clin Invest* **2**:384, 1972.
70. McNeil LW, McKee LC Jr, Lorber D, et al: The endocrine manifestations of hemochromatosis. *Am J Med Sci* **285**:7, 1983.
71. Stacpoole PW, Interlandi JW, Nicholson WE, et al: Isolated ACTH deficiency: A heterogeneous disorder. *Medicine (Baltimore)* **61**:13, 1982.
72. Steinberg A, Shechter FR, Segal HI: True pituitary unitropic deficiency. *J Clin Endocrinol (Metab)* **14**:1519, 1954.
73. Satterfield RG, Williamson HO: Isolated ACTH deficiency and pregnancy. *Obstet Gynecol* **48**:693, 1976.
74. Martin MM, Martin ALA: Hypoglycemia due to isolated ACTH deficiency. *South Med J* **62**:1539, 1969.
75. Major P, Kuchel O, Boucher R, et al: Selective hypopituitarism with severe hyponatremia and secondary hyporeninism. *J Clin Endocrinol Metab* **46**:15, 1978.
76. Merriam GR, Baer L: Adrenocorticotropin deficiency: correction of hyponatremia and hypoaldosteronism with chronic glucocorticoid therapy. *J Clin Endocrinol Metab* **50**:19, 1980.
77. Abboud CF: Laboratory diagnosis of hypopituitarism. *Mayo Clin Proc* **61**:35, 1986.
78. Snyder PJ, Jacobs LS, Rabello MM, et al: Diagnostic value of thyrotropin-releasing hormone in pituitary and hypothalamic disease: Assessment of thyrotrophin and prolactin secretion in 100 patients. *Ann Intern Med* **81**:751, 1974.
79. Faglia G, Beck-Peccoz P, Ferrari C, et al: Plasma thyrotropin response to thyrotropin-releasing hormone in patients with pituitary and hypothalamic disorders. *J Clin Endocrinol Metab* **37**:595, 1973.
80. Wollesen F, Swerdloff RS, Odell WD: LH and FSH responses to luteinizing-releasing hormone in normal, adult human males. *Metabolism* **25**:845, 1976.

81. Wentz AC. Clinical applications of luteinizing hormone-releasing hormone. *Fertil Steril* **28:**901, 1977.
82. Mortimer CH, Besser GM, McNeilly AS, et al: Luteinizing hormone and follicle stimulating hormone-releasing hormone test in patients with hypothalamic-pituitary gonadal dysfunction. *Br Med J* **4:73, 1973.**
83. Laron Z, Keret R, Bauman B, et al: Differential diagnosis between hypothalamic and pituitary hGH deficiency with the aid of synthetic GH-RH 1–44. *Clin Endocrinol (Oxf)* **21:**9, 1984.
84. Rogol AD, Blizzard RM, Johanson AJ, et al: Growth hormone release in response to human pancreatic tumor growth hormone-releasing hormone-40 in children with short stature. *J Clin Endocrinol Metab* **59:**580, 1984.
85. Orth DN, Jackson RV, DeChermey GS, et al: Effect of synthetic ovine corticotropin-releasing factor: Dose response of plasma adrenocorticotropin and cortisol. *J Clin Invest* **71:**587, 1983.

16

Tumors of the Pituitary

Introduction

Pituitary tumors represent 10% of all intracranial neoplasms. Routine examination of pituitary glands at autopsy has disclosed that the prevalence rate of microadenomas may be as high as 25%.[1,2] The diagnosis and management of pituitary tumors have been revolutionized by five developments in the recent past—the emergence of high-resolution computed tomography (CT), the development of sensitive radioimmunoassays (RIAs) for measurement of adenohypophyseal hormones, improved histological staining techniques, expertise in transsphenoidal microsurgery, and the availability of adjunctive drug therapies. These developments have contributed significantly to the early detection and better management of patients with pituitary tumors. These tumors, considered benign in terms of their nonmetastatic proclivity, often demonstrate a locally aggressive behavior, resulting in compromised sight and life. The morbidity caused by pituitary tumors relates to both the mass effect as well as the hormonal effects secondary to the tumor; the latter can be due to both hypersecretion of hormones by the tumor as well as deficient secretion of hormones by the normal portion of the gland compressed by the tumor. These morbid consequences can be prevented by instituting therapy early in the course of the disease.

Classification

Pituitary tumors are generally classified on the basis of their anatomical and histological characteristics. Thus, anatomically, these tumors can be classified as microadenomas (<10 mm) or macroadenomas (>10 mm), depending on size. Macroadenomas can be further classified as enclosed or invasive macroadenomas, depending on whether these tumors are contained within the osteoaponeural confines of the sella turcica. Invasive macroadenomas are further classified into locally invasive (merely suprasellar extension) versus diffuse invasive varieties, the latter reflecting supra-, para-, and even subsellar invasion (Fig. 36).

Histologically, the traditional classification of pituitary tumors into acidophilic, basophilic, and chromophobe adenomas is inadequate and highly unsatisfactory for several reasons. First, this classification does not provide any information regarding hormonal secretion. Second, it is now well estab-

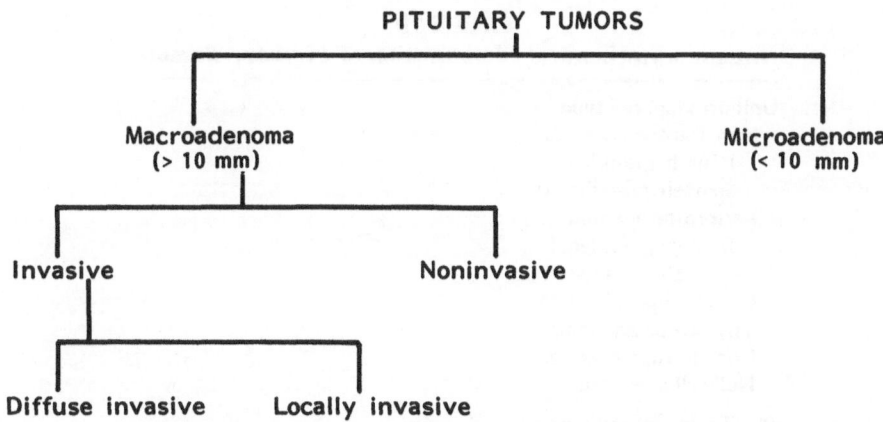

PITUITARY TUMORS

Macroadenoma (> 10 mm) — Microadenoma (< 10 mm)

Macroadenoma: Invasive — Noninvasive

Invasive: Diffuse invasive — Locally invasive

FIGURE 36. Anatomic classification of pituitary tumors.

lished that the chromophobe adenoma, traditionally assumed to be nonsecretory in the old classification, is in fact capable of secreting peptide hormones, as well as glycoproteins. Third, and most importantly, the conventional classification may not even unequivocally establish the pituitary origin of the tumor, since a chromophobic appearance can be seen in several other tumors as well. Finally, information regarding hormone storage, cytogenesis, and the structure–function relationship cannot be obtained by the staining characteristics of the cytoplasm. The new classification of pituitary tumors is based on immunocytological and ultrastructural features of the tumor cell. The introduction of the immunoperoxidase technique and its application to pituitary tumors permits accurate demonstration of the cell origin of these tumors. The procedure is based on the use of specific antibodies that seek out their respective antigens (hormones) within the secretory granules of the tumor cells. A positive immunostaining for one or more adenohypophyseal hormones in the tumor cell would establish the adenohypophyseal origin of the tumor conclusively. This information is crucial, since several tumors in the region of the sella can mimic the clinical behavior and radiological appearance of pituitary tumors. With the use of immunoperoxidase staining, it is possible to identify the five major cell types of the anterior pituitary: the somatotropes, the corticotropes, the lactotropes, the thyrotropes, and the gonadotropes. Thus, the six major types of pituitary tumors are the somatotrope adenomas, the corticotrope adenomas, the lactotrope adenomas, the thyrotrope adenomas, the gonadotrope adenomas, and the null-cell adenomas (no hormone-containing cells). The major classification is further fine-tuned by the introduction of subtypes based on ultrastructural characteristics. Thus, somatotrope and lactotrope adenomas are further categorized into densely granulated and sparsely granulated adenomas. The identification of specific hormonal content within pituitary tumors has facilitated the addition of a third facet to the classification, based on embryological derivation, i.e., plurihormonal tumors. Within this category are included tumors that demonstrate pluripotential secretory activity, such as mixed-cell adenomas, acidophil

TABLE 79
Immunocytochemical Classification of Pituitary Tumors

Unihormonal cell type
  Somatotrope adenoma
    Densely granulated
    Sparsely granulated
  Lactotrope adenoma
    Densely granulated
    Sparsely granulated
  Corticotrope adenoma
  Thyrotrope adenoma
  Gonadotrope adenoma
  Null-cell adenoma

Plurihormonal cell type
  Mixed somatotrope and lactotrope adenoma
  Acidophil stem cell adenoma
  Mammosomatotropic adenoma

stem cell adenomas, and mammosomatotropic ademonas. This classification, proposed by Dr. Kalman Kovacs[3] is currently accepted as the most accurate and descriptive classification of pituitary tumors (Table 79). It is apparent that the introduction of the immunoperoxidase staining technique to pituitary tumor tissue has completely revolutionized the concept of tumor histology.[4–7]

Somatotrope Adenoma (Somatotropinoma)

As the underlying cause of acromegaly and gigantism, somatrope adenomas originate in the GH-producing somatotropes. These cells constitute approximately 50% of the adenohypophyseal cell population. Somatotropes primarily populate the lateral wings of the pars distalis.[8] Scattered somatotropes can also be found in the median wedge. The three characteristics of the normal somatotropes are the presence of evenly electron-dense granules (250–500 nm), prominent rough endoplasmic reticulum, and positive immunoperoxidase staining for GH. Several of these normal characteristics are extended to somatotrope adenomas. The tumor is usually located in one of the lateral wings and is often well demarcated. The two major types of somatotrope adenomas are the densely granulated adenoma[9–11] and the sparsely granulated variety.[12–14] Of these, the appearance of the densely granulated variety bears a striking resemblance to normal somatotropes but reflects an increased populace. The secretory granules are numerous, mostly round, measuring 300–600 nm, and occasionally showing "giant" granules as large as 2000 nm. The densely granulated adenoma stains with acid dyes and represents the classic eosinophilic adenoma in the conventional classification. By contrast, the sparsely granulated somatotrope adenoma represents the chromophobic variety on conventional stains, but, with immunoperoxidase stains, GH can be located to the sparse secretory granules. It is noteworthy that in

TABLE 80
Two Types of Somatotropinoma

Feature	Densely granulated somatotrope adenoma	Sparsely granulated somatotrope adenoma
Cell of origin	Somatotropes	Somatotropes
Staining with conventional stains	Acidophilic	Chromophobic
Immunoperoxidase staining with GH antisera	Positive	Positive
Secretory granule size	300–600 nm	100–300 nm
Nucleus of cells	Regular	Irregular
Similarity to normal somatotropes	Preserved	Less differentiated
Characteristic ultrastructural feature	Well-developed rough endoplasmic reticulum	Fibrous bodies composed of microfilaments
Clinical presentation with acromegaly	+	+
Course	Slow growing; nonaggressive	Fast growing; invasive

contrast to the densely granulated variety, these tumors are composed of cells that differ sharply, in appearance, from the normal somatotropes. The chief differences are the presence of irregular nuclei, sparse secretory granules that are smaller in size, and striking fibrous bodies composed of microfilaments. Clinically, the sparsely granulated somatotrope adenomas tend to be more aggressive, and are fast growing. The features of both varieties of somato-tropinomas are outlined in Table 80. In clinical practice, both varieties are encountered with equal frequency.

Lactotrope Adenoma (Prolactinoma)

Lactotrope adenomas are responsible for causing the clinical syndrome of galactorrhea, amenorrhea, and infertility in females, and impotence in males. These tumors originate from the lactotropes, which represent approximately 10–25% of the pituitary cell population. The pars distalis is the favored location of these cells, clusters of cells usually visible at the posterior aspect of the pars distalis. One notable feature of normal lactotropes is the differential staining of these cells with prolactin antisera: some cells are densely granulated and show intense immunopositivity, reflected as diffuse brown deposits throughout the cytoplasm while other lactotropes are only mildly immunopositive, demonstrating a ringlike immunopositive reaction with PRL antisera, especially discernible over the Golgi complex. Accordingly, prolactinomas can be densely granulated, or sparsely granulated. The densely granulated prolactinoma presents a striking histological picture on immunope-roxidase staining with PRL antisera, demonstrating diffuse intense brown deposits throughout the cytoplasm. The densely granulated prolactinoma represents the classic eosinophilic adenoma in the conventional classification. The sparsely granulated prolactinoma is chromophobic and may pose minor difficulties in recognition by the immunoperoxidase technique, the immu-nopositivity often being limited to a crescentic or ringlike area in the Golgi complex.

Mixed-Cell Adenoma

Although somatotropinomas and prolactinomas represent tumors orig-inating from monomorphous secretory cells adhering to the "one-cell, one-hormone" concept, mixed adenomas can occur. True mixed adenomas are bimorphous and bihormonal, containing two separate and distinct cell types. Immunoperoxidase stains identify both cell types coexisting side by side in the same tumor, with either dense or sparse granulation. These tumors cause hypersecretion of both GH and prolactin.

In addition to the true mixed-cell adenoma, two other tumors are char-acterized by immunopositivity to both GH and PRL—the acidophil stem cell adenomas and the mammosomatotropic adenoma. Acidophil stem cell adenomas[6,15–17] are tumors that originate in the common precursor cell, the acidophil stem cell, from which both the somatotropes and the lactotropes are derived. The acidophil stem cell adenoma may appear chromophobic or mildly acidophilic with traditional stains. The remarkable aspect, on immu-noperoxidase staining, is the localization of both GH and PRL to the same cell. The tumor cells contain small secretory granules, usually sparse, with poorly developed rough endoplasmic reticulum. One of the features that highlights the electron microscopic appearance of the acidophil stem cell adenoma is the characteristic presence of giant mitochondria.

The clinical manifestations of acidophil stem cell adenoma are dual: hy-persecretion of PRL and of GH. Three clinical observations are pertinent in regard to these acidophil stem cell adenomas. First, while hyperprolactinemia (often with amenorrhea, galactorrhea, and impotence) is a very frequent fea-ture, acromegaly or even elevated serum levels of GH are seen less frequently. Second, these tumors tend to be aggressive, often invading the sphenoid sinus. Third, these tumors are often large, with rapid growth rates. The general behavior of these tumors, coupled with their staining properties on light microscopy, is strikingly reminiscent of the all-too-familiar chromophobe ad-enoma.

Mammosomatotrope adenomas[18] are believed to be the mature counter-parts of the acidophil stem cell adenoma. This is reflected by their slow growth rate, less aggressive behavior, and tendency to show numerous densely gran-ulated monomorphous cells. Also, in contrast to acidophil stem cell adenomas, these tumors result more frequently in acromegaly, while the PRL levels in the serum are only mildly or moderately elevated.

The combined occurrence of GH and PRL hypersecretion by the same tumor may be related to the fact that PRL and GH are derived from the same ancestral gene.[19–21] Molecular cloning techniques used to study gene expression have indicated substantive homology in the sequences of cDNA (the DNA complementary to mRNA) of PRL and GH. It appears that PRL and GH are derived from duplication of the same precursor gene.

Corticotrope Adenoma

Corticotrope adenomas underlie the pathophysiology of Cushing's disease, the Nelson syndrome, and endorphinomas. In the normal pituitary gland, corticotropes constitute 10–20% of pituicytes. These cells characteristically appear oval or angular and are found mostly in the median wedge of the pars distalis. Corticotropes can be identified by their immunopositivity for ACTH, β-LPH, and endorphins. It is generally believed that all corticotropic cells react with antisera against all POMC-derived peptides.[22,23] However, dissociation between the presence of β-endorphin and other POMC peptides has been demonstrated in normal tissue,[24] as well as in corticotrope adenomas.[25,26] This dissociation is exemplified in corticotrope adenomas that demonstrate immunopositivity for only β-endorphins, but not for ACTH. Trouillas et al.[27] described a patient with a pituitary corticotrope adenoma composed of tumor cells that reacted only with β-endorphin antiserum and was associated with hypogonadism and hypocortisolemia.

Corticotrope adenomas express considerable heterogeneity. The classic corticotrope adenomas are small, readily stain basophilic, and are positive for periodic acid-Schiff (PAS) as well as lead hematoxylin. They are densely granular and demonstrate intense immunopositivity to ACTH/β-LPH. Less commonly, the corticotrope adenomas appear chromophobic and are sparsely granulated. Such tumors tend to be larger and more aggressive in behavior. A third subtype of corticotrope adenomas are those originating in the intermediate lobe.[28] These tumors are in a class by themselves and are recognized as a distinct subtype of tumors that produce Cushing's disease. The five characteristics of intermediate lobe corticotrope adenomas are resistance to dexamethasone suppression, response to dopamine agonists such as bromocriptine, frequent association with hyperprolactinemia, presence of argyrophilic nerve fibers coursing through the tumor suggestive of a neural origin, and poor cure rate with surgery. Perturbations in dopaminergic regulation of corticotropes may play some role in the development of intermediate lobe tumors.

A fourth subset of corticotrope adenoma is referred to as silent corticotrope adenoma.[29] These tumors may appear basophilic or chromophobic by conventional stains but contain secretory granules that demonstrate intense immunopositivity for ACTH, as well as several parts of the proopiomelanocortin molecule. Yet these tumors do not cause hypersecretion of ACTH. Thus, patients with these adenomas remain eucortisolemic. The reason for this paradox is unclear.

Thyrotrope Adenoma

These adenomas represent an important cause for TSH-mediated hyperthyroidism. These tumors are rare and probably account for no more than 1% of all pituitary tumors. Hill et al.[30] found only four cases of TSH-secreting adenomas in 545 surgically operated cases. In another report, Ipse et al.[31] found only a single thyrotrope adenoma among 312 pituitary tumors. Approximately 50 cases of adenomatous TSH-mediated hyperthyroidism have been described in the literature. Interestingly, 30–45% of these cases had been associated with concomitant hypersecretion of GH or prolactin.[32] The classic hormonal hallmark of thyrotrope adenomas is hyperthyroidism with an inappropriately elevated TSH level.

Thyrotrope adenomas originate from thyrotropes—medium to large sized, angular, or polyhedral cells with characteristic long cystoplasmic processes. These cells contain fine granules that are immunopositive for TSH thyrotropes, constituting 10% of the adenohypophyseal cell population. Morphologically, thyrotope adenomas are recognized by their immunopositivity to TSH and appear chromophobic by conventional stains. Occasionally, thyrotrope adenomas result from chronic untreated or undertreated primary hypothyroidism.

Gonadotrope Adenoma

Until the mid-1970s, gonadotrope cell adenomas were thought to be extremely rare and to result from chronic hyperplasia of gonadotropes, secondary to chronic primary hypogonadism.[33–36] Primary gonadotrope adenomas have emerged as a distinct clinicopathological entity only in the past decade due partly to the availability of the immunoperoxidase staining techniques and partly to routine preoperative measurement of FSH, LH, and glycoprotein subunits in all patients with pituitary tumors. Trouillas et al.[37] estimated the incidence of gonadotrope adenomas in 230 excised pituitary tumors to be approximately 3.5%. A similar figure, 4.1% of 728 adenomas, was reported by Horvath and Kovacs.[38] These adenomas originate from gonadotropes, cells that are randomly distributed in the central wedge and the lateral wings of the gland. They account for approximately 10% of the total cell population. Immunoperoxidase stain localizes both FSH and LH to the same cell. Electron microscopic studies have revealed that the tumor cells of the gonadotrope adenoma differ morphologically from the normal gonadotropes. The adenomatous cells are smaller and more angular and have a striking increase in microtubules. The secretory granules, which are smaller than normal, demonstrate immunopositive staining for β-FSH, and to a lesser extent β-LH. Horvath and Kovacs[38] observed the presence of a unique honeycomb-like Golgi apparatus in gonadotrope adenomas from females, but not in males. The gonadotrope adenoma appears chromophobic by conventional stains.

The presentation of gonadotrope adenoma is remarkably similar to that of the conventional chromophobe adenoma—middle-aged, normally virilized, eugonadal males, manifesting the parasellar compression effects of an enormous pituitary tumor that appears chromophobic on gross histology. The hormonal features that characterize these adenomas are discussed in the section on hormonal diagnosis.

Null-Cell Adenoma

As the name implies, the tumor cells of the null-cell adenoma fail to show immunopositivity with any of the adenohypophyseal hormones. Clinically, there is no evidence of hypersecretory syndromes, and hormonally they do not show elevated levels of peptide hormones or glycoprotein hormones. The exception is the prolactin level, which can be elevated secondary to impairment in the production or transport of dopamine, caused by suprasellar extension of tumor. These tumors appear chromophobic and PAS negative on conventional staining. Ultrastructurally, the tumor cells do show cytoplasmic secretory granules, which characteristically fail to demonstrate immunopositivity for all adenohypophyseal hormones, their subunits as well as their precursors. These tumors manifest as large intrasellar masses that cause headache and visual-field defect and that compromise pituitary reserve. They represent the true chromophobe adenomas by the conventional classification.

Clinical Features

The manifestations of pituitary tumors fall into three categories: those caused by the mass effect of an enlarging tumor, those caused by hypersecretion of adenohypophyseal hormones, and those due to hyposecretion of these hormones. Each of these, however, can be caused by nontumorous conditions involving the pituitary gland.

Mass Effect of Tumor

The initial presentation of pituitary tumors is frequently characterized by pressure effects caused by the tumors. The most frequent symptom experienced by patients with pituitary tumor is headache. However, there are no specific features of these headaches that would permit identification. Caused by stretching of the diaphragma, occasionally headaches may be totally absent or insignificant. Pituitary tumors can remain asymptomatic for years, until they begin to expand the sella and extend to the suprasellar, parasellar, and infrasellar regions. Suprasellar extension is the most frequent extension, owing to the relative resilience of the diaphragma of sella. Invasive pituitary tumors can also cause destruction of the floor and extend inferiorly to involve the sphenoid sinus. The consequences of expanding pituitary lesions (i.e., the mass effect) are several and include the development of chiasmal syndromes, cranial nerve involvement, compression of the stalk, compression of aqueduct of Sylvius, vasopressin deficiency, destruction of normal pituitary tissue, and sphenoid sinus invasion.

Chiasmal Syndromes

The optic chiasm forms the most important superior relationship to the pituitary gland, lying just above the diaphragma sella. The presence, type, and severity of the chiasmal syndrome that occurs from suprasellar extension of the tumor depend on whether the chiasm is prefixed, postfixed, or mid-positional. In approximately 80% of normal persons, the optic chiasm is located over the center of the posterior two-thirds of the sella (postfixed); in 12% of cases, the chiasm sits directly at the center of the sella (mid-positional); and in the remainder (8%) the chiasm is found anteriorly[39] (prefixed). The optic chiasm is composed of fibers derived from both optic nerves at their point of decussation. At the optic chiasm, the fibers from the inner nasal half of each retina (representing the temporal fields of each eye) decussate, while fibers from the outer temporal half of the retina (representing the nasal fields of each eye) remain on the same side. The optic tracts on each side are formed after the decussation in the chiasm. Therefore, each optic tract contains fibers from the outer half of the retina on the same side as well as those from the inner half of the retina on the opposite side. Each optic tract passes posteriorly, terminating at the lateral geniculate body of the same side. A few fibers of the optic tract ("pregeniculate fibers) project to the superior colliculus. The optic radiation originates from the lateral geniculate body, moves through the posterior limb of the internal capsule and projects posteriorly to the occipital cortex, at the calcarine area.

The chiasmal syndrome results when the optic chiasm is compressed from below by the superiorly expanding solid tumor mass. Obviously, the most vulnerable fibers are the lower fibers of the chiasm, which consist predominantly of the decussated fibers from the nasal halves of retinae on both sides (usually serving the upper temporal fields bilaterally). This is the reason for the development of bitemporal hemianopia, the field defect most commonly associated with chiasmal compression. However, the chiasmal compression syndrome consists of several distinct visual-field defects, of which bitemporal hemianopia is but one. There are five well-recognized field cuts associated with the chiasmal compression syndrome:[40] (1) bitemporal hemianopic scotoma (Fig. 37), (2) bitemporal hemianopia, (3) bitemporal hemianopia with a unilateral central scotoma, (4) central scotoma with a contralateral superior quadrantic temporal field cut, and (5) homonymous hemianopia plus central scotoma.

The earliest sign of chiasmal compression is the development of bitemporal hemianopic scotoma. This variety of field cut is difficult, if not impossible, to detect clinically, but should be suspected when the patient has bitemporal hemianopia with decreased vision in one eye. The demonstration of bitemporal hemianopic scotoma implies the involvement of the chiasm plus

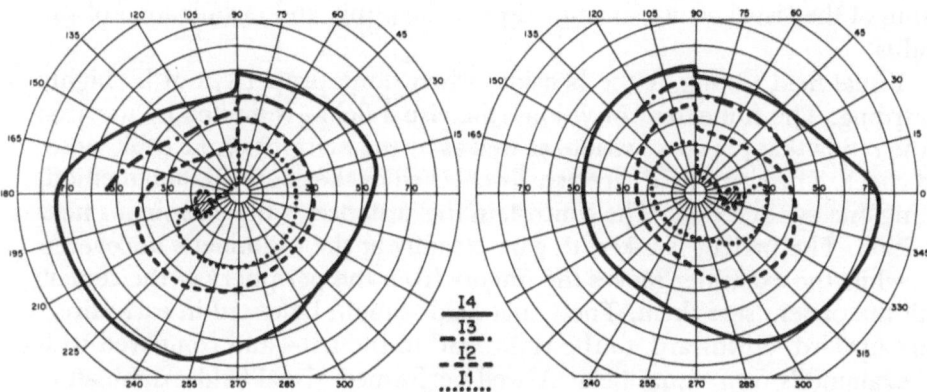

FIGURE 37. Bitemporal superior field defect caused by an enlarging pituitary tumor.

the optic nerve on one side. The second lesion, and the one usually detected by the confrontation method at the patient's bedside, is bitemporal hemianopia. The third and fourth types of field cuts—central scotoma associated with either a superior quadrantic temporal field cut, or bitemporal hemianopia—result from compression at the junction of the optic nerve and the chiasm (junctional scotoma). The lesion compromises the ipsilateral optic nerve and the fibers from the inferior nasal retina of the opposite eye after having decussated. Thus, the patient will demonstrate decreased central vision in one eye and a superior quadratic temporal scotoma on the opposite side. Finally, when the optic tract is compressed, classic homonymous hemianopia will develop.

The relative positional alignments between the expanding mass and the optic apparatus (optic nerve, chiasm, and optic tract) will determine the type and severity of lesions. If the chiasm is significantly prefixed—moved forward—and the tumor is posterior to the chiasm, optic tract syndromes dominate the picture, giving rise to classic homonymous hemianopia. If the chiasm is strictly midline and becomes entrapped by the tumor, the brunt of compression falls on the chiasm, with the development of the conventional field cuts associated with pituitary tumors—superior quadratic temporal field cuts and bitemporal hemianopia. In this regard, it should be realized that after decussation, the posterior fibers form a bundle called the papulomacular bundle. Compression of this bundle will result in loss of central color vision, an early sign of chiasmal compression. When the optic chiasm is significantly posterior, and the tumor is relatively anterior, the brunt of compression falls on the ipsilateral optic nerve, giving rise to decreased vision, sometimes to the point of blindness, as well as junctional scotomas (field cuts involving decreased central vision in one eye and superior quadratic temporal defects on the contralateral side.

The clinical examination of patients with suspected pituitary tumors must be performed carefully, with an understanding of the regional anatomy of the suprasellar area. The examination should include testing of visual fields, testing of the visual acuity, examination of the pupil, and examination of the fundus.

Visual-field testing at the bedside, when done properly, can be highly rewarding. The full extent of vision—the visual field—is limited by the area of the retina that perceives vision, as well as by the margins of the orbit, nose, and cheek. The visual fields are usually checked by the confrontation method, testing one eye at a time. The control, in this instance, is the examiner's field of vision. The patient is asked to gaze steadily at the examiner's uncovered eye, while the examiner moves the fingers from the periphery to the center, until the patient sees them. The visual fields should be tested in each direction—upward, downward, to the right, and to the left—and compared with the examiner's own visual field. A well-performed visual-field examination correlates rather well with findings on perimetry. In one study of 27 patients with pituitary tumors, only in four were the defects missed by confrontation and documented by perimetry.[41] While the confrontation test can grossly detect the presence of visual-field cuts, perimetry is required in all patients to document the degree of these defects, as well as to serve as a baseline record for future comparisons.

In addition to conventional testing of visual fields by the confrontation method, it is essential to perform the red desaturation test in all patients with pituitary lesions. The red pin test outlines the central visual field. The test can be done by holding a pinhead up within the patient's field of vision, moving it gradually to the center. Central scotomas can be perceived in this manner, since the intensity of the red color is reduced in the central part of the field, permitting the detection of less severe forms of central scotomas. It should be remembered that the physiological blind spot is situated in the temporal side of the central point of fixation in the visual field. It corresponds to the entrance point of optic nerve into the retina. This point should always be identified (based on comparisons with the examiner's blind spot), in order to properly interpret the presence of central scotomas.

Visual acuity must be tested in all patients suspected of having pituitary disease. In many patients with pituitary tumors the diagnosis was initially suspected by ophthalmologist. Any decrease in visual acuity should be reason enough to suspect an underlying pituitary etiology. Decreased visual acuity can be a result of optic nerve involvement, papilledema, or both.

Examination of the pupil and papillary reflexes can provide useful information in patients with pituitary tumors. The demonstration of the Marcus Gunn pupil is suggestive of partial afferent involvement and is associated with irritation or partial compression of the optic nerve. The Marcus Gunn pupil is said to exist when the pupillary constriction is brisk and well sustained in one eye, but when the light is immediately shone into the other eye, this pupil may slowly dilate, indicating that its reaction to the consensually mediated light reflex is more active than its response to direct light applied after the consensual reflex. Another abnormality seen in the pupils is oculomotor nerve involvement by the tumor, but this is an extremely rare manifestation of pituitary tumors, generally occurring when there is apoplexy in the tumor.

Finally, examination of the fundus is vital to detect papilledema or optic atrophy secondary to chronic optic nerve involvement. A new concept has been emphasized by Burde[40] in terms of predicting the presence of a field cut by the use of the ophthalmoscope alone. The nerve fiber bundles of the retina can be visualized using the red-free filters on the opthalmoscope.[42] Large nerve fiber bundles enter the optic disc in a well-defined topographical pattern. Thus, large accurate bundles that enter the superior and inferior aspect of the temporal part of the disc serve the nasal visual fields, while the fibers that enter the nasal part of the disc serve the temporal visual fields. When temporal hemianopia is caused by compression of the chiasm, at a location distant to the retina, there is loss of all nerve fibers around the disc, with the exception of the large accurate bundles that serve the nasal fields and enter the disc at its temporal aspect. Preservation of these accurate bundles alone imparts "bow-tie" atrophy that is horizontally oriented. The demonstration of such an atrophy, seen through the ophthalmoscope, indicates bitemporal hemianopia, hence chiasmal compression. Recognition of such a lesion requires familiarity with the visualization of nerve bundles around the disc.

The importance of careful examination of the eyes in patients with pituitary tumors is illustrated by the fact that these tumors exhibit abnormalities in sellar polytomography in nearly all instances and abnormalities in plain films of the skull in approximately 40% of cases,[43,44] when field cuts are detected.

Cranial Nerve Involvement

While involvement of the optic nerve is a frequent phenomenon with superiorly extending pituitary tumors, involvement of other cranial nerves is distinctly unusual. Involvement of the optic nerve can result in the development of central scotomas, the Marcus Gunn pupil, decreased visual acuity, and even partial or complete blindness. Involvement of the third, fourth, and sixth cranial nerves by pituitary tumors rarely occurs when the cavernous sinus is invaded. This phenomenon is usually produced by pituitary apoplexy rather than by the tumor per se.

Compression of the Stalk

Compression of the pituitary stalk by a superiorly expanding pituitary tumor results in an auto-stalk-section effect, partially or completely interrupting transport of the hypothalamic factors to the pituitary. The dual consequences of this phenomenon are hyperprolactinemia (due to loss of hypothalamic inhibition) and secondary hypopituitarism (due to loss of hypothalamic drive). Hormonal testing in this situation will reveal mild to modest hyperprolactinemia with low levels of other adenohypophyseal hormones that respond to the administration of their respective releasing factors.

Compression of the Aqueduct of Sylvius

When the enlarging tumor, in its upward sojourn, compresses the aqueduct of Sylvius, the CSF traffic is interrupted, resulting in hydrocephalus. As a result, increased intracranial tension as well as bilateral papilledema may develop. The process may be so gradual that the only symptom experienced by the patient may be some intensification in the headache. Nausea and vomiting are unusual but may occur.

Vasopressin Deficiency

When the pituitary tumor extends above the sella, it may interrupt the neural connections above the median eminence or may actually destroy the supraoptic and paraventricular hypothalamic nuclei. This phenomenon results in central diabetes insipidus, often complete and sometimes permanent. It is well known that the severity of the polyuria caused by diabetes insipidus can be masked by the concomitant development of pituitary hormone deficiency, particularly ACTH, and to a lesser extent TSH. It is believed that glucocorticoid deficiency (resulting from ACTH deficiency) results in defective free water clearance. Since the expression of vasopression deficiency rests on adequate free water excretion, glucocorticoid deficiency results in ameliorating or even masking the polyuria of diabetes insipidus. It is believed that glucocorticoid deficiency directly increases water permeability of the distal and collecting tubules,[45] with the result of counteracting the polyuria of vasopressin deficiency. When the glucocorticoid deficiency is corrected by the administration of cortisone, the polyuria reappears, often with the full-blown picture of diabetes insipidus.[46]

Destruction of Normal Pituitary Tissue

With progressive expansion, the pituitary tumor compromises the hormonal reserve of the normal pituitary cells. Pituitary tumors represent the most common cause of pituitary hypofunction. The incidence of demonstrating suboptimal hormone reserve (laboratory hypopituitarism) far exceeds the incidence of clinically overt hypopituitarism, which is a later occurrence.

Sphenoid Sinus Invasion

Involvement of the sphenoid sinuses by pituitary tumors is a feature seen with the invasive tumors. The invasion of the sphenoid sinus can be so complete that intranasal presentation of pituitary adenomas is a well-documented entity.[47] In general, pituitary tumors that have become large enough to invade the sphenoid sinus pose therapeutic problems, seldom permitting complete surgical removal.

Hypersecretory Syndromes

Several hypersecretory syndromes are associated with pituitary tumors. Prolactin is the most frequently hypersecreted hormone by these tumors, followed by hypersecretion of GH, ACTH, TSH, gonadotropins, and endorphins. In addition, hypersecretion of glycoprotein subunits is a frequent phenomenon in pituitary tumors. Such a phenomenon has no clinical consequences. But hypersecretion of the other adenohypophyseal hormones confers serious consequences to the patient. Hypersecretion of GH results in acromegaly or gigantism and is potentially the most dangerous. Hypersecretion of ACTH causes Cushing's disease. Hyperprolactinemia, the most common hypersecretory syndrome results in menstrual irregularities, galactorrhea, and infertility in females, and impotence in males. Hypersecretion of TSH by pituitary tumors is an important but rare cause of hyperthyroidism. Finally, gonadotropin hypersecretion by tumors can paradoxically result in hypogonadism in males. It is important to recognize that the hypersecretory syndrome may be the first clue to the presence of a large underlying tumor.

PRL Hypersecretion

The concept that PRL hypersecretion can result in a clinical syndrome dates back to before the introduction of the PRL RIA; in 1954, Forbes et al.[48] described the first association among amenorrhea-galactorrhea, pituitary tumor, and PRL excess. Currently, PRL hypersecretion has the status of being the most frequently encountered pituitary disorder. The presenting features of PRL excess in females and males are separately reviewed. The spectrum of symptoms that can be associated with hyperprolactinemia is outlined in Table 81.

In females, the most frequent symptoms are related to menstruation. Oligomenorrhea is the most common complaint in women with PRL excess. Boyd et al.[49] indicated that approximately 20% of patients develop symptoms during the postpartum period, 25% develop symptoms after discontinuing oral contraceptive agents, and 5% of patients may present with primary amenorrhea with normal secondary sexual characteristics. Severe oligomenorrhea and amenorrhea are more frequent when galactorrhea is present. In one report by Hardy,[50] 90% of woman presented with a combination of galactorrhea and amenorrhea. The incidence of galactorrhea in hyperprolactinemic women is 30–80%.[51,52] It is a well-noted observation that galactorrhea is not an accurate reflector of the degree, or even the presence, of hyperprolactinemia. Galactorrhea may be present with perfectly normal basal PRL levels and absent despite markedly elevated PRL levels.[53] Infertility is a major concern in 65% of women with hyperprolactinemia. In addition to menstrual irregularities, galactorrhea, and infertility, decreased libido and dyspareunia are experienced in approximately 30% of hyperprolactinemic women. Less commonly, headache, visual symptoms, and dermatological symptoms may be experienced.

TABLE 81
Clinical Features of Hyperprolactinemia

Females
    Menstrual abnormalities
        Oligomenorrhea
        Amenorrhea
        Menorrhagia
    Infertility; decreased luteal phase
    Galactorrhea
    Decreased libido
    Dyspareunia
    Headache
    Weight gain
    Dermatological symtoms
        Acne
        Seborrhea
        Dry skin
        Hirsutism
    Visual symptoms

Males
    Headaches
    Visual-field defects
    Impotence
    Galactorrhea
    Hypopituitarism
    Infertility

The age difference in the presentation between men and women harboring prolactinomas is noteworthy. In women, hyperprolactinemia is a condition seen mostly between the ages of 20 and 35, although the disorder can be encountered at any age. The mean age, in most series, is 28–30. By contrast, male cases of prolactinomas come to attention in their fifth or sixth decade. In one series of 55 men with prolactinoma reported by Hardy,[52] the mean age for males with prolactinoma was younger, at 39 years of age. Morphologically, the tumor is much larger in males, often causing significant compressive effects. Thus, the presentation of prolactinoma in males is dominated by the pressure effects caused by the tumor. Impotence is the leading symptom in men with PRL excess.[54] The incidence of hyperprolactinemia as a cause of male impotence has been reported to range from 5 to 25%.[54–58] This may reflect an overestimation of the problem owing to the highly selected population of patients. Boyd et al.[49] noted that of 100 males screened for impotence, only three had elevated PRL levels. In addition to impotence, males with hyperprolactinemia may present with infertility,[49,58] galactorrhea, or signs of decreased virilization such as reduced beard growth. It should be noted that galactorrhea is present in only 20–30% of males with hyperprolactinemia.[54,58,59] In addition to the effects of hyperprolactinemia, the most important symptoms that bring the patient to the physician are those related to the mass effect of tumor (e.g., headaches, chiasmal compression, deficient

TABLE 82
Hyperprolactinemia in Males versus Females

Features	Females with hyperprolactinemia	Males with hyperprolactinemia
Age of presentation	20–35 (mean 28)	40–60 (mean 50)
Headaches	Relatively uncommon	Very common
Chiasmal compression	Uncommmon	Common
Tumor size	Mostly microadenoma	Macroadenomas, often invasive
Coexistent hypopituitarism	Extremely rare	Very frequent
Galactorrhea	Common	Rare
Prolactin level	Mild to moderate elevation	Marked elevation often >400 ng

pituitary hormone reserve). In summary, prolactinomas in males are characterized by symptoms related to perisellar compression, impotence, varying degrees of hypopituitarism, lower incidence of galactorrhea, and higher circulating PRL levels (Table 82).

In a patient with pituitary tumor, the presence of hyperprolactinemia per se should not invoke the diagnosis of prolactinoma (Table 83). A pituitary tumor that extends superiorly can lead to PRL hypersecretion by the auto-stalk secretion effect. This occurs because of interruption in the production or transport of dopamine—the physiological hypothalamic prolactin-inhibitory factor. While the level of PRL in the circulation is generally only modestly

TABLE 83
Hyperprolactinemia in Pituitary Tumors

Prolactin-secreting adenomas
  Microadenoma
  Macroadenoma
  Invasive macroadenoma
Intrasellar tumors with suprasellar extension
  Pressure on infundibulum
  Auto-stalk-section effect
  Hypothalamic involvement (median eminence)
Plurihormonal hypersecretion
  Plurihormonal secretion by same cell (e.g., GH)
  Plurihormonal secretion by different cells (e.g., TSH, GH, ACTH, and endorphins)
Intrinsic hypothalamic disease
  Suprasellar tumors
    Craniopharyngiomas
    Dysgerminomas
    Hamartomas
  Granulomatous disease mimicking tumor
    Sarcoidosis
    Histiocytosis
    Eosinophilic granuloma
  Metastatic disease

elevated in such cases, considerable overlap exists in PRL levels of patients with prolactinomas and those with the auto-stalk section effect of tumors. Definitive proof that PRL hypersecretion is caused by a prolactinoma depends on demonstration of prolactin immunopositivity in the tumor cells.

In addition to these mechanisms, hyperprolactinemia may be the marker of lesions that are completely suprasellar. Thus, suprasellar tumors, granulomatous disease (sarcoidosis), and metastatic disease to the hypothalamus may manifest with hyperprolactinemia. This is usually, but not invariably, associated with diabetes insipidus.

Finally, it should be noted that PRL-secreting adenomas may coexist with other hypersecretory tumors of the pituitary. Thus, PRL hypersecretion may coexist with tumors secreting GH, TSH, ACTH, and endorphins. Interestingly, there are reports in the literature describing PRL hypersecretion (and galactorrhea) as the sole manifestation of acromegaly[60] and of TSH-mediated hyperthyroidism.[61] The presence of hyperprolactinemia in ACTH-dependent Cushing's disease is especially relevant to intermediate lobe tumors that secrete ACTH. This disorder is particularly associated with hyperprolactinemia: It is believed that disturbances in dopamine regulation may underlie the excessive secretion of both ACTH and PRL seen in this disorder. This is supported by the observation that dopamine agonists, when administered to patients with this disorder, cause an impressive decline in both the ACTH and PRL levels.

GH Hypersecretion

The clinical effects of chronic GH hypersecretion caused by pituitary tumors results in acromegaly and gigantism—spectacular disorders involving somatic growth. The clinical features of acromegaly are discussed in Chapter 3 (summarized in Table 19).

ACTH Hypersecretion

It is estimated that in approximately 50–80% of patients with Cushing syndrome, the etiology is ACTH-dependent Cushing's disease.[62,63] Solitary adenomas are present in approximately 80% of patients with pituitary-dependent Cushing's disease,[64,65] most of these are microadenomas and are not always visualized by CT of the sella. A minority of patients have macroadenomas.[66] The clinical features of hypercortisolism in general, and Cushing's disease in particular are discussed in chapter 9.

In addition to Cushing's disease, ACTH hypersecretion is also encountered in Nelson syndrome.[67,68] Characteristically, the Nelson syndrome develops after bilateral adrenalectomy for Cushing's disease. The original tumor in the pituitary gland continues to grow, often rapidly, and often to extremely large sizes, causing perisellar invasion. Clinically, the two symptoms experienced by patients with Nelson's syndrome are hyperpigmentation (due to the markedly elevated β-LPH levels) and visual-field defects caused by the upwardly enlarging tumor. The diagnostic triad of Nelson's syndrome is (1) marked elevation of ACTH and β-LPH levels, (2) the presence of a large intrasellar mass, and (3) evidence of perisellar invasion.

Endorphin Hypersecretion

The clinical features of endrophin-secreting tumors are yet ill defined. The diagnosis of endorphinoma is often made only when the adenoma tissue is examined by immunoperoxidase staining techniques. Trouillas et al.[27] described a β-endorphinoma in a 43-yr-old man who presented with impotence and a pituitary tumor associated with hyperprolactinemia, low cortisol level, and normal LH and FSH levels that did not significantly respond to LRH. After removal of the adenoma, normalization of hormone dynamics was noted, along with clinical improvement. It is notable that endorphin hypersecretion in these patients occurs in the absence of ACTH hypersecretion.

TSH Hypersecretion

Thyrotrope adenomas are rare but represent an important etiology for TSH-induced hyperthyroidism. Since the original report by Jailer and Holub[69] in 1960, several reports and reviews have documented the nature of hyperthyroidism that results from TSH-secreting pituitary adenomas.[70–81] The general characteristics of TSH-secreting pituitary tumors are as follows:

1. These tumors are rare, accounting for less than 1% of all pituitary tumors.
2. These tumors occur with equal frequency in males and females, in contrast to the female preponderance of Graves' disease.
3. The clinical expression of these tumors is hyperthyroidism.
4. Like Graves' hyperthyroidism, patients with TSH-induced hyperthyroidism demonstrate thyromegaly, and varying degrees of thyrotoxicosis. But in contrast to Graves' disease, there are not signs of infiltrative ophthalmopathy or dermopathy. This is corroborated by the demonstration that thyroid-stimulating immmunoglobulins (TSI) are not detected in the sera of patients with adenomatous TSH-induced hyperthyroidism.[82] Occasionally, however, unilateral proptosis has been reported to occur in association with TSH-secreting tumors of the pituitary gland. Yovos et al.[83] described a unique case of a 17-yr-old girl who presented with hyperthyroidism, goiter, unilateral exophthalmos, and an elevated TSH level. Computed tomography revealed a large pituitary tumor invading the orbit.
5. Most TSH-secreting adenomas are large, often causing visual-field defects due to chiasmatic compression. In reviewing the literature of TSH-secreting pituitary tumors reported between 1958 and 1983, Ridgway[84] found visual-field defects in 20 of 29 patients tested and suprasellar invasion in 19 of 23 patients.

6. It is noteworthy that in nearly one-half of patients with TSH-secreting adenomas, a second hypersecretory syndrome can be recognized. The most common accompaniment is PRL hypersecretion,[85–88] followed by hypersecretion of GH[89–91] and gonadotropins.[92]

7. The two most important disorders to be differentiated from TSH-secreting adenomas[93–96] are Graves' hyperthyroidism and nontumorous TSH-induced hyperthyroidism secondary to peripheral resistance to thyroid hormones.

8. The diagnostic triad that established the diagnosis of TSH-induced hyperthyroidism consists of (a) inappropriately elevated TSH levels in the presence of elevated thyroidal hormone levels, (b) increased ratio of α-TSH/molar TSH ratio, and (c) failure of TSH or its subunits to respond to TRH stimulation of T3 suppression.

Gonadotropin Hypersecretion

Until the mid-1970s, it had been assumed that gonadotropin-secreting adenomas usually occur as a consequence of longstanding primary hypogonadal states. With the advent of immunoperoxidase staining techniques and the availability of glycoprotein subunit assays, several instances of gonadotropin cell tumors have been recognized in the recent literature.[97–107] The largest number of patients, 16 in all, with gonadotropin-secreting adenomas was reported in two separate reports by Synder and co-workers.[105,106] Much has been learned regarding the clinical presentation, hormonal characteristics, and morphological behavior of these tumors.

The clinical characteristics of gonadotrope cell adenomas have been reasonably well characterized. These tumors occur predominantly in middle-aged men. In a recent review, Snyder[108] pointed out that of the 59 patients reported in the literature with gonadotropin-secreting adenomas, only four were females. Very rarely, the disorder has been described in children. Faggiano et al.[109] reported a 4½-yr-old boy with precocious puberty caused by a pituitary adenoma that secreted LH (and PRL). Aside from the characteristic predilection to affect middle-aged and elderly men, the most notable feature is the large size of the tumor at the time of diagnosis, with a tendency to extend outside the sella. In more than 75% of patients, the presenting feature of these tumors is visual impairment of varying degrees.

Endocrinologically, patients with gonadotropin-secreting adenomas do not generally show any signs of gonadal dysfunction. These patients have a normal history of pubertal development, sexual activity, and fertility. This is in sharp contrast to patients who develop secondary gonadotropinomas consequent to longstanding primary gonadal failure such as Turner's or Klinefelter's syndromes. On physical examination, patients with gonadotropin-secreting tumors appear normally virilized with normal testicular dimensions and hair growth. Occasionally, these patients may have mild secondary hypogonadism because the tumor may secrete suboptimal amounts of intact LH. This, coupled with compression of normal gonadotropes by the large tumor, can result in hypogonadotropic hypogonadism.

The presentation and behavior of gonadotropin-secreting adenomas would appear to resemble that of chromophobe adenoma. The large size, often with suprasellar extension, the lack of clinical hypersecretory syndromes or overt hypopituitarism, and the chromophobic appearance with routine conventional stains bring the similarities even closer. It is therefore possible that tumors apparently labeled as nonsecretory may in fact be gonadotropinomas. Beckers et al.[110] retrospectively reviewed 40 patients with an apparently nonsecretory tumor operated between 1971 and 1981. They noted that 6 men of 40 had elevated FSH levels preoperatively, and all had complained of sexual impotence. Tumor tissue obtained during operation stained positively for gonadotropins. The authors concluded that 6 of a series of 40 patients had primary gonadotropinomas. Thus, immunoperoxidase staining techniques may confirm the possibility that gonadotropinomas occur more commonly than was previously estimated.

Hypopituitarism

Pituitary hypofunction in general, and tumorous hypopituitarism in particular, is discussed in Chapter 15. Three brief comments bear emphasis. First, since pituitary tumors are the most frequent cause of hypopituitarism, every patient with clinical or laboratory evidence of hypopituitarism requires radiological evaluation of the sella, preferably by CT scan. Second, while the sequence of hormone loss generally follows a pattern—in descending order of disappearance, GH, LH, and FSH, TSH, ACTH, and PRL—such a pattern is not invariable. Third, loss of hormonal reserve need not be permanent. Removal of the tumor may result in relief of pressure on the normal population of cells, the stalk, or even the hypothalamus. Therefore, careful pre- and postoperative hormonal evaluation is necessary in patients with pituitary tumors. The only exception to this principle is patients with microadenomas, which seldom if ever cause trophic hormone deficiency.

Diagnostic Studies

As in the discussion of the clinical features, the diagnostic studies in patients with pituitary tumors are viewed from three vantage points: tests directed to detect the mass lesion, tests aimed at detecting hormonal hypersecretion, and tests that disclose trophic hormone deficiency.

Radiological Evaluation

Anatomic Evaluation

Topographical evaluation of pituitary tumors is best attained by the use of CT of the sellar region and the the surrounding structures.[111] The application of fourth-generation high-resolution CT has obviated the use of polytomography, angiography, and pneumoencephalography in the diagnosis of pituitary adenomas. The impact of nuclear magnetic resonance (NMR) imaging for pituitary tumors is still being evaluated. Several general comments apply to the radiological evaluation of pituitary tumors.

1. Microadenomas (tumors under 10 mm in size) do not produce alterations evident in lateral skull radiographs.
2. An enlarged sella turcica can result from numerous cases besides a pituitary tumor (Table 84). Conversely, a tumor in the pituitary gland need not necessarily cause sellar enlargement.
3. Variations in the shape of the sella, as well as qualitative changes in anatomic details can be a reflection of an intrasellar process, or may indicate normal variants. Thus, sellar enlargement with a double floor contour and thinning of the walls with erosion can be seen in association with pituitary tumors.[112] However, as has been aptly pointed out by Swanson and du Boulay,[113] such findings can be encountered as normal variants in patients with no evidence of pituitary tumors.
4. The most common disorder that needs to be differentiated from the enlarged sella caused by pituitary tumors is the primary empty sella syndrome, a totally benign entity. The distinction between tumors and primary empty sella syndrome cannot be made on the basis of plain radiographs of the skull alone.
5. Pluridiectional polytomography is helpful in demonstrating microadenomas.[114-116] At some institutions, polytomography continues to be used as an adjunct to CT in the detection of microadenomas, while at others it has been replaced entirely by high-resolution CT.
6. Pneumoencephalography, a laborious and uncomfortable procedure, also has been replaced by the newer models of CT scanners. When contrast enhancement of the cerebrospinal fluid (CSF) is desired, metrizamide cisternography with CT studies is the preferred procedure.
7. Carotid angiography is no longer used to detect the presence or extent of pituitary tumors. Angiography is still used preoperatively, however, to obtain information regarding the vascular anatomy, as well as for exclusion of coexistent aneurysms. This assumes importance in light of the report that the incidence of aneurysms associated with pituitary adenomas is very high (7.4%).[117]

With the above overview, the test employed in the evaluation of pituitary tumors is high-resolution CT using intravenous contrast medium. In addition, all patients with pituitary tumors must undergo visual-field testing by the Goldmann perimetry.

The optimal method for visualizing the sellar region is direct coronal CT scans, supplemented by sagittal reconstructed views (Figs. 38A and B, 39, and 40). Most pituitary tumors are nonenhancing or minimally enhancing lesions in comparison to normal pituitary tissue. Therefore, some radiologists prefer not to obtain nonenhanced scans routinely and recommend the administration of a bolus of iodinated contrast material to enhance the normal pituitary tissue. This procedure will highlight the poorly enhancing pituitary tumor

TABLE 84
Causes of an Enlarged Sella Turcica

Pituitary tumors
Empty sella syndrome
Aneurysm
Meningioma of the tuberculum sellae
Craniopharyngioma
Cysts (arachinoid, dermoid, epidermoid)
Hypophysitis (granulomatous and lymphocytic)
Hypothalamic glioma
Pituitary abscess
Metastases
Miscellaneous
  Chordoma
  Rathke cleft cyst
  Eosinophilic granuloma
  Teratoma
  Mucocele of the sphenoid sinus
  Tuberculoma

with greater precision. Nonenhanced scans can be added later if distinction between calcification and hypervascularity is desired. The CT appearance of microadenomas and macroadenomas is discussed separately.

Generally, a microadenoma tends to appear as a low-density area within a normal sized gland.[118] If this finding is associated with thinning and depression of the bony floor contiguous to the microadenoma, the diagnosis is virtually certain. Three important comments are pertinent when a microadenomalike picture is seen by CT. First, the classic appearance of a microadenoma—low-density area—can be mimicked by small cysts in the pituitary, a not infrequent condition as judged by autopsy studies. Muhr et al.[119] performed a parallel study of the roentgenographic anatomy of the sella and the histopathology of the pituitary gland in 205 autopsy subjects and found pituitary cysts, 1–6 mm in diameter with fibrotic walls, in 8.7% of the autopsied specimens. Thus, neuroepithelial, Rathke cleft, and arachnoid cycts will resemble the appearance of microadenomas on CT scan. The presence of contiguous bony changes would argue strongly in favor of microadenomas. Second, it has been abundantly documented that silent microadenomas exist in an impressive population of normal subjects, the incidence ranging from 8.5 to 27%.[120–123] The presence of such an incidental adenoma might be conducive to mistakenly conferring a causal role to the coexistent clinical problem, such as PRL excess or ACTH excess. Third, microadenomas may occasionally demonstrate contrast enhancing, as has been most notably documented in the case of prolactinomas.[124,125]

The false-positive results associated with CT of the pituitary can be quite significant, sometimes as high as 20%; when nonhomogeneous enhancement of the gland is seen, the resemblance to microadenoma is so close as to be mistaken for an abnormality. Reliable signs that point strongly to the presence

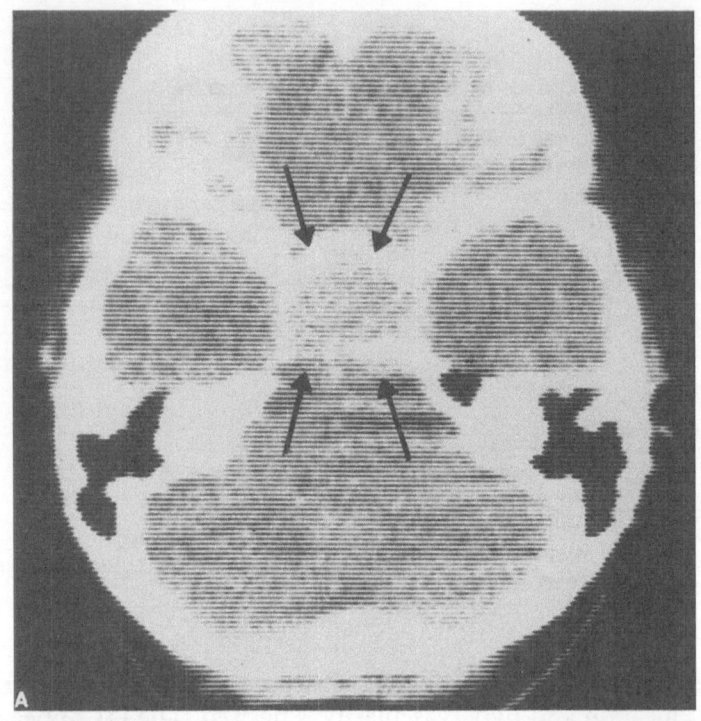

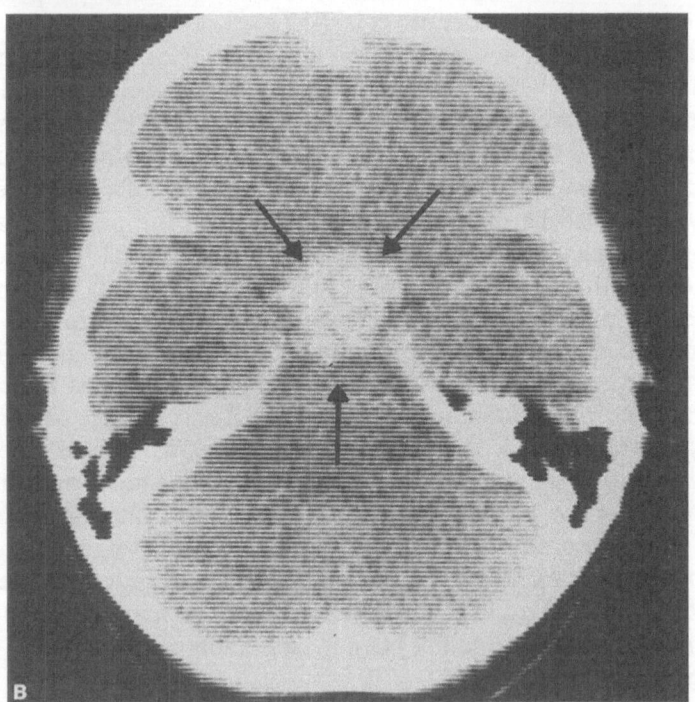

FIGURE 38. (A) Contrast-enhanced CT scan demonstrating enlargement of the sella turcica (arrows) with contrast-enhancing tumor within it. (B) Scan obtained 13 mm cephalad to (A). (From Wolpert.[215])

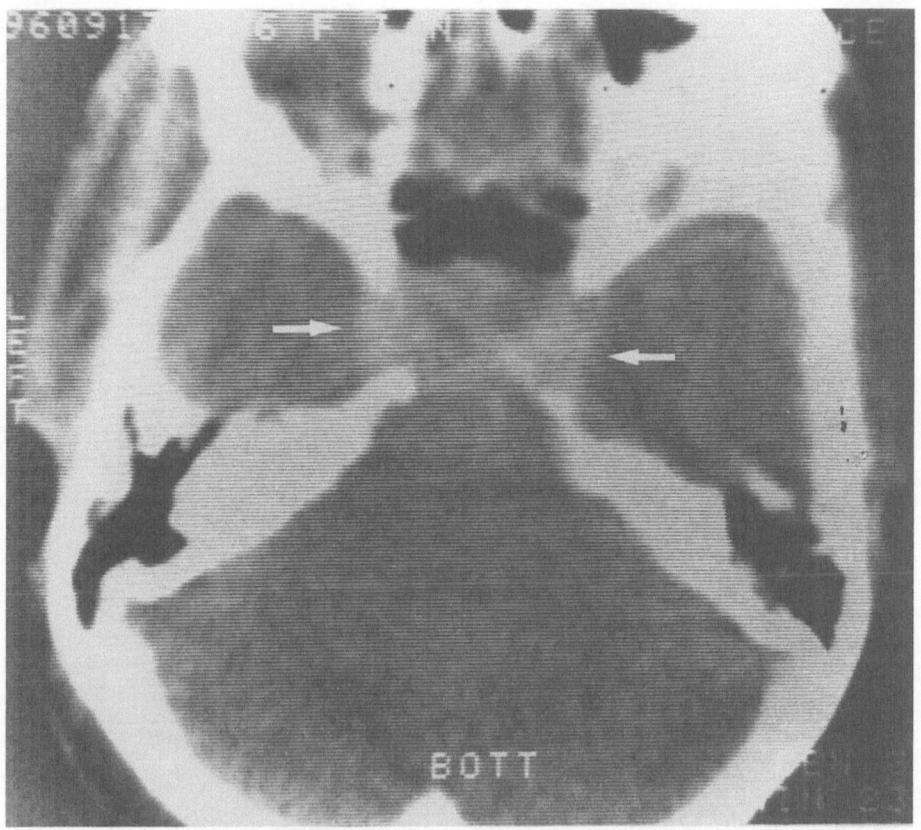

FIGURE 39. CT scan demonstrating large intrasellar tumor with involvement of the cavernous sinuses on both sides. (From Post.[216])

of microadenoma are a gland height of 10 mm or more, asymmetrical location of the lesion in the gland, bulging of the superior margin of the gland, and abnormalities in the bone contiguous to the presumed microadenoma. When a suspicious small hypodense lesion is seen in association with some of these abovementioned findings, the diagnosis of microadenoma is more tenable.

Macroadenomas, in contrast to microadenomas, are invariably detected by CT. In addition to documentation of its intrasellar location, CT outlines the margins, provides information regarding tumor consistency, and delineates the extent of invasion by the tumor. Thus, the presence and degree of suprasellar, intrasellar, and perisellar invasion can be accurately defined by CT. Since these tumors are large, high detail scanning is not necessary; the usual recommendation is to obtain 4-mm axial cuts. The bony margins of the sella will often demonstrate expansion, erosion, destruction. Macroadenomas generally reveal homogeneous contrast enhancement.[126] However, considerable individual variations exist owing to the presence of degenerative, necrotic, and hemorrhagic changes that occur within large adenomas. When

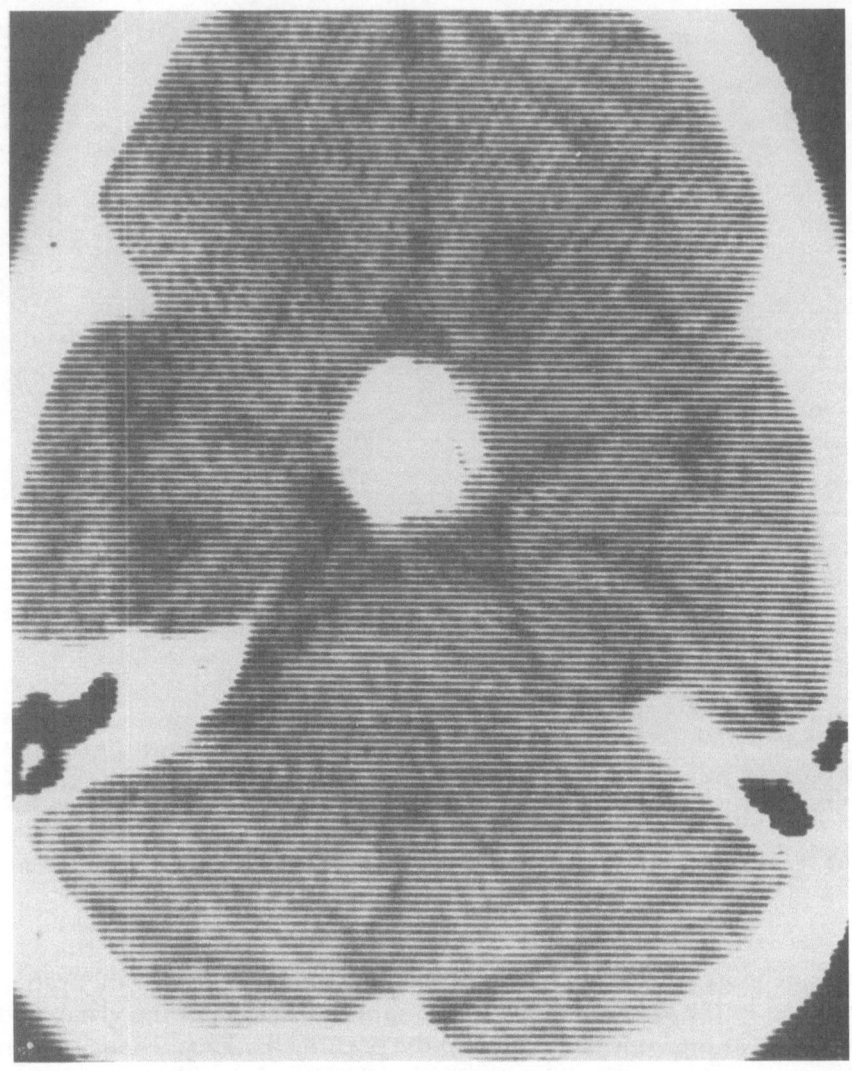

FIGURE 40. CT scan of the suprasellar area demonstrating a high-density lesion. Diagnosis: pituitary tumor with suprasellar extension. (From Post.[216])

ring enhancement is seen, the possibility of hemorrhagic infarction (apoplexy) should be considered. However, tumors without infarction also occasionally show ring enhancement. The CT appearance of apoplexy within the tumor is highly variable; the hemorrhagic areas may appear as a high-density lesion in the preinfusion films, or the classic appearance of the ring and core lesion may be seen after contrast enhancement. Quite frequently, an area of mottled mixed density may be seen within the tumor.[127,128]

The most significant aspect of CT is its ability to define the extent of the tumor. The scans should be obtained commencing at the level of the sphenoid sinus and continued to the level of the suprasellar cisterns. In the preinfusion films, the density of the tumor will be the same as, or higher than, the density of brain tissue. When necessary, the accurate delineation of the optic chiasm and the adjacent vascular structures can be obtained by the intracisternal administration of metrizamide.[129] This procedure will also provide excellent

TABLE 85
Radiology of Sellar and Suprasellar Tumors

Disorder	Plain films of the skull	CT
Microadenomas[118]	Normal	Hypodense lesions (<10 mm)
Macroadenomas[130]	Enlarged sella often with erosions and destruction of floor	Large, often contrast-enhancing lesions in the sella, frequently with suprasellar extension
Aneurysm of internal carotid[131-133]	Enlarged sella	Unremarkable; dynamic CT may reveal the clot and lumen of the aneurysm
Apoplexy[134]	Enlarged sella, often with destruction of walls	Peripheral ringlike enhancement after contrast
Empty sella syndrome[135,136]	Enlarged sella usually, but not invariably, symmetrical	Continuation of CSF density into the sella; presence of infundibulum sign
Meningioma[137,138]	Enlarged sella; hyperostosis of adjacent bone evident in 50%	Hyperostosis of adjacent bone present; parasellar location; calcification may be seen in nonenhanced films. Enhancement after contrast
Craniopharyngioma[139,140]	Enlarged sella, suprasellar calcification in 80% of young patients and 50% of older patients	Intra- or suprasellar tumor usually hypodense, cystic; floccular or curvilinear calcification present; enhancement variable
Germinoma[141]	Enlarged sella	Suprasellar; midline mass; cystic; solid or mixed; variable enhancement
Hypothalamic glioma[142]	Enlarged sella	Suprasellar lesion, often indents the suprasellar cistern, and the third ventricle; often small; isodense tumor with moderate enhancement
Chordoma[143]	Sellar enlargement; heavy calcification. Extensive destruction of sella	Not consistent enough to be regarded as diagnostic
Eosinophilic granuloma[144]	Enlarged sella turcica, cystic lesions in the mandible or other facial bones	Sellar or parasellar mass

detail regarding suprasellar extension of pituitary tumors, as well as vital information regarding pressure on the ventricular cisterns.

Table 85 illustrates the diverse radiological findings encountered in sellar suprasellar tumors.[118,130–144]

Visual-Field Testing

Visual-field testing is mandatory in all patients presenting with sellar and suprasellar tumors. Target screen perimetry and Goldmann perimetry are conventional methods to determine the extent of visual fields. It should be realized that even with the Goldmann perimeter, which contains a small mirror with a centrally located fixation device, small hemianopic scotomas can be missed unless another fixation device is employed to carefully investigate the central area. Records of visual field testing are extremely helpful in prospectively documenting changes after any treatment modality.

Hormonal Evaluation

The five hypersecretory syndromes associated with pituitary tumors are hypersecretion of prolactin, GH, ACTH, TSH, and gonadotropins. In addition, free subunits can be secreted in excessive quantities by pituitary tumors.

PRL Hypersecretion

Hyperprolactinemia represents the most common hypersecretory disorder associated with pituitary tumors.[145] Measurement of basal prolactin is often the only test needed to establish hormonal hypersecretion. This disorder is discussed in Chapter 14.

GH Hypersecretion

Acromegaly or gigantism caused by hypersecretion of GH can be readily established by measurement of basal GH or somatomedin levels as well as by demonstration of abnormal GH dynamics to various agents (glucose, TRH, and L-dopa). The hormonal evaluation of acromegaly is discussed in Chapter 3.

ACTH Hypersecretion

ACTH hypersecretion by pituitary tumors is usually secondary to microadenomas. The hormonal features of pituitary-dependent hypercortisolism are preservation of suppression to high-dose dexamethasone. preservation of a normal response of ACTH and 11-deoxycortisol to metyrapone administration, and normal or minimally elevated basal plasma ACTH levels. When all the three of these criteria are concordant, the diagnosis of ACTH-mediated Cushing's disease is straightforward. However, considerable variation in the hormonal dynamics can pose a vexing problem in some patients with pituitary-dependent Cushing's disease. These difficulties, as well as the use of newly emerging studies (CRH test, and selective venous sampling of the inferior petrosal sinus) are discussed in Chapter 9.

TSH Hypersecretion

Hyperthyroidism caused by adenomatous TSH hypersecretion is discussed in Chapter 6. The hormonal criteria for establishing the diagnosis of TSH hypersecretion by the thyrotrope adenoma are clinical and biochemical hyperthyroidism associated with inappropriately detectable or elevated TSH level in the plasma, disproportionate elevation of α-TSH subunits, resulting in a ratio of α-TSH/TSH molar greater than unity,[146] absence of alteration in α-TSH, β-TSH, or intact TSH levels following the administration of TRH or T3.[147]

A high proportion of patients with TSH-secreting adenomas demonstrate an association with coexistent hypersecretory syndromes. These usually involve hypersecretion of PRL, GH, and gonadotropins.[85-88] TSH hypersecretion by adenomas must be differentiated from the TSH-mediated hyperthyroidism associated with partial peripheral resistance (see Chapter 6).

FSH and LH Hypersecretion

Hypersecretion of gonadotropins by pituitary adenomas is of three types: those that secrete FSH alone, those that secrete both FSH and LH, and those that secrete only LH. The most common variety is hypersecretion of intact FSH. This can be demonstrated by gel-filtration chromatography wherein the majority of FSH can be eluted as intact FSH rather than as its subunits. In some cases, hypersecretion of intact FSH is also accompanied by secretion of excessive quantities of FSH subunits, both β-FSH and α-FSH. Occasionally, hypersecretion of β-LH or even intact LH is associated as well. Thus, the characteristic marker of gonadotropin-secreting tumor is high levels of FSH in the circulation. Since this closely resembles the values seen in postmenopausal women, these tumors naturally tend to be missed when they occur in postmenopausal women. Gonadotropin-secreting tumors, however, predominantly occur in middle-aged or elderly men. Table 86 outlines the characteristics of gonadotropin-secreting adenomas.

The sex steroid milieu in patients with gonadotropin-secreting tumors is highly variable. A large number of these patients demonstrate subnormal testosterone levels, without an elevation in the level of intact LH, implying defective secretion of the intact hormone. The fact that the testosterone level of such patients respond to administration of human chorionic gonadotropin (HCG) implies that the mechanism for hypotestosteronemia is central, i.e., impaired secretion of intact LH by the normal pituitary tissue compressed by the tumor. Less commonly, some patients with gonadotropin-secreting tumors may demonstrate supranormal testosterone level due to hypersecretion of intact LH by the tumor.

A third characteristic of the gonadotropin elevation seen in association

TABLE 86
Gonadotropin-Secreting Tumors

Clinical
    Uncommon hypersecretory tumors
    (3–4% in autopsy, and surgical series)
    Affects middle-aged or elderly males
    No past history of gonadal failure
    Presenting symptoms are those of an invasive intrasellar mass
Hormonal
    ↑ FSH with or without
        ↑ α-FSH or ↑ β-FSH
        ↑ LH (intact)
        ↓ β-LH
    Testosterone levels normal, low, or high
    Paradoxical FSH response to TRH
    Normal testosterone response to hCG
    Varying degres of compromise in pituitary reserve
Immunohistochemical
    Immunopositivity to FSH and sometimes to LH
Morphological behavior
    Dispersed adenoma cells actively secrete FSH

with gonadotropinomas is a paradoxical increase in FSH and or LH following the administration of TRH,[148] a finding not observed in postmenopausal women. This behavior resembles the paradoxical increase in GH level in acromegalics following intravenous administration of TRH and may imply loss of receptor specificity of tumor cells.

Comparison of the in vivo and in vitro behavior of gonadotropin-secreting tumors generally reveals concordance. It is of interest that even "pure" α-subunit-secreting adenomas in vivo, secrete large amounts of intact FSH when the tumor cells are dispersed in culture.[149]

Gonadotropin-secreting pituitary adenoma must be differentiated from secondary pituitary tumors that develop in patients with longstanding primary hypogonadal states, such as Turner's or Klinefelter's syndrome. Table 87 outlines useful features in making that distinction.

Subunit Hypersecretion

The concept that pituitary adenomas can secrete free α-subunits has been advanced and documented since 1976, when Kourides et al.[150] described patients with pituitary tumors that secreted PRL and α-subunits. The impact of this phenomenon is threefold: (1) it confirms the notion that the so-called nonsecretory chromophobe adenomas are capable of secreting pure α-subunits, (2) measurement of these subunits could serve as a marker for pituitary tumors, and (3) subunit measurement could be used as a parameter to assess therapeutic efficacy.

Following the original description, MacFarlane et al.[151] demonstrated elevated α-subunit levels in 18 of 99 patients with pituitary adenomas. Most

TABLE 87
Primary versus Secondary Gonadotropinomas

Features	Gonadotropin-secreting tumors (gonadotropinomas)	Secondary gonadotropin-secreting tumors
Clinical		
History of longstanding primary hypogonadism	Absent	Present (Turner's or Klinefelter's)
Manifestations of hypogonadism	None	Present (e.g., small testes, eunuchoidal proportions)
Visual defects and other suprasellar compressive effects	Very common; represents the major mode of presentation	Unusual
Laboratory		
FSH level	Elevated, sometimes to a marked extent	Elevated
LH level	May be elevated	Often elevated
α-FSH level	May be elevated	May be elevated
β-LH level	May be elevated	Not elevated
Testosterone level	Normal, low, or supranormal	Very low
Testosterone response to hCG administration	Normal rise	No rise at all
FSH response to TRH	Paradoxical increase may be seen	Not seen

of these patients with α-subunit secretion also demonstrated concomitant hypersecretion of PRL or GH, but none had hypersecretion of intact glycoproteins. Reports of α-subunit hypersecretion in association with TSH hypersecretion[81] and gonadotropin hypersecretion[152] have been well documented in the literature. The measurement of subunit levels in plasma was soon added to the diagnostic armamentarium of pituitary tumors. The syndrome of monotropic α-hypersecretion (i.e., in the absence of hypersecretion of other trophic hormones) was described by Ridgway et al.[153] These investigators reported two patients with large, chromophobe adenomas, previously thought to be nonsecretory, and demonstrated isolated hypersecretion of free α-subunits associated with partial panhypopituitarism resulting from tumor invasion. Further dynamic studies revealed subunit autonomy (i.e., failure of the subunit level to change following the administration of TRH, LRH, T4, or sex steroids). Characterization of the free subunit in the serum and the tumor homogenate by gel chromatography revealed that the fraction eluted only to the α-subunit. Immunologically, these subunits cross-reacted with LH but displayed absolutely no biological activity.

The glycoprotein hormones of the pituitary contain both an α- and a β-subunit.[154] The α-subunit is identical to that for TSH, LH, and FSH. Theα-subunits can be readily measured in normal serum, while β-subunits are undetectable. Several conditions are characterized be elevated an α-subunit level in the serum, the two most common being primary hypogonadism[155] and primary hypothyroidism.[156] However, the increased α-subunit level in these situations is responsive to the administration of LRH and TRH, respectively. Rarely, ectopic secretion of α-subunits has been associated with nonpituitary neoplasms.[157] The α-subunit level in such a setting is not affected by administration of hypothalamic-releasing factors.

α-Subunit hypersecretion by pituitary tumors is characterized by several features:

1. This phenomenon is more common in males.
2. α-Subunit hypersecretion is associated with large tumors. Symptoms of chiasmal compression, headaches, and hypopituitarism are usually present.
3. α-Subunit hypersecretion can be an isolated phenomenon or can be associated with hypersecretion of intact glycoproteins or peptide hormones.
4. α-Subunit hypersecretion by tumors is not affected by administration of hypothalamic releasing factors, thyroid hormones, or sex steroids.
5. α-Subunit levels decline following surgery or radiation.

Thus, the measurement of α-subunit may serve as yet another diagnostic tool in the evaluation of pituitary tumors.

Pituitary Reserve

The hypopituitarism that ensues as a result of tumor encroachment on the normal pituitary tissue is discussed in Chapter 15. It will suffice to underscore the concept that the incidence of impaired hormone reserve in macroadenomas is high. The combination of an enlarged sella with perfectly normal reserve testing of pituitary function decreases the likelihood of tumor. However, several space-occupying intrasellar lesions can impair pituitary function; these conditions should be considered in the differential diagnosis.

Differential Diagnosis

The differential diagnosis of macroadenomas revolves around conditions that cause sellar enlargement as shown radiographically (see Table 85). Five in particular need to be considered: suprasellar and parasellar tumors, primary empty sella syndrome, hypophysitis, aneurysms, and pituitary cysts. Less commonly, pituitary abscess, metastatic disease, and infiltrative disease, and secondary pituitary tumors need to be considered. The availability of high-resolution CT has considerably aided in the differential diagnosis of sellar and parasellar lesions.

Suprasellar and Parasellar Tumors

The most important conditions in this category are craniopharyngioma, germ cell tumors, parasellar meningioma, and hypothalamic tumors, the prototype of which is glioma. These tumors can very closely mimic the clinical behavior of pituitary tumors with suprasellar extension. The clinical, radiological, and hormonal aspects of these tumors are discussed separately.

Primary Empty Sella Syndrome

This condition, caused by arachnoid herniation into the sella, due to an inherent weakness in the diaphragma, is an extremely frequent cause (20–40%) of sellar enlargement detected by radiographs. This entity is usually seen in middle-aged females who are often obese and hypertensive. The two hallmarks of this condition are the absence of symptoms and the absence of tropic hormone deficiencies. The diagnosis can be established by high-resolution CT; the *sine qua non* of primary empty sella syndrome is the demonstration of CSF within the enlarged sella that also contains a flattened displaced small pituitary. This condition is fully discussed in Chapter 18.

Hypophysitis

Two forms of hypophysitis can affect the pituitary gland: lymphocytic and granulomatous (giant cell type). Both are rare, both can present as mass lesions within the sella, and both can result in varying degrees of hypopituitarism as well as suprasellar extension. In fact, most patients with hypophysitis are initially considered to have pituitary tumors. While lymphocytic hypophysitis shows a strong predilection for females, and bears a temporal relation to pregnancy or postpartum, granulomatous hypophysitis can occur at any age and affects both sexes equally. There are no specific radiological features seen by CT that permit identification of these disorders. They enhance variably and mimic the appearance of a pituitary tumor. In most cases the diagnosis is arrived at by biopsy or by exploration. Hypophysitis is discussed in Chapter 15.

Aneurysms

Protrusion of an internal carotid artery aneurysm into the sella can result in both an enlarged sella as well as in adenohypophyseal hormone deficiency. White and Ballantine[158] estimated that pituitary involvement occurs in approximately 2% of all intracranial aneurysms. A high incidence (7%) of aneurysms coexisting with pituitary tumors has been suggested by the study of Wakai et al.[117] There are no clinical clues that help delineate the silent aneurysm from a pituitary tumor. Three radiological clues, when present, should raise the suspicion of an aneurysm: (1) the bone destruction caused by the aneurysm is asymmetrical, since aneurysms that invade the sella generally arise from the infraclinoidal portion of the internal carotid artery; (2) the presence of abnormal calcification, especially curvilinear, is a strongly suggestive sign; and (3) there is often demonstrable widening of the superior orbital fissure, due to eccentric bone destruction by the aneurysm. It is crucial to recognize the presence of aneurysms preoperatively, to avoid rupture during surgery. The only definitive study that established the presence of an aneurysm is angiography. Even a negative angiogram is not foolproof, since a clot-filled aneurysm can be missed by the study. Dynamic CT may be helpful in demonstrating the lumen as well as the clot, following the rapid injection of contrast.[132,133] The clinical presentation of an aneurysm that has ruptured into the sellar cavity is identical to that of pituitary apoplexy. The presentation is characterized by dramatic onset and progressive expansion of the sellar contents (see Chapter 17).

Pituitary Cysts

Several types of cystic lesion can occur within the pituitary gland. Most of these are developmental in origin. Thus, arachnoid cysts, epidermoid cysts, Rathke cleft cysts, and dermoid cysts can present with sellar enlargement. The CT-enhanced study would reveal the cystic nature of these lesions. The major differential diagnosis of pituitary cysts is intrasellar craniopharyngioma.

Pituitary Abscess

This disorder is an extremely rare cause of the development of an intrasellar mass that enlarges the sella. Most reports of pituitary abscesses are from neurosurgical literature.[159–161] Underlying causes (e.g., cysts, tumor, cavernous sinus thrombophlebitis, purulent meningitis, and sphenoid sinusitis) are present in most of these patients. However, abscesses of the pituitary can occur in the absence of any overt focus of infection. The clinical presentation can be acute (mimicking that of apoplexy) or subacute (with headaches, visual-field cuts, and low-grade fever). The microorganisms found in these patients are diverse: gram-positive, gram-negative, anaerobes and even fungi, especially mucomycosis and aspergillosis. Involvement of the sphenoid sinus by the fungus *Aspergillus*, can result in a progressive relentless infection that can invade and involve the pituitary gland. Fuchs et al.[162] described a 48-year-old woman who had constant bifrontal headaches for 3 months, with rapid development of diploplia, third nerve palsy, and dizziness, 2 weeks prior to hospitalization. A lateral skull radiograph revealed an enlarged sella with destruction of the floor and involvement of the sphenoid sinus. Computed tomography revealed a large intrasellar mass with involvement of the sphenoid sinus inferiorly, the ethmoid sinuses posteriorly, and extending superiorly into the base of the brain. On exploration, a large pituitary abscess was found, cultures of which identified *Aspergillus*. Despite its rare occurrence, it is important to keep abscesses in mind as a cause of pituitary masses, since early intervention can prevent the invasive sequelae as well as the development of hypopituitarism.

Metastatic Disease

The involvement of the pituitary gland by metastases from malignant disease of the breast, GI tract, and kidneys as well as by hematopoietic malignancies is discussed in Chapter 15. In general, pituitary involvement by metastatic disease is seen in the presence of obvious dissemination of the primary malignancy.

Infiltrative Disease

A tumorlike presentation can be encountered when the pituitary gland is involved by granulomatous (histiocytosis X, eosinophilic granuloma) processes, sarcoidosis, tuberculosis, and even syphilitic gummata. These, however, are extremely unusual, and the diagnosis can be established with certainty only by exploring the sella. One possible exception, however, needs to be mentioned. Neuroendocrine sequelae can occur following tuberculous meningitis. Postmeningitic hypopituitarism with suprasellar calcification can closely mimic suprasellar tumors. Haslam et al.[163] described a patient with diabetes insipidus, growth retardation, hypogonadism, and suprasellar calcifications, years after successful treatment for meningitis. Similarly, Sherman et al.[164] reported a 17-yr-old boy who developed diabetes insipidus and GH and gonadotropin deficiencies 6 years after treatment for meningitis. Striking suprasellar calcifications were found on X-ray examination. These cases illustrate the importance of obtaining a past history of meningitis in patients presenting with a picture resembling suprasellar tumors.

Secondary Pituitary Tumors

Rarely, pituitary tumors develop as a consequence of longstanding primary target organ failure. This is seen particularly in association with primary hypothyroidism and primary hypogonadism. Although several reports in the early 1960s had speculated on the development of secondary pituitary enlargement in primary hypothyroidism, two reports in 1969, one by Leiba et al.[165] and one by Patel and Kilpatrick,[166] clearly documented this association. This phenomenon may not be a rarity. Lawrence et al.[167] reported eight patients in a single series who displayed pituitary enlargement subsequent to the development of primary hypothyroidism. These investigators made the interesting observation that the hypothyroidism in such patients need not be profound or longstanding.

In addition to radiographically demonstrable abnormalities, these patients may demonstrate abnormal dynamics of unrelated pituitary hormones, such as GH, PRL, and even gonadotropins. Indeed, the syndrome of precocious puberty and galactorrhea in girls with longstanding primary hypothyroidism, originally described by Van Wyck and Grumbach,[168] is prototypical of this phenomenon at its extreme. In patients who develop pituitary thyrotrope hyperplasia (or even adenomas) secondary to primary hypothyroidism, the elevations in basal TSH level would be obvious. This is associated with elevations in both, the α- and β-subunits of TSH, which appropriately respond to the administration of TRH as well as T3. These features, coupled with the obvious hypothyroid status of the patient, will help in differentiating this condition from primary TSH-secreting adenomas.

A similar situation exists between primary gonadal failure and the development of secondary gonadotropinomas. It is well known that patients with Klinefelter's and Turner's syndromes have a heightened tendency for the development of pituitary tumors.[169,170] The distinguishing features between primary gonadotropin-secreting tumors and the secondary tumors that develop in hypogonadal patients are outlined in Table 87.

Natural History and Complications

Macroadenomas of the pituitary cause considerable morbidity by virtue of their location. While the individual behavior of these tumors is highly variable, progressive enlargement eventually results in loss of pituitary function, loss of vision, and even loss of life. The incidence of apoplexy (spontaneous bleeding) in association with a macroadenoma has been variably reported, ranging from 2 to 28% (see Chapter 17). The consequences of pituitary apoplexy are considerable, with sudden loss of sight and life.

By contrast, microadenomas behave in a vastly different manner. The potential for growth and invasiveness of microadenomas is so low that these tumors should be regarded as an entirely separate entity. The natural history of microadenomas in general, and of prolactinomas in particular, has been studied by several workers. Microadenomas do not appear to progress naturally into macroadenomas. Koppelman et al.[171] studied the course of untreated hyperprolactinemia, chiefly caused by microadenomas, in 25 women. The mean interval from onset of symptoms to reevaluation was 11.3 years. No patient was judged to have worsened clinically, and indeed many improved by clinical as well as hormonal criteria. Progression of the radiological abnormality was not in only one patient. Menses had resumed in an impressive number (32%) of patients without any form of therapy. Koppelman and co-workers also noted that the greatest degree of clinical improvement tended to occur in patients whose symptoms were estrogen related. These data indicate that most patients with microadenomas demonstrate a stable, and clearly benign course. Similar data were presented by Weiss et al.,[172] who followed 27 patients with microprolactinomas for 6 years; their data suggest that only 3 of their 27 patients showed any significant growth of tumor. March et al.,[173] in a 3- to 20-yr longitudinal evaluation of 43 symptomatic patients with prolactinomas, found progressive tumor growth in only two cases. These studies strongly suggest that most microadenomas remain as such for long periods of time. These data have also indicated that conservative treatment does indeed have a place in the management of microprolactinomas. However, the same cannot be said for microadenomas that hypersecrete ACTH and GH, since the adverse effects of untreated hypercortisolism and hypersomatotropism are considerable.

Treatment

Treatment for hypersecretory microadenomas that cause acromegaly, Cushing's disease, and hyperprolactinemia is discussed in Chapters 3, 9, and 14, respectively. The lack of uniformity in defining cures as well as in documenting recurrences has generated problems in the assessment of therapeutic efficacy. However, some general observations are reasonably accepted.[174]

1. The remission rate for ACTH-secreting microadenomas following transsphenoidal microsurgery is approximately 80–90%, when the adenoma is found.
2. The remission rate for acromegaly caused by microadenomas is 40–80% following transsphenoidal microsurgery.
3. The remission rate for microprolactinomas following primary surgery is 67–91%, averaging 74%.

Transsphenoidal microadenectomy carries an extremely low, almost neglible, mortality rate (0.27%). The morbidity of the procedure is also low; the usual catalogue of complications includes CSF rhinorrhea, meningitis, pituitary insufficiency, and occasionally visual loss.

Surgical Removal

The treatment for macroadenomas associated with or without hypersecretory syndromes is predominantly surgical. The results of surgery for secretory macroadenomas are disappointing. The remission rates for secretory macroadenomas causing hyperprolactinemia, acromegaly, and Cushing's disease are, respectively, 30%, 33%, and 61%.[174] The results are even more discouraging when the tumor is invasive. In such circumstances, the best that can be hoped for is a debulking procedure to remove as much of the tumor as possible, in order to avoid compression syndromes. Postoperative radiation therapy, perhaps coupled with the appropriate adjunctive drug therapy, is recommended in such instances. Ciric et al.[175] reported the results of transsphenoidal surgery for pituitary macroadenomas. Their data revealed that gross total removal without demonstrable evidence of residual tumor tissue was attainable in approximately 41% of cases. They also confirmed the low endocrine cure rates for hypersecretory macroadenomas. The notable aspect of their study was the relatively low (12%) incidence of tumor recurrence rate, which was further reduced when radiotherapy was administered post-operatively. The role of dopamine-agonist drugs as the sole modality of therapy for macroprolactinomas is under investigation, with optimistic initial reports (see Chapter 14).

The major problems with pituitary surgery for macroadenomas are the complications. Aside from the generic complications associated with surgery per se, the main problems pertain to the visual and hypothalamic domains. The rate for major complications is 5–14%. The incidence of postoperative hypopituitarism following transsphenoidal surgery has been reported by Nelson et al.[176] They studied 84 patients with macroadenomas operated by the transsphenoidal route and showed that, of those patients who had normal anterior pituitary function preoperatively, 88% retained normal function after surgery. In patients who already had impairment of pituitary function, deterioration occurred in 33%. Their data suggest that the risk of developing hypopituitarism was greater in patients with larger tumors, and in those with already compromised reserve. In a small number of patients, pituitary reserve improves following surgical removal of the adenoma. Arafah et al.[177] studied the effect of removal of nonfunctioning adenomas in patients who already had impaired function; these workers were able to demonstrate significant improvement in several of the 11 patients studied. This phenomenon is likely if the adenoma is responsible for impaired reserve by compressing the pituitary stalk.

Radiation Therapy

The treatment of nonsecretory macroadenomas with conventional radiation therapy has not found wide application, since these tumors are generally not as radiosensitive as their hypersecretory counterparts. Besides, the potential for visual loss and other suprasellar pressure phenomena precludes the use of a modality of therapy that takes considerable time to work, if indeed it does.

Bromocriptine Therapy

The use of bromocriptine as the sole mode of therapy for nonsecretory macroadenomas is highly controversial and not accepted by most authorities. While the effect of this drug in reducing the size of prolactinomas, and even somatotropinomas, is well documented, such is not the case for nonsecretory macroadenomas. There have been reports of apparent reduction in size of nonsecretory macroadenomas treated with bromocriptine.[178,179] However, this therapy cannot be recommended for a life-threatening condition when other modalities of proven efficacy are available. Therefore, bromocriptine is not an alternative for the conventional therapies used to treat nonsecretory macroadenomas.

Suprasellar Tumors

Suprasellar tumors, as the term implies, are tumors that have their origin above the diaphragma sella. Their importance lies in the fact that these tumors are located in an extemely strategic location; their expansion has profound neuroendocrine consequences. Suprasellar tumors lie in close proximity to the hypothalamus, the optic chiasm, and the ventricular system. If one views the base of the brain from below, the crucial nature of this strategic area will become evident. The hypothalamus is limited anteriorly by the optic chiasm and posteriorly by the mammilary bodies and interpeduncular fossa. The posterior communicating arteries form the lateral borders of the hypothalamus. Inferiorly, the floor of the hypothalamus is formed by the suprasellar cistern. Superiorly, the roof of the hypothalamus is formed by the third ventricle and the thalamus. The hypothalamus assumes a funnel-like shape inferiorly as it forms the pituitary stalk. The stalk is closely surrounded by the lateral walls of the third ventricle and its infundibular recess. It is therefore obvious that an expanding tumor above the sella can play havoc by damaging several vital structures.

Although many tumors can occur in the suprasellar region, the two most frequent ones are craniopharyngioma and germ cell tumors. Meningioma, hypothalamic glioma, optic glioma, hamartoma, infundibuloma, and chordoma are less frequently encountered and require histological confirmation for their diagnosis.

Craniopharyngioma

Craniopharyngioma arises from the remnants of the Rathke's pouch, an evagination of the primitive stomodeum from which the anterior pituitary develops. The cells from which this tumor is derived are epithelial in origin. Embryological remnants of the Rathke's pouch can be found to be scattered in the infundibular region, in the suprasellar area, and within the anterior pituitary. The term *Rathke's pouch cyst* was replaced by Frazier and Alpers,[180] who called it craniopharyngioma.

Incidence and Histopathology

Craniopharyngiomas account for 2–4% of all intracranial tumors, regardless of age. In children, these tumors account for 8.2–13% of all intracranial tumors[181–183] and represent the third most common tumor in childhood. Macroscopically, these tumors can be cystic, solid, or mixed. Histologically, most of these tumors are cystic, lined by mature squamous epithelium. While the traditional classification of craniopharyngioma by pathologists into squamous and adamantinomous forms is a well-accepted notion, transitional forms have been described.[184] The tumor is well differentiated, benign, and often well capsulated. The squamous cells that make up the tumor are arranged as strands or sheets separated by cystic pools. The epithelial cells demonstrate irregular nuclei, with peripherally clumped chromatin. One striking histological feature is that the cytoplasm of tumor cells extends out of the cells as microvilli that link themselves with microvilli of adjacent cells. This impressive feature can be carried to its extreme in certain craniopharyngiomas that demonstrate finger-shaped projections that extend and become attached to surrounding structures with reactive gliosis. This poses a major problem for the neurosurgeon by precluding complete resection of the tumor. It is this local invasiveness that imparts to craniopharyngiomas an aggressiveness uncharacteristic of a benign neoplasm. Surgical experience indicates that, although encapsulated, the capsule of craniopharyngioma is unusually adherent to neural and vascular structures.

Although the histological characteristics of craniopharyngioma are well defined, occasionally the distinction from suprasellar epidermoid cysts can be difficult. In some cases, the tumor contains considerable amounts of keratin, or keratohyaline granules, and calcium deposits can be seen in larger cyst cavities. Reactive astrocytes and phagocytes can be seen lying adjacent to the tumorous squamous epithelial cells. The distinction between craniopharyngioma and pituitary adenoma can be readily made by immunoperoxidase staining; cells of craniopharyngioma do not contain hormone secreting granules, and hence show no immunopositivity with a panel of adenohypophyseal hormone antisera. Yet, occasionally difficulties may arise in differentiating cystic, truly nonsecretory chromobhobe adenomas from craniopharyngiomas.

Clinical Features

Craniopharyngiomas are always symptomatic. The clinical effects of these tumors are a direct result of tumor expansion and compression of surrounding structures. To the clinician who has cared for patients with craniopharyngioma, this tumor conjures vivid images of pain, suffering, and horror—the baby with relentlessly progressive hydrocephalus, the child who fails to grow and thrive, the adolescent waiting for puberty that never arrives, and patients of any age who develop blindness from this tumor. Unfortunately, the disease is not often recognized in the early phases, and many teenage children have been treated for behavioral problems, until their vision is irrevocably lost.

The clinical effects of craniopharyngioma depend on the location of the tumor and the degree of compression exerted on the surrounding structures. The four major phenomena that occur are compression of the ventricular system, the optic chiasm, the hypothalamus, and the pituitary.

Compression of the Ventricles.   Increased intracranial tension is encountered when craniopharyngiomas compress and obstruct the third ventricle. This is a particularly likely event when the optic chiasm is prefixed; thus, the tumor can expand only posteriorly into the floor of the third ventricle and the interpeduncular fossa. Occasionally, very large tumors can expand laterally into the Sylvian fissure. Headache, papilledema, and optic atrophy result.

Pressure on the Optic Chiasm.   Compression of optic nerves and/or optic chiasm results in visual-field defects, decreased visual acuity, and partial or complete blindness. Optic nerve compression is particularly prevalent when the chiasm is postfixed or in the normal position.

Compression of the Hypothalamus.   Hypothalamic involvement by cranio-pharyngioma results in the development of diabetes insipidus. In comparison with germinoma (another common suprasellar tumor), the occurrence of vasopressin deficiency is a relatively delayed development in patients with craniopharyngioma. The presence of diabetes insipidus adversely affects the outcome after surgery, since complete excision of the tumor in these cases is quite difficult. In addition to vasopressin deficiency, craniopharyngiomas can cause deficiencies of hypothalamic releasing factors by dual mechanisms: impairment in production or transport. In one series of 42 children with cran-iopharyngioma,[185] growth retardation was present in 53%, and GH deficiency was documented in 72% prior to treatment. Hormonal testing of pituitary hormone reserve reveals abnormalities in the vast majority of patients with craniopharyngioma.[186] Thus, impaired reserve of LH, FSH, TSH, and to a lesser extent ACTH is encountered. The abnormalities detected can resemble a primary pituitary type of deficiency pattern or a hypothalamic pattern. The latter, a common feature of any suprasellar lesion, is usually due to interruption of the pituitary stalk and is accompanied by hyperprolactinemia.

Compression of the Pituitary Gland.   This phenomenon is particularly evi-dent in the relatively rare entity. Destruction of the normal pituitary tissue by the expanding intrasellar tumor results in hypopituitarism and an enlarged sella turcica. In some instances, the tumor can destroy the dorsum sella and invade the tissues inferiorly. In a remarkably unique and resounding auto-biographical report, Dr. Maier[187] gave a detailed description of a cranio-pharyngioma that penetrated the sphenoid sinus and eroded into the naso-pharynx, with drainage of cystic parts of the tumor intermittently through the nose over a period of 30 years. The tumor ultimately found its way to the oropharynx. The only aspect of this case that was more remarkable than the tumor itself was how this physician triumphed over the aggressive tumor and excelled in his professional career as a student, surgeon, and educator. Such heartening reports punctuate the otherwise gloomy outlook for invasive craniopharyngioma.

The craniopharyngiomas of adulthood are less frequent than the child-hood form of tumor. The clinical features are identical to those associated with childhood craniopharyngiomas. The tumor in adults shows less of a tendency for calcification, an important radiological hallmark of the tumor in children.

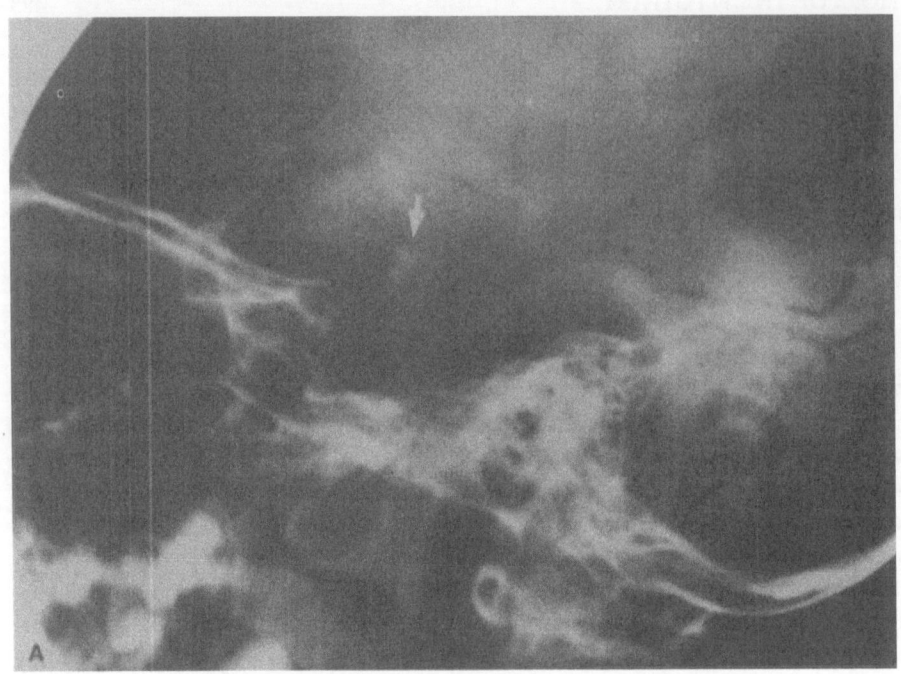

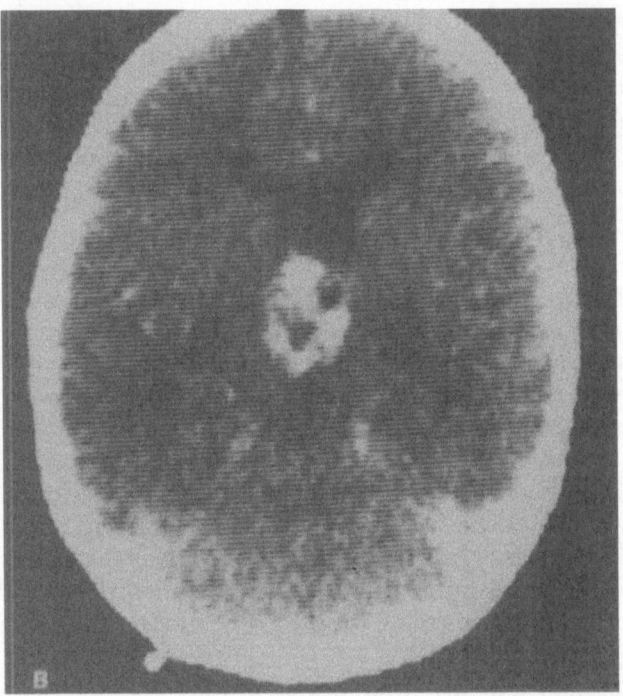

FIGURE 41. (A) Lateral film of the skull demonstrating loss of dorsum sella with suprasellar calcification (arrow). (B) Contrast-enhanced horizontal CT showing a calcified, enhancing tumor invading the third ventricle. (C) Contrast-enhanced coronal CT showing suprasellar extension of the tumor (arrows). (From Post.[216])

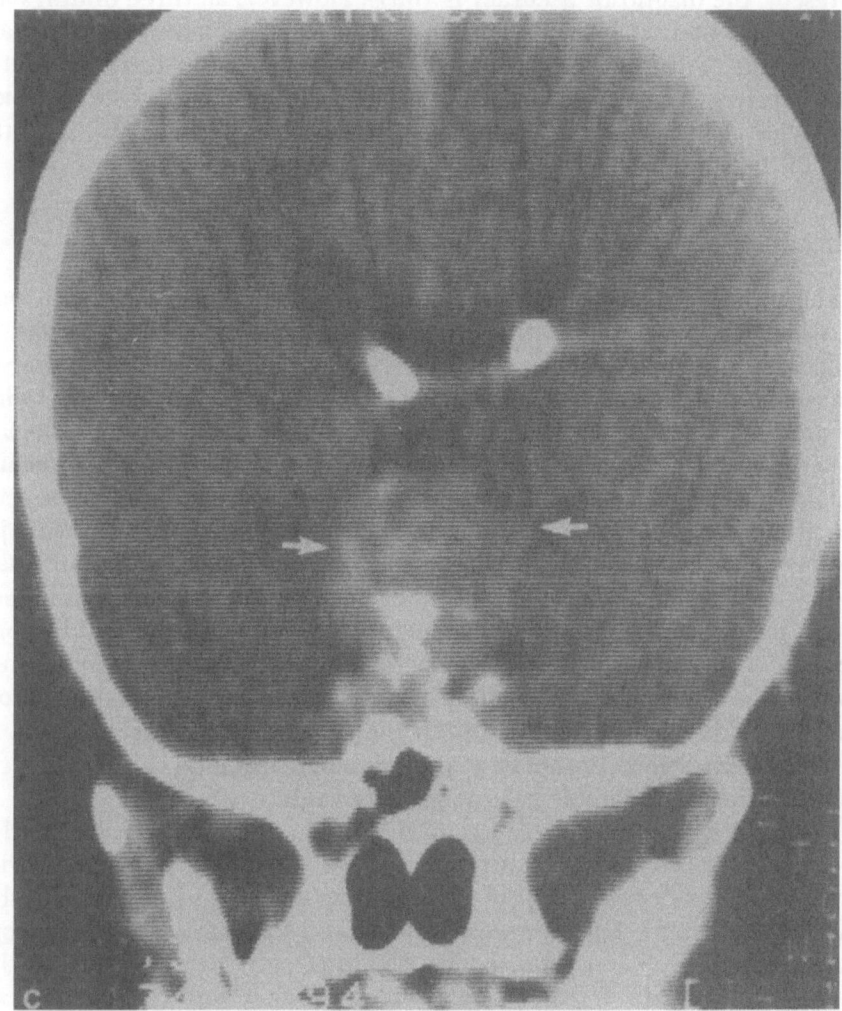

FIGURE 41. *(Continued)*

Radiological Diagnosis

Radiological clues for the presence of a craniopharyngioma are often provided first by plain films of the skull. The presence of suprasellar calcification is a highly suggestive sign of craniopharyngiomas; calcification is present in approximately 80% of children and in 45–50% of adults with craniopharyngioma. When the calcification is dense and floccular, it is even more suggestive. Enlargement of the sella and signs of increased intracranial tension may be evident by plain X-ray films.

The CT diagnosis of craniopharyngioma rests on demonstrating three characteristics associated with this suprasellar tumor: contrast enhancement, presence of cysts, and presence of calcification. When all three characteristics are present, the diagnostic accuracy is 100%. However, all three findings are noted in only about 75% of the patients. Craniopharyngiomas generally are of low density, with partial or complete enhancement after contrast. Metrizamide cisternography can be useful in delineating the extent of compression of the ventricles. Pneumoencephalography is seldom resorted to, with the advent of high-resolution CT (Fig. 41A–C).

Treatment

The best treatment for craniopharyngioma is total excision, when possible, when it can be accomplished without causing significant mortality or morbidity. Unfortunately, most craniopharyngiomas are large, at the time of diagnosis, and have involved surrounding structures. In such a setting, aiming for total surgical removal cannot be attained without disastrous neurological sequelae. Even after so-called complete resection, recurrences are not uncommon. Favorable results with surgery alone have been reported by several groups,[188–195] with variable rates of mortality and morbidity. One of the major problems encountered during surgery is the extensive reactive gliosis associated with the tumor. The advent of superior surgical techniques has indeed lowered the mortality considerably, but many workers believe that complete removal of tumor is illusory. Some of the factors that adversely affect the outcome are the presence of intracranial hypertension, altered mentation, diabetes insipidus, hypothalamic compression, and chiasmal compression. In these cases, the surgeon cannot be expected to perform total curative surgery without causing considerable neurological damage.

During the past two decades, radiotherapy has become well accepted as a supplement to conservative surgery. In two separately published series, Kramer et al.[196,197] noted improved long-term survival in adults and children when radiation therapy was supplemented with conservative surgery. Similar results have been reported by Bloom et al.,[198] who also noted that a 3-year recurrent-free period following combined therapy for patients with craniopharyngioma augured well. An increasing body of literature has confirmed the benefits of supplemental full-course radiation therapy for craniopharyngiomas.[185,199–201]

In some series, the reduction in the recurrence rate has been dramatic; The adverse effects of radiation therapy are acceptably low. Danoff et al.[202] studied 19 patients (children and adolescents) following surgery and supervoltage radiation; they demonstrated no additional endocrine deficiencies that could be directly attributable to radiation; 16% had developed a recurrence, despite supplemental radiotherapy, and 79% had minimal or no neurological disability. The consensus of opinion among experts is that post operative irradiation after subtotal surgical excision is superior to surgery alone. The quality of life following combined therapy also seems to be better than after radical aggressive surgery. Shilito[203] recommended that craniopharyngiomas be explored by the subfrontal route in order to assess operability, the minimal acceptable morbidity being unilateral anosmia and transient diabetes insipidus. If total removal is not possible, the optic system should be decompressed, following which radiation therapy should be instituted. Thus, even though the tumor is benign, the type of surgery performed is palliative. Finally, the postoperative intratumoral injection of bleomycin has been reported.[204] This is based on the histological fact that craniopharyngiomas are of squamous cell origin.

Suprasellar Germ Cell Tumors

Incidence and Histology

Germ cell tumors account for approximately 0.5–2.0% of all primary intracranial tumors.[205] The term germ cell tumor comprises several histological types, the most frequent of which is the germinoma (Table 88). These tumors are derived from the embryonic germ cell. Germinomas are derived from the primordial germ cell and teratomas are derived from the pluripotential stem cell, while endodermal sinus tumors are derived from yolk sac endoderm, and choricarcinomas originate from the trophoblast. All types of germ cell tumors are malignant.

Germ cell tumors in the CNS are located in the midline, either in the pineal gland, or at the suprasellar cistern. In general, most germinomas arise in the suprasellar cistern, while the other varieties of germ cell tumors tend to occur more frequently in the pineal gland. Germ cell tumors primarily affect children of both sexes, with a slight predominance in males. Pineal involvement is more common in males.

Clinical Features

The clinical presentation of germ cell tumors depends on the location of the tumor. The largest collected series of intracranial germ cell tumors is from Memorial Sloan-Kettering Center, consisting of 389 patients reported by Jennings et al.[206] The main presentation of suprasellar germinomas is characterized by the triad of chiasmal field defects, diabetes insipidus, and other

TABLE 88
Intracranial Germ Cell Tumors

Tumor	Percent
Germinoma	65
Teratoma	18
Embryonal carcinoma	5
Endodeimal sinus tumor	7
Choriocarcinoma	5

signs of hypothalamic–pituitary dysfunction. The latter includes growth retardation, delayed puberty, and other trophic hormone deficiencies of the
pituitary gland. The pressure effects on the ventricular system can dominate
the clinical picture, resulting in increased intracranial tension, and neurological syndromes (Table 89). In general, diabetes insipidus, and hypothalamic
pituitary dysfunction is more prevalent with suprasellar germinomas, than
with the nongerminomatous germ cell tumors.

Germ cell tumors may present with metastatic manifestations. Thus, spinal
cord compression syndromes, as well as pulmonary or bony lesions, may be
the initial manifestation. Germ cell tumors metastasize through the ventricular
and subarachnoid pathways as well as by the systemic route.

Diagnosis

Diagnostic studies useful in the delineation of germ cell tumors are
the CT study, cisternography, and, when indicated, myelography. Abnormal
CSF cytology may be seen in as many as 60%. Elevated levels of β-HCG, or
α-fetoprotein are seen in a high percentage of patients with nongerminomatous germ cell tumors, and to a lesser extent in patients with germinomas.
In general, the less mature, and poorly differentiated germ cell tumors tend
to be associated with increased levels of biomarkers.

TABLE 89
Clinical Manifestations of Germ Cell Tumors

Suprasellar germinomas
  Chiasmal compression
  Hypothalamic compression
    Diabetes insipidus
    Hypopituitarism
  Obstruction to CSF flow
    Increased ICT
    Hydrocephalus
  Neurological syndromes
    Parinaud's sign
    Pyramidal tract signs
    Extrapyramidal signs
    Seizures
  Psychiatric syndromes
    Dementia
    Psychosis

Nongerminomatous germ cell tumors
  Pineal region mass
  Sexual precocity
  Neurological syndromes (same as germ cell tumors)
  Increased intracranial pressure
  Hypothalamic dysfunction

Treatment

Treatment for germ cell tumors is controversial, with opinions ranging from radiation alone[207-210] (since these tumors are rather radiosensitive), surgery combined with radiation[211,212] and a combination of surgery, radiation, and chemotherapy.[213] Unfortunately, there have been no large prospective studies to determine the role of chemotherapy alone or in combination. Selection of the appropriate treatment for germ cell tumors is further complicated by the fact that a higher incidence of spinal cord metastasis is seen in patients who undergo biopsy for purposes of diagnostic and histological classification. The approach of initial radiation, followed by surgical intervention if the mass fails to regress, has been recommended by Japanese investigators.[208,209,214]

References

1. Costello RT: Subclinical adenoma of the pituitary gland. *Am J Pathol* **12**:205, 1936.
2. Kernohan JW, Sayre GP: Tumors of the pituitary gland and infundibulum. In *Atlas of Tumor Pathology*, First Series, Fascicle 36. Armed Forces Institute of Pathology, Washington, DC, 1956.
3. Kovacs K: Light and electron microscopic pathology of pituitary tumors: Immunohistochemistry. In Black PM, Zervas NT, Ridgway EC, et al. (eds): *Secretory Tumors of the Pituitary Gland. Progress in Endocrine Research and Therapy*, Vol. 1. Raven, New York, 1984, p. 365.
4. Horvath E, Kovacs K: Pathology of the pituitary gland. In Ezrin C, Horvath E, Kaufman B, Kovacs K, Weiss MH (eds): *Pituitary Diseases*. CRC Press, Boca Raton, FL, 1980, p. 1.
5. Kovacs K, Horvath E: Pituitary adenomas: Pathologic aspects. In Tolis G, Labrie F, Martin JB, Naftolin F (eds): *Clinical Neuroendocrinology: A Pathophysiological Approach*. Raven, New York, 1979, p. 367.
6. Kovacs K, Horvath E: Pathology of pituitary adenomas. In Givens JR (ed): *Hormone Secreting Pituitary Tumors*. Year Book, Chicago, 1982, p. 97.
7. Kovacs K, Horvath E, Ezrin C: Pituitary adenomas. *Pathol Annu* 12(2):341, 1977.
8. Phifer RF, Spicer SS, Hennigar GG: Histochemical reactivity and staining properties of functionally defined cell types in the human adenohypophysis. *Am J Pathol* **73**:569, 1973.
9. Horvath E, Kovacs K: Ultrastructural classification of pituitary adenomas. *Can J Neurol Sci* **3**:9, 1976.
10. Halmi NS, Duello T: "Acidophilic" pituitary tumor. A reappraisal with differential staining and immunocytochemical techniques. *Arch Pathol Lab Med* **100**:346, 1976.
11. Kovacs K, Horvath E, Killinger DW, et al: Growth hormone producing pituitary adenoma with giant secretory granules. *Acta Neuropathol* **46**:239, 1979.
12. Robert F: Electron microscopy of human pituitary tumors. In Tindall GT, Collins WF (eds): *Clinical Management of Pituitary Disorders*. Raven, New York, 1979, p. 113.
13. Cardell RR Jr, Knighton RS: The cytology of a human pituitary tumor. An electron microscopic study. *Trans Am Microsc Soc* **85**:58, 1966.
14. Roy S: Cytoplasmic filamentous masses in chromophobe adenoma of the human pituitary gland. *J Pathol Lab Med* **125**:151, 1978.
15. Horvath E, Kovacs K: Pathology of the pituitary gland. In Ezrin C, Horvath E, Kaufman B, Kovacs K, Weiss MH (eds): *Pituitary Diseases*. CRC Press, Boca Raton, FL, 1980, p. 1.
16. Horvath E, Kovacs K, Singer W, et al: Acidophil stem cell adenoma of the human pituitary. *Arch Pathol Lab Med* **101**:594, 1977.
17. Horvath E, Kovacs K, Singer W, et al: Acidophil stem cell adenoma of the human pituitary. Clinico-pathological analysis of 15 cases. *Cancer* **47**:761, 1981.

18. Horvath E, Kovacs K, Killinger DW, et al: Mammosomatotroph cell adenoma of the human pituitary. In *Proceedings of the Thirty-eighth Annual Meeting of the Electron Microscopy Society of America, 1980*, p. 726 (abst.).

19. Miller WL, Eberhardt NL: Structure and evolution of the growth hormone gene family. *Endoc Rev* 4:97, 1983.

20. Martial JA, Hallewell RA, Baxter JD, et al: Human growth hormone: Complementary DNA cloning and expression in bacteria. *Science* 205:602, 1979.

21. Niall HD, Hogan ML, Sayer R, et al: Sequences of pituitary and placental lactogenic and growth hormones: Evolution from a primordial peptide by gene duplication. *Proc Natl Acad Sci USA* 68:866, 1971.

22. Guillemin R, Vargo T, Rossier J, et al: β-Endorphin and adrenocorticotropin are secreted concomitantly by the pituitary gland. *Science* 197:1367, 1977.

23. Wilkes MM, Watkins WB, Stewart RD, et al: Localization and quantitation of β-endorphin in human brain and pituitary. *Neuroendocrinology* 30:113, 1980.

24. Mendelsohn G, d'Agostino R, Eggleston JC, et al: Distribution of β-endorphin immuno-reactivity in normal human pituitary. *J Clin Invest* 63:1297, 1979.

25. Wilson MG, Nicholson WE, Holscher MA, et al: Proopiolipomelanocortin peptides in normal pituitary, pituitary tumor, and plasma of normal and Cushing's horses. *Endocrinology* 110:941, 1982.

26. Martin R, Cetin Y, Fehm HL, et al: Multiple cellular forms of corticotrophs in surgically removed pituitary adenomas and periadenomatous tissue in Cushing's disease. *Am J Pathol* 106:332, 1982.

27. Trouillas J, Girod C, Sassolas G, et al: A human β-endorphin pituitary adenoma. *J Clin Endocrinol Metab* 58:242, 1984.

28. Lamberts SWJ, Delange SA, Stefanko S: Adrenocorticotropin-secreting pituitary adenomas originate from the anterior or the intermediate lobe in Cushing's disease: Differences in the regulation of hormone secretion. *J Clin Endocrinol Metab* 54:286, 1982.

29. Horvath E, Kovacs K, Killinger DW, et al: Silent corticotropic adenomas of the human pituitary gland. A histologic, immunocytologic, and ultrastructural study. *Am J Pathol* 98:617, 1980.

30. Hill SA, Falko JM, Wilson CB, et al: Thyrotropin-producing pituitary adenomas. *J Neurosurg* 57:515, 1982.

31. Ipse G, Ryan N, Kovacs K, et al: Calcium deposition in human pituitary adenomas studied by histology, electron microscopy, electron diffraction and x-ray spectometry. *Exp Pathol (Jena)* 18:377, 1980.

32. Smallridge RC, Smith CE: Hyperthyroidism due to thyrotropin-secreting pituitary tumors: Diagnostic and therapeutic considerations. *Arch Intern Med* 143:503, 1983.

33. Caughey JE: The etiology of pituitary tumors: The role of hypogonadism and hypothyroidism. *Aust Ann Med* 6:93, 1957.

34. Gordon SJ, Moses AM: Multiple endocrine organ refractoriness to trophic hormone stimulation. A patient with an enlarged sella turcica and increased FSH secretion. *Ann Intern Med* 63:313, 1965.

35. Bower BF: Pituitary enlargement secondary to untreated primary hypogonadism. *Arch Intern Med* 69:107, 1968.

36. Burke CW, Marshall JC: Ovarian failure with probable pituitary tumor. *Proc R Soc Med* 64:1066, 1971.

37. Trouillas J, Girod C, Sassolas G, et al: Human pituitary gonadotropic adenoma; histological, immunocytochemical, and ultrastructural and hormonal studies in eight cases. *J Pathol* 135:315, 1981.

38. Horvath E, Kovacs K: Gonadotroph adenomas of the human pituitary: Sex-related fine-structural dichotomy. *Am J Pathol* 117:429, 1984.

39. Smith JL (ed): *Neuro-ophthalmology Symposium of the University of Miami and the Bascom Palmer Eye Institute—The Human Optic Chiasm: A Neuroanatomical Review of Current Concepts, Recent Investigations, and Unsolved Problems*. Charles C Thomas, Springfield, IL, 1964, p. 64.

40. Ludmerer KM, Kissane JM: Visual impairment, pituitary dysfunction, and hilar adenopathy in a young man. *Am J Med* **80**:259, 1986.

41. Weisberg LA, Zimmerman EA, Frantz AG: Diagnosis and evaluation of patients with an enlarged sella turcica. *Am J Med* **61**:590, 1976.

42. Hoyt WF, Frisen L, Newman NM: Fundoscopy of nerve fiber layer defects in glaucoma. *Invest Ophthalmol* **12**:814, 1973.

43. Wilson P, Falconer MA: Patterns of visual failure with pituitary tumors. *Br J Ophthalmol* **52**:94, 1978.

44. Knight CL, Hoyt WF, Wilson CB: Syndrome of incipient prechiasmal optic nerve compression. *Arch Ophthalmol* **87**:1, 1972.

45. Kleeman CR, Czaczkes JW, Cutler R: Mechanisms of impaired water excretion in adrenal and pituitary insufficiency. IV. Antidiuretic hormone in primary and secondary adrenal insufficiency. *J Clin Invest* **43**:1641, 1964.

46. Robson JS, Lambie AT: Cortisone-induced polyuria following hypophysectomy. *Am J Med* **26**:769, 1959.

47. Lessard ML, Attia EL, Baxter JD, et al: Intranasal presentation of a pituitary adenoma. *J Otolaryngol* **14**(4):251, 1985.

48. Forbes AP, Henneman PH, Griswold GC, et al: Syndrome characterized by galactorrhea, amenorrhea, and low urinary FSH: comparison with acromegaly and normal lactation. *J Clin Endocrinol* **14**:265, 1954.

49. Boyd AE III, Hamilton D, Murray BG, et al: Medical management of prolactinomas: II. In Black PM, Zervas NT, Ridgway EC, et al (eds): *Secretory Tumors of the Pituitary Gland. Progress in Endocrine Research and Therapy*, Vol 1. Raven, New York, 1984, p. 65.

50. Hardy J: Transsphenoidal microsurgery of prolactinomas. In: Black PM, Zervas NT, Ridgway EC, et al. (eds): *Secretory Tumors of the Pituitary Gland. Progress in Endocrine Research and Therapy*, Vol 1. Raven, New York, 1984, p.73.

51. Franks S, Murray MA, Jequier AM, et al: Incidence and significance of hyperprolactinemia in women with amenorrhea. *Clin Endocrinol (Oxf)* **4**:597, 1975.

52. Hardy J, Beauregard H, Robert F: Prolactin-secreting pituitary adenomas: Transsphenoidal microsurgical treatment. In Robyn C, Harter M (eds): *Progress in Prolactin Physiology and Pathology*. Elsevier, New York, 1978, p. 361.

53. Kleinberg DL, Noel GL, Frantz AG: Galactorrhea: 235 cases including 48 with pituitary tumors. *N Engl J Med* **296**:589, 1977.

54. Carter JN, Tyson JE, Tolis G, et al: Prolactin-secreting tumors and hypogonadism in 22 men. *N Engl J Med* **299**:847, 1978.

55. Ambrosi B, Rosella B, Travaglini P, et al: Study of the effect of bromocriptine on sexual impotence. *Clin Endocrinol (Oxf)* **7**:417, 1977.

56. Skrabanek P, McDonald D, deValera E, et al: Plasma prolactin in amenorrhea, infertility and other disorders: A retrospective study of 608 patients. *Ir J Med Sci* **149**:236, 1980.

57. Spark RF, White RA, Connolly PB: Impotence is not always psychogenic. *JAMA* **243**:750, 1980.

58. Segal S, Yaffee H, Laufer N, et al: Male hyperprolactinemia: Effects on fertility. *Fertil Steril* **32**:556, 1979.

59. Thorner MO, Edwards CRW, Hanker JP, et al: Prolactin and gonadotropin interaction in the male. In Troen P, Nankin H (eds): *The Testis in Normal and Infertile Men*. Raven, New York, 1977, p. 351.

60. Tourniaire J, Trouillas J, Chalender D, et al: Somatotropic adenoma manifested by glactorrhea without acromegaly. *J Clin Endocrinol Metab* **61**:451, 1985.

61. Scanlon MF, Howells S, Peters JR, et al: Hyperprolactinaemia, amenorrhoea and galactorrhoea due to a pituitary thyrotroph adenoma. *Clin Endocrinol (Oxf)* **23**:35, 1985.

62. Lagerquist LG, Meikle AW, West CD, et al: Cushing's disease with cure by resection of a pituitary adenoma: Evidence against a primary hypothalamic defect. *Am J Med* **57**:826, 1974.

63. Salassa RM, Kearns TP, Kernohan JW, et al: Pituitary tumors in patients with Cushing's syndrome. *J Clin Endocrinol Metab* **19**:1523, 1959.

64. Boggan JE, Tyrrell JB, Wilson CB: Transphenoidal microsurgical management of Cushing's disease: Report of 100 cases. *J Neurosurg* **59**:195, 1983.

65. Tyrrell JB, Brooks RM, Fitzgerald PA, et al: Cushing's disease: Selective transsphenoidal resection of pituitary microadenomas. *N Engl J Med* **298**:753, 1978.

66. Findling JW, Tyrrell JB: The anterior pituitary. In Greenspan FS, Forsham PH (eds): *Basic and Clinical Endocrinology*. Lange Medical Publications, Los Altos, CA, 1983, p. 38.

67. Nelson DH, Meakin JW, Thorn GW: ACTH-producing pituitary tumors following adrenalectomy for Cushing's syndrome. *Ann Intern Med* **52**:560, 1960.

68. Moore TJ, Dluhy RG, Williams GH, et al: Nelson's syndrome: Frequency, prognosis, and effect of prior irradiation. *Ann Intern Med* **85**:731, 1976.

69. Jailer JW, Holub DA: Remission of Graves' disease following radiotherapy of a pituitary neoplasm. *Am J Med* **28**:497, 1960.

70. Hamilton CR, Adams LC, Maloof F: Hyperthyroidism due to thyrotropin-producing pituitary chromophobe adenoma. *N Engl J Med* **283**:1077, 1970.

71. Afrasiabi A, Valenta L, Gwinup G: A TSH secreting pituitary tumour causing hyperthyroidism: Presentation of a case and review of the literature. *Acta Endocrinol (Copenh)* **192**:448, 1979.

72. Barbarino A, DeMarinis L, Anile C, et al: Normal pituitary function and reserve after selective transsphenoidal removal of a thyrotropin-producing pituitary adenoma. *Metabolism* **29**:739, 1980.

73. Baylis PH: Case of hyperthyroidism due to a chromophobe adenoma. *Clin Endocrinol (Oxf)* **5**:145, 1976.

74. Cooper DS, Ridgway EC, Maloof F: Unusual types of hyperthyroidism. *Clin Endocrinol Metab* **7**:199, 1978.

75. Cravioto H, Fukaya T, Zimmerman EA, et al: Immunohistochemical and electron-microscopic studies of functional and non-functional pituitary adenomas including one TSH secreting tumor in a thyrotoxic patient. *Acta Neuropathol* **53**:281, 1981.

76. Faglia G, Ferrari C, Neri V, et al: High plasma thyrotropin levels in two patients with pituitary tumour. *Acta Endocrinol (Copenh)* **69**:649, 1972.

77. Filetti S, Rapoport B, Aron DC, et al: TSH and TSH-subunit production by human thyrotrophic tumour cells in monolayer culture. *Acta Endocrinol (Cophen)* **99**:224, 1982.

78. Gharib H, Carpenter PC, Scheithauer BW, et al: The spectrum of inappropriate pituitary thyrotropin secretion associated with hyperthyroidism. *Mayo Clin Proc* **57**:556, 1982.

79. Hill SA, Falko JM, Wilson CB, et al: Thyrotropin-producing pituitary adenomas. *J Neurosurg* **57**:515, 1982.

80. Jackson IMD: Hyperthyroidism in a patient with a pituitary chromophobe adenoma. *J Clin Endocrinol Metab* **25**:491, 1965.

81. Kourides IA, Ridgway EC, Weintraub BD, et al: Thyrotropin-induced hyperthyroidism: Use of alpha and beta subunit levels to identify patients with pituitary tumors. *J Clin Endocrinol Metab* **45**:534, 1977.

82. Kourides IA, Pekonen F, Weintraub BD: Absence of thyroid-binding immunoglobulins in patients with thyrotropin-mediated hyperthyroidism. *J Clin Endocrinol Metab* **51**:271, 1980.

83. Yovos JG, Falko JM, O'Dorisio TM, et al: Thyrotoxicosis and a thyrotropin-secreting pituitary tumor causing unilateral exophthalmos. *J Clin Endocrinol Metab* **53**:338, 1981.

84. Ridgway EC: Glycoprotein hormone production by pituitary tumors. In Black PM, Zervas NT, Ridgway EC, et al (eds): *Secretory Tumors of the Pituitary Gland. Progress in Endocrine Research Therapy*, Vol. 1. Raven, New York, 1984, p. 343.

85. Benoit R, Pearson-Murphy BE, Robert F, et al: Hyperthyroidism due to a pituitary TSH secreting tumour with amenorrhoea-galactorrhoea. *Clin Endocrinol (Oxf)* **12**:11, 1980.

86. Duello TM, Halmi NS: Pituitary adenoma producing thyrotropin and prolactin, an immunocytochemical and electron microscopic study. *Virchows Arch [Pathol Amat]* **376**:255, 1977.

87. Horn K, Erhardt F, Fahlbusch R, et al: Recurrent goiter, hyperthyroidism, galactorrhea and amenorrhea due to a thyrotropin and prolactin-producing pituitary tumor. *J Clin Endocrinol Metab* **43**:137, 1976.

88. Werner S: Human pituitary adenomas with hypersecretion of TSH and prolactin. *Horm Metab Res* **11**:452, 1979.

89. Carlson HE, Linfoot JA, Braunstein GD, et al: Hyperthyroidism and acromegaly due to a thyrotropin- and growth hormone-secreting pituitary tumor. *Am J Med* **74**:915, 1983.

90. Lamberg B-A, Ripatti J, Gordin A, et al: Chromophobe pituitary adenoma with acromegaly and TSH-induced hyperthyroidism associated with parathyroid adenoma. *Acta Endocrinol (Copenh)* **60**:157, 1969.

91. Takano K, Kogawa M, Tsushima T, et al: A TSH secreting pituitary tumor accompanied by high stature: Presentation of a case and review of the literature. *Endocrinol Jpn* **28**:215, 1981.

92. Koide Y, Kugai N, Kimura S, et al: A case of pituitary adenoma with possible simultaneous secretion of thyrotropin and follicle-stimulating hormone. *J Clin Endocrinol Metab* **54**:397, 1982.

93. Reschini E, Giustina G, Cantalamessa, et al: Hyperthyroidism with elevated plasma TSH levels and pituitary tumor: Study with somatostatin. *J Clin Endocrinol Metab* **43**:924, 1976.

94. Smallridge RC, Smith CE: Hyperthyroidism due to thyrotropin-secreting pituitary tumors: Diagnostic and therapeutic considerations. *Arch Intern Med* **143**:503, 1983.

95. Tolis G, Bird C, Bertrand G, et al: Pituitary hyperthyroidism. Case report and review of the literature. *Am J Med* **64**:177, 1978.

96. Smith CE, Smallridge RC, Dimond RC, et al: Hyperthyroidism due to a thyrotropin-secreting pituitary adenoma. *Arch Intern Med* **142**:1709, 1982.

97. Woolf PD, Schenk EA: An FSH-producing pituitary tumor in a patient with hypogonadism. *J Clin Endocrinol Metab* **38**:561, 1974.

98. Snyder PJ, Sterling FH: Hypersecretion of LH and FSH by a pituitary adenoma. *J Clin Endocrinol Metab* **42**:544, 1976.

99. Kovacs K, Horvath E, VanLoon GR, et al: Pituitary adenomas associated with elevated blood follicle-stimulating hormone levels: A histologic, immunocytologic, and electron microscopic study of two cases. *Fertil Steril* **29**:622, 1978.

100. DeMura R, Kubo O, DeMura H, et al: FSH and LH secreting pituitary adenoma. *J Clin Endocrinol Metab* **45**:653, 1977.

101. Friend JN, Judge DM, Sherman BM, et al: FSH-secreting pituitary adenomas: Stimulation and suppression studies in two patients. *J Clin Endocrinol Metab* **43**:650, 1976.

102. Cunningham GR, Huckins C: An FSH and prolactin-secreting pituitary tumor: Pituitary dynamics and testicular histology. *J Clin Endocrinol Metab* **44**:248, 1977.

103. Nicolis GL, Modhi G, Gabrilove JL: Gonadotropin producing pituitary adenomas. A case report and review of the literature. *Mt Sinai J Med* **49**:297, 1982.

104. Peterson RE, Kourides IA, Horwith M, et al: Leuteinizing hormone- and α-subunit-secreting pituitary tumor: Positive feedback of estrogen. *J Clin Endocrinol Metab* **52**:692, 1981.

105. Snyder PJ, Sterling FH: Hypersecretion of LH and FSH by a pituitary adenoma. *J Clin Endocrinol Metab* **42**:544, 1976.

106. Snyder PJ, Johnson J, Muzyka R: Abnormal secretion of glycoprotein α-subunit and follicle-stimulating hormone (FSH) β-subunit in men with pituitary adenomas and FSH hypersecretion. *J Clin Endocrinol Metab* **51**:579, 1980.

107. Harris RI, Schatz NJ, Gennarelli T, et al: Follicle-stimulating hormone-secreting pituitary adenomas: Correlation of reduction of adenoma size with reduction of hormonal hypersecretion after transsphenoidal surgery. *J Clin Endocrinol Metab* **56**:1288, 1983.

108. Snyder PJ: Gonadotroph cell adenomas of the pituitary. *Endocrin Rev* **6**:552, 1985.

109. Faggiano M, Criscuolo T, Perrone I, et al: Sexual precocity in a boy due to hypersecretion of LH and prolactin by a pituitary adenoma. *Acta Endocrinol (Copenh)* **102**:167, 1983.

110. Beckers A, Stevenaert A, Mashiter K, et al: Follicle-stimulating hormone-secreting pituitary adenomas. *J Clin Endocrinol Metab* **61**:525, 1985.

111. Wolpert SM: The radiology of pituitary adenomas. *Semin Roentgenol* **19**:53, 1984.
112. McLachlan MSF, Wright AD, Doyle FH: Plain film and tomographic assessment of the pituitary fossa in 140 acromegalic patients. *Br J Radiol* **43**:360, 1970.
113. Swanson HA, du Boulay G: Borderline variants of the normal pituitary fossa. *Br J Radiol* **48**:366, 1975.
114. Vezina JL, Sutton TJ: Prolactin-secreting pituitary microadenomas: Roentgenologic diagnosis. *AJR* **120**:46, 1974.
115. Geehr RB, Allen WE, Rothman SLG, et al: Pluridirectional tomography in the evaluation of pituitary tumors. *AJR* **130**:105, 1978.
116. Robertson WD, Newton TH: Radiologic assessment of pituitary microadenomas. *AJR* **131**:489, 1978.
117. Wakai S, Fukushima T, Furihata T, et al: Association of cerebral aneurysm with pituitary adenoma. *Surg Neurol* **12**:503, 1979.
118. Syvertsen A, Houghton VM, Williams AL, et al: The computed tomographic appearance of the normal pituitary gland and pituitary microadenomas. *Radiology* **89**:389, 1979.
119. Muhr C, Bergström K, Grimelius L, et al: A parallel study of the roentgen anatomy of the sella turcica and the histopathology of the pituitary gland in 205 autopsy specimens. *Neuroradiology* **21**:55, 1981.
120. Costello RT: Subclinical adenoma of the pituitary gland. *Am J Pathol* **12**:205, 1936.
121. Parent AD, Bebin J, Smith RR: Incidental pituitary adenomas. *J Neurosurg* **54**:228, 1981.
122. Kovacs K, Ryan N, Horvath E, et al: Pituitary adenomas in old age. *J Gerontol* **35**:16, 1980.
123. Burrow GN, Wortzman G, Rewcastle NB, et al: Microadenomas of the pituitary and abnormal sellar tomograms in an unselected autopsy series. *N Engl J Med* **304**:156, 1981.
124. Gardeur D, Naidich TP, Metzger J: CT analysis of intrasellar pituitary adenomas with emphasis on patterns of contrast enhancement. *Neuroradiology* **20**:241, 1981.
125. Sakoda K, Mukada K, Yonezwa M, et al: CT scan of pituitary adenomas. *Neuroradiology* **20**:249, 1981.
126. Kuuliala I: Computed axial tomography of pituitary adenomas. *Clin Radiol* **32**:259, 1981.
127. Ebersold MJ, Law ER, Scheithauer BW, et al: Pituitary apoplexy treated by transsphenoidal surgery. A clinicopathological and immunocytochemical study. *J Neurosurg* **58**:315, 1983.
128. David NJ, Gargano FP, Glaser JS: Pituitary apoplexy in clinical perspective. *Neurophthalmology* **8**:140, 1975.
129. Drayer BP, Rosenbaum AE, Kennerdell JS, et al: Computed tomographic diagnosis of suprasellar masses by intrathecal enhancement. *Radiology* **123**:339, 1977.
130. Molitch ME: Nonsecretory adenomas. In Post KD, Jackson IMD, Reichlin S (eds): *The Pituitary Adenoma.* Plenum, New York, 1980, p. 151.
131. Cartlidge NEF, Shaw DA: Intrasellar aneurysm with subarachnoid hemorrhage and hypopituitarism. *J Neurosurg* **36**:640, 1972.
132. Lombardi G, Passerini A, Migliavacca F: Intracavernous aneurysms of the internal carotid artery. *AJR* **89**:361, 1963.
133. Pinto RS, Cohen WA, Kricheff II, et al: Giant intracranial aneurysms: Rapid sequential computed tomography, *AJNR* **3**:495, 1982.
134. Fitz-Patrick D, Tolis G, McGarry EE, et al: Pituitary apoplexy. The importance of skull roentgenograms and computerized tomography in diagnosis. *JAMA* **244**:59, 1980.
135. Haughton VM, Rosenbaum AE, Williams AL, et al: Recognizing the empty cella by CT: The infundibulum sign. *AJNR* **1**:527, 1980.
136. Smaltino F, Bernini FP, Muras J: Computed tomography for diagnosis of empty sella associated with enhancing pituitary microadenomas. *J Comput Assist Tomogr* **4**:592, 1980.
137. Naidich TP, Pinto RS, Kushner MJ, et al: Evaluation of sellar and parasellar masses by computed tomography. *Radiology* **120**:91, 1976.
138. Numaguchi Y, Kishikawa T, Ikeda J, et al: Neuroradiological manifestations of suprasellar pituitary adenomas, meningiomas and craniopharyngiomas. *Neuroradiology* **21**:67, 1981.
139. Banna M: Craniopharyngioma: Based on 160 cases. *Br J Radiol* **49**:206, 1976.
140. Fitz CR, Wortzman G, Harwood-Nash DC, et al: Computed tomography in craniopharyngiomas. *Radiology* **127**:687, 1978.

141. Sklar CA, Grumbach MM, Kaplan SL, et al: Hormonal and metabolic abnormalities associated with central nervous system germinoma in children and adolescents and the effect of therapy: Report of 10 patients. *J Clin Endocrinol Metab* **52**:9, 1981.

142. Post KD, Kasdon DL: Sellar and parasellar lesions mimicking adnomas. In Post KD, Jackson IMD, Reichlin S (eds): *The Pituitary Adenoma.* Plenum, New York, 159–210, 1980.

143. Firooznia H, Pinto RS, Lin JP, et al: Chordoma: Radiologic evaluation of 20 cases. *AJR* **127**:797, 1976.

144. Goodman RH, Post KD, Molitch ME, et al: Eosinophilic granuloma mimicking a pituitary tumor. *Neurosurgery* **5**:723, 1979.

145. Frantz AG: Prolactin. *N Engl J Med* **298**:201, 1978.

146. Weintraub BD, Gershengorn MC, Kourides IA: Inappropriate secretion of thyroid-stimulating hormone. *Ann Intern Med* **95**:339, 1981.

147. Kourides IA, Weintraub BD, Ridgway EC, et al: Pituitary secretion of free alpha and beta subunit of human thyrotropin in patients with thyroid disorders. *J Clin Endocrinol Metab* **40**:872, 1975.

148. Snyder PJ, Muzyka R, Johnson J, et al: Thyrotropin-releasing hormone provokes abnormal follicle-stimulating hormone (FSH) and luteinizing hormone responses in men who have pituitary adenomas and FSH hypersecretion. *J Clin Endocrinol Metab* **51**:744, 1980.

149. Snyder PJ, Bashey HM, Phillips JL, et al: Comparison of hormonal secretory behavior of gonadotroph cell adenomas in vivo and in culture. *J Clin Endocrinol Metab* **61**:1061, 1985.

150. Kourides IA, Weintraub BD, Rosen SW, et al: Secretion of alpha subunit of glycoprotein hormones by pituitary adenomas. *J Clin Endocrinol Metab* **43**:97, 1976.

151. MacFarlane IA, Beardwell CG, Shalet SM, et al: Glycoprotein hormone alpha subunit secretion by pituitary adenomas: Influence of external irradiation. *Clin Endocrinol (Oxf)* **13**:215, 1980.

152. Peterson RE, Kourides IA, Horwith M, et al: Luteinizing hormone and α-subunit secreting pituitary tumor: Positive feedback of estrogen. *J Clin Endocrinol Metab* **52**:692, 1981.

153. Ridgway EC, Klibanski A, Ladenson PW, et al: Pure alpha-secreting pituitary adenomas. *N Engl J Med* **304**:1254, 1981.

154. Pierce JG: Eli Lilly Lecture: The subunits of pituitary thyrotropin—their relationship to other glycoprotein hormones. *Endocrinology* **89**:1331, 1971.

155. Kourides IA, Weintraub BD, Ridgway EC, et al: Increase in the beta subunit of human TSH in hypothyroid serum after thyrotropin releasing hormone. *J Clin Endocrinol Metab* **37**:836, 1973.

156. Kourides IA, Weintraub BD, Ridgway EC, et al: Pituitary secretion of free alpha and beta subunit of human thyrotropin in patients with thyroid disorders. *J Clin Endocrinol Metab* **40**:872, 1975.

157. Rosen SW, Weintraub BD: Ectopic production of the isolated alpha subunit of the glycoprotein hormones: A quantitative marker in certain cases of cancer. *N Engl J Med* **290**:1441, 1974.

158. White JC, Ballantine HJ: Intrasellar aneurysms simulating hypophyseal tumors. *J Neurosurg* **18**:34, 1961.

159. Dominque JN, Wilson CB: Pituitary abscesses: Report of seven cases and review of the literature. *J Neurosurg* **46**:601, 1977.

160. Rudwan MA: Pituitary abscess. *Neuroradiology* **12**:243, 1977.

161. Lindholm J, Rasmussen P, Korsgaard B: Intrasellar or pituitary abscess. *J Neurosurg* **38**:616, 1973.

162. Fuchs HA, Evans RM, Gregg CR: Invasive aspergillosis of the sphenoid sinus manifested as a pituitary tumor. *South Med J* **78**:1365, 1985.

163. Haslam RHA, Winternitz WW, Howieson J: Selective hypopituitarism following tuberculous meningitis. *Am J Dis Child* **118**:903, 1969.

164. Sherman BM, Gorden P, di Chiro G: Postmeningitic selective hypopituitarism with suprasellar calcification. *Arch Intern Med* **128**:600, 1971.

165. Leiba S, Landau B, Ber A: Target gland insufficiency and pituitary tumors. *Acta Endocrinol (Copenh)* **60**:112, 1969.

166. Patel YC, Kilpatrick JA: Pituitary enlargement with longstanding myxoedema. *N Z Med J* **70:**21, 1969.

167. Lawrence AM, Wilber JF, Hagen TC: The pituitary and primary hypothyroidism. Enlargement and unusual growth hormone secretory responses. *Arch Intern Med* **132:**327, 1973.

168. Van Wyck JJ, Grumbach MM: Syndrome of precocious menstruation and galactorrhea in juvenile hypothyroidism: An example of hormonal overlap in pituitary feedback. *J Pediatr* **57:**416, 1960.

169. Burt AS, et al: Kleinfelter's syndrome: Report of an autopsy, with particular reference to the histology and histochemistry of the endocrine glands. *J Clin Endocrinol Metab* **14:**719, 1954.

170. Kelley LW Jr: Ovarian dwarfism with pituitary tumor. *J Clin Endocrinol Metab* **23:**50, 1963.

171. Koppelman MCS, Jaffe MJ, Rieth KG, et al: Hyperprolactinemia, amenorrhea, and galactorrhea: A retrospective assessment of twenty-five cases. *Ann Intern Med* **100:**115, 1984.

172. Weiss MH, Teal J, Gott P, et al: Natural history of microprolactinomas: Six-year follow-up. *Neurosurgery* **12:**180, 1983.

173. March CM, Kletzky OA, Davajan V, et al: Longitudinal evaluation of patients with untreated prolactin-secreting pituitary adenomas. *Am J Obstet Gynecol* **139:**835, 1981.

174. Zervas NT: Surgical results for pituitary adenomas: Results of an international survey. In Black PM, Zervas NT, Ridgway EC, et al. (eds): *Secretory Tumors of the Pituitary Gland. Progress in Endocrine Research and Therapy.* Vol. 1. Raven, New York, 1984, p. 377.

175. Ciric I, Mikhael M, Stafford T, et al: Transsphenoidal microsurgery of pituitary macroadenomas with long-term follow-up results. *J Neurosurg* **59:**395, 1983.

176. Nelson AT, Tucker SG Jr, Becker DP: Residual anterior pituitary function following transsphenoidal resection of pituitary macroadenomas. *J Neurosurg* **61:**577, 1984.

177. Arafah BM, Brodkey JS, Manni A, et al: Recovery of pituitary function following surgical removal of large nonfunctioning pituitary adenomas. *Clin Endocrinol (Oxf)* **17:**213, 1982.

178. Wollesen F, Andersen T, Karle A: Size reduction of extrasellar pituitary tumors during bromocriptine treatment. *Ann Intern Med* **96:**281, 1982.

179. McGregor AM, Scanlon MF, Hall R, et al: Effects of bromocriptine on pituitary tumour size. *Br Med J* **2:**700, 1979.

180. Frazier CH, Alpers BJ: Adamantinoma of the craniopharyngeal duct. *Arch Neurol Psychiatry* **26:**905, 1931.

181. Koos WT, Miller MH: *Intracranial Tumors of Infants and Children.* Thieme, Stuttgart, 1971.

182. Banna M, Honre RD, Stanley P, et al: Craniopharyngioma in children. *J Pediatr* **85:**781, 1973.

183. Matson DD, Crigler FJ: Management of craniopharyngioma in childhood. *J Neurosurg* **30:**377, 1969.

184. Petto CK, DeGirolami U, Earle KM: Craniopharyngiomas: A clinical and pathological review. *Cancer* **37:**1944, 1976.

185. Thomsett MJ, Conte FA, Kaplan SL, et al: Endocrine and neurologic outcome in childhood craniopharyngioma: Review of effect of treatment in 42 patients. *J Pediatr* **97:**728, 1980.

186. Jenkins JS, Gilbert CJ, Ang V: Hypothalamic pituitary function in patients with craniopharyngioma. *J Clin Endocrinol Metab* **43:**394, 1976.

187. Maier HC: Craniopharyngioma with erosion and drainage into the nasopharynx: An autobiological case report. *J Neurosurg* **62:**132, 1985.

188. Katz EL: Late results of radical excision of craniopharyngiomas in children. *J Neurosurg* **42:**86, 1975.

189. Sweet WH: Radical surgical treatment of craniopharyngioma. *Clin Neurosurg* **23:**52–79, 1976.

190. Hoffman HJ, Hendrick B, Humphreys RP, et al: Management of craniopharyngioma in children. *J Neurosurg* **47:**218, 1977.

191. Symon L, Sprich W: Radical excision of craniopharyngioma: Results in 20 patients. *J Neurosurg* **62:**174, 1985.

192. Carmel PW, Antunes JL, Chang CH: Craniopharyngiomas in children. *Neurosurgery* **11:**382, 1982.

193. Matson DD, Crigler JF Jr: Management of craniopharyngioma in childhood. *J Neurosurg* **30**:377, 1969.
194. Humphreys RP, Hoffman HJ, Hendrick EB: A long-term postoperative follow-up in craniopharyngioma. *Childs Brain* **5**:530, 1979.
195. Kahn EA, Gosch HH, Seeger JF, et al: Forty-five years experience with the craniopharyngiomas. *Surg Neurol* **1**:5, 1973.
196. Kramer S, McKissock W, Concannon JP: Craniopharyngiomas: Treatment by combined surgery and radiation therapy. *J Neurosurg* **18**:217, 1961.
197. Kramer S, Southard M, Mansfield CM: Radiotherapy in the management of craniopharyngiomas: Further experiences and late results. *AJR* **103**:44, 1968.
198. Bloom HJG: Combined modality therapy for intracranial tumors. *Cancer* **35**:111, 1975.
199. Manaka S, Teramoto A, Takakura K: The efficacy of radiotherapy for craniopharyngioma. *J Neurosurg* **62**:648, 1985.
200. Richmond IL, Wara WM, Wilson CB: Role of radiation therapy in the management of craniopharyngiomas in children. *Neurosurgery* **6**:513, 1980.
201. Sweet WH: Recurrent craniopharyngiomas: Therapeutic alternatives. *Clin Neurosurg* **23**:52, 1976.
202. Danoff BE, Cowchock FS, Kramer S: Childhood craniopharyngioma: Survival, local control, endocrine and neurologic function following radiotherapy. *Int J Radiat Oncol Biol Phys* **9**:171, 1983.
203. Shilito J Jr: The treatment of craniopharyngiomas of childhood. In Morley W (ed): *Current Controversies in Neurosurgery.* WB Saunders, Philadelphia, 1976, p. 332.
204. Takahashi H, Nakazawa S, Shimura T: Evaluation of postoperative intratumoral injection of bleomycin for craniopharyngioma in children. *J Neurosurg* **62**:120, 1985.
205. Walsh JW: Suprasellar germinomas. In Wilkins RH (ed): *Neurosurgery* (in press).
206. Jennings MT, Gelman R, Hochberg F: Intracranial germ-cell tumors: Natural history and pathogenesis. *J Neurosurg* **63**:155, 1985.
207. Sung DI, Harasiadis L, Chang CH: Midline pineal tumors and suprasellar germinomas: Highly curable by irradiation. *Radiology* **128**:745, 1978.
208. Griffin BR, Griffin TW, Tong DYK, et al: Pineal region tumors: Results of radiation therapy and indications for elective spinal irradiation. *Int J Radiat Oncol Biol Phys* **7**:605, 1981.
209. Handa H, Yamashita J: Summary: Current treatment of pineal tumors. *Surg Neurol* **16**:279, 1981.
210. Takeuchi J, Handa H, Nagata I: Suprasellar germinoma. *J Neurosurg* **49**:41, 1978.
211. Sano K, Matsutani M: Pinealoma (germinoma) treated by direct surgery and postoperative irradiation. A long-term follow-up. *Childs Brain* **8**:81, 1981.
212. Jooma R, Kendall BE: Diagnosis and management of pineal tumors. *J Neurosurg* **58**:654, 1983.
213. Logothetis CJ, Samuels ML, Trindade A: The management of brain metastases in germ cell tumors. *Cancer* **49**:12, 1982.
214. Wara WM, Jenkin RDT, Evans A, et al: Tumors of the pineal and suprasellar region. Children's Cancer Study Group Treatment results 1960–1975. *Cancer* **43**:698, 1979.
215. Wolpert SM: The Radiology of Pituitary Adenomas: An Update. In Post KD, Jackson IM, Reichlin S (eds): *The Pituitary Adenoma,* Plenum Press, New York, 1980, pp. 297–320.
216. Post KD, Kasdon DL: Sellar and Parasellar Lesions Mimicking Adenoma. In Post KD, Jackson IM, Reichlin S (eds): *The Pituitary Adenoma,* Plenum Press, New York, 1980, pp. 159–218.

17

Pituitary Apoplexy

Pituitary apoplexy refers to the sudden expansion in the size of a preexisting pituitary tumor, with the consequence of rapidly developing pressure effects. The term is derived from the Greek words, *apo* (meaning "from") and *plessin* ("to strike"). Pituitary apoplexy is usually the result of sudden bleeding into a tumor located in a closed but flexible space, an expanded sella turcica. Although Bleibtreu[1] has been credited with describing the entity in 1905 in German literature, the condition had been recognized as early as 1898, by Dr. Pearce Bailey,[2] an eminent neurologist from New York. The term *pituitary apoplexy* was coined in 1950 by Brougham et al.,[3] who reported five patients who experienced a sudden and fatal expansion of their pituitary tumors. The term *apoplexy* effectively conveyed the dark and sinister side of the clinical spectrum. Before discussing the clinical and diagnostic features of pituitary apoplexy, a few general comments are applicable:

1. Pituitary apoplexy is a clinical concept.[4–6] In the purest sense, it implies not merely bleeding into a tumor, but the development of rapid pressure effects on the sellar, suprasellar, and parasellar structures. The term *pituitary apoplexy* should be applied only when clinical evidence of perisellar compression or meningeal irritation have developed fol-

lowing the bleeding. Unfortunately, reports in the recent literature have somewhat trivialized this concept by fostering the tendency to include all cases of bleeding, even minimal grades, as apoplexy. Review of surgical material has clearly shown that asymptomatic bleeding into pituitary adenomas occurs more frequently than previously supposed, but to the clinician these episodes may not reflect pituitary apoplexy as we know it.

2. While hemorrhage into the anterior pituitary can occur from several causes, compressive phenomena do not occur unless the bleeding occurs in an underlying tumor. For instance, hemorrhage into a non-tumorous adenohypophysis can occur in association with postpartum shock, diabetes, hemolytic crisis, or increased intracranial pressure, but the bleeding in such cases is limited to the sella whose boundaries are intact. Pressure effects on the perisellar structures are therefore not evident. By contrast, when the boundaries of the sella turcica are destroyed by an underlying tumor and the barriers are broken, the blood within the sella can no longer be contained within; the combined mass effect of the blood clot and the tumor together cause the compressive effect on the perisellar structures. When the tumor is large and the bleeding massive, the result is no less than a sellar explosion.

3. Recognition of pituitary apoplexy requires a high index of suspicion, since timely intervention can be either life saving or sight saving, or both. The clinical resemblance of pituitary apoplexy to diverse neurological disorders, ranging from meningitis or subarachnoid hemorrhage to ruptured aneurysm or even a stroke, can result in dangerous misdiagnoses, if the condition if not kept in mind. The problem is further compounded by the fact that the apoplectic episode may be the first clue to the presence of an underlying tumor.

4. It has now become evident that pituitary apoplexy displays a wide spectrum of clinical and radiological features, the recognition of which is crucial for establishing the diagnosis. Thus, the clinician, radiologist and neurosurgeon play a major role in the diagnosis and treatment of apoplexy.

Incidence

The reported incidence of apoplexy occurring in pituitary tumors is highly variable. The problem in the past was the criteria used to define hemorrhage. When surgical material is reviewed for the presence of old or new blood in the tumor specimen, the incidence is obviously higher than the calculated incidence of hemorrhage occurring during the life of a pituitary adenoma. Wakai et al.[7] evaluated the occurrence of hemorrhage in pituitary adenomas by examining 560 consecutive patients operated for pituitary tumors. These investigators demonstrated the presence of hematoma or old bloody fluid within tumor tissue in 16.1% of surgical specimens. They made the interesting observation that nearly one-half (7.5%) of those patients had suffered asymptomatic hemorrhage. Similarly, Mohr and Hardy[8] reviewed the records of 664 patients who underwent surgery for pituitary tumors and found hemorrhagic and/or necrotic degeneration in 63 (9.5%) patients. Symptoms of classic pituitary apoplexy were present only in 4 of the 63 patients. Clearly, there is a major dichotomy between the incidence of demonstrating hemorrhage in pituitary tumors and the incidence of such hemorrhage manifesting itself clinically. It should be noted that, in general, there is a distinct and definable increase in the risk of hemorrhage occurring in all intracranial tumors. Drake and McGee[9] evaluated 236 patients and reported a 7.6% incidence of bleeding in intracranial tumors, a finding supported by others.[10] Pituitary tumors are believed to have a higher tendency for bleeding in comparison with other intracranial tumors. The reported incidence of hemorrhage occurring in pituitary tumors ranges between 1.5 and 27.7%.[7,8,11–13] It is estimated that clinically manifested hemorrhage into the pituitary tumor probable occurs in 5% of patients with pituitary tumors.[4,11]

It is generally believed that apoplexy tends to occur more frequently in patients harboring large tumors, especially those characterized by hypersecretion of anterior pituitary hormones. Analysis of the cell types of pituitary tumors demonstrates that apoplexy can occur in association with all types of pituitary tumors. Several studies have suggested that pituitary apoplexy is seen more frequently in association with tumors causing acromegaly.[1,3,14–20] Lawrence et al.[20] collected and reviewed the development of pituitary hemorrhage in 32 patients harboring growth hormone (GH)-secreting tumors. The incidence of pituitary apoplexy occurring in chromophobe adenomas ranges from 1% to as high as 10%.[21,22] In the study reported by Mohr and Hardy,[8] the most prevalent histological type was prolactin-secreting adenomas. Pituitary apoplexy has also been described in association with adenomas secreting ACTH/β-LPH,[8] particularly in the clinical setting of Nelson's syndrome. In one recent report,[7] the incidence of apoplexy was not significantly different among patients with hypersecretory and nonsecretory tumors. The rarity of pituitary apoplexy in patients with pituitary-dependent Cushing's disease has been reemphasized in this report. Rarely, pituitary apoplexy has been reported to occur in patients with craniopharynagioma.[23]

Pathophysiology

The pathophysiological mechanisms underlying the development of apoplexy have not been clearly elucidated. Review of the pathological material obtained from patients with apoplexy has left considerable doubt regarding the primary versus secondary nature of the bleeding episode. It is uncertain whether the hemorrhagic necrosis is a consequence of a primary bleeding episode within a vascular tumor or whether the bleeding was a secondary response to ischemic necrosis occurring in the tumor. Rovit et al.[24] suggested that the peculiar vascular supply of the pituitary makes the latter more probable. The pituitary gland is supplied arterially by the superior and inferior hypophyseal arteries. Arborization of these vessels with their contralateral counterparts supplies the neural hypophysis and lower infundibular stem. The hypophyseal portal system is responsible for supplying the anterior lobe with vessels derived from the capillary network originating from the superior and inferior hypophyseal arteries. Rovit et al.[24] proposed that when an expanding pituitary tumor extends superiorly, it stretches the diaphragma sella and insinuates itself through the diaphragmatic notch. When the tumor impacts at this critical location, the result may be acute compression of the infundibular stem, a phenomenon that can completely cut off the supply to the entire anterior pituitary. This may result in ischemic necrosis and secondary bleeding into the tumor.

Whereas the ischemic theory has gained popularity, Cardoso and Peterson[25] point out that occlusion of the hypophyseal portal vessels in animals or humans leads to mostly ischemic infarction and not to hemorrhage. It is debatable whether the vascular anatomy of tumors is different, predisposing to a hemorrhagic episode. Also, pathological correlation for the ischemic theory has not been borne out, except in two reports of occlusion of hypophyseal arteries in conjunction with apoplexy.[22,26] The old concept that tumors may also outgrow their arterial blood supply and develop ischemia has been challenged by the fact that apoplexy does develop in small and medium tumors as well.

The lack of data to support a predominantly ischemic mechanism to account for the pathophysiology of apoplexy has led to the belief that perhaps local factors within the sella play a permissive role. The association between pituitary adenomas and coincidental aneurysm has been reported to be as high as 7.4%.[27] This fact, coupled with the well-known tendency of pituitary tumors to bleed, may be responsible for an inherent local bleeding diathesis within the pituitary tumor.

Unclear as the predisposing factors are, several precipitating factors have been recognized (Table 90). Although none has been shown to be a consistent

---

TABLE 90
Factors That May Precipitate Apoplexy

Radiation
Trauma (head injury)
Drug therapy
   Estrogen
   Anticoagulants
   Bromocriptine
Postprocedural
   Angiography
   Pneumoencephalography
   Myelography
Post-tussive (severe coughing or sneezing)
Diabetes mellitus/diabetic ketoacidosis

offender, the frequency of these factors in association with the hemorrhage has impressed neurosurgeons. Of these factors, radiation therapy has been the most controversial. Several observers have noted that radiation therapy to pituitary tumors may be followed, within hours or days, by the development of apoplexy.[11,28–30] The development of increased vascularity following radiation therapy may enhance the susceptibility to apoplexy. However, the rarity with which apoplexy follows radiation, a commonly employed procedure, casts some doubt on a causal relationship between the two. The same problem is encountered with evaluating the role of drugs such as estrogens,[25] bromocriptine,[7] and anticoagulants,[8,31,32] although the case for anticoagulant therapy is slightly stronger than for the other two drugs. The association of apoplexy with other precipitating factors such as trauma,[33] after procedures,[11] and severe cough,[34] is reasonably accepted. The relationship between apoplexy and diabetes is an intriguing one.[15,30,35] The role of diabetic vascular disease in predisposing the patient with pituitary adenoma to an apoplectic episode is unclear.

Clinical Features

The clinical spectrum of pituitary apoplexy is highly variable, ranging from the classic fulminant catastrophe to milder versions of an attack. The full-blown expression is described first, because it is this variety that is fatal if untreated, making recognition crucial.[25]

The clinical expression of the classic pituitary apoplexy is characterized by five features:

1. Dramatic onset with excruciating headache, often with vomiting
2. Rapid development of visual impairment (e.g., decreased acuity, restriction of fields)
3. Ophthalmoplegia, characterized by palsies of the III, IV, VI, and even the V cranial nerves
4. Signs of meningeal irritation
5. Alteration in sensorium

When any of the above occurs in a patient known to harbor a pituitary adenoma, the diagnosis is plain as day. However, it has been repeatedly emphasized in the literature that in more than one-half of patients who suffer an apoplectic episode, the hemorrhagic event was the first clue pointing to a pituitary tumor.

The clinical features of pituitary apoplexy can be discussed in terms of the four major effects that occur as a consequence of the mass effect: features due to the intrasellar process, features secondary to suprasellar pressure, the consequences of lateral extension of the expanding mass, and finally the endocrine disturbances associated with apoplexy.

Intrasellar Process

The most immediate reflection of bleeding into the tumor is the development of headache, which is excruciating. The onset is often described as lightning like. The location of the headache can be retroorbital, frontal, or frontotemporal. Although headaches are quite frequently experienced by patients harboring pituitary tumors, the character, intensity, and progression of the pain is distinctly different when apoplexy develops. Pressure on the diaphragma sella with stretching of that membrane is thought to be the reason for the headache. Headache is present in virtually all patients with pituitary apoplexy.

Sudden Upward Enlargement of Mass

Suprasellar compression is the most frequent sequel of pituitary apoplexy and results from the suddenly swollen tumor mass herniating superiorly through the aperture in the diaphragma sella. Depending on the severity of the hemorrhage, the size of the bleeding tumor mass, and the size of the aperture in the diaphragma, the suprasellar pressure effects can be mild, moderate, or severe. The mildest form of suprasellar compression is the development of visual-field cuts. With more compression, visual acuity decreases, with progressive development of bitemporal hemianopsia, homonymous hemianopsia, and eventually total blindness. Early optic nerve involvement (afferent) can be detected by the presence of the Marcus Gunn pupil in these patients. Visual disturbances are present in an overwhelming majority of patients who suffer pituitary apoplexy, but there have been reports of apoplexy occurring in the absence of visual symptoms.[7,36–39] Visual impairment is generally associated with impaired consciousness.

In addition to causing compression of the optic nerve, optic chiasm, and optic tracts, sudden suprasellar enlargement can compress four other structures within the brain, the hypothalamus, the diencephalon, the brain stem, and the middle cerebral artery, each of which deserves brief mention. Hypothalamic compression may result in the development of serious physiological and hormonal sequelae, such as disturbances in temperature regulation, cardiac and respiratory disturbances, hypotension, and clouding of consciousness. Hormonally, central diabetes insipidus may develop as a consequence of loss of vasopressin synthesized by the supraoptic and paraventricular nuclei.[31,40,41] Hypothalamic compression can result in loss of releasing factors, which can lead to the development of hypothalamic hypopituitarism. Lawrence et al.[20] described two patients with pituitary apoplexy in whom hypothalamic deficiencies of releasing factors developed, leading to hypopituitarism.

Diencephalic compression can be seen even with moderate degrees of suprasellar compression. Alteration in consciousness is seen in 30–60% of patients with apoplexy. In severe cases, compression of the brain stem may result in stupor, coma, labored respiration, dilated pupils, and even death.

With severe upward extension, even the middle cerebral artery may be compressed, resulting in an ischemic stroke.[42]

Lateral Extension of Mass

The main targets for lateral compression are the cavernous sinuses on one or both sides. The brunt of the pressure effects fall on the oculomotor nerves, resulting in varying degrees of external ophthalmoplegia. Involvement of the third and sixth cranial nerves is the most frequent phenomenon, resulting in diplopia, ptosis, and pupillary abnormalities.[29,32,36,38] The presence of oculomotor nerve palsies in a patient with an enlarged sella should clearly point to apoplexy, owing to the rarity of cranial nerve involvement in patients with uncomplicated pituitary adenomas. The oculomotor nerve involvement is the most reversible of the pressure effects following decompression. Extreme lateral extension can result in compression of the internal carotid artery, a rare complication. Infrequently, the central retinal vein may be occluded, resulting in the development of ipsilateral proliferative retinopathy at a later date.[43] This may contribute to, or compound, the preexisting visual compromise. Table 91 outlines the multifaceted effects of pituitary apoplexy.

In addition to pressure-related effects, blood can extravasate into the ventricular system, resulting in meningeal irritation. As a result, the clinical picture can resemble aseptic meningitis, subarachnoid hemorrhage, or even purulent meningitis—when the necrotic material irritates the meningeal system.

TABLE 91
Pituitary Apoplexy: Clinical Manifestations

Pressure effects	Clinical expression
Intrasellar	Headaches
Suprasellar	
Optic nerves, chiasm, and tracts	Field cuts (bitemporal and homonymous hemianopsia); decreased acuity; Marcus Gunn pupil
Hypothalamus	Diabetes insipidus; hypothalamic hypopituitarism; cardiorespiratory and temperature dysregulation; altered sensorium
Diencephalon	Altered sensorium
Middle cerebral artery	Ischemic stroke
Parasellar	
Cavernous sinus compression	Cavernous sinus syndrome (proptosis)
Oculomotor compression	III, IV, V, VI cranial nerve palsy; external ophthalmoplegia
Internal carotid artery	Ischemic (hemispherical) lesion
Infrasellar	Epistaxis; CSF rhinorrhea
Meningeal irritation	Meningitic picture

Endocrine Effects

The degree of endocrine damage that can be associated with pituitary apoplexy is variable; it depends on four factors: massiveness of the hemorrhage, preexisting hormonal reserve of the adenohypophyseal hormones, degree of stalk involvement, and presence of coexistent hypothalamic damage. The classic concept that apoplexy wipes out all pituitary function has to be reviewed in light of data that support the development of partial hypopituitarism, as well as deficiencies of hypothalamic releasing factors. The two most frequent endocrinopathies that result immediately after an apoplectic episode are secondary adrenal failure caused by acute ACTH deficiency and diabetes insipidus due to vasopressin deficiency, the latter more often being transient.[44] Even when the ACTH deficiency is not clinically overt, its occurrence must be presumed and treated in patients with apoplexy. The transient nature of the diabetes insipidus may reflect neurohypophyseal sparing, since its blood supply is derived differently than the adenohypophysis. Less frequently, apoplexy may be associated with the syndrome of inappropriate secretion of ADH (SIADH).[36,45]

Recent literature,[7,8,23] based on examination of surgical specimens obtained from patients harboring pituitary tumors, has clearly disclosed the presence of subclinical and even asymptomatic expressions of pituitary hemorrhage. The incidence of asymptomatic hemorrhage appears to be greatest in the third decade.[7] The symptoms of minor attacks consists of moderate headache, nausea, vomiting, dizziness, or vertigo. It is debatable whether such incidences should be referred to as true apoplexy.

Differential Diagnosis

Pituitary apoplexy needs to be differentiated from ruptured aneurysm, meningitis, and abscess of the pituitary gland. Aneurysmal rupture can resemble apoplexy since headache, ocular signs and altered sensorium are common to both disorders. This distinction between these two conditions is extremely critical, but can be well nigh impossible on clinical grounds alone. As early as 1877, Bramwell[46] described an internal carotid artery aneurysm protruding into the sella and resulting in rupture with subarachnoid hemorrhage. The presence of intermittent ophthalmoplegia and severe intermittent headaches can be associated with aneurysms. Furthermore, in reviewing the literature, White and Ballantine[47] suggest that 2% of all intracranial aneurysms involve the pituitary gland. Aneurysms of the internal carotid artery that are embedded in the pituitary fossa are usually atherosclerotic, occurring in older persons, although congenital aneurysms in younger patients can also be involved; one clue that may help in suspecting an aneurysm, on the basis of the skull roentgenogram, is the fact that aneurysms usually arise from the infraclinoidal portion of the internal carotid artery, hence tend to erode the sella in an asymmetrical fashion.[48] Arteriography is the only method for excluding an aneurysm.

TABLE 92
When to Think "Apoplexy"

Sudden, dramatic symptomatology in a patient known to harbor
  a pituitary tumor
Triad of severe headache, altered sensorium and field cuts in
  any patient
Rapid evolution of hypopituitarism
In the differential diagnosis of
  Meningitis
  Cavernous syndrome
  Subarachnoid hemorrhage

Meningitis is an important differential diagnosis of apoplexy. The combination of headache, neck rigidity, cranial nerve palsies, and altered sensorium coupled with the fever seen in some cases of pituitary apoplexy can be easily mistaken for aseptic or purulent meningitis. The meningeal reaction caused by the extravasation of blood or necrotic tissue can be further compounded by superimposed bacterial infection. The spinal fluid in pituitary apoplexy may reveal pleocytosis, further simulating meningitis.

Rarely, a pituitary abscess[49,50] can mimic apoplexy. An abscess within the pituitary is rare, and may occur in conjunction with an overt focus of infection (meningitis, infected pituitary tumor or cyst) or in the absence of an obvious source of infection. The clinical syndrome caused by a pituitary abscess is also characterized by a rapidly expanding sellar mass: fever, visual defects, headache, and even cranial nerve palsies. The offending pathogens are usually anerobic, a fact that may have contributed to the failure to recover the organism from cerebrospinal fluid (CSF) in the early reports.

Table 92 outlines the clinical settings that should alert the physician to suspect apoplexy. If the question "Could this be due to pituitary apoplexy?" is asked of every patient with altered sensorium and neurological deficit, the diagnosis would be less often missed.

Diagnostic Studies

Two diagnostic studies invaluable in establishing the diagnosis of pituitary apoplexy are plain radiographs of the skull and computed tomography (CT). Since pituitary adenomas underlie apoplexy in nearly all instances, the combination of plain skull roentgenograms and CT will be abnormal in virtually all cases of apoplexy.

The abnormalities seen in the skull roentgenograms are those caused by space-occupying lesions within the sella turcica. Thus, an enlarged sella is seen in approximately 90–95% of patients. Other abnormalities, such as erosion, double floor, and destruction of the floor or walls of the sella, may be associated with the enlarged sella turcica. Rarely, the appearance of the sella turcica can be completely normal.[29]

The classic appearance of pituitary apoplexy on CT is the ringlike peripheral enhancement in the postcontrast tomogram. The central core area often represents the necrotic center. Recent hemorrhage appears as a round, high-density lesion, without much enhancement.[11,51,52] The mixed density tumor can be recognized in most cases. It has been emphasized in recent literature that accurate interpretation of the CT scan for the diagnosis of apoplexy requires knowledge of the time course of the hemorrhage. Serial CT scans may sometimes be more revealing than a single tomographic study done at the onset of illness. The presence of a mixed lesion, i.e., high-density fluid level within a homogeneous density, is a highly suggestive sign of apoplexy.[11]

Prognosis

Pituitary apoplexy, in its fully evolved form, carries an extremely high (70–80%) mortality rate.[3,30,53] When the bleeding is not massive, varying degrees of spontaneous resolution can occur.[7,23,34,37] The reversal of sequelae depends on their severity. The visual impairment caused by apoplexy often tends to be permanent. Oculomotor nerve paralyses tend to recover rather well. Varying degrees of endocrine deficiency, ranging from compromised reserve detectable only by provocative testing to overt panhypopituitarism, can occur as a result of pituitary apoplexy. While spontaneous cures of hypersecretory syndromes have been reported following apoplexy,[15,18,54] recurrence of the hypersecretory state following apoplexy has also been reported.[16]

Treatment

The prevailing consensus among neurosurgeons is that all patients with pituitary apoplexy need to be "decompressed," preferably by the transsphenoidal route.[6,25,37] It has been pointed out, however, that spontaneous recovery does occur in at least some of these patients[23,41,55]; therefore, a role for conservative therapy in carefully selected patients has been suggested.

Indeed, recovery from apoplexy with conservative treatment has been documented in several reports.[21,29,34,54,56] Conservative treatment implies maintenance of fluid and electrolyte status, administration of steroids, and close observation of the patients. The development of progressive visual deterioration, neurological signs, or impairment in consciousness would be reasons for intervention. Several factors argue against the routine advocation of conservative treatment for apoplexy. First, there are no predictive parameters that permit selection of patients. The unpredictability of the outcome in patients with apoplexy is well recognized. Delay in decompression can result in unnecessary loss of vision, especially in patients fortunate enough to have had a timely diagnosis. Second, the morbidity of transsphenoidal surgery is so low, and its efficacy in relieving the pressure effects so satisfactory, that delaying the procedure is not cost effective. This would entail a certain degree of overutilization of indications for such surgery, but for the most part the successful outcome argues in favor of surgery. Third, the surgical procedure also permits treatment of the underlying disorder, the pituitary tumor. Finally, the advocation of conservative treatment for a disorder as deadly as apoplexy minimizes its impact, after having established the concept that the situation is analogous to a "walking time bomb."

Timely transsphenoidal surgery has greatly influenced the mortality rate of apoplexy, a condition once thought to be uniformly fatal. In a review of the operative mortality in a total of 105 patients from 35 reports published in the literature since 1970, Cardoso and Peterson[25] found the mortality rate to be only 6.7%. The procedure should be considered on an emergency basis in every patient with the fully evolved clinical syndrome. Preoperative CT is essential, not only for the diagnosis of apoplexy, but also to decide the feasibility of operation by the transsphenoidal route. Tumors with large suprasellar extension or patients with nonaerated sphenoid sinuses are ideally suited for the transcranial route of decompression. Transsphenoidal surgical decompression is particularly easy in patients with apoplexy, in whom the contents can be literally suctioned off.

Another neurosurgical procedure successfully employed in the treatment of apoplexy is stereotactic transsphenoidal needle aspiration. Since the original description of the procedure by Berti et al.,[57] several neurosurgeons have successfully treated patients with apoplexy by sterotactic aspiration.[58–60]

Regardless of the procedure used to decompress the sellar pressure, the patient with apoplexy should receive glucocorticoid coverage and close attention to fluid and electrolyte problems, especially those related to vasopressin deficiency or excess. Improvement in the medical and endocrine management coupled with timely decompression has greatly contributed to the improved mortality rate in patients with pituitary apoplexy.

References

1. Bleibtreu L: Ein fall von Akromegalic (Zerstorung der Hypophysis durch Blutung). *MMW* **52:**2079, 1905.
2. Bailey P: Pathological report of a case of acromegaly with especial reference to the lesions in the hypophysis cerebri and in the thyroid gland; and of a case of hemorrhage into the pituitary. *Philadelphia Med J* **1:**789, 1898.
3. Brougham M, Heusner AP, Adams RD: Acute degenerative changes in adenomas of the pituitary body—with special reference to pituitary apoplexy. *J Neurosurg* **7:**421, 1950.
4. Conomy JP, Ferguson JH, Brodkey JS, et al: Spontaneous infarction in pituitary tumors: Neurologic and therapeutic aspects. *Neurology (NY)* **25:**580, 1975.
5. David NJ, Gargano FP, Glaser JS: Pituitary apoplexy in clinical perspective. *Neuroophthalmology* **8:**140, 1975.
6. Epstein S, Pimstone BL, De Villiers JC, et al: Pituitary apoplexy in five patients with pituitary tumours. *Br Med J* **2:**267, 1971.
7. Wakai S, Fukushima T, Teramoto A, et al: Pituitary apoplexy: Its incidence and clinical significance. *J Neurosurg* **55:**187, 1981.

8. Mohr G, Hardy J: Hemorrhage, necrosis, and apoplexy in pituitary adenomas. *Surg Neurol* **18**:181, 1982.

9. Drake CG, McGee D: Apoplexy associated with brain tumors. *Can Med Assoc J* **84**:303, 1961.

10. Glass B, Abbott KH: Subarachnoid hemorrhage consequent to intracranial tumors. *Arch Neurol Psychiatry* **73**:369, 1955.

11. Weisberg LA: Pituitary apoplexy. Association of degenerative change in pituitary adenoma with radiotherapy and detection by cerebral computed tomography. *Am J Med* **63**:109, 1977.

12. Mohanty S, Tandon PN, Banerji AK, et al: Haemorrhage into pituitary adenomas. *J Neurol Neurosurg Psychiatry* **40**:987, 1977.

13. Poppen JL: Changing concepts in the treatment of pituitary adenomas. *Bull NY Acad Med* **39**:21, 1963.

14. Jacobi JD, Fishman LM, Daroff RB: Pituitary apoplexy in acromegaly followed by partial pituitary insufficiency. *Arch Intern Med* **134**:559, 1974.

15. Taylor AL, Finster JL, Raskin P, et al: Pituitary apoplexy in acromegaly. *J Clin Endocrinol Metab* **28**:1784, 1968.

16. Werner PL, Shah JH, Kukreja SC, et al: Recurrence of acromegaly after pituitary apoplexy. *JAMA* **247**:2816, 1982.

17. Rigolosi RS, Schwartz E, Glick SM: Occurrence of growth-hormone deficiency in acromegaly as a result of pituitary apoplexy. *N Engl J Med* **279**:362, 1968.

18. Dunn PJ, Donald RA, Espiner EA: Regression of acromegaly following pituitary apoplexy. *Aust NZ J Med* **5**:369, 1975.

19. Pieper WJ, Ryan RJ: Pituitary apoplexy in a patient with acromegaly. *Ann Intern Med* **55**:478, 1961.

20. Lawrence AM: Hypothalamic hypopituitarism after pituitary apoplexy in acromegaly. *Arch Intern Med* **137**:1134, 1977.

21. List CF, Williams JR, Balyeat GW: Vascular lesions in pituitary adenomas. *J Neurosurg* **9**:177, 1952.

22. Locke S, Tyler HR: Pituitary apoplexy. Report of two cases, with pathological verification. *Am J Med* **30**:643, 1961.

23. Lloyd MH, Belchetz PE: The clinical features and management of pituitary apoplexy. *Postgrad Med J* **53**:82, 1977.

24. Rovit RL, Fein JM: Pituitary apoplexy: A review and reappraisal. *J Neurosurg* **37**:280, 1972.

25. Cardoso ER, Peterson EW: Pituitary apoplexy: A review. *Neurosurgery* **14**:363, 1984.

26. Sussman EB, Porro RS: Pituitary apoplexy: The role of atheromatous emboli. *Stroke* **5**:318, 1974.

27. Wakai S, Fukushima T, Furihata T, et al: Association of cerebral aneurysm with pituitary adenoma. *Surg Neurol* **12**:503, 1979.

28. Sosman MC: The roentgen therapy of pituitary adenomas. *JAMA* **113**:1282, 1939.

29. Wright RL, Ojemann RG, Drew JH: Hemorrhage into pituitary adenomata. *Arch Neurol* **12**:326, 1965.

30. Uihilein A, Balfour WM, Donovan PT: Acute hemorrhage into pituitary adenomas. *J Neurosurg* **14**:140, 1957.

31. Nourizadeh AR, Pitts FW: Hemorrhage into pituitary adenoma during anticoagulant therapy. *JAMA* **193**:623, 1965.

32. Poisson M, Van Effentere R, Marshaly R: Pituitary apoplexy with retraction nystagmus. *Ann Neurol* **7**:286, 1980.

33. Van Wagenen WP: Haemorrhage into a pituitary tumor following trauma. *Ann Surg* **95**:625, 1932.

34. Dawson BH, Kothandaram P: Acute massive infarction of pituitary adenomas. A study of five patients. *J Neurosurg* **37**:275, 1972.

35. Gurling KJ: Diabetic coma and pituitary necrosis in an acromegalic patients: A case report. *Diabetes* **4**:138, 1955.

36. Conomy JP, Ferguson JH, Brodkey JS, et al: Spontaneous infarction in pituitary tumors: Neurologic and therapeutic aspects. *Neurology (NY)* **25**:580, 1975.

37. Ebersold MJ, Laws ER Jr, Scheithauer BW, et al: Pituitary apoplexy treated by transsphenoidal surgery: A clinicopathological and immunocytochemical study. *J Neurosurg* **58**:315, 1983.

38. Robinson JL: Sudden blindness with pituitary tumors: Report of three cases. *J Neurosurg* **36**:83, 1972.

39. Symon L, Monhanty S: Haemorrhage in pituitary tumours. *Acta Neurochir (Wien)* **65**:41, 1982.

40. Gebel P: Pituitary apoplexy: Review of literature with a report of an unusual case associated with diabetes insipidus. *Milit Med* **127**:753, 1962.

41. Pelkonen R, Kuusisto A, Salmi J, et al: Pituitary function after pituitary apoplexy. *Am J Med* **65**:773, 1978.

42. Dastur HM, Pandya SK: Haemorrhagic adenomas of the pituitary gland: Their clinical and radiological presentations and treatment. *Neurosurg India* **19**:4, 1971.

43. Petersen P, Lindholm J: Pituitary apoplexy, the Houssay phenomenon, and accelerated proliferative retinopathy. *Am J Med* **79**:385, 1985.

44. Veldhuis JD, Hammond JM: Endocrine function after spontaneous infarction of the human pituitary: Report, review, and reappraisal. *Endocr Rev* **1**:100, 1980.

45. Cooperman D, Malarkey WB: Pituitary apoplexy. *Heart Lung* **7**:450, 1978.

46. Bramwell B: Two enormous intra-cranial aneurysms. *Edinb Med J* **32**:911, 1887.

47. White JC, Ballantine HJ: Intrasellar aneurysms simulating hypophyseal tumors. *J Neurosurg* **18**:34, 1961.

48. Cryer PE, Kissane JM: Headaches, visual loss and an enlarged sella turcica. *Am J Med* **67**:665, 1979.

49. Dominque JN, Wilson CB: Pituitary abscesses: Report of seven cases and review of the literature. *J Neurosurg* **46**:601, 1977.

50. Lindholm J, Rasmussen P, Korsgaard B: Intrasellar or pituitary abscess. *J Neurosurg* **38**:616, 1973.

51. Post MJD, David NJ, Glaser JS, et al: Pituitary apoplexy: Diagnosis by computed tomography. *Radiology* **134**:665, 1980.

52. Fitz-Patrick D, Tolis G, McGarry EE, et al: Pituitary apoplexy: the importance of skull roentgenograms and computerized tomography in diagnosis. *JAMA* **244**:59, 1980.

53. Kalyanaraman UP: Clinically asymptomatic pituitary adenoma manifesting as pituitary apoplexy and fatal third-ventricular hemorrhage. *Hum Pathol* **13**:1141, 1982.

54. McLaren EH, Keet PC: Regression of the Forbes-Albright syndrome after pituitary apoplexy. *Br Med J* **2**:314, 1973.

55. Sachdev Y, Evered DC, Hall R: Spontaneous pituitary necrosis. *Br Med J* **1**:942, 1976.

56. Fountain EM, Baird WC, Poppen JL: Pituitary apoplexy: A report of three cases with recovery. *Lahey Clin Bull* **7**:117, 1951.

57. Berti G, Heisey WG, Dohn DF: Pituitary apoplexy treated by stereotactic transsphenoidal aspiration. *Cleve Clin Q* **41**:163, 1974.

58. Zervas NT, Mendelson G: Treatment of acute haemorrhage of pituitary tumours. *Lancet* **1**:604, 1975.

59. Goodman JM, Gilson M, Shapiro B: Pituitary apoplexy—a cause of sudden blindness. *J Indiana State Med Assoc* **66**:320, 1973.

60. Kosary IZ, Braham J, Tadmor R, et al: Trans-sphenoidal surgical approach in pituitary apoplexy. *Neurochirurgia (Stuttg)* **19**:55, 1976.

18

The Empty Sella Syndrome

Introduction

The term *empty sella syndrome* denotes the presence of an enlarged sella containing a remodeled pituitary gland flattened against its posterior and inferior walls. The designation "empty" is a misnomer, since the sella is filled partially by the compressed pituitary gland and partially by cerebrospinal fluid (CSF). The condition results from extension or herniation of the arachnoid membrane into the pituitary fossa, due to an inherent weakness of the diaphragma sella. This weakness, probably congenital, permits herniation of a small arachnoid diverticulum filled with CSF into the fossa. As the diverticulum gradually enlarges, it occupies more volume within the sella, pushing the pituitary to a corner and thus giving the illusion of an empty sella turcica. Since the early description of this entity in the 1960s,[1] the empty sella syndrome has been recognized with increasing frequency. The prevalence rate of empty sella syndrome in adults is estimated at approximately 6%.[2] Minor degrees of empty sella syndrome may not cause sufficient distortion or enlargement of the sella to be detected radiologically. Thus, the reported prevalence rate of radiologically evident empty sella syndrome underestimates the true incidence of this entity. Zull and Falko[3] estimated that the empty sella syndrome accounts for 35% of radiologically demonstrable sellar enlargements. Thus, the empty sella syndrome runs a close second to pituitary tumors as a cause for sellar enlargement. The importance of recognizing empty sella syndrome resides in its superficial resemblance to pituitary tumors. In some cases, the roentgenographic and hormonal manifestations of the empty sella syndrome can be almost identical to those of pituitary tumors.

Pathogenesis

The term *primary empty sella syndrome* refers to the de novo development of the syndrome as a consequence of congenital weakness in the diaphragma sella. By contrast, secondary empty sella syndrome results following surgery, radiation, or vascular catastrophes involving the pituitary gland. The most common reason for the development of secondary empty sella syndrome is operative removal of the sellar contents (sellar cleanout). Less commonly, the secondary empty sella syndrome develops after conventional radiation to the pituitary gland or yittrium implantation. Rarely, an empty sella syndrome can be the consequence of spontaneous hemorrhage within a tumorous pituitary[4,5] (pituitary apoplexy) or postpartum necrosis[6] (Sheehan's syndrome). The secondary nature of the empty sella syndrome in these cases is readily apparent by history. The etiology and pathogenesis of the primary empty sella syndrome is more complex. The basic defect that permits the development of primary empty sella syndrome is the anatomical weakness within the diaphragma sella. The diaphragma sella is the portion of the dura mater that forms the roof of the sella turcica. Extending between the clinoidal attachments of the tentorium cerebelli and the tuberculum sellae, the diaphragma contains a small central fenestra for the passage of the pituitary stalk (infundibular opening). The competency of the normal diaphragma is so good that it does not permit the arachnoid mater to enter the fossa.[7] Studies performed on autopsy material have shown, however, that in as many as 20% of normal subjects, the central opening in the diaphragma sella is incompetent, permitting herniation of the arachnoid mater into the sella.[8]

While it is believed that the prerequisite for the development of the primary empty sella syndrome is an anatomic defect in the diaphragma permitting arachnoid herniation, other factors must be necessary for perpetuation of the process. One such factor that may play a role in the formation of the primary empty sella syndrome is increased CSF pressure. Since the arachnoid diverticulum can transmit pressure changes from the ventricular system to the pituitary fossa, it is possible that increased CSF pressure may be conducive to the progressive enlargement of the diverticulum. Indeed, an increased association has been reported between the development of primary empty sella syndrome and conditions characterized by increased CSF pressure. Thus, hydrocephalics,[9–12] pseudotumor cerebri,[13] intracranial tumors,[14,15] and, to a lesser extent congestive cardiac failure and the pickwickian hypoventilation syndrome, have been reportedly associated with primary empty sella syndrome.

The stages in the development of an empty sella syndrome have been

TABLE 93
Anatomical Consequences of a Defect in Diaphragma Sella

Extension of arachnoid diverticulum through the diaphragm
Compression of the pituitary gland
Enlargement of the sellar walls
Thinning of sellar walls
Erosion of the sellar floor
Communication with sphenoid sinus
Prolapse of optic tracts, optic nerves, or chiasm

---

outlined by Randall.[16] The first step is the herniation of arachnoid membrane into the pituitary fossa, resulting in an arachnoid diverticulum filled with CSF. The gradual expansion of the diverticulum results in formation of a cystlike structure within the sella, resulting in compression of the pituitary gland. The compressed gland is pushed toward the inferoposterior corner of the sella, where it is flattened against the wall. Meanwhile, the sellar walls undergo thinning and expansion to accommodate the enlarging diverticulum. Randall[16] pointed out that while the remodeling of the sella is usually symmetrical, occasionally it can be impressively asymmetrical. As the condition progresses, in extreme cases the floor of the sella gives in, eroding under the pressure, and resulting in communications with the sphenoid sinus. In longstanding severe cases, the negative pressure created within the sella can lead to prolapse of the optic chiasm and optic nerve into the sella turcica. Thus, what started out as a small arachnoid diverticulum can gradually culminate in disastrous consequences. These anatomical consequences are outlined in Table 93.

Primary empty sella syndrome is predominantly seen in females, who tend to be obese, multiparous, and often hypertensive.[17] The reason for the striking female preponderance in this entity is not well understood. It is unclear whether the effects of pregnancy on the pituitary gland play any role in the development of primary empty sella syndrome.[18–20] Although the peak incidence of detecting primary empty sella syndrome is in the fourth decade, the condition has been described in both children[21–24] and adolescents.[25,26]

Clinical Features

Judging from the frequency of this entity in autopsy studies, primary empty sella syndrome remains asymptomatic in most persons. The most frequent mode of detection is by serendipity, when skull radiographs are obtained for unrelated purposes (e.g., headache, following head trauma). It is interesting to note that approximately 10% of patients with empty sella syndrome complain of headaches that tend to be nonspecific without any characteristic localization. Visual-field symptoms are rare in primary empty sella syndrome[18,27] and, when encountered, are usually associated with increased intracranial pressure. Visual-field disturbances, however, have been reported in the absence of raised intracranial pressure.[28-30] Bursztyn et al.[31] documented a case of empty sella syndrome with herniation of the optic chiasm. When visual-field cuts and headaches complicate primary empty sella syndrome, the similarity to pituitary tumors is heightened. Rarely, CSF rhinorrhea may be seen in association with empty sella syndrome. This phenomenon places the patient at risk of developing retrograde meningitis.

Endocrine Function

Pituitary function testing in patients with primary empty sella syndrome reveals four categories of patients: (1) those with entirely normal pituitary function; (2) those with subtle perturbations in pituitary hormone reserve, but without clinical consequences; (3) those with associated syndromes of trophic hormone deficiency; and (4) those with associated syndromes of trophic hormone excess. Most patients with primary empty sella syndrome have normal pituitary function.[18,32] Snyder et al.,[33] Ridgway et al.,[34] and Faglia et al.[35] demonstrated normal thyroid-stimulating hormone (TSH) and prolactin responses to the administration of thyrotropin-releasing-hormone (TRH) in patients with primary empty sella syndrome. The GH and cortisol responses to hypoglycemia are generally well preserved, as is the response of leuteinizing hormone (LH) and follicle-stimulating hormone (FSH) to administration of leuteinizing hormone-releasing hormone (LRH).[35] Occasionally, however, blunted GH or cortisol responses, or both, may be encountered in patients with primary empty sella syndrome, but these are of little or no clinical consequence. Since the overwhelming majority of patients with primary empty sella syndrome demonstrate preservation of pituitary function, while the overwhelming majority of patients with pituitary tumors do not,[33,36] determining pituitary reserve serves as a good procedure for the differentiation. Thus, for all practical and clinical purposes, it can be assumed that the functional integrity of the pituitary gland is preserved in patients with primary empty sella syndrome. In fact, histological and immunocytological studies of the diminutive pituitary gland of empty sella syndrome demonstrate remarkable normalcy. Bergeron et al.[37] demonstrated the presence of adequate numbers of all six types of pituicytes in the pituitary glands of patients with empty sella syndrome. The morphological normalcy of the compressed, and perhaps architecturally distorted, pituitary has also been confirmed by Gharib et al.,[38] who performed histological examination of pituitary tissue obtained by transsphenoidal approach in eight patients with primary empty sella syndrome.

When clinically overt trophic hormone deficiencies are seen in association with empty sella syndrome, the possibility of pituitary infarction, spontaneous or otherwise, needs to be considered. Such an event can be associated with pituitary tumors or with Sheehan's syndrome. Fleckman et al.[6] described 11 patients with Sheehan's syndrome, with varying degrees of hypopituitarism and a partially empty sella as evaluated by CT. They postulated that postpartum pituitary necrosis may be followed by the development of an empty sella of normal size. The notion that at least some cases of empty sella syndrome are derived by necrosis of pituitary tumors has been proposed.[39] The development of primary empty sella in patients with pituitary tumors was studied by Lindholm et al.[40] These workers evaluated the sella turcica in 44 patients with documented pituitary tumors and showed that in 20 of these patients, the tumor had partially "vanished," leaving behind a partially empty sella as seen by CT. Approximately 50% of these patients had in fact experienced symptoms of apoplexy. There were no differences in the degree or frequency of trophic hormone deficiencies in both groups. This report supports the concept that pituitary apoplexy may be followed by the development of empty sella syndrome.[41] The frequency of such a phenomenon following apoplexy is unknown. The demonstration of multiple hormone deficiencies in association with empty sella syndrome should suggest an underlying vascular etiology in the development of the empty sella. In the strict sense of the term, these cases represent secondary empty sella. Rarely, clinically significant unitrophic or selective trophic hormone deficiency may be associated with primary empty sella syndrome. The occurrence of selective ACTH deficiency in association with primary empty sella syndrome has been amply documented in the literature.[42,43] Rarely, primary empty sella syndrome has been reported in association with primary amenorrhea. Farber et al.[25] describe the association of primary empty sella syndrome in an 18-year-old phenotypic female with short stature, primary amenorrhea, and deficient secretion of GH and gonadotropins. The causal relationship between primary empty sella and hypogonadotropism, however, has not been established.

The fourth type of endocrine profile that may be encountered in patients with primary empty sella syndrome is characterized by hypersecretion of anterior pituitary hormones. While acromegaly and Cushing's disease can develop in patients with primary empty sella syndrome, the most frequently encountered abnormality is hyperprolactinemia.[44,45] The incidence of galactorrhea or hyperprolactinemia, or both, in primary empty sella syndrome ranges between 2.5 and 8%. Oligomenorrhea and amenorrhea have been reported to be associated with hyperprolactinemia–primary empty sella syndrome complex.[46,47] There appear to be three subsets of patients in the hyperprolactinemia–empty sella syndrome complex: (1) patients with primary empty sella syndrome with hyperprolactinemia caused by a coexistent prolactin-secreting microadenoma in the flattened pituitary, (2) patients with primary empty sella syndrome with nontumorous hyperprolactinemia caused by hypothalamic dysfunction, and (3) patients with primary empty sella syndrome and coexistent idiopathic galactorrhea with or without basal hyperprolactinemia.

The coexistence of PRL-secreting microadenomas in primary empty sella syndrome has been well documented in the literature.[48–54] These reports highlight the fact that the anatomically distorted, flattened pituitary gland is not immune from developing microadenomata. The diagnostic difficulty lies in the fact that hormonal and radiological procedures may not be totally helpful in detecting the presence of these neoplasms. Even with the high-resolution capacity of fourth-generation CT, it may be difficult to outline a microadenoma occurring in a flattened, distorted, and inferoposteriorly displaced pituitary gland.

Recent reports in the literature have suggested the existence of a subset of patients with nontumorous hyperprolactinemia in association with primary empty sella syndrome.[55–58] One mechanism proposed to explain such an occurrence is the presence of coexistent hypothalamic dysfunction in these patients. Futterweit et al.[59] studied 19 patients with primary empty sella syndrome by evaluating their serum prolactin response to sequential administration of TRH and chlorpromazine (CPZ). As compared with normal subjects, as well as those with idiopathic galactorrhea, patients with primary empty sella showed impaired prolactin responses to CPZ, with preservation of response to TRH. These data are suggestive of an underlying hypothalamic defect in prolactin regulation. It has been supposed that the elevated CSF pressure often seen in patients with empty sella syndrome, when associated with a low anterior end of the third ventricle, can result in compression of the hypothalamus. Anatomical distortion of the infundibular stalk can also theoretically affect dopamine transport, resulting in hyperprolactinemia. Regardless of the mechanism proposed, the message from these reports is clear: the presence of hyperprolactinemia in a patient with an enlarged sella should not lead to the assumption that a pituitary tumor underlies the hyperprolactinema. Careful radiological as well as hormonal studies should be undertaken. The demonstration of an absent prolactin response to TRH, associated with a well-delineated tumor by CT, should point to a tumor. If the CT study does not reveal a tumor but shows only the primary empty sella, careful follow-up is recommended. In some instances, the only definitive method to exclude a coexistent microadenoma is by transsphenoidal exploration. Gharib et al.[38] evaluated 11 hyperprolactinemic patients with primary empty sella syndrome and intact pituitary reserve; of the eight patients who underwent transsphenoidal exploratory surgery, tumor was found only in one. This study emphasizes the fact that the frequency with which nontumorous hyperprolactinemia is encountered in primary empty sella syndrome far exceeds that of prolactin-secreting microadenomas associated with the entity.

Finally, idiopathic galactorrhea can occur in patients with primary empty sella syndrome, with or without basal hyperprolactinemia. These patients often demonstrate integrity of hypothalamic control of prolactin secretion, as well as negative evidence of microadenoma by computed tomography.

Radiological Studies

Plain Films of the Skull and Polytomography

The characteristics of the sella in the empty sella syndrome have been outlined by several workers.[16–18,60] Typically, the sella shows symmetrical enlargement. The dorsum sella is normally curved, with proper approximation of the clinoid processes. Erosive changes are usually absent or minimal. Occasionally, however, the changes in the sella turcica may be impressive enough to mimic those caused by a pituitary tumor. Thus, deformity of the sella, moderate to severe erosive changes, and even the "double floor sign" can be encountered in patients with primary empty sella syndrome. Since the frequency of asymmetrical sellar remodeling can be considerable, it is impossible on the basis of skull radiographs alone, to differentiate this entity from pituitary tumors. A moderately enlarged sella with symmetrical, smooth, uneroded contours is highly suggestive, but not diagnostic, of empty sella syndrome. The diagnosis can be established conclusively only with the aid of CT, metrizamide cisternography, or pneumoencephalography.

Computed Tomography

Currently, the fourth-generation CT scanners provide a noninvasive method for establishing the diagnosis of empty sella syndrome. The appearance of the sellar area in axial and coronal views in a CT study obtained after

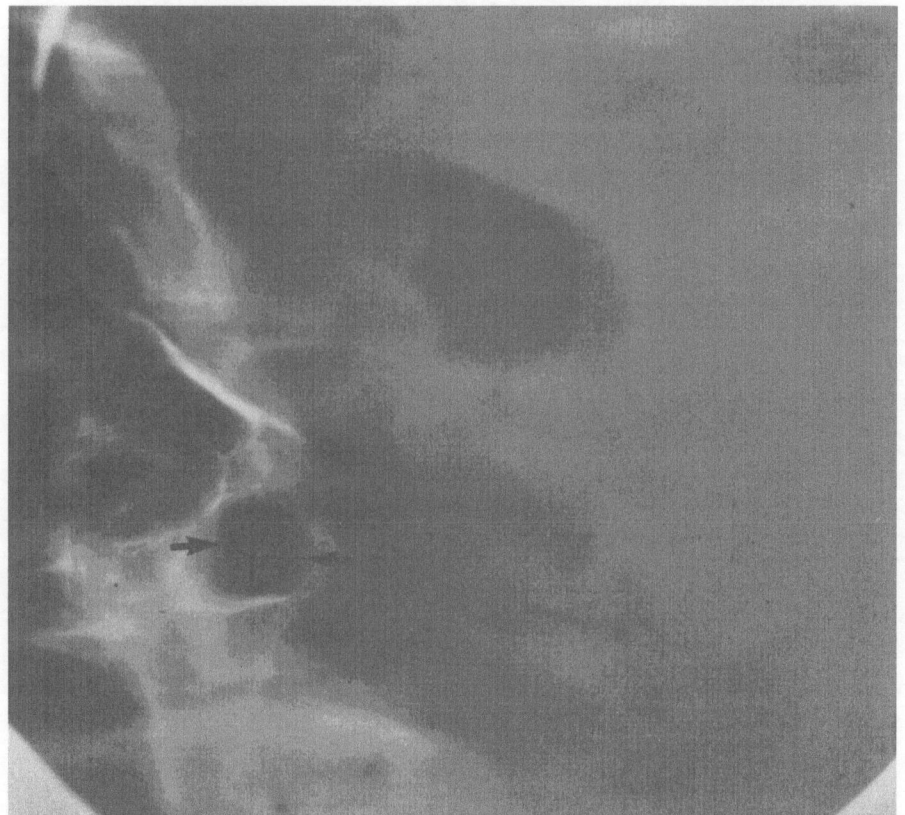

FIGURE 42.   Pneumoencephalography demonstrating air within the empty sella. (From Post.[66])

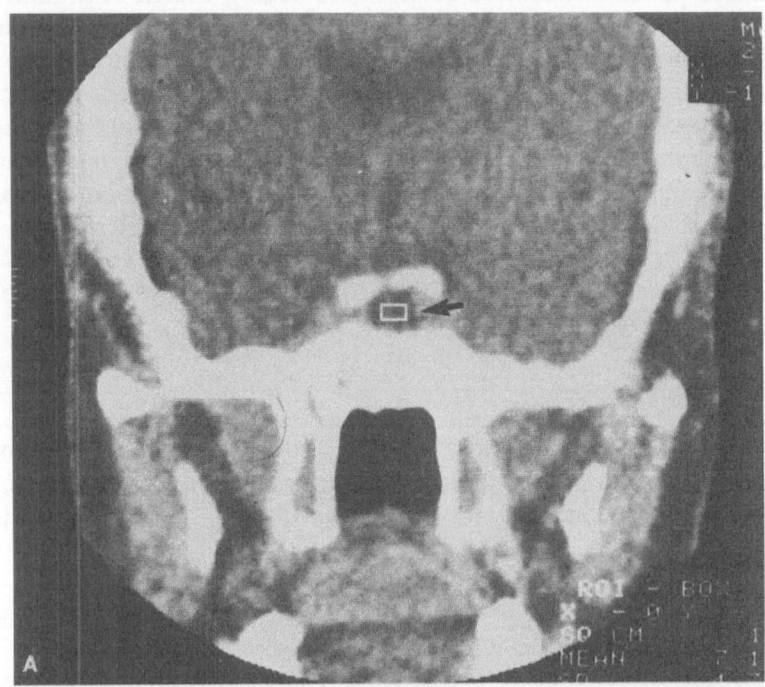

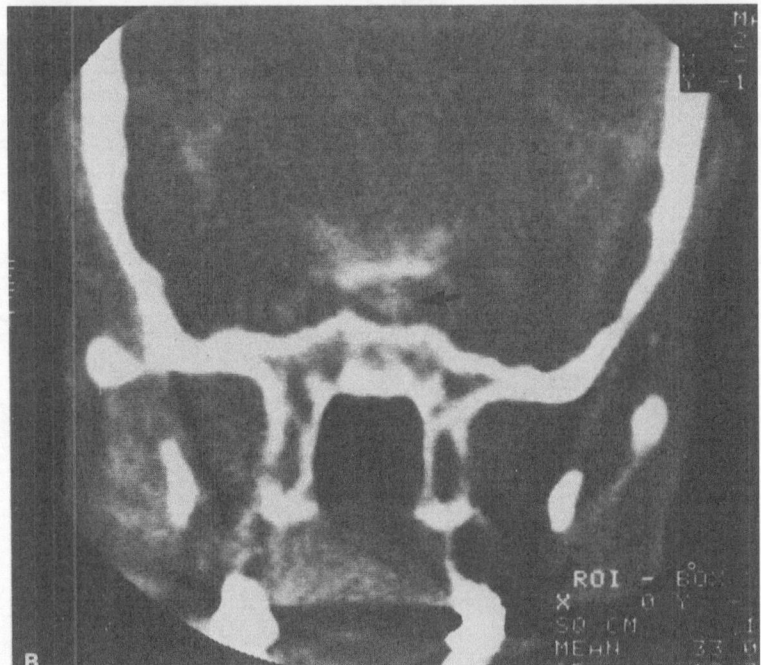

FIGURE 43. (A) Coronal CT scan demonstrating a hypodense area within the sella (arrow). (B) Coronal CT scan after metrizamide demonstrating contrast medium within the sella (arrow). (From Post.[66])

intravenous administration of contrast material is typified by the following signs (Figs. 42 and 43):

1. The infundibulum sign,[61] or the appearance of a normal infundibulum (i.e., following the injection of the contrast material—the pituitary stalk seen and identified as extending inferiorly to the flattened pituitary at the floor of the sella

2. The presence of a low attenuation density caused by CSF that can be traced from the suprasellar cistern all the way to an infradiaphragmatic location within the sella

3. The demonstration of a low-lying optic chiasm, which may dip below the level of the diaphragma sella

These findings, best appreciated in coronal views, when present together, are diagnostic of primary empty sella syndrome. The CT study has obviated the need for performing pneumoencephalography, an invasive procedure aimed at demonstrating air entry into the sella.

Occasionally, as when the CT study is inconclusive, metrizamide-enhanced CT may be required to document the empty sella.[62,63] Metrizamide, an iodinated contrast material, is intracisternally introduced and the CT study performed. Metrizamide is contraindicated in patients with iodine sensitivity and is poorly tolerated in patients with severe cardiovascular disease, asthma, epilepsy, as well as in patients on neuroleptic drugs. In such patients, small amounts of air can be used for cisternal enhancement. The safety and efficacy of air cisternography (air-enhanced CT) in delineating empty sella syndrome, when metrizamide cannot be used, has been amply documented in radiology literature.[61,64,65] These procedures—metrizamide or air-enhanced CT, or both—have almost obviated the conventional pneumoencephalographic studies that once were the only standard means for documenting the empty sella syndrome. Carotid arteriography plays no role in the diagnosis of empty sella syndrome.

Differential Diagnosis

Primary empty sella syndrome constitutes an important differential diagnosis of an enlarged sella turcica. Thus, the most significant differential diagnosis revolves around intrasellar pituitary tumors. Primary empty sella syndrome accounts for approximately 35% of enlarged sella turcicas seen by X-ray films of the skull, second only to pituitary tumors. The natural history, prognosis, and treatment between the two conditions is so different that exclusion of empty sella syndrome is a crucial concern in the workup of every patient with an enlarged sella turcica. The diagnostic triad for empty sella syndrome is the asymptomatic nature of presentation, the symmetrical contour of the enlargement, and the well-preserved pituitary hormone reserve (Table 94). The diagnosis can be confirmed only by CT. The advent of high-resolution CT has obviated the need for measurement of adenohypophyseal

TABLE 94
Similarities and Differences between Pituitary Tumors and
Empty Sella Syndrome

Features	Pituitary tumor	Empty sella syndrome
Clinical		
Headaches	Common, but not invariable	10% incidence of nonspecific headaches
Visual-field cuts	Often present	May rarely be encountered
CSF rhinorrhea	Uncommon	A rare complication
Hormonal deficiency syndromes	Occur frequently	Not a feature
Hormonal hypersecreting syndromes	May occur	Unusual, except for galactorrhea
Hormonal		
Trophic hormone reserve	Usually impaired	Usually preserved
Radiological		
Plain radiography, tomography		
Sellar size	Enlarged, asymmetrical	Enlarged, often symmetrical
Erosions	Present	Can be present
CT	Delineates tumor	Demonstrates low attenuation density within sella continuous with CSF; positive infundibular sign
Suprasellar lesions	May be present	Absent

hormones in CSF to distinguish between tumor and empty sella syndrome.[17] The concomitant occurrence of microadenoma in the flattened pituitary of empty sella syndrome can pose a major diagnostic challenge, sometimes resolved only by exploration.

The enlarged sella caused by arachnoid cyst, increased intracranial pressure, and aqueductal stenosis can present with almost identical radiological findings. In fact, the mechanism underlying the sellar enlargement in these instances is somewhat analogous to that of primary empty sella syndrome. The distinction, in some cases, may be well nigh impossible.

Natural History and Complications

The natural history of empty sella syndrome is characterized by its impressively benign and uneventful outcome. The neurological and endocrine status of patients with empty sella syndrome continues to remain unaltered. In extreme cases, prolapse of optic chiasm, optic nerve, or optic tracts into the sella may cause visual-field cuts. Similarly, CSF rhinorrhea and increased intracranial pressure can result when the arachnoid herniation is severe. These complications are rare, however.

Treatment

Uncomplicated primary empty sella syndrome requires no treatment. Reparative surgery is required only when the disorder is complicated by the development of intracranial hypertension, CSF rhinorrhea, and visual-field defects. Progressive increase in the intracranial pressure can be corrected by a shunt. CSF rhinorrhea is best managed by repair of the floor of the sella by the transsphenoidal approach. Finally, prolapse of the optic nerve, chiasm, or the optic tracts can be managed by corrective surgery involving reinforcement of the roof of the sella. This is done by restoring the optic apparatus to its original position and preventing prolapse by packing the interior of the sella with muscle and bone tissue. Such treatment is quite effective in preventing further deterioration in visual-field cuts.[16]

References

1. Kaufman B: The "empty" sella syndrome—a manifestation of the intrasellar subarachnoid space. *Radiology* **90**:931, 1968.
2. Kaufman B, Chamberlain WB Jr: The ubiquitous "empty" sella turcica. *Acta Radiol (Diagn) (Stockh)* **13**:413, 1972.
3. Zull DN, Falko JM: Metrizamide cisternography in the investigation of the empty sella syndrome. *Arch Intern Med* **141**(4):487, 1981.
4. Login I, Santen RJ: Empty sella syndrome: Sequela of the spontaneous remission of acromegaly. *Arch Intern Med* **135**:1519, 1975.
5. Drury M, O'Loughlin S, Sweeney E: Houssay phenomena in a diabetic. *Br Med J* **2**:709, 1970.
6. Fleckman AM, Schubart UK, Danziger A, et al: Empty sella of normal size in Sheehan's syndrome. *Am J Med* **75**:585, 1983.
7. Gray H, Goss CM: *Anatomy of the Human Body.* Lea & Febiger, Philadelphia, 1959.
8. Bergland RM, Ray BS, Torack RM: Anatomical variations in the pituitary gland and adjacent structures in 225 human autopsy cases. *J Neurosurg* **28**:93, 1968.
9. Mortara R, Norell H: Consequences of a deficient sellar diaphragma. *J Neurosurg* **32**:565, 1970.
10. Kaufman HH: Nontraumatic cerebrospinal fluid rhinorrhea. *Arch Neurol* **21**:59, 1969.
11. Kaufman B, Pearson OH, Chamberlin WB: Radiographic features of intrasellar masses and progressive, asymmetrical, nontumorous enlargements of the sella turcica, the 'empty' sella. In Ross GT, Kohler PO (eds): *Diagnosis and Treatment of Pituitary Tumors.* Excerpta Medica, Amsterdam, 1973, p. 100.
12. Bjerre P, Lindholm J: Adult hydrocephalus and the empty sella. *Acta Neurol Scand* **70**(3):201, 1984.
13. Muhlendahl KE, Bradac GB: Empty sella syndrome in a boy with mucopolysaccharidosis type VI (Maroteaux-Lamy). *Helv Pediatr Acta* **30**:185, 1975.
14. Banerjee T, Neagher JN: Foster Kennedy syndrome, aqueductal stenosis and empty sella. *Am Surg* **40**:552, 1974.
15. du Boulay GH, El Gammal T: The classification, clinical value and mechanism of sella turcica changes in raised intracranial pressure. *Br J Radiol* **39**:421, 1966.

16. Randall RV: Empty sella syndrome *Comp Ther* **10**(4):57, 1984.

17. Jordan RM, Kendall JW, Kerber CW: The primary empty sella syndrome: Analysis of the clinical characteristics, radiographic features, pituitary function and cerebrospinal fluid adenohypophysial hormone concentrations. *Am J Med* **62**:569, 1977.

18. Neelon FA, Goree JA, Lebovitz HE: The primary empty sella: Clinical and radiographic characteristics and endocrine function. *Medicine (Baltimore)* **52**:73, 1973.

19. Zatz IM, Janon EA, Newton TH: The enlarged sella and the intrasellar cistern. *Radiology* **93**:1085, 1969.

20. Obrador S: The empty sella and some related syndromes. *J Neurosurg* **36**:162, 1972.

21. Wilkinson IA, Duck SC, Gager WE, et al: Empty sella syndrome. *Am J Dis Child* **136**(3):245, 1982.

22. Costigan DC, Daneman D, Harwood-Nash D, et al: The "empty sella" in childhood. *Clin Pediatr* **23**:437, 1983.

23. Onur K, Lala V, Zimmer J, et al: The primary empty sella syndrome in child. *J Pediatr* **90**:425, 1977.

24. Von Petrykowski W, Reinwein H, Ostertag D, et al: Asymptomatic primary empty sella in a 14-year-old girl: Comparison of computer tomography and nuclear magnetic resonance imaging. *Horm Res* **22**:58, 1985.

25. Farber M, Turksoy RN, Rodgers J: The primary empty sella syndrome. *Obstet Gynecol* **49**(Suppl 1):6, 1977.

26. Raiti S, Albrink MJ, MacLaren NK, et al: Empty sella syndrome secondary to intrasellar cyst in adolescence. *Am J Dis Child* **130**:1009, 1976.

27. Lee WM, Adams JE: The empty sella syndrome. *J Neurosurg* **28**:351, 1968.

28. Buckman MT, Husain M, Carlow TJ, et al: Primary empty sella syndrome with visual field defects. *Am J Med* **61**:124, 1976.

29. Dahlstrom R, Acers TE: Chiasmatic arachnoiditis and empty sella: Report and discussion of a case. *Ann Ophthalmol* **7**:73, 1975.

30. Shinoda Y, Ohnishi Y, Abe M, et al: Empty sella syndrome with visual field disturbance. *Jpn J Ophthalmol* **27**:248, 1983.

31. Bursztyn EM, Lavyne MH, Hisen M: Empty sella syndrome with intrasellar herniation of the optic chiasm. *AJNR* **4**:167, 1983.

32. Brisman R, Hughes JO, Holub DA: Endocrine function in nineteen patients with empty sella syndrome. *J Clin Endocrinol Metab* **34**:570, 1972.

33. Snyder PJ, Jacobs LS, Rabello MM, et al: Diagnostic value of thyrotropin-releasing hormone in pituitary hypothalamic diseases. *Ann Intern Med* **81**:751, 1974.

34. Ridgway EC, Kourides IA, Kliman B, et al: Thyrotropin and prolactin pituitary reserve in the empty sella syndrome. *J Clin Endocrinol Metab* **41**:953, 1975.

35. Faglia G, Beck-Peccoz P, Ambrosi B, et al: The empty sella syndrome. *Lancet* **1**:149, 1973.

36. Nieman EA, Landon J, Wynn V: Endocrine function in patients with untreated chromophobe adenomas. *Q J Med* **36**:356, 1967.

37. Bergeron C, Kovacs K, Bilbao JM: Primary empty sella: A histologic and immunocytologic study. *Arch Intern Med* **139**:248, 1979.

38. Gharib H, Frey HM, Laws ER Jr, et al: Coexistent primary empty sella syndrome and hyperprolactinemia: Report of 11 cases. *Arch Intern Med* **143**:1383, 1983.

39. Bjerre P, Glydensted G, Rhshede J, et al: The empty sella and pituitary adenoma: a theory on the causal relationship. *Acta Neurol Scand* **66**:82, 1982.

40. Lindholm J, Bjerre P, Rhshede J, et al: Pituitary function in patients with evidence of spontaneous disappearance of a pituitary adenoma. *Clin Endocrinol (Oxf)* **18**:599, 1983.

41. Veldhuis JD, Hammond JM: Endocrine function after spontaneous infarction of the human pituitary: Report, review, and reappraisal. *Endocr Rev* **1**:100, 1980.

42. Stacpoole PW, Kandell TW, Fisher WR: Primary empty sella, hyperprolactinemia, and isolated ACTH deficiency after postpartum hemorrhage. *Am J Med* **74**:905, 1983.

43. Stephens WP, Goddard KJ, Laing I, et al: Isolated adrenocorticotrophin deficiency and empty sella associated with hypothyroidism. *Clin Endocrinol (Oxf)* **22**:771, 1985.

44. Bryner JR, El Gammal T, Acker JD, et al: Intrasellar subarachnoid herniation or empty sella associated with galactorrhea. *Obstet Gynecol* **51**:198, 1978.
45. Brismar K, Efendic S: Pituitary function in the empty sella syndrome *Neuroendocrinology* **32**:70, 1981.
46. Haney AF, Kramer RS, Weibe RH, et al: Hypothalamic-pituitary function and radiographic evaluation of women with hyperprolactinemia and an empty sella turcica. *Am J Obstet Gynecol* **134**:917, 1979.
47. Futterweit W: Galactorrhea, amenorrhea, hyperprolactinemia and pseudotumor cerebri in a patient with primary empty sella syndrome: Case report with review of the literature. *Mt Sinai J Med* **49**:514, 1982.
48. Schaison G, Metzger J: The primary empty sella: An endocrine study on 12 cases. *Acta Endocrinol (Kbh)* **83**:483, 1976.
49. Dominque JN, Wing SD, Wilson CB: Coexisting pituitary adenomas and partially empty sellas. *J Neurosurg* **48**:23, 1978.
50. Swanson JA, Sherman BM, Van Gilder JC, et al: Coexistent empty sella and prolactin-secreting microadenoma. *Obstet Gynecol* **53**:258, 1979.
51. Archer DF, Maroon JC, DuBois PJ: Galactorrhea, amenorrhea, hyperprolactinemia, and an empty sella. *Obstet Gynecol* **52**(Suppl):23, 1978.
52. Kleinberg DL, Noel GL, Frantz AG: Galactorrhea: A study of 235 cases including 48 with pituitary tumors. *N Engl J Med* **296**:589, 1977.
53. Jones JR, DeHempel PAC, Kemmann E, et al: Galactorrhea and amenorrhea in a patient with an empty sella. *Obstet Gynecol* **49**:S9, 1977.
54. Valenta LJ, Sostrin RD, Eisenberg H, et al: Diagnosis of pituitary tumors by hormone assays and computerized tomography. *Am J Med* **72**:861, 1982.
55. Bar RS, Mazzaferri EL, Malarkey WB: Primary empty sella, galactorrhea, hyperprolactinemia and renal tubular acidosis. *Am J Med* **59**:863, 1975.
56. Badawy SZA, Nusbaum ML, Omar M: Hypothalamic-pituitary evaluation in patients with galactorrhea-amenorrhea and hyperprolactinemia. *Obstet Gynecol* **55**:1, 1980.
57. Hsu TH, Shapiro JR, Tyson JE, et al: Hyperprolactinemia associated with empty sella syndrome. *JAMA* **235**:2002, 1976.
58. Gates RB, Friesen H, Samaan NA: Inappropriate lactation and amenorrhoea: Pathological and diagnostic considerations. *Acta Endocrinol (Kbh)* **72**:101, 1973.
59. Futterweit W, Smith H Jr, Holt JE: Dissociation of serum prolactin response to sequential thyrotropin-releasing hormone and chlorpromazine stimulation in patients with primary empty sella syndrome. *Fertil Steril* **42**:573, 1984.
60. Hodgson SF, Randall RV, Laws ER: Empty sella syndrome. In Youmans JR (ed): *Neurological Surgery*. 2nd Ed. WB Saunders, Philadelphia, 1982, p. 3170.
61. Haughton VM, Rosenbaum AE, et al: Recognizing the empty sella by CT: The infundibulum sign. *AJR* **136**:293, 1981.
62. Gross CE, Binet EF, Esquerra JV: Metrizamide cisternography in the evaluation of pituitary adenomas and the empty sella syndrome. *J Neurosurg* **50**:472, 1979.
63. Ketonen L, Kuuliala I: Diagnosis of primary empty sella syndrome by computed tomography. *Ann Clin Res* **11**:125, 1979.
64. Penley MW, Pribram HFV: Diagnosis of empty sella with small amounts of air at computed tomography. *Surg Neurol* **14**:296, 1980.
65. Al-Mefty O, Fox JL, Reese MS: The empty sella: Air-enhanced CT. *Radiology* **155**:253, 1985.
66. Post KD, Kasdon DL: Sellar and Parasellar Lesions Mimicking Adenoma. In Post KD, Jackson IM, Reichlin S (eds): *The Pituitary Adenoma*, Plenum Press, New York, 1980, pp. 159–218.

19

Arginine Vasopressin

Introduction

Arginine vasopressin (AVP) is the antidiuretic principle secreted by the hypothalamus and stored in the posterior pituitary. The control of AVP secretion and its role in regulating water balance form the basis of this discussion.

Secretion of AVP

Arginine vasopressin is synthesized in highly specialized neurosecretory cells located within the supraoptic and paraventricular nuclei of the hypothalamus. Much of our current knowledge regarding the biosynthesis of AVP has been provided by Sachs et al.,[1,2] who used the radioisotope labeling model of hormone biosynthesis to elucidate the synthesis of AVP. The supraoptic nucleus is the major site of AVP synthesis, which is initiated on the ribosomes. It has been proposed that AVP is derived from a precursor form, provasopressin, which like other prohormones is biologically inactive. In vitro isotopic labeling experiments indicate that puromycin inhibits the biosynthesis of the labeled precursor without affecting the release of hormone from its precursor. The secretion of AVP is paralleled by the secretion of the protein neurophysin. This protein, also known as Van Dyke protein, is thought to originate from the same common precursor of AVP. Using supraoptic neurons maintained in culture, Sachs et al.[3] showed that AVP and its specific neurophysin are derived from a single precursor, provasopressin. The situation is thus somewhat analogous to the derivation of insulin and C peptide from the single-chain precursor proinsulin.[4] The similarity is heightened by the demonstration that release of neurophysin into the circulation is regulated by the same factors that provoke release of AVP. Thus, dehydration and hypertonic saline affect the release of neurophysin into the circulation.[5,6] The neurophysin for AVP (vasopressin-neurophysin) has no antidiuretic properties but plays an important role in the binding and intraneuronal transport of the hormone.

Arginine vasopressin represents the antidiuretic principle seen in most mammals. By contrast, the antidiuretic principle in animals of the suborder suina is lysine vasopressin (LVP). The presence of a basic amino acid in the penultimate position of the side chain confers antidiuretic properties to the octapeptide AVP. The chemical structure of AVP is shown in Fig. 44. All antidiuretic principles (AVP, LVP, oxytocin, and vasotocin) contain a sulfhydryl bond between the two cysteine residues at positions 1 and 6. The differences are in subsitiations at positions 3 and 8.

The synthetic pathway for secretion of vasopressin involves multiple stages. First, the hormone is synthesized as a prohormone at the ribosomes of the magnocellular neurons. Next, conversion of the prohormone into hormone probably takes place at the Golgi apparatus. The neurosecretory granule, containing both vasopressin and its neurophysin, is then transported along the supraoptical–hypophyseal tracts as osmotically inactive granules. Finally, the posterior pituitary serves as the storage reservoir for both AVP and its neurophysin.

The synthesis and transport of oxytocin are identical to those of AVP, except that oxytocin has its own neurophysin. The concept that AVP and

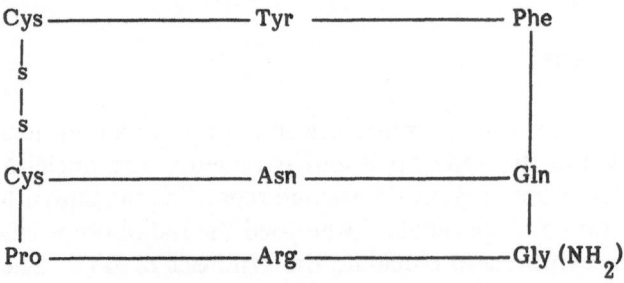

FIGURE 44. Structure of vasopressin.

oxytocin have separate neurophysins is strongly supported by experimental observations. The availability of specific immunoperoxidase staining techniques using antibodies to AVP, oxytocin, and neurophysins has permitted specific identification of their respective secretory cells. Thus, the supraoptic and paraventricular nuclei contain magnocellular neurons that secrete either AVP or oxytocin. The individuality of these cells is demonstrated by immunopositivity to AVP (and its neurophysin) or oxytocin (and its neurophysin).[7,8] Stimuli that preferentially release AVP release only its neurophysin, and vice versa. Pathological processes can affect one set of magnocellular neurons without affecting the other. The classic illustration of such a phenomenon is exemplified by the congenital DI in the Brattleboro rat, in which a total lack of neurons that secrete AVP and its neurophysin can be encountered, with preservation of oxytocin- (and its neurophysin) containing neurons.[9] The terms *neurophysin I* and *neurophysin II* are applied to denote vasopressin- and oxytocin-associated neurophysins, respectively.

Release of AVP

The release of AVP into the circulation begins with depolarization impulses that originate in the supraoptic nuclei of the hypothalamus and that extend progressively down the axons, terminating in the posterior pituitary. Electron microscopic studies[10–12] have demonstrated that the neurosecretory granules discharge the hormone, with its neurophysin, by a process of exocytosis. This process is initiated by the fusion of the secretory granule with the plasma membrane, followed by the development of a "window" in the membrane that communicates with the extracellular space. The hormone is released by a process of extrusion. A similar sequence of events occurs when the posterior pituitary gland is stimulated. The process of hormone extrusion appears to be calcium dependent. As early as 1964, Douglas and Poisner[13] demonstrated that release of vasopressin from the neural lobe is calcium dependent. These workers advanced the notion that AVP release is mediated by stimulus–secretion coupling, influenced by the efflux of calcium ions across the plasma cell membrane of axons.

Experiments in animals to define the role of the posterior pituitary gland as a storage organ have yielded important information. Sachs et al.[14] showed that the initial response to bleeding (hypovolemia) is characterized by rapid discharge of AVP into the circulation. It is generally accepted that approximately 10–20% of the hormonal content of the posterior pituitary is in a readily releasable pool. If the stimulus for AVP release continues to be applied with the same intensity, the neurohypophysis will still respond, but at a significantly reduced rate. This attenuation in response is partly pituitary in origin, because direct electrical stimulation of the posterior pituitary following bleeding results in a significantly reduced rate of hormone release.

TABLE 95
Stimuli for AVP Release

Osmotic
    Hypertonicity of body fluids
Nonosmotic
    Receptors in the aortic arch and left atrial wall
    Temperature
    Renin-angiotensin system
    Sympathetic nervous system
    Hypoxia, hypercapnia
    Chemicals

Regulation of AVP Release

The secretion and release of AVP are mediated by osmotic and nonosmotic stimuli. The osmotic stimulus—hypertonicity of plasma—mediates the release of AVP by osmoreceptors located at the hypothalamus. Several nonosmotic stimuli participate in the release of AVP; these include volume and baroreceptors located in the left atrial wall and aortic arch, temperature, the renin–angiotensin system, the sympathetic system, hypoxia or hypercapnia, and chemicals (Table 95).

Osmotic Regulation

It is now well recognized that osmoregulation of vasopressin release is mediated by osmoreceptors. Work done by Hayward[15] has led to the recognition of two types of osmoreceptors in the hypothalamus: the osmoreceptors of Verney and the osmoreceptors of Sawyer. The osmoreceptors of Verney are cells that respond exclusively to hyperosmolality, whereas the osmoreceptors of Sawyer respond not only to changes in osmolality but also to noxious stimuli. The osmoreceptors are anatomically discrete cells, with axons projecting to the magnocellular neurons within the supraoptic and paraventricular nuclei. These magnocellular neurons represent the neuroendocrine secretory cells that secrete AVP. Naturally, then, the magnocellular neurons also receive projections from the osmoreceptors of Sawyer as well as from afferents related to behavioral stimuli. The osmoreceptors, probably located in the anteroventral aspect of the third ventricle, are exquisitely sensitive to changes in the osmolality of the blood perfusing the receptors. As originally pointed out in the epoch-making Croonian lecture delivered by Verney in 1947,[16] even minor changes, as little as a 2% increase in plasma osmolality, result in significant antidiuresis. It is important to recognize that the osmoreceptors respond predominantly to sodium-mediated rises in osmolality. When hyperosmolality is induced by the administration of hypertonic urea solution, or hypertonic glucose, no antidiuresis occurs, indicating that the osmoreceptors lie outside the blood-brain barrier and are freely permeable to both urea and glucose.[17,18] The search for sodium-sensitive receptors in the brain has met with conflicting reports. Experiments performed by intraventricular infusions of hypertonic saline have supported the belief that sodium receptors exist in the floor of the third ventricle and that they play an important role in the modulation of AVP release.[19] Such a notion, however, has been contraindicated by others. Thrasher et al.[20] infused several solutions intravenously in conscious dogs and demonstrated that although comparable rises in plasma and cerebrospinal fluid (CSF) osmolality were induced by several solutions, only hypertonic saline and sucrose resulted in significant rises of AVP. The relative unimportance of sodium receptors was pointed out by these workers by the demonstration that infusion of sucrose results in AVP release despite a fall in the serum sodium concentrations. The consensus of opinions favors the notion that sodium receptors, wherever located, do not possess an overriding influence over the osmoreceptors.

The availability of an immunoassay for AVP has permitted evaluation of the relationship between rising osmolality and AVP release.[21-24] In ambulatory subjects in a state of normal hydration, the osmolality of the plasma is 275–280 mOsm/kg, and the plasma AVP levels range between 0.5 and 1.5 pg/ml. The osmotic threshold appears to be 280 mOsm/kg, since a rise in plasma osmolality above this level is attended by a brisk rise in plasma AVP level. When the changes in plasma osmolality are plotted against the urine/plasma osmolar ratio, a sigmoidal curve is usually obtained. Moore[25] convincingly demonstrated this sigmoidal pattern by studying the relationship between hormone levels in blood and the U/P osmolar ratio in 203 pairs of data. The implication is that small changes in the plasma concentrations of the hormone are accompanied by relatively large and impressive changes in the U/P osmolar ratio. Thus, the greatest water conservation by the tubules is attained by the release of only modest amounts of AVP. This ability of the renal collecting tubules to conserve water maximally at AVP levels far below maximal secretory rate is a reflection of the efficiency of the countercurrent system of the kidney. The importance of osmotic regulatory mechanisms for AVP release can be overridden by volume-mediated (nonosmotic) signals during volume contraction.

Nonosmotic Stimuli

Volume and Baroreceptors

The most important nonosmotic stimuli for release of AVP are the volume receptors located in the left atrium and the baroreceptors located in the carotid sinus and the aortic arch.

Volume Receptors in the Left Atrium.   It is well known that erect posture increases antidiuresis. In addition, maneuvers that affect intrathoracic pressure also affect AVP levels.[26,27] The receptors in the left atrial wall are exquisitely sensitive to changes in the left atrial pressure. The afferent pathways from the left atrium travel via the vagus nerve. Any condition that leads to an increase in the left atrial pressure distends the left atrial wall, thereby increasing the parasympathetic drive and suppressing AVP release. Any condition that results in a fall in left atrial pressure decreases the parasympathetic drive and increases the release of AVP. Three clinical observations relate to this mechanism. First, in patients on positive-pressure ventilatory assistance, the negative intrathoracic pressure leads to a drop in left atrial pressure, resulting in an increase in AVP release; similarly, patients on negative-pressure breathing develop polyuria, which can be abolished by the administration of pitressin.[28] Second, in patients with paroxysmal atrial tachycardia (PAT), increased left atrial pressure develops, leading to distention in the left atrial wall and resulting in suppression of AVP and polyuria. Third, depletion of ECF volume is sensed by the left atrial wall (even in early stages, prior to the development of hypotension), leading to a decline in left atrial pressure and causing AVP release. The mediation of these phenomena is assumed to be via the vagus.

Baroreceptors in the Aortic Arch and Carotid Sinus. These receptors are sensitive to changes in arterial blood pressure. A drop in the arterial blood pressure stimulates these receptors and leads to the release of AVP. The afferent pathways for this reflex arc are the vagus and glossopharyngeal nerves. This is the mechanism whereby volume depletion and hypotension (from dehydration, hemorrhage, shock, etc.) result in an increased release of AVP in an attempt to replete volume by water conservation.[29]

Thus, it appears that the two most important stimuli for the release of AVP are hypertonicity of plasma (increased sodium) and hypotension (from volume depletion). Of these two stimuli, the most sensitive one is the osmotic stimulus. A 1–2% change in osmolality will affect AVP release, while 5–10% change in blood volume is necessary to stimulate AVP release. The relative role of blood osmolality and volume depletion have been examined in the rat by Dunn et al.[22] They have demonstrated the greater sensitivity for osmotic release of AVP and greater potency for the nonosmotic release of AVP. Once stimulated by the nonosmotic pathway, the response of AVP occurs in an exponential fashion, accounting for the greater response. Usually the osmoreceptors and the baroreceptor pathways work in concert. Very rarely, disease states in the hypothalamus may destroy the osmotic pathway, but the baroreceptor pathway might be intact. Patients labeled as having essential hypernatremia or reset osmostats might in fact have such a syndrome. These patients are unable to release AVP in response to increasing plasma osmolality but are unable to do so when the blood pressure is lowered. A hypothalamic lesion is usually in the background, having destroyed the osmotic receptor (and often the thirst center), but leaving intact the baroreceptor pathway that travels through the parasympathetic system, medulla, and hypothalamus. This explains why these patients are not able to concentrate the urine despite hypernatremia but are able to do so when they become dehydrated enough to become hypotensive (see Chapter 20).

The Renin–Angiotensin System

Angiotensin II has three major actions on water metabolism (1) dipsogenic action (i.e., it stimulates thirst), (2) stimulation of AVP secretion, and (3) provocation of AVP release from the posterior pituitary.

The dipsogenic action of angiotensin II has been documented experimentally by the injection of this substance directly into the ventricles, as well by the intravenous route.[30,31] While these experimental data in animals are impressive, it is not clear whether angiotensin II has a substantive effect on the drinking habits of humans. Similarly, the effects of angiotensin infusion on vasopressin release are also convincing experimentally.[32,33] Since angiotensin II has to be given in rather large doses to effect AVP release in animals, it is doubtful whether it plays any role in the physiological regulation of AVP release.

Temperature and AVP Release

Physiologically, an increase in body temperature stimulates the release of AVP as well as the thirst center. When thermal electrodes placed in the preoptic area of the dog are stimulated, a brisk release of AVP is observed.[34] In normal subjects exposed to 50°C for 2 hr, a four- to fivefold increase in the plasma AVP level may be seen.[21] The effect on AVP release is partly due to the stimulatory effect of the warm blood perfusing the supraoptic and paraventricular nuclei and is partly due to the peripheral vasodilatation, which stimulates nonosmotic AVP release.

Adrenergic Modulation

It is well recognized that adrenergic stimulation affects AVP release (stress, fear, and pain). Berl et al.[35,36] demonstrated that α-adrenergic stimulation with norepinephrine suppresses AVP release, while β-adrenergic stimulation with isoproterenol stimulates vasopressin release.[37,38] It is unclear whether the effects of β-adrenergic stimulation are mediated exclusively by central mechanisms or mediated by the baroreceptors. The phenomenon is further complicated by the fact that catecholamines possess an added intrarenal diuretic effect.[38]

Chemicals and AVP Release

Several chemicals and drugs stimulate AVP release (Table 96). A direct effect has been presumed for most of these drugs, of which nicotine is the prototype. Analgesics, barbiturates, cyclophosphamide, and vincristine probably mediate antidiuretic action via AVP release. In the case of chlorpropamide and clofibrate, the peripheral mechanisms are more important in causing antidiuresis, although a central mechanism has not been ruled out. Alcohol and diphenylhydantoin are well-recognized examples of drugs that inhibit AVP release.

TABLE 96.
Chemical Factors That Affect
AVP Release

Stimulate AVP release	Suppress AVP release
Nicotine	Alcohol
Analgesics	Diphenylhydantoin
Barbiturates	
Cyclophosphamide	
Vincristine	
Chlorpropamide[a]	
Clofibrate[a]	

[a] Also enhances peripheral action of arginine vasopressin
   (AVP).

In summary, the rate of release of AVP at a given moment is dependent on the algebraic sum of the stimulating and inhibiting influences. Even the most potent of influences on AVP release—hyperosmolality—can be modified by the presence of other stimuli. The single most important factor that mediates osmoreceptor function is the state of cellular hydration. When the cells are dehydrated, a phenomenon sensed by osmoreceptors with as little as a 2% rise in tonicity of body fluids, AVP is released; when the cells are rehydrated, AVP release is suppressed. The mechanism of AVP release is closely synchronized with the functioning of the thirst center.

Thirst and Its Regulation

The thirst centers are located in the hypothalamus, in close proximity to the supraoptic and paraventricular nuclei. Thirst plays an important role in the day-to-day regulation of water metabolism. Thirst is a cortical sensation and is dependent on the preservation of consciousness. The most important stimulus for thirst is hypertonicity of body fluids. Increases by as little as 1–2% in the plasma osmolality stimulate the thirst center.

Hypertonicity stimulates both AVP release and the thirst center. Figure 45 outlines the concerted efforts of both these centers in counteracting hypertonicity. This is referred to as the water-repletion reaction, aimed at restoring renal conservation of water. Destruction of the thirst center experimentally, or by pathological processes, results in hypodipsia or adipsia. This would lead to dangerous elevations in serum sodium levels and serum osmolality. The presence of an intact thirst center is crucial in maintaining water metabolism of patients with diabetes insipidus. This is reflected by the fact that these patients develop severe dehydration when thirst sensation is suppressed by loss of consciousness. Given the proximity between the centers for thirst and those for AVP secretion, it is indeed surprising that the former is destroyed only rarely by diseases that invade the centers for AVP secretion.

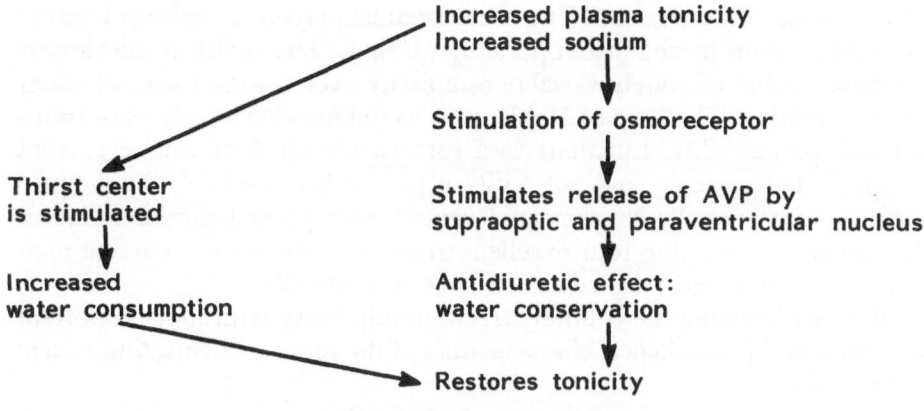

FIGURE 45. Water-repletion reaction.

The regulatory mechanisms that stimulate and suppress the thirst center are analogous to those that control vasopressin. Thus, the two major forces that stimulate thirst are hypertonicity of body fluids and depletion of the vascular and extracellular volume. The baroreceptors and volume receptors that stimulate AVP release probably mediate stimulation of thirst as well. The consistency with which hypovolemic states stimulate thirst as well as the renin-angiotensin system has led to the evaluation of the role of angiotensin in stimulating thirst. The thirst reflex of hypotensive rats can be abolished by nephrectomy and restored by the intracranial administration of angiotensin II.[39-41] The evidence for angiotensin-mediated thirst is more pronounced in animals than in humans.

The recent addition to the list of substances that stimulate thirst is prostaglandin E (PGE).[42,43] Intraventricular infusion or perfusion of PGE is associated with stimulation of both thirst and AVP release. Furthermore, PGE enhances the response of AVP to such stimuli as angiotensin II and hypertonic saline.[43,44]

Circulation and Metabolism

Arginine vasopressin circulates in the plasma, mostly in an unbound form. The hormone is rapidly cleared from the circulation and is degraded at multiple sites—the liver, brain, and kidneys. The biological half-life of vasopressin is approximately 30 min.[45] The short half-life of AVP readily permits dynamic assessment of hormonal perturbations in response to provocative and suppressive influences.

Action

Since the action of AVP is to produce concentration of urine, it is essential to focus on the mechanisms, both AVP mediated as well as non-AVP mediated, involved in this phenomenon. The renal medulla plays a crucial role because of the counterflow mechanisms operating within it. The ability of the kidneys to produce urine of widely variable osmolarity over a wide range of water intake is amazing. The loop of Henle, with its unique characteristics of transport and permeability, functions as a remarkably efficient countercurrent multiplier. This system—and not AVP—is primarily responsible for the production of a maximally concentrated urine under physiological conditions. The reader is referred to four excellent treatises on the countercurrent multiplier theory of urinary concentration (see refs. 46–49).

Before discussing the countercurrent multiplier system, some important anatomical and physiological characteristics of the renal concentrating system bear mention:

1. In the normal adult, approximately 100 ml of glomerular filtrate is presented to the proximal tubule every minute of which two-thirds to three-fourths is reabsorbed.

2. The loops of Henle, especially the long loops, are specialized structures that participate in the concentration of urine. Three portions of the loop are involved in this process: the thin descending limb, the thin ascending limb, and the thick ascending limb.

3. The thin descending limb is very permeable to water but impermeable to sodium ions and urea. The thin ascending limb is impermeable to water but very permeable to sodium ions and somewhat permeable to urea. The thick ascending limb is also impermeable to water, while being very permeable to sodium and chloride ions and impermeable to urea.

4. The distal tubules and the collecting ducts, in contrast to the loop of Henle, depend on AVP for water permeability.

5. The countercurrent multiplier system works on the premise that the loop of Henle provides the countercurrent flow, with marked differences in permeability between the limbs carrying fluid in opposite directions. The basic energy source is provided by the active transport of NaCl out of the thick ascending limb into the medullary interstitium.

In order to understand the mechanism of the countercurrent system, first a simplistic, basic view is presented. This is followed by recent modifications. The concept of the countercurrent system is used to illustrate the process of urine concentration. The renal medullary interstitium is an extremely important structure that, along with AVP, regulates this process. The countercurrent system is visualized as resembling a fluid stream bent back on itself like a hairpin, with fluid flowing in opposite directions but very close to each other. This system is buried in the substance of the renal medullary interstitium, which provides the tonicity needed for the functioning of this countercurrent system. Another anatomical fact that has enormous significance is that the proximal and distal convuluted tubules are in the cortex of the kidney, whereas Henle's loop with both limbs and the collecting tubules are in the medullary interstitium (Fig. 46).

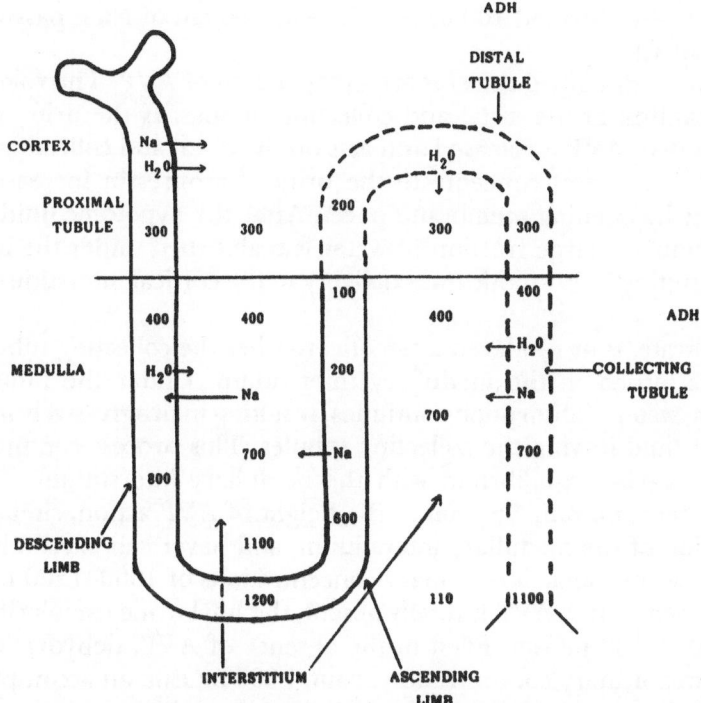

FIGURE 46. The descending limb of the loop of Henle is freely permeable to water, whereas the adjoining medulla is relatively concentrated. Hence, the filtrate progressively becomes concentrated. In the ascending limb, sodium is removed from the tubule without a simultaneous efflux of water. Thus, the filtrate reaching the distal tubule has an osmolarity of 100 mOsm. Vasopressin increases water permeability at the distal and collecting tubules.

The glomerular filtrate is isosmotic to the plasma with an osmolality of 280–300 mOsm/kg. The proximal tubule neither concentrates nor dilutes the urine but performs the phenomenal task of reabsorbing almost 120 liters of filtrate per day. It delivers 40 liters of isotonic filtrate to the descending limb of the loop of Henle. The loop of Henle is closely intertwined with blood vessels of the vasa recta, an important anatomical consideration for the events of the countercurrent system that subsequently take place.

Two facts govern the events occurring at the descending limb of the loop of Henle: (1) the descending limb is freely permeable to water, and (2) the adjoining medulla is relatively concentrated. Therefore, as the filtrate moves down along this limb, it becomes progressively concentrated, because the water exits from the descending limb into the concentrated medulla. Maximal concentrations are reached at the hairpin bend.

The concentrated filtrate reaches the ascending limb. This structure actively removes Na and Cl from the tubule into the medullary interstitium but is impermeable to backdiffusion of water; i.e., Na leaves the ascending limb into the interstitium to increase its solute content without a simultaneous efflux of water. As the dilute filtrate moves up, it reaches the distal convoluted tubule with an osmolarity around 100 mOsm. The movement of Na is passive to the movement of Cl.

The events described thus far are independent of AVP. The role of AVP becomes manifest at the distal and collecting tubules. If the urine needs to be concentrated, AVP is released and acts on the distal and collecting tubules to reabsorb water and concentrate the urine. Vasopressin increases water permeability by opening membrane pores. After the hypotonic fluid reaches the distal tubule, a large fraction of water is reabsorbed under the influence of AVP, resulting in isosmotic fluid delivery to the cortical interstitium (~300 mOsm).

The filtrate, thus rendered isosmotic, reaches the collecting tubule. This structure is buried in the medullary interstitium. Under the influence of vasopressin, water reabsorption continues, resulting in progressively increased amounts of fluid leaving the collecting tubules. This process continues until the filtrate reaches equilibrium with the medullary interstitium. The final urine can therefore only be even at the height of AVP action, similar to the concentration of the medullary interstitium, and never can exceed it. At the height of vasopressin action, urinary concentrations of 1000–1200 mOsm/kg can be attained. When AVP is totally absent, the final urine osmolarity ranges between 50 to 100 mOsm. Even in the absence of AVP, dehydration alone can result in a urinary concentration around 300 mOsm, an accomplishment achieved entirely by the hypertonic medullary interstitium.

In general terms, then, the medullary countercurrent system forms the basis for excreting urines of widely variable concentrations, but it is AVP that modulates the fine tuning. If water homeostasis demands a concentrated urine, AVP is released and can concentrate the final urine. By contrast, if the homeostasis requires a dilute urine, AVP is suppressed, and the final urine remains hypoosmolar.

There is considerable controversy as to the relative roles of active and passive steps involved in the countercurrent multiplier system. Two spatially distinct steps are believed to exist—the active step in the outer medulla and the passive step in the inner medulla.[50] The first step in the countercurrent multiplier system is the efflux of NaCl from the water-permeable thick ascending limbs. While active efflux of Na is the major contributor to the hypertonicity of the medullary interstitium, passive transport of urea from the papillary collecting ducts also contributes to the medullary hypertonicity. This is especially so with ADH-induced antidiuresis.

The modifications to the countercurrent multiplier system have been provided by the models of both Stephenson[51,52] and Kokko and Rector.[50] These models highlight the fact that the active transport of NaCl is confined to the thick ascending limb and represent the local driving force in the outer medulla. The chloride pump in the thick ascending limb (the outer medulla) pumps out the Cl and Na into the medullary interstitium, with the formation of hypotonic fluid. This fluid, although hypotonic in terms of NaCl, is rich in urea, since the thick ascending limb is impermeable to urea. The hypotonic urea-rich fluid reaches the distal convoluted tubule and the collecting ducts in the outer medullary and cortical regions. These segments are also impermeable to urea but are freely permeable to water (under the influence of AVP) and salt. As the fluid moves progressively down into the collecting ducts of the inner medulla, these segments are freely permeable to the urea (under the influence of AVP). The passive passage of urea into the medullary interstitium adds to the hypertonicity, which is now prepared to draw fluid from the descending limb of the loop of Henle, and the whole process restarts. Thus, the countercurrent multiplier system is an amazing example of economy and recycling, where no event is a waste, and all phenomena are geared toward providing a hypertonic interstitium to concentrate the filtrate (Fig. 46).

It is essential to understand that urinary concentration can be attained, often to an impressive degree, even in the absence of AVP. This is especially so when the glomerular filtration rate (GFR) is reduced. In animal studies, this is illustrated by the fact that the Brattleboro rat, an animal with congenital lack of vasopressin, can increase urinary osmolality by threefold when the GRF is reduced by partial aortic clamping.[53,54] Similarly, progressive solute diuresis (osmotic diuresis) can modify urinary osmolality regardless of the presence or absence of AVP.[55]

Cellular Basis of AVP Action

Arginine vasopressin acts on the distal and collecting tubules and promotes water reabsorption. The first step in the expression of AVP action is binding of the hormone to specific vasopressin receptors located at the basolateral membrane of responsive cells.[56] The tissue receptors for vasopressin are classified into V_1 receptors located in the smooth muscle and V_2 receptors located in the renal epithelial cells. The property of stimulating adenylate cyclase in response to AVP is limited to the V_2-type receptors.[57]

The second step, following hormone receptor binding, is the activation of adenylate cyclase, in response to AVP. Work reported by Jard et al.[58,59] supports the notion that even with low rates of receptor occupancy by AVP, and suboptimal activation of adenylate cyclase, the hormone can elicit a full antidiuretic response. Activation of adenylate cyclase stimulates conversion of ATP to cyclic adenosine monophosphate (3'5'-cAMP). The degradation of this cyclic nucleotide is regulated by the intracellular concentration of phosphodiesterases.

The third step in the action of AVP involves the activation of protein kinases within the cell by cAMP.[60,61] The protein kinases within the cell regulate the phosphorylation of specific cell proteins; these proteins play a major role in the formation of certain organelles involved in the expression of AVP action.

The fourth step involves the formation of intracellular microtubules or microfilaments. These intracellular organelles are somehow involved in expressing the hydroosmotic effect of vasopressin.[62] AVP increases the water permeability of the distal and collecting tubules by increasing the number of pores in the membrane. The method of water movement induced by AVP is by diffusion. By maximizing water permeability, AVP allows water to move down its concentration gradient, shown in Fig. 46. The hypotonic fluid emerging from the ascending limb of the loop of Henle approaches osmotic equilibrium with the isotonic cortical interstitium. As a result of the action of AVP, a large fraction of water delivered to this segment is reabsorbed. As the isotonic fluid reaches the medullary collecting duct, also under the influence of the AVP, the fluid becomes more concentrated, reaching a concentration in equilibrium with that of the medullary interstitium.

The following conditions may override the effect of AVP. Each one is very important when considering the failure of AVP to act.

1. *Significant reduction in the nephron population:* Obviously, loss of nephrons and interstitium would limit the urinary concentrating ability.
2. *Increased excretion of solute (glucose or urea):* This condition can impair the ability to concentrate urine—"solute diuresis."
3. *Dilute medullary interstitium:* This would limit the concentration of the final urine. Even with AVP, the final urine can only be as concentrated as the medullary interstitium. A hypotonic interstitium can be encountered under the following conditions: compulsive water drinking, renal disease, sickle cell disease, and Sjögren's syndrome.
4. *Antagonists to the action of AVP:* These can minimize or neutralize the action of AVP. Included are such conditions as hypercalcemia and hypokalemia and a variety of drugs, notably lithium and demeclocycline.

Clinical Disorders Involving AVP

Diabetes insipidus is prototypical of the disorder that results from inadequate secretion of vasopressin (central) or from impairment of its peripheral action (nephrogenic). The syndrome of inappropriate secretion of ADH (SIADH) results from sustained secretion of AVP despite a hypotonic plasma. The syndrome of reset osmostats can manifest at extreme ends of the spectrum. When the osmostats are set at a higher threshold, the osmoregulatory problem is characterized by the release of AVP only when the serum osmolality reaches very high levels. These patients manifest with polyuria, dehydration, and hypernatemia. By contrast, when the osmostats are reset at a lower level, the osmoregulatory problem is characterized by continued secretion of AVP, even when the serum is hypotonic. These patients present with hyponatremia.

Loss of thirst (hypodipsic or adipsic) syndromes present with polyuria, dehydration, and hypernatemia. Adipsia can be caused by tumors or can be primary (essential hypernatremia).

Psychogenic polydipsia results from excessive and chronic ingestion of water. The resulting water diuresis is caused by dilution of the medullary interstitium as well as by suppression of endogenous release of vasopressin due to chronic water imbibing.

References

1. Sachs H, Portanova R, Haller EW, et al: Cellular processes concerned with vasopressin biosynthesis, storage, and release. In Stutinsky F (ed): *Neurosecretion. Fourth International Symposium on Neurosecretion.* Springer-Verlag, Berlin, 1967, p. 146.
2. Sachs H: Biosynthesis and release of vasopressin. *Am J Med* **42**:687, 1967.
3. Sachs H, Goodman R, Osinchak J, et al: Supraoptic neurosecretory neurons of the guinea pig in organ culture. Biosynthesis of vasopressin and neurophysin. *Proc Natl Acad Sci USA* **68**:2782, 1971.
4. Buford GD, Jones CW, Pickering BT: Tentative identification of a vasopressin-neurophysin and an oxytocin-neurophysin in the rat. *Biochem J* **124**:809, 1971.
5. Cheng KW, Friesen HG: Physiological factors regulating secretion of neurophysin. *Metabolism* **19**:876, 1970.
6. Cheng KW, Friesen HG: A radioimmunoassay for vasopressin binding protein, neurophysin. *Endocrinology* **88**:608, 1971.
7. Zimmerman EA, Robinson AG: Hypothalamic neurons secreting vasopressin and neurophysin. *Kidney Int* **10**:12, 1976.
8. Zimmerman EA: Localization of hypothalamic hormones by immunocytochemical techniques. In Martini L, Ganong WF (eds): *Frontiers in Neuroendocrinology.* Vol. 4. Raven, New York, 1976.
9. Sunde DA, Sokol HW: Quantification of rat neurophysins by polyacrylamide gel electrophoresis: Application to the rat with hereditary hypothalamic diabetes insipidus. *Ann NY Acad Sci* **248**:345, 1975.
10. Nagasawa J, Douglas WW, Schultz RA: Ultrastructural evidence of secretion by exocytosis and of "synaptic vesicle" formation in posterior pituitary gland. *Nature (Lond)* **227**:407, 1970.
11. Nagasawa J, Douglas WW, Schultz RA: Micropinocytotic origin of coated and smooth microvesicles ("synaptic vesicles") in neurosecretory terminals of posterior pituitary glands demonstrated by incorporation of horseradish peroxidase. *Nature (Lond)* **232**:341, 1971.
12. Douglas WW, Nagasawa J, Schultz RA: Coated microvesicles in neurosecretory terminals of posterior pituitary glands shed their coats to become smooth "synaptic" vesicles. *Nature (Lond)* **232**:340, 1971.
13. Douglas WW, Poisner AM: Stimulus–secretion coupling in a neurosecretory organ. The role of calcium in the release of vasopressin from the neurohypophysis. *J Physiol (Lond)* **172**:1, 1964.
14. Sachs H, Share L, Osinchak J, et al: Capacity of the neurohypophysis to release vasopressin. *Endocrinology* **81**:755, 1967.
15. Hayward JN: Neural control of the posterior pituitary. *Annu Rev Physiol* **37**:191, 1975.
16. Verney EB: The antidiuretic hormone and the factors which determine its release. *Proc R Soc Lond* **135**:25, 1947.
17. Thrasher TN: Osmoreceptor mediation of thirst and vasopressin secretion in the dog. *Fed Proc* **41**:2528, 1982.
18. McKinley MJ, Denton DA, Weisinger RS: Sensors for antidiuresis and thirst—osmoreceptors or CSF sodium detectors? *Brain Res* **141**:89, 1978.
19. Andersson B, Olsson K: Evidence for periventricular sodium-sensitive receptors of importance in the regulation of ADH secretion. In Moses AM, Sharle L (eds): *Neurohypophysis.* Karger, Basel, 1977, p. 118.
20. Thrasher TN, Brown CJ, Keil LC, et al: Thirst and vasopressin release in the dog: An osmoreceptor or sodium receptor mechanism? *Am J Physiol* **238**:R333, 1980.
21. Segar WE, Moore WW: The regulation of antidiuretic hormone release in man. I. Effects of change in position and ambient temperature on blood ADH levels. *J Clin Invest* **47**:2143, 1968.
22. Dunn FL, Brennan JT, Nelson AE, et al: The role of blood osmolality and volume in regulating vasopressin secretion in the rat. *J Clin Invest* **52**:3212, 1973.
23. Robertson GL, Shelton RL, Athar S: The osmoregulation of vasopressin. *Kidney Int* **10**:25, 1976.

24. Culpepper RM, Herbert SC, Andreoli TE: Nephrogenic diabetes insipidus. In Stanbury JB, Wyngarden JB, Fredrickson DS, et al. (eds): *The Metabolic Basis of Inherited Disease*. 5th Ed. McGraw-Hill, New York, 1983, p. 1867.

25. Moore WW: Antidiuretic hormone levels in normal subjects. *Fed Proc* **30**:1387, 1971.

26. Gauer OH, Henry JP, Sieker HO, et al: The effect of negative pressure breathing on urine flow. *J Clin Invest* **33**:287, 1954.

27. White WA, Bergland RM: Experimental inappropriate ADH secretion caused by positive pressure respirators. *J Neurol Surg* **36**:608, 1972.

28. Murdaugh HV, Sieker HO, Manfredi F: Effect of altered intrathoracic pressure on renal hemodynamics, electrolyte excretion and water clearance. *J Clin Invest* **38**:834, 1959.

29. Weinstein H, Berne RM, Sachs H: Vasopressin in blood: Effect of hemorrhage. *Endocrinology* **66**:712, 1960.

30. Reid IA: Actions of angiotensin II on the brain: Mechanisms and physiologic role. *Am J Physiol* **246**:F533, 1984.

31. Ramsay DJ: Effects of circulating angiotensin II on the brain. In Ganong WF, Martini L (eds): *Frontiers in Neuroendocrinology*. Raven, New York, 1982, p. 263.

32. Reid IA, Brooks VL, Rudolph CD, et al: Analysis of the actions of angiotensin on the central nervous system of conscious dogs. *Am J Physiol* **243**:R82, 1982.

33. Mouw D, Bonjour J, Malvin RL, et al: Central action of angiotensin in stimulating ADH release. *Am J Physiol* **220**:239, 1971.

34. Szczepanska-Sadowska E: Plasma ADH increase and thirst suppression elicited by preoptic heating in the dog. *Am J Physiol* **226**:155, 1974.

35. Berl T, Cadnapaphornchi P, Harbottoe JA, et al: Mechanism of stimulation of vasopressin release during beta adrenergic stimulation with isoproterenol. *J Clin Invest* **53**:857, 1974.

36. Berl T, Cadnapaphornchi P, Harbottoe JA, et al: Mechanism of suppression of vasopressin during alpha adrenergic stimulation with norepinephrine. *J Clin Invest* **53**:219, 1974.

37. Schrier RW, Liberman R, Ufferman RC: Mechanism of antidiuretic effect of beta-adrenergic stimulation. *J Clin Invest* **51**:97, 1972.

38. Klein LA, Lieberman B, Laks M, et al: Interrelated effects of antidiuretic hormone and adrenergic drugs on water metabolism. *Am J Physiol* **221**:1657, 1971.

39. Epstein AN, Fitzsimons JT, Roots BJ: Drinking induced by injection of angiotensin into the brain of the rat. *J Physiol (Lond)* **210**:457, 1970.

40. Fitzsimons JT: The effect on drinking of peptide precursors and of shorter chain peptide fragments of angiotensin II injected into the rat's diencephalon. *J Physiol (Lond)* **214**:295, 1971.

41. Severs WB, Severs AED: Effects of angiotensin on the central nervous system. *Pharmacol Rev* **415**:448, 1973.

42. Leksell LG: Influence of PGE on cerebral mechanisms involved in control of fluid balance. *Acta Physiol Scand* **98**:85, 1976.

43. Andersson B, Leksell LG: Effects on fluid balance of intraventricular infusions of prostaglandin E_1. *Acta Physiol Scand* **93**:286, 1975.

44. Yamamoto M, Share L, Shade RE: Vasopressin release during ventriculocisternal perfusion with prostaglandin E_2. *J Endocrinol* **71**:325, 1976.

45. Bauman G, Dingman JF: Distribution, blood transport and degradation of antidiuretic hormone in man. *J Clin Invest* **57**:1109, 1976.

46. Rector FC Jr: Renal concentrating mechanisms. In Andreoli TE, Grantham JJ, Rector FC Jr (eds): *Disturbances in Body Fluid Osmolality*. American Physiology Society, Bethesda, MD, 1977, p. 179.

47. Jamison RL, Maflly RH: The urinary concentrating mechanism. *N Engl J Med* **295**:1059, 1976.

48. Tisher CC: Anatomy of the kidney. In Brenner BM, Rector FC Jr (eds): *The Kidney*. WB Saunders, Philadelphia, 1976, p. 280.

49. Weitzman RE, Kleeman CR: The clinical physiology of water metabolism. Part II. Renal mechanisms for urinary concentration; diabetes insipidus. *West J Med* **131**:486, 1979.

50. Kokko JP, Rector FC Jr: Countercurrent multiplication system without active transport in inner medulla. *Kidney Int* **2**:214, 1972.
51. Stephenson JL: Concentration of urine in a central core model of the renal counterflow system. *Kidney Int* **2**:85, 1972.
52. Stephenson JL: Concentrating engines and the kidney. I and II. *Biophys J* **13**:512, 1973.
53. Gellai M, Edwards BR, Valtin H: Urinary concentrating, ability during dehydration in the absence of vasopressin. *Am J Physiol* **237**:F100, 1979.
54. Edwards BR, Gallai M, Valtin H: Concentration of urine in the absence of ADH with minimal or no decrease in GFR. *Am J Physiol* **239**:F84, 1980.
55. Gennari FJ, Kassirer JP: Osmotic diuresis. *N Engl J Med* **291**:714, 1974.
56. Walter R, Clark WS, Mehta PK, et al: Conformational considerations of vasopressin as a guide to development of biological probes and therapeutic agents. In Andreoli TE, Grantham JJ, Rector FC Jr (eds): *Disturbances in Body Fluid Osmolality*. American Physiology Society, Bethesda, MD, 1977, p. 1.
57. Sawyer WH, Manning M: Effective antagonists of the antidiuretic action of vasopressin in rats. *Ann NY Acad Sci* **394**:464, 1982.
58. Jard S, Roy C, Barth T, et al: Antidiuretic hormone-sensitive kidney adenylate cyclase. *Adv Cyclic Nucleotide Res* **5**:31, 1975.
59. Jard S, Bockaert J: Stimulus–response coupling in neurohypophyseal peptide target cells. *Physiol Rev* **55**:489, 1975.
60. Dousa TP, Valtin H: Cellular actions of vasopressin in the mammalian kidney. *Kidney Int* **10**:55, 1976.
61. Andreoli TE, Schafer JA: Mass transport across cell membranes. The effects of antidiuretic hormone on water and solute flows in epithelia. In Knobil E, Sonnenschein RR, Edelman IS (eds): *Annual Review of Physiology*. Vol. 38, Annual Reviews, Palo Alto, CA, 1977, p. 451.
62. Taylor A: Role of microtubules and microfilaments in the action of vasopressin. In Andreoli TE, Grantham JJ, Rector FC Jr (eds): *Disturbances in Body Fluid Osmolality*. American Physiology Society, Bethesda, MD, 1977, p. 97.

20

Diabetes Insipidus

Introduction

The diabetes insipidus syndrome is characterized by the inability to concentrate urine or conserve water. The term *insipidus* reflects the pale, almost colorless waterlike urine excreted by patients in the fully expressed syndrome. Diabetes insipidus results from either impaired synthesis and release of vasopressin (central diabetes insipidus, neurogenic diabetes insipidus, vasopressin-sensitive diabetes insipidus) or peripheral resistance to the actions of vasopressin at the level of its target organ, the renal tubules (nephrogenic diabetes insipidus). Regardless of the type, the resulting disturbances in water metabolism are identical, with the continuous production of extremely large volumes of very dilute urine. Diabetes insipidus can be partial or complete, temporary or permanent.

Etiology of Central Diabetes Insipidus

The etiology of central diabetes insipidus is considered first (Table 97). The largest single category in the causation of diabetes insipidus is the idiopathic variety, which accounts for 32–50% of cases. It should be emphasized that in nearly one-half of cases of central diabetes insipidus, an underlying etiology will be discovered if carefully sought. The diagnosis of idiopathic diabetes insipidus therefore be made only after an exhaustive search has failed to disclose a secondary etiology.

Idiopathic diabetes insipidus can be either familial or nonfamilial. The hereditary form is inherited either as an autosomal dominant trait with incomplete penetrance or as an X-linked recessive trait. The basic defect in the

TABLE 97
Etiology of Central Diabetes Insipidus

Idiopathic
  Familial
  Nonfamilial
Posthypophysectomy
  Transient
  Permanent
Tumorous
  Suprasellar tumors and cysts
      Germinoma
      Craniopharyngioma
      Hamartoma
  Sellar tumors that extend superiorly
      Pituitary adenomas
      Stalk compression
  Metastatic disease to hypothalamic pituitary area
      Breast
      Lymphoma leukemia
      Colon
Traumatic
  Basal skull fracture
Granuloma and granulomatous disease
  Eosinophilic granuloma
  Hand-Schüller-Christian disease
  Sarcoidosis
  Granulomatous hypophysitis
  Wegener granulomatosis
Cerebrovascular catastrophe
  Intraventricular hemorrhage
  Pituitary apoplexy
  Aneurysm rupture
  Cerebrovascular accident (CVA)
  Postpartum bleeding
Infection
  Meningitis
  Encephalitis

idiopathic variety is believed to be a decrease in the number of secretory neurons of the supraoptic and paraventricular nuclei. Braverman et al.[1] reported the autopsy findings in a patient with central diabetes insipidus, and demonstrated an impressive reduction in the number of nerve cells in the hypothalamic supraoptic and paraventricular nuclei, a striking decrease in the secretory granules (Nissl granules), increased gliosis, and a diminutive posterior pituitary gland. Although arginine vasopressin (AVP) and oxytocin are synthesized by the same nuclei, it is interesting to note that patients with idiopathic central diabetes insipidus do not always lose oxytocin as well. This is illustrated by the report by Legros and Crabbe,[2] who studied five members from a single family of hereditary central diabetes insipidus, and showed absence of AVP and its neurophysin in the serum and urine, while oxytocin and its neurophysin were sometimes present. The reason for this dichotomy is unknown. More recently, Nagai et al.[3] described the autopsy findings in a patient who had a lifelong history of hereditary central diabetes insipidus, and had died of a myocardial infarction. At autopsy, there was no atrophy of the supraoptic and paraventricular nuclei. Immunohistochemically, vasopressin was demonstrable in the nerve cells of the supraoptic nucleus. The paraventricular nuclear cells, however, showed very little immunopositivity for AVP.

Hypophysectomy is an extremely common reason for transient and sometimes permanent diabetes insipidus. Experimentally, when the neurohypophyseal system is injured, the resultant diabetes insipidus is classically triphasic:[4,5] (1) an acute phase of polyuria, polydipsia; (2) an interphase of normalcy characterized by a tendency to excrete a less than maximally dilute urine; and (3) permanent diabetes insipidus, if the stalk is sectioned above the median eminence. The mechanism of the first phase is probably akin to spinal shock, with failure to discharge AVP from the hypothalamus and the posterior lobe. The second phase is characterized by a slow leak of AVP from the hypothalamus and the posterior lobe. This phase shares similarities with the SIADH syndrome, for example, the tendency for hyponatremia and water intoxication, when fluid loads are administered. The final phase evolves when all the stored hormone is depleted and no more supply can reach the posterior lobe. Three important observations pertain to postneurosurgical diabetes insipidus. First, permanent diabetes insipidus does not develop if the posterior lobe is removed or if the stalk is sectioned below the median eminence. Second, the characteristic triphasic response seen with experimental stalk section may not be always encountered with surgery; for instance, the normal interphase will not be seen if the posterior lobe or the median eminence is removed. Finally, during the postoperative period, it may be difficult to distinguish the polyuria of mild diabetes insipidus from the polyuria of vigorous postoperative fluid administration.

Diabetes insipidus secondary to tumors represents the most important category of etiologies, owing to their strategic location and the potential pressure effects. Suprasellar tumors are the most frequent intracranial neoplasms that present with diabetes insipidus. Thus, craniopharyngiomas, germ cell tumors, optic and hypothalamic gliomas, parasellar meningioma, and a whole variety of other hypothalamic tumors can result in the development of diabetes insipidus.[6-9] The mechanism for such a phenomenon involves destruction of either the supraoptic and paraventricular nuclei or the neuronal pathways above the median eminence. Diabetes insipidus is a particularly early manifestation of germ cell tumors, especially the germinomatous variety that originates in the suprasellar cistern. Craniopharyngiomas are the most common cause of tumor-related diabetes insipidus in children. This is especially so in patients with craniopharyngiomas occurring in association with a prefixed optic chiasm, which directs the growth of tumor in the posteroinferior direction (Chapter 16).

Metastatic disease is an important etiology for diabetes insipidus occurring for the first time in older patients. Kimmel et al.[10] recently reported that diabetes insipidus can be the first manifestation of systemic cancer. In their series of 11 patients with the initial presentation of diabetes insipidus, the most common malignant disease was small cell cancer of the lung. Computed tomography (CT) of the sellar area demonstrated detectable abnormalities in one-half of patients. The study underscores the well-known notion that metastatic disease should be suspected if no other cause for diabetes insipidus is evident. The most common metastatic disease associated with diabetes insipidus is metastatic breast carcinoma,[11] followed by bronchogenic neoplasms, hematogenous malignancies and neoplasms of the gastrointestinal (GI) tract. In general, when hypothalamic metastases have occurred, there is evidence of generalized dissemination of the malignancy. Occasionally, however, hypothalamic-pituitary involvement by metastatic disease can be a solitary event.[12] Such a phenomenon can be encountered with GI neoplasms and breast cancer.

The diabetes insipidus associated with granulomatous disease is particularly evident in patients with Hand-Schüller-Christian disease[13-19] and sarcoidosis.[20-23] Diabetes insipidus, in both conditions, is usually associated with other perturbations in hypothalamic-pituitary dysfunction. This is most commonly represented by hyperprolactinemia, and varying degrees of decreased secretion of the other adenohypophyseal hormones. Growth hormone (GH) deficiency, manifested as growth retardation in children, is the most common endocrinopathy seen in association with the diabetes insipidus of Hand-Schüller-Christian disease,[18,19] while partial panhypopituitarism is the endocrinopathy associated with the diabetes insipidus caused by sarcoidosis. Lytic lesions in the skull bones are less frequently encountered in sarcoidosis.

Other causes of central diabetes insipidus are even rarer. Two in particular bear relevance: pituitary apoplexy and Sheehan's syndrome. Both conditions represent acute vascular catastrophes involving hemorrhagic or ischemic necrosis of the pituitary gland. Varying degrees of diabetes insipidus has been reported in the acute phase following pituitary apoplexy.[24-28] Since the posterior pituitary has a different blood supply than that of the anterior pituitary, it is usually spared from the apoplectic damage. Therefore, the development of diabetes insipidus in these cases, which is quite often permanent, must be on the basis of damage to the hypothalamus or the median eminence. Since the infundibulum is often "choked" in pituitary apoplexy, the possibility of median eminence damage is quite appealing. However, convincing autopsy proof for such a phenomenon is lacking.

The association of diabetes insipidus with Sheehan's postpartum hemorrhage is also well known.[29-34] Approximately 30 cases of diabetes insipidus with postpartum hemorrhage have been reported. Recently, Schwartz and Leddy[35] described yet another case of diabetes insipidus associated with Sheehan hypopituitarism, pointing out an important clinical dictum, that adenohypophyseal insufficiency may mask the symptomatology of diabetes insipidus. The reason for the development of diabetes insipidus in Sheehan's syndrome has been elucidated fairly well. Histopathological demonstration of vascular damage to the hypothalamus by postpartum hemorrhage has been well documented in the literature.[36,37]

Nephrogenic Diabetes Insipidus

Congenital nephrogenic diabetes insipidus is predominantly seen in boys and is inherited as an X-linked disorder. It is believed that the gene for nephrogenic diabetes insipidus was introduced into the North American continent by the Ulster clan from Scotland, who landed in 1761 in Halifax, Nova Scotia, aboard the carrier The Hopewell.[38] The disorder is a consequence of failure to generate cAMP in the renal tubular cell in response to vasopressin.

The most common causes of acquired nephrogenic diabetes insipidus are metabolic abnormalities (hypokalemia, hypercalcemia) and drugs that inhibit the action of vasopressin (lithium carbonate, demeclocycline). Any form of interstitial nephropathy (e.g., analgesic, lead, pychonephritis), can affect tubular function and may result in suboptimal urinary concentration in response to endogenous and exogenous vasopressin. A recent addition to the causes of nephrogenic diabetes insipidus is the transient vasopressin resistance seen in the latter part of pregnancy. Barron et al.[39] described three women who had transient diabetes insipidus during late pregnancy and early puerperium. The collective findings in all three included polyuria, polydipsia, hypotonic urine despite substantial hypertonic body fluids, measurable AVP levels, and failure of urine osmolality to respond to vasopressin administration. Regardless of the mechanisms underlying this phenomenon, this condition is truly transitory. Table 98 outlines the diverse causes of nephrogenic diabetes insipidus.

Clinical Features

The two salient symptoms of any form of diabetes insipidus are polyuria and polydipsia. The physical examination is usually negative; the presence of associated features depends on the underlying etiology.

TABLE 98
Etiology of Nephrogenic Diabetes Insipidus

Congenital
  X-linked
Acquired
  Hypercalcemia
  Hypokalemia
  Chronic renal disease–interstitial nephropathy
  Sickle cell disease
  Amyloidosis
  Multiple myeloma
  Sjögren's syndrome
Transient
  Drugs
    Lithium carbonate
    Demeclocycline
  Pregnancy

Polyuria

In complete diabetes insipidus, the urine output can exceed 10 liters/day. The onset is acute in central idiopathic diabetes insipidus, patients often being able to recall the exact day the polyuria started. Nocturia is invariably present. The severity of the polyuria poses restrictions on the patient's life style, often to a frustrating degree. The description of the urine is often characteristic: pale, dilute, waterlike urine. As long as the patient is conscious, with an intact thirst center, and has access to water, overt signs of dehydration do not develop. When the polyuria is severe, secondary bladder distention and hydronephrosis may supervene. Occasionally, the polyuria in diabetes insipidus may not be impressive. It should be realized that the development of hypoadrenalism tends to mask or improve the diabetes insipidus.[40] This is because of the marked decrease in free water clearance that develops secondary to glucocorticoid deficiency. Upon replacement with glucocorticoids, the polyuria of diabetes insipidus becomes apparent and is unmasked.[41]

Polydipsia

Increased thirst invariably accompanies the polyuria of diabetes insipidus as a compensatory mechanism to balance the fluid losses in the urine. One situation in which thirst may be absent is when central diabetes insipidus is caused by a suprasellar tumor that has also encroached on and destroyed the thirst center in the hypothalamus. Such a combination is obviously disastrous, as such patients rapidly become dehydrated. Yet despite the proximity of the thirst center to the supraoptic and paraventricular nuclei, tumors that cause diabetes insipidus seldom destroy the thirst center.

Associated Features

Associated features depend on the underlying etiology and are mostly encountered when the diabetes insipidus is of central origin.

1. Signs of chiasmal compression are indicative of a suprasellar tumor causing diabetes insipidus.
2. Optic atrophy or papilledema with central diabetes insipidus points increased intracranial tension, usually secondary to tumors that obstruct the flow of cerebrospinal fluid (CSF).
3. The demonstration of the Parinaud sign, impairment of upward gaze, in a patient with diabetes insipidus is highly suggestive of a pinealoma (dysgerminoma).
4. The combination of diabetes insipidus and precocious puberty again suggests a suprasellar tumor as the etiology for both.
5. The development of diabetes insipidus in a patient with a past or present history of breast cancer or colorectal cancer is indicative of metastatic disease to the median eminence or the hypothalamus.
6. The features of clinical hypopituitarism in a patient with diabetes insipidus suggest a suprasellar tumor extending inferiorly or an intrasellar tumor extending upward.
7. The relationship between central diabetes insipidus and memory deficits is an intriguing and controversial one. Danguir[42] demonstrated the existence of paradoxical sleep deficits and memory deficits in the Brattleboro rat, the animal model for central diabetes insipidus. Laczi et al.[43] administered a vasopressin analogue to 13 patients with central diabetes insipidus and demonstrated consistent facilitation of short-term memory, as well as some improvement in long-term memory. These changes were unrelated to changes in renal handling of water, as the analogue was devoid of antidiuretic properties.

Diagnostic Studies

The standard test used in the distinction of the various types of diabetes insipidus is the water-deprivation test. The assay for vasopressin assumes an adjunctive role in the diagnosis. Additional tests are employed, depending on whether the diabetes insipidus is central or nephrogenic in origin.

Water-Deprivation Test

The principle of this test is to evaluate the release of vasopressin, the antidiuretic hormone (ADH), in response to a potent physiological stimulus, such as rendering the plasma hypertonic by dehydration. The test, usually performed in the morning, begins by obtaining of the basal parameters: weight, blood, pressure, pulse rate, urine output, and serum and urine osmolality.

Water is withheld, and the response of the urine osmolality and urine output is evaluated on an hourly basis. The serum osmolality should be measured at the beginning and end of the test. As the water deprivation proceeds, the patient with true diabetes insipidus becomes desperate for water and will obtain water by any means unless closely supervised. The end point of fluid deprivation can be any of the following: (1) three consecutive urine samples that are nearly identical (<30 mOsm difference), (2) clinical evidence of volume depletion, or (3) loss in body weight exceeding 3–5% of the weight at the start of the study. The second part of the water-deprivation test involves evaluating the urinary concentrating ability following the injection of 5 units of aqueous vasopressin.

The normal subject, following water deprivation, shows a rapid increase in urinary osmolality, reaching isomotic levels within hours. By 4–6 hr, the urinary osmolality reaches 800–1000 mOsm, with a prompt drop in urine flow. These effects result from the physiological release of vasopressin in response to dehydration. After maximal concentration has been attained, as demonstrated by a plateau in the urine osmolality, further injection of aqueous vasopressin fails to increase the urinary concentration significantly. In most normal subjects, the urinary osmolality either declines slightly or rises by less than 9% after aqueous vasopressin. Malnourished and debilitated patients may show a suboptimal rise in urine osmolality (400–800 mOsm/kg) following dehydration.

The patient with complete central diabetes insipidus responds in a characteristic manner to water deprivation: to begin with, these patients start with a very low urinary osmolality (50–100 mOsm) and hardly attain an isosmotic range by dehydration; however, following aqueous vasopressin administration, there is a dramatic increase in the urinary osmolality, often exceeding 800–1000 mOsm, and at least 50% greater than the baseline.

The patient with nephrogenic diabetes insipidus behaves similarly to the patient with central diabetes insipidus during the first part of the test; however, there is a characteristic failure to respond to vasopressin during the second part of the test, highlighting the target organ resistance of the renal tubules to vasopresssin.

Partial central diabetes insipidus is characterized by a slow, steady, but inadequate increase in the urinary osmolality. In fact, the end point in these patients takes a frustratingly long time to be reached. Following the administration of aqueous vasopressin, there is an increase in the urine osmolality by 9% or more over the baseline.[44] This contrasts sharply with the normal individual, who attains maximal concentration with dehydration, following which vasopressin injection causes little if any change.

Patients with psychogenic polydipsia, a disorder characterized by compulsive water drinking, can be regarded as having two separate problems. First, because of chronic water ingestion, there is suppression of endogenous vasopressin, leading to sluggishness in responding to water deprivation; the first part of the test resembles partial central diabetes insipidus. Second, the renal medullary interstitium of these patients has been rendered hypotonic

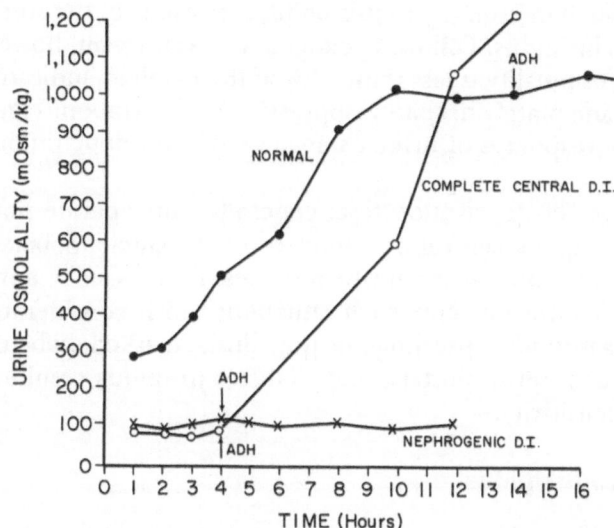

FIGURE 47. The water-deprivation test in complete central diabetes insipidus (DI) demonstrates no urinary concentration with water deprivation but normal concentration following antidiuretic hormone (ADH). The water-deprivation test in nephrogenic diabetes insipidus resembles complete central diabetes insipidus during the first part of the test. However, the disease is characterized by failure to respond to ADH administration.

(dilute) as a result of chronic water imbibing; therefore, there is impairment even in responding to exogenous aqueous vasopressin. Thus, the second part of the test demonstrates an element of vasopressin resistance. The patient with psychogenic polydipsia starts with a urine osmolality of 200–300 mOsm and then shows a gradual but inadequate concentrating ability, reaching a

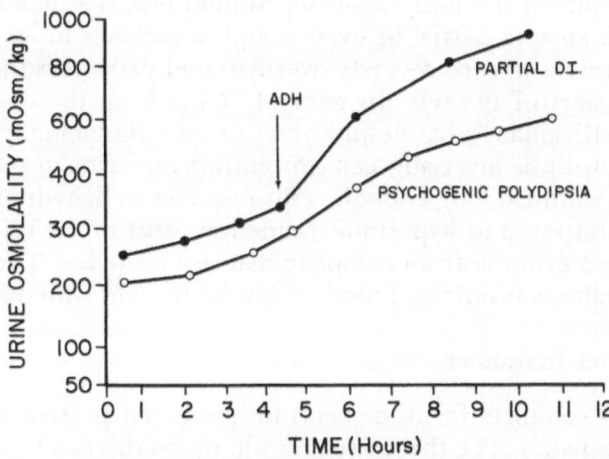

FIGURE 48. The results of the water-deprivation test in patients with partial diabetes insipidus (DI) may resemble those of patients with psychogenic polydipsia (see text).

plateau of 400–600 mOsm. Up to this point, the results of the study resemble partial diabetes insipidus. Following exogenous vasopressin, however, the increase in urine osmolality is less than 20% of the baseline, indicating inability to concentrate adequately despite vasopressin administration. Figures 47 and 48 illustrate the responses of urine osmolality to water deprivation in various disease states.

While the water-deprivation tests generally can separate normals from patients with complete central or complete nephrogenic diabetes insipidus, there are several circumstances in which the test results can be nondiagnostic, equivocal, and confusing. Five such situations merit consideration; partial central diabetes insipidus, psychogenic polydipsia, masked diabetes insipidus, the syndrome of reset osmostats, and diabetes insipidus resulting from osmoreceptor insensitivity.

Partial Diabetes Insipidus

Diabetes insipidus is not an all-or-none disorder. Miller et al.[44] demonstrated that patients with partial diabetes insipidus, after attaining maximal but subnormal urinary concentrations by dehydration, show further rise in urinary concentrations following pitressin. This rise in osmolality (9–25%) is not seen in normals or in patients with psychogenic polydipsia.

Psychogenic Polydipsia

The dual problems caused by chronic, excessive water intoxication are suppression of endogenous AVP secretion and dilution of the medullary interstitium. These patients must first excrete the positive water load before they can begin to concentrate their urine in response to dehydration. In some patients who are profoundly overhydrated, it may take more than 24 hr before the urine even becomes isosmotic. When such a patient is evaluated on the basis of the results of the 8-hr water-deprivation test, the data may show a striking resemblance to partial or even complete diabetes insipidus. An important clue present in such severely overhydrated patients is the hypotonic plasma at the start of the test. In general, it has been the observation of Weitzman and Kleeman[45] that despite years of water imbibing, patients with psychogenic polydipsia invariably can concentrate the urine to at least isotonicity if water is withheld long enough. This response to dehydration coupled with the AVP response to hypertonic saline can assist in the differentiation of psychogenic polydipsia from complete diabetes insipidus. The distinction from partial diabetes insipidus, however, can be fraught with difficulties.

Masked Diabetes Insipidus

Even in its complete form, diabetes insipidus can present with absence of significant polyuria. The three situations in which this can be encountered are decreased GFR, depletion of salt, and combined anterior and posterior pituitary failure. When the GFR of patients with complete diabetes insipidus is severely reduced (as in volume depletion and shock), the polyuria can ameliorate to a surprising extent. In these circumstances, backdiffusion of water occurs even in the absence of AVP, resulting in urinary volumes of less than 3 liters/day and urinary osmolalities near or above plasma osmolality. Another important condition that can ameliorate the voluminous water excretion of diabetes insipidus is severe salt depletion. This, however, is unusually encountered in patients with diabetes insipidus. A more frequent entity that masks diabetes insipidus is the presence of concomitant glucocorticoid deficiency, as in patients with combined anterior and posterior pituitary insufficiency. In such patients, the free water clearance is so reduced that the lack of AVP matters little, in terms of expression of the polyuria. When glucocorticoid therapy is reinstituted, polyuria promptly, and often dramatically, reappears. It is unclear whether the decrease in free water clearance is due to a direct effect of lack of glucocorticoids or to an indirect effect of water and salt depletion associated with glucocorticoid deficiency.

Reset Osmoreceptors—Essential Hypernatremia

This relatively rare entity is characterized by normally operating central mechanisms for vasopressin release, albeit at much higher levels of osmoregulation.[46] These patients characteristically present with features of diabetes insipidus but are able to concentrate the urine normally after prolonged dehydration. The feature that sets these patients apart is that vasopressin release occurs only when the serum osmolality rises to remarkably high levels (>290 mOsm) in response to protracted dehydration. It is believed that the osmoreceptors in these patients are reset and consequently recognize and respond to hyperosmolality only when the levels rise to very high degrees. The serum sodium levels are usually elevated (essential hypernatremia), reflecting the hyperosmolar state. The term essential hypernatremia denotes a rare syndrome caused by resetting of the osmostats; it is characterized by disturbed osmoreceptor function, hence a chronically elevated serum sodium.[46–49] There is usually a well-defined underlying etiology, such as trauma, tumors in the hypothalamic region, and granulomatous disease (histiocytosis). Patients with essential hypernatremia demonstrate several facets suggestive of hypothalamic involvement. The two most commonly described phenomena are obesity and defective or absent thirst sensation. The association between essential hypernatremia and anterior hypophyseal deficiency syndrome has been described in the literature.[49] Yet these patients are different from those with central diabetes insipidus, since the basic problem is not one of AVP deficiency, but represents a resetting of the osmostats (i.e., normal-shaped release curve of AVP), but one that occurs at a higher osmotic threshold. This theory of reset osmostats was based on indirect result obtained by evaluating the urine osmolalities following hypertonic saline infusions. Gill et al.[50] were the first to document resetting of osmostats in the essential hypernatremia syndrome, by direct measurement of vasopressin. These workers described a 24-yr-old, short, obese woman with neurological and behavioral changes attributable to hypernatremia. Studies of osmoregulation by water loading and hypertonic saline infusion documented elevation of the threshold for AVP release, which had a normal slope after release. In addition to abnormal osmoregulation, the patient also showed suboptimal AVP release to baroregulation. Two additional remarkable features of this case were the presence of associated deficiencies in GH, gonadotropins, and prolactin and entirely normal radiology of the sellar and suprasellar region by CT and nuclear magnetic resonance (NMR) scans.

Two additional patients were described by Dunger et al.,[51] who presented with hypernatremia, lack of thirst, osmoreceptor dysfunction, precocious puberty, and aggressive behavior. The random AVP levels in both patients were inappropriately low, to the degree of hyperosmolarity. Hypertonic saline infusion resulted in a rise of AVP in one but not the other patient. Insulin and hypotension resulted in release of AVP in both boys, suggesting the presence of a selective defect in osmoreceptor function. No abnormalities were demonstrable by CT scan.

Although adipsic hypernatremia can be essential and isolated,[50–52] several etiologies should be excluded.[53–60] Table 99 outlines the several underlying causes of adipsic hypernatremia.

The syndrome of reset osmoreceptors can be suspected by the triad of a diabetes insipidus-like clinical picture, normal response to protracted dehydration, and extremely high serum osmolality. In contrast to central diabetes insipidus, serum AVP levels can be shown to rise in response to raising plasma osmolality. The condition can be confirmed by measurement of osmotic threshold. A modified version of the conventional Hickey-Hare test[61] was developed by Moses and Streeten[62] to assist in determining the osmotic threshold. The protocol described by these workers is followed by many for quantification of the responses in free water clearance to gradually increasing

TABLE 99
Adipsic Hypernatremia

Essential
  No underlying cause
Secondary to intracranial lesions
  Congenital
      Agenesis of corpus callosum
      Microcephaly
  Vascular
      Cerebral artery aneurysm
      Cerebrovascular accidents
  Granulomatous disease
      Histiocytosis
  Tumors
      Pineal germinomas
      Craniopharyngiomas
      Gliomas
  Postsurgery for craniopharyngiomas

plasma osmolality. The test involves the administration of 5% NaCl solution by an infusion pump at 0.05 ml/kg per min after the collection of six to eight approximately equal 15-min urine samples. Urine osmolality and the serum osmolality are calculated every 15 min while the water load is maintained. The end point is reached when the infusion has continued for 2 hr or 30–45 min after an abrupt drop in urine flow has occurred, or when symptoms supervene. The free water clearance is calculated in each sample to determine whether a fall has occurred.[63] The release of AVP coincides with a fall in the free water clearance by more than 2 SD below the mean free water clearance in the control period. This is correlated with the contemperaneously obtained plasma osmalility, which will be rather high in patients with reset osmostats.

Osmoreceptor Insensitivity

Rarely a patient with the classic clinical picture of central diabetes insipidus and the characteristic response in the water deprivation test may show a response to nonosmotic stimuli, for instance, nicotine. This suggests that while incapable of releasing AVP in response to osmotic challenge, the patient is indeed capable of responding to nonosmotic stimuli. The response to nicotine is evaluated by asking the patient to rapidly smoke 3–4 cigarettes. This constitutes an intense nonosmotic challenge to AVP release, especially in the nonsmoker. However, it is difficult in some cases to separate the effects of nicotine on the neurohypophyseal system from its effects on the renal hemodynamics. Furthermore, the nausea, vomiting, and decreased blood pressure seen in some patients with nicotine in the bloodstream may stimulate AVP through baroreceptor-mediated mechanisms.

The existence of a subset of patients with central diabetes insipidus who demonstrate qualitative abnormalities in AVP secretion is well documented. DeRubertis et al.[64] described the results of detailed studies in one such patient who showed little or no response of AVP to osmotic stimuli but with well-preserved AVP responses to baroreceptor and nonvolume-mediated stimuli. The presentation of these patients with polyuria resembles the classic variety of central diabetes insipidus where AVP production is lost or impaired. Hypernatremia is an often present clue in patients with qualitative abnormalities in AVP release.[65]

Plasma AVP Levels

Measurement of AVP levels performed in conjunction with the water-deprivation test can provide a clear separation among normals, patients with central diabetes insipidus, and those with nephrogenic diabetes insipidus. More importantly, the assay provides a means to differentiate patients with partial diabetes insipidus from those with psychogenic polydipsia. When a sensitive assay for AVP is used, patients with central diabetes insipidus fail to demonstrate a rise in AVP levels following water deprivation.[66,67] Zerbe and Robertson[68] documented the value of measuring AVP levels in conjunction with the standard water-deprivation test. They compared the diagnoses provided by a standard water-deprivation test with those obtained by radioimmunoassay (RIA) of vasopressin. While the assay confirmed the diagnosis of central complete diabetes insipidus in all cases, two patients presumed to have partial central diabetes insipidus, based on indirect testing, proved to have primary polydipsia and nephrogenic diabetes insipidus by RIA. Although advocating the use of vasopressin measurements in the evaluation of polyuric patients, Zerbe and Robertson also pointed out the variability in specificity and sensitivity between different assays in use for measurement of AVP and the lack of reference standards.

Topographical Tests

If central diabetes insipidus is diagnosed, the single most crucial test is evaluation of the suprasellar and sellar area by CT. Exclusion of tumors is a prerequisite for the diagnosis of idiopathic diabetes insipidus. A bone survey to detect lytic lesion may be essential if Hand-Schüller-Christian disease is suspected.

Differential Diagnosis

The diagnosis of central diabetes insipidus is secure when it can be demonstrated that major stimuli such as water deprivation or hypertonic saline infusion fail to increase AVP release. The first step in the approach to a patient with polyuria is to quantitate the volume of urine excreted and to qualitate the degree of dilution. This can be done by collecting and measuring the 24-hr urine output and measuring basal (random) urine osmolality. It should be remembered that diabetes insipidus can exist with urine outputs below 3 liters/day. If the overnight urine osmolality exceeds 800 mOsm/kg, diabetes insipidus is unlikely.

The second step in a polyuric patient is to exclude some very common causes for poluria: glycosuria, hypercalcemia, hypokalemia, drug intake (e.g., particularly diuretics, high doses of lithium, dichlormethyl tetracyline), excessive fluid and saline admininistration (particularly during the postoperative setting), and renal disease (Table 100).

The third step is the performance of the water-deprivation test to obtain information regarding two phenomena: the ability of the hypothalamus to release AVP in response to osmotic stimuli (dehydration) and the ability of the renal interstitum to concentrate urine with endogenous and exogenous vasopressin. The combination of sensitive AVP assays to the conventional water-deprivation test may be of enormous assistance in separating partial forms of central diabetes insipidus from partial forms of nephrogenic diabetes insipidus, as well as in separating psychogenic polydipsia from pure diabetes insipidus.

The patient should be carefully questioned about thirst. While patients with both diabetes insipidus, as well as psychogenic polydipsia, have an in-

TABLE 100
Polyuric Syndromes from Renal Problems

Renal tubular unresponsiveness
  Nephrogenic (diabetes insipidus, congenital)
Chronic renal diseases
  Chronic pyelonephritis
  Analgesic nephropathy
  Lead nephropathy
  Uric acid nephropathy
  Interstitial nephropathy
  Medullary cystic disease
  Polycystic kidneys
  Bilateral hydronephrosis
Diuretic phase of acute tubular necrosis
Pronounced hyperreninemic states

creased thirst mechanism, the latter consume water far out of proportion to their polyuria. The three facets of psychogenic polydipsia are the slightly hypotonic plasma, the ability to concentrate the urine eventually to at least isosmotic levels with protracted water restriction, and the presence of underlying psychiatric abnormalities. These patients consume enormous quantities of water for self-satisfaction. (The author knew of a memorable case in which the patient would "quench" her thirst by stuffing a fully open garden hose into her mouth!) Occasionally, however, psychiatric abnormalities need not be present, as in primary polidipsic syndromes. Mellinger and Zafar[69] described a 24-year-old woman with polyuria and polydipsia since infancy, and normal urinary concentration following water deprivation. The patient's disorder was interpreted as being an abnormal thirst center, wherein the thirst stimulus was not inhibited by normal plasma osmolality. Interestingly, although the release of endogenous vasopressin was normal, treatment with exogenous pitressin relieved her symptoms without resulting in water intoxication. This differs sharply from patients with psychogenic polydipsia who develop dangerous hyponatremia from water intoxication when pitressin is administered. Thus, it appears that the polydipsic syndromes are heterogeneous, and the existence of primary polidipsia is well supported in the literature. Occasionally, polydipsia syndromes can result from intracranial tumors.

The fourth step in the evaluation consists of exclusion of intracranial pathology when central diabetes insipidus is demonstrated. Exclusion of primary or secondary neoplastic lesions in the sella or suprasellar region is mandatory in all patients with central diabetes insipidus.

Complications

The most dangerous complication of complete diabetes insipidus is the development of volume depletion and dehydration. This occurs when access to water is denied, as when the patient is unconscious. Bladder distention and even hydronephrosis can complicate severe disease.

Treatment

The treatment for acute diabetes insipidus with dehydration is liberal fluid replacement coupled with the use of a short-acting vasopressin preparation such as aqueous vasopressin. The usual dosage of aqueous pitressin is 5 to 10 units every 4–6 hr. The advantage of the aqueous preparation is its short half-life, which makes water intoxication a less likely complication with treatment. The drug should be used with extreme caution in patients with coronary heart disease, owing to its vasoconstrictive potential.

For chronic central diabetes insipidus, wherein the symptoms are distressing and interfere with the patient's life style, replacement therapy with 1-desamino-8-D-arginine vasopressin (dDAVP) or with a long-acting preparation such as pitressin tannate in oil is indicated. Of the two, dDAVP is superior for a variety of reasons.[70–73] First, the drug can be conveniently administered intranasally twice a day. Second, the drug is highly potent; doses as small as 2.5 µg twice a day provide effective relief from polyuria. Third, the drug has a relatively longer duration of action, permitting BID dosage. Fourth, side effects such as vasoconstriction, hypotension, and particularly, water intoxication are seldom encountered with dDAVP. Finally, the development of resistance to the drug with continued use is rare with dDAVP. Thus, therapy with dDAVP is a most effective, convenient, and safe modality. The only drawback is the expense, since the drug is costly.

dDAVP (desmopressin) is supplied in a 2.5-ml bottle containing 100 µg/ml preserved desmopressin. The bottle is used as a dropper to load the calibrated rhinyl catheters. Some skill and practice is required in using these catheters to accurately measure the intranasal dose of the drug. Loss of hormone may occur if part of the dose runs out of the catheter when raising it to the nose. The availability of parenteral preparations of dDAVP has provided a safe means of acute replacement in patients with coronary disease. The action of dDAVP can be prolonged when drugs such as chlorpropamide, clofibrate, and indomethacin are used concurrently.[73]

Pitressin tannate in oil is a long-acting vasopressin preparation. Its duration and potency are unpredictable, however, especially with long-term treatment. The particular danger with long-term use is the vasoconstrictive effect on arteries, especially the coronaries. Other undesirable effects include water intoxication and the development of resistance to the drug. A practical point that should be remembered is that the "hormone" is contained in a brown speck at the bottom of the ampule and needs to be gently hand warmed prior to injection. (If this not done properly, the patient would be injecting peanut oil, which obviously does not possess any antidiuretic properties).

Treatment of partial central diabetes insipidus, when necessary, can be successfully achieved by the use of chlorpropamide. The drug is effective only in vasopressin-sensitive diabetes insipidus, and depends on the availability of some endogenous AVP. Thus, chlorpropamide will not be effective in patients with nephrogenic diabetes insipidus, or in those with complete central diabetes insipidus. Several studies have confirmed the beneficial effects of chlorpropamide in the treatment of partial central diabetes insipidus.[74-77] It is believed that chlorpropamide potentiates the antidiuretic action of AVP.[78] In vitro experiments[79] have supported the notion that the drug shares with vasopressin at least one common site of action and that this site may be proximal to the cAMP generation within the receptor cell. Other mechanisms of action have been ascribed to explain the antidiuretic effect of chlorpropamide. Inhibition of adenosine 3'5'-cyclic monophosphate phosphodiesterase,[80] as well as inhibition of PGE within tissues,[81] have been proposed to explain the vasopressin-potentiating effect of chlorpropamide. The average dose of chlorpropamide required to treat partial central diabetes insipidus ranges between 250 and 500 mg of the drug per day. The adverse effects of the drug are mainly twofold: the development of hypoglycemia and hyponatremia especially when water is indiscriminately consumed (water intoxication).

Other drugs, such as clofibrate[82] and carbamazepine,[83] have been used as alternatives to chlorpropamide, but with less success. Carbamazepine, in addition to peripheral augmentation of vasopressin action, may have a central effect in enhancing release of AVP. Regardless of the therapy used, patients with diabetes insipidus should be well warned about the possibility of water intoxication.

The treatment of nephrogenic diabetes insipidus is extremely difficult. The use of thiazide diuretics for amelioration of the polyuria in nephrogenic diabetes insipidus was first proposed by Crawford and Kennedy.[84] The mechanism of action of the thiazide diuretic is rather nonspecific. The diuretic induces a mild negative salt balance and causes a reduction of GFR, which results in augmented reabsorption of isosmotic fluid in the proximal nephron. As a consequence, there is decreased delivery of fluid to the water impermeable segments which in turn leads to a reduced urinary volume. The oral intake of sodium should be restricted, since this potentiates the antidiuretic effectiveness of the thiazide diuretic used in the treatment of nephrogenic diabetes insipidus. More recently, Battle et al.[85] showed that the diuretic amiloride ameliorates the polyuria of lithium-induced nephrogenic diabetes insipidus.

References

1. Braverman LE, Mancini JP, McGoldrick DPM: Hereditary idiopathic diabetes insipidus—A case report with autopsy findings. *Ann Intern Med* **63**:504, 1965.
2. Legros JJ, Crabbe J: Serum neurophysins in familial central diabetes insipidus. *J Clin Endocrinol Metab* **47**:1065, 1978.
3. Nagai I, Li CH, Hsieh SM, et al: Two cases of hereditary diabetes insipidus, with an autopsy finding in one. *Acta Endocrinol (Copenh)* **105**:318, 1984.

4. Laszlo FA, de Wied D: Antidiuretic hormone content of the hypothalamoneurohypophyseal system and urinary excretion of antidiuretic hormone in rats during the development of diabetes insipidus after lesions in the pituitary stalk. *J Endocrinol* **36**:125, 1966.

5. Kovacs K, Laszlo FA, David MA: The antidiuretic phase of water metabolism in rats after lesions of the pituitary stalk. II. The role of the antidiuretic hormone. *J Endocrinol* **25**:397, 1962.

6. Petto CK, DeGirolami U, Earle KM: Craniopharyngiomas: A clinical and pathological review. *Cancer* **37**:1944, 1976.

7. Thomsett MJ, Conte FA, Kaplan SL, et al: Endocrine and neurologic outcome in childhood craniopharyngioma: Review of effect of treatment in 42 patients. *J Pediatr* **97**:728, 1980.

8. Jennings MT, Gelman R, Hochberg F: Intracranial germ-cell tumors: Natural history and pathogenesis. *J Neurosurg* **63**:155, 1985.

9. Suprasellar tumors in children: A review of clinical manifestations and managements. *Cancer* **50**:1420, 1982.

10. Kimmel DW, O'Neill BP: Systemic cancer presenting as diabetes insipidus. *Cancer* **52**:2355, 1983.

11. Yap HY, Tashima CK, Blumenschein MD, et al: Diabetes insipidus and breast cancer. *Arch Intern Med* **139**:1009, 1979.

12. Case Records of the Massachusetts General Hospital: Case 33-1983. *N Engl J Med* **309**:418, 1983.

13. Braunstein GD, Kohler PO: Pituitary function in Hand-Schüller-Christian disease: Evidence for deficient growth-hormone release in patients with short stature. *N Engl J Med* **286**:1225, 1972.

14. Rowland RS: Christian's syndrome and lipoid cell hyperplasias of the reticulo-endothelial system. *Ann Intern Med* **2**:1277, 1929.

15. Avioli LV, Lasersohn JT, Lopresti JM: Histiocytosis X (Schüller-Christian disease): A clinico-pathological survey, review of ten patients and the results of prednisone therapy. *Medicine (Baltimore)* **42**:119, 1963.

16. Avery ME, McAfee JG, Guild HG: The course and prognosis of reticuloendotheliosis (eosinophilic granuloma, Schüller-Christian disease and Letterer-Siwe disease): A study of forty cases. *Am J Med* **22**:636, 1957.

17. Pressman BD, Waldron RL, II, Wood EH: Histiocytosis-X of the hypothalamus. *Br. J. Radiol* **48**:176, 1975.

18. Zinkham WH: Multifocal eosinophilic granuloma—natural history, etiology and management. *Am J Med* **60**:457, 1976.

19. Rothman JG, Snyder PJ, Utiger RD, et al: Hypothalamic endocrinopathy in Hand-Schüller-Christian disease. *Ann Intern Med* **88**:512, 1978.

20. Herring AB, Urich H: Sarcoidosis of the central nervous system. *J Neurol Sci* **9**:405, 1969.

21. Pennell WH: Boeck's sarcoid with involvement of the central nervous system. *Arch Neurol* **66**:728, 1951.

22. Vesely DL, Maldonodo A, Levey GS: Partial hypopituitarism and possible hypothalamic involvement in sarcoidosis: Report of a case and review of the literature. *Am J Med* **62**:425, 1977.

23. Stuart CA, Neelon FA, Lebovitz HE: Hypothalamic insufficiency: The cause of hypopituitarism in sarcoidosis. *Ann Intern Med* **88**:589, 1978.

24. Pelkonen R, Kuusisto A, Salmi J, et al: Pituitary function after pituitary apoplexy. *Am J Med* **65**:773, 1978.

25. Veldhuis JD, Hammond JM: Endocrine function after spontaneous infarction of the human pituitary: Report, review, and reappraisal. **Endocr Rev** **1**:100, 1980.

26. Conomy JP, Ferguson JH, Brodkey JS: Spontaneous infarction in pituitary tumors: Neurologic and therapeutic aspects. *Neurology (NY)* **25**:580, 1975.

27. Gebel P: Pituitary apoplexy: Review of literature with a report of an unusual case associated with diabetes insipidus. *Milit Med* **127**:753, 1962.

28. Cardoso ER, Peterson EW: Pituitary apoplexy: A review. *Neurosurgery* **14**:363, 1984.

29. Merker E, Futterweit W: Postpartum amenorrhea, diabetes insipidus and galactorrhea: Report of a case and review of the literature. *Am J Med* **56**:554, 1974.

30. Doxiades T, Tiliakos M: Diabetes insipidus in association with postpartum hypopituitarism *Br Med J* **1**:23, 1956.

31. Evans HW: Sheehan's syndrome with diabetes insipidus. *Am J Med* **28**:648, 1960.

32. Ahn CS, Kim DS: Sheehan's syndrome associated with diabetes insipidus. *Lancet* **2**:1045, 1964.

33. Aguilo F, Vega LA, Haddock L, et al: Diabetes insipidus syndrome in hypopituitarism of pregnancy. *Acta Endocrinol (Kbh)* **60**(suppl):137-H, 1969.

34. Spain AW, Geoghehan F: Diabetes insipidus in association with postpartum pituitary necrosis. *J Obstet Gynaecol Br Commonw* **53**:223, 1946.

35. Schwartz AR, Leddy AL: Recognition of diabetes insipidus in postpartum hypopituitarism. *Obstet Gynecol* **59**:394, 1982.

36. Sheehan HL, Whitehead R: The neurohypophysis in post-partum hypopituitarism. *J Pathol* **85**:145, 1963.

37. Whitehead R: The hypothalamus in post-partum hypopituitarism. *J Pathol* **86**:55, 1963.

38. Bode HH, Crawford JD: Nephrogenic diabetes insipidus in North America: The Hopewell hypothesis. *N Engl J Med* **280**:750, 1969.

39. Barron WM, Cohen LH, Ulland LA, et al: Transient vasopressin-resistant diabetes insipidus of pregnancy. *N Engl J Med* **310**:442, 1984.

40. Kleeman CR, Czaczkes JW, Cutler R: Mechanisms of impaired water excretion in adrenal and pituitary insufficiency. IV. Antidiuretic hormone in primary and secondary adrenal insufficiency. *J Clin Invest* **43**:1641, 1964.

41. Robson JS, Lambie AT: Cortisone-induced polyuria following hypophysectomy. *Am J Med* **26**:769, 1959.

42. Danguir J: Sleep deficits in rats with hereditary diabetes insipidus. *Nature (Lond)* **304**:163, 1983.

43. Laczi F, van Ree JM, Wagner A, et al: Effects of desglycinamide-arginine-vasopressin (DG-AVP) on memory processes in diabetes insipidus patients and non-diabetic subjects. *Acta Endocrinol (Copenh)* **102**:205, 1983.

44. Miller M. Dalakos T, Moses AM, et al: Recognition of partial defects in antidiuretic hormone secretion. *Ann Intern Med* **73**:721, 1970.

45. Weitzman RE, Kleeman CR: The clinical physiology of water metabolism. Part II. Renal mechanisms for urinary concentration; diabetes insipidus. *West J Med* **131**:486, 1979.

46. Halter JB, Goldberg AP, Robertson GL, et al: Selective osmoreceptor dysfunction in the syndrome of chronic hypernatremia. *J Clin Endocrinol Metab* **44**:609, 1977.

47. Avioli LV, Earley LE, Kashima HK: Chronic and sustained hypernatremia, absence of thirst, diabetes insipidus, and adrenocorticotropin insufficiency resulting from widespread destruction of the hypothalamus. *Ann Intern Med* **56**:131, 1982.

48. De Rubertis FR: Michelis MF, Davis BB: Essential hypernatremia. *Arch Intern Med* **134**:889, 1974.

49. Brezis M, Weiler-Ravell D: Hypernatremia, hypodipsia and partial diabetes insipidus: A model for defective osmoregulation. *Am J Med Sci* **279**:37, 1980.

50. Gill G, Baylis P, Burn J: A case of 'essential' hypernatraemia due to resetting of the osmostat. *Clin Endocrinol (Oxf)* **22**:545, 1985.

51. Dunger DB, Lightman S, Williams M, et al: Lack of thirst, osmoreceptor dysfunction, early puberty and abnormally aggressive behaviour in two boys. *Clin Endocrinol (Oxf)* **22**:469, 1985.

52. Alford FP, Scoggins BA, Wharton C: Symptomatic normovolemic essential hypertension. *Am J Med* **54**:359, 1973.

53. Leaf A, Mamby AR, Ramussen H, et al: Some hormonal aspects of water extraction in man. *J Clin Invest* **31**:914, 1952.

54. Kastin AJ, Lipsett MB, Ommaya AK, et al: Asymptomatic hypernatremia, physiological and clinical study. *Am J Med* **38**:306, 1965.

55. Zazgornik J, Jellinger K, Waldhausal W, et al: Excessive hypernatraemia and hyperosmolality associated with a germinoma in the hypothalamic and pituitary regions. *Eur Neurol* **12**:38, 1974.

56. Mahoney JH, Goodman AD: Hypernatremia due to hypodysplasia and elevated threshold for vasopressin release. *N Engl J Med* **279**:1191, 1968.

57. Pleasure D, Goldberg M: Neurogenic hypernatremia. *Arch Neurol* **15**:78, 1966.
58. Schaffblass S, Robertson GL, Rosenfield RL, et al: Chronic hypernatremia from a congenital defect in osmoregulation of thirst and vasopressin. *J Pediatr* **102**:703, 1983.
59. Bannister P, Sheridan P, Penney MD: Chronic reset osmoreceptor response, agenesis of the corpus callosum and hypothalamic cyst. *J Pediatr* **104**:97, 1984.
60. Wise BL: Neurogenic hyperosomolality (hypernatremia). *Neurology (NY)* **12**:453, 1962.
61. Hickey RC, Hare K: The renal excretion of chloride and water in diabetes insipidus. *J Clin Invest* **23**:768, 1944.
62. Moses AM, Streeten DHP: Differentiation of polyuric states by measurement of responses to changes in plasma osmolality induced by hypertonic saline infusions. *Am J Med* **42**:368, 1967.
63. Moses AM, Notman DD: Diabetes insipidus and syndrome of inappropriate antidiuretic hormone secretion (SIADH). *Adv Intern Med* **27**:73, 1982.
64. DeRubertis FR, Michelis MF, Beck N, et al: "Essential" hypernatremia due to ineffective osmotic and intact volume regulation of vasopressin secretion. *J Clin Invest* **50**:97, 1971.
65. Goldberg M, Weinstein G, Adesman J, et al: Asymptomatic hypovolemic hypernatremia, a variant of essential hypernatremia. *Am J Med* **43**:804, 1967.
66. Robertson GL: Immunoassay of plasma vasopressin in man. *Proc Natl Acad Sci USA* **66**:1298, 1970.
67. Beardwell CG: Radioimmunoassay of arginine vasopressin in human plasma. *J Clin Endocrinol Metab* **33**:254, 1971.
68. Zerbe RL, Robertson GL: A comparison of plasma vasopressin measurements with a standard indirect test in the differential diagnosis of polyuria. *N Engl J Med* **305**:1539, 1981.
69. Mellinger RC, Zafar MS: Primary polydipsia: Syndrome of inappropriate thirst. *Arch Intern Med* **143**:1249, 1983.
70. Vavra G, Machova A, Holecek V: Effect of synthetic analogue of vasopressin in animals and in patients with diabetes insipidus. *Lancet* **1**:948, 1968.
71. Robinson AG: DDAVP in the treatment of diabetes insipidus. *N Engl J Med* **294**:507, 1976.
72. Lee WP, Lippe BM, La Franchi SH, et al: Vasopressin analog DDAVP in the treatment of diabetes insipidus. *Am J Dis Child* **130**:166, 1976.
73. Richardson DW, Robinson AG: Desmopressin. *Ann Intern Med* **103**:228, 1985.
74. Kunstadter RH, Cabana EC, Oh W: Treatment of vasopressin-sensitive diabetes insipidus with chlorpropamide. *Am J Dis Child* **117**:436, 1969.
75. Vallet HL, Prasad M, Goldbloom RB: Chlorpropamide treatment of diabetes insipidus in children. *Pediatrics* **45**:246, 1970.
76. Webster B, Bain J: Antidiuretic effect and complications of chlorpropamide therapy in diabetes insipidus. *J Clin Endocrinol Metab* **30**:215, 1970.
77. Miller M, Moses AM: Mechanism of chlorpropamide action in diabetes insipidus. *J Clin Endocrinol Metab* **30**:488, 1970.
78. Berndt WO, Miller M, Kettyle WM, et al: Potentiation of the antidiuretic effect of vasopressin by chlorpropamide. *Endocrinology* **86**:1028, 1970.
79. Ingelfinger JR, Hays RA: Evidence that chlorpropamide and vasopressin share a common site of action. *J Clin Endocrinol Metab* **29**:738, 1969.
80. Brooker G, Fichman M: Chlorpropamide and tolbutamide inhibition of adenosine 3'5'-cyclic monophosphate phosphodiesterase. *Biochem Biophys Res Commun* **42**:824, 1971.
81. Zusman RM, Keiser HR, Handler JS: A hypothesis for the molecular mechanism of action of chlorpropamide in the treatment of diabetes mellitus and diabetes insipidus. *Fed Proc* **36**:2728, 1977.
82. Moses AM, Howanitz J, Van Gemert M, et al: Clofibrate induced antidiuresis. *J Clin Invest* **52**:535, 1973.
83. Kimura T, Matsui K, Sato T, et al: Mechanism of carbamazepine (Tegretol) induced antidiuresis: Evidence for release of antidiuretic hormone and impaired excretion of a water load. *J Clin Endocrinol Metab* **38**:356, 1974.
84. Crawford JD, Kennedy G: Chlorothiazide in diabetes insipidus. *Nature (Lond)* **183**:891, 1959.
85. Batlle DC, von Riotte AB, Gaviria M, et al: Amelioration of polyuria by amiloride in patients receiving long-term lithium therapy. *N Engl J Med* **312**:408, 1985.

21

Syndrome of Inappropriate Secretion of ADH

Introduction

Since the original description by Bartter and Schwartz,[1-3] the syndrome of inappropriate ADH secretion (SIADH) has progressed from a rare syndrome to one of the most frequent metabolic disorders encountered in hospitalized patients. The diverse diseases that cause this syndrome, the clinical features, and the criteria for establishing the diagnosis, as well as the therapeutic options for this disorder are discussed in this chapter.

TABLE 101
Response to Hypotonic Plasma in the Normal State and in SIADH

Normal	SIADH
Hypotonic plasma	Hyponatremia/hypotonic plasma
↓	
Prompt suppression of vasopressin (ADH)	Failure of vasopressin suppression (or increased vasopressin levels)
↓	↓
Prompt suppression of water reabsorption in distal and collecting tubule	Continuous water reabsorption at the distal and collecting tubule
↓	↓
Dilution of urine due to increased water excretion	Urine relatively concentrated
↓	
Conservation of solute	Continuous natriuresis
↓	↓
Excretion of dilute urine	Excretion of less than maximally dilute urine
↓	↓
Restoration of plasma osmolality to normal	Persistance of low plasma osmolality

Pathophysiology

The excess AVP in SIADH can be secreted by either the hypothalamus or an ectopic source. Regardless of the source, the pathophysiology of SIADH is identical. The effects of excess secretion of AVP are dual: decreased free water clearance because of increased reabsorption of water by the distal and collecting tubules and hypotonicity of the plasma as a result of dilution. The hallmark of SIADH is nonsuppressible AVP secretion despite hypotonicity of plasma. This is in contrast to the normal situation, in which prompt suppression of AVP is the response to hypotonicity. With the continuous secretion of AVP, dilutional hyponatremia occurs, analogous to water intoxication. This hyponatremia is compounded by natriuresis, a paradox, in light of the fact that the serum Na^+ is already low. Table 101 contrasts the physiological response to hypotonicity in normal patients and in patients with SIADH. The specific mechanisms underlying the pathophysiology of each facet of SIADH are discussed separately.

Etiology

Several disorders are related etiologically to the development of SIADH. Despite the variety of disorders that cause SIADH, most cases result from malignant disease or infectious disease of the CNS or lungs. Table 102 illus-

TABLE 102
Etiology of SIADH

Malignancy
   Bronchogenic carcinoma
   Pancreatic carcinoma
   GI neoplasms
   Hodgkin's and non-Hodgkin's lymphoma
   Thymoma
Nonmalignant pulmonary disease
   Tuberculosis
   Bacterial and viral pneumonia
   Lung abscess
   Empyema
   Fungal diseases
CNS disease
   Head injury
   Subarachnoid and subdural hemorrhage
   Meningitis, encephalitis
   Brain abscess
Drug induced
   Chlorpropamide
   Carbamazepine
   Clofibrate
   Vincristine
   Cyclophosphamide
   Tricyclic antidepressants
Idiopathic
Miscellaneous
   Acute psychosis
   Acute intermittent porphyria
   Positive-pressure ventilatory assistance
   Mitral commissurotomy
   Chronic obstructive lung disease
   Posttranssphenoidal surgery
   Pituitary apoplexy

trates the numerous etiologies of SIADH. When no etiology for SIADH is discerned, the diagnosis of idiopathic SIADH can be made, a disorder especially prevalent in older patients. The development of the syndrome secondary to neoplastic and infectious etiologies, as well as its occurrence related to drugs, central release of AVP, and the idiopathic form are discussed below.

SIADH and Neoplastic Disease

One of the most common etiologies underlying the development of SIADH is bronchogenic carcinoma, especially the oat cell variety. As early as 1963, Amatruda et al.[4] demonstrated bioassayable ADH-like activity in extracts of oat cell carcinoma. Employing purer techniques for measurement of vasopressin activity, several workers[5,6] demonstrated vasopressin in the extracts of tumor tissue of patients with bronchogenic carcinoma. Cheng et al.[8] and

TABLE 103
Neoplasms Associated with SIADH

Bronchogenic carcinoma[11]
Pancreatic carcinoma[1]
Carcinoma of duodenum[1]
Bladder tumors[12]
Hematological malignancies[13,14]
  Hodgkin's disease
  Lymphosarcoma
  Reticulum cell sarcoma
Ewing's sarcoma[15]
Thymoma[16]

Hamilton et al.[9] showed independently that vasopressin-associated neurophysin can be measured in the blood and from tumor extracts of patients with lung cancer, particularly the oat-cell variety. Furthermore, George et al.[7] demonstrated that bronchogenic tumor tissue grown in tissue culture is capable of synthesizing vasopressin. Clearly, ectopic secretion of vasopressin or vasopressinlike material by tumorous lung tissue is one of the most common causes of SIADH. This is not surprising, since it is currently well accepted that an extremely high proportion of tumor extracts, especially lung tumors, contain detectable protein hormone precursors, or hormones per se. It is also recognized that a dichotomy exists between the incidence of finding elevated hormonal levels in the circulation of patients with neoplastic disease and the evidence of the clinical expression of the respective hormonal excess. For example, Odell and Wolfsen[10] demonstrated increased vasopressin concentrations in the blood of 41% of patients with lung cancer, but only a fraction manifested the clinical syndrome of SIADH. This statement should be tempered, however, by the fact that Comis et al.[11] found varying degrees of impairment in excretion of a water load in approximately two-thirds of patients with small cell anaplastic lung cancer.

In addition to bronchogenic carcinoma, several other neoplasms have been reported to be associated with SIADH, the two most frequent ones being carcinoma of the pancreas and of the duodenum.[1,11–16] Table 103 outlines the numerous neoplasias associated with SIADH. Baylin and Mendelsohn[17] proposed that the secretion of hormones by tumor cells represents the persistence of a function normally present in a more primitive cell that has been arrested in the process of its differentiation.

SIADH and Infectious Disease

Several infections are recognized as being associated with secretion of AVP or of AVP-like material. The ability of nontumorous but tuberculous lung tissue to secrete an ADH-like substance was documented by Vorherr et al.[18] There appears to be considerable diversity in the nature and type of pulmonary infections associated with SIADH,[19] including pneumonias (bacterial and viral), lung abscess, empyema, and aspergillosis. In addition to chest infections, a variety of infections involving the neurological system may be associated with SIADH. Thus, encephalitis, meningitis (viral, purulent, and tuberculous), brain abscess, and Guillain-Barré syndrome have been recognized as underlying conditions for the development of hyponatremia secondary to SIADH.

SIADH and Drugs

Iatrogenic SIADH represents an important category of patients with SIADH. The most common example of this phenomenon is the development of hyponatremia with the chronic use of chlorpropamide. Moses and Notman[20] indicated that the estimated incidence of chlorpropamide-induced hyponatremia is approximately 4% and is more likely to occur in patients with congestive heart failure or cirrhosis of the liver. Similarly, the use of carbamazepine has been reported to cause SIADH.[21] The mechanism of SIADH in both entities reflects the peripheral effects of these drugs in enhancing the action of endogenous vasopressin. SIADH is a well-recognized complication of chemotherapy with vincristine[22,23] and cyclophosphamide[24,25] and more recently with the use of cis-dichlorodiamine platinum (CPPD).[16] Another group of drugs related to SIADH are psychotropic agents. Antidepressant therapy with tricyclic agents is a well-recognized setting for the development of SIADH. Thus, amitryptyline, desipramine, fluphenazine, and thiothixene may all be associated with SIADH-like syndrome. The development of SIADH in patients being treated for psychiatric disorders must take into account the fact that several psychiatric disorders may be associated with SIADH. Dubovsky et al.[26] described SIADH in a patient during an acute psychotic reaction on no psychotropic drugs. The syndrome resolved with improvement of the acute psychotic state. Furthermore, Kramer and Drake[27] reported a relationship among acute psychosis, water ingestion, and SIADH. Patients with this triad may present with bizarre neurological syndromes due to profound hyponatremia. The major psychiatric disorders reported to underlie SIADH are schizophrenia and delusional thought disorders.[28,29] Confusion exists in sorting out the real etiology of hyponatremia in these patients, as several patients with such an episode are placed on haloperidol, a drug linked with producing SIADH.[30–32]

Iatrogenic SIADH can also result from pituitary surgery, or the use of vasopressin analogues. Cusik et al.[33] described three patients who developed the clinical and biochemical characteristics of SIADH following transsphenoidal surgery. The mechanism is probably related to leakage of AVP from the damaged neurohypophyseal tissue. The development of SIADH following the use of dDAVP has been reported.[34] This is more in the nature of ADH intoxication rather than true SIADH.

TABLE 104
SIADH from Excessive Endogenous AVP Secretion

Head trauma
Posttranssphenoidal surgery
Cerebrovascular hemorrhage
Pituitary apoplexy
Infections of CNS
Positive-pressure breathing
Postmitral commissurotomy
Chronic obstructive lung disease
Acute psychosis
Acute intermittent porphyria
Drugs (e.g., chemotherapeutic agents, morphine)
Chronic diuretic therapy with hypokalemia

SIADH Caused by Excess Central Release of AVP

SIADH can result from excessive secretion of AVP from the hypothalamus itself. Table 104 outlines the several situations in which excessive secretion of endogenous AVP underlies the etiology of SIADH. The most prototypical of these is the SIADH associated with head trauma. The tradional view that patients with intracranial injury demonstrated cerebral wasting was replaced in the 1970s by the concept that sustained hypersecretion of AVP release caused the salt wasting of head trauma and other related intracranial disorders. Recently, it seems that the cycle has come full circle: Nelson et al.[35] measured intravascular blood volumes in 12 neurosurgical patients with classic SIADH and showed a decreased intravascular volume in 10 of 12 patients. Those with decreased intravascular volume demonstrated significant reductions in red blood cell mass, plasma volume, and total blood volume compared with normal controls, as well as control neurosurgical patients. The blood loss was not excessive in the neurosurgical patients. Interesting as this study is, it does not clarify the role of endogenous AVP in the development of this syndrome in patients with head trauma.

Idiopathic SIADH

Occasionally, SIADH occurs in the absence of an identifiable cause.[36-38] This entity, termed idiopathic SIADH, had not been fully documented until Goldstein et al.[39] reported the syndrome in an 88-year-old man who had no associated disease. Upon reviewing 10 other cases in the literature, Goldstein and co-workers concluded that idiopathic SIADH clearly does exist as an entity and is likely to be encountered in older patients. The reason for such an occurrence is unclear. Hyponatremia is an extremely common problem in hospitalized patients. Anderson et al.[40] prospectively analyzed the epidemiology of hyponatremia in hospitalized patients and found a daily incidence rate of 0.97% and a prevalence rate of 2.48%. In this study, nonosmotic secretion of AVP played a major role in the pathogenesis of hyponatremia. The single largest group in this series was the group of hospitalized patients with normovolemic hyponatremia, a feature prototypical of SIADH. Whether some of these patients represented transient idiopathic SIADH is unclear. Since the hyponatremia of most of these patients developed during the hospital stay, the impact of fluids, drugs, and the stress of hospitalization cannot be ignored in the causation of the AVP-mediated hyponatremia.

Clinical Features

The features of SIADH result from the effects of hyponatremia. The clinical features of SIADH evolve slowly and are characterized by fatigue, muscle weakness, and dizziness. As the disorder progresses, muscle twitching, alteration in behavior, and drowsiness supervene. When serum sodium declines below 120 mEq/liter, stupor, convulsions, and coma occur. The development of metabolic encephalopathy depends on the severity of the hyponatremia as well as the rapidity with which the syndrome evolved. Edema or dehydration are characteristically àbsent. Patients with SIADH are euvolemic, showing no signs of edema or dehydration.

Diagnostic Studies

The criteria for the diagnosis of SIADH are (1) hypotonicity of plasma; (2) hyponatremia; (3) less than maximally dilute urine; (4) natriuresis despite hyponatremia; and (5) the exclusion of hepatic, renal, thyroid, and adrenal dysfunction. Each of these criteria needs elucidation.

Hyponatremia/Hypotonicity

Hyponatremia is the initial clue that often raises the suspicion of SIADH. The depression in the serum sodium level can be mild, moderate, or severe. The decrease in the plasma osmolality is a direct consequence of the hyponatremia; changes in the plasma osmolality closely parallel the changes in serum sodium. The two reasons for the hyponatremia are the dilutional factor caused by excessive water reabsorption by the collecting tubules under the influence of AVP and the loss of sodium by natriuresis.

Since hyponatremia is the marker of SIADH, it is important to exclude pseudohyponatremia as a cause of the lowered level of measured sodium. This can be easily attained by simultaneously measuring the plasma osmolality. Generally, hyponatremia is indeed associated with hypoosmolar plasma. There are two exceptions wherein a normal plasma osmolality may be seen in a hyponatremic patient; hyponatremia caused by severe hyperlipidemia (or paraproteinemia) and hyperglycemia and is referred to as factitious or pseudo-hyponatremia. It is important to understand the principle behind this phenomenon: 93–95% of plasma is aqueous and 5–7% is nonaqueous, due to the presence of a variety of macromolecules (e.g., large proteins or lipids). Sodium is confined to the aqueous phase only and is measured and expressed as milliequivalents per unit volume of plasma divided by the fraction of plasma normally occupied by water (normally 93–95%). When hyperlipidemia or significant paraproteinemia exists, the aqueous portion of plasma may be considerably reduced because of the presence of these nonpolar substances.[41,42] Thus, the sodium levels measured are obviously in a reduced aqueous phase, giving rise to a low sodium determination, but the plasma osmolality remains normal. When this combination is seen, the plasma should be observed for lactescence and protein electrophoresis done.

The other mechanism for lowering serum sodium without a concomitant lowering of plasma osmolality is in the presence of excessive extracellular solute, i.e., hyperglycemia or mannitol administration. The mechanism of hyponatremia caused by excessive glucose begins with the accumulation of glucose in the extracellular fluid, which contributes to the extracellular osmolality. This creates an osmolar gradient, permitting movement of water from the cells into the extracellular fluid. The water keeps expanding the extracellular space, resulting in dilution of this space. Therefore, the sodium is distributed in a larger volume, and its concentration becomes progressively lower as the glucose becomes progressively higher. The rule-of-thumb calculation used to be that for every 100-mg increase in the glucose (above a normal of 100 mg), the sodium declined by a factor of 2.8 mEq/liter; adjustments were made accordingly in reading the "true" sodium. Katz[43] indicated that the classic calculation may be erroneous and has provided data with which to adapt a new formula, i.e., for each 100-mg increment of glucose there is an expected decrease of only 1.6 mEq/liter sodium. Obviously, the plasma osmolality in these patients is normal or high despite the "lowered" sodium level.

Less Than Maximally Dilute Urine

The demonstration of a less than maximally dilute urine in the presence of a hypotonic plasma constitutes one of the most important criteria for the diagnosis. The reasons for this phenomenon are the continuous water reabsorption at the distal and collecting tubules and the continuous natriuresis, both of which are conducive to excretion of a concentrated urine. In general, the patient with SIADH demonstrates a urine osmolality greater than the plasma osmolality. It should be emphasized that urine/plasma osmolality ratio of >1 is valid only when the patient has a hypotonic plasma. For example, any normal individual who has been deprived of water overnight will have a urine/plasma osmolarity ratio of >1. The serum osmolality in these normal subjects is not low.

Occasionally, a severely solute-restricted patient with SIADH may demonstrate a urine osmolality below the plasma osmolality, but the inappropriateness, i.e., the less than maximally dilute nature, will be evident. For instance, a severely salt-restricted patient with SIADH may have a plasma osmolality of 250 mOsm and a simultaneous urine osmolarity of 200 mOsm. Although below the plasma level, the urine osmolality in this situation is higher than one would expect for the degree of hypotonicity of the plasma. In normal situations, a plasma tonicity of 250 mOsm is associated with a very dilute urine, not exceeding 100 mOsm. Thus, in comparing the urine and plasma osmolalities in the diagnosis of SIADH, the question is not whether the urine osmolality is greater than that of the plasma, but whether the urine is less than maximally dilute. This criterion, however, occurs in several other situations as well (see the section Differential Diagnosis). Continued natriuresis despite hyponatremia is a result of expansion of ECF and suppression of the renin-aldosterone system.

Natriuresis

The exact mechanism for the sodium loss in SIADH is unclear but is most probably from chronic expansion of extracellular fluid (ECF) volume. Renal wasting of sodium chloride is a documented phenomenon i n normal volunteers infused with saline or water.[44–46] Volume expansion clearly inhibits the reabsorption of sodium and chloride by the proximal renal tubules. This natriuretic effect is probably a reflection of suppression in renin and aldosterone levels consequent to volume expansion. Severe salt restriction will often correct the urinary sodium loss by stimulating the renin–angiotensin–aldosterone system with a resultant increase in sodium reabsorption.

Exclusion of Hepatic, Renal, Thyroid, and Adrenal Dysfunction

The exclusion of hepatic, cardiac, renal, and adrenal dysfunction, as well as volume depletion, is an essential criterion for the diagnosis of SIADH. Several conditions are characterized by dilutional hyponatremia, decreased free water clearance, and production of a less than maximally dilute urine in the presence of a hypotonic plasma. Cirrhosis, nephrotic syndrome, congestive heart failure, and volume depletion are all characterized by a reduction in effective circulating blood volume. A decrease in renal flow or GFR is associated with production of urine that is more concentrated than the plasma even in the absence of AVP. The sequence for this occurrence is as follows:

1. A decrease in the effective circulating blood volume results in a decreased GFR. Del Greco and de Wardener[47] documented that even transient reductions in GFR are attended by production of urines that are less than maximally dilute. That such a phenomenon can take place in the absence of AVP, is suggested by observations that patients with diabetes insipidus can produce a hypertonic urine (even in the absence of AVP) when the GFR is reduced.[48]

2. The decrease in GFR stimulates the renin–angiotensin–aldosterone system, which causes increased proximal tubular reabsorption of sodium. The net result is a reduced delivery of filtrate to the diluting segment of the nephron, i.e., the ascending limb of the loop of Henle. The diluting process in this region depends on movement of sodium from the ascending limb into the medullary interstitium, thereby rendering the fluid progressively more dilute. Berliner and Davidson,[49] in 1957, proposed the now classic concept that reduction in the delivery of filtrate to the diluting segments of nephron is associated with production of a urine that is more concentrated relative to the plasma. This is partly because reduced delivery of filtrate from the proximal tubule limits the quantity of dilute fluid that can be generated by the ascending limb. In addition, the decreased filtrate is conducive to a decreased flow rate through the distal tubule and collecting ducts, an area somewhat permeable to backdiffusion of water into the medullary interstitium.

3. Most importantly, the decrease in the effective circulating blood volume stimulates the baroreceptors with the appropriate release of AVP. The AVP further compounds the events at the collecting tubules, resulting in further concentration of urine. The net result is the production of a less than maximally dilute urine.

Thus, the diagnosis of SIADH cannot be made in the presence of conditions that reduce the effective circulating blood volume. The two reasons that preclude the diagnosis of SIADH under these circumstances are because these conditions can produce an identical abnormality in the handling of water by the renal tubules and because these conditions elicit an appropriate response in AVP release mediated by the nonosmotic pathways of AVP release.

The exclusion of adrenocortical insufficiency is an important criterion that needs to be satisfied. The abnormalities in urinary dilution seen in association with adrenocortical insufficiency occur as a consequence depletion of salt, water, and volume, leading to a decrease in GFR. Any decrease in GFR, regardless of the cause and regardless of whether AVP is present, can result in the production of a less than maximally dilute urine.

The exclusion of hypothyroidism is a less important, but well accepted, criterion for the diagnosis of SIADH. Chronic, and often severe, hypothyroidism per se can result in the dual occurrences of enhanced tubular reabsorption of sodium at the proximal tubule and decreased reabsorption at the distal tubule.[50,51] This may be compounded by other mechanisms, such as resetting of the osmostats, decreased half-life of AVP due to the hypothyroid state, and the presence of relative glucocorticoid deficiency. In any event, the hyponatremia of hypothyroidism bears a close relationship to the clinical severity of the hypothyroidism.

Differential Diagnosis

The combination of hyponatremia, natriuresis, and the production of a less than maximally dilute urine in the presence of a hypotonic plasma is not unique to SIADH. Volume depletion, ascites, nephrotic syndrome, hepatic failure, renal disease, and congestive heart failure are common situations in which such a combination is frequently encountered. These can be excluded on the basis of clinical examination and routine laboratory tests. The definitive diagnosis of SIADH largely depends on the exclusion of other conditions. The lack of edema, hypovolemia, or dehydration provides important supportive clinical clues to the diagnosis of SIADH. The presence of natriuresis greater than 20 mEq/liter is a strong supportive biochemical clue. In addition, the low levels of BUN, creatinine, and uric acid serve as mildly supportive biochemical clues. The measurement of AVP levels in the plasma has not found a strong place in the clinical setting of SIADH. This is partly due to the lack of ready availability as well as the lack of uniform standards for the assay.

Measurement of vasopressin levels in the plasma, while corroborating the inappropriateness of the hormone in relationship to the plasma hypoosmolality, would not be very helpful in separating other conditions that enter into the differential diagnosis. It has been emphasized that evaluation of the relationship between plasma osmolality and plasma AVP levels might be useful in difficult cases,[38] but most workers agree that many hyponatremic conditions other than classic SIADH do indeed have abnormal relationships between the hormone level and plasma osmolality.[20] Therefore, measurement of vasopressin levels does not substantially add to the original criteria of Bartter and Schwartz in diagnosing the entity. The AVP assay is both difficult and expensive, and the patient can often be evaluated, diagnosed, and treated without obtaining levels of AVP in the plasma.

The performance of water-loading tests in the diagnosis of SIADH assumes a very low priority. The test is based on the principle that when 20 ml/kg of water is orally administered, normal persons excrete more than 65% of the load within 4 hr, with concomitant dilution of urine. While it is true that all patients with SIADH will have an abnormal water loading test, i.e., they are unable to excrete 65% of the water load and the urine remains concentrated, so will patients with other conditions that usually enter into the differential diagnosis of SIADH. Thus, abnormal water-loading tests are not uniquely diagnostic to SIADH. The one reported exception are patients with so-called sick cell syndrome, which causes hyponatremia and hypoosmolality of plasma,[52] but a normal ability to excrete a water load. It must be cautioned that when water loading is done in a patient with hyponatremia ($\leq$125 mEq/liter) severe water intoxication may develop, resulting in seizures and neurological manifestations.

Although the clinical diagnosis of SIADH can generally be made with a reasonable degree of certainty, three situations deserve special emphasis in the differential diagnosis: adrenal insufficiency, chronic diuretic use, and the sick cell syndrome.

Adrenal Failure

Chronic adrenal failure and SIADH share several features in common. Both disorders are characterized by an impairment in urinary diluting ability. Patients with either disease have difficulty in excreting a water load and demonstrate a decrease in free water clearance. Natriuresis and hyponatremia with a hypotonic plasma are seen in both entities. Lung cancer or tuberculosis can be responsible for either of the two entities, i.e., may secrete ADH and result in SIADH or may rarely involve both adrenals and cause adrenocortical failure. The differentation may be quite difficult at times. The following findings may be of help. First, in most patients with adrenal failure, potassium levels tend to be high. When the adrenal deficiency is severe enough to render the patient hyponatremic, generally potassium retention would also be present. The value of serum potassium levels is especially relevant in light of the observation that several studies have reported mild potassium wasting in the syndrome of SIADH.[53,54] Thus, patients with SIADH tend to have normal to slightly low serum potassium levels, while hyperkalemia is evident in two-thirds of patients with adrenal failure. Second, volume depletion is often present in patients with adrenocortical failure. Third, the low serum cortisol levels may help in suggesting adrenocortical failure. When in doubt, evaluation of adrenal reserve by the administration of ACTH should be performed to exclude Addison's disease.

It is believed the abnormalities in urinary dilution in adrenocortical failure are a result of salt and water depletion that leads to a decrease in GFR. This results in increased reabsorption of water in the proximal tubule (vasopressin independent), reducing the delivery of filtrate to the diluting segments of the nephron. Consequently, the urine is less than maximally dilute. The natriuresis from the mineralocorticoid deficiency further contributes to urinary concentration. There are several lines of evidence to support the above theory:

1. Ufferman and Schrier[55] studied the diluting capacity of the nephrons of adrenalectomized dogs. Their observations indicated that the free water clearance was normal up to 4 days without any steroids as long as the animals were fed abundant NaCl. If salt and water depletion were allowed to occur, the animals rapidly became hyponatremic with a decrease in free water clearance and urine that was less than maximally dilute.

2. In patients with Addison's disease, restoration of water diuresis can be demonstrated simply by expanding the ECF volume by normal saline.[56]

3. A direct effect of glucocorticoids has also been suggested because of the striking improvement in free water clearance following hydrocortisone therapy in patients with adrenal insufficiency.[57]

4. Finally, an impressive line of evidence to suggest that the impaired diluting ability seen in adrenocortical failure is non-ADH dependent comes from experiments in the Brattleboro rats. These rats are a special strain and are predisposed to diabetes insidipus (lack of ADH). When these rats with diabetes insipidus were adrenalectomized they became unable to maximally dilute urine, implying adrenal insufficiency as the cause of the phenomenon and ruling out ADH mediation.[58]

SIADH and Diuretic Therapy

Fichman et al.[59] were among the first to report the occurrence of severe hyponatremia (<120 mEq) in patients receiving chronic diuretic therapy in the absence of dehydration or alterations in the creatinine clearances. The consistent biochemical changes encountered were hypokalemic alkalosis and elevated circulating ADH levels. The development of the syndrome in conjunction with marked decreases in the exchangeable potassium pool suggested that hypokalemia may play a role in the genesis of the syndrome. This is supported by the observation that the hyponatremia promptly responds to discontinuation of diuretic and potassium replenishment. It is unclear whether these cases represent true hypersecretion of ADH or an appropriate release of ADH in response to subclinical volume depletion.

Sick Cell Syndrome

The hyponatremia of sick cell syndrome is a loosely defined entity. Observed in chronically debilitated patients, it is believed to occur as a result of shifting sodium ions from the extracellular to the intracellular compartment; this contrasts with the situation in dilutional hyponatremia, in which water excretion is defective. Flear and Singh[60] have suggested that the hypoosmolality seen in the chronically ill patient may be secondary to the loss of organic intracellular osmoles (products of cellular metabolism) resulting from the cachectic, malnourished state. The hyponatremia in these patients is asymptomatic and stable and is restored to normal when the underlying disease is treated. This entity may represent a true example of resetting of the osmostat, as has been suggested by DeFronzo et al.,[52] who demonstrated that following a water load, four patients with hyponatremia from sick cell syndrome demonstrated normal excretion of dilute urine.

Treatment

The treatment of SIADH depends on the degree of hyponatremia, the presence of cerebral dysfunction, the response to prior therapies, and to some extent the underlying etiology. The mainstay of treatment for chronic SIADH is fluid restriction. When fluid restriction fails or the response to such therapy reaches a plateau, drug therapy with demeclocycline or lithium carbonate can be instituted. When the hyponatremia is severe, symptomatic, and life threatening, aggressive therapy with hypertonic saline is one option. In patients with marked volume expansion who excrete all the infused saline, the use of furosemide combined with replacement of urinary losses is another therapeutic strategy.

Fluid Restriction

Fluid restriction is the mainstay for correcting mild to modest hyponatremia due to SIADH,[61] usually 600–800 ml/day. With this intake, the patient should soon develop negative fluid balance owing to the continued insensible losses; the sodium gradually increases as the water intoxication is treated by simple water restriction. A good indication that the water restriction is working is to demonstrate weight loss. As weight decreases, the plasma osmolality and the serum sodium rise. This treatment usually increases the serum osmolality to normal within 7–10 days. Unfortunately, the practical problem of maintaining a fluid-restricted regimen undermines the efficiency of this therapy on a chronic basis. Most patients with chronic SIADH eventually require additional therapy, usually demeclocycline.

Drug Therapy: Demeclocycline

Demeclocycline is a useful adjunct to fluid restriction.[62] This drug interferes with the action of vasopressin at the level of the distal and collecting tubules. The drug is believed to interfere in the steps proximal and distal to cyclic adenosine monophosphate (cAMP) generation. Several studies have established the efficacy of demeclocycline and its superiority over lithium, another drug with similar properties.

Forrest et al.[63] compared the use of demeclocycline with that of lithium in the treatment of SIADH. They noted that demeclocycline restored serum sodium concentration to normal within 5–14 days, allowing unrestricted water intake in all the 10 patients studied. In contrast to lithium, no patient failed to respond to demeclocycline. Thus, demeclocycline is clearly superior to lithium in treatment of SIADH. Besides, the adverse effects of lithium, particularly CNS symptoms, make it a less desirable choice. Demeclocycline can be used regardless of whether the etiology of SIADH is ectopic or eutopic hypersecretion of the hormone. Similarly gratifying results have been reported by DeTroyer,[64] who used demeclocycline successfully in seven patients with SIADH.

Demeclocycline blocks the action of vasopressin regardless of whether it is secreted by the hypothalamus or secreted ectopically by tumor. In most patients, the electrolyte abnormality is restored to normal within 5–14 days. It is given in doses of 900–1200 mg/day. The occasional side effects include the occurrence of renal failure and bacterial superinfection. In general, the side effects are minimal, and the drug can be used safely, provided that there is adequate follow-up and monitoring of renal function. Demeclocycline should be used with caution in patients with cirrhosis[65] and heart failure.[66]

Hypertonic Saline

In patients with profound hyponatremia (serum sodium below 115 mEq/liter) or in those with neurological symptoms the initial treatment of choice is the administration of hypertonic saline. Careful evaluation of the cardiopulmonary status is essential prior to instituting such therapy, to avoid the development of pulmonary edema. It has been suggested that rapid correction of hyponatremia may result in the development of central pontine myelinolysis, a fatal condition.[67,68] The validity of such a suggestion has to be weighed against the dangers of not correcting severe hyponatremia with reasonable rapidity. The relationship between the use of hypertonic saline and the development of central pontine myelinolysis has been examined by several workers, without the emergence of a consensus.[69-71] It appears that central pontine myelinolysis can occur in association with numerous factors often found in hyponatremic patients (e.g., alcohol, malnutrition, hepatic encephalopathy). Furthermore, experimental evidence suggests that untreated severe hyponatremia per se can contribute to demyelinating lesions, having nothing to do with the rapidity of correction. Since severe hyponatremia is associated with an extremely high mortality and morbidity (permanent neurological damage), rapid correction to mildly hyponatremic levels is advocated. Ayus et al.[72] suggested that when the serum sodium level is more than 105 mEq/liter, it can be corrected to a value of 125–130 mEq/liter. However, when the serum sodium levels are lower than 105 mEq/liter, it is not safe to raise the serum sodium in excess of 20 mEq/liter over the baseline. The problem of developing central pontine myelinolysis with reasonably rapid correction of hyponatremia is overridden by the high mortality of untreated severe hyponatremia.

Furosemide Diuresis with Electrolyte Replacement

Hantman et al.[73] treated five patients with severe hyponatremia caused by SIADH with the combined used of furosemide and meticulous replacement of electrolyte losses. This therapy is based on the fact that patients with severe hyponatremia and volume expansion may excrete the administered saline load and may therefore fail to respond to hypertonic saline. In such a setting, furosemide produces a negative water balance rapidly, permitting retention of subsequently or comtemporaneously administered hypertonic saline. The negative water balance desired to raise the osmolality to 270 can be calculated using the formulas outlined in Table 105. The duration of furosemide administration depends on the rate by which the serum sodium rises. In the study reported by Hantman et al.,[73] when furosemide therapy was combined with replacement of Na and K intravenously, serum sodium could be raised to 130

TABLE 105
Calculation of Desired Negative Fluid Balance

1. Total body water = wt in kg $\times$ 60 or 70%
   (TBW)
2. Total body solute = TBW $\times$ plasma osmolality
   (in mOsm)
3. $\dfrac{\text{Total body solute}}{x \text{ (liters)}} = \dfrac{\text{desired plasma osmolality}}{1 \text{ liter}}$

4. Desired neg water balance = TBW $-x$

^a From Hantman et al.[75]

mEq/liter within 6–8 hr. These investigators also reported rapid improvement in the CNS symptoms of these patients. It should be pointed out that such therapy can be undertaken only when the urinary losses of electrolytes can be measured and replaced on an hourly basis.

References

1. Bartter FC, Schwartz WB: The syndrome of inappropriate secretion of antidiuretic hormone. *Am J Med* **42:**790, 1967.
2. Schwartz, WB, Bennett W, Curelop S, et al: A syndrome of renal sodium loss and hyponatremia probably resulting from inappropriate secretion of antidiuretic hormone. *Am J Med* **23:**529, 1957.
3. Schwartz, WB, Tassel D, Barter FC: Further observations on hyponatremia and renal sodium loss probably resulting from inappropriate secretion of antidiuretic hormone. *N Engl J Med* **262:**743, 1960.
4. Amatruda TT, Mulrow PJ, Gallagher JC, et al: Carcinoma of the lung with inappropriate antidiuresis: Demonstration of antidiuretic hormone-like activity in tumor extract. *N Engl J Med* **269:**544, 1963.
5. DeSousa RC, Delaere J, Berde B: Inappropriate secretion of vasopressin. *Lancet* **1:**436, 1965.
6. Vorherr H, Massry SG, Utiger RD, et al: Antidiuretic principle in malignant tumor extracts from patients with inappropriate ADH syndrome. *J Clin Endocrinol* **28:**162, 1968.
7. George JM, Capen CC, Phillips AS: Biosynthesis of vasopressin in vitro and ultrastructure of a bronchogenic carcinoma: Patient with the syndrome of inappropriate secretion of antidiuretic hormone. *J Clin Invest* **51:**141, 1972.
8. Cheng KW, Friesen HG: Studies of human neurophysin by radioimmunoassay. *J Clin Endocrinol Metab* **36:**553, 1973.
9. Hamilton BPM, Upton GV, Amatruda TT Jr: Evidence for the presence of neurophysin in tumors producing the syndrome of inappropriate antidiuresis. *J Clin Endocrinol Metab* **35:**764, 1972.
10. Odell WD, Wolfsen AR: Hormones from tumors: Are they ubiquitous? *Am J Med* **68:**317, 1980.
11. Comis RL, Miller M, Ginsberg SJ: Abnormalities in water homeostasis in small cell anaplastic lung cancer. *Cancer* **45:**2414, 1980.
12. Kaye SB, Ross EJ: Inappropriate anti-diuretic hormone (ADH) secretion in association with carcinoma of bladder. *Postgrad Med J* **53:**274, 1977.
13. Cassileth PA, Trotman BW: Inappropriate antidiuretic hormone in Hodgkin's disease. *Am J Med Sci* **265:**233, 1973.
14. Miller R, Ashkar FS, Rudzinski DJ: Inappropriate secretion of antidiuretic hormone in reticulum cell sarcoma. *South Med J* **64:**763, 1971.

15. Zimbler H, Robertson GL, Bartter FC, et al: Ewing's sarcoma as a cause of the syndrome of inappropriate secretion of antidiuretic hormone. *J Clin Endocrinol Metab* **41**:390, 1975.

16. Levin L, Sealy, Barron J: Syndrome of inappropriate antidiuretic hormone secretion following cis-dichlorodiammineplatinum II in a patient with malignant thymoma. *Cancer* **50**:2279, 1982.

17. Baylin SB, Mendelsohn G: Ectopic (inappropriate) hormone production by tumors: Mechanisms involved and the biological and clinical implications. *Endocr Rev* **1**:45, 1980.

18. Vorherr H, Massry SG, Fallet R, et al: Antidiuretic principle in tuberculous lung tissue of a patient with pulmonary tuberculosis and hyponatremia. *Ann Intern Med* **72**:383, 1970.

19. Spanos A, Spry CJ: Inappropriate secretion of antidiuretic hormone with chronic chest infections. *Br Med J* **3**:785, 1974.

20. Moses AM, Notman DD: Diabetes insipidus and syndrome of inappropriate antidiuretic hormone secretion (SIADH). *Adv Intern Med* **27**:73, 1982.

21. Sordillo P, Sagransky DM, Mercado R, et al: Carbamazepine-induced syndrome of inappropriate antidiuretic hormone secretion. *Arch Intern Med* **138**:299, 1978.

22. Stuart MJ, Cuaso C, Miller M, et al: Syndrome of recurrent increased secretion of antidiuretic hormone following multiple doses of vincristine. *Blood* **45**:315, 1975.

23. Cutting HO: Inappropriate secretion of antidiuretic hormone secondary to vincristine therapy. *Am J Med* **51**:269, 1971.

24. Steele TH, Serpick AA, Block JB: Antidiuretic response to cyclophosphamide in man. *J Pharmacol Exp Ther* **185**:245, 1973.

25. DeFronzo RA, Braine H, Colvin OM, et al: Water intoxication in man after cyclophosphamide therapy: Time course and relation to drug activation. *Ann Intern Med* **78**:861, 1973.

26. Dubovsky SL, Grabon S, Berl T, et al: Syndrome of inappropriate secretion of antidiuretic hormone with exacerbated psychosis. *Ann Intern Med* **79**:551, 1973.

27. Kramer DS, Drake ME Jr: Acute psychosis, polydipsia, and inappropriate secretion of antidiuretic hormone. *Am J Med* **75**:712, 1983.

28. Raskind MA, Orenstein H, Christopher TG: Acute psychosis, increased water ingestion, and inappropriate antidiuretic hormone secretion. *Am J Psychiatry* **132**:907, 1975.

29. Fowler RC, Kronfol ZA, Perry PJ: Water intoxication, psychosis, and inappropriate secretion of antidiuretic hormone. *Arch Gen Psychiatry* **34**:1097, 1977.

30. Husband C, Mai FM, Carruthers G: Syndrome of inappropriate secretion of antidiuretic hormone in a patient treated with haloperidol. *Can J Psychiatry* **26**:196, 1981.

31. Vincent FM: Inappropriate ADH secretion. (Letter.) *Arch Neurol* **34**:725, 1977.

32. Peck P, Schenkman L: Haloperidol-induced syndrome of inappropriate secretion of antidiuretic hormone. *Clin Pharmacol Ther* **26**:442, 1979.

33. Cusick JF, Hagen TC, Findling JW: Inappropriate secretion of antidiuretic hormone after transsphenoidal surgery for pituitary tumors. *N Engl J Med* **311**:36, 1984.

34. Valkusz Z, Laczi F, Laszlo FA: Development of Schwartz-Bartter syndrome after administration of chlorpropamide and 1-Deamino-8-D-arginine vasopressin. *Endokrinologie* **79**:345, 1982.

35. Nelson PB, Seif SM, Maroon JC, et al: Hyponatremia in intracranial disease: Perhaps not the syndrome of inappropriate secretion of antidiuretic hormone (SIADH). *J Neurosurg* **55**:938, 1981.

36. Hostetter TH, Martinez-Maldonado M: Syndromes of ADH excess and deficiency. *Min Elect Metab* **5**:159, 1981.

37. Moses AM, Miller M, Streeten DHP: Pathophysiologic and pharmacologic alterations in the release and action of ADH. *Metabolism* **25**:697, 1976.

38. Zerbe R, Stropes L, Robertson G: Vasopressin function in the syndrome of inappropriate antidiuresis. *Annu Rev Med* **31**:315, 1980.

39. Goldstein CS, Braunstein S, Goldfarb S: Idiopathic syndrome of inappropriate antidiuretic hormone secretion possibly related to advanced age. *Ann Intern Med* **99**:185, 1983.

40. Anderson RJ, Chung H, Kluge R: Hyponatremia: A prospective analysis of its epidemiology and the pathogenetic role of vasopressin. *Ann Intern Med* **102**:164, 1985.

41. Albrink MJ, et al: The displacement of serum water by the lipids of hyperlipemic serum: A new method for the rapid determination of serum water. *J Clin Invest* **34**:1483, 1955.

42. Tarail R, et al: Misleading reductions of serum sodium and chloride associated with hyper-proteinemia in patients with multiple myeloma. *Proc Soc Exp Biol Med* **110**:145, 1962.

43. Katz MA: Hyperglycemia induced hyponatremia—Calculation of expected serum sodium depression. *N Engl J Med* **289**:843, 1973.

44. Dirks JH, Cirksena WJ, Berliner RW: The effect of saline infusion on sodium reabsorption by the proximal tubule of the dog. *J Clin Invest* **44**:1160, 1965.

45. Martino JA, Earley LE: The effects of infusion of water on renal hemodynamics and the tubular reabsorption of sodium. *J Clin Invest* **46**:1229, 1967.

46. Alexander EA, Doner DW, Auld B, et al: Tubular reabsorption of sodium during acute and chronic volume expansion in man. *J Clin Invest* **51**:2370, 1972.

47. Del Greco F, de Wardener HE: The effect of urine osmolality of a transient reduction in glomerular filtration rate and solute output during a 'water' diuresis. *J Physiol (Lond)* **131**:307, 1956.

48. Kleeman CR, Maxwell MH, Rockney R: Production of hypertonic urine in humans in the probable absence of antidiuretic hormone (ADH). *Proc Soc Exp Biol Med* **96**:189, 1957.

49. Berliner RW, Davidson DG: Production of hypertonic urine in the absence of pituitary antidiuretic hormone. *J Clin Invest* **36**:1416, 1957.

50. Discala VA, Kinney MJ: Effect of myxedema on the renal diluting and concentrating mechanism. *Am J Med* **50**:325, 1971.

51. Goldberg M, Reivich M: Studies on mechanism of hyponatremia and impaired water excretion in myxedema. *Ann Intern Med* **56**:120, 1962.

52. DeFronzo RA, Goldberg M, Agus ZS: Normal diluting capacity in hyponatremic patients. *Ann Intern Med* **84**:538, 1976.

53. Lowance DC, Garfinkel HB, Mattern WD, et al: The effect of chronic hypotonic volume expansion on the renal regulation of acid–base equilibrium. *J Clin Invest* **51**:2928, 1972.

54. Jones NF, Barraclough MA, Forsling ML, et al: Inappropriate production of vasopressin, potassium deficiency and cerebrovascular disease. *Am J Med* **45**:474, 1968.

55. Ufferman RC, Schrier RW: Importance of sodium intake and mineralocorticoid hormone in the impaired water excretion in adrenal insufficiency. *J Clin Invest* **51**:1639, 1972.

56. Gill JR Jr, Gann DS, Bartter FC: Restoration of water diuresis in Addisonian patients by expansion of the volume of extracellular fluid. *J Clin Invest* **41**:1078, 1962.

57. Agus ZS, Goldberg M: Role of antidiuretic hormone in the abnormal water diuresis of anterior hypopituitarism in man. *J Clin Invest* **50**:1478, 1971.

58. Green H, Harrington AR, Valtin H: On the role of antidiuretic hormone in the inhibition of acute water diuresis in adrenal insufficiency and the effect of gluco- and mineralocorticoids in reversing the inhibition. *J Clin Invest* **49**:1724, 1970.

59. Fichman MP, Vorherr H, Kleeman CR, et al: Diuretic-induced hyponatremia. *Ann Intern Med* **75**:853, 1971.

60. Flear DTG, Singh CM: Hyponatremia and sick cells. *Br J Anesth* **45**:976, 1973.

61. Harrington JT, Cohen JJ: Clinical disorders of urine concentration and dilution. *Arch Intern Med* **131**:810, 1973.

62. Padfield PL, Morton JJ, Hodsman GP: Demeclocycline in the treatment of the syndrome of inappropriate antidiuretic hormone release: With measurement of plasma ADH. *Postgrad Med J* **54**:623, 1978.

63. Forrest JN, Cox M, Hong C, et al: Superiority of demeclocycline over lithium in the treatment of chronic syndrome of inappropriate secretion of antidiuretic hormone. *N Engl J Med* **298**:173, 1978.

64. DeTroyer A: Demeclocycline. Treatment for syndrome of inappropriate antidiuretic hormone secretion. *JAMA* **237**:2723, 1977.

65. Miller PD, Linas SL, Schrier RW: Plasma demeclocycline levels and nephrotoxicity: Correlation in hyponatremic cirrhotic patients. *JAMA* **243**:2513, 1980.

66. Zegers de Beyl D, Naeije R, de Troyer A: Demeclocycline treatment of water retention in congestive heart failure. *Br Med J* **1**:760, 1978.

67. Tomlinson BE, Pierides AM, Bradley WG: Central pontine myelinolysis: Two cases with associated electrolyte disturbance. *Q J Med* **179**:373, 1976.

68. Kleinschmidt-deMasters BK, Norenberg MD: Rapid correction of hyponatremia causes demyelination: Relation to central pontine myelinolysis. *Science* **211**:1068, 1981.
69. Norenberg MD, Leslie KO, Robertson AS: Association between rise in serum sodium and central pontine myelinolysis. *Ann Neurol* **11**:128, 1982.
70. Arieff AI: Permanent neurological disability from hyponatremia in healthy women undergoing elective surgery. *Kidney Int* **27**:132, 1985 (abst.).
71. Ayus JC, Krothapalli RK, Arieff AI, et al: Overcorrection rather than rapid correction induces central pontine myelinolysis in patients with severe hyponatremia. *Kidney Int* **27**:132, 1985 (abst).
72. Ayus JC, Krothapalli RK, Arieff AI: Changing concepts in treatment of severe symptomatic hyponatremia: Rapid correction and possible relation to central pontine myelinolysis. *Am J Med* **78**:897, 1985.
73. Hantman D, Rossier B, Zohlman R, et al: Rapid correction of hyponatremia in the syndrome of inappropriate secretion of antidiuretic hormone: An alternative treatment to hypertonic saline. *Ann Intern Med* **78**:870, 1973.

Index